Vol. 2

Essential Calculus-based

PHYSICS

Study Guide Workbook

Volume 2: Electricity and Magnetism

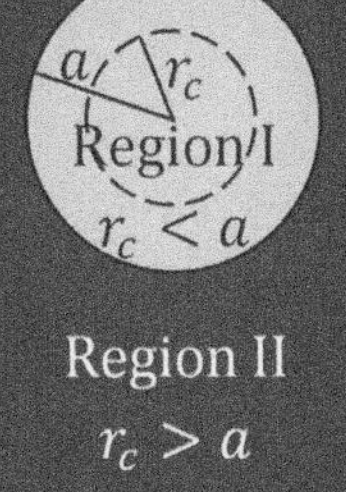

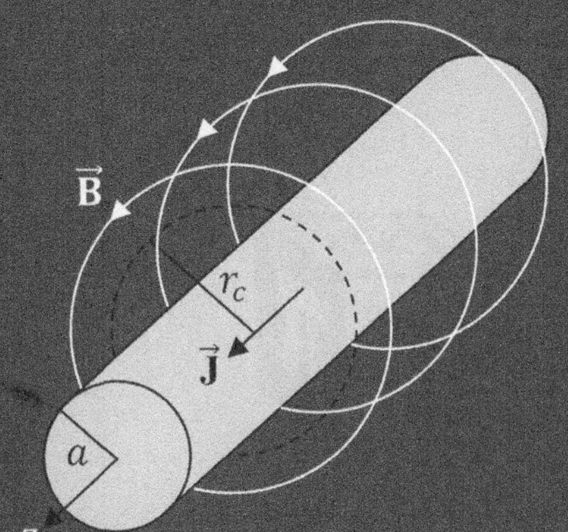

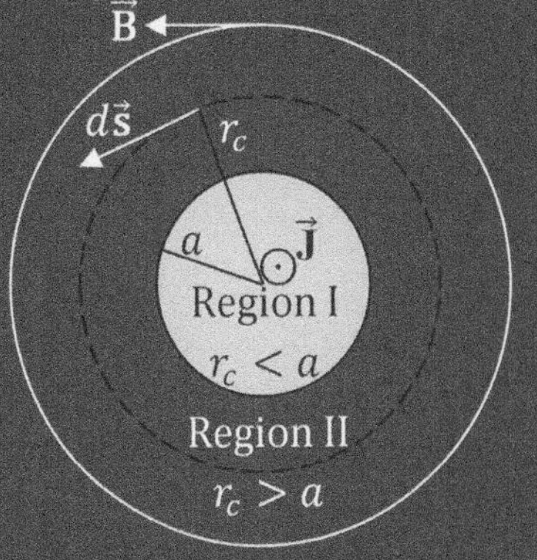

$$\oint_C \vec{\mathbf{B}} \cdot d\vec{\mathbf{s}} = \mu_0 I_{enc}$$

Chris McMullen, Ph.D.

Includes Answers!

Essential Calculus-based Physics Study Guide Workbook
Volume 2: Electricity and Magnetism
Learn Physics with Calculus Step-by-Step

Chris McMullen, Ph.D.
Physics Instructor
Northwestern State University of Louisiana

www.monkeyphysicsblog.wordpress.com
www.improveyourmathfluency.com
www.chrismcmullen.wordpress.com

Zishka Publishing

ISBN: 978-1-941691-11-3

Textbooks > Science > Physics
Study Guides > Workbooks> Science

CONTENTS

EXPECTATIONS

Prerequisites:

- As this is the second volume of the series, the student should already have studied the material from first-semester physics, including uniform acceleration, vector addition, applications of Newton's second law, conservation of energy, Hooke's law, and rotation.
- The student should have learned basic calculus skills, including how to find the derivative of a polynomial or trig function and how to integrate over a polynomial or trig function. This book reviews essential integration skills in Chapter 6.
- The student should know some basic algebra skills, including how to combine like terms, how to isolate an unknown, how to solve the quadratic equation, and how to apply the method of substitution. Needed algebra skills were reviewed in Volume 1.
- The student should have prior exposure to trigonometry. Essential trigonometry skills were reviewed in Volume 1.

Use:

- This book is intended to serve as a supplement for students who are attending physics lectures, reading a physics textbook, or reviewing physics fundamentals.
- The goal is to help students quickly find the most essential material.

Concepts:

- Each chapter reviews relevant definitions, concepts, laws, or equations needed to understand how to solve the problems.
- This book does not provide a comprehensive review of every concept from physics, but does cover most physics concepts that are involved in solving problems.

Strategies:

- Each chapter describes the problem-solving strategy needed to solve the problems at the end of the chapter.
- This book covers the kinds of fundamental problems which are commonly found in standard physics textbooks.

Help:

- Every chapter includes representative examples with step-by-step solutions and explanations. These examples should serve as a guide to help students solve similar problems at the end of each chapter.
- Each problem includes the main answer(s) on the same page as the question. At the **back of the book**, you can find hints, intermediate answers, directions to help walk you through the steps of each solution, and explanations regarding common issues that students encounter when solving the problems. It's very much like having your own **physics tutor** at the back of the book to help you solve each problem.

INTRODUCTION

The goal of this study guide workbook is to provide practice and help carrying out essential problem-solving strategies that are standard in electricity and magnetism. The aim here is not to overwhelm the student with comprehensive coverage of every type of problem, but to focus on the main strategies and techniques with which most physics students struggle.

This workbook is not intended to serve as a substitute for lectures or for a textbook, but is rather intended to serve as a valuable supplement. Each chapter includes a concise review of the essential information, a handy outline of the problem-solving strategies, and examples which show step-by-step how to carry out the procedure. This is not intended to teach the material, but is designed to serve as a time-saving review for students who have already been exposed to the material in class or in a textbook. Students who would like more examples or a more thorough introduction to the material should review their lecture notes or read their textbooks.

Every exercise in this study guide workbook applies the same strategy which is solved step-by-step in at least one example within the chapter. Study the examples and then follow them closely in order to complete the exercises. Many of the exercises are broken down into parts to help guide the student through the exercises. Each exercise tabulates the corresponding answers on the same page. Students can find additional help in the hints section at the back of the book, which provides hints, answers to intermediate steps, directions to walk students through every solution, and explanations regarding issues that students commonly ask about.

Every problem in this book can be solved without the aid of a calculator. You may use a calculator if you wish, though it is a valuable skill to be able to perform basic math without relying on a calculator.

The mathematics and physics concepts are not two completely separate entities. The equations speak the concepts. Let the equations guide your reasoning.

— Chris McMullen, Ph.D.

1 COULOMB'S LAW

Relevant Terminology

Electric charge – a fundamental property of a particle that causes the particle to experience a force in the presence of an electric field. (An electrically neutral particle has no charge and thus experiences no force in the presence of an electric field.)
Electric force – the push or pull that one charged particle exerts on another. Oppositely charged particles attract, whereas like charges (both positive or both negative) repel.
Coulomb's constant – the constant of proportionality in Coulomb's law (see below).

Coulomb's Law

According to Coulomb's law, any two objects with charge attract or repel one another with an **electrical force** that is directly proportional to each **charge** and inversely proportional to the square of the separation between the two charges:

$$F_e = k\frac{|q_1||q_2|}{R^2}$$

The absolute values around each charge indicate that the magnitude of the force is positive. Note the subscript on F_e: It's F sub e (**<u>not</u>** F times e). The subscript serves to distinguish electric force (F_e) from other kinds of forces, such as gravitational force (F_g). The proportionality constant in Coulomb's law is called **Coulomb's constant** (k):

$$k = 8.99 \times 10^9 \ \frac{\text{N}\cdot\text{m}^2}{\text{C}^2} \approx 9.0 \times 10^9 \ \frac{\text{N}\cdot\text{m}^2}{\text{C}^2}$$

In this book, we will round Coulomb's constant to $9.0 \times 10^9 \ \frac{\text{N}\cdot\text{m}^2}{\text{C}^2}$ such that the problems may be solved without using a calculator. (This rounding is good to 1 part in 900.)

Symbols and SI Units

Symbol	Name	SI Units
F_e	electric force	N
q	charge	C
R	separation	m
k	Coulomb's constant	$\frac{\text{N}\cdot\text{m}^2}{\text{C}^2}$ or $\frac{\text{kg}\cdot\text{m}^3}{\text{C}^2\cdot\text{s}^2}$

Notes Regarding Units

The SI units of Coulomb's constant (k) follow by solving for k in Coulomb's law:

$$k = \frac{F_e R^2}{q_1 q_2}$$

The SI units of k equal $\frac{\text{N}\cdot\text{m}^2}{\text{C}^2}$ because these are the SI units of $\frac{F_e R^2}{q_1 q_2}$. This follows since the SI unit of electric force (F_e) is the Newton (N), the SI unit of charge (q) is the Coulomb (C), and the SI unit of separation (R) is the meter (m). Recall from first-semester physics that a Newton is equivalent to:

$$1 \text{ N} = 1 \frac{\text{kg}\cdot\text{m}}{\text{s}^2}$$

Plugging this into $\frac{\text{N}\cdot\text{m}^2}{\text{C}^2}$, the SI units of k can alternatively be expressed as $\frac{\text{kg}\cdot\text{m}^3}{\text{C}^2\cdot\text{s}^2}$.

Essential Concepts

The matter around us is composed of different types of atoms. Each atom consists of protons and neutrons in its nucleus, surrounded by electrons.

- **Protons** have **positive** electric charge.
- Neutrons are electrically neutral.
- **Electrons** have **negative** electric charge.

Whether two charges attract or repel depends on their relative signs:

- **Opposite** charges **attract**. For example, electrons are attracted to protons.
- **Like** charges **repel**. For example, two electrons repel. Similarly, two protons repel.

The charge of an object depends on how many protons and electrons it has:

- If the object has more protons than electrons (meaning that the object has lost electrons), the object has positive charge.
- If the object has more electrons than protons (meaning that the object has gained electrons), the object has negative charge.
- If the object has the same number of protons as electrons, the object is electrically neutral. Its net charge is zero.

(Atoms tend to gain or lose valence electrons from their outer shells. It's not easy to gain or lose protons since they are tightly bound inside the nucleus of the atom. One way for objects to become electrically charged is through rubbing, such as rubbing glass with fur.)

Some materials tend to be good conductors of electricity; others are good insulators.

- Charges flow readily through a **conductor**. Most metals are good conductors.
- Charges tend not to flow through an **insulator**. Glass and wood are good insulators.

When two charged objects touch (or are connected by a conductor), charge can be transferred from one object to the other. See the second example in this chapter.

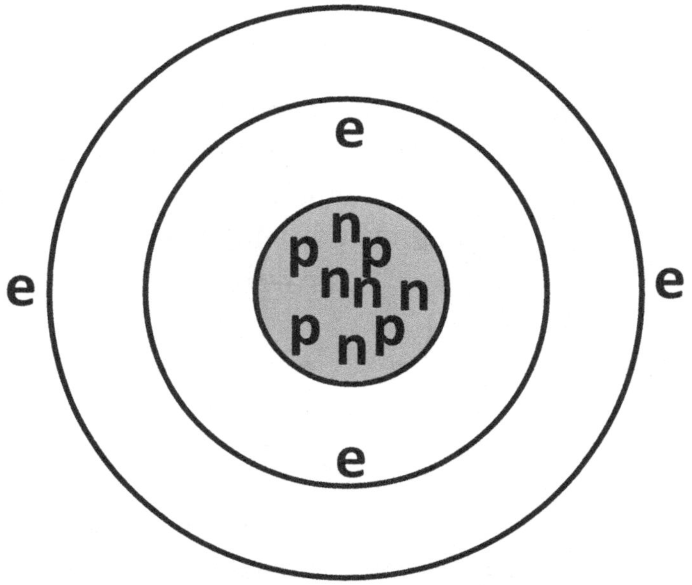

Metric Prefixes

Since a Coulomb (C) is a very large amount of charge, we often use the following metric prefixes when working with electric charge.

Prefix	Name	Power of 10
m	milli	10^{-3}
μ	micro	10^{-6}
n	nano	10^{-9}
p	pico	10^{-12}

Note: The symbol μ is the lowercase Greek letter mu. When it is used as a metric prefix, it is called micro. For example, 32 μC is called 32 microCoulombs.

Algebra with Powers

It may be helpful to recall the following rules of algebra relating to powers:

$$x^a x^b = x^{a+b} \quad , \quad \frac{x^a}{x^b} = x^{a-b} \quad , \quad x^{-a} = \frac{1}{x^a} \quad , \quad \frac{1}{x^{-a}} = x^a$$

$$x^0 = 1 \quad , \quad (x^a)^b = x^{ab} \quad , \quad (ax)^b = a^b x^b$$

Coulomb's Law Strategy

How you solve a problem involving Coulomb's depends on which kind of problem it is:

- In this chapter, we will focus on the simplest problems, which involve two charged objects attracting or repelling one another. For simple problems like these, plug the known values into the following equation and solve for the unknown quantity.

$$F_e = k\frac{|q_1||q_2|}{R^2}$$

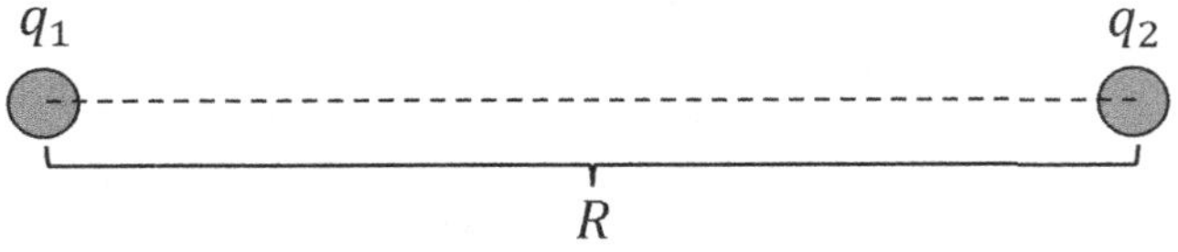

  Look at the units and wording to determine which symbols you know.
  - A value in Coulombs (C) is electric charge, q.
  - A value in meters (m) is likely related to the separation, R.
  - A value in N is a force, such as electric force, F_e.
  - You should know Coulomb's constant: $k = 9.0 \times 10^9 \ \frac{\text{N}\cdot\text{m}^2}{\text{C}^2}$.

  If two charged objects touch or are connected by a **conductor**, in a fraction of a second the excess charge will redistribute and the system will attain static equilibrium. The two charges will then be equal: The new charge, q, will equal $q = \frac{q_1+q_2}{2}$. Coulomb's law then reduces to $F_e = k\frac{q^2}{R^2}$.
- If a problem gives you three or more charges, apply the technique of vector addition, as illustrated in Chapter 3.
- If a problem involves other forces, like tension in a cord, apply Newton's second law, as illustrated in Chapter 5.
- If a problem involves an **electric field**, E, (not to be confused with electric force, F_e, or electric charge, q), see Chapter 2, 3, or 5, depending on the nature of the problem.

Inverse-square Laws

Coulomb's law and Newton's law of gravity are examples of inverse-square laws: Each of these laws features a factor of $\frac{1}{R^2}$. (Newton's law of gravity was discussed in Volume 1.)

$$F_e = k\frac{|q_1||q_2|}{R^2} \quad , \quad F_g = G\frac{m_1 m_2}{R^2}$$

Coulomb's law has a similar structure to Newton's law of gravity: Both force laws involve a proportionality constant, a product of sources ($|q_1||q_2|$ or $m_1 m_2$), and $\frac{1}{R^2}$.

Elementary Charge

Protons have a charge equal to 1.60×10^{-19} C (to three significant figures). We call this elementary charge and give it the symbol e. Electrons have the same charge, except for being negative. Thus, protons have charge $+e$, while electrons have charge $-e$. When a macroscopic object is charged, its charge will be a multiple of e, since all objects are made up of protons, neutrons, and electrons. If you need to use the charge of a proton or electron to solve a problem, use the value of e below.

Elementary Charge

$$e = 1.60 \times 10^{-19} \text{ C}$$

Example: A small strand of monkey fur has a net charge of 4.0 µC while a small piece of glass has a net charge of −5.0 µC. The strand of fur is 3.0 m from the piece of glass. What is the electric force between the monkey fur and the piece of glass?

Make a list of the known quantities:

- The strand of fur has a charge of $q_1 = 4.0$ µC.
- The piece of glass has a charge of $q_2 = -5.0$ µC.
- The separation between them is $R = 3.0$ m.
- Coulomb's constant is $k = 9.0 \times 10^9 \frac{\text{N}\cdot\text{m}^2}{\text{C}^2}$.

Convert the charges from microCoulombs (µC) to Coulombs (C). Recall that the metric prefix micro (µ) stands for one millionth: $\mu = 10^{-6}$.

$$q_1 = 4.0 \text{ µC} = 4.0 \times 10^{-6} \text{ C}$$
$$q_2 = -5.0 \text{ µC} = -5.0 \times 10^{-6} \text{ C}$$

Plug these values into Coulomb's law. It's convenient to suppress units until the end in order to avoid clutter. Note that the absolute value of −5 is +5.

$$F_e = k\frac{|q_1||q_2|}{R^2} = (9 \times 10^9)\frac{|4 \times 10^{-6}||-5 \times 10^{-6}|}{(3)^2} = (9 \times 10^9)\frac{(4 \times 10^{-6})(5 \times 10^{-6})}{(3)^2}$$

If not using a calculator, it's convenient to separate the powers:

$$F_e = \frac{(9)(4)(5)}{(3)^2} \times 10^9 10^{-6} 10^{-6} = 20 \times 10^{-3} = 0.020 \text{ N}$$

Note that $10^9 10^{-6} 10^{-6} = 10^{9-6-6} = 10^{9-12} = 10^{-3}$ according to the rule $x^m x^n = x^{m+n}$. The answer is $F_e = 0.020$ N, which could also be expressed as 20×10^{-3} N, 2.0×10^{-2} N, or 20 mN (meaning milliNewtons, where the prefix milli, m, stands for 10^{-3}).

Example: A metal banana-shaped earring has a net charge of $-3.0\ \mu\text{C}$ while a metal monkey-shaped earring has a net charge of $7.0\ \mu\text{C}$. The two earrings are brought together, touching one another for a few seconds, after which the earrings are placed 6.0 m apart. What is the electric force between the earrings when they are placed 6.0 m apart?

The "trick" to this problem is to realize that charge is transferred from one object to the other when they touch. The excess charge splits evenly between the two earrings. The excess charge (or the net charge) equals $q_{net} = -3.0\ \mu\text{C} + 7.0\ \mu\text{C} = 4.0\mu\text{C}$. Half of this charge will reside on each earring after contact is made: $q = \frac{q_{net}}{2} = \frac{4.0\ \mu\text{C}}{2} = 2.0\ \mu\text{C}$. You could obtain the same answer via the following formula:

$$q = \frac{q_1 + q_2}{2} = \frac{-3.0\ \mu\text{C} + 7.0\ \mu\text{C}}{2} = \frac{4.0\ \mu\text{C}}{2} = 2.0\ \mu\text{C}$$

Convert the charge from microCoulombs (μC) to Coulombs (C). Recall that the metric prefix micro (μ) stands for one millionth: $\mu = 10^{-6}$.

$$q = 2.0\ \mu\text{C} = 2.0 \times 10^{-6}\ \text{C}$$

Set the two charges equal to one another in Coulomb's law.

$$F_e = k\frac{q^2}{R^2} = (9 \times 10^9)\frac{(2 \times 10^{-6})^2}{(6)^2} = (9 \times 10^9)\frac{(2)^2(10^{-6})^2}{(6)^2} = (9 \times 10^9)\frac{(2)^2(10^{-12})}{(6)^2}$$

Note that $(2 \times 10^{-6})^2 = (2)^2(10^{-6})^2$ according to the rule $(xy)^2 = x^2y^2$ and note that $(10^{-6})^2 = 10^{-12}$ according to the rule $(x^m)^n = x^{mn}$. If not using a calculator, it's convenient to separate the powers:

$$F_e = \frac{(9)(2)^2}{(6)^2} \times 10^9 10^{-12} = 1.0 \times 10^{-3}\ \text{N}$$

Note that $10^9 10^{-12} = 10^{9-12} = 10^{-3}$ according to the rule $x^m x^{-n} = x^{m-n}$. The answer is $F_e = 0.0010$ N, which could also be expressed as 1.0×10^{-3} N or 1.0 mN (meaning milliNewtons, where the prefix milli, m, stands for 10^{-3}).

1. A coin with a monkey's face has a net charge of $-8.0\ \mu C$ while a coin with a monkey's tail has a net charge of $-3.0\ \mu C$. The coins are separated by 2.0 m. What is the electric force between the two coins? Is the force attractive or repulsive?

2. A small glass rod and a small strand of monkey fur are each electrically neutral initially. When a monkey rubs the glass rod with the monkey fur, a charge of 800 nC is transferred between them. The monkey places the strand of fur 20 cm from the glass rod. What is the electric force between the fur and the rod? Is the force attractive or repulsive?

Want help? Check the hints section at the back of the book.

Answers: 0.054 N (repulsive), 0.144 N (attractive)

3. A metal banana-shaped earring has a net charge of $-2.0\ \mu C$ while a metal apple-shaped earring has a net charge of $8.0\ \mu C$. The earrings are 3.0 m apart.

(A) What is the electric force between the two earrings? Is it attractive or repulsive?

(B) The two earrings are brought together, touching one another for a few seconds, after which the earrings are once again placed 3.0 m apart. What is the electric force between the two earrings now? Is it attractive or repulsive?

Want help? Check the hints section at the back of the book.

Answers: 0.016 N (attractive), 0.0090 N (repulsive)

2 ELECTRIC FIELD

Relevant Terminology

Electric charge – a fundamental property of a particle that causes the particle to experience a force in the presence of an electric field. (An electrically neutral particle has no charge and thus experiences no force in the presence of an electric field.)
Electric force – the push or pull that one charged particle exerts on another. Oppositely charged particles attract, whereas like charges (both positive or both negative) repel.
Electric field – force per unit charge.

Electric Field Equations

Electric force ($\vec{\mathbf{F}}_e$) equals charge (q) times electric field ($\vec{\mathbf{E}}$). This equation is always true, but is only useful when you wish to relate electric force to electric field.

$$\vec{\mathbf{F}}_e = q\vec{\mathbf{E}}$$

If you want to find the electric field created by a single **pointlike** charge, use the following equation, where R is the distance from the pointlike charge. The absolute values represent that the magnitude (E) of the electric field vector ($\vec{\mathbf{E}}$) is always positive.

$$E = \frac{k|q|}{R^2}$$

To find the electric field created by a system of pointlike charges, see the strategy outlined in Chapter 3. To find the electric field created by a continuous distribution of charge (such as a plate or sphere with charge on it), see Chapters 7-8.

Symbols and SI Units

Symbol	Name	SI Units
E	electric field	N/C or V/m
F_e	electric force	N
q	charge	C
R	distance from the charge	m
k	Coulomb's constant	$\frac{\text{N}\cdot\text{m}^2}{\text{C}^2}$ or $\frac{\text{kg}\cdot\text{m}^3}{\text{C}^2\cdot\text{s}^2}$

Notes Regarding Units

The SI units of electric field (E) follow by solving for E in the equation $\vec{\mathbf{F}}_e = q\vec{\mathbf{E}}$:

$$\vec{\mathbf{E}} = \frac{\vec{\mathbf{F}}_e}{q}$$

The SI units of E equal $\frac{\text{N}}{\text{C}}$ because force (F_e) is measured in Newtons (N) and charge (q) is measured in Coulombs (C). Since a Newton is equivalent to $1 \text{ N} = 1 \frac{\text{kg}\cdot\text{m}}{\text{s}^2}$, the SI units of electric field could be expressed as $\frac{\text{kg}\cdot\text{m}}{\text{C}\cdot\text{s}^2}$. In Chapter 10, we will learn that the electric field between two parallel plates equals $E = \frac{\Delta V}{d}$, where potential difference (ΔV) is measured in Volts (V) and the distance between the plates (d) is measured in meters (m). Therefore, yet another way to express electric field is $\frac{\text{V}}{\text{m}}$. It's convenient to use $\frac{\text{N}}{\text{C}}$ for the units of electric field when working with force, and to use $\frac{\text{V}}{\text{m}}$ for the units of electric field when working with electric potential.

Essential Concepts

Every charged particle creates an electric field around it. The formula $E = \frac{k|q|}{R^2}$ expresses that the electric field is stronger in the region of space close to the charged particle and gets weaker farther away from the particle (that is, as the distance R increases, electric field decreases by a factor of $\frac{1}{R^2}$).

When there are two or more charged particles, each charge creates its own electric field, and the net electric field is found through vector addition (as we will see in Chapter 3).

Electric field has different values at different locations in space (just like gravitational field is stronger near earth's surface and noticeably weaker if you go halfway to the moon).

Once you know the value of the electric field at a given point in space, if a charged particle is placed at that same point (that is, the point where you know the value of the electric field), then you can use the equation $\vec{\mathbf{F}}_e = q\vec{\mathbf{E}}$ to determine what force would be exerted on that charged particle. We call such a particle a **test charge**.

The direction of the electric field is based on the concept of a test charge. To determine the direction of the electric field at a particular point, imagine placing a **positive** test charge at that point. The direction of the electric field is the same as the direction of the electric force that would be exerted on that positive test charge.

Electric Field Strategy

How you solve a problem involving electric field depends on which kind of problem it is:

- If you want to find the electric field created by a single pointlike charge, use the following equation.

$$E = \frac{k|q|}{R^2}$$

  R is the distance from the pointlike charge to the point where the problem asks you to find the electric field. If the problem gives you the coordinates (x_1, y_1) of the charge and the coordinates (x_2, y_2) of the point where you need to find the electric field, apply the distance formula to find R.

$$R = \sqrt{(x_2 - x_1)^2 + (y_2 - y_1)^2}$$

- If you need to relate the magnitude of the electric field (E) to the magnitude of the electric force (F_e), use the following equation. The absolute values are present because the magnitudes (E and F_e) of the vectors must be positive.

$$F_e = |q|E$$

- If a problem gives you three or more charges, apply the technique of vector addition, as illustrated in Chapter 3.
- If a problem involves other forces, like tension in a cord, apply Newton's second law, as illustrated in Chapter 5.
- If a problem involves finding the electric field created by a continuous distribution of charge (such as a plate or cylinder), see Chapters 7-8.

Important Distinctions

The first step toward mastering electric field problems is to study the terminology: You need to be able to distinguish between electric charge (q), electric field (E), and electric force (F_e). Read the problems carefully, memorize which symbol is used for each quantity, and look at the units to help distinguish between them:

- The SI unit of electric **charge** (q) is the Coulomb (C).
- The SI units of electric **field** (E) can be expressed as $\frac{\text{N}}{\text{C}}$ or $\frac{\text{V}}{\text{m}}$.
- The SI unit of electric **force** (F_e) is the Newton (N).

The second step is to learn which equations to use in which context.

- Use the equation $E = \frac{k|q|}{R^2}$ when you want to find the electric field **created by a pointlike charge** at a particular point in space.
- Use the equation $F_e = |q|E$ to find the force that would be exerted on a pointlike charge **in the presence of an (external) electric field.**

It may be helpful to remember that a pointlike charge doesn't exert a force on itself. The electric field created by one charge can, however, exert a force on a different charge.

Coulomb's Law and Electric Field

Consider the two pointlike charges illustrated below. The left charge, q_1, creates an electric field everywhere in space, including the location of the right charge, q_2. We could use the formula $E_1 = \frac{k|q_1|}{R^2}$ to find the magnitude of q_1's electric field at the location of q_2. We could then find the electric force exerted on q_2 using the equation $F_e = |q_2|E_1$. If we combine these two equations together, we get Coulomb's law, as shown below.

$$F_e = |q_2|E_1 = |q_2|\left(\frac{k|q_1|}{R^2}\right) = k\frac{|q_1||q_2|}{R^2}$$

Similarly, we could use the formula $E_2 = \frac{k|q_2|}{R^2}$ to find the magnitude of q_2's electric field at the location of q_1, and then we could then find the electric force exerted on q_1 using the equation $F_e = |q_1|E_2$. We would obtain the same result, $F_e = k\frac{|q_1||q_2|}{R^2}$. This should come as no surprise, since Newton's third law of motion states that the force that q_1 exerts on q_2 is equal in magnitude and opposite in direction to the force that q_2 exerts on q_1.

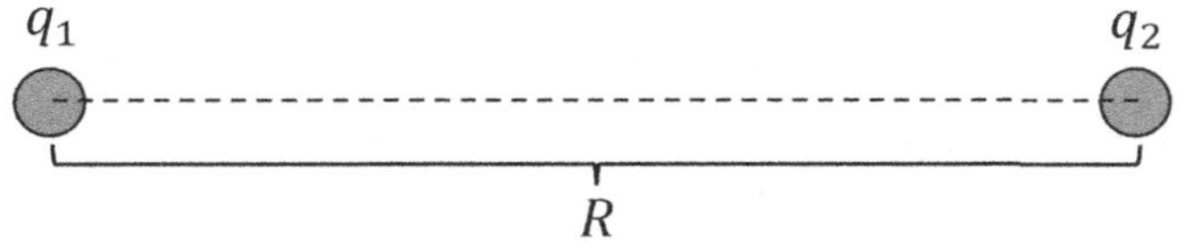

Example: A monkey's earring has a net charge of $-400\ \mu\text{C}$. The earring is in the presence of an external electric field with a magnitude of 5,000 N/C. What is the magnitude of the electric force exerted on the earring?

Make a list of the known quantities and the desired unknown:

- The earring has a charge of $q = -400\ \mu\text{C} = -400 \times 10^{-6}\ \text{C} = -4.0 \times 10^{-4}\ \text{C}$. Recall that the metric prefix micro (μ) stands for 10^{-6}. Note that $100 \times 10^{-6} = 10^{-4}$.
- The electric field has a magnitude of $E = 5{,}000$ N/C.
- We are looking for the magnitude of the electric force, F_e.

Based on the list above, we should use the following equation.

$$F_e = |q|E = |-4.0 \times 10^{-4}| \times 5000 = (4.0 \times 10^{-4}) \times 5000 = 20{,}000 \times 10^{-4} = 2.0\ \text{N}$$

The magnitude of a vector is always positive (that's why we took the absolute value of the charge). The electric force exerted on the earring is $F_e = 2.0$ N.

Note that we used the equation $F_e = |q|E$ because we were finding the force exerted on a charge in the presence of an external electric field. We didn't use the equation $E = \frac{k|q|}{R^2}$ because the problem didn't require us to find an electric field created by the charge. (The problem didn't state what created the electric field, and it doesn't matter as it isn't relevant to solving the problem.)

Example: A small strand of monkey fur has a net charge of 50 μC. (A) Find the magnitude of the electric field a distance of 3.0 m from the strand of fur. (B) If a small strand of lemur fur with a net charge of 20 μC is placed 3.0 m from the strand of monkey fur, what force will be exerted on the strand of lemur fur?

(A) Make a list of the known quantities and the desired unknown:

- The strand of fur has a charge of $q = 50\ \mu\text{C} = 50 \times 10^{-6}\ \text{C} = 5.0 \times 10^{-5}\ \text{C}$. Recall that the metric prefix micro (μ) stands for 10^{-6}. Note that $10 \times 10^{-6} = 10^{-5}$.
- We wish to find the electric field at a distance of $R = 3.0$ m from the strand of fur.
- We also know that Coulomb's constant is $k = 9.0 \times 10^9\ \frac{\text{N}\cdot\text{m}^2}{\text{C}^2}$.

Based on the list above, we should use the following equation.

$$E = \frac{k|q|}{R^2} = \frac{(9 \times 10^9)|5 \times 10^{-5}|}{(3)^2} = \frac{(9)(5)}{(3)^2} \times 10^9 10^{-5} = 5.0 \times 10^4\ \text{N/C}$$

Note that $10^9 10^{-5} = 10^{9-5} = 10^4$ according to the rule $x^m x^{-n} = x^{m-n}$. The answer is $E = 5.0 \times 10^4$ N/C, which could also be expressed as 50,000 N/C.

(B) Use the result from part (A) with the following equation.

$$F_e = |q_2|E = |20 \times 10^{-6}| \times 5.0 \times 10^4 = 100 \times 10^{-2} = 1.0\ \text{N}$$

The electric force exerted on the strand of lemur fur is $F_e = 1.0$ N.

Example: A gorilla-shaped earring, shown as a dot (●) below, with a charge of 600 μC lies at the origin. What is the magnitude of the electric field at the point (3.0 m, 4.0 m), which is marked as a star (★) below?

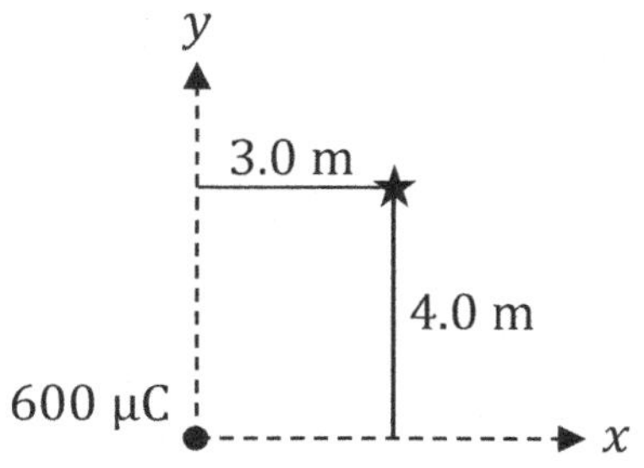

First, we need to find the distance between the charge and the point where we're trying to find the electric field. Apply the distance formula.

$$R = \sqrt{(x_2 - x_1)^2 + (y_2 - y_1)^2} = \sqrt{(3-0)^2 + (4-0)^2} = \sqrt{9+16} = \sqrt{25} = 5.0\ \text{m}$$

Next, use the equation for the electric field created by a pointlike object.

$$E = \frac{k|q|}{R^2} = \frac{(9 \times 10^9)|600 \times 10^{-6}|}{(5)^2} = 2.16 \times 10^5\ \text{N/C}$$

Note that $10^9 10^{-6} = 10^{9-6} = 10^3$ and $10^3 10^2 = 10^5$ according to the rule $x^m x^n = x^{m+n}$. The answer is $E = 2.16 \times 10^5$ N/C, which could also be expressed as 216,000 N/C.

4. A small furball from a monkey has a net charge of $-300\ \mu\text{C}$. The furball is in the presence of an external electric field with a magnitude of 80,000 N/C. What is the magnitude of the electric force exerted on the furball?

5. A gorilla's earring has a net charge of $800\ \mu\text{C}$. Find the magnitude of the electric field a distance of 2.0 m from the earring.

Want help? Check the hints section at the back of the book.

Answers: 24 N, 1.8×10^6 N/C

6. A monkey's earring experiences an electric force of 12 N in the presence of an external electric field of 30,000 N/C. The electric force is opposite to the electric field. What is the net charge of the earring?

7. A small strand of monkey fur has a net charge of 80 μC. Where does the electric field created by the strand of fur equal 20,000 N/C?

Want help? Check the hints section at the back of the book.

Answers: −400 μC, 6.0 m

8. A pear-shaped earring, shown as a dot (●) below, with a charge of 30 μC lies at the point (3.0 m, 6.0 m).

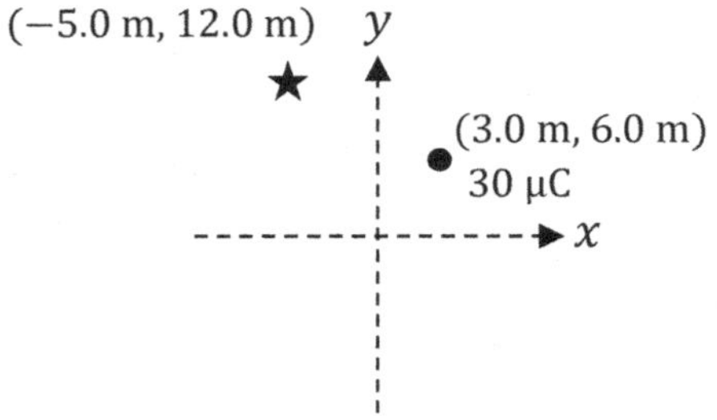

(A) What is the magnitude of the electric field at the point (−5.0 m, 12.0 m), which is marked as a star (★) above?

(B) If a small strand of chimpanzee fur with a net charge of 500 μC is placed at the point (−5.0 m, 12.0 m), marked with a star (★), what force will be exerted on the strand of fur?

Want help? Check the hints section at the back of the book.

Answers: 2700 N/C, 1.35 N

3 SUPERPOSITION OF ELECTRIC FIELDS

Essential Concepts

When two or more charges create two or more electric fields, the net electric field can be found through the principle of superposition. What this means is to add the electric field vectors together using the strategy of **vector addition**. Similarly, when there are three or more charges and you want to find the net electric force exerted on one of the charges, the net force can be found using vector addition.

Superposition Equations

Depending on the question, either find the magnitude of the electric **field** created by each charge at a specified point <u>or</u> find the magnitude of the electric **force** exerted on a specified charge by each of the other charges.

$$E_1 = \frac{k|q_1|}{R_1^2} \quad , \quad E_2 = \frac{k|q_2|}{R_2^2} \quad , \quad E_3 = \frac{k|q_3|}{R_3^2} \quad \cdots$$

$$F_1 = \frac{k|q_1||q|}{R_1^2} \quad , \quad F_2 = \frac{k|q_2||q|}{R_2^2} \quad , \quad F_3 = \frac{k|q_3||q|}{R_3^2} \quad \cdots$$

Use the distance formula to determine each of the R's.

$$R_i = \sqrt{(x_2 - x_1)^2 + (y_2 - y_1)^2}$$

Apply trigonometry to find the direction of each electric field or electric force vector.

$$\theta_i = \tan^{-1}\left(\frac{\Delta y}{\Delta x}\right)$$

Add the electric field or electric force vectors together using the vector addition strategy. Either use E's or F's, depending on what the questions asks for. Apply trig to determine the components of the given vectors.

$$E_{1x} = E_1 \cos\theta_1 \quad , \quad E_{1y} = E_1 \sin\theta_1 \quad , \quad E_{2x} = E_2 \cos\theta_2 \quad , \quad E_{2y} = E_2 \sin\theta_2 \quad \cdots$$

$$F_{1x} = F_1 \cos\theta_1 \quad , \quad F_{1y} = F_1 \sin\theta_1 \quad , \quad F_{2x} = F_2 \cos\theta_2 \quad , \quad F_{2y} = F_2 \sin\theta_2 \quad \cdots$$

Combine the respective components together to find the components of the resultant vector. Either use E's or F's (not both).

$$E_x = E_{1x} + E_{2x} + \cdots + E_{Nx} \quad , \quad E_y = E_{1y} + E_{2y} + \cdots + E_{Ny}$$

$$F_x = F_{1x} + F_{2x} + \cdots + F_{Nx} \quad , \quad F_y = F_{1y} + F_{2y} + \cdots + F_{Ny}$$

Use the Pythagorean theorem to find the magnitude of the net electric field or net electric force. Use an inverse tangent to determine the direction of the resultant vector.

$$E = \sqrt{E_x^2 + E_y^2} \quad , \quad \theta_E = \tan^{-1}\left(\frac{E_y}{E_x}\right)$$

$$F_e = \sqrt{F_x^2 + F_y^2} \quad , \quad \theta_F = \tan^{-1}\left(\frac{F_y}{F_x}\right)$$

Symbols and SI Units

Symbol	Name	Units
E	magnitude of electric field	N/C or V/m
F_e	magnitude of electric force	N
θ	direction of $\vec{\mathbf{E}}$ or $\vec{\mathbf{F}}_e$	°
q	charge	C
R	distance from the charge	m
d	distance between two charges	m
k	Coulomb's constant	$\frac{\text{N}\cdot\text{m}^2}{\text{C}^2}$ or $\frac{\text{kg}\cdot\text{m}^3}{\text{C}^2\cdot\text{s}^2}$

Important Distinction

There are two similar kinds of superposition problems:

- One kind gives you two or more pointlike charges and asks you to find the net electric **field** at a particular point in space (usually where there is no charge).
- The other kind gives you three or more pointlike charges and asks you to find the net electric **force** exerted on one of the charges.

When you are finding electric **field**, work with E's. When you are finding electric **force**, work with F's. Note that the equation for the electric field created by a pointlike charge has a single charge, $E = \frac{k|q|}{R^2}$, whereas the equation for the electric force exerted on one charge by another has a pair of charges, $F_e = \frac{k|q_1||q|}{R^2}$. Use the absolute value of the charge in the formulas $E = \frac{k|q|}{R^2}$ and $F_e = \frac{k|q_1||q|}{R^2}$, since the magnitude of a vector is always positive. The sign of the charge instead factors into the direction of the vector (see the first step of the strategy on the next page).

With electric **field**, you work with one charge at a time: $E = \frac{k|q|}{R^2}$. With electric **force**, you work with pairs of charges: $F_e = \frac{k|q_1||q|}{R^2}$.

If you need to find net electric **force**, you could first find the net electric **field** at the location of the specified charge, and once you're finished you could apply the equation $\vec{\mathbf{F}}_{net} = q\vec{\mathbf{E}}_{net}$. In this case, $F_{net} = |q|E_{net}$ and $\theta_F = \theta_E$ if $q > 0$ whereas $\theta_F = \theta_E + 180°$ if $q < 0$.

Superposition Strategy

Either of the two kinds of problems described below are solved with a similar strategy.

- Given two or more pointlike charges, find the net electric field at a specified point.
- Given three or more pointlike charges, find the net electric force on one charge.

To find the net electric **field**, work with E's. To find the net electric **force**, work with F's.

1. Begin by drawing a sketch of the individual electric **fields** created by each charge at the specified point <u>**or**</u> the electric **forces** exerted on the specified charge.
   - For electric **field**, imagine a positive "test" charge at the specified point and draw arrows based on how the positive "test" charge would be pushed by each of the given charges. Label these $\vec{\mathbf{E}}_1$, $\vec{\mathbf{E}}_2$, etc.
   - For electric **force**, consider whether each of the charges attracts or repels the specified charge. Opposite charges attract, whereas like charges repel. Label these $\vec{\mathbf{F}}_1$, $\vec{\mathbf{F}}_2$, etc.
2. Use the distance formula to determine the distance from each charge to the point specified (for electric **field**) <u>**or**</u> to the specified charge (for electric **force**).
$$R_i = \sqrt{\Delta x^2 + \Delta y^2}$$
3. Apply trig to determine the direction of each vector counterclockwise from the $+x$-axis. First, get the reference angle using the rise (Δy) and run (Δx) with the formula below. Then use geometry to find the angle counterclockwise from the $+x$-axis.
$$\theta_{ref} = \tan^{-1}\left|\frac{\Delta y}{\Delta x}\right|$$
4. Either find the magnitude of the electric **field** created by each pointlike charge at the specified point <u>**or**</u> find the magnitude of the electric **force** exerted by each of the other charges on the specified charge.
$$E_1 = \frac{k|q_1|}{R_1^2} \quad , \quad E_2 = \frac{k|q_2|}{R_2^2} \quad , \quad E_3 = \frac{k|q_3|}{R_3^2} \quad \cdots$$
$$F_1 = \frac{k|q_1||q|}{R_1^2} \quad , \quad F_2 = \frac{k|q_2||q|}{R_2^2} \quad , \quad F_3 = \frac{k|q_3||q|}{R_3^2} \quad \cdots$$
5. Apply trig to determine the components of the given vectors.
$$E_{1x} = E_1 \cos\theta_1 \quad , \quad E_{1y} = E_1 \sin\theta_1 \quad , \quad E_{2x} = E_2 \cos\theta_2 \quad , \quad E_{2y} = E_2 \sin\theta_2 \quad \cdots$$
$$F_{1x} = F_1 \cos\theta_1 \quad , \quad F_{1y} = F_1 \sin\theta_1 \quad , \quad F_{2x} = F_2 \cos\theta_2 \quad , \quad F_{2y} = F_2 \sin\theta_2 \quad \cdots$$
6. Add the respective components together to find the components of the resultant.
$$E_x = E_{1x} + E_{2x} + \cdots + E_{Nx} \quad , \quad E_y = E_{1y} + E_{2y} + \cdots + E_{Ny}$$
$$F_x = F_{1x} + F_{2x} + \cdots + F_{Nx} \quad , \quad F_y = F_{1y} + F_{2y} + \cdots + F_{Ny}$$
7. Use the Pythagorean theorem to find the magnitude of the resultant vector. Use an inverse tangent to determine the direction of the resultant vector.
$$E = \sqrt{E_x^2 + E_y^2} \quad , \quad \theta_E = \tan^{-1}\left(\frac{E_y}{E_x}\right) \quad \textbf{or} \quad F = \sqrt{F_x^2 + F_y^2} \quad , \quad \theta_F = \tan^{-1}\left(\frac{F_y}{F_x}\right)$$

Example: A strand of monkey fur with a charge of 4.0 μC lies at the point ($\sqrt{3}$ m, 1.0 m). A strand of gorilla fur with a charge of −16.0 μC lies at the point ($-\sqrt{3}$ m, 1.0 m). What is the magnitude of the net electric field at the point ($\sqrt{3}$ m, −1.0 m)?

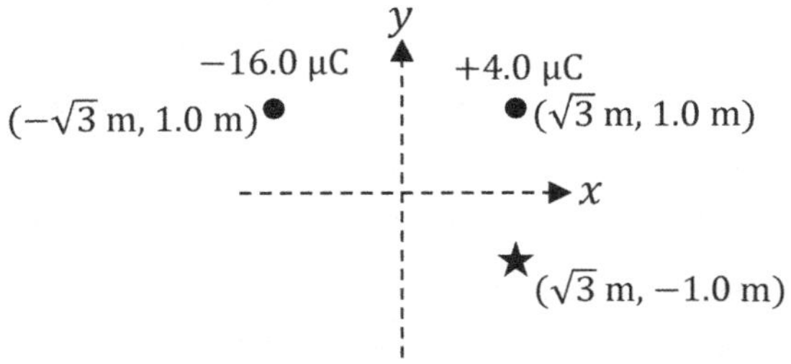

Begin by sketching the electric field vectors created by each of the charges. In order to do this, imagine a positive "test" charge at the point ($\sqrt{3}$ m, −1.0 m), marked by a star above. We call the monkey fur charge 1 ($q_1 = 4.0$ μC) and the gorilla fur charge 2 ($q_2 = -16.0$ μC).

- A positive "test" charge at ($\sqrt{3}$ m, −1.0 m) would be repelled by $q_1 = 4.0$ μC. Thus, we draw $\vec{\mathbf{E}}_1$ directly away from $q_1 = 4.0$ μC (straight down).
- A positive "test" charge at ($\sqrt{3}$ m, −1.0 m) would be attracted to $q_2 = -16.0$ μC. Thus, we draw $\vec{\mathbf{E}}_2$ towards $q_2 = -16.0$ μC (up and to the left).

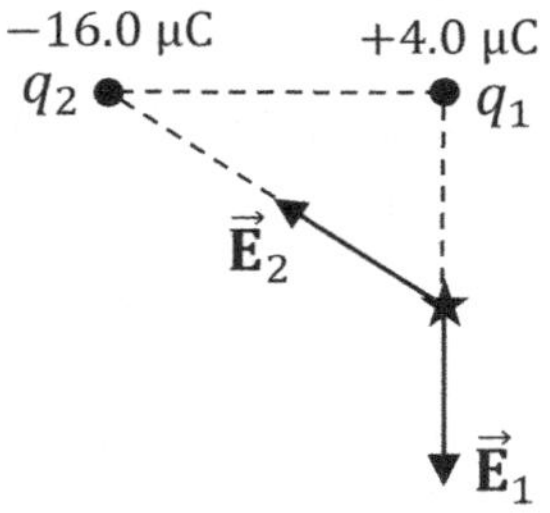

Find the distance between each charge (shown as a dot above) and the point (shown as a star) where we're trying to find the net electric field. Apply the distance formula.

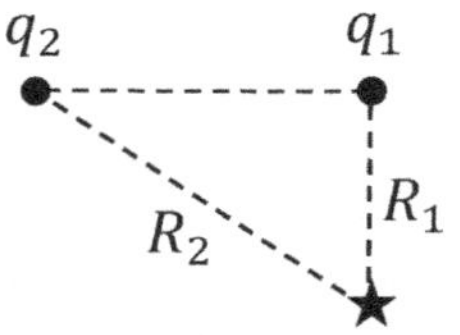

$$R_1 = \sqrt{\Delta x_1^2 + \Delta y_1^2} = \sqrt{\left(\sqrt{3} - \sqrt{3}\right)^2 + (-1 - 1)^2} = \sqrt{0^2 + (-2)^2} = \sqrt{4} = 2.0 \text{ m}$$

$$R_2 = \sqrt{\Delta x_2^2 + \Delta y_2^2} = \sqrt{\left[\sqrt{3} - \left(-\sqrt{3}\right)\right]^2 + (-1 - 1)^2} = \sqrt{\left(2\sqrt{3}\right)^2 + (-2)^2} = \sqrt{16} = 4.0 \text{ m}$$

Note that $\sqrt{12} = \sqrt{(4)(3)} = \sqrt{4}\sqrt{3} = 2\sqrt{3}$. In the diagram above, q_1 is 2.0 m above the star (that's why $R_1 = 2.0$ m), while R_2 is the hypotenuse of the right triangle. Since the top side is $2\sqrt{3}$ m wide and the right side is 2.0 m high, the Pythagorean theorem can be used to find the hypotenuse: $\sqrt{\left(2\sqrt{3}\right)^2 + 2^2} = \sqrt{12 + 4} = \sqrt{16} = 4.0$ m. That's why $R_2 = 4.0$ m.

Next, find the direction of $\vec{\mathbf{E}}_1$ and $\vec{\mathbf{E}}_2$. Since $\vec{\mathbf{E}}_1$ points straight down, $\theta_1 = 270°$. (Recall from trig that 0° points along $+x$, 90° points along $+y$, 180° points along $-x$, and 270° points along $-y$.) The angle θ_2 points in Quadrant II (up and to the left). To determine the precise angle, first find the reference angle from the right triangle. The reference angle is the smallest angle with the positive or negative x-axis.

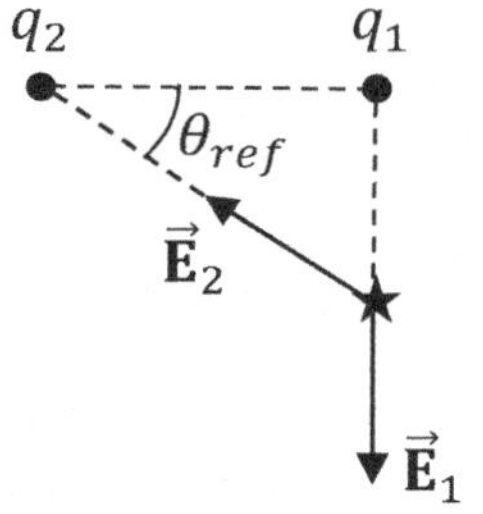

Apply trig to determine the reference angle, using the rise (Δy) and the run (Δx).

$$\theta_{ref} = \tan^{-1}\left|\frac{\Delta y}{\Delta x}\right| = \tan^{-1}\left(\frac{2}{2\sqrt{3}}\right) = \tan^{-1}\left(\frac{1}{\sqrt{3}}\right) = \tan^{-1}\left(\frac{\sqrt{3}}{3}\right) = 30°$$

Note that $\frac{1}{\sqrt{3}} = \frac{1}{\sqrt{3}}\frac{\sqrt{3}}{\sqrt{3}} = \frac{\sqrt{3}}{3}$ since $\sqrt{3}\sqrt{3} = 3$. Relate the reference angle to θ_2 using geometry. In Quadrant II, the angle counterclockwise from the $+x$-axis equals 180° minus the reference angle. (Recall that trig basics were reviewed in Volume 1 of this series.)

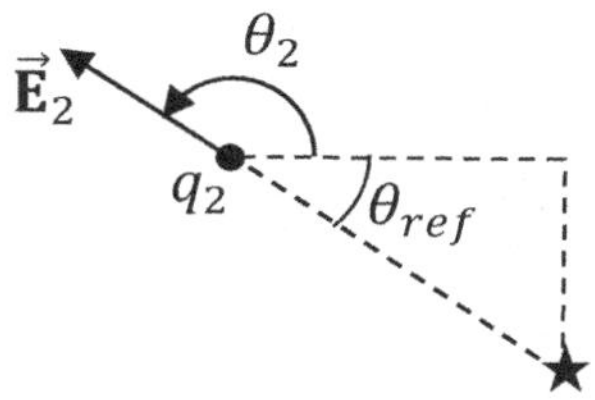

$$\theta_2 = 180° - \theta_{ref} = 180° - 30° = 150°$$

Find the magnitude of the electric field created by each pointlike charge at the specified point. First convert the charges from µC to C: $q_1 = 4.0 \times 10^{-6}$ C and $q_2 = -16.0 \times 10^{-6}$ C. Use the values $R_1 = 2.0$ m and $R_2 = 4.0$ m, which we found previously.

$$E_1 = \frac{k|q_1|}{R_1^2} = \frac{(9 \times 10^9)|4 \times 10^{-6}|}{(2)^2} = \frac{(9)(4)}{(2)^2} \times 10^9 10^{-6} = 9.0 \times 10^3 \text{ N/C}$$

$$E_2 = \frac{k|q_2|}{R_2^2} = \frac{(9 \times 10^9)|-16 \times 10^{-6}|}{(4)^2} = \frac{(9)(16)}{(4)^2} \times 10^9 10^{-6} = 9.0 \times 10^3 \text{ N/C}$$

Note that $10^9 10^{-6} = 10^{9-6} = 10^3$ according to the rule $x^m x^{-n} = x^{m-n}$. Also note that the minus sign from $q_2 = -16.0 \times 10^{-6}$ C vanished with the absolute values (because the magnitude of the electric field vector can't be negative). Following is a summary of what we know thus far.

- $q_1 = 4.0 \times 10^{-6}$ C, $R_1 = 2.0$ m, $\theta_1 = 270°$, and $E_1 = 9.0 \times 10^3$ N/C.
- $q_2 = -16.0 \times 10^{-6}$ C, $R_2 = 4.0$ m, $\theta_2 = 150°$, and $E_2 = 9.0 \times 10^3$ N/C.

We are now prepared to add the electric field vectors. We will use the magnitudes and directions ($E_1 = 9.0 \times 10^3$ N/C, $E_2 = 9.0 \times 10^3$ N/C, $\theta_1 = 270°$, and $\theta_2 = 150°$) of the two electric field vectors to determine the magnitude and direction (E and θ_E) of the resultant vector. The first step of vector addition is to find the components of the given vectors.

$$E_{1x} = E_1 \cos\theta_1 = (9 \times 10^3)\cos(270°) = (9 \times 10^3)(0) = 0$$

$$E_{1y} = E_1 \sin\theta_1 = (9 \times 10^3)\sin(270°) = (9 \times 10^3)(-1) = -9.0 \times 10^3 \text{ N/C}$$

$$E_{2x} = E_2 \cos\theta_2 = (9 \times 10^3)\cos(150°) = (9 \times 10^3)\left(-\frac{\sqrt{3}}{2}\right) = -\frac{9\sqrt{3}}{2} \times 10^3 \text{ N/C}$$

$$E_{2y} = E_2 \sin\theta_2 = (9 \times 10^3)\sin(150°) = (9 \times 10^3)\left(\frac{1}{2}\right) = \frac{9}{2} \times 10^3 \text{ N/C}$$

The second step of vector addition is to add the respective components together.

$$E_x = E_{1x} + E_{2x} = 0 + \left(-\frac{9\sqrt{3}}{2} \times 10^3\right) = -\frac{9\sqrt{3}}{2} \times 10^3 \text{ N/C}$$

$$E_y = E_{1y} + E_{2y} = -9 \times 10^3 + \frac{9}{2} \times 10^3 = -\frac{9}{2} \times 10^3 \text{ N/C}$$

The final step of vector addition is to apply the Pythagorean theorem and inverse tangent to determine the magnitude and direction of the resultant vector.

$$E = \sqrt{E_x^2 + E_y^2} = \sqrt{\left(-\frac{9\sqrt{3}}{2} \times 10^3\right)^2 + \left(-\frac{9}{2} \times 10^3\right)^2} = \left(\frac{9}{2} \times 10^3\right)\sqrt{\left(-\sqrt{3}\right)^2 + (-1)^2}$$

We factored out $\left(\frac{9}{2} \times 10^3\right)$ to make the arithmetic simpler. The minus signs disappear since they are squared. For example, $(-1)^2 = +1$.

$$E = \left(\frac{9}{2} \times 10^3\right)\sqrt{3 + 1} = \left(\frac{9}{2} \times 10^3\right)\sqrt{4} = \left(\frac{9}{2} \times 10^3\right)(2) = 9.0 \times 10^3 \text{ N/C}$$

$$\theta_E = \tan^{-1}\left(\frac{E_y}{E_x}\right) = \tan^{-1}\left(\frac{-\frac{9}{2} \times 10^3}{-\frac{9\sqrt{3}}{2} \times 10^3}\right) = \tan^{-1}\left(\frac{1}{\sqrt{3}}\right) = \tan^{-1}\left(\frac{\sqrt{3}}{3}\right)$$

Note that $-\frac{9}{2} \times 10^3$ cancels out (divide both the numerator and denominator by this). Also note that $\frac{1}{\sqrt{3}} = \frac{1}{\sqrt{3}}\frac{\sqrt{3}}{\sqrt{3}} = \frac{\sqrt{3}}{3}$ since $\sqrt{3}\sqrt{3} = 3$. The reference angle for the answer is 30° since $\tan 30° = \frac{\sqrt{3}}{3}$. However, this isn't the answer because the answer doesn't lie in Quadrant I. Since $E_x < 0$ and $E_y < 0$ (find these values above), the answer lies in Quadrant III. Apply trig to determine θ_E from the reference angle: In Quadrant III, add 180° to the reference angle.

$$\theta_E = 180° + \theta_{ref} = 180° + 30° = 210°$$

The final answer is that the magnitude of the net electric field at the specified point equals $E = 9.0 \times 10^3$ N/C and its direction is $\theta_E = 210°$.

Example: A monkey places three charges on the corners of a right triangle, as illustrated below. Determine the magnitude and direction of the net electric force exerted on q_3.

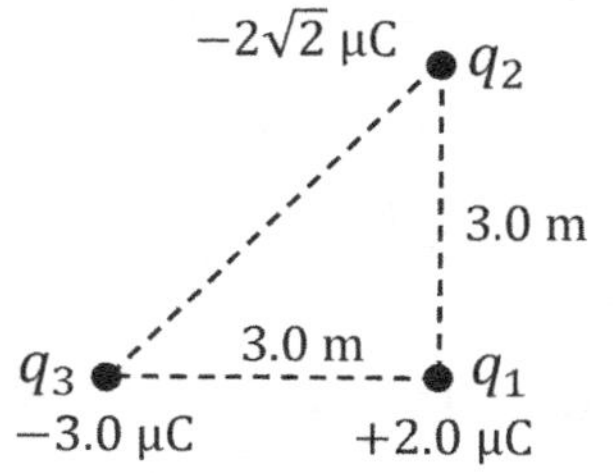

Begin by sketching the two electric forces exerted on q_3.

- Charge q_3 is attracted to q_1 because they have opposite signs. Thus, we draw $\vec{\mathbf{F}}_1$ towards q_1 (to the right).
- Charge q_3 is repelled by q_2 because they are both negative. Thus, we draw $\vec{\mathbf{F}}_2$ directly away from q_2 (down and to the left).

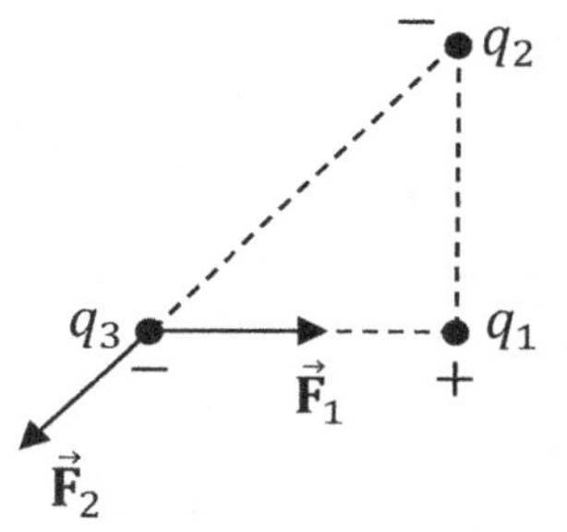

Find the distance between each charge. Note that $R_1 = 3.0$ m (the width of the triangle). Apply the distance formula to find R_2.

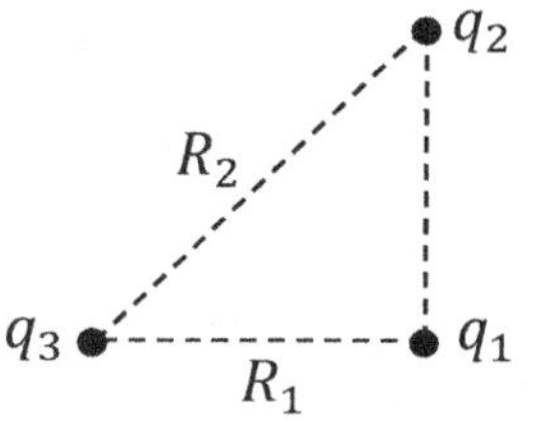

$$R_2 = \sqrt{\Delta x_2^2 + \Delta y_2^2} = \sqrt{3^2 + 3^2} = \sqrt{9+9} = \sqrt{18} = \sqrt{(9)(2)} = \sqrt{9}\sqrt{2} = 3\sqrt{2} \text{ m}$$

Next, find the direction of $\vec{\mathbf{F}}_1$ and $\vec{\mathbf{F}}_2$. Since $\vec{\mathbf{F}}_1$ points to the right, $\theta_1 = 0°$. (Recall from trig that 0° points along $+x$.) The angle θ_2 points in Quadrant III (down and to the left). To determine the precise angle, first find the reference angle from the right triangle. The reference angle is the smallest angle with the positive or negative x-axis. Apply trig to determine the reference angle, using the rise (Δy) and the run (Δx).

$$\theta_{ref} = \tan^{-1}\left|\frac{\Delta y}{\Delta x}\right| = \tan^{-1}\left(\frac{3}{3}\right) = \tan^{-1}(1) = 45°$$

Relate the reference angle to θ_2 using geometry. In Quadrant III, the angle counterclockwise from the $+x$-axis equals 180° plus the reference angle.

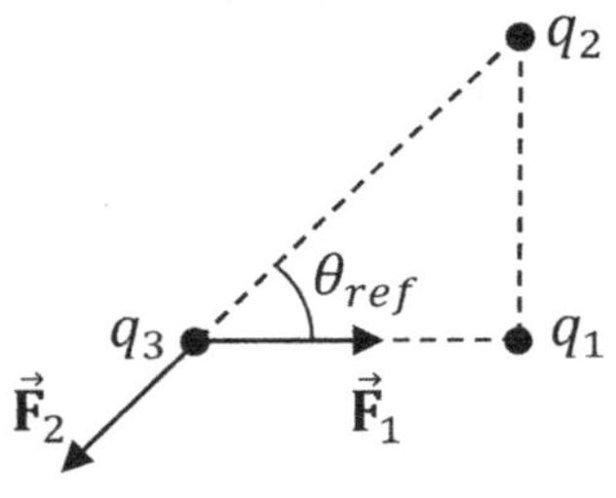

$$\theta_2 = 180° + \theta_{ref} = 180° + 45° = 225°$$

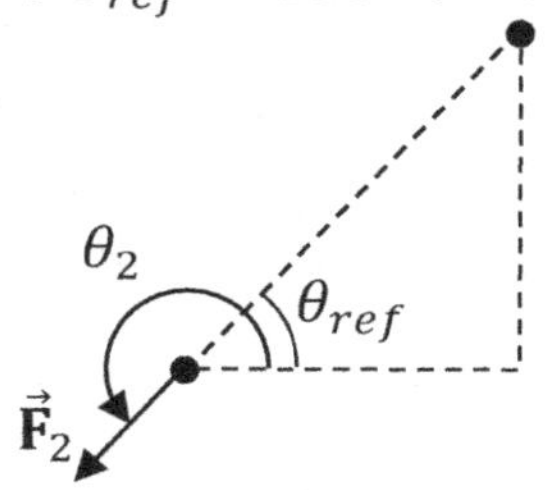

Find the magnitude of the electric force exerted on q_3 by each of the other charges. First convert the charges to SI units: $q_1 = 2.0 \times 10^{-6}$ C, $q_2 = -2\sqrt{2} \times 10^{-6}$ C, and $q_3 = -3.0 \times 10^{-6}$ C. Use the values $R_1 = 3.0$ m and $R_2 = 3\sqrt{2}$ m, which we found previously. Note that q_3 appears in both formulas below because we are finding the net force exerted on q_3.

$$F_1 = \frac{k|q_1||q_3|}{R_1^2} = \frac{(9 \times 10^9)|2 \times 10^{-6}||-3 \times 10^{-6}|}{(3)^2} = 6.0 \times 10^{-3} \text{ N}$$

$$F_2 = \frac{k|q_2||q_3|}{R_2^2} = \frac{(9 \times 10^9)\left|-2\sqrt{2} \times 10^{-6}\right||-3 \times 10^{-6}|}{\left(3\sqrt{2}\right)^2} = 3\sqrt{2} \times 10^{-3} \text{ N}$$

Note that $10^9 10^{-6} 10^{-6} = 10^{9-6-6} = 10^{-3}$ according to the rule $x^m x^n = x^{m+n}$, and that $\left(3\sqrt{2}\right)^2 = (3)^2\left(\sqrt{2}\right)^2 = (9)(2) = 18$. Also note that the minus signs vanished with the absolute values (because the magnitude of the force can't be negative). Following is a summary of what we know thus far.

- $q_1 = 2.0 \times 10^{-6}$ C, $R_1 = 3.0$ m, $\theta_1 = 0°$, and $F_1 = 6.0 \times 10^{-3}$ N.
- $q_2 = -2\sqrt{2} \times 10^{-6}$ C, $R_2 = 3\sqrt{2}$ m, $\theta_2 = 225°$, and $F_2 = 3\sqrt{2} \times 10^{-3}$ N.
- $q_3 = -3.0 \times 10^{-6}$ C.

We are now prepared to add the force vectors. We will use the magnitudes and directions ($F_1 = 6.0 \times 10^{-3}$ N, $F_2 = 3\sqrt{2} \times 10^{-3}$ N, $\theta_1 = 0°$, and $\theta_2 = 225°$) of the two force vectors to determine the magnitude and direction (F and θ_F) of the resultant vector. The first step of vector addition is to find the components of the given vectors.

$$F_{1x} = F_1 \cos\theta_1 = (6 \times 10^{-3})\cos(0°) = (6 \times 10^{-3})(1) = 6.0 \times 10^{-3} \text{ N}$$

$$F_{1y} = F_1 \sin\theta_1 = (6 \times 10^{-3})\sin(0°) = (6 \times 10^{-3})(0) = 0$$

$$F_{2x} = F_2 \cos\theta_2 = \left(3\sqrt{2} \times 10^{-3}\right)\cos(225°) = \left(3\sqrt{2} \times 10^{-3}\right)\left(-\frac{\sqrt{2}}{2}\right) = -3.0 \times 10^{-3} \text{ N}$$

$$F_{2y} = F_2 \sin\theta_2 = \left(3\sqrt{2} \times 10^{-3}\right)\sin(225°) = \left(3\sqrt{2} \times 10^{-3}\right)\left(-\frac{\sqrt{2}}{2}\right) = -3.0 \times 10^{-3} \text{ N}$$

Note that $(3\sqrt{2})\left(-\frac{\sqrt{2}}{2}\right) = -\frac{(3)(\sqrt{2})(\sqrt{2})}{2} = -\frac{(3)(2)}{2} = -\frac{6}{2} = -3$ since $\sqrt{2}\sqrt{2} = 2$. The second step of vector addition is to add the respective components together.

$$F_x = F_{1x} + F_{2x} = 6 \times 10^{-3} + (-3 \times 10^{-3}) = 3.0 \times 10^{-3} \text{ N}$$

$$F_y = F_{1y} + F_{2y} = 0 + (-3 \times 10^{-3}) = -3.0 \times 10^{-3} \text{ N}$$

The final step of vector addition is to apply the Pythagorean theorem and inverse tangent to determine the magnitude and direction of the resultant vector.

$$F = \sqrt{F_x^2 + F_y^2} = \sqrt{(3 \times 10^{-3})^2 + (-3 \times 10^{-3})^2} = (3 \times 10^{-3})\sqrt{(1)^2 + (-1)^2}$$

We factored out (3×10^{-3}) to make the arithmetic simpler. The minus sign disappears since it is squared: $(-1)^2 = +1$.

$$F = (3 \times 10^{-3})\sqrt{1+1} = (3 \times 10^{-3})\sqrt{2} = 3\sqrt{2} \times 10^{-3} \text{ N}$$

$$\theta_F = \tan^{-1}\left(\frac{F_y}{F_x}\right) = \tan^{-1}\left(\frac{-3 \times 10^{-3}}{3 \times 10^{-3}}\right) = \tan^{-1}(-1)$$

The reference angle for the answer is 45° since $\tan 45° = 1$. However, this isn't the answer because the answer doesn't lie in Quadrant I. Since $F_x > 0$ and $F_y < 0$ (find these values above), the answer lies in Quadrant IV. Apply trig to determine θ_F from the reference angle: In Quadrant IV, subtract the reference angle from 360°.

$$\theta_F = 360° - \theta_{ref} = 360° - 45° = 315°$$

The final answer is that the magnitude of the net electric force exerted on q_3 equals $F = 3\sqrt{2} \times 10^{-3}$ N and its direction is $\theta_F = 315°$.

9. A monkey-shaped earring with a charge of 6.0 μC lies at the point (0, 2.0 m). A banana-shaped earring with a charge of −6.0 μC lies at the point (0, −2.0 m). Your goal is to find the magnitude of the net electric field at the point (2.0 m, 0).

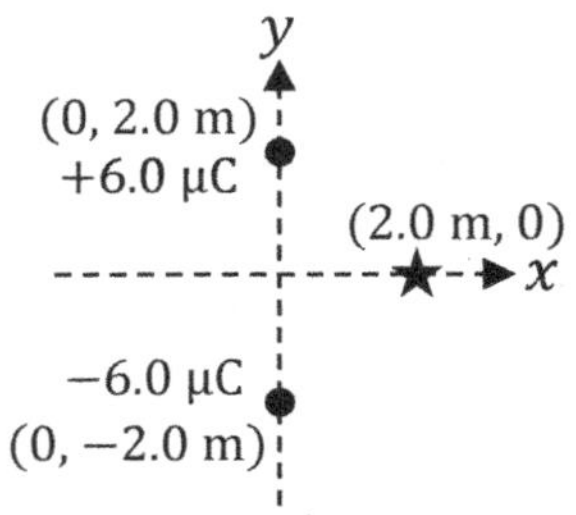

(A) Sketch the electric field vectors created by the two charges at the point (2.0 m, 0). Label $\vec{\mathbf{E}}_1$, $\vec{\mathbf{E}}_2$, θ_1, and θ_2. Label the angles counterclockwise from the $+x$-axis.

(B) Apply the distance formula to determine R_1 and R_2.

(C) Apply trig to find the reference angles for each electric field vector.

(D) Use your answers to part (C) and part (A) to determine θ_1 and θ_2 counterclockwise from the $+x$-axis. Be sure that their Quadrants match your sketch from part (A).

(E) Determine the magnitudes (E_1 and E_2) of the two electric field vectors.

Note: You're not finished yet. This problem is **continued** on the next page.

(F) Determine the components (E_{1x}, E_{1y}, E_{2x}, and E_{2y}) of the electric field vectors.

(G) Determine the components (E_x and E_y) of the resultant vector.

(H) Determine the magnitude (E) of the net electric field at the point (2.0 m, 0).

(I) Determine the direction (θ_E) of the net electric field at the point (2.0 m, 0).

Want help or intermediate answers? Check the hints section at the back of the book.

Answers to (H) and (I): $6750\sqrt{2}$ N/C, 270°

10. A monkey places three charges on the vertices of an equilateral triangle, as illustrated below. Your goal is to determine the magnitude and direction of the net electric force exerted on q_3.

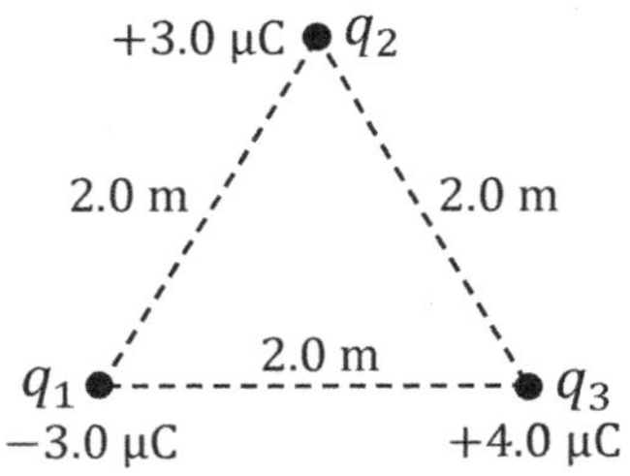

(A) Sketch the forces that q_1 and q_2 exert on q_3. Label $\vec{\mathbf{F}}_1$, $\vec{\mathbf{F}}_2$, θ_1, and θ_2. Label the angles counterclockwise from the $+x$-axis. (Choose $+x$ to point right and $+y$ to point up.)

(B) Determine the distances R_1 and R_2.

(C) Find the reference angles for each force vector.

(D) Use your answers to part (C) and part (A) to determine θ_1 and θ_2 counterclockwise from the $+x$-axis. Be sure that their Quadrants match your sketch from part (A).

(E) Determine the magnitudes (F_1 and F_2) of the two force vectors.

Note: You're not finished yet. This problem is **continued** on the next page.

(F) Determine the components (F_{1x}, F_{1y}, F_{2x}, and F_{2y}) of the force vectors.

(G) Determine the components (F_x and F_y) of the resultant vector.

(H) Determine the magnitude (F) of the net electric force exerted on q_3.

(I) Determine the direction (θ_F) of the net electric force exerted on q_3.

Want help or intermediate answers? Check the hints section at the back of the book.

Answers to (H) and (I): 27 mN, 240°

Strategy to Find the Point Where the Net Electric Field Equals Zero

If a problem gives you two pointlike charges and asks you where the net electric field is zero, follow these steps. This strategy is illustrated in the example that follows.

1. Define three distinct regions (as shown in the following example):
   - One region lies between the two pointlike charges.
   - Two of the regions are outside of the two pointlike charges.
2. Draw a "test" point in each of the three regions defined in the previous step. Sketch the electric fields created by each pointlike charge at each "test" point (so there will be three pairs of electric fields in your diagram). To do this, imagine placing a positive "test" charge at each "test" point. Ask yourself which way each charge would push (or pull) on the positive "test" charge.
3. In which region could the two electric fields have opposite direction and also have equal magnitude?
   - If the two charges have the same sign, the answer is the region between the two charges.
   - If the two charges have opposite sign, the answer is the region outside of the two charges, on the side with whichever charge has the smallest value of $|q|$.
4. Set the magnitudes of the two electric fields at the "test" point equal to one another.
$$E_1 = E_2$$
$$\frac{k|q_1|}{R_1^2} = \frac{k|q_2|}{R_2^2}$$
Divide both sides by k. Cross-multiply. Plug in the values of the charges.
5. You have one equation with two unknowns. The unknowns are R_1 and R_2. Get the second equation from the picture that you drew in Steps 1-3. Label R_1 and R_2 (the distances from the "test" point to each pointlike charge) for the correct region. Call d the distance between the two pointlike charges.
   - If the two charges have the same sign, the equation will be:
$$R_1 + R_2 = d$$
   - If the two charges have opposite sign, the equation will be one of these, depending upon which charge has the greater value of $|q|$:
$$R_1 = R_2 + d$$
$$R_2 = R_1 + d$$
6. Isolate R_1 or R_2 in the previous equation (if one isn't already isolated). Substitute this expression into the simplified equation from Step 4. Plug in the value of d.
7. Squareroot both sides of the equation. Carry out the algebra to solve for R_1 and R_2. Use these answers to determine the location where the net electric field equals zero.

Example: A monkey places two small spheres a distance of 5.0 m apart, as shown below. The left sphere has a charge of $-4.0\ \mu C$, while the right sphere has a charge of $-9.0\ \mu C$. Find the point where the net electric field equals zero.

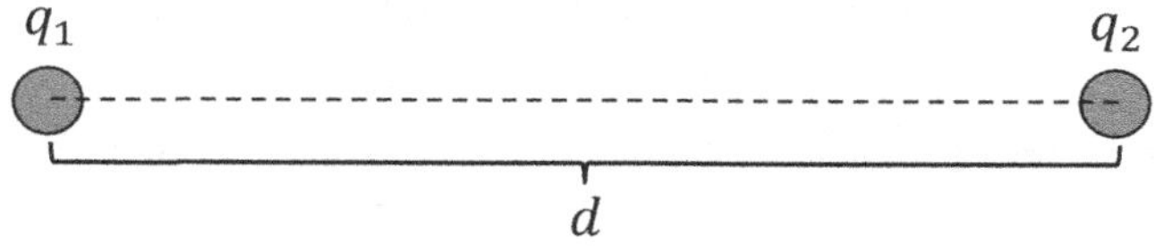

Consider each of the three regions shown below. Imagine placing a positive "test" charge where you see the three stars (★) in the diagram below. Since q_1 and q_2 are both negative, the positive "test" charge would be attracted to both q_1 and q_2. Draw the electric fields towards q_1 and q_2 in each region.

I. Region I is left of q_1. $\vec{\mathbf{E}}_1$ and $\vec{\mathbf{E}}_2$ both point to the right. They won't cancel here.

II. Region II is between q_1 and q_2. $\vec{\mathbf{E}}_1$ points left, while $\vec{\mathbf{E}}_2$ points right. They can cancel out in Region II.

III. Region III is right of q_2. $\vec{\mathbf{E}}_1$ and $\vec{\mathbf{E}}_2$ both point left. They won't cancel here.

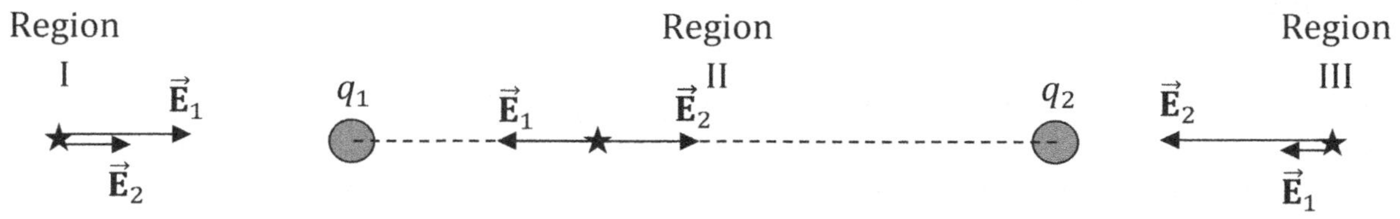

We know that the answer lies in Region II, since that is the only place where the electric fields could cancel out. In order to find out exactly where in Region II the net electric field equals zero, set the magnitudes of the electric fields equal to one another.

$$E_1 = E_2$$

$$\frac{k|q_1|}{R_1^2} = \frac{k|q_2|}{R_2^2}$$

Divide both sides by k (it will cancel out) and **cross-multiply**.

$$|q_1|R_2^2 = |q_2|R_1^2$$

Plug in the values of the charges.

$$|-4.0 \times 10^{-6}|R_2^2 = |-9.0 \times 10^{-6}|R_1^2$$

Divide both sides by 10^{-6} (it will cancel out). The minus signs disappear when you apply the absolute values.

$$4R_2^2 = 9R_1^2$$

Since we have two unknowns (R_1 and R_2), we need a second equation. We can get the second equation by studying the diagram below.

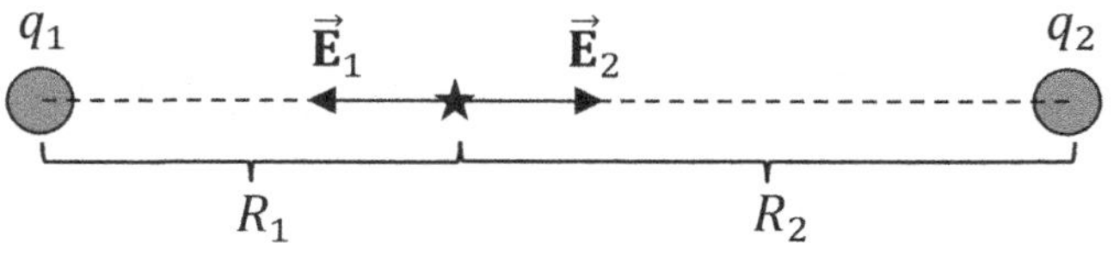

$$R_1 + R_2 = d$$

Recall that d is the distance between the two charges. Isolate R_2 in the previous equation.

$$R_2 = d - R_1$$

Substitute the previous equation into the equation $4R_2^2 = 9R_1^2$, which we found earlier.

$$4(d - R_1)^2 = 9R_1^2$$

Squareroot both sides of the equation. (This trick lets you avoid the quadratic equation.)

$$\sqrt{4(d - R_1)^2} = \sqrt{9R_1^2}$$

Recall the rule from algebra that $\sqrt{xy} = \sqrt{x}\sqrt{y}$.

$$\sqrt{4}\sqrt{(d - R_1)^2} = \sqrt{9}\sqrt{R_1^2}$$

$$2\sqrt{(d - R_1)^2} = 3\sqrt{R_1^2}$$

Also recall from algebra that $\sqrt{x^2} = x$.

$$2(d - R_1) = 3R_1$$

Distribute the 2.

$$2d - 2R_1 = 3R_1$$

Add $2R_1$ to both sides of the equation.

$$2d = 5R_1$$

Divide both sides of the equation by 5. Recall that the distance between the charges was given in the problem: $d = 5.0$ m.

$$R_1 = \frac{2d}{5} = \frac{(2)(5)}{5} = \frac{10}{5} = 2.0 \text{ m}$$

The net electric field is zero in Region II, a distance of $R_1 = 2.0$ m from the left charge (and therefore a distance of $R_2 = 3.0$ m from the right charge, since $R_1 + R_2 = d = 5.0$ m).

11. A monkey places two small spheres a distance of 9.0 m apart, as shown below. The left sphere has a charge of 25 μC, while the right sphere has a charge of 16 μC. Find the point where the net electric field equals zero.

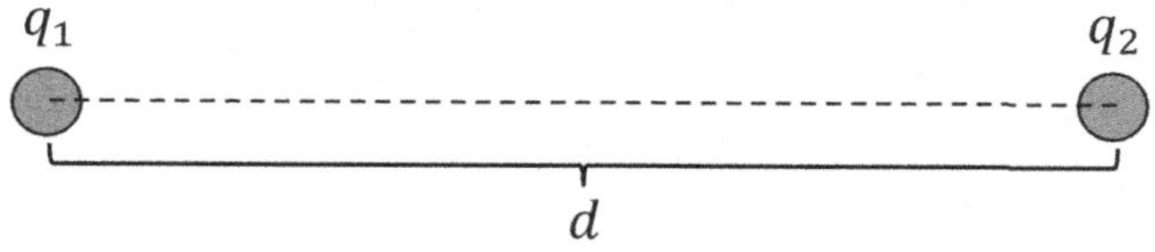

Want help? Check the hints section at the back of the book.

Answer: between the two spheres, 5.0 m from the left sphere (4.0 m from the right sphere)

12. A monkey places two small spheres a distance of 4.0 m apart, as shown below. The left sphere has a charge of 2.0 μC, while the right sphere has a charge of −8.0 μC. Find the point where the net electric field equals zero.

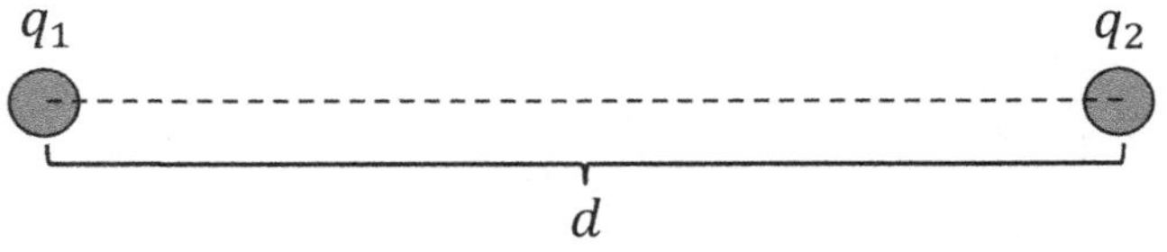

Want help? Check the hints section at the back of the book.

Answer: left of the left sphere, a distance of 4.0 m from the left sphere (8.0 m from the right sphere)

4 ELECTRIC FIELD MAPPING

Relevant Terminology

Lines of force – a map of electric field lines showing how a positive "test" charge at any point on the map would be pushed, depending on the location of the "test" charge.

Electric field lines – the same as lines of force. Since $\vec{\mathbf{F}}_e = q\vec{\mathbf{E}}$, if a positive "test" charge q is placed at a particular point in an electric field map, the electric field lines show which way the "test" charge would be pushed. In this way, electric field lines serve as lines of force.

Equipotential surface – a set of points (lying on a surface) that have the same electric potential. The change in **electric potential** is proportional to the **work done** moving a charge between two points. **No work is done moving a charge along an equipotential surface**, since electric potential is the same throughout the surface. We will explore electric potential more fully in Chapter 9.

Essential Concepts

Given a set of charged objects, an electric field map shows electric field lines (also called lines of force) and equipotential surfaces.

- The **lines of force** show which way a positive "test" charge would be pushed at any point on the diagram. It would be pushed along a tangent line to the line of force.
- The **equipotential surfaces** show surfaces where electric potential is constant.

The lines of force and equipotential surfaces really exist everywhere in space, but we draw a fixed number of lines of force and equipotential surfaces in an electric field map to help visualize the electrostatics. When drawing an electric field map, it is important to keep the following concepts in mind:

- Lines of force travel from positive charges to negative charges, or in the direction of decreasing electric potential (that is, from higher potential to lower potential).
- Lines of force are **perpendicular** to equipotential surfaces.
- Two lines of force **can't intersect**. Otherwise, a "test" charge at the point of intersection would be pushed in two different directions at once.
- Two different equipotential surfaces **can't intersect**. Otherwise, the point of intersection would be multi-valued.
- Lines of force are **perpendicular** to charged conductors.

When interpreting an electric field map, keep the following concepts in mind:

- The magnitude of the electric field is stronger where the electric field lines are more dense and weaker where the electric field lines are less dense.
- Electric field lines are very dense around charged objects.
- The value of electric potential is constant along an equipotential surface.

Electric Field Mapping Equations

In first-year physics courses, we mostly analyze electric field maps conceptually, using equations mostly as a guide.

For example, consider the equation for the electric field created by a single pointlike charge.

$$E = \frac{k|q|}{R^2}$$

The above equation shows that electric field is stronger near a charged object (where R is smaller), and weaker further away from the charged object (where R is larger). It also shows that a larger value of $|q|$ produces a stronger electric field. The effect of R is greater than the effect of $|q|$, since R is squared.

If there are two or more charged objects, we find the net electric field at a particular point in the diagram through superposition (vector addition), as discussed in Chapter 3. In an electric field map, we do this conceptually by joining the vectors tip-to-tail: The tip of one vector is joined to the tail of the other vector. The net electric field, $\vec{\mathbf{E}}_{net}$, is the **resultant** of the two vectors joined together tip-to-tail.

$$\vec{\mathbf{E}}_{net} = \vec{\mathbf{E}}_1 + \vec{\mathbf{E}}_2$$

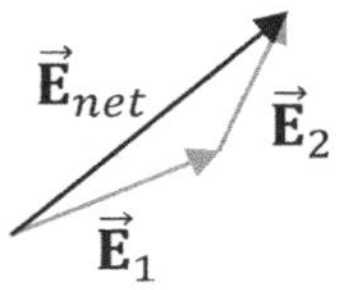

The equation $\vec{\mathbf{F}}_e = q\vec{\mathbf{E}}$ is helpful for interpreting an electric field map. It shows that if a positive "test" charge q is placed at a particular point on the map, the "test" charge would be pushed along a tangent line to the line of force (or electric field line).

One equation that we use computationally, rather than conceptually, when interpreting an electric field map is the following equation for approximating the electric field at a specified point on an electric field map.

$$E \approx \left|\frac{\Delta V}{\Delta R}\right|$$

The way to use this equation is to draw a straight line connecting two equipotential surfaces, such that the straight line passes through the desired point and is, on average, roughly perpendicular to each equipotential. Then measure the length of this line, ΔR. To find the value of ΔV, simply subtract the values of electric potential for each equipotential.

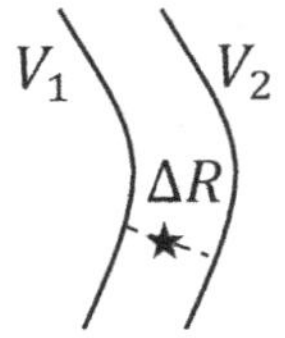

Symbols and SI Units

Symbol	Name	SI Units
E	magnitude of electric field	N/C or V/m
q	charge	C
k	Coulomb's constant	$\frac{\text{N}\cdot\text{m}^2}{\text{C}^2}$ or $\frac{\text{kg}\cdot\text{m}^3}{\text{C}^2\cdot\text{s}^2}$
R	distance from the charge	m
F_e	magnitude of electric force	N
V	electric potential	V
ΔV	potential difference	V
ΔR	distance between two points	m

Strategy for Conceptually Drawing Net Electric Field

A prerequisite to drawing and interpreting an electric field map is to be able to apply the principle of superposition visually. If a problem gives you two or more pointlike charges and asks you to sketch the electric field vector at a particular point, follow this strategy.

1. Imagine placing a positive "test" charge at the specified point.
2. Draw an electric field vector at the location of the "test" charge for each of the pointlike charges given in the problem.
   - For a positive charge, draw an arrow directly away from the positive charge, since a positive "test" charge would be repelled by the positive charge.
   - For a negative charge, draw an arrow towards the negative charge, since a positive "test" charge would be attracted to the negative charge.
3. The lengths of the arrows that you draw should be proportional to the magnitudes of the electric field vectors. The closer the "test" charge is to a charged object, the greater the magnitude of the electric field. If any of the charged objects have more charge than other charged objects, this also increases the magnitude of the electric field.
4. Apply the principle of superposition to draw the net electric field vector. Join the arrows from Step 2 together **tip-to-tail** in order to form the resultant vector, as illustrated in the examples that follow (pay special attention to the last example).

Example: Sketch the electric field at points A, B, and C for the isolated positive sphere shown below.

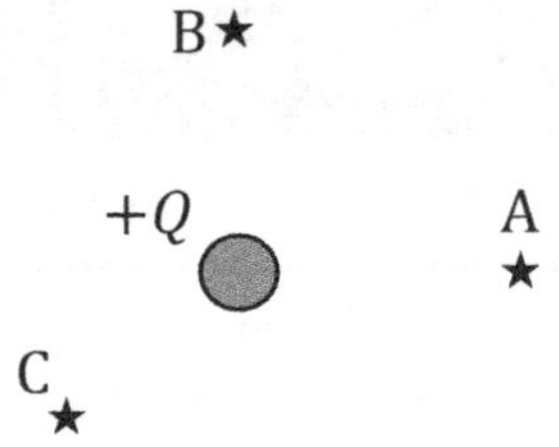

If a positive "test" charge were placed at points A, B, or C, it would be repelled by the positive sphere. Draw the electric field directly away from the positive sphere.

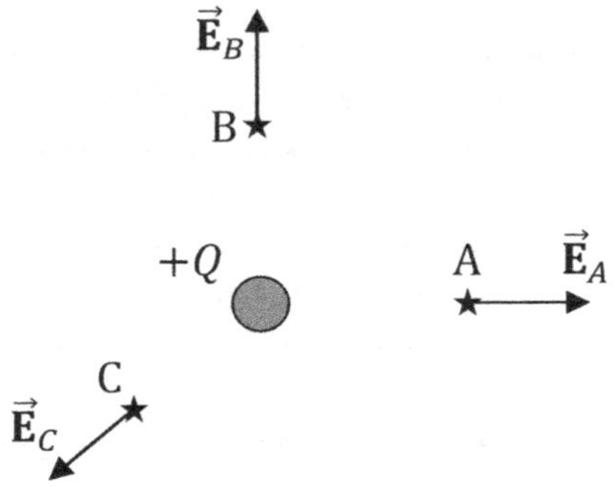

Example: Sketch the electric field at points D, E, and F for the isolated negative sphere shown below.

If a positive "test" charge were placed at points D, E, or F, it would be attracted to the negative sphere. Draw the electric field towards the negative sphere. Furthermore, the electric field gets weaker as a function of $\frac{1}{R^2}$ as the points get further from the charged sphere. Therefore, the arrow should be longest at point D and shortest at point F.

Example: Sketch the electric field at point G for the pair of equal but oppositely charged spheres shown below. (This configuration is called an **electric dipole**.)

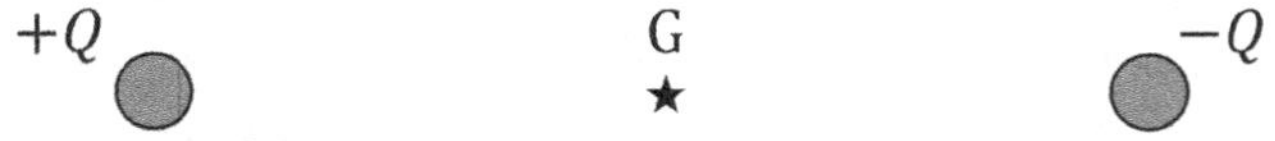

If a positive "test" charge were placed at point G, it would be repelled by the positive sphere and attracted to negative sphere. First draw two separate electric fields (one for each sphere) and then draw the resultant of these two vectors for the net electric field.

The electric fields **<u>don't</u>** cancel: They both point right. Contrast this with the next example.

Example: Sketch the electric field at point H for the pair of positive spheres shown below.

If a positive "test" charge were placed at point H, it would be repelled by each positive sphere. First draw two separate electric fields (one for each sphere) and then draw the resultant of these two vectors for the net electric field.

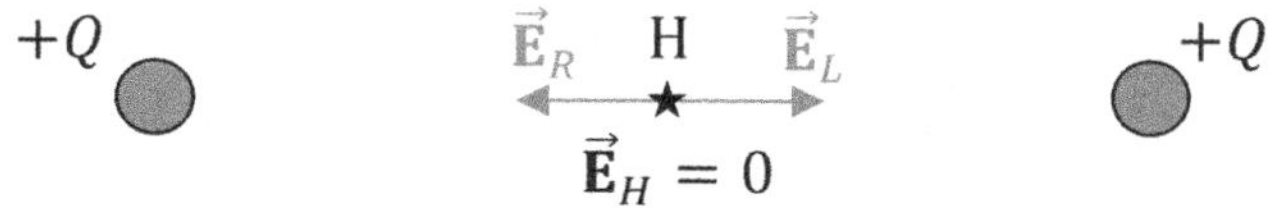

The two electric fields cancel out. The net electric field at point H is zero: $\vec{\mathbf{E}}_H = 0$.
It is instructive to compare this example with the previous example.
Example: Sketch the electric field at points I and J for the pair of equal but oppositely charged spheres shown below.

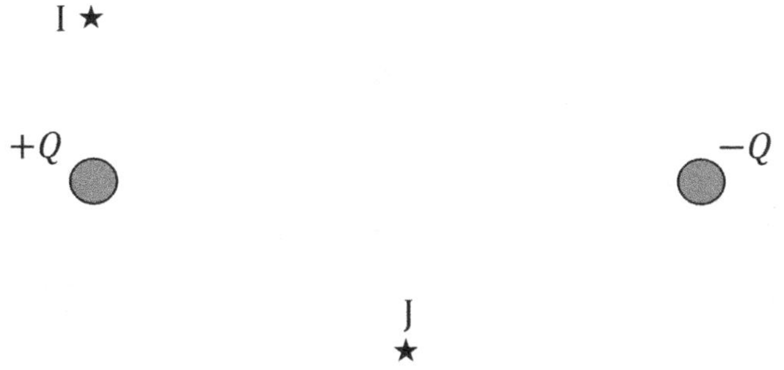

If a positive "test" charge were placed at point I or J, it would be repelled by the positive sphere and attracted to the negative sphere. First draw two separate electric fields (one for each sphere) and then draw the resultant of these two vectors for the net electric field. Move one of the arrows (we moved $\vec{\mathbf{E}}_R$) to join the two vectors ($\vec{\mathbf{E}}_L$ and $\vec{\mathbf{E}}_R$) **tip-to-tail**. The resultant vector, $\vec{\mathbf{E}}_I$, which is the net electric field at the specified point, begins at the tail of $\vec{\mathbf{E}}_L$ and ends at the tip of $\vec{\mathbf{E}}_R$, as shown below. Note how the vertical components cancel at J.

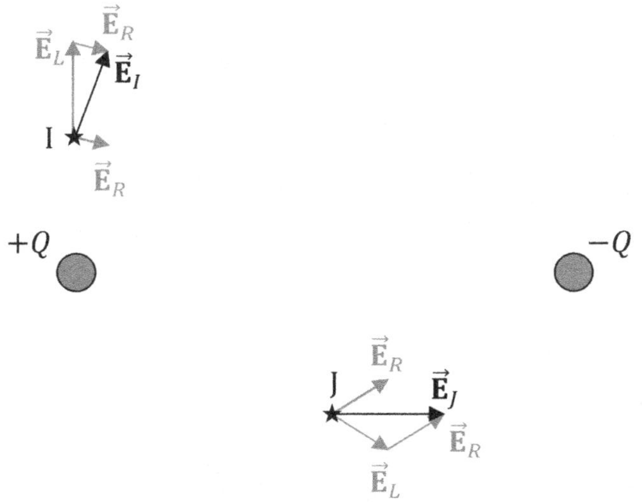

13. Sketch the electric field at points A, B, and C for the isolated positive sphere shown below.

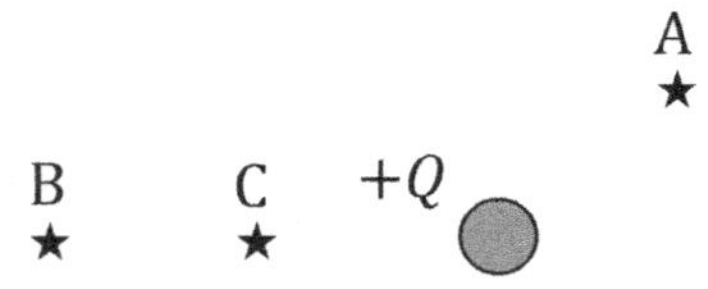

14. Sketch the electric field at points D, E, and F for the isolated negative sphere shown below.

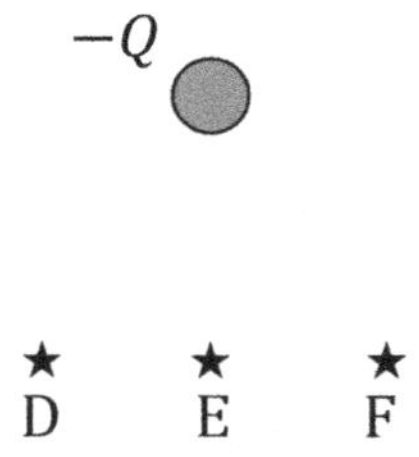

15. Sketch the electric field at points G, H, I, J, K, and M for the pair of equal but oppositely charged spheres shown below. (We skipped L in order to avoid possible confusion in the answer key, since we use $\vec{\mathbf{E}}_L$ for the electric field created by the left charge.)

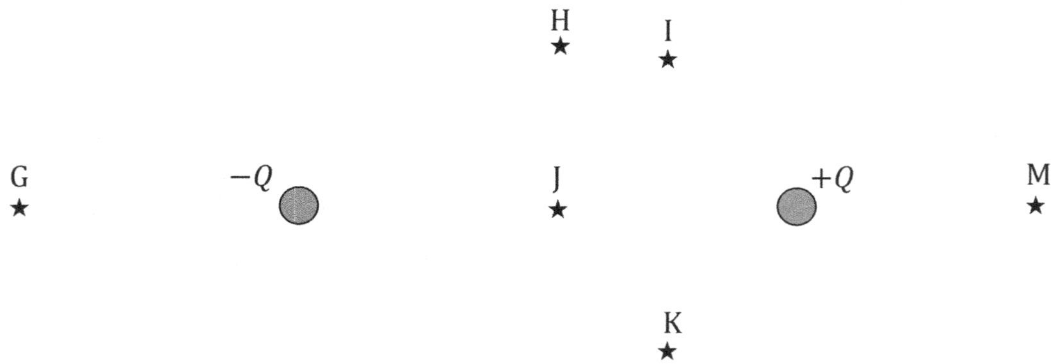

Want help? The problems from Chapter 4 are fully solved in the back of the book.

16. Sketch the electric field at points N, O, P, S, T, and U for the pair of negative spheres shown below. (We skipped Q in order to avoid possible confusion with the charge. We also skipped R in order to avoid possible confusion in the answer key, since we use $\vec{\mathbf{E}}_R$ for the electric field created by the right charge.)

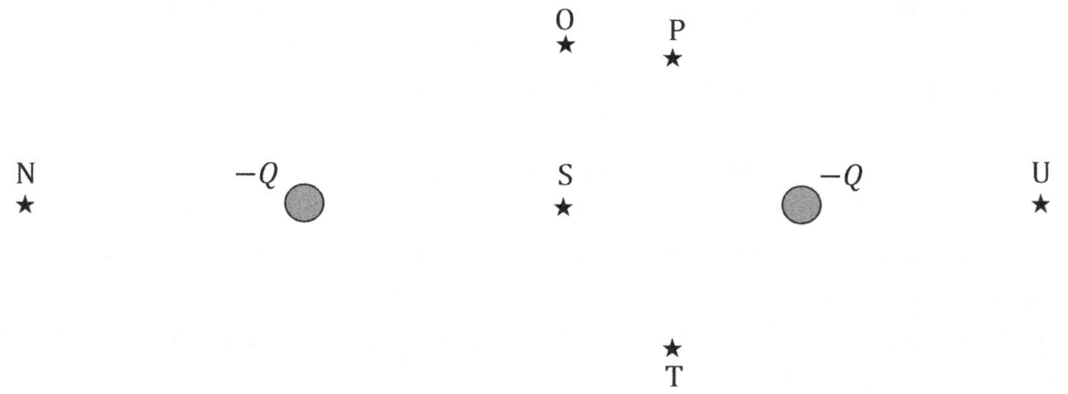

17. Sketch the electric field at point V for the three charged spheres (two are negative, while the bottom left is positive) shown below, which form an equilateral triangle.

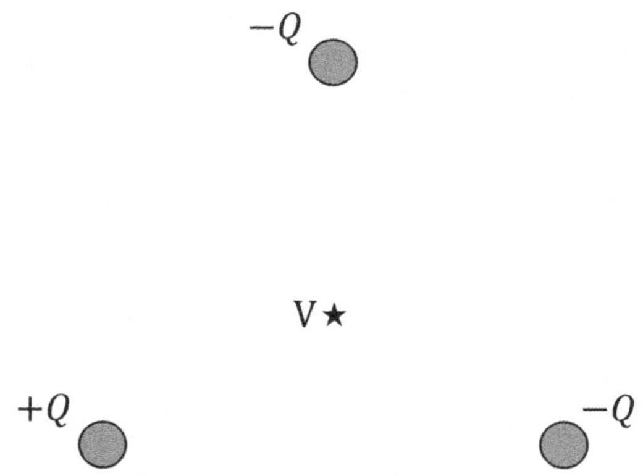

Want help? The problems from Chapter 4 are fully solved in the back of the book.

Strategy for Drawing an Electric Field Map

To draw an electric field map for a system of charged objects, follow these steps:

1. The lines of force (or electric field lines) should travel **perpendicular** to the surface of a charged object (called an **electrode**) when they meet the charged object.
2. Lines of force **come out of positive charges and go into negative charges**. Put arrows on your lines of force to show the direction of the electric field along each line.
3. **Near a pointlike charge**, the lines of force should be **approximately radial** (like the spokes of a bicycle wheel). If there are multiple charges in the picture, the lines won't be perfectly radial, but will curve somewhat, and will appear less radial as they go further away from one pointlike charge towards a region where other charges will carry significant influence.
4. If there are multiple charges and if all of the charges have the same value of $|q|$, the same number of lines of force should go out of (or into) each charge. If any of the charges has a higher value of $|q|$ than the others, there should be more lines of force going out of (or into) that charge. **The charge is proportional to the number of lines of force going out of (or into) it.**
5. Draw smooth lines of force from positive charges to negative charges. Note that some lines may go off the page. Also not that although we call them "lines of force," they are generally "curves" and seldom are straight "lines."
6. Make sure that every point on your lines of force satisfies the **superposition principle** applied in the previous examples and problems. The **tangent line** must be along the direction of the **net electric field** at any given point. Choose a variety of points in different regions of your map and apply the superposition principle at each point in order to help guide you as to how to draw the lines of force.
7. Two lines of force **can't intersect**. When you believe that your electric field map is complete, make sure that your lines of force don't cross one another anywhere on your map.
8. To represent equipotential surfaces, draw smooth curves that pass perpendicularly through your lines of force. Make sure that your equipotentials are **perpendicular** to your lines of force at every point of intersection. (An equipotential will intersect several lines of force. This doesn't contradict point 7, which says that two lines of force can't intersect one another.) Although equipotentials are really surfaces, they appear to be curves when drawn on a sheet of paper: They are surfaces in 3D space, but your paper is just a 2D cross section of 3D space.
9. Two equipotential surfaces representing two different values of electric potential (meaning that they have different values in Volts) won't cross.

Example: Sketch an electric field map for the single isolated positive charge shown below.

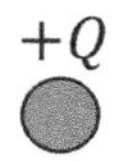

No matter where you might place a positive "test" charge in the above diagram, the "test" charge would be repelled by the positive charge $+Q$. Therefore, the lines of force radiate outward away from the positive charge (since it is completely isolated, meaning that there aren't any other charges around). The equipotential surfaces must be perpendicular to the lines of force: In this case, the equipotential surfaces are concentric spheres centered around the positive charge. Although they are drawn as circles below, they are really spheres that extend in 3D space in front and behind the plane of the paper.

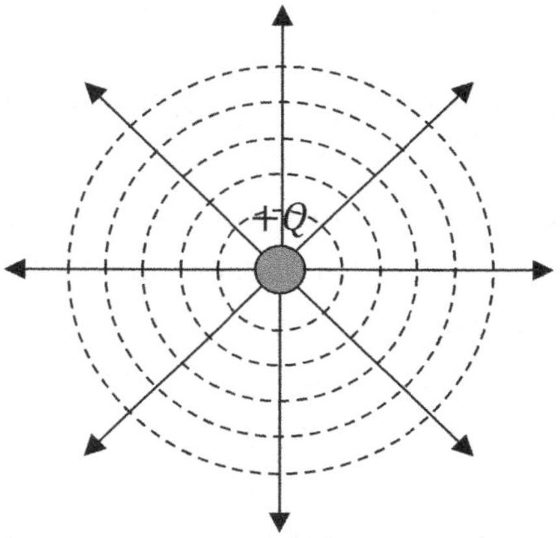

Example: Sketch an electric field map for the single isolated negative charge shown below.

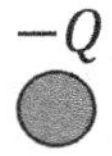

This electric field map is nearly identical to the previous electric field map: The only difference is that the lines radiate inward rather than outward, since a positive "test" charge would be attracted to the negative charge $-Q$.

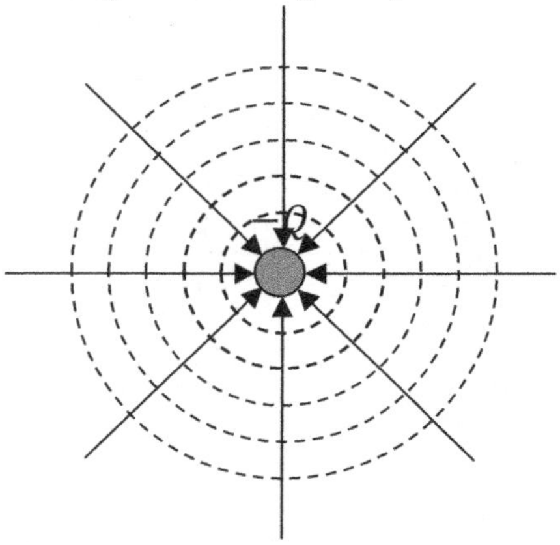

Example: Sketch an electric field map for the equal but opposite charges (called an electric dipole) shown below.

Make this map one step at a time:

- First, sample the net electric field in a variety of locations using the superposition strategy that we applied in the previous examples. If you review the previous examples, you will see that a couple of them featured the electric dipole:
    - The net electric field points to the **right** in the **center** of the diagram (where a positive "test" charge would be repelled by $+Q$ and attracted to $-Q$).
    - The net electric field points to the **right** anywhere along the **vertical line** that **bisects** the diagram, as the vertical components will cancel out there.
    - The net electric field points to the left along the horizontal line to the left of $+Q$ or to the right of $-Q$, but points to the right between the two charges.
    - The net electric field is somewhat **radial** (like the spokes of a bicycle wheel) **near either charge**, where the closer charge has the dominant effect.
- Beginning with the above features, draw smooth curves that leave $+Q$ and head into $-Q$ (except where the lines go beyond the scope of the diagram). The lines aren't perfectly radial near either charge, but curve so as to leave $+Q$ and reach $-Q$.
- Check several points in different regions: The **tangent** line at any point on a line of force should match the direction of the net electric field from **superposition**.
- Draw smooth curves for the equipotential surfaces. Wherever an equipotential intersects a line of force, the two curves must be perpendicular to one another.

In the diagram below, we used solid (—) lines for the lines of force (electric field lines) with arrows (→) indicating direction, and dashed lines (----) for the equipotentials.

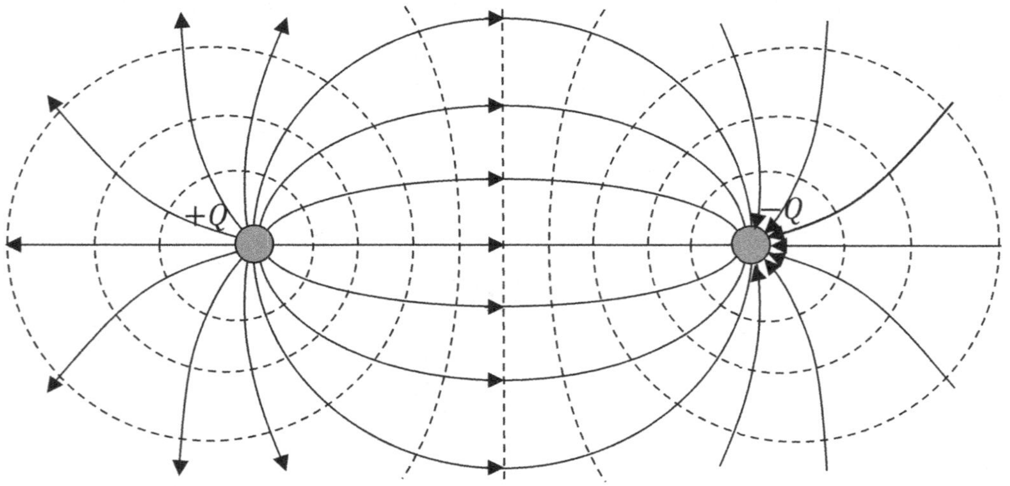

18. Sketch an electric field map for the two positive charges shown below.

(A) First just draw the lines of force.

(B) Now add the equipotential surfaces.

Want help? The problems from Chapter 4 are fully solved in the back of the book.

Strategy for Interpreting an Electric Field Map

When analyzing an electric field map, there are a variety of concepts that may apply, depending upon the nature of the question:

- To find the location of a pointlike charge on an electric field map, look for places where electric field lines converge or diverge. They diverge from a positive point-charge and converge towards a negative point-charge.
- If a question asks you which way a "test" charge would be pushed, find the direction of the tangent to the line of force at the specified point. The tangent to the line of force tells you which way a positive "test" charge would be pushed; if a negative "test" charge is used, the direction will be opposite.
- To find the value of electric potential at a specified point, look at the values of the electric potentials labeled for the equipotential surfaces. If the specified point is between two electric potentials, look at the two equipotentials surrounding it and estimate a value between them.
- To find the potential difference between two points (call them A and B), read off the electric potential at each point and subtract the two values: $\Delta V = V_B - V_A$.
- To estimate the magnitude of the electric field at a specified point, apply the following equation.

$$E \approx \left|\frac{\Delta V}{\Delta R}\right|$$

  First draw a straight line connecting two equipotential surfaces, such that the straight line passes through the desired point and is, on average, roughly perpendicular to each equipotential. (See the example below.) Then measure the length of this line, ΔR. To find the value of ΔV, simply subtract the values of electric potential for each equipotential.

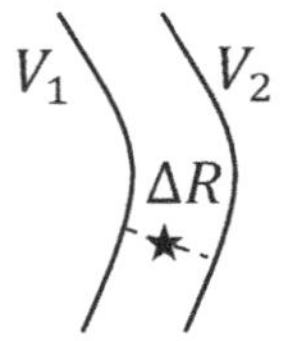

- If a question asks you about the strength of the electric field at different points, electric field is stronger where the lines of force are more dense and it is weaker where the lines of force are less dense.
- If a question gives you a diagram of equipotential surfaces and asks you to draw the lines of force (or vice-versa), draw smooth curves that pass perpendicularly through the equipotentials, from positive to negative.

Example: Consider the electric field map drawn below.

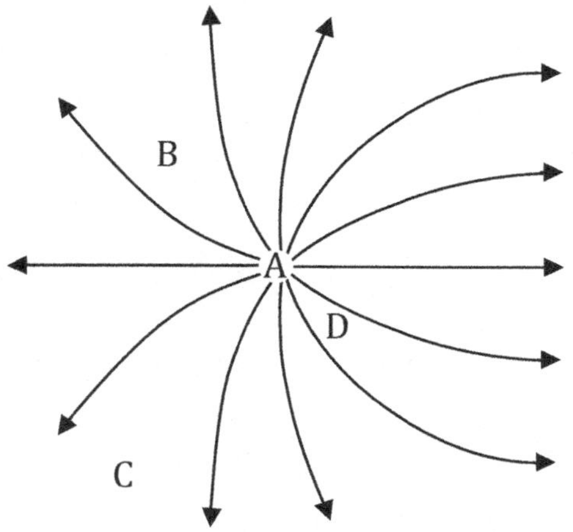

(A) At which point(s) is there a pointlike charge? Is the charge positive or negative?
There is a charge at point A where the lines of force **diverge** from. The charge is positive because the lines of force go away from point A.
(B) Rank the electric field strength at points B, C, and D.
Between these 3 points, the electric field is stronger at point D, where the lines of force are **more dense**, and weaker at point C, where the lines of force are less dense: $E_D > E_B > E_C$.

Example: Consider the map of equipotentials drawn below.

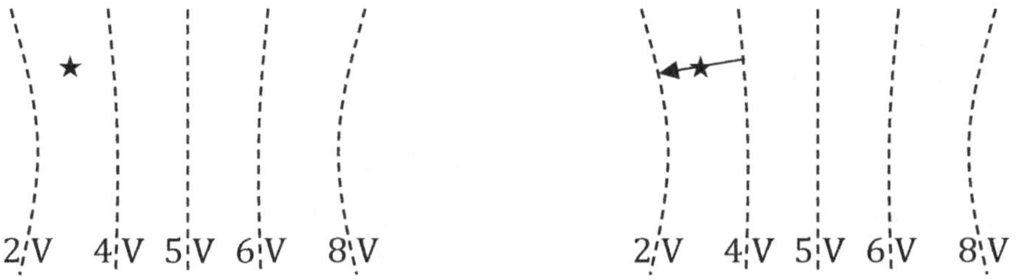

(A) Estimate the magnitude of the electric field at the star (★) in the diagram above.
First note that the above diagram shows equipotentials, <u>**not**</u> lines of force. Draw a straight line through the star (★) from higher voltage to lower voltage that is approximately perpendicular (on average) to the two equipotentials on either side of point X. Subtract the values of electric potential (in Volts) for these two equipotentials to get the potential difference: $\Delta V = 4 - 2 = 2$ V. Measure the length of the line with a ruler: $\Delta R = 1.0$ cm. Convert R to meters: $R = 0.010$ m. Apply the following formula. Recall that a Volt per meter $\left(\frac{\text{V}}{\text{m}}\right)$ equals a Newton per Coulomb $\left(\frac{\text{N}}{\text{C}}\right)$.

$$E \approx \left|\frac{\Delta V}{\Delta R}\right| = \frac{2}{0.01} = 200\,\frac{\text{V}}{\text{m}}$$

(B) If a positive "test" charge were placed at the star (★), which way would it be pushed?
It would be pushed along the line of force, which is the straight line that we drew through the star (★) in the diagram above.

19. Consider the map of equipotentials drawn below. Note that the diagram below shows equipotentials, **not** lines of force.

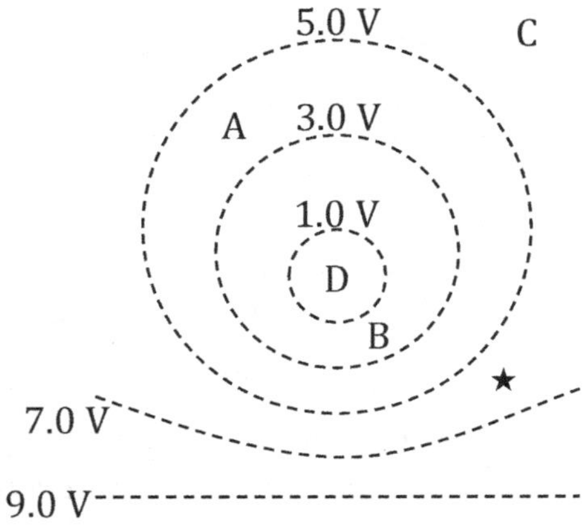

(A) Sketch the lines of force for the diagram above.

(B) At which point(s) is there a pointlike charge? Is the charge positive or negative?

(C) Rank the electric field strength at points A, B, and C.

(D) Estimate the magnitude of the electric field at the star (★) in the diagram above.

(E) If a negative "test" charge were placed at the star (★), which way would it be pushed?

Want help? The problems from Chapter 4 are fully solved in the back of the book.

5 ELECTROSTATIC EQUILIBRIUM

Relevant Terminology

Electrostatic equilibrium – when a system of charged objects is in static equilibrium (meaning that the system is motionless).
Electric charge – a fundamental property of a particle that causes the particle to experience a force in the presence of an electric field. (An electrically neutral particle has no charge and thus experiences no force in the presence of an electric field.)
Electric force – the push or pull that one charged particle exerts on another. Oppositely charged particles attract, whereas like charges (both positive or both negative) repel.
Electric field – force per unit charge.

Essential Concepts

When a system of charged objects is in electrostatic equilibrium, we can apply Newton's second law with the components of the acceleration set equal to zero (since an object doesn't have acceleration when it remains in static equilibrium).

Electrostatic Equilibrium Equations

Many electrostatic equilibrium problems are an application of Newton's second law. Recall that Newton's second law was covered in Volume 1 of this series. To apply Newton's second law, draw a free-body diagram (FBD) for each object, and sum the components of the forces acting on each object. In electrostatic equilibrium, set $a_x = 0$ and $a_y = 0$.

$$\sum F_x = ma_x \quad , \quad \sum F_y = ma_y$$

When an object with charge q is in the presence of an external electric field ($\vec{\mathbf{E}}$), the object experiences an electric force ($\vec{\mathbf{F}}_e$), which is parallel to $\vec{\mathbf{E}}$ if $q > 0$ and opposite to $\vec{\mathbf{E}}$ if $q < 0$.

$$F_e = |q|E$$

If there are two (or more) charged objects in the problem, apply Coulomb's law.

$$F_e = \frac{k|q_1||q_2|}{R^2}$$

Any object in the presence of a gravitational field experiences **weight** (W), which equals mass (m) times gravitational acceleration (g). Weight pulls straight down.

$$W = mg$$

When a stationary object is in contact with a surface, the force of **static friction** (f_s) is less than or equal to the coefficient of static friction (μ_s) times normal force (N).

$$f_s \leq \mu_s N$$

Symbols and SI Units

Symbol	Name	SI Units
E	magnitude of electric field	N/C or V/m
F_e	magnitude of electric force	N
q	charge	C
R	distance between two charges	m
k	Coulomb's constant	$\frac{\text{N}\cdot\text{m}^2}{\text{C}^2}$ or $\frac{\text{kg}\cdot\text{m}^3}{\text{C}^2\cdot\text{s}^2}$
m	mass	kg
mg	weight	N
N	normal force	N
f	friction force	N
μ	coefficient of friction	unitless
T	tension	N
a_x	x-component of acceleration	m/s^2
a_y	y-component of acceleration	m/s^2

Important Distinction

It's important to distinguish between electric charge (q), electric field (E), and electric force (F_e). Read the problems carefully, memorize which symbol is used for each quantity, and look at the units to help distinguish between them:

- The SI unit of electric **charge** (q) is the Coulomb (C).
- The SI units of electric **field** (E) can be expressed as $\frac{\text{N}}{\text{C}}$ or $\frac{\text{V}}{\text{m}}$.
- The SI unit of electric **force** (F_e) is the Newton (N).

It's also important to learn when to use which equation for electric force.

- Use the equation $F_e = |q|E$ for a charged object that is in the presence of an external electric field.
- Use the equation $F_e = \frac{k|q_1||q_2|}{R^2}$ when there are two or more charges in a problem.

Electrostatic Equilibrium Strategy

To apply Newton's second law to a system in electrostatic equilibrium, follow these steps.

1. Draw a free-body diagram (FBD) for each object. Label the forces acting on each object. Consider each of the following forces:
   - Every object has **weight** (mg). Draw mg **straight down**. If there are multiple objects, distinguish their masses with subscripts: $m_1 g, m_2 g$, etc.
   - Does the problem mention an external electric field? If so, any charged object in the problem will experience an **electric force** ($|q|E$), which is parallel to $\vec{\mathbf{E}}$ if $q > 0$ and opposite to $\vec{\mathbf{E}}$ if $q < 0$.
   - Are there two or more charged objects in the problem? If so, each pair will experience **electric forces** $\left(\frac{k|q_1||q_2|}{R^2}\right)$ according to Coulomb's law. The forces are attractive for opposite charges and repulsive for like charges.
   - Is the object in contact with a surface? If it is, draw **normal** force (N) **perpendicular to the surface**. If there are two or more normal forces in the problem, use N_1, N_2, etc.
   - If the object is in contact with a surface, there will also be a **friction** force (f). Draw the friction force **opposite to the potential velocity**. If there is more than one friction force, use f_1, f_2, etc.
   - Is the object connected to a cord, rope, thread, or string? If so, there will be a tension (T) **along the cord**. If two objects are connected to one cord, draw **equal and oppositely directed** forces acting on the two objects in accordance with Newton's third law. If there are two separate cords, then there will be two different pairs of tensions (T_1 and T_2), one pair for each cord.
   - Does the problem describe or involve any other forces, such as a monkey's pull (P)? If so, draw and label these forces.
2. Label the $+x$- and $+y$-axes in each FBD.
3. Write Newton's second law in component form for each object. The components of acceleration (a_x and a_y) equal zero in electrostatic equilibrium.

$$\sum F_{1x} = 0 \quad , \quad \sum F_{1y} = 0 \quad , \quad \sum F_{2x} = 0 \quad , \quad \sum F_{2y} = 0$$

4. Rewrite the left-hand side of each sum in terms of the x- and y-components of the forces acting on each object. Consider each force one at a time. Ask yourself if the force lies on an axis:
   - If a force lies on a positive or negative coordinate axis, the force only goes in that sum (x or y) with no trig.
   - If a force lies in the middle of a Quadrant, the force goes in both the x- and y-sums using trig. One component will involve cosine, while the other will involve sine. Whichever axis is adjacent to the angle gets the cosine.

5. Check the signs of each term in your sum. If the force has a component pointing along the $+x$-axis, it should be positive in the x-sum, but if it has a component pointing along the $-x$-axis, it should be negative in the x-sum. Apply similar reasoning for the y-sum.
6. Apply algebra to solve for the desired unknown(s).

Example: A monkey dangles two charged banana-shaped earrings from two separate threads, as illustrated below. Once electrostatic equilibrium is attained, the two earrings have the same height and are separated by a distance of 3.0 m. Each earring has a mass of $\frac{\sqrt{3}}{25}$ kg. The earrings carry equal and opposite charge.

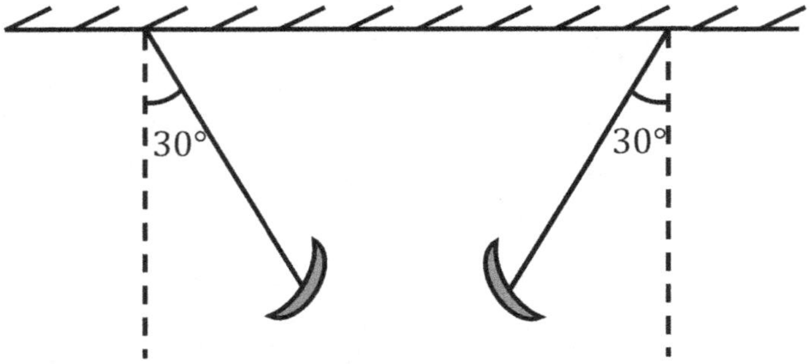

(A) What is the tension in each thread?
Draw and label a FBD for each earring.

- Each earring has weight (mg) pulling straight down.
- Tension (T) pulls along each thread.
- The two earrings attract one another with an electric force (F_e) via Coulomb's law. The electric force is attractive because the earrings have opposite charge.

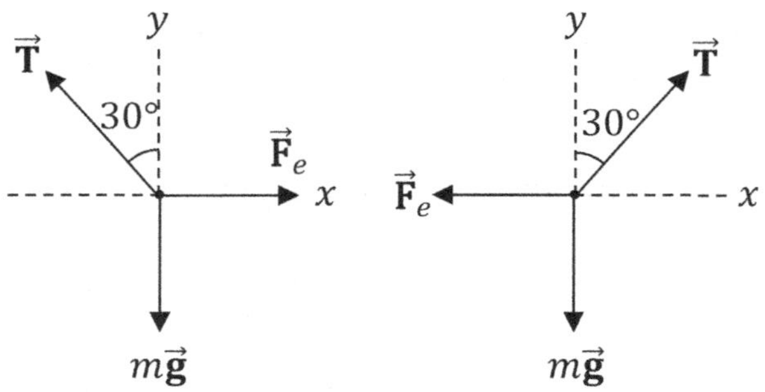

Apply Newton's second law. In electrostatic equilibrium, $a_x = 0$ and $a_y = 0$.

- Since tension doesn't lie on an axis, T appears in both the x- and y-sums with trig. In the FBD, since y is adjacent to 30°, cosine appears in the y-sum.
- Since the electric force is horizontal, F_e appears only in the x-sums with no trig.
- Since weight is vertical, mg appears only in the y-sums with no trig.

$$\sum F_{1x} = ma_x \quad , \quad \sum F_{1y} = ma_y \quad , \quad \sum F_{2x} = ma_x \quad , \quad \sum F_{2y} = ma_y$$

$$F_e - T\sin 30° = 0 \quad , \quad T\cos 30° - mg = 0 \quad , \quad T\sin 30° - F_e = 0 \quad , \quad T\cos 30° - mg = 0$$

We can solve for tension in the equation from the y-sums.

$$T\cos 30° - mg = 0$$

Add weight (mg) to both sides of the equation.

$$T\cos 30° = mg$$

Divide both sides of the equation by $\cos 30°$.

$$T = \frac{mg}{\cos 30°}$$

In this book, we will round $g = 9.81 \text{ m/s}^2$ to $g \approx 10 \text{ m/s}^2$ in order to show you how to obtain an approximate answer without using a calculator. (Feel free to use a calculator if you wish. It's a valuable skill to be able to estimate an answer without a calculator, which is the reason we will round 9.81 to 10.)

$$T = \frac{\left(\frac{\sqrt{3}}{25}\right)(9.81)}{\cos 30°} \approx \frac{\left(\frac{\sqrt{3}}{25}\right)(10)}{\frac{\sqrt{3}}{2}}$$

To divide by a fraction, multiply by its **reciprocal**. Note that the reciprocal of $\frac{\sqrt{3}}{2}$ is $\frac{2}{\sqrt{3}}$.

$$T \approx \left(\frac{\sqrt{3}}{25}\right)(10)\frac{2}{\sqrt{3}} = \frac{20\sqrt{3}}{25\sqrt{3}} = \frac{20}{25} = \frac{4}{5} \text{ N}$$

Note that $\frac{20}{25}$ reduces to $\frac{4}{5}$ if you divide the numerator and denominator both by 5. The answer is $T \approx \frac{4}{5}$ N.

(B) What is the charge of each earring?

First, solve for the electric force in the equation from the x-sums from part (A).

$$F_e - T\sin 30° = 0$$

Add $T\sin 30°$ to both sides of the equation.

$$F_e = T\sin 30°$$

Plug in the tension that we found in part (A).

$$F_e \approx \frac{4}{5}\sin 30° = \frac{4}{5}\left(\frac{1}{2}\right) = \frac{2}{5} \text{ N}$$

Now apply Coulomb's law (Chapter 1).

$$F_e = k\frac{|q_1||q_2|}{R^2}$$

Since the charges are equal in value and opposite in sign, either $q_1 = q$ and $q_2 = -q$ or vice-versa. It doesn't matter which, since the magnitude of the electric force involves absolute values (magnitudes of vectors are always positive). Note that $qq = q^2$.

$$F_e = k\frac{q^2}{R^2}$$

Multiply both sides of the equation by R^2.

$$F_e R^2 = kq^2$$

Divide both sides of the equation by Coulomb's constant.

$$\frac{F_e R^2}{k} = q^2$$

Squareroot both sides of the equation. Note that $\sqrt{R^2} = R$. Recall that $k = 9.0 \times 10^9 \frac{\text{N·m}^2}{\text{C}^2}$.

$$q = \sqrt{\frac{F_e R^2}{k}} = R\sqrt{\frac{F_e}{k}}$$

$$q \approx 3\sqrt{\frac{\frac{2}{5}}{9 \times 10^9}}$$

Note that $\frac{2}{5}$ can be expressed as 4×10^{-1} since $\frac{2}{5} = 0.4$ and $4 \times 10^{-1} = 0.4$.

$$q \approx 3\sqrt{\frac{4 \times 10^{-1}}{9 \times 10^9}}$$

Apply the rule $\frac{x^{-m}}{x^n} = x^{-m-n}$.

$$q \approx 3\sqrt{\frac{4 \times 10^{-1-9}}{9}} = 3\sqrt{\frac{4 \times 10^{-10}}{9}}$$

Apply the rule $\sqrt{\frac{xy}{z}} = \frac{\sqrt{x}\sqrt{y}}{\sqrt{z}}$.

$$q \approx 3\frac{\sqrt{4}\sqrt{10^{-10}}}{\sqrt{9}} = \frac{(3)(2)(10^{-5})}{3} = 2.0 \times 10^{-5}\ \text{C}$$

Note that $\sqrt{10^{-10}} = 10^{-5}$ since $(10^{-5})^2 = 10^{-10}$. The answer is $q \approx 2.0 \times 10^{-5}$ C, which can also be expressed as $q \approx 20$ µC using the metric prefix micro ($\mu = 10^{-6}$). One charge is $+20$ µC, while the other charge is -20 µC. (Without more information, there is no way to determine which one is positive and which one is negative.)

Example: The 80-g banana-shaped earring illustrated below is suspended in midair in electrostatic equilibrium. The earring is connected to the ceiling by a thread that makes a 60° angle with the vertical. The earring has a charge of +200 μC and there is a uniform electric field directed horizontally to the right. Find the magnitude of the electric field.

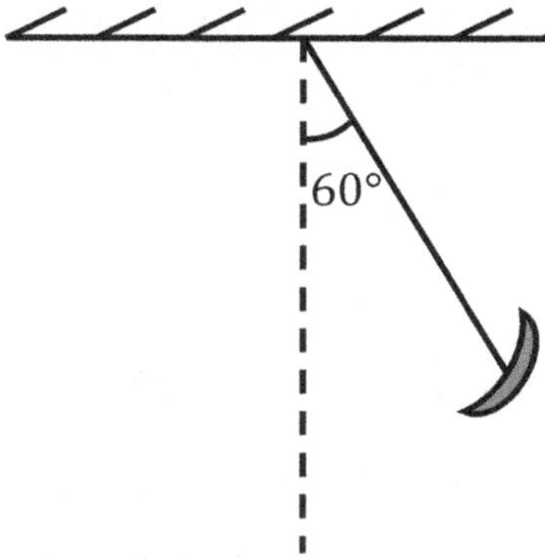

Draw and label a FBD for the earring.

- The earring has weight (mg) pulling straight down.
- Tension (T) pulls along the thread.
- The horizontal electric field ($\vec{\mathbf{E}}$) exerts an electric force ($\vec{\mathbf{F}}_e = q\vec{\mathbf{E}}$) on the charge, which is parallel to the electric field since the charge is positive.

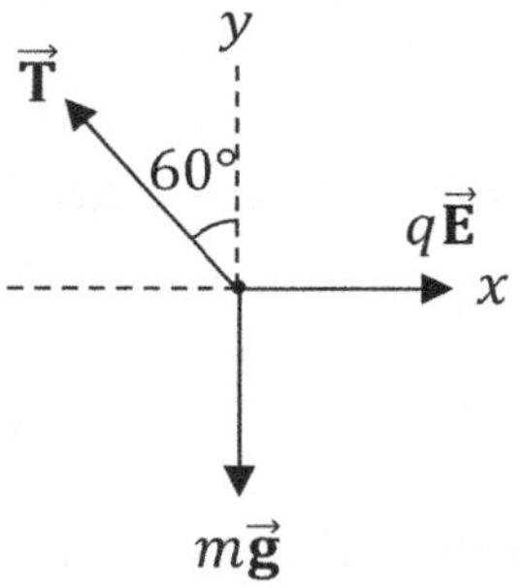

Apply Newton's second law. In electrostatic equilibrium, $a_x = 0$ and $a_y = 0$.

- Since tension doesn't lie on an axis, T appears in both the x- and y-sums with trig. In the FBD, since y is adjacent to 60°, cosine appears in the y-sum.
- Since the electric force is horizontal, $|q|E$ appears in only the x-sum with no trig. Recall that $F_e = |q|E$ for a charge in an external electric field.
- Since weight is vertical, mg appears only in the y-sum with no trig.

$$\sum F_x = ma_x \quad , \quad \sum F_y = ma_y$$

$$|q|E - T\sin 60° = 0 \quad , \quad T\cos 60° - mg = 0$$

We can solve for magnitude of the electric field (E) in the equation from the x-sum.

$$|q|E - T\sin 60° = 0$$

$$|q|E = T\sin 60°$$

Divide both sides of the equation by the charge.

$$E = \frac{T \sin 60°}{|q|}$$

If we knew the tension, we could plug it into the above equation in order to find the electric field. However, the problem didn't give us the tension (and it's **not** equal to weight). Fortunately, we can solve for tension using the equation from the y-sum from Newton's second law.

$$T \cos 60° - mg = 0$$
$$T \cos 60° = mg$$
$$T = \frac{mg}{\cos 60°}$$

Convert the mass from grams (g) to kilograms (kg): $m = 80 \text{ g} = 0.08 \text{ kg} = \frac{2}{25}$ kg. (Note that $0.08 = \frac{2}{25}$. Try it on your calculator, if needed.)

$$T = \frac{\left(\frac{2}{25}\right)(9.81)}{\cos 60°} \approx \frac{\left(\frac{2}{25}\right)(10)}{\frac{1}{2}}$$

Note that we rounded 9.81 to 10. To divide by a fraction, multiply by its **reciprocal**. The reciprocal of $\frac{1}{2}$ is 2.

$$T \approx \left(\frac{2}{25}\right)(10)(2) = \frac{40}{25} = \frac{8}{5} \text{ N}$$

We can plug this in for tension in the equation for electric field that we had found before. Convert the charge from microCoulombs to Coulombs: $q = 200\ \mu\text{C} = 2.00 \times 10^{-4}$ C.

$$E = \frac{T \sin 60°}{|q|} \approx \frac{\left(\frac{8}{5}\right)\sin 60°}{2 \times 10^{-4}} = \frac{\left(\frac{8}{5}\right)\left(\frac{\sqrt{3}}{2}\right)}{2 \times 10^{-4}} = \frac{2\sqrt{3}}{5} \times 10^{4}$$

Note that $\frac{8\sqrt{3}}{5(2)} = \frac{8\sqrt{3}}{10} = \frac{4\sqrt{3}}{5}$. If you divide this by 2, you get $\frac{2\sqrt{3}}{5}$. Then when you divide by 10^{-4}, you multiply by 10^4 according to the rule $\frac{1}{x^{-n}} = x^n$.

$$E \approx \frac{2\sqrt{3}}{5} \times 10^4 = \frac{2\sqrt{3}}{5} \times 10 \times 10^3 = 2\sqrt{3} \times 2 \times 10^3 = 4\sqrt{3} \times 10^3 = 4000\sqrt{3}\ \frac{\text{N}}{\text{C}}$$

Here, we used the fact that $10^4 = 10 \times 10^3$. Note that $\frac{10}{5} = 2$ and that $10^3 = 1000$. The answer is that the magnitude of the electric field is $E \approx 4000\sqrt{3}\ \frac{\text{N}}{\text{C}}$.

20. A poor monkey spends his last nickel on two magic bananas. When he dangles the two $9\sqrt{3}$-kg (which is heavy because the magic bananas are made out of metal) bananas from cords, they "magically" spread apart as shown below. Actually, it's because they have equal electric charge.

(A) Draw a FBD for each object (two in all). Label each force and the x- and y-coordinates.

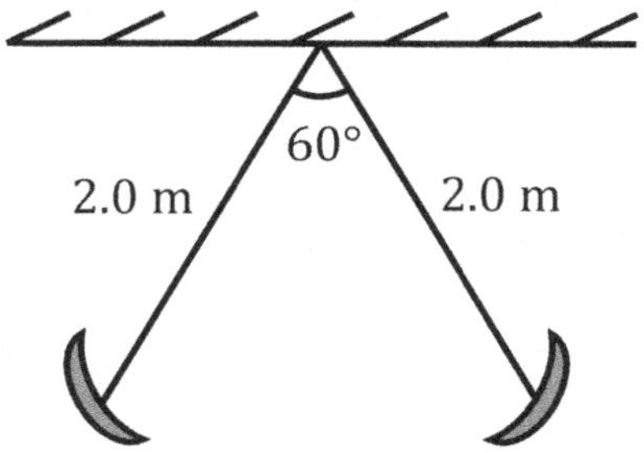

(B) Write the x- and y-sums for the forces acting on each object. There will be four sums. On the line immediately below each sum, rewrite the left-hand side in terms of the forces.

(C) What is the tension in each cord?

(D) What is the charge of each banana?

Want help? Check the hints section at the back of the book.

Answers: 180 N, 200 μC

21. The 60-g banana-shaped earring illustrated below is suspended in midair in electrostatic equilibrium. The earring is connected to the floor by a thread that makes a 60° angle with the vertical. The earring has a charge of +300 μC and there is a uniform electric field directed 60° above the horizontal.

(A) Draw a FBD for the earring. Label each force and the x- and y-coordinates.

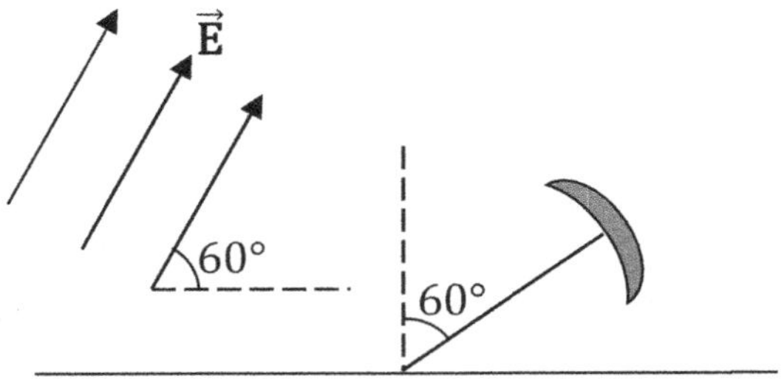

(B) Write the x- and y-sums for the forces acting on the earring. There will be two sums. On the line immediately below each sum, rewrite the left-hand side in terms of the forces.

(C) Determine the magnitude of the electric field.

(D) Determine the tension in the thread.

Want help? Check the hints section at the back of the book.

Answers: $2000\sqrt{3}$ N/C, $\frac{3}{5}$ N

6 INTEGRATION ESSENTIALS

Do you already know how to integrate over polar coordinates, spherical coordinates, and cylindrical coordinates? If not, you will have something to learn in this chapter. Although this chapter starts out easy, it will grow progressively harder. We will cover these topics:

- review of basic integrals (pages 65-71)
- trigonometric substitutions (pages 72-76)
- double and triple integrals (pages 77-82)
- 2D polar coordinates (pages 83-90)
- cylindrical coordinates (pages 83-90)
- spherical coordinates (pages 83-90)

Quick Review of Calculus with Polynomials

To find the **derivative** of a polynomial term of the form ax^b, bring the exponent (b) down and reduce the exponent by 1 (from b to $b-1$).

$$\frac{d}{dx}(ax^b) = bax^{b-1}$$

To find the **anti-derivative** of a polynomial term of the form ax^b, raise the exponent by 1 (from b to $b+1$) and divide by the new exponent ($b+1$), provided that $b \neq -1$. (We will review the case where $b = -1$ in Chapter 17.)

$$\int ax^b\,dx = \frac{ax^{b+1}}{b+1} \qquad (b \neq -1)$$

In the above indefinite integral, there is also a constant of integration ($+c$). However, in physics, we generally perform definite integrals rather than indefinite integrals, in which case the constant of integration will cancel out. In a **definite integral**, the anti-derivative is evaluated at the upper and lower limits of the integral and the two results are subtracted.

$$\int_{x=x_i}^{x_f} ax^b\,dx = \left[\frac{ax^{b+1}}{b+1}\right]_{x=x_i}^{x_f} = \frac{ax_f^{b+1}}{b+1} - \frac{ax_i^{b+1}}{b+1}$$

For a polynomial expression with **multiple terms**, apply calculus to each term individually.

$$\frac{d}{dx}(y_1 + y_2 + \cdots + y_N) = \frac{dy_1}{dx} + \frac{dy_2}{dx} + \cdots + \frac{dy_N}{dx}$$

$$\int (y_1 + y_2 + \cdots + y_N)\,dx = \int y_1\,dx + \int y_2\,dx + \cdots + \int y_N\,dx$$

If you see a squareroot, recall from algebra that you can express it as an exponent. For example, $\sqrt{x} = x^{1/2}$ and $\sqrt{cx} = (cx)^{1/2} = c^{1/2}x^{1/2}$. If a variable appears in a denominator, it may be helpful to apply the rule $\frac{1}{x^n} = x^{-n}$.

Example: Consider the function $y = 8x^3$.

(A) Evaluate $\frac{dy}{dx}$ when $x = 2$.

First take the derivative (before plugging in the numerical value for x). The exponent (3) comes down to multiply the coefficient (8), and the exponent is reduced by 1 (from 3 to 2).

$$\frac{dy}{dx} = \frac{d}{dx}(8x^3) = (3)(8)x^{3-1} = 24x^2$$

The derivative of $8x^3$ with respect to x is $24x^2$. Now plug in the specified value of x.

$$\left.\frac{dy}{dx}\right|_{x=2} = 24(2)^2 = 24(4) = 96$$

The answer to the problem is 96.

(B) Evaluate $\int_{x=1}^{2} y\,dx$.

Find the anti-derivative of $8x^3$. The exponent is raised by 1 (from 3 to 4). Also divide by the new exponent (4).

$$\int_{x=1}^{2} y\,dx = \int_{x=1}^{2} 8x^3\,dx = \left[\frac{8x^{3+1}}{3+1}\right]_{x=1}^{2} = \left[\frac{8x^4}{4}\right]_{x=1}^{2} = [2x^4]_{x=1}^{2}$$

The anti-derivative of $8x^3$ is $2x^4$, but we're not finished yet. The notation $[2x^4]_{x=1}^{2}$ means to evaluate the function $2x^4$ when $x = 2$, then evaluate the function $2x^4$ when $x = 1$, and finally to subtract the two results: $2(2)^4 - 2(1)^4$.

$$\int_{x=1}^{2} y\,dx = \int_{x=1}^{2} 8x^3\,dx = [2x^4]_{x=1}^{2} = 2(2)^4 - 2(1)^4 = 2(16) - 2(1) = 32 - 2 = 30$$

The answer to the definite integral is 30.

Example: Perform the following definite integral: $\int_{x=4}^{8} \frac{dx}{x^2}$.

Since x appears in the denominator, we use the rule $\frac{1}{x^n} = x^{-n}$ to write $\frac{1}{x^2}$ as x^{-2}. Compare x^{-2} with ax^b to see that $a = 1$ and $b = -2$. The anti-derivative is $\frac{ax^{b+1}}{b+1}$.

$$\int_{x=4}^{8} \frac{dx}{x^2} = \int_{x=4}^{8} x^{-2}\,dx = \left[\frac{x^{-2+1}}{-2+1}\right]_{x=4}^{8} = \left[\frac{x^{-1}}{-1}\right]_{x=4}^{8} = [-x^{-1}]_{x=4}^{8} = \left[-\frac{1}{x}\right]_{x=4}^{8}$$

Note that $-2 + 1 = -1$ and that $\frac{1}{-1} = -1$. In the last step, we used the rule that $x^{-1} = \frac{1}{x}$.

The notation $\left[-\frac{1}{x}\right]_{x=4}^{8}$ means to evaluate the function $-\frac{1}{x}$ when $x = 8$, then evaluate the function $-\frac{1}{x}$ when $x = 4$, and finally to subtract the two results: $-\frac{1}{8} - \left(-\frac{1}{4}\right)$.

$$\int_{x=4}^{8} \frac{dx}{x^2} = \left[-\frac{1}{x}\right]_{x=4}^{8} = -\frac{1}{8} - \left(-\frac{1}{4}\right) = -\frac{1}{8} + \frac{1}{4} = -\frac{1}{8} + \frac{2}{8} = \frac{-1+2}{8} = \frac{1}{8}$$

Recall that two minus signs make a plus sign. The answer to the definite integral is $\frac{1}{8}$.

Example: Perform the following definite integral: $\int_{x=4}^{9} \frac{dx}{\sqrt{x}}$.

The trick to this problem is to recall the rules from algebra that $\sqrt{x} = x^{1/2}$ and $x^{-1} = \frac{1}{x}$. Combine these rules together to write $\frac{1}{\sqrt{x}} = x^{-1/2}$. Compare $x^{-1/2}$ with ax^b to see that $a = 1$ and $b = -\frac{1}{2}$ (since a coefficient of 1 is implied when you don't see a coefficient). The anti-derivative is $\frac{ax^{b+1}}{b+1}$.

$$\int_{x=4}^{9} \frac{dx}{\sqrt{x}} = \int_{x=4}^{9} x^{-1/2}\,dx = \left[\frac{x^{-1/2+1}}{-\frac{1}{2}+1}\right]_{x=4}^{9} = \left[\frac{x^{1/2}}{\frac{1}{2}}\right]_{x=4}^{9} = \left[2x^{1/2}\right]_{x=4}^{9} = \left[2\sqrt{x}\right]_{x=4}^{9}$$

Note that $-\frac{1}{2} + 1 = -\frac{1}{2} + \frac{2}{2} = \frac{-1+2}{2} = \frac{1}{2}$ (this step involves finding a **common denominator**). Also note that $\frac{1}{1/2} = 2$ (this step involves multiplying by the **reciprocal**). In the last step, we applied the rule that $\sqrt{x} = x^{1/2}$. The notation $\left[2\sqrt{x}\right]_{x=4}^{9}$ means to evaluate the function $2x^{1/2}$ when $x = 9$, then evaluate the function $2\sqrt{x}$ when $x = 4$, and finally to subtract the two results: $2\sqrt{9} - 2\sqrt{4}$.

$$\int_{x=4}^{9} \frac{dx}{\sqrt{x}} = \left[2\sqrt{x}\right]_{x=4}^{9} = 2\sqrt{9} - 2\sqrt{4} = 2(3) - 2(2) = 6 - 4 = 2$$

The answer to the definite integral is 2.

Example: Perform the following definite integral: $\int_{x=0}^{3} (6x^2 - 4)\,dx$.

First find the anti-derivative of each term:

- The first term is $6x^2$. Compare $6x^2$ with ax^b to see that $a = 6$ and $b = 2$.

$$\int 6x^2\,dx = \frac{6x^{2+1}}{2+1} = \frac{6x^3}{3} = 2x^3$$

- The second term is -4. Compare -4 with ax^b to see that $a = -4$ and $b = 0$ (recall from algebra that $x^0 = 1$).

$$\int -4\,dx = \frac{-4x^{0+1}}{0+1} = -\frac{4x^1}{1} = -4x$$

Add these two anti-derivatives together and evaluate the definite integral over the limits:

$$\int_{x=0}^{3} (6x^2 - 4)\,dx = [2x^3 - 4x]_{x=0}^{3} = [2(3)^3 - 4(3)] - [2(0)^3 - 4(0)] = 54 - 12 = 42$$

22. Perform the following definite integral.

$$\int_{x=2}^{3} 6x^2\, dx =$$

Answer: 38

23. Perform the following definite integral.

$$\int_{x=1}^{2} \frac{8\, dx}{x^3} =$$

Want help? Check the hints section at the back of the book.

Answer: 3

24. Perform the following definite integral.

$$\int_{x=9}^{16} 3\sqrt{x}\,dx =$$

Answer: 74

25. Perform the following definite integral.

$$\int_{x=1}^{3} (8x^3 - 6x)\,dx =$$

Want help? Check the hints section at the back of the book.

Answer: 136

Quick Review of Calculus with Trig Functions

The first **derivatives** of the basic trig functions are:

$$\frac{d}{d\theta}\sin\theta = \cos\theta \quad , \quad \frac{d}{d\theta}\cos\theta = -\sin\theta \quad , \quad \frac{d}{d\theta}\tan\theta = \sec^2\theta$$

$$\frac{d}{d\theta}\sec\theta = \sec\theta\tan\theta \quad , \quad \frac{d}{d\theta}\csc\theta = -\csc\theta\cot\theta \quad , \quad \frac{d}{d\theta}\cot\theta = -\csc^2\theta$$

Recall the secant ($\sec\theta$), cosecant ($\csc\theta$), and cotangent ($\cot\theta$) functions:

$$\sec\theta = \frac{1}{\cos\theta} \quad , \quad \csc\theta = \frac{1}{\sin\theta} \quad , \quad \cot\theta = \frac{1}{\tan\theta}$$

Note that the co's **<u>don't</u>** go together: Secant goes with cosine, but cosecant goes with sine.

The **integrals** of the basic trig functions are:

$$\int \sin\theta\, d\theta = -\cos\theta \quad , \quad \int \cos\theta\, d\theta = \sin\theta \quad , \quad \int \tan\theta\, d\theta = \ln|\sec\theta|$$

$$\int \sec\theta\, d\theta = \ln|\sec\theta + \tan\theta| \quad , \quad \int \csc\theta\, d\theta = -\ln|\csc\theta + \cot\theta| \quad , \quad \int \cot\theta\, d\theta = \ln|\sin\theta|$$

Example: Evaluate $\frac{d}{d\theta}\tan\theta$ at $\theta = 30°$.

First perform the derivative (before plugging in the numerical value for the angle).

$$\frac{d}{d\theta}\tan\theta = \sec^2\theta$$

Now plug in the specified angle.

$$\sec^2 30° = \frac{1}{\cos^2 30°} = \frac{1}{\left(\frac{\sqrt{3}}{2}\right)^2} = \frac{1}{\left(\frac{3}{4}\right)} = \frac{4}{3}$$

Note that $\left(\sqrt{3}\right)^2 = \sqrt{3}\sqrt{3} = 3$. In the last step, to divide by $\frac{3}{4}$, multiply by its **reciprocal** $\left(\frac{4}{3}\right)$.

Example: Perform the following definite integral: $\int_{\theta=0°}^{90°} \sin\theta\, d\theta$.

First find the anti-derivative of $\sin\theta$:

$$\int_{\theta=0°}^{90°} \sin\theta\, d\theta = [-\cos\theta]_{\theta=0°}^{90°} = -\cos 90° - (-\cos 0°) = -0 - (-1) = 1$$

Note that the two minus signs make a plus sign. The answer to the definite integral is 1.

26. Perform the following definite integral.

$$\int_{\theta=30°}^{90°} \cos\theta \, d\theta =$$

Answer: $\frac{1}{2}$

27. Perform the following definite integral.

$$\int_{\theta=45°}^{135°} \sin\theta \, d\theta =$$

Want help? Check the hints section at the back of the book.

Answer: $\sqrt{2}$

Strategy for Integrating by Substitution

To integrate via a substitution, follow these steps:

1. Visualize a substitution of the form $x = x(u)$, meaning that the old variable x is a function of the new variable u, which will transform the integral from $\int_{x=x_0}^{x} f(x)\,dx$ to the form $\int_{u=u_0}^{u} g(u)\,du$ in such a way that you can perform the new integral (in terms of the variable u) with a known technique. Two common substitutions are:
    - **Polynomial** substitutions. Example: $\int (6x+4)^3\,dx$. Try $u = 6x + 4$.
    - **Trig** substitutions. This is common with quadratic functions in squareroots. The idea is to make a substitution that collapses two terms down to a single term through the trig identities $\sin^2 u + \cos^2 u = 1$ or $\tan^2 u + 1 = \sec^2 u$.
        - Example: $\int \sqrt{a^2 - x^2}\,dx$. Try $x = a \sin u$ because $1 - \sin^2 u = \cos^2 u$.
        - Example: $\int \sqrt{x^2 + a^2}\,dx$. Try $x = a \tan u$ because $\tan^2 u + 1 = \sec^2 u$.
        - Example: $\int \sqrt{x^2 - a^2}\,dx$. Try $x = a \sec u$ because $\sec^2 u - 1 = \tan^2 u$.
2. Implicitly differentiate the function $x = x(u)$ in order to write du in terms of dx. On one side, take a derivative with respect to u and multiply by du, and on the other side, take a derivative with respect to x and multiply by dx.
    - Example: $u = 3x^2$. Beginning with $\frac{d}{du}(u) = 1$ and $\frac{d}{dx}(3x^2) = 6x$, we obtain $du = 6x\,dx$.
    - Example: $x = a \sin u$. Beginning with $\frac{d}{dx}(x) = 1$ and $\frac{d}{du}(a \sin u) = a \cos u$, we obtain $dx = a \cos u\,du$. (Unlike the previous example, in this case we wouldn't need to do any algebra to solve for dx.)
3. Solve for dx from the equation in Step 2.
4. Determine the new lower and upper limits of integration for the variable u which correspond to the old limits of integration for the variable x. Use the equation from Step 1 to determine the new limits from the old ones.
5. Make **three** substitutions in the original integral:
    - First replace x with the function of u from Step 1.
    - Next replace dx with the equation from Step 3. Don't forget the du.
    - Replace the old limits of integration with the new ones for u.
6. You should now be able to do the new integral in terms of u.
7. Carry out the new integral over the variable u. If you're using a trigonometric substitution, the following trig identities may come in handy.

$$\sin^2 u + \cos^2 u = 1 \quad , \quad \tan^2 u + 1 = \sec^2 u \quad , \quad 1 + \cot^2 u = \csc^2 u$$

$$\cos^2 u = \frac{1 + \cos 2u}{2} \quad , \quad \sin^2 u = \frac{1 - \cos 2u}{2}$$

Example: Perform the following definite integral: $\int_{x=1}^{2}(2x-1)^4\,dx$.

We make the substitution $u = 2x - 1$. Take an implicit derivative of $u = 2x - 1$:

- On the left-hand side: $\frac{d}{du}(u) = 1$. Multiply by du to get $1\,du = du$.
- On the right-hand side: $\frac{d}{dx}(2x-1) = 2$. Multiply by dx to get $2\,dx$.

The implicit derivative of $u = 2x - 1$ is therefore $du = 2\,dx$. Solve for dx to get $dx = \frac{du}{2}$.

$$u = 2x - 1 \quad , \quad dx = \frac{du}{2}$$

Now we must adjust the limits. Plug each limit into $u = 2x - 1$:

$$u(x=1) = 2(1) - 1 = 2 - 1 = 1$$
$$u(x=2) = 2(2) - 1 = 4 - 1 = 3$$

The integral becomes:

$$\int_{x=1}^{2}(2x-1)^4\,dx = \int_{u=1}^{3} u^4 \frac{du}{2} = \frac{1}{2}\int_{u=1}^{3} u^4\,du$$

Now we can integrate over u.

$$\frac{1}{2}\int_{u=1}^{3} u^4\,du = \frac{1}{2}\left[\frac{u^{4+1}}{4+1}\right]_{u=1}^{3} = \frac{1}{2}\left[\frac{u^5}{5}\right]_{u=1}^{3}$$

$$= \frac{1}{2}\frac{(3)^5}{5} - \frac{1}{2}\frac{(1)^5}{5} = \frac{243}{10} - \frac{1}{10} = \frac{242}{10} = \frac{121}{5}$$

Both the old and new integrals equal $\frac{121}{5}$.

$$\int_{x=1}^{2}(2x-1)^4\,dx = \frac{121}{5}$$

$$\frac{1}{2}\int_{u=1}^{3} u^4\,du = \frac{121}{5}$$

The answer to this example is $\frac{121}{5}$.

Example: Perform the following definite integral:

$$\int_{x=0}^{2} \frac{dx}{\sqrt{x^2+4}}$$

Write $x^2 + 4$ as $x^2 + 2^2$. Now it looks like $x^2 + a^2$ with $a = 2$. Following the suggestion in the strategy, we make the substitution $x = 2\tan u$. Take an implicit derivative:

- On the left-hand side: $\frac{d}{dx}(x) = 1$. Multiply by dx to get $1\,dx = dx$.
- On the right-hand side: $\frac{d}{du}(2\tan u) = 2\sec^2 u$. Multiply by du to get $2\sec^2 u\ du$.

The implicit derivative of $x = 2\tan u$ is therefore $dx = 2\sec^2 u\ du$. We will make the following pair of substitutions in the original integral:

$$x = 2\tan u \quad , \quad dx = 2\sec^2 u\ du$$

Now we must adjust the limits. Solve for u to obtain $u = \tan^{-1}\left(\frac{x}{2}\right)$:

$$u(x=0) = \tan^{-1}\left(\frac{0}{2}\right) = 0°$$

$$u(x=2) = \tan^{-1}\left(\frac{2}{2}\right) = \tan^{-1}(1) = 45°$$

The integral becomes:

$$\int_{x=0}^{2} \frac{dx}{\sqrt{x^2+4}} = \int_{u=0°}^{45°} \frac{2\sec^2 u\ du}{\sqrt{(2\tan u)^2+4}} = \int_{u=0°}^{45°} \frac{2\sec^2 u\ du}{\sqrt{4\tan^2 u+4}}$$

Factor out the 4 to write $\sqrt{4\tan^2 u + 4} = \sqrt{4(\tan^2 u + 1)} = 2\sqrt{\tan^2 u + 1}$

$$\int_{x=0}^{2} \frac{dx}{\sqrt{x^2+4}} = \int_{u=0°}^{45°} \frac{2\sec^2 u\ du}{2\sqrt{\tan^2 u+1}}$$

Use the trig identity $\tan^2 u + 1 = \sec^2 u$ to replace $\sqrt{\tan^2 u + 1}$ with $\sqrt{\sec^2 u}$.

$$\int_{x=0}^{2} \frac{dx}{\sqrt{x^2+4}} = \int_{u=0°}^{45°} \frac{\sec^2 u\ du}{\sqrt{\sec^2 u}} = \int_{u=0°}^{45°} \frac{\sec^2 u\ du}{\sec u} = \int_{u=0°}^{45°} \sec u\, du$$

Look up the integral of secant (on page 70).

$$\int_{u=0°}^{45°} \sec u\, du = [\ln|\sec u + \tan u|]_{u=0°}^{45°} = \ln|\sec 45° + \tan 45°| - \ln|\sec 0° + \tan 0°|$$

$$\int_{u=0°}^{45°} \sec u\, du = \ln\left|\sqrt{2}+1\right| - \ln|1+0| = \ln\left|\sqrt{2}+1\right| - \ln|1| = \ln\left|\sqrt{2}+1\right| - 0 = \ln\left|\sqrt{2}+1\right|$$

We used the logarithm identity $\ln(1) = 0$. The answer to the definite integral is $\ln\left|\sqrt{2}+1\right|$.

28. Perform the following definite integral.

$$\int_{x=0}^{8} \left(\frac{x}{4}+1\right)^3 dx =$$

Answer: 80

29. Perform the following definite integral.

$$\int_{x=1}^{5} 3\sqrt{2x-1}\, dx =$$

Want help? Check the hints section at the back of the book.

Answer: 26

30. Perform the following definite integral.

$$\int_{x=0}^{3} 4\sqrt{9 - x^2}\, dx =$$

Answer: 9π

31. Perform the following definite integral.

$$\int_{x=0}^{4} \frac{dx}{\sqrt{16 + x^2}} =$$

Want help? Check the hints section at the back of the book.

Answer: $\ln\left|\sqrt{2} + 1\right|$

Strategy for Performing Multiple Integrals

To perform a **double integral** or a **triple integral**, follow these steps:

1. Feel free to reverse the order of the differentials. For example, it doesn't really matter whether you write $\iint f(x,y)\,dxdy$ or $\iint f(x,y)\,dydx$. What matters (if there is a variable limit) is which integral you do first. The order of the differentials does not tell you which integral to do first. Look at the limits for this (see Step 2).
2. If any of the integrals has a variable limit, you must perform that integration first. The following examples will help you decide which integral to do first.
   - In the example below, the upper y-limit is a function of x. Since the y-limit has a variable, you must integrate over y before integrating over x.

$$\int_{x=0}^{1}\int_{y=0}^{x^2} f(x,y)\,dxdy = \int_{x=0}^{1}\left(\int_{y=0}^{x^2} f(x,y)\,dy\right)dx$$

   - In the example below, the upper x-limit is a function of y. Since the x-limit has a variable, you must integrate over x before integrating over y.

$$\int_{x=0}^{4y}\int_{y=-8}^{8} f(x,y)\,dxdy = \int_{y=-8}^{8}\left(\int_{x=0}^{4y} f(x,y)\,dx\right)dy$$

   - In the example below, all of the limits are constants. In this case, you can do the integrals in any order.

$$\int_{x=0}^{2}\int_{y=-1}^{1} f(x,y)\,dxdy = \int_{x=0}^{2}\left(\int_{y=-1}^{1} f(x,y)\,dy\right)dx = \int_{y=-1}^{1}\left(\int_{x=0}^{2} f(x,y)\,dx\right)dy$$

   Note: In the above equations, on the right-hand side the order of the differentials does matter, as the integrals have been separated (that is, they were reorganized to show which integral will be performed first).
3. When integrating over one variable, treat the other independent variables as constants. For example:
   - When integrating over x, treat y and z the same way as you would treat any other constants.
   - When integrating over y, treat x and z the same way as you would treat any other constants.
   - When integrating over z, treat x and y the same way as you would treat any other constants.
4. Evaluate the first integral over its limits before performing the second integral.
5. A triple integral works the same way as a double integral, except that you do it in three stages, one integral at a time.

The following examples illustrate how to perform a double or triple integral.

Example: Perform the following double integral:

$$\int_{x=0}^{2}\int_{y=0}^{x^2} x^3 y \, dxdy$$

Since the upper y-limit is a function of x, we must carry out the y-integration first.

$$\int_{x=0}^{2}\int_{y=0}^{x^2} x^3 y \, dxdy = \int_{x=0}^{2}\left(\int_{y=0}^{x^2} x^3 y \, dy\right)dx$$

When integrating over y, we treat x like any other constant. Therefore, we can factor x^3 out of the y integral (but be careful not to pull x^3 out of the x integral). We're applying the same concept as $\int cf(y)\,dy = c\int f(y)\,dy$.

$$\int_{x=0}^{2}\left(\int_{y=0}^{x^2} x^3 y \, dy\right)dx = \int_{x=0}^{2} x^3\left(\int_{y=0}^{x^2} y \, dy\right)dx$$

To help make this clear, we will carry out the complete definite integral over y in parentheses before proceeding.

$$\int_{x=0}^{2} x^3\left(\int_{y=0}^{x^2} y \, dy\right)dx = \int_{x=0}^{2} x^3\left(\left[\frac{y^2}{2}\right]_{y=0}^{y=x^2}\right)dx$$

$$= \int_{x=0}^{2} x^3\left(\frac{(x^2)^4}{2} - \frac{(0)^4}{2}\right)dx = \int_{x=0}^{2} x^3\left(\frac{x^4}{2}\right)dx = \int_{x=0}^{2} \frac{x^7}{2}dx$$

Now we have a single integral over x.

$$\int_{x=0}^{2} \frac{x^7}{2}dx = \frac{1}{2}\left[\frac{x^{7+1}}{7+1}\right]_{x=0}^{x=2} = \frac{1}{2}\left[\frac{x^8}{8}\right]_{x=0}^{x=2} = \frac{2^8}{16} - \frac{0^8}{16} = \frac{256}{16} - 0 = 16$$

Example: Perform the following double integral:

$$\int_{x=0}^{4}\int_{y=1}^{2} xy \, dxdy$$

Since all of the limits are constants, in this example we are free to integrate in any order.

$$\int_{x=0}^{4}\int_{y=1}^{2} xy \, dxdy = \int_{x=0}^{4}\left(\int_{y=1}^{2} xy \, dy\right)dx$$

When integrating over y, we treat x like a constant. We can factor x out of the y integral.

$$\int_{x=0}^{4} x\left(\int_{y=1}^{2} y \, dy\right)dx = \int_{x=0}^{4} x\left(\left[\frac{y^2}{2}\right]_{y=1}^{y=2}\right)dx = \int_{x=0}^{4} x\left(\frac{2^2}{2} - \frac{1^2}{2}\right)dx$$

$$\int_{x=0}^{4} x\left(2 - \frac{1}{2}\right)dx = \int_{x=0}^{4} \frac{3x}{2}dx = \frac{3}{2}\int_{x=0}^{4} x \, dx = \frac{3}{2}\left[\frac{x^2}{2}\right]_{x=0}^{x=4} = \frac{3}{2}\frac{(4)^2}{2} - \frac{3}{2}\frac{(0)^2}{2} = 12$$

Example: Perform the following double integral:

$$\int_{x=y^2}^{9y^2}\int_{y=0}^{5} 3y\sqrt{x}\,dxdy$$

Since the x-limits are functions of y, we must carry out the x-integration first.

$$\int_{x=y^2}^{9y^2}\int_{y=0}^{5} 3y\sqrt{x}\,dxdy = \int_{y=0}^{5}\left(\int_{x=y^2}^{9y^2} 3y\sqrt{x}\,dx\right)dy$$

When integrating over x, we treat y like a constant. We can factor $3y$ out of the x integral.

$$\int_{y=0}^{5}\left(\int_{x=y^2}^{9y^2} 3y\sqrt{x}\,dx\right)dy = \int_{y=0}^{5} 3y\left(\int_{x=y^2}^{9y^2} \sqrt{x}\,dx\right)dy$$

First we write $\frac{1}{\sqrt{x}}$ as $x^{1/2}$. We then compare $x^{1/2}$ to ax^b to see that $a = 1$ and $b = \frac{1}{2}$. Since the anti-derivative of ax^b is $\frac{ax^{b+1}}{b+1}$, it follows that the anti-derivative of $x^{1/2}$ is $\frac{(1)x^{1/2+1}}{1/2+1} = \frac{x^{3/2}}{3/2} = \frac{2x^{3/2}}{3}$ (to divide by $\frac{3}{2}$, multiply by its **reciprocal**). Therefore, $\int \sqrt{x}\,dx = \frac{2x^{3/2}}{3}$.

$$\int_{y=0}^{5} 3y\left(\int_{x=y^2}^{9y^2} \sqrt{x}\,dx\right)dy = \int_{y=0}^{5} 3y\left(\left[\frac{2x^{3/2}}{3}\right]_{x=y^2}^{x=9y^2}\right)dy$$

Evaluate the expression $\frac{2x^{3/2}}{3}$ from $x = y^2$ to $x = 9y^2$.

$$\int_{y=0}^{5} 3y\left(\left[\frac{2x^{3/2}}{3}\right]_{x=y^2}^{x=9y^2}\right)dy = \int_{y=0}^{5} 3y\left(\frac{2(9y^2)^{3/2}}{3} - \frac{2(y^2)^{3/2}}{3}\right)dy$$

Note that $(y^2)^{3/2} = y^3$ and $9^{3/2} = \left(\sqrt{9}\right)^3 = 3^3 = 27$.

$$\int_{y=0}^{5} 3y\left(\frac{2(9y^2)^{3/2}}{3} - \frac{2(y^2)^{3/2}}{3}\right)dy = \int_{y=0}^{5} 3y\left(\frac{2(27y^3)}{3} - \frac{2y^3}{3}\right)dy$$

$$= \int_{y=0}^{5} 3y\left(\frac{54y^3}{3} - \frac{2y^3}{3}\right)dy = \int_{y=0}^{5} 3y\left(\frac{54y^3 - 2y^3}{3}\right)dy = \int_{y=0}^{5} 3y\left(\frac{52y^3}{3}\right)dy = \int_{y=0}^{5} 52y^4\,dy$$

Now we have a single integral over y.

$$\int_{y=0}^{5} 52y^4\,dy = 52\left[\frac{y^{4+1}}{4+1}\right]_{y=0}^{y=5} = 52\left[\frac{y^5}{5}\right]_{y=0}^{y=5} = \frac{52(5)^5}{5} - \frac{52(0)^5}{5} = 32{,}500$$

Example: Perform the following triple integral:

$$\int_{x=1}^{2}\int_{y=0}^{x^2}\int_{z=0}^{y} 8xy\,dx\,dy\,dz$$

Since the y- and z-limits both involve variables, we must perform the y and z integrals first. Furthermore, in this example we must perform the z integral **before** performing the y integral. Why? Because there is a y in the upper limit of the z integral, we won't be able to integrate over y until we perform the z-integration. Thus we begin with the z-integration.

$$\int_{x=1}^{2}\int_{y=0}^{x^2}\left(\int_{z=0}^{y} 8xy\,dz\right)dy\,dx$$

When integrating over z, we treat the independent variables x and y like constants. We can factor $8xy$ out of the z integral. We're left with a trivial integral: $\int dz = z$.

$$\int_{x=1}^{2}\int_{y=0}^{x^2} 8xy\left(\int_{z=0}^{y} dz\right)dy\,dx = \int_{x=1}^{2}\int_{y=0}^{x^2} 8xy\left([z]_{z=0}^{z=y}\right)dy\,dx = \int_{x=1}^{2}\int_{y=0}^{x^2} 8xy(y-0)\,dy\,dx$$

This simplifies to:

$$\int_{x=1}^{2}\int_{y=0}^{x^2} 8xy(y)\,dy\,dx = \int_{x=1}^{2}\int_{y=0}^{x^2} 8xy^2\,dy\,dx$$

Now we have a double integral similar to the previous examples. We perform the y-integration next because there is a variable in the upper limit of the y integral.

$$\int_{x=1}^{2}\int_{y=0}^{x^2} 8xy^2\,dy\,dx = \int_{x=1}^{2}\left(\int_{y=0}^{x^2} 8xy^2\,dy\right)dx$$

We factor out the $8x$ since we treat the independent variable x as a constant when integrating over y.

$$\int_{x=1}^{2} 8x\left(\int_{y=0}^{x^2} y^2\,dy\right)dx = \int_{x=1}^{2} 8x\left(\left[\frac{y^{2+1}}{2+1}\right]_{y=0}^{x^2}\right)dx = \int_{x=1}^{2} 8x\left(\left[\frac{y^3}{3}\right]_{y=0}^{x^2}\right)dx$$

$$\int_{x=1}^{2} 8x\left(\left[\frac{y^3}{3}\right]_{y=0}^{x^2}\right)dx = \int_{x=1}^{2} 8x\left(\frac{(x^2)^3}{3}-\frac{(0)^3}{3}\right)dx = \int_{x=1}^{2} 8x\left(\frac{x^6}{3}\right)dx = \int_{x=1}^{2}\frac{8x^7}{3}dx$$

In the last two steps, we applied the rules $(x^m)^n = x^{mn}$ and $x^m x^n = x^{m+n}$.

$$\int_{x=1}^{2}\frac{8x^7}{3}dx = \left[\frac{8x^{7+1}}{(3)(7+1)}\right]_{x=1}^{2} = \left[\frac{8x^8}{24}\right]_{x=1}^{2} = \left[\frac{x^8}{3}\right]_{x=1}^{2} = \frac{2^8}{3}-\frac{1^8}{3} = \frac{256}{3}-\frac{1}{3} = \frac{255}{3} = 85$$

32. Perform the following double integral.

$$\int_{x=0}^{3} \int_{y=0}^{x} 10xy^2 \, dxdy =$$

Answer: 162

33. Perform the following double integral.

$$\int_{x=0}^{y} \int_{y=0}^{3} 10xy^2 \, dxdy =$$

Want help? Check the hints section at the back of the book.

Answer: 243

34. Perform the following double integral.

$$\int_{x=0}^{2}\int_{y=0}^{3} x^3y^2\,dxdy =$$

Answer: 36

35. Perform the following triple integral.

$$\int_{x=0}^{4}\int_{y=0}^{x}\int_{z=0}^{y} 6z\,dx\,dy\,dz =$$

Want help? Check the hints section at the back of the book.

Answer: 64

Coordinate Systems

Physics problems are often easier to solve if you work with a coordinate system well-suited to the geometry. We will see examples of this in Chapters 7 and 8 (and later in this book).

- **Cartesian coordinates**. A problem featuring straight edges and flat sides (such as a triangle or cube) is often easier to solve using Cartesian coordinates (also called rectangular coordinates).

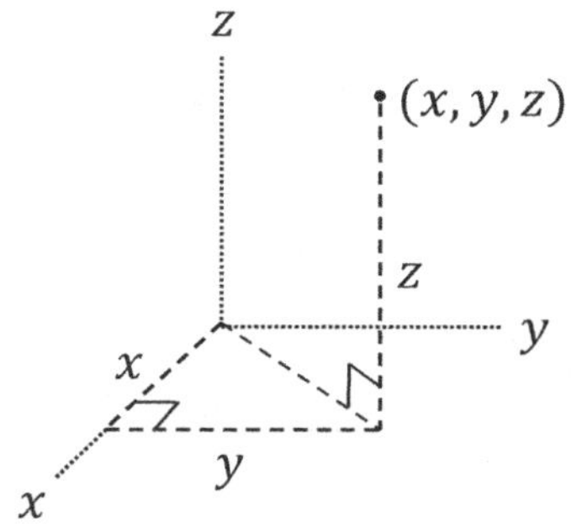

  - x is measured along the x-axis. In the diagram above, x is out of the page.
  - y is measured along the y-axis. In the diagram above, y is to the right.
  - z is measured along the z-axis. In the diagram above, z is up.
- **Polar coordinates**. A problem featuring a circular shape (such as a disc, ring, or circular arc) is often easier to solve using 2D polar coordinates:

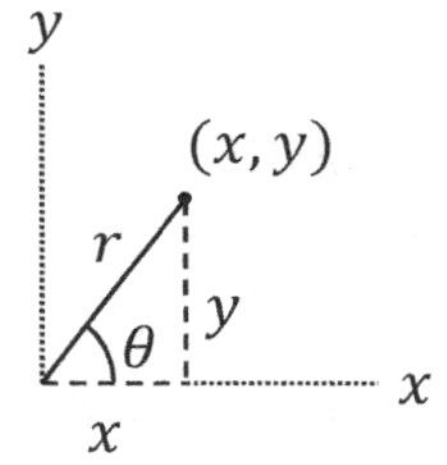

  - r extends radially outward from the origin.
  - θ is measured counterclockwise from the $+x$-axis.
- **Cylindrical coordinates**. A problem featuring cylindrical symmetry (such as a cylinder, cone, or helix) is often easier to solve using cylindrical coordinates:

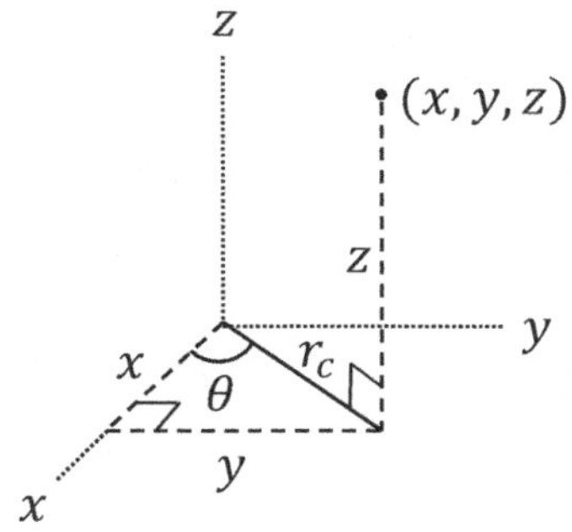

- r_c extends horizontally outward from the z-axis (**not** from the origin).
- θ is measured counterclockwise from the $+x$-axis.
- z is measured along the z-axis. In the previous diagram, z is up.

Cylindrical coordinates are merely a combination of 2D polar coordinates and the z-coordinate.

- **Spherical coordinates**. A problem featuring a spherical shape (such as a solid sphere or a hemispherical shell) is often easier to solve using spherical coordinates.

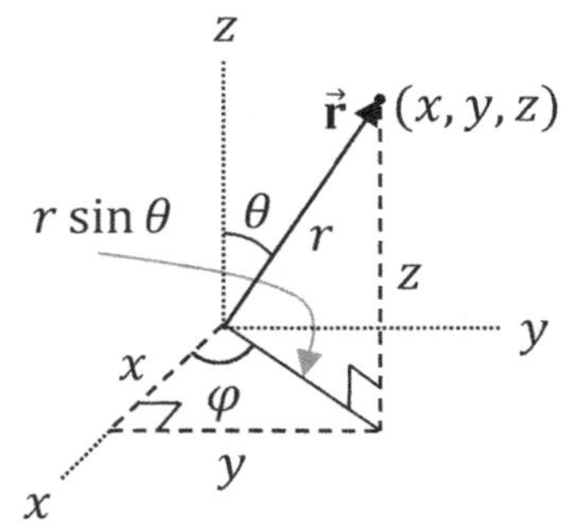

- r extends radially outward from the origin.
- θ is measured down from the $+z$-axis.
- φ is measured counterclockwise from the $+x$-axis after projecting down onto the xy plane. The symbol φ is the lowercase Greek letter phi.

Important Distinctions

Note that the r of spherical coordinates is different from the r_c of cylindrical coordinates: r extends outward from the origin, whereas r_c extends horizontally from the z-axis.

The roles of the angles θ and φ are often swapped in physics textbooks compared to math textbooks. Therefore, if you study physics and vector calculus simultaneously, for example, you may have an additional challenge of trying to keep these two angles straight.

In physics, setting φ constant creates a great vertical circle called a longitude, whereas setting θ constant creates a horizontal circle called a latitude. As shown below, except for the equator, all of the latitudes are smaller than the longitudes.

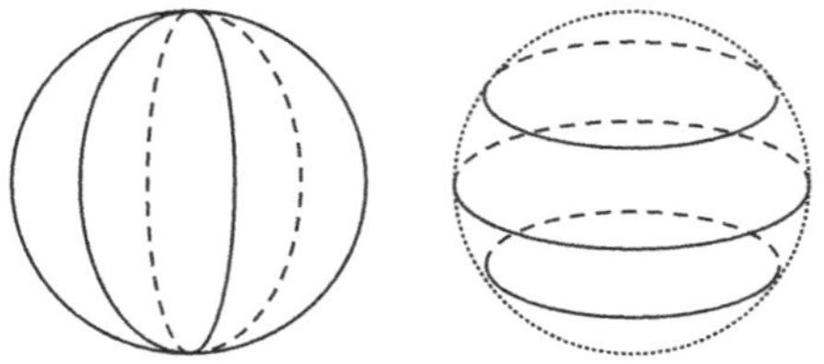

longitude
s

latitude
s

Relations between Different Coordinate Systems

The 2D polar coordinates r and θ are related to the Cartesian coordinates x and y by:

$$x = r\cos\theta \quad , \quad y = r\sin\theta$$

$$r = \sqrt{x^2 + y^2} \quad , \quad \theta = \tan^{-1}\left(\frac{y}{x}\right)$$

The cylindrical coordinates r_c, θ, and z are related to the Cartesian coordinates x, y, and z by (where z is the same in both coordinate systems):

$$x = r_c\cos\theta \quad , \quad y = r_c\sin\theta$$

$$r_c = \sqrt{x^2 + y^2} \quad , \quad \theta = \tan^{-1}\left(\frac{y}{x}\right)$$

The spherical coordinates r, θ, and φ are related to the Cartesian coordinates x, y, and z by:

$$x = r\sin\theta\cos\varphi \quad , \quad y = r\sin\theta\sin\varphi \quad , \quad z = r\cos\theta$$

$$r = \sqrt{x^2 + y^2 + z^2} \quad , \quad \theta = \cos^{-1}\left(\frac{z}{r}\right) \quad , \quad \varphi = \tan^{-1}\left(\frac{y}{x}\right)$$

Differential Elements

The differential arc length is represented in general by ds. If you integrate over ds, you get a finite length of arc: $s = \int ds$. The arc length s is finite, whereas ds is infinitesimal.

- For a straight line parallel to the x-axis, use $ds = dx$. For a straight line parallel to the y-axis, use $ds = dy$. For a straight line parallel to the z-axis, use $ds = dz$.
- For a circular arc of radius a, in 2D polar coordinates use $ds = ad\theta$. **Notation**: Lowercase a is a constant (the radius), whereas lowercase r is a variable.

The differential area element is represented in general by dA. If you integrate over dA, you get surface area: $A = \int dA$. This will be done as a double integral in the examples.

- For a flat surface bounded entirely by straight lines (like a triangle or rectangle) lying in the xy plane, use $dA = dxdy$.
- If one of the sides of a flat surface is circular (like a solid semicircle or like a slice of pizza), in 2D polar coordinates use $dA = rdrd\theta$.
- If the surface is spherical (like a thin hemispherical shell) with radius a, in spherical coordinates use $dA = a^2\sin\theta\, d\theta d\varphi$. Recall that a is a constant (unlike r).

The differential volume element is represented in general by dV. If you integrate over dV, you get volume: $V = \int dV$. This will be done as a triple integral in the examples.

- For a solid consisting of flat sides and straight edges (like a cube), use $dV = dxdydz$.
- For a cylindrical solid (like a cylinder or cone), use $dV = r_c dr_c d\theta dz$.
- For a spherical solid (like a solid hemisphere), use $dV = r^2\sin\theta\, drd\theta d\varphi$.

Strategy for Integrating over Arc Length, Surface Area, or Volume

To integrate over arc length, surface area, or volume, follow these steps:

1. First make the appropriate substitution for the differential element:
   - For a straight line parallel to the x-axis, $ds = dx$.
   - For a straight line parallel to the y-axis, $ds = dy$.
   - For a straight line parallel to the z-axis, $ds = dz$.
   - For a circular arc of radius a, $ds = ad\theta$.
   - For a solid polygon like a rectangle or triangle, $dA = dxdy$.
   - For a solid semicircle or pie slice, $dA = rdrd\theta$.
   - For a very thin spherical shell of radius a, $dA = a^2 \sin\theta \, d\theta d\varphi$.
   - For a solid polyhedron like a cube $dV = dxdydz$.
   - For a solid cylinder or cone, $dV = r_c dr_c d\theta dz$.
   - For a portion of a solid sphere like a hemisphere, $dV = r^2 \sin\theta \, drd\theta d\varphi$.
2. An integral over ds is a single integral, an integral over dA is a double integral, and an integral over dV is a volume integral.
3. Set the limits of each integration variable that map out the region of integration, as illustrated in the examples.
4. Perform the integral using techniques from earlier in this chapter.

Example: Find the arc length around the curved path of a semicircle by integrating over arc length.

For a circular arc, $ds = ad\theta$. This entails a single integral over θ. For a **semicircle** (not a full circle), θ varies form 0 to π radians.

$$C = \int ds = \int_{\theta=0}^{\pi} ad\theta$$

When integrating over a circular arc length, the radius a is constant. (In contrast, when integrating over a surface area, with $dA = rdrd\theta$, the symbol r is a variable.)

$$C = a \int_{\theta=0}^{\pi} d\theta = a[\theta]_{\theta=0}^{\pi} = a(\pi - 0) = \pi a$$

The result is the usual equation for one-half the circumference of a circle (since this example involves a semicircle, **<u>not</u>** a full circle).

Example: Find the area of the triangle illustrated below using a double integral.

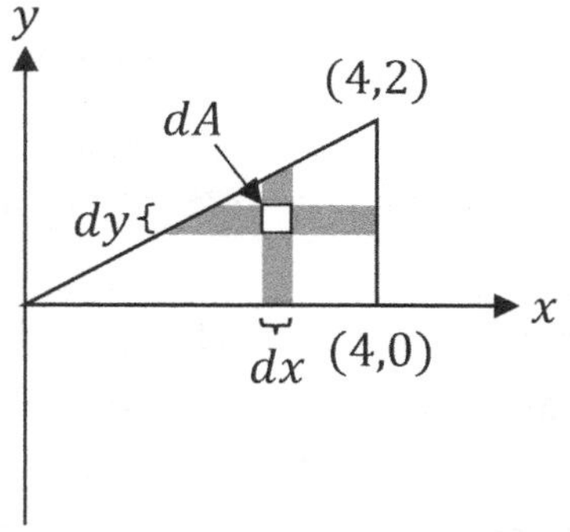

For a triangle, use $dA = dxdy$. This requires integrating over both x and y. We **don't** want to let x vary from 0 to 4 and let y vary from 0 to 2 because that would give us a rectangle instead of a triangle. If we let x vary from 0 to 4, observe that for a given value of x, y varies from 0 to the hypotenuse of the triangle, which is less than 2 (except when x reaches its upper limit).

To find the proper upper limit for y, we need the equation of the line that serves as the triangle's hypotenuse. That line has a slope equal to $\frac{y_2-y_1}{x_2-x_1} = \frac{2-0}{4-0} = \frac{1}{2}$ and a y-intercept of 0. Since the general equation for a line is $y = mx + b$, the equation for this line is $y = \frac{1}{2}x + 0$. For a given value of x, y will vary from 0 to $y = \frac{1}{2}x$, where the upper limit came from the equation for the line (of the hypotenuse). See the vertical gray band in the previous diagram. Now we have the integration limits.

$$A = \int dA = \int_{x=0}^{4} \int_{y=0}^{\frac{x}{2}} dydx$$

We must perform the y-integration first because it has x in its upper limit.

$$A = \int_{x=0}^{4} \left(\int_{y=0}^{\frac{x}{2}} dy \right) dx = \int_{x=0}^{4} \left([y]_{y=0}^{y=\frac{x}{2}} \right) dx = \int_{x=0}^{4} \left(\frac{x}{2} - 0 \right) dx = \int_{x=0}^{4} \frac{x}{2} dx$$

$$A = \int_{x=0}^{4} \frac{x}{2} dx = \left[\frac{x^2}{4} \right]_{x=0}^{x=4} = \frac{4^2}{4} - \frac{0^2}{4} = \frac{16}{4} = 4$$

Of course, we don't need calculus to find the area of a triangle. We could just use the formula one-half base times height: $A = \frac{1}{2}bh = \frac{1}{2}(4)(2) = 4$. However, there are some integrals that can only be done with calculus. (On a similar note, while area can be calculated as a single integral, there are some physics integrals that can only be done as double or triple integrals, so multi-integration is a necessary skill.)

Example: Find the area of a solid semicircle using a double integral.

For a solid semicircle, use $dA = rdrd\theta$. This requires integrating over both r and θ.

Unlike in the previous example, r and θ each have constant limits:

- $0 \leq r \leq a$
- $0 \leq \theta \leq \pi$ (not 2π, since this is a **semicircle**, not a full circle)

Note that a is a constant (the radius of the circle), whereas r is a variable of integration. For any value of θ, the variable r ranges between 0 and a. Contrast this with the previous example where the upper limit of y depended on the value of x.

Substitute $dA = rdrd\theta$ into the area integral.

$$A = \int dA = \int_{r=0}^{a} \int_{\theta=0}^{\pi} rdrd\theta$$

Since all of the limits are constant, we can do these integrals in any order.

$$A = \int_{r=0}^{a} \left(\int_{\theta=0}^{\pi} rd\theta \right) dr$$

When integrating over θ, treat the independent variable r as a constant. This means that you can pull r out of the θ integral (but be careful not to pull r out of the r integral).

$$A = \int_{r=0}^{a} r \left(\int_{\theta=0}^{\pi} d\theta \right) dr = \int_{r=0}^{a} r[\theta]_{\theta=0}^{\pi}\, dr = \int_{r=0}^{a} r(\pi - 0)\, dr = \int_{r=0}^{a} \pi r\, dr$$

$$A = \int_{r=0}^{a} \pi r\, dr = \pi \int_{r=0}^{a} r\, dr = \pi \left[\frac{r^2}{2}\right]_{r=0}^{a} = \pi \left(\frac{a^2}{2} - \frac{0^2}{2}\right) = \frac{\pi a^2}{2}$$

As expected, the area of a semicircle is $A = \frac{\pi a^2}{2}$ (one-half the area of a circle, since this example involves a semicircle, **<u>not</u>** a full circle).

36. Find the area of the triangle illustrated below using a double integral.

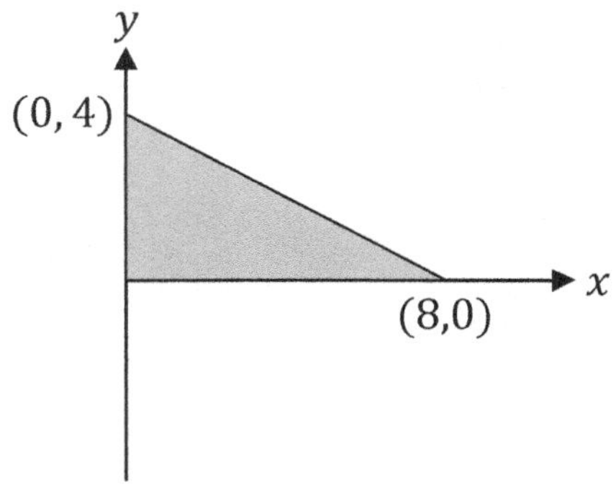

Want help? Check the hints section at the back of the book.

Answer: 16

37. A thick circular ring has an inner radius of 2.0 m and an outer radius of 4.0 m. Find the area of this ring using a double integral.

Want help? Check the hints section at the back of the book.

Answer: 12π m^2

7 ELECTRIC FIELD INTEGRALS

Relevant Terminology

Electric charge – a fundamental property of a particle that causes the particle to experience a force in the presence of an electric field. (An electrically neutral particle has no charge and thus experiences no force in the presence of an electric field.)
Electric field – force per unit charge.
Field point – the point where you are trying to calculate the electric field. The field point is ordinarily specified in the problem. There often is **not** a charge at the field point.
Source – the charge that is creating the electric field.
Conductor – a material through which electrons are able to flow readily. Metals tend to be good conductors of electricity.
Insulator – a material through which electrons are **not** able to flow readily. Wood and glass are examples of good insulators.

Charge Densities

We use three different kinds of charge densities when performing electric field integrals, depending upon the geometry.

- Linear charge density λ (lambda) applies to charge that is distributed along a line (like a long, thin rod) or an arc (like a wire bent into a semicircle).
- Surface charge density σ (sigma) applies to charge that is distributed over a surface (like a metal sheet or like the surface of a sphere).
- Volume charge density ρ (rho) applies to charge that is distributed over a volume (like charge that is distributed throughout the volume of a sphere, instead of just on the sphere's surface).

For a conductor in electrostatic equilibrium, the charge will only reside on the surface. For a charged insulator, it's possible to have the charge spread throughout its volume. That's why you wouldn't use ρ for a conducting sphere, but why you would use ρ for a spherical insulator with charge spread throughout its volume.

Notation

Unfortunately, not all electricity and magnetism books adopt the same notation. Some textbooks may use different symbols for λ, σ, and ρ, and also for ds, dA, and dV. If you're also reading another book, be sure to compare notation carefully (this workbook has a handy chart in each chapter defining each symbol). For example, a few books use $d\ell$ for arc length (instead of ds) and dS (uppercase) for surface area (instead of dA).

Symbols and SI Units

Symbol	Name	SI Units
dq	differential charge element	C
Q	total charge of the object	C
$\vec{\mathbf{E}}$	electric field	N/C or V/m
k	Coulomb's constant	$\frac{\text{N}\cdot\text{m}^2}{\text{C}^2}$ or $\frac{\text{kg}\cdot\text{m}^3}{\text{C}^2\cdot\text{s}^2}$
$\vec{\mathbf{R}}$	a vector from each dq to the field point	m
$\hat{\mathbf{R}}$	a unit vector from each dq toward the field point	unitless
R	the distance from each dq to the field point	m
x, y, z	Cartesian coordinates of dq	m, m, m
$\hat{\mathbf{x}}, \hat{\mathbf{y}}, \hat{\mathbf{z}}$	unit vectors along the $+x$-, $+y$-, $+z$-axes	unitless
r, θ	2D polar coordinates of dq	m, rad
r_c, θ, z	cylindrical coordinates of dq	m, rad, m
r, θ, φ	spherical coordinates of dq	m, rad, rad
$\hat{\mathbf{r}}, \hat{\boldsymbol{\theta}}, \hat{\boldsymbol{\varphi}}$	unit vectors along spherical coordinate axes	unitless
$\hat{\mathbf{r}}_c$	a unit vector pointing away from the $+z$-axis	unitless
λ	linear charge density (for an arc)	C/m
σ	surface charge density (for an area)	C/m^2
ρ	volume charge density	C/m^3
ds	differential arc length	m
dA	differential area element	m^2
dV	differential volume element	m^3

Note: The symbols λ, σ, and ρ are the lowercase Greek letters lambda, sigma, and rho.

Strategy for Performing the Electric Field Integral

To find the electric field created by a continuous distribution of charge (called the **source**) at a specified point (called the **field point**), follow these steps:

1. Draw the charged object. We will call Q the total charge of the object. The total charge Q is composed of an infinite number of infinitesimal charges dq. Visualize integrating over every dq that makes up the charged object.
2. Draw and label a representative dq somewhere within the object. **Don't** draw dq at the origin or on an axis (unless the object is a rod and every point of the rod lies on an axis, in which case you have no choice). Draw an arrow from dq (the **source**) to the **field point** (the point where the problem asks you to find the electric field). Label this displacement as $\vec{\mathbf{R}}$. See the diagrams in the examples that follow.
3. Begin with the electric field integral.
$$\vec{\mathbf{E}} = k \int \frac{\hat{\mathbf{R}}}{R^2}\, dq$$
Interpret these symbols as follows:
   - $\vec{\mathbf{E}}$ is the electric field at the field point.
   - k is Coulomb's constant: $k = 9.0 \times 10^9\ \frac{\text{N·m}^2}{\text{C}^2}$.
   - $\hat{\mathbf{R}}$ is a unit vector pointing from each dq to the field point. In most of the problems, $\hat{\mathbf{R}}$ is a variable and may **not** come out of the integral.
   - R is the distance from each dq to the field point. In most of the problems, R is a variable and may **not** come out of the integral.
   - dq represents each infinitesimal charge element that makes up the object.
4. Study your diagram to express R in terms of suitable coordinates and to express $\hat{\mathbf{R}}$ in terms of suitable unit vectors.
   - For a line or an object with all straight edges (like a polygon), work with Cartesian coordinates (x, y, z) and unit vectors $(\hat{\mathbf{x}}, \hat{\mathbf{y}}, \hat{\mathbf{z}})$. Recall that $\hat{\mathbf{x}}$ points along $+x$, $\hat{\mathbf{y}}$ points along $+y$, and $\hat{\mathbf{z}}$ points along $+z$.
   - For a circular shape, work with 2D polar coordinates (r, θ) and unit vectors $(\hat{\mathbf{r}}, \hat{\boldsymbol{\theta}})$. Recall that $\hat{\mathbf{r}}$ points outward from the origin.
   $$\hat{\mathbf{r}} = \hat{\mathbf{x}} \cos\theta + \hat{\mathbf{y}} \sin\theta$$
   - For a spherical shape, work with spherical coordinates (r, θ, φ) and unit vectors $(\hat{\mathbf{r}}, \hat{\boldsymbol{\theta}}, \hat{\boldsymbol{\varphi}})$. Recall that $\hat{\mathbf{r}}$ points outward from the origin.
   $$\hat{\mathbf{r}} = \hat{\mathbf{x}} \cos\varphi \sin\theta + \hat{\mathbf{y}} \sin\varphi \sin\theta + \hat{\mathbf{z}} \cos\theta$$
   - For a cylinder or cone, work with cylindrical coordinates (r_c, θ, φ) and unit vectors $(\hat{\mathbf{r}}_c, \hat{\boldsymbol{\theta}}, \hat{\mathbf{z}})$. Recall that $\hat{\mathbf{r}}_c$ and $\hat{\boldsymbol{\theta}}$ are the same as the 2D polar unit vectors and $\hat{\mathbf{z}}$ points along the $+z$-axis.
   $$\hat{\mathbf{r}}_c = \hat{\mathbf{x}} \cos\theta + \hat{\mathbf{y}} \sin\theta$$

5. For a problem where the unit vector $\hat{\mathbf{R}}$ (which is the direction of $\vec{\mathbf{R}}$) doesn't seem easy to express in terms of other unit vectors, it may be easier to express $\vec{\mathbf{R}}$ in terms of unit vectors and then divide by its magnitude: $\hat{\mathbf{R}} = \frac{\vec{\mathbf{R}}}{R}$. See the second example.
6. Make one of the following substitutions for dq, depending on the geometry:
   - $dq = \lambda ds$ for an arc length (like a rod or circular arc).
   - $dq = \sigma dA$ for a surface area (like a triangle, disc, or thin spherical shell).
   - $dq = \rho dV$ for a 3D solid (like a solid cube or a solid hemisphere).
7. Choose the appropriate coordinate system and make a substitution for ds, dA, or dV from Step 6 using the strategy from Chapter 6 (on page 86):
   - For a straight line parallel to the x-axis, $ds = dx$.
   - For a straight line parallel to the y-axis, $ds = dy$.
   - For a straight line parallel to the z-axis, $ds = dz$.
   - For a circular arc of radius a, $ds = a d\theta$.
   - For a solid polygon like a rectangle or triangle, $dA = dxdy$.
   - For a solid semicircle (**<u>not</u>** a circular arc) or pie slice, $dA = rdrd\theta$.
   - For a very thin spherical shell of radius a, $dA = a^2 \sin\theta \, d\theta d\varphi$.
   - For a solid polyhedron like a cube, $dV = dxdydz$.
   - For a solid cylinder or cone, $dV = r_c dr_c d\theta dz$.
   - For a portion of a solid sphere like a hemisphere, $dV = r^2 \sin\theta \, drd\theta d\varphi$.
8. Is the density uniform or non-uniform?
   - If the density is uniform, you can pull λ, σ, or ρ out of the integral.
   - If the density is non-uniform, leave λ, σ, or ρ in the integral.
9. If you have Cartesian coordinates in your integrand, but are integrating over polar, cylindrical, or spherical coordinates (or vice-versa), use the following substitutions, as needed, to put all of your coordinates in the same system.

$$x = r\cos\theta \quad , \quad y = r\sin\theta \quad \text{(2D polar)}$$
$$x = r_c\cos\theta \quad , \quad y = r_c\sin\theta \quad \text{(cylindrical)}$$
$$x = r\sin\theta\cos\varphi \quad , \quad y = r\sin\theta\sin\varphi \quad , \quad z = r\cos\theta \quad \text{(spherical)}$$

10. An integral over ds is a single integral, an integral over dA is a double integral, and an integral over dV is a triple integral. Set the limits of each integration variable that map out the region of integration, as illustrated in Chapter 6. Perform the integral using techniques from Chapter 6.
11. Perform the following integral to determine the total charge (Q) of the object:

$$Q = \int dq$$

    Make the same substitutions as you made in Steps 6-10. Use the same limits of integration as you used in Step 10.
12. When you finish with Step 11, substitute your expression for Q into the result from your original electric field integral (Step 10). Simplify the resulting expression.

Example: A rod has non-uniform charge density $\lambda = \beta x^2$, where β is a constant, and endpoints at $(L, 0)$ and $(2L, 0)$, where L is a constant, as illustrated below. The positively charged rod has total charge Q. Derive an equation for the electric field at the origin in terms of k, Q, L, and appropriate unit vectors.

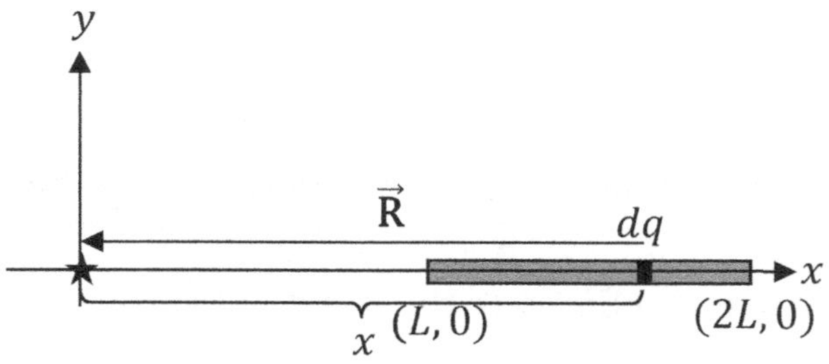

Begin with a labeled diagram. Draw a representative dq. Draw $\vec{\mathbf{R}}$ from the source, dq, to the field point $(0,0)$. When we perform the integration, we effectively integrate over every dq that makes up the rod. Start the math with the electric field integral.

$$\vec{\mathbf{E}} = k \int \frac{\hat{\mathbf{R}}}{R^2} dq$$

Examine the picture above:

- The vector $\vec{\mathbf{R}}$ has a different length for each dq that makes up the rod. Therefore, its magnitude, R, can't come out of the integral. Note that $R = x$, since x is the distance between the source (each dq) and the field point (the origin).
- Observe that $\vec{\mathbf{R}}$ points in the $-x$-direction. Therefore, $\hat{\mathbf{R}} = -\hat{\mathbf{x}}$.

Substitute the expressions for R and $\hat{\mathbf{R}}$ into the electric field integral.

$$\vec{\mathbf{E}} = k \int \frac{\hat{\mathbf{R}}}{R^2} dq = k \int \frac{-\hat{\mathbf{x}}}{x^2} dq$$

For a thin rod, we write $dq = \lambda ds$ (Step 6 on page 94). Note that $-\hat{\mathbf{x}}$ is a constant and may come out of the integral, since $\hat{\mathbf{x}}$ is exactly one unit long and always points along the $+x$-axis. In contrast, the symbol x is <u>**not**</u> a constant and may <u>**not**</u> come out of the integral.

$$\vec{\mathbf{E}} = -k\hat{\mathbf{x}} \int \frac{1}{x^2} \lambda ds$$

Since the rod has non-uniform charge density, we may <u>**not**</u> pull λ out of the integral. Instead, make the substitution $\lambda = \beta x^2$ using the expression given in the problem.

$$\vec{\mathbf{E}} = -k\hat{\mathbf{x}} \int \frac{1}{x^2} (\beta x^2) ds$$

The constant β may come out of the integral. Note that x^2 cancels out: $\frac{x^2}{x^2} = 1$.

$$\vec{\mathbf{E}} = -k\beta\hat{\mathbf{x}} \int ds$$

For a rod lying along the x-axis, we work with Cartesian coordinates and write the differential arc length as $ds = dx$. The limits of integration correspond to the length of the rod: $L \le x \le 2L$ (the endpoints of the rod).

$$\vec{\mathbf{E}} = -k\beta\hat{\mathbf{x}} \int_{x=L}^{2L} dx$$

This integral is trivial: The anti-derivative is x.

$$\vec{\mathbf{E}} = -k\beta\hat{\mathbf{x}}[x]_{x=L}^{2L}$$

Evaluate $[x]_{x=L}^{2L}$ over the limits.

$$\vec{\mathbf{E}} = -k\beta\hat{\mathbf{x}}(2L - L)$$

Note that $2L - L = L$. (Two lengths minus one length equals one length.)

$$\vec{\mathbf{E}} = -k\beta L\hat{\mathbf{x}}$$

We're not finished yet. We must eliminate the constant β from our answer. The way to do this is to integrate over dq to find the total charge of the rod, Q. This integral is convenient since we use the same substitutions from before.

$$Q = \int dq = \int \lambda\, ds = \int (\beta x^2)\, ds = \beta \int x^2\, ds = \beta \int_{x=L}^{2L} x^2\, dx$$

$$Q = \beta \left[\frac{x^3}{3}\right]_{x=L}^{2L} = \beta\left(\frac{(2L)^3}{3} - \frac{L^3}{3}\right) = \beta\left(\frac{8L^3}{3} - \frac{L^3}{3}\right) = \beta\left(\frac{8L^3 - L^3}{3}\right) = \beta\frac{7L^3}{3}$$

Note that $(2L)^3 = 2^3L^3 = 8L^3$ and that $8L^3 - L^3 = 7L^3$. Solve for β in terms of Q: Multiply both sides of the equation by 3 and divide both sides by $7L^3$.

$$\beta = \frac{3Q}{7L^3}$$

Substitute this expression for β into our previous expression for $\vec{\mathbf{E}}$.

$$\vec{\mathbf{E}} = -k\beta L\hat{\mathbf{x}} = -k\left(\frac{3Q}{7L^3}\right)L\hat{\mathbf{x}} = -\frac{3kQ}{7L^2}\hat{\mathbf{x}}$$

Note that $\frac{L}{L^3} = \frac{1}{L^2}$. The electric field at the origin is $\vec{\mathbf{E}} = -\frac{3kQ}{7L^2}\hat{\mathbf{x}}$. It has a magnitude of $E = \frac{3kQ}{7L^2}$ and a direction of $\hat{\mathbf{E}} = -\hat{\mathbf{x}}$.

Tip: Check the units of your answer for consistency, as this can help to catch mistakes. Your final expression for electric field should include the units of k times the unit of charge divided by the unit of length squared. Our answer, $E = \frac{3kQ}{7L^2}$, meets this criteria.

Example: A semi-infinite* rod lies on the y-axis with one end at the origin, as illustrated below. The positively charged rod has uniform charge density λ. Derive an equation for the electric field at the point $(a, 0)$, where a is a constant, in terms of k, λ, a, and appropriate unit vectors.

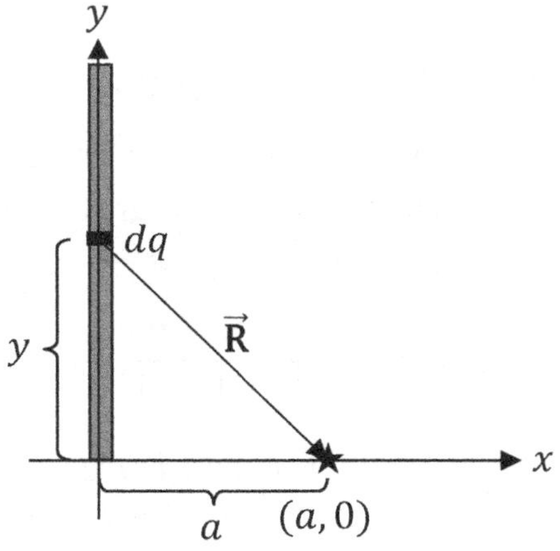

Begin with a labeled diagram. Draw a representative dq. Draw $\vec{\mathbf{R}}$ from the source, dq, to the field point $(a, 0)$. When we perform the integration, we effectively integrate over every dq that makes up the rod. Start the math with the electric field integral.

$$\vec{\mathbf{E}} = k \int \frac{\hat{\mathbf{R}}}{R^2} dq$$

Examine the picture above:

- The vector $\vec{\mathbf{R}}$ has a different length for each dq that makes up the rod. Therefore, its magnitude, R, can't come out of the integral. What we can do is express R in terms of the constant a and the variable y using the Pythagorean theorem: $R = \sqrt{a^2 + y^2}$.
- Similarly, $\vec{\mathbf{R}}$ points in a different direction for each dq that makes up the rod, so its direction, $\hat{\mathbf{R}}$, can't come out of the integral. Note that $\vec{\mathbf{R}}$ extends a units to the right (along $\hat{\mathbf{x}}$) and y units downward (along $-\hat{\mathbf{y}}$). Therefore, $\vec{\mathbf{R}} = a\,\hat{\mathbf{x}} - y\,\hat{\mathbf{y}}$. Find the unit vector $\hat{\mathbf{R}}$ by dividing $\vec{\mathbf{R}}$ by R.†

$$\hat{\mathbf{R}} = \frac{\vec{\mathbf{R}}}{R} = \frac{a\,\hat{\mathbf{x}} - y\,\hat{\mathbf{y}}}{\sqrt{a^2 + y^2}}$$

Substitute the expressions for R and $\hat{\mathbf{R}}$ into the electric field integral.

$$\vec{\mathbf{E}} = k \int \frac{\hat{\mathbf{R}}}{R^2} dq = k \int \hat{\mathbf{R}} \frac{1}{R^2} dq = k \int \frac{a\,\hat{\mathbf{x}} - y\,\hat{\mathbf{y}}}{\sqrt{a^2 + y^2}} \frac{1}{a^2 + y^2} dq = k \int \frac{a\,\hat{\mathbf{x}} - y\,\hat{\mathbf{y}}}{(a^2 + y^2)^{3/2}} dq$$

Note that $R^2 = \left(\sqrt{a^2 + y^2}\right)^2 = a^2 + y^2$ and that $\sqrt{a^2 + y^2}(a^2 + y^2) = (a^2 + y^2)^{3/2}$, where

---

* Although a rod can't really be infinite, a rod can be approximately infinite, provided that the distance a is small compared to the length of the rod, L. That is, $a \ll L$ (where $\ll$ means "much less than").

† The vector $\vec{\mathbf{R}}$ equals its magnitude (R) times its direction ($\hat{\mathbf{R}}$): $\vec{\mathbf{R}} = R\hat{\mathbf{R}}$. Solving for $\hat{\mathbf{R}}$, we get $\hat{\mathbf{R}} = \frac{\vec{\mathbf{R}}}{R}$.

we applied the rule from algebra that $\sqrt{x}\,x = x^{1/2}x^1 = x^{1/2+1} = x^{3/2}$.
For a thin rod, we write $dq = \lambda ds$ (Step 6 on page 94).

$$\vec{\mathbf{E}} = k \int \frac{a\,\hat{\mathbf{x}} - y\,\hat{\mathbf{y}}}{(a^2 + y^2)^{3/2}} \lambda ds$$

Since the rod has uniform charge density, we may pull λ out of the integral.

$$\vec{\mathbf{E}} = k\lambda \int \frac{a\,\hat{\mathbf{x}} - y\,\hat{\mathbf{y}}}{(a^2 + y^2)^{3/2}} ds$$

For a rod lying along the y-axis, we work with Cartesian coordinates and write the differential arc length as $ds = dy$. The limits of integration correspond to the length of the rod: $0 \leq y < \infty$. (The semi-infinite rod extends to $y \to \infty$.)

$$\vec{\mathbf{E}} = k\lambda \int_{y=0}^{\infty} \frac{a\,\hat{\mathbf{x}} - y\,\hat{\mathbf{y}}}{(a^2 + y^2)^{3/2}} dy$$

Split this into two separate integrals using the rule $\int (f_1 + f_2)\,dy = \int f_1\,dy + \int f_2\,dy$.

$$\vec{\mathbf{E}} = k\lambda \int_{y=0}^{\infty} \frac{a\,\hat{\mathbf{x}}}{(a^2 + y^2)^{3/2}} dy - k\lambda \int_{y=0}^{\infty} \frac{y\,\hat{\mathbf{y}}}{(a^2 + y^2)^{3/2}} dy$$

Note that the constant a and the unit vectors $\hat{\mathbf{x}}$ and $\hat{\mathbf{y}}$ can come out of the integrals.

$$\vec{\mathbf{E}} = k\lambda a\hat{\mathbf{x}} \int_{y=0}^{\infty} \frac{1}{(a^2 + y^2)^{3/2}} dy - k\lambda\hat{\mathbf{y}} \int_{y=0}^{\infty} \frac{y}{(a^2 + y^2)^{3/2}} dy$$

These integrals can be performed via the following trigonometric substitution:

$$y = a \tan\theta$$
$$dy = a \sec^2\theta \, d\theta$$

Solving for θ, we get $\theta = \tan^{-1}\left(\frac{y}{a}\right)$, which shows that the new limits of integration are from $\theta = \tan^{-1}(0) = 0°$ to $\theta = \tan^{-1}(\infty) = 90°$ (since $\tan\theta$ approaches infinity in the limit that θ approaches 90°). Note that the denominators of the integrals simplify as follows, using the trig identity $1 + \tan^2\theta = \sec^2\theta$:

$$(a^2 + y^2)^{3/2} = [a^2 + (a\tan\theta)^2]^{3/2} = [a^2(1 + \tan^2\theta)]^{3/2} = (a^2\sec^2\theta)^{3/2} = a^3\sec^3\theta$$

In the last step, we applied the rule from algebra that $(a^2x^2)^{3/2} = (a^2)^{3/2}(x^2)^{3/2} = a^3x^3$.
Substitute the above expressions for y, dy, and $(a^2 + y^2)^{3/2}$ into the previous integrals.

$$\vec{\mathbf{E}} = k\lambda a\hat{\mathbf{x}} \int_{\theta=0°}^{90°} \frac{1}{a^3\sec^3\theta}(a\sec^2\theta\,d\theta) - k\lambda\hat{\mathbf{y}} \int_{\theta=0°}^{90°} \frac{a\tan\theta}{a^3\sec^3\theta}(a\sec^2\theta\,d\theta)$$

$$\vec{\mathbf{E}} = k\lambda a\hat{\mathbf{x}} \int_{\theta=0°}^{90°} \frac{d\theta}{a^2\sec\theta} - k\lambda\hat{\mathbf{y}} \int_{\theta=0°}^{90°} \frac{\tan\theta\,d\theta}{a\sec\theta}$$

Recall from trig that $\sec\theta = \frac{1}{\cos\theta}$ and $\tan\theta = \frac{\sin\theta}{\cos\theta}$. Therefore, it follows that:

$$\frac{1}{\sec\theta} = \cos\theta \quad \text{and} \quad \frac{\tan\theta}{\sec\theta} = \frac{\sin\theta}{\cos\theta} \div \frac{1}{\cos\theta} = \frac{\sin\theta}{\cos\theta} \times \frac{\cos\theta}{1} = \sin\theta$$

To divide by a fraction, multiply by its **reciprocal**. The electric field integral becomes:

$$\vec{\mathbf{E}} = k\lambda a\hat{\mathbf{x}} \int_{\theta=0^\circ}^{90^\circ} \frac{\cos\theta}{a^2} d\theta - k\lambda\hat{\mathbf{y}} \int_{\theta=0^\circ}^{90^\circ} \frac{\sin\theta}{a} d\theta$$

We can pull the constant a out of these integrals. Note that $a\left(\frac{1}{a^2}\right) = \frac{1}{a}$.

$$\vec{\mathbf{E}} = \frac{k\lambda\hat{\mathbf{x}}}{a} \int_{\theta=0^\circ}^{90^\circ} \cos\theta \, d\theta - \frac{k\lambda\hat{\mathbf{y}}}{a} \int_{\theta=0^\circ}^{90^\circ} \sin\theta \, d\theta = \frac{k\lambda\hat{\mathbf{x}}}{a} [\sin\theta]_{\theta=0^\circ}^{90^\circ} - \frac{k\lambda\hat{\mathbf{y}}}{a} [-\cos\theta]_{\theta=0^\circ}^{90^\circ}$$

$$\vec{\mathbf{E}} = \frac{k\lambda\hat{\mathbf{x}}}{a}(\sin 90^\circ - \sin 0^\circ) - \frac{k\lambda\hat{\mathbf{y}}}{a}(-\cos 90^\circ + \cos 0^\circ) = \frac{k\lambda\hat{\mathbf{x}}}{a}(1 - 0) - \frac{k\lambda\hat{\mathbf{y}}}{a}(-0 + 1)$$

$$\vec{\mathbf{E}} = \frac{k\lambda\hat{\mathbf{x}}}{a} - \frac{k\lambda\hat{\mathbf{y}}}{a} = \frac{k\lambda}{a}(\hat{\mathbf{x}} - \hat{\mathbf{y}})$$

The instructions in this problem permit us to include λ in our answer, so it's not necessary to integrate $Q = \int dq$. Contrast this with the previous problem, where the instructions wanted Q in our answer instead of β. The electric field at the point $(a, 0)$ is $\vec{\mathbf{E}} = \frac{k\lambda}{a}(\hat{\mathbf{x}} - \hat{\mathbf{y}})$.

Example: A thin wire is bent into the shape of the semicircle illustrated below. The radius of the semicircle is denoted by the symbol a and the negatively charged wire has total charge $-Q$ and uniform charge density. Derive an equation for the electric field at the origin in terms of k, Q, a, and appropriate unit vectors.

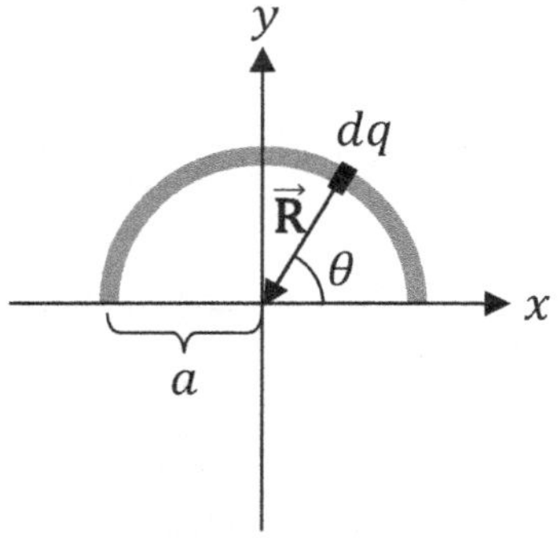

Begin with a labeled diagram. Draw a representative dq: Since this is a semicircular wire (and **not** a solid semicircle – that is, it's **not** half of a solid disc), dq must lie on the circumference and not inside the semicircle. Draw $\vec{\mathbf{R}}$ from the source, dq, to the field point (at the origin). When we perform the integration, we effectively integrate over every dq that makes up the semicircle. Start the math with the electric field integral.

$$\vec{\mathbf{E}} = k\int \frac{\hat{\mathbf{R}}}{R^2}dq$$

Examine the picture above:

- The vector $\vec{\mathbf{R}}$ has the same length for each dq that makes up the semicircle. Its magnitude, R, equals the radius of the semicircle: $R = a$. (Note that this would **not** be the case for a thick ring or a solid semicircle.)
- However, $\vec{\mathbf{R}}$ points in a different direction for each dq that makes up the semicircle, so its direction, $\hat{\mathbf{R}}$, can't come out of the integral. Note that $\hat{\mathbf{R}} = -\hat{\mathbf{r}}$ because the $\hat{\mathbf{r}}$ of 2D polar coordinates is **outward**, whereas $\hat{\mathbf{R}}$ in the picture above is **inward**.

Substitute the expressions $R = a$ and $\hat{\mathbf{R}} = -\hat{\mathbf{r}}$ into the electric field integral.

$$\vec{\mathbf{E}} = k\int \frac{\hat{\mathbf{R}}}{R^2}dq = k\int \frac{-\hat{\mathbf{r}}}{a^2}dq$$

We may pull $-\frac{1}{a^2}$ out of the integral, since the radius (a) is constant.

$$\vec{\mathbf{E}} = -\frac{k}{a^2}\int \hat{\mathbf{r}}\,dq$$

For a thin semicircular wire, we write $dq = \lambda ds$ (Step 6 on page 94).

$$\vec{\mathbf{E}} = -\frac{k}{a^2}\int \hat{\mathbf{r}}\,\lambda ds$$

Since the semicircle has uniform charge density, we may pull λ out of the integral.

$$\vec{\mathbf{E}} = -\frac{k\lambda}{a^2}\int \hat{\mathbf{r}}\,ds$$

For a semicircle, we work with 2D polar coordinates and write the differential arc length as $ds = a\,d\theta$. (It's the same formula as $ds = r d\theta$, where the radius of the semicircle is $r = a$.) The limits of integration are from $\theta = 0$ to $\theta = \pi$ **radians** (for half a circle).

$$\vec{\mathbf{E}} = -\frac{k\lambda}{a^2} \int_{\theta=0}^{\pi} \hat{\mathbf{r}}\,(a\,d\theta)$$

Since the radius (a) of the semicircle is constant, we may pull it out of the integral. Note that $\frac{1}{a^2}(a) = \frac{a}{a^2} = \frac{1}{a}$.

$$\vec{\mathbf{E}} = -\frac{k\lambda}{a} \int_{\theta=0}^{\pi} \hat{\mathbf{r}}\,d\theta$$

Since $\hat{\mathbf{r}}$ points in a different direction for each value of θ, it is not constant and therefore may **not** come out of the integral. What we need to do is express $\hat{\mathbf{r}}$ in terms of θ. You can find such an equation on page 93. For this problem, we need the expression for $\hat{\mathbf{r}}$ in 2D polar coordinates (not the $\hat{\mathbf{r}}$ of spherical coordinates).‡

$$\hat{\mathbf{r}} = \hat{\mathbf{x}}\cos\theta + \hat{\mathbf{y}}\sin\theta$$

Substitute this expression into the electric field integral.

$$\vec{\mathbf{E}} = -\frac{k\lambda}{a} \int_{\theta=0}^{\pi} (\hat{\mathbf{x}}\cos\theta + \hat{\mathbf{y}}\sin\theta)\,d\theta$$

Separate this integral into two integrals – one for each term.

$$\vec{\mathbf{E}} = -\frac{k\lambda}{a} \int_{\theta=0}^{\pi} \hat{\mathbf{x}}\cos\theta\,d\theta - \frac{k\lambda}{a} \int_{\theta=0}^{\pi} \hat{\mathbf{y}}\sin\theta\,d\theta$$

Unlike the 2D polar unit vector $\hat{\mathbf{r}}$, the Cartesian unit vectors $\hat{\mathbf{x}}$ and $\hat{\mathbf{y}}$ are constants (for example, $\hat{\mathbf{x}}$ always points one unit along the $+x$-axis). Therefore, we may factor $\hat{\mathbf{x}}$ and $\hat{\mathbf{y}}$ out of the integrals.

$$\vec{\mathbf{E}} = -\frac{k\lambda}{a}\hat{\mathbf{x}} \int_{\theta=0}^{\pi} \cos\theta\,d\theta - \frac{k\lambda}{a}\hat{\mathbf{y}} \int_{\theta=0}^{\pi} \sin\theta\,d\theta$$

$$\vec{\mathbf{E}} = -\frac{k\lambda}{a}\hat{\mathbf{x}}[\sin\theta]_{\theta=0}^{\pi} - \frac{k\lambda}{a}\hat{\mathbf{y}}[-\cos\theta]_{\theta=0}^{\pi}$$

$$\vec{\mathbf{E}} = -\frac{k\lambda}{a}\hat{\mathbf{x}}(\sin\pi - \sin 0) - \frac{k\lambda}{a}\hat{\mathbf{y}}(-\cos\pi + \cos 0)$$

‡ Here is where this equation comes from. First, any 2D vector can be expressed in terms of unit vectors in the form $\vec{\mathbf{A}} = A_x\hat{\mathbf{x}} + A_y\hat{\mathbf{y}}$. Next, the position vector is special in that its x- and y-components are the x- and y-coordinates: $\vec{\mathbf{r}} = x\hat{\mathbf{x}} + y\hat{\mathbf{y}}$. An alternative way to express the position vector is $\vec{\mathbf{r}} = r\hat{\mathbf{r}}$. Going x units along $\hat{\mathbf{x}}$ and then going y units along $\hat{\mathbf{y}}$ is equivalent to going r units outward along $\hat{\mathbf{r}}$. (See the 2D polar coordinate diagram from Chapter 6 on page 83, and note that $\hat{\mathbf{r}}$ is along the hypotenuse of the right triangle.) Set these two equations equal to one another: $x\hat{\mathbf{x}} + y\hat{\mathbf{y}} = r\hat{\mathbf{r}}$. Divide both sides of the equation by r: $\hat{\mathbf{r}} = \frac{x}{r}\hat{\mathbf{x}} + \frac{y}{r}\hat{\mathbf{y}}$. Recall that $x = r\cos\theta$ and $y = r\sin\theta$, such that $\frac{x}{r} = \cos\theta$ and $\frac{y}{r} = \sin\theta$. Therefore, $\hat{\mathbf{r}} = \hat{\mathbf{x}}\cos\theta + \hat{\mathbf{y}}\sin\theta$.

$$\vec{\mathbf{E}} = -\frac{k\lambda}{a}\hat{\mathbf{x}}(0-0) - \frac{k\lambda}{a}\hat{\mathbf{y}}(1+1) = -\frac{k\lambda}{a}\hat{\mathbf{y}}(2) = -\frac{2k\lambda}{a}\hat{\mathbf{y}}$$

Note that π radians $= 180$ degrees. We're not finished yet. We must eliminate the charge density λ from our answer. The way to do this is to integrate over dq to find the total charge of the semicircle, $-Q$. This integral involves the same substitutions from before. Note that the integral equals $-Q$ because the problem states that the semicircle is **negative** and has a total charge of $-Q$.

$$-Q = \int dq = \int \lambda\, ds = \lambda \int ds = \lambda \int a\, d\theta = a\lambda \int_{\theta=0}^{\pi} d\theta = a\lambda[\theta]_{\theta=0}^{\pi} = a\lambda(\pi - 0) = \pi a\lambda$$

Multiply both sides of the equation by -1 in order to solve for Q. In our notation, the symbol Q is positive, such that we can write the total charge of the semicircle as $-Q$, which is negative. Also, the symbol λ is negative (so that the two minus signs in the equation below make Q positive, such that $-Q$ is negative).

$$Q = -\pi a\lambda$$

The signs get a little tricky with a negatively charged object. However, there is a simpler alternative: You could instead solve the problem for a positively charged object, and simply change the direction of the final answer at the end of the solution to deal with the negative charge. (We didn't do it that way, but we could have. You would get the same final answer.)

We **<u>must</u>** use **radians** (instead of degrees) for the limits of θ (unless **<u>every</u>** θ appears in the argument of a trig function, which is **<u>not</u>** the case here). Solve for λ in terms of Q: Divide both sides of the equation by πa.

$$\lambda = \frac{-Q}{\pi a}$$

Substitute this expression for λ into our previous expression for $\vec{\mathbf{E}}$.

$$\vec{\mathbf{E}} = -\frac{2k\lambda}{a}\hat{\mathbf{y}} = -\frac{2k}{a}\left(\frac{-Q}{\pi a}\right)\hat{\mathbf{y}} = \frac{2kQ}{\pi a^2}\hat{\mathbf{y}}$$

The electric field at the origin is $\vec{\mathbf{E}} = \frac{2kQ}{\pi a^2}\hat{\mathbf{y}}$. It has a magnitude of $E = \frac{2kQ}{\pi a^2}$ and a direction of $\hat{\mathbf{E}} = +\hat{\mathbf{y}}$. If you study the picture, by symmetry it should make sense that the electric field points straight **up** at the origin, along the $+y$-axis: A positive "test" charge placed at the origin would be attracted upward towards the negatively charged semicircle. Note that the horizontal component of the electric field cancels out.

Example: A uniformly charged solid disc lies in the xy plane, centered about the origin as illustrated below. The radius of the disc is denoted by the symbol a and the positively charged disc has total charge Q. Derive an equation for the electric field at the point $(0, 0, p)$ in terms of k, Q, a, p, and appropriate unit vectors.

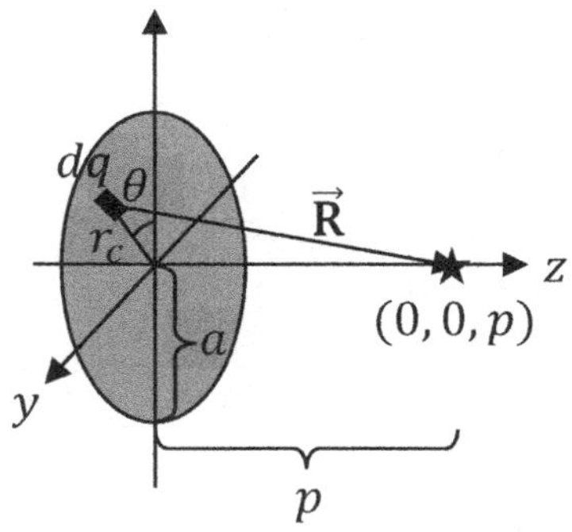

Begin with a labeled diagram. Draw a representative dq: Since this is a solid disc (unlike the previous example), most of the dq's lie within the area of the disc, so we drew dq inside the disc (not on its circumference). Draw $\vec{\mathbf{R}}$ from the source, dq, to the field point $(0, 0, p)$. When we perform the integration, we effectively integrate over every dq that makes up the solid disc. Start the math with the electric field integral.

$$\vec{\mathbf{E}} = k \int \frac{\hat{\mathbf{R}}}{R^2} dq$$

Examine the picture above:

- The vector $\vec{\mathbf{R}}$ has a different length for each dq that makes up the solid disc. Therefore, its magnitude, R, can't come out of the integral. What we can do is express R in terms of the variable r_c and the constant p using the Pythagorean theorem: $R = \sqrt{r_c^2 + p^2}$.
- Similarly, $\vec{\mathbf{R}}$ points in a different direction for each dq that makes up the solid disc, so its direction, $\hat{\mathbf{R}}$, can't come out of the integral. Note that $\vec{\mathbf{R}}$ extends r_c units towards the origin (along $-\hat{\mathbf{r}}_c$) and p units along the $+z$-axis (along $\hat{\mathbf{z}}$). Therefore, $\vec{\mathbf{R}} = -r_c\,\hat{\mathbf{r}}_c + p\,\hat{\mathbf{z}}$. Find the unit vector $\hat{\mathbf{R}}$ by dividing $\vec{\mathbf{R}}$ by R.

$$\hat{\mathbf{R}} = \frac{\vec{\mathbf{R}}}{R} = \frac{-r_c\,\hat{\mathbf{r}}_c + p\,\hat{\mathbf{z}}}{\sqrt{r_c^2 + p^2}}$$

Substitute the expressions for R and $\hat{\mathbf{R}}$ into the electric field integral. Note that $R^2 = \left(\sqrt{r_c^2 + p^2}\right)^2 = r_c^2 + p^2$ and that $\sqrt{r_c^2 + p^2}(r_c^2 + p^2) = (r_c^2 + p^2)^{1/2+1} = (r_c^2 + p^2)^{3/2}$.

$$\vec{\mathbf{E}} = k \int \frac{\hat{\mathbf{R}}}{R^2} dq = k \int \hat{\mathbf{R}} \frac{1}{R^2} dq = k \int \frac{-r_c\,\hat{\mathbf{r}}_c + p\,\hat{\mathbf{z}}}{\sqrt{r_c^2 + p^2}} \frac{1}{r_c^2 + p^2} dq = k \int \frac{-r_c\,\hat{\mathbf{r}}_c + p\,\hat{\mathbf{z}}}{(r_c^2 + p^2)^{3/2}} dq$$

For a solid disc, we write $dq = \sigma dA$ (Step 6 on page 94).

$$\vec{\mathbf{E}} = k \int \frac{-r_c\,\hat{\mathbf{r}}_c + p\,\hat{\mathbf{z}}}{(r_c^2 + p^2)^{3/2}} (\sigma dA)$$

Since the solid disc has uniform charge density, we may pull σ out of the integral.

$$\vec{\mathbf{E}} = k\sigma \int \frac{-r_c\,\hat{\mathbf{r}}_c + p\,\hat{\mathbf{z}}}{(r_c^2 + p^2)^{3/2}} dA$$

For a solid disc, we write the differential area element as $dA = r_c dr_c d\theta$ (Step 7 on page 94), and since the field point at $(0, 0, p)$ does not lie in the xy plane, we work with cylindrical coordinates (which simply adds the z-coordinate to the 2D polar coordinates). The limits of integration are from $r_c = 0$ to $r_c = a$ and $\theta = 0$ to $\theta = 2\pi$ **radians**.

$$\vec{\mathbf{E}} = k\sigma \int_{r_c=0}^{a} \int_{\theta=0}^{2\pi} \frac{-r_c\,\hat{\mathbf{r}}_c + p\,\hat{\mathbf{z}}}{(r_c^2 + p^2)^{3/2}} (r_c dr_c d\theta)$$

Split the integral into two parts, noting that $\frac{-r_c\,\hat{\mathbf{r}}_c + p\,\hat{\mathbf{z}}}{(r_c^2+p^2)^{3/2}} = \frac{-r_c\,\hat{\mathbf{r}}_c}{(r_c^2+p^2)^{3/2}} + \frac{p\,\hat{\mathbf{z}}}{(r_c^2+p^2)^{3/2}}$.

$$\vec{\mathbf{E}} = k\sigma \int_{r_c=0}^{a} \int_{\theta=0}^{2\pi} \frac{-r_c\hat{\mathbf{r}}_c r_c dr_c d\theta}{(r_c^2 + p^2)^{3/2}} + k\sigma \int_{r_c=0}^{a} \int_{\theta=0}^{2\pi} \frac{p\hat{\mathbf{z}} r_c dr_c d\theta}{(r_c^2 + p^2)^{3/2}}$$

Since p and $\hat{\mathbf{z}}$ are constants (since $\hat{\mathbf{z}}$ is one unit long and points along the $+z$-axis), we may pull them out of the second integral. However, r_c and $\hat{\mathbf{r}}_c$ are **not** constants, as they differ for each dq that makes up the solid disc, so r_c and $\hat{\mathbf{r}}_c$ may **not** come out of the integrals.

$$\vec{\mathbf{E}} = k\sigma \int_{r_c=0}^{a} \int_{\theta=0}^{2\pi} \frac{-r_c\hat{\mathbf{r}}_c r_c dr_c d\theta}{(r_c^2 + p^2)^{3/2}} + k\sigma p\hat{\mathbf{z}} \int_{r_c=0}^{a} \int_{\theta=0}^{2\pi} \frac{r_c dr_c d\theta}{(r_c^2 + p^2)^{3/2}}$$

Study the diagram at the beginning of this example. It should be clear from symmetry that a positive test charge placed at the field point $(0, 0, p)$ would be pushed directly to the right, along the $+z$-axis. Therefore, the first integral, which contains $\hat{\mathbf{r}}_c$, will equal zero (though you can do the extra work if you like), and we only need to perform the second integral.

$$\vec{\mathbf{E}} = k\sigma p\hat{\mathbf{z}} \int_{r_c=0}^{a} \int_{\theta=0}^{2\pi} \frac{r_c dr_c d\theta}{(r_c^2 + p^2)^{3/2}}$$

When we integrate over θ, we treat the independent variable r_c as a constant. In fact, nothing in the integrand depends on the variable θ, so we can factor everything out of the θ-integration.

$$\vec{\mathbf{E}} = k\sigma p\hat{\mathbf{z}} \int_{r_c=0}^{a} \frac{r_c dr_c}{(r_c^2 + p^2)^{3/2}} \int_{\theta=0}^{2\pi} d\theta = k\sigma p\hat{\mathbf{z}} \int_{r_c=0}^{a} \frac{r_c dr_c}{(r_c^2 + p^2)^{3/2}} [\theta]_{\theta=0}^{2\pi}$$

$$\vec{\mathbf{E}} = k\sigma p\hat{\mathbf{z}} \int_{r_c=0}^{a} \frac{r_c dr_c}{(r_c^2 + p^2)^{3/2}} (2\pi - 0) = 2\pi k\sigma p\hat{\mathbf{z}} \int_{r_c=0}^{a} \frac{r_c dr_c}{(r_c^2 + p^2)^{3/2}}$$

This integral can be performed via the following trigonometric substitution:

$$r_c = p \tan\psi$$
$$dr_c = p \sec^2\psi\, d\psi$$

Solving for ψ, we get $\theta = \tan^{-1}\left(\frac{r_c}{p}\right)$, which shows that the new limits of integration are from $\psi = \tan^{-1}(0) = 0°$ to $\psi = \tan^{-1}\left(\frac{a}{p}\right)$, which for now we will simply call ψ_{max}. Note that the denominator of the integral simplifies as follows, using the trig identity $\tan^2\psi + 1 = \sec^2\psi$:

$$(r_c^2 + p^2)^{3/2} = [(p\tan\psi)^2 + p^2]^{3/2} = [p^2(\tan^2\psi + 1)]^{3/2} = (p^2\sec^2\psi)^{3/2} = p^3\sec^3\psi$$

In the last step, we applied the rule from algebra that $(p^2x^2)^{3/2} = (p^2)^{3/2}(x^2)^{3/2} = p^3x^3$. Substitute the above expressions for r_c, dr_c, and $(r_c^2 + p^2)^{3/2}$ into the previous integral.

$$\vec{\mathbf{E}} = 2\pi k\sigma p\hat{\mathbf{z}} \int_{r_c=0}^{a} \frac{r_c dr_c}{(r_c^2 + p^2)^{3/2}} = 2\pi k\sigma p\hat{\mathbf{z}} \int_{\psi=0°}^{\psi_{max}} \frac{(p\tan\psi)(p\sec^2\psi\, d\psi)}{p^3\sec^3\psi}$$

Note that $\frac{p^2}{p^3} = \frac{1}{p}$ and $\frac{\sec^2\psi}{\sec^3\psi} = \frac{1}{\sec\psi}$.

$$\vec{\mathbf{E}} = 2\pi k\sigma p\hat{\mathbf{z}} \int_{\psi=0°}^{\psi_{max}} \frac{\tan\psi\, d\psi}{p\sec\psi}$$

The p from outside the integral cancels with the $\frac{1}{p}$ in the integrand.

$$\vec{\mathbf{E}} = 2\pi k\sigma\hat{\mathbf{z}} \int_{\psi=0°}^{\psi_{max}} \frac{\tan\psi\, d\psi}{\sec\psi}$$

Recall from trig that $\sec\psi = \frac{1}{\cos\psi}$ and $\tan\psi = \frac{\sin\psi}{\cos\psi}$. Therefore, it follows that:

$$\frac{\tan\psi}{\sec\psi} = \frac{\sin\psi}{\cos\psi} \div \frac{1}{\cos\psi} = \frac{\sin\psi}{\cos\psi} \times \frac{\cos\psi}{1} = \sin\psi$$

To divide by a fraction, multiply by its **reciprocal**. The electric field integral becomes:

$$\vec{\mathbf{E}} = 2\pi k\sigma\hat{\mathbf{z}} \int_{\psi=0°}^{\psi_{max}} \sin\psi\, d\psi = 2\pi k\sigma\hat{\mathbf{z}}[-\cos\psi]_{\psi=0°}^{\psi_{max}} = 2\pi k\sigma\hat{\mathbf{z}}(-\cos\psi_{max} + \cos 0°)$$

Recall that what we called ψ_{max} is given by $\psi_{max} = \tan^{-1}\left(\frac{a}{p}\right)$, such that $\cos\psi_{max}$ is the complicated looking expression $\cos\left(\tan^{-1}\left(\frac{a}{p}\right)\right)$. When you find yourself taking the cosine of an inverse tangent, you can find a simpler way to write it by drawing a right triangle and applying the Pythagorean theorem. Since $\psi_{max} = \tan^{-1}\left(\frac{a}{p}\right)$, it follows that $\tan\psi_{max} = \frac{a}{p}$. We can make a right triangle from this: Since the tangent of ψ_{max} equals the opposite over the adjacent, we draw a right triangle with a opposite and p adjacent to ψ_{max}.

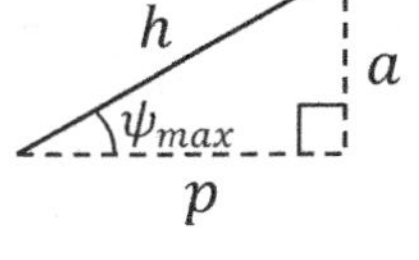

Find the hypotenuse, h, of the right triangle from the Pythagorean theorem.

$$h = \sqrt{p^2 + a^2}$$

Now we can write an expression for the cosine of ψ_{max}. It equals the adjacent (p) over the hypotenuse ($h = \sqrt{p^2 + a^2}$).

$$\cos \psi_{max} = \frac{p}{h} = \frac{p}{\sqrt{p^2 + a^2}}$$

Substitute this expression into the previous equation for electric field.

$$\vec{\mathbf{E}} = 2\pi k\sigma\hat{\mathbf{z}}(-\cos \psi_{max} + \cos 0\,°) = 2\pi k\sigma\hat{\mathbf{z}}\left(-\frac{p}{\sqrt{p^2 + a^2}} + 1\right)$$

To add fractions, make a **common denominator**.

$$\vec{\mathbf{E}} = 2\pi k\sigma\hat{\mathbf{z}}\left(-\frac{p}{\sqrt{p^2 + a^2}} + \frac{\sqrt{p^2 + a^2}}{\sqrt{p^2 + a^2}}\right) = 2\pi k\sigma\hat{\mathbf{z}}\left(\frac{-p + \sqrt{p^2 + a^2}}{\sqrt{p^2 + a^2}}\right)$$

We're not finished yet because we need to eliminate the charge density σ from our answer. The way to do this is to integrate over dq to find the total charge of the disc, Q. This integral involves the same substitutions from before.

$$Q = \int dq = \int \sigma\, dA = \sigma \int dA = \sigma \int_{r_c=0}^{a} \int_{\theta=0}^{2\pi} r_c dr_c d\theta = \sigma \int_{r_c=0}^{a} r_c dr_c \int_{\theta=0}^{2\pi} d\theta$$

$$Q = \sigma \left[\frac{r_c^2}{2}\right]_{r_c=0}^{a} [\theta]_{\theta=0}^{2\pi} = \sigma\left(\frac{a^2}{2} - \frac{0^2}{2}\right)(2\pi - 0) = \sigma\left(\frac{a^2}{2}\right)(2\pi) = \pi\sigma a^2$$

Solve for σ in terms of Q: Divide both sides of the equation by πa^2.

$$\sigma = \frac{Q}{\pi a^2}$$

Substitute this expression for σ into our previous expression for $\vec{\mathbf{E}}$.

$$\vec{\mathbf{E}} = 2\pi k\sigma\hat{\mathbf{z}}\left(\frac{-p + \sqrt{p^2 + a^2}}{\sqrt{p^2 + a^2}}\right) = 2\pi k\left(\frac{Q}{\pi a^2}\right)\hat{\mathbf{z}}\left(\frac{-p + \sqrt{p^2 + a^2}}{\sqrt{p^2 + a^2}}\right)$$

$$\vec{\mathbf{E}} = \frac{2kQ\hat{\mathbf{z}}}{a^2}\left(\frac{-p + \sqrt{p^2 + a^2}}{\sqrt{p^2 + a^2}}\right)$$

There is an alternative way to write this. First note that $-p + \sqrt{p^2 + a^2} = \sqrt{p^2 + a^2} - p$.

$$\vec{\mathbf{E}} = \frac{2kQ\hat{\mathbf{z}}}{a^2}\left(\frac{\sqrt{p^2 + a^2} - p}{\sqrt{p^2 + a^2}}\right)$$

Now apply the rule from algebra that $\frac{a+b}{c} = \frac{a}{c} + \frac{b}{c}$.

$$\vec{\mathbf{E}} = \frac{2kQ\hat{\mathbf{z}}}{a^2}\left(\frac{\sqrt{p^2 + a^2}}{\sqrt{p^2 + a^2}} + \frac{-p}{\sqrt{p^2 + a^2}}\right) = \frac{2kQ\hat{\mathbf{z}}}{a^2}\left(1 + \frac{-p}{\sqrt{p^2 + a^2}}\right)$$

38. A uniformly charged rod of length L lies on the y-axis with one end at the origin, as illustrated below. The positively charged rod has total charge Q. Derive an equation for the electric field at the point $(0, -a)$, where a is a constant, in terms of k, Q, a, L, and appropriate unit vectors.

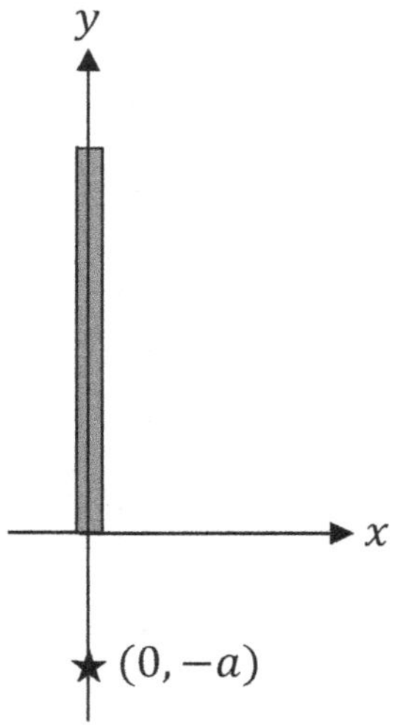

Want help? Check the hints section at the back of the book.

Answer: $\vec{\mathbf{E}} = -\frac{kQ}{a(L+a)}\hat{\mathbf{y}}$

39. A rod has non-uniform charge density $\lambda = \beta|x|$ (note the absolute values, which make λ nonnegative), where β is a constant, and endpoints at $\left(-\frac{L}{2}, 0\right)$ and $\left(\frac{L}{2}, 0\right)$, where L is a constant, as illustrated below. The positively charged rod has total charge Q. Derive an equation for the electric field at the point $(0, p)$, where p is a constant, in terms of k, Q, L, p, and appropriate unit vectors.

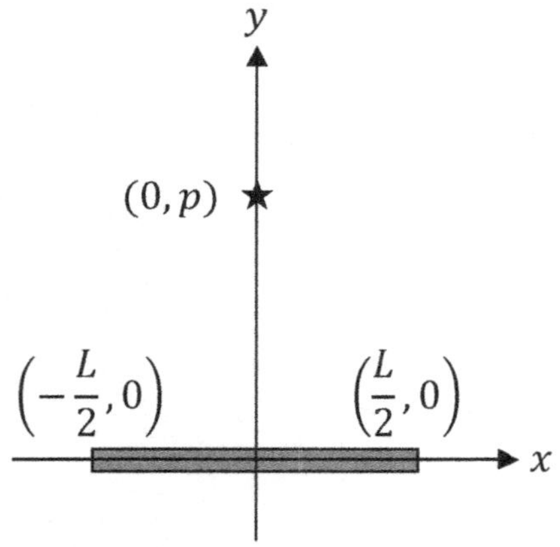

Want help? Check the hints section at the back of the book.

Answer: $\vec{\mathbf{E}} = \frac{8kQ}{L^2}\hat{\mathbf{y}}\left(1 - \frac{p}{\sqrt{\frac{L^2}{4}+p^2}}\right)$

40. Two thin wires are bent into the shape of semicircles, as illustrated below. Each semicircle has the same radius (a) and is uniformly charged. The top semicircle has charge $+Q$, while the bottom semicircle has charge $-Q$. Derive an equation for the electric field at the origin in terms of k, Q, a, and appropriate unit vectors.

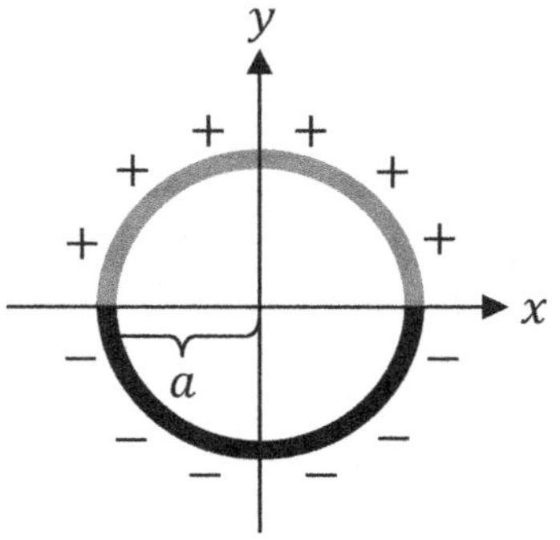

Want help? Check the hints section at the back of the book.

Answer: $\vec{\mathbf{E}} = -\frac{4kQ}{\pi a^2}\hat{\mathbf{y}}$

41. A very thin uniformly charged ring lies in the xy plane, centered about the origin as illustrated below. The radius of the ring is denoted by the symbol a and the positively charged ring has total charge Q. Derive an equation for the electric field at the point $(0, 0, p)$ in terms of k, Q, a, p, and appropriate unit vectors.

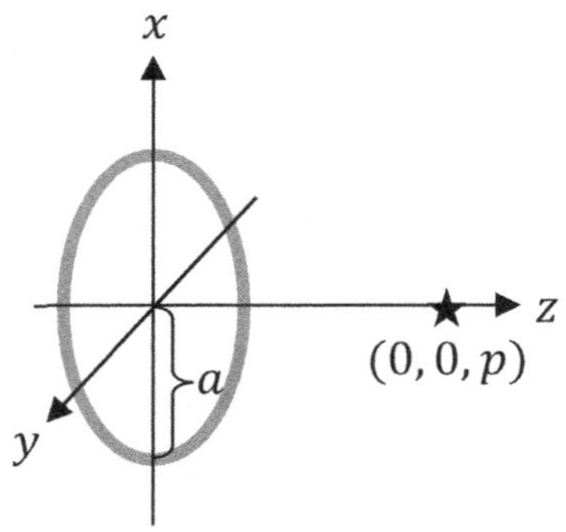

Want help? Check the hints section at the back of the book.

Answer: $\vec{\mathbf{E}} = \frac{kQp}{(p^2+a^2)^{3/2}}\hat{\mathbf{z}}$

42. A solid disc with non-uniform charge density $\sigma = \beta r_c$, where β is a constant, lies in the xy plane, centered about the origin as illustrated below. The radius of the disc is denoted by the symbol a and the positively charged disc has total charge Q. Derive an equation for the electric field at the point $(0, 0, p)$ in terms of k, Q, a, p, and appropriate unit vectors.

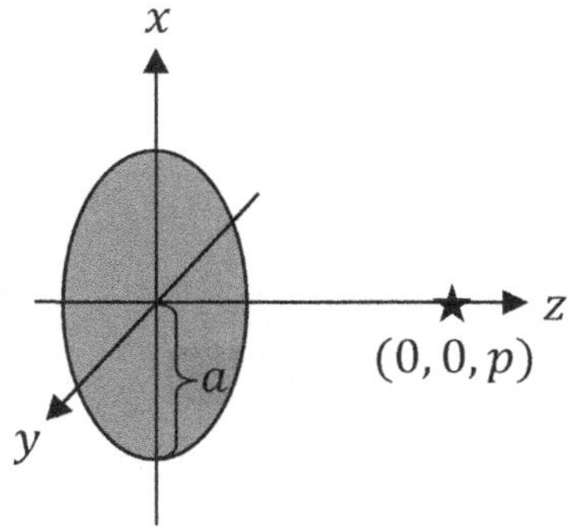

Want help? Check the hints section at the back of the book.

Answer: $\vec{\mathbf{E}} = \frac{3kQp\hat{\mathbf{z}}}{a^3}\left[\ln\left|\frac{a+\sqrt{a^2+p^2}}{p}\right| - \frac{a}{\sqrt{a^2+p^2}}\right]$

43. A thick semicircular ring with non-uniform charge density $\sigma = \beta r$, where β is a constant, lies in the xy plane, centered about the origin as illustrated below. The thick ring has inner radius $\frac{a}{2}$, outer radius a, and the positively charged ring has total charge Q. Derive an equation for the electric field at the origin in terms of k, Q, a and appropriate unit vectors.

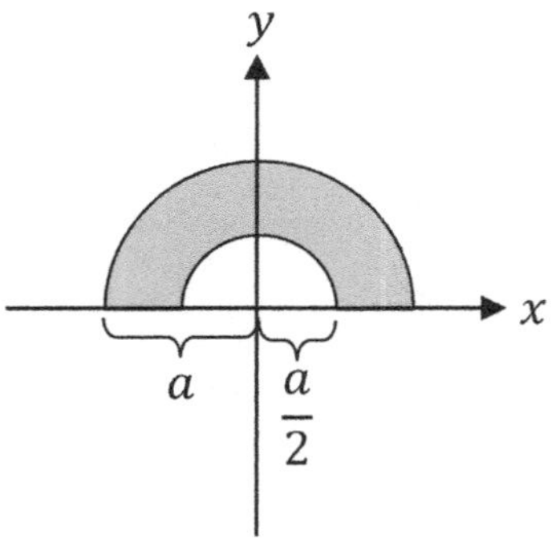

Want help? Check the hints section at the back of the book.

Answer: $\vec{\mathbf{E}} = -\frac{24kQ}{7\pi a^2}\hat{\mathbf{y}}$

8 GAUSS'S LAW

Relevant Terminology

Electric charge – a fundamental property of a particle that causes the particle to experience a force in the presence of an electric field. (An electrically neutral particle has no charge and thus experiences no force in the presence of an electric field.)
Electric field – force per unit charge.
Electric flux – a measure of the relative number of electric field lines that pass through a surface.
Net flux – the electric flux through a **closed** surface.
Open surface – a surface that doesn't entirely enclose a volume. An ice-cream cone is an example of an open surface because an ant could crawl out of it.
Closed surface – a surface that completely encloses a volume. A sealed box is an example of a closed surface because an ant could be trapped inside of it.
Conductor – a material through which electrons are able to flow readily. Metals tend to be good conductors of electricity.
Insulator – a material through which electrons are <u>**not**</u> able to flow readily. Wood and glass are examples of good insulators.

Gauss's Law

Electric flux (Φ_e) is a measure of the relative number of electric field lines passing through a surface. (Electric field lines were discussed in Chapter 4.) The equation for electric flux involves the scalar product between electric field ($\vec{\mathbf{E}}$) and the differential area element ($d\vec{\mathbf{A}}$), where the direction of $d\vec{\mathbf{A}}$ is **perpendicular to the surface**. The symbol S below stands for the surface over which area is integrated.

$$\Phi_e = \int_S \vec{\mathbf{E}} \cdot d\vec{\mathbf{A}}$$

Recall the scalar product from Volume 1, where θ is the angle between $\vec{\mathbf{E}}$ and $d\vec{\mathbf{A}}$. Note that $\vec{\mathbf{E}} \cdot d\vec{\mathbf{A}}$ is maximum when $\theta = 0°$ and zero when $\theta = 90°$.

$$\vec{\mathbf{E}} \cdot d\vec{\mathbf{A}} = E \cos\theta \, dA$$

According to **Gauss's law**, the **net** electric flux (Φ_e^{net}) through a **closed** surface (often called a Gaussian surface) is proportional to the charge enclosed (q_{enc}) by the surface.

$$\oint_S \vec{\mathbf{E}} \cdot d\vec{\mathbf{A}} = \frac{q_{enc}}{\epsilon_0}$$

The symbol $\oint$ is called a **closed** integral: It represents that the integral is over a **closed** surface. The constant ϵ_0 is called the **permittivity of free space**.

Symbols and SI Units

Symbol	Name	SI Units
dq	differential charge element	C
q_{enc}	the charge enclosed by the Gaussian surface	C
Q	total charge of the object	C
$\vec{\mathbf{E}}$	electric field	N/C or V/m
Φ_e	electric flux	$\frac{N \cdot m^2}{C}$ or $\frac{kg \cdot m^3}{C \cdot s^2}$
Φ_e^{net}	the net electric flux through a closed surface	$\frac{N \cdot m^2}{C}$ or $\frac{kg \cdot m^3}{C \cdot s^2}$
ϵ_0	permittivity of free space	$\frac{C^2}{N \cdot m^2}$ or $\frac{C^2 \cdot s^2}{kg \cdot m^3}$
x, y, z	Cartesian coordinates of dq	m, m, m
$\hat{\mathbf{x}}, \hat{\mathbf{y}}, \hat{\mathbf{z}}$	unit vectors along the $+x$-, $+y$-, $+z$-axes	unitless
r, θ	2D polar coordinates of dq	m, rad
r_c, θ, z	cylindrical coordinates of dq	m, rad, m
r, θ, φ	spherical coordinates of dq	m, rad, rad
$\hat{\mathbf{r}}, \hat{\boldsymbol{\theta}}, \hat{\boldsymbol{\varphi}}$	unit vectors along spherical coordinate axes	unitless
$\hat{\mathbf{r}}_c$	a unit vector pointing away from the $+z$-axis	unitless
λ	linear charge density (for an arc)	C/m
σ	surface charge density (for an area)	C/m^2
ρ	volume charge density	C/m^3
ds	differential arc length	m
dA	differential area element	m^2
dV	differential volume element	m^3

Note: The symbol Φ is the uppercase Greek letter phi* and ϵ is epsilon.

* Cool physics note: If you rotate uppercase phi (Φ) sideways, it resembles Darth Vader's™ spaceship.

Essential Concepts

The net flux passing through a closed surface is proportional to the number of electric field lines passing through the surface, and according to Gauss's law this is proportional to the net charge enclosed by the surface:

- If more electric field lines go out of the surface than come into the surface, the net flux through the surface is **positive**. This indicates that there is a net positive charge inside of the closed surface. In the example below, electric field lines are heading out of the spherical surface (represented by the dashed circle), but none are heading into the surface. According to Gauss's law, there must be a net positive charge inside of the closed surface.

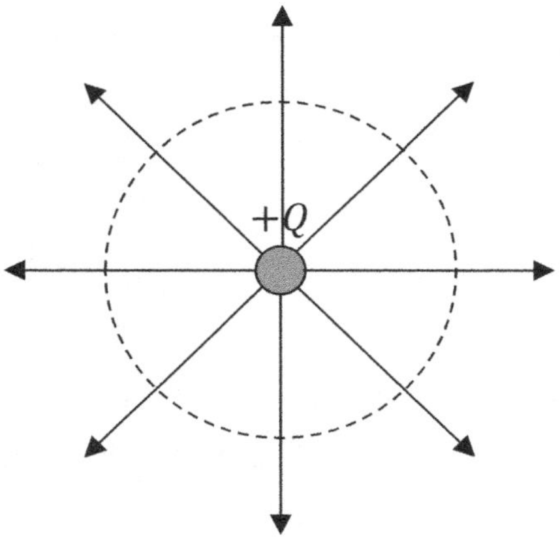

- If fewer electric field lines go out of the surface than come into the surface, the net flux through the surface is n**egative**. This indicates that there is a net negative charge inside of the closed surface. In the example below, electric field lines are heading into the spherical surface (represented by the dashed circle), but none are heading out of the surface. According to Gauss's law, there must be a net negative charge inside of the closed surface.

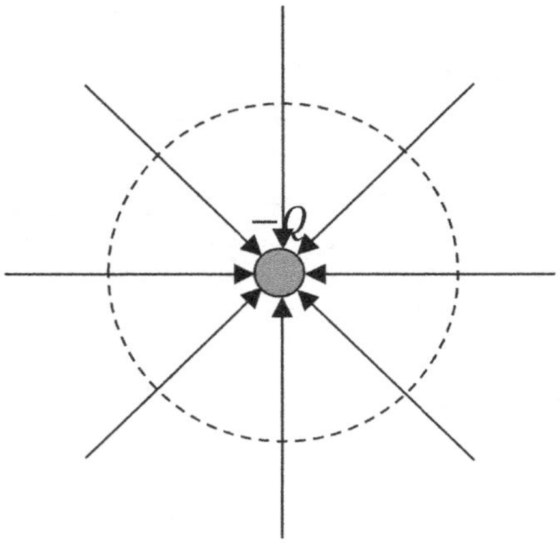

- If the same number of electric field lines go out of the surface as come into the surface, the net flux through the surface is zero. This indicates that the net charge inside of the closed surface is **zero**. In the example below, for every electric field line that heads out of the elliptical surface (represented by the dashed ellipse) there is another electric field line which heads back into the surface. The net charge inside of the closed surface is zero (the positive and negative charges cancel out).

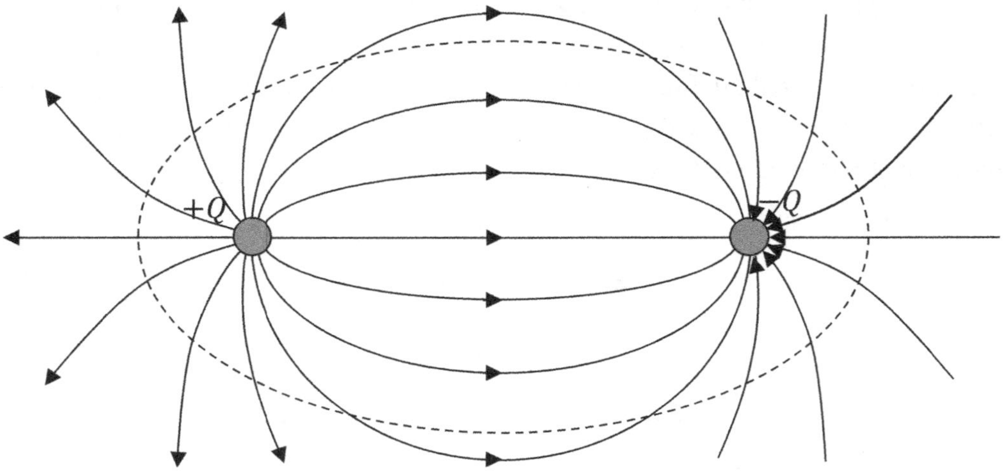

The example below includes 4 different closed surfaces:

- The net flux through closed surface A is positive. It encloses a positive charge.
- The net flux through closed surface B is zero. The positive and negative charges inside cancel out (the net charge enclosed is zero).
- The net flux through closed surface C is zero. There aren't any charges enclosed.
- The net flux through closed surface D is negative. It encloses a negative charge.

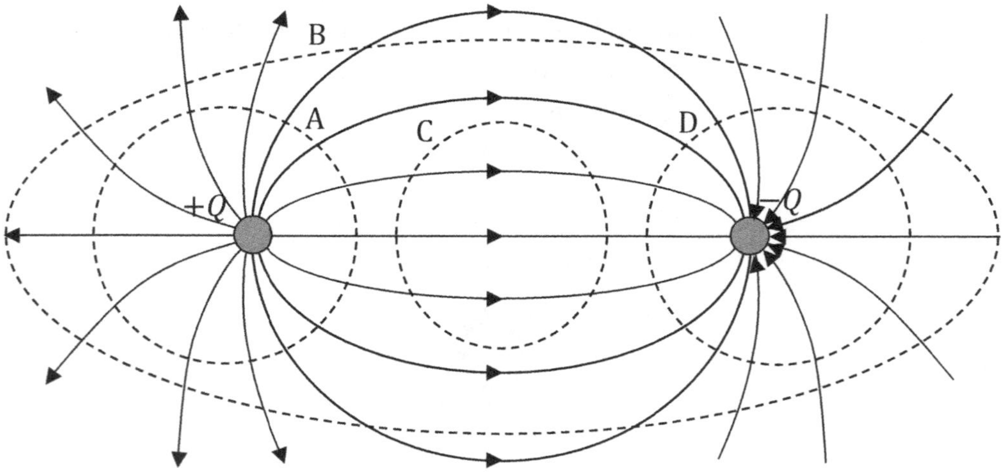

Strategy for Applying Gauss's Law

If there is enough symmetry in an electric field problem for it to be practical to take advantage of Gauss's law, follow these steps:

1. Sketch the electric field lines (or lines of force) for the problem following the technique discussed in Chapter 4.
2. Look at these electric field lines. Try to visualize a closed surface (like a sphere, cylinder, or cube) called a **Gaussian surface** for which the electric field lines would always be parallel to, anti-parallel to, or perpendicular to (or a combination of these) the surface no matter which part of the surface they pass through. When the electric field lines are parallel or anti-parallel to the surface, you want the magnitude of the electric field to be **constant** over that part of the surface. These features make the left-hand side of Gauss's law very easy to compute. The closed surface must also enclose some of the electric charge.
3. Study the examples that follow. Most Gauss's law problems have a geometry that is very similar to one of these examples (an infinite line, an infinite plane, an infinitely long cylinder, or a sphere). The Gaussian surfaces that we draw in the examples are basically the same as most of the Gaussian surfaces encountered in the problems. This makes Steps 1-2 very easy.
4. Write down the formula for Gauss's law.
$$\oint_S \vec{\mathbf{E}} \cdot d\vec{\mathbf{A}} = \frac{q_{enc}}{\epsilon_0}$$
If the surface is a sphere, there will just be one integral. If the surface is a cylinder, break the closed integral up into three open integrals: one for the body and one for each end (see the examples).
5. Simplify the left-hand side of Gauss's law. Recall that the direction of $d\vec{\mathbf{A}}$ is **perpendicular** to the surface.
   - If $\vec{\mathbf{E}}$ is parallel to $d\vec{\mathbf{A}}$, then $\vec{\mathbf{E}} \cdot d\vec{\mathbf{A}} = EdA$.
   - If $\vec{\mathbf{E}}$ is anti-parallel to $d\vec{\mathbf{A}}$, then $\vec{\mathbf{E}} \cdot d\vec{\mathbf{A}} = -EdA$.
   - If $\vec{\mathbf{E}}$ is perpendicular to $d\vec{\mathbf{A}}$, then $\vec{\mathbf{E}} \cdot d\vec{\mathbf{A}} = 0$.
6. If by symmetry the magnitude of the electric field (E) is constant over any part of the surface, $\int_S E\,dA = E\int_S dA = EA$ for that part of the surface. If you choose your Gaussian surface wisely in Step 2, you will either get EA or zero for each part of the closed surface, such that Gauss's law simplifies to:
$$E_1A_1 + E_2A_2 + \cdots + E_NA_N = \frac{q_{enc}}{\epsilon_0}$$
Here, E_1A_1 is the flux through one part of the surface, E_2A_2 is the flux through another part of the surface, and so on. For a sphere, there is just one term. For a cylinder, there are three terms (though one or two terms may be zero).

7. Replace each area with the appropriate expression, depending upon the geometry.
   - The surface area of a sphere is $A = 4\pi r^2$.
   - The area of the end of a right-circular cylinder is $A = \pi r_c^2$.
   - The surface area of the body of a right-circular cylinder is $A = 2\pi r_c L$.

   Gauss's law problems with spheres or cylinders generally involve two different radii: The radius of the Gaussian sphere or cylinder is r or r_c, whereas the radius of a charged sphere or cylinder is a different symbol (which we will usually call a in this book, but may be called R in other textbooks – we are using a to avoid possible confusion between lowercase and uppercase r's).
8. Isolate the electric field in your simplified equation from Gauss's law. (This should be a simple algebra exercise.)
9. Consider each region in the problem. There will ordinarily be at least two regions. One region may be inside of a charged object and another region may be outside of the charged object, for example. We will label the regions with Roman numerals (I, II, III, IV, V, etc.).
10. Determine the net charge enclosed in each region. In a given region, if the Gaussian surface encloses the entire charged object, the enclosed charge (q_{enc}) will equal the total charge of the object (Q). However, if the Gaussian surface encloses only a fraction of the charged object, you will need to integrate in order to find the charge enclosed, applying the technique from Chapter 7 (Step 11 on page 94).
$$q_{enc} = \int dq$$
    Note that the above integral is over the Gaussian surface for the given region.
11. For each region, substitute the charge enclosed (from Step 10) into the simplified expression for the electric field (from Step 8).

The Permittivity of Free Space

The constant ϵ_0 in Gauss's law is called the permittivity of **free space** (meaning **vacuum**, a region completely devoid of matter). The permittivity of free space (ϵ_0) is related to Coulomb's constant $\left(k = 9.0 \times 10^9 \ \frac{\text{N·m}^2}{\text{C}^2}\right)$:
$$\epsilon_0 = \frac{1}{4\pi k}$$
From this equation, we see that ϵ_0's units are the reciprocal of k's units. Since the SI units of k are $\frac{\text{N·m}^2}{\text{C}^2}$, it follows that the SI units of ϵ_0 are $\frac{\text{C}^2}{\text{N·m}^2}$. If you plug the numerical value for k $\left(9.0 \times 10^9 \ \frac{\text{N·m}^2}{\text{C}^2}\right)$ into the above formula, you will find that $\epsilon_0 = 8.8 \times 10^{-12} \ \frac{\text{C}^2}{\text{N·m}^2}$. However, in this book we will often write $\epsilon_0 = \frac{10^{-9}}{36\pi} \frac{\text{C}^2}{\text{N·m}^2}$ in order to avoid the need of a calculator.

Example: A very thin, infinitely† large sheet of charge lies in the xy plane. The positively charged sheet has uniform charge density σ. Derive an expression for the electric field on either side of the infinite sheet.

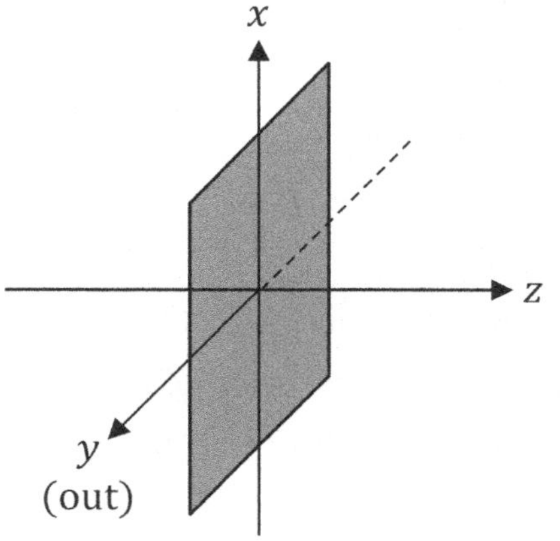

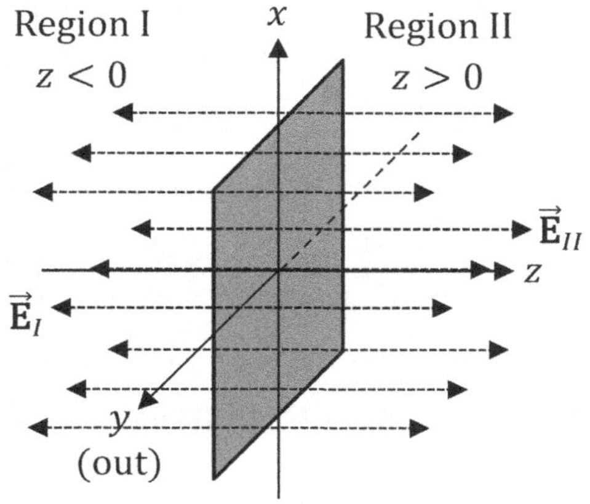

First sketch the electric field lines for the infinite sheet of positive charge. If a positive test charge were placed to the right of the sheet, it would be repelled to the right. If a positive test charge were placed to the left of the sheet, it would be reprelled to the left. Therefore, the electric field lines are directed away from the sheet, as shown above on the right. We choose a right-circular cylinder to serve as our Gaussian surface, as illustrated below. The reasons for this choice are:

- $\vec{\mathbf{E}}$ is parallel to $d\vec{\mathbf{A}}$ at the ends of the cylinder (both are horizontal).
- $\vec{\mathbf{E}}$ is perpendicular to $d\vec{\mathbf{A}}$ over the body of the cylinder ($\vec{\mathbf{E}}$ is horizontal while $d\vec{\mathbf{A}}$ is not).
- The magnitude of $\vec{\mathbf{E}}$ is constant over either end of the cylinder, since every point on the end is equidistant from the infinite sheet.

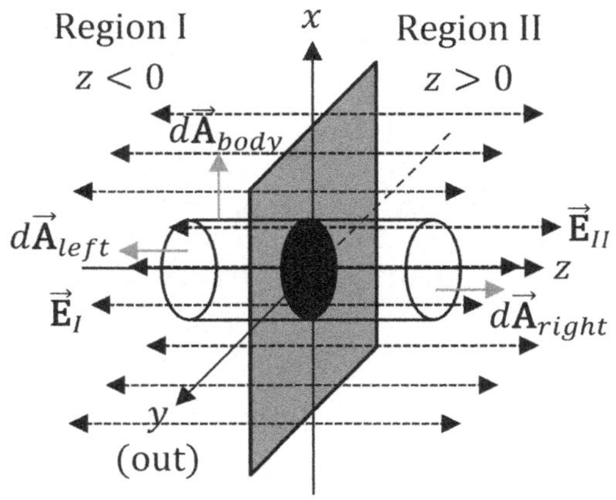

† In practice, this result for the "infinite" sheet applies to a finite sheet when you're calculating the electric field at a distance from the sheet that is very small compared to the dimensions of the sheet and which is not near the edges of the sheet. A common example encountered in the laboratory is the parallel-plate capacitor (Chapter 12), for which the separation between the plates is very small compared to the size of the plates.

Write the formula for Gauss's law.

$$\oint_S \vec{\mathbf{E}} \cdot d\vec{\mathbf{A}} = \frac{q_{enc}}{\epsilon_0}$$

The net flux on the left-hand side of the equation involves integrating over the complete surface of the Gaussian cylinder. The surface of the cylinder includes a left end, a body, and a right end.

$$\int_{left} \vec{\mathbf{E}} \cdot d\vec{\mathbf{A}} + \int_{body} \vec{\mathbf{E}} \cdot d\vec{\mathbf{A}} + \int_{right} \vec{\mathbf{E}} \cdot d\vec{\mathbf{A}} = \frac{q_{enc}}{\epsilon_0}$$

Recall (from Volume 1) that the scalar product can be expressed as $\vec{\mathbf{E}} \cdot d\vec{\mathbf{A}} = E \cos\theta \, dA$, where θ is the angle between $\vec{\mathbf{E}}$ and $d\vec{\mathbf{A}}$. Also recall that the direction of $d\vec{\mathbf{A}}$ is perpendicular to the surface. Study the direction of $\vec{\mathbf{E}}$ and $d\vec{\mathbf{A}}$ at each end and the body of the cylinder in the previous diagram.

- For the ends, $\theta = 0°$ because $\vec{\mathbf{E}}$ and $d\vec{\mathbf{A}}$ either both point right or both point left.
- For the body, $\theta = 90°$ because $\vec{\mathbf{E}}$ and $d\vec{\mathbf{A}}$ are perpendicular.

$$\int_{left} E \cos 0° \, dA + \int_{body} E \cos 90° \, dA + \int_{right} E \cos 0° \, dA = \frac{q_{enc}}{\epsilon_0}$$

Recall from trig that $\cos 0° = 1$ and $\cos 90° = 0$.

$$\int_{left} E \, dA + 0 + \int_{right} E \, dA = \frac{q_{enc}}{\epsilon_0}$$

Over either end, the magnitude of the electric field is constant, since every point on one end is equidistant from the infinite sheet. Therefore, we may pull E out of the integrals. (We choose our Gaussian surface to be centered about the infinite sheet such that the value of E must be the same[‡] at both ends.)

$$E \int_{left} dA + E \int_{right} dA = \frac{q_{enc}}{\epsilon_0}$$

The remaining integrals are trivial: $\int dA = A$.

$$EA_{left} + EA_{right} = \frac{q_{enc}}{\epsilon_0}$$

The two ends have the same area (the area of a circle): $A_{left} = A_{right} = A_{end}$.

$$2EA_{end} = \frac{q_{enc}}{\epsilon_0}$$

Isolate the magnitude of the electric field by dividing both sides of the equation by $2A_{end}$.

$$E = \frac{q_{enc}}{2\epsilon_0 A_{end}}$$

[‡] Once we reach our final answer, we will see that this doesn't matter: It turns out that the electric field is independent of the distance from the infinite charged sheet.

Now we need to determine how much charge is enclosed by the Gaussian surface. We find this the same way that we calculated Q in the previous chapter.

$$q_{enc} = \int dq$$

This integral isn't over the Gaussian surface itself: Rather, this integral is over the region of the sheet enclosed by the Gaussian surface. This region is the circle shaded in black on the previous diagram. For a sheet of charge, we write $dq = \sigma dA$ (Step 6 on page 94).

$$q_{enc} = \int \sigma \, dA$$

Since the sheet has uniform charge density, we may pull σ out of the integral.

$$q_{enc} = \sigma \int dA = \sigma A_{end}$$

This area, which is the area of the sheet enclosed by the Gaussian cylinder, and which is shown as a black circle on the previous diagram, is the same as the area of either end of the cylinder. Substitute this expression for the charge enclosed into the previous equation for electric field.

$$E = \frac{q_{enc}}{2\epsilon_0 A_{end}} = \frac{\sigma A_{end}}{2\epsilon_0 A_{end}} = \frac{\sigma}{2\epsilon_0}$$

The magnitude of the electric field is $E = \frac{\sigma}{2\epsilon_0}$, which is a constant. Thus, the electric field created by an infinite sheet of charge is uniform. We can use a unit vector to include the direction of the electric field with our answer: $\vec{\mathbf{E}} = \frac{\sigma}{2\epsilon_0}\hat{\mathbf{z}}$ for $z > 0$ (to the right of the sheet) and $\vec{\mathbf{E}} = -\frac{\sigma}{2\epsilon_0}\hat{\mathbf{z}}$ for $z < 0$ (to the left of the sheet), since $\hat{\mathbf{z}}$ points one unit along the $+z$-axis. There is a clever way to combine both results into a single equation: We can simply write $\vec{\mathbf{E}} = \frac{\sigma}{2\epsilon_0}\frac{z}{|z|}\hat{\mathbf{z}}$, since $\frac{z}{|z|} = +1$ if $z > 0$ and $\frac{z}{|z|} = -1$ if $z < 0$.

Example: An infinite slab is like a very thick infinite sheet: It's basically a rectangular box with two infinite dimensions and one finite thickness. The infinite nonconducting slab with thickness T shown below is parallel to the xy plane, centered about $z = 0$, and has uniform positive charge density ρ. Derive an expression for the electric field in each region.

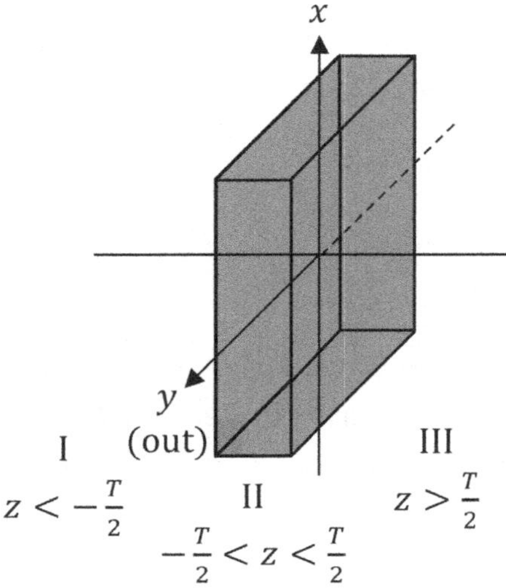

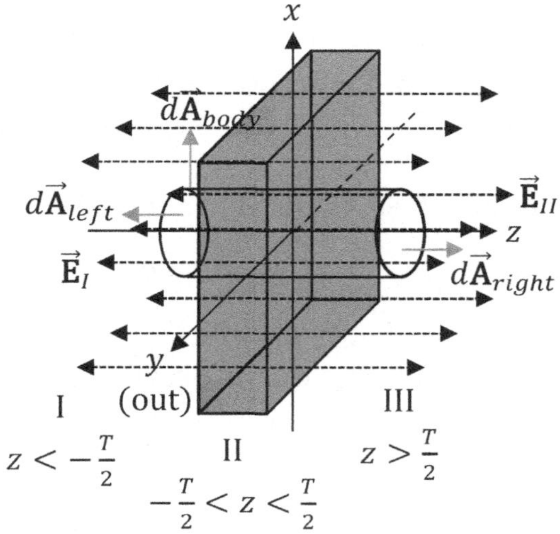

This problem is similar to the previous example. The difference is that the infinite slab has thickness, unlike the infinite sheet. The electric field lines look the same, and we choose the same Gaussian surface: a right-circular cylinder. The math will also start out the same with Gauss's law. To save time, we'll simply repeat the steps that are identical to the previous example, and pick up from where this solution deviates from the previous one. It would be a good exercise to see if you can understand each step (if not, review the previous example for the explanation).

$$\oint_S \vec{\mathbf{E}} \cdot d\vec{\mathbf{A}} = \frac{q_{enc}}{\epsilon_0}$$

$$\int_{left} \vec{\mathbf{E}} \cdot d\vec{\mathbf{A}} + \int_{body} \vec{\mathbf{E}} \cdot d\vec{\mathbf{A}} + \int_{right} \vec{\mathbf{E}} \cdot d\vec{\mathbf{A}} = \frac{q_{enc}}{\epsilon_0}$$

$$\int_{left} E \cos 0° \, dA + \int_{body} E \cos 90° \, dA + \int_{right} E \cos 0° \, dA = \frac{q_{enc}}{\epsilon_0}$$

$$\int_{left} E \, dA + 0 + \int_{right} E \, dA = \frac{q_{enc}}{\epsilon_0}$$

$$E \int_{left} dA + E \int_{right} dA = \frac{q_{enc}}{\epsilon_0}$$

$$2EA_{end} = \frac{q_{enc}}{\epsilon_0}$$

$$E = \frac{q_{enc}}{2\epsilon_0 A_{end}}$$

What's different now is the charge enclosed by the Gaussian surface. Since the slab has thickness, the Gaussian surface encloses a volume of charge, so we use $dq = \rho dV$ (Step 6 on page 94) instead of σdA in the integral to find q_{enc}.

$$q_{enc} = \int dq = \int \rho \, dV$$

Since the slab has uniform charge density, we may pull ρ out of the integral.

$$q_{enc} = \rho \int dV = \rho V_{enc}$$

Here, V_{enc} is the volume of the slab enclosed by the Gaussian surface. There are two cases to consider:

- The Gaussian surface could be longer than the thickness of the slab. This will help us find the electric field in regions I and III (see the regions labeled below).
- The Gaussian surface could be shorter than the thickness of the slab. This will help us find the electric field in region II.

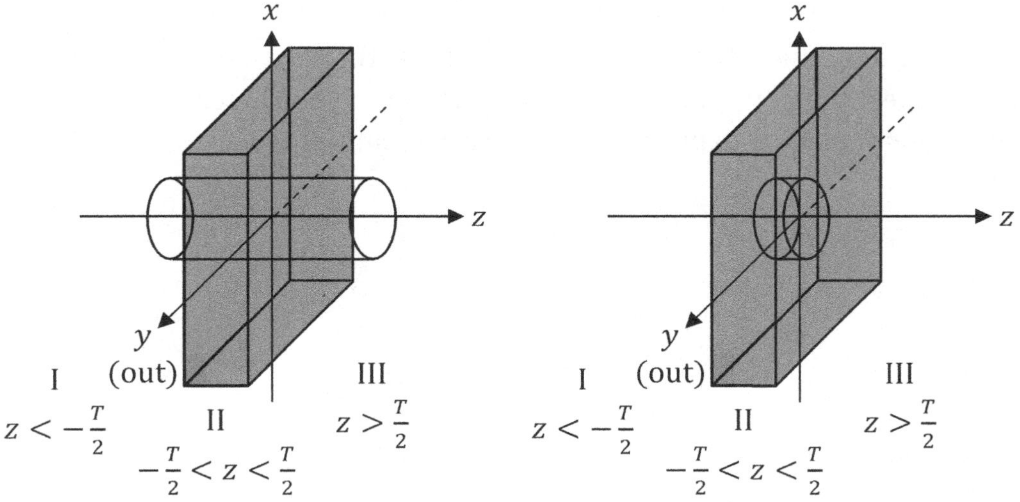

<u>**Regions I and III**</u>: $z < -\frac{T}{2}$ and $z > \frac{T}{2}$.

When the Gaussian cylinder is longer than the thickness of the slab, the volume of charge enclosed equals the intersection of the cylinder and the slab: It is a cylinder with a length equal to the thickness of the slab. The volume of this cylinder equals the thickness of the slab times the area of the circular end.

$$V_{enc} = TA_{end}$$

In this case, the charge enclosed is:

$$q_{enc} = \rho V_{enc} = \rho T A_{end}$$

Substitute this expression into the previous equation for electric field.

$$E = \frac{q_{enc}}{2\epsilon_0 A_{end}} = \frac{\rho T A_{end}}{2\epsilon_0 A_{end}} = \frac{\rho T}{2\epsilon_0}$$

The answer will be different in region II.

<u>Region II</u>: $-\frac{T}{2} < z < \frac{T}{2}$.

When the Gaussian cylinder is shorter than the thickness of the slab, the volume of charge enclosed equals the volume of the Gaussian cylinder. The length of the Gaussian cylinder is $2|z|$ (since the Gaussian cylinder extends from $-z$ to $+z$), where $-\frac{T}{2} < z < \frac{T}{2}$. The shorter the Gaussian cylinder, the less charge it encloses. The volume of the Gaussian cylinder equals $2|z|$ times the area of the circular end.

$$V_{enc} = 2|z|A_{end}$$

In this case, the charge enclosed is:

$$q_{enc} = \rho V_{enc} = \rho 2|z|A_{end}$$

Substitute this expression into the previous equation for electric field.

$$E = \frac{q_{enc}}{2\epsilon_0 A_{end}} = \frac{\rho 2|z|A_{end}}{2\epsilon_0 A_{end}} = \frac{\rho|z|}{\epsilon_0}$$

Note that the two expressions (for the different regions) agree at the boundary: That is, if you plug in $z = \frac{T}{2}$ into the expression for the electric field in region II, you get the same answer as in regions I and III. It is true in general that the electric field is continuous across a boundary (**except** when working with a **conductor**): If you remember this, you can use it to help check your answers to problems that involve multiple regions.

Example: A solid spherical insulator[§] centered about the origin with positive charge Q has radius a and uniform charge density ρ. Derive an expression for the electric field both inside and outside of the sphere.

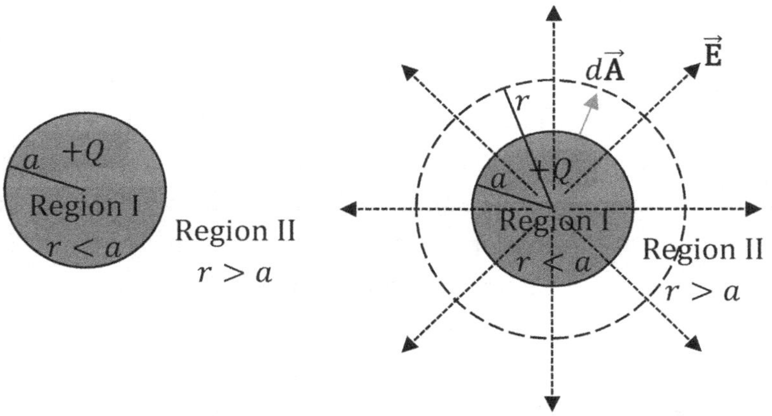

First sketch the electric field lines for the positive sphere. Regardless of where a positive test charge might be placed, it would be repelled directly away from the center of the sphere. Therefore, the electric field lines radiate outward, as shown above on the right (but realize that the field lines really radiate outward in three dimensions, and we are really working with spheres, not circles). We choose our Gaussian surface to be a sphere (shown as a dashed circle above) concentric with the charged sphere such that $\vec{\mathbf{E}}$ and $d\vec{\mathbf{A}}$ will be parallel and the magnitude of $\vec{\mathbf{E}}$ will be constant over the Gaussian surface (since every point on the Gaussian sphere is equidistant from the center of the positive sphere). Write the formula for Gauss's law.

$$\oint_S \vec{\mathbf{E}} \cdot d\vec{\mathbf{A}} = \frac{q_{enc}}{\epsilon_0}$$

The scalar product is $\vec{\mathbf{E}} \cdot d\vec{\mathbf{A}} = E \cos\theta \, dA$, and $\theta = 0°$ since the electric field lines radiate outward, perpendicular to the surface (and therefore parallel to $d\vec{\mathbf{A}}$, which is always perpendicular to the surface).

$$\oint_S E \cos 0° \, dA = \frac{q_{enc}}{\epsilon_0}$$

Recall from trig that $\cos 0° = 1$.

$$\oint_S E \, dA = \frac{q_{enc}}{\epsilon_0}$$

The magnitude of the electric field is constant over the Gaussian sphere, since every point on the sphere is equidistant from the center of the positive sphere. Therefore, we may pull E out of the integral.

[§] If it were a conductor, all of the excess charge would move to the surface due to Gauss's law. We'll discuss why that's the case as part of the following example (see the example with the conducting shell).

$$E \oint_S dA = \frac{q_{enc}}{\epsilon_0}$$

This area integral is over the surface of the Gaussian sphere of radius r, where r is a variable because the electric field depends on how close or far the field point is away from the charged sphere. For a sphere, we work with spherical coordinates (Chapter 6) and write $dA = r^2 \sin\theta \, d\theta d\varphi$. This is a purely angular integration (over the two spherical angles θ and φ), and we treat r as a constant in this integral because every point on the surface of the Gaussian sphere has the same value of r. For a sphere, φ varies from 0 to 2π while θ varies from 0 to π: φ sweeps out a horizontal circle, while a given value of θ sweeps out a cone and once θ reaches π the cone traces out a complete sphere (this was discussed in Volume 1, Chapter 32, on page 348).

$$A = \oint_S dA = \int_{\theta=0}^{\pi} \int_{\varphi=0}^{2\pi} r^2 \sin\theta \, d\theta \, d\varphi = r^2 \int_{\theta=0}^{\pi} \int_{\varphi=0}^{2\pi} \sin\theta \, d\theta \, d\varphi$$

When integrating over φ, we treat the independent variable θ as a constant. Therefore, we may pull θ out of the φ-integration.

$$A = r^2 \int_{\theta=0}^{\pi} \sin\theta \left(\int_{\varphi=0}^{2\pi} d\varphi \right) d\theta = r^2 \int_{\theta=0}^{\pi} \sin\theta \, [\varphi]_{\varphi=0}^{2\pi} \, d\theta = r^2 \int_{\theta=0}^{\pi} \sin\theta \, (2\pi) \, d\theta$$

$$A = 2\pi r^2 \int_{\theta=0}^{\pi} \sin\theta \, d\theta = 2\pi r^2 [-\cos\theta]_{\theta=0}^{\pi} = 2\pi r^2(-\cos\pi + \cos 0) = 2\pi r^2(1+1) = 4\pi r^2$$

Although we could have simply looked up the formula for the surface area of a sphere and found that it was $A = 4\pi r^2$, there are some problems where you need to know how to perform the integration. Substitute this expression for surface area into the previous equation for electric field.

$$E \oint_S dA = EA = E 4\pi r^2 = \frac{q_{enc}}{\epsilon_0}$$

Isolate the magnitude of the electric field by dividing both sides of the equation by $4\pi r^2$.

$$E = \frac{q_{enc}}{4\pi\epsilon_0 r^2}$$

Now we need to determine how much charge is enclosed by the Gaussian surface. For a solid charged sphere, we write $dq = \rho dV$ (Step 6 on page 94). Since the sphere has uniform charge density, we may pull ρ out of the integral.

$$q_{enc} = \int dq = \int \rho \, dV = \rho \int dV$$

For a sphere, we work with spherical coordinates and write the differential volume element as $dV = r^2 \sin\theta \, dr d\theta d\varphi$ (see Chapter 6). Since we are now integrating over volume (not surface area), r is a variable and we will have a triple integral.

We must consider two different regions:

- The Gaussian sphere could be smaller than the charged sphere. This will help us find the electric field in region I.
- The Gaussian sphere could be larger than the charged sphere. This will help us find the electric field in region II.

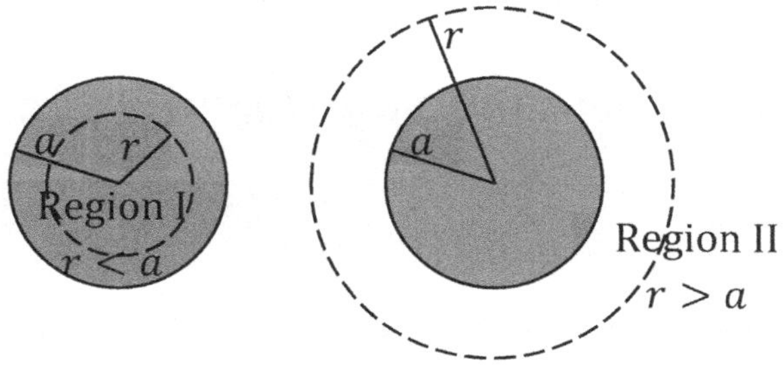

Region I: $r < a$.

Inside of the charged sphere, only a fraction of the sphere's charge is enclosed by the Gaussian sphere. In this region, the upper limit of the r-integration is the variable r: The larger the Gaussian sphere, the more charge it encloses, up to a maximum radius of a (the radius of the charged sphere).

$$q_{enc} = \rho \int dV = \rho \int_{r=0}^{r} \int_{\theta=0}^{\pi} \int_{\varphi=0}^{2\pi} r^2 \sin\theta \, dr \, d\theta \, d\varphi$$

This integral is separable:

$$q_{enc} = \rho \int_{r=0}^{r} r^2 \, dr \int_{\theta=0}^{\pi} \sin\theta \, d\theta \int_{\varphi=0}^{2\pi} d\varphi = \rho \left[\frac{r^3}{3}\right]_{r=0}^{r} [-\cos\theta]_{\theta=0}^{\pi} [\varphi]_{\varphi=0}^{2\pi}$$

$$q_{enc} = \rho\left(\frac{r^3}{3}\right)(-\cos\pi + \cos 0)(2\pi) = \rho\left(\frac{r^3}{3}\right)(1+1)(2\pi) = \frac{4\pi\rho r^3}{3}$$

You should recognize that $\frac{4\pi r^3}{3}$ is the volume of a sphere. For a uniformly charged sphere, charge equals ρ times volume. (For a non-uniform sphere, you would need to integrate: Then you couldn't just multiply ρ times volume.) Substitute this expression for the charge enclosed into the previous equation for electric field.

$$E = \frac{q_{enc}}{4\pi\epsilon_0 r^2} = \frac{4\pi\rho r^3}{3} \div 4\pi\epsilon_0 r^2 = \frac{4\pi\rho r^3}{3} \times \frac{1}{4\pi\epsilon_0 r^2} = \frac{\rho r}{3\epsilon_0}$$

Since the electric field points outward (away from the center of the sphere), we can include a direction with the electric field by adding on the spherical unit vector $\hat{\mathbf{r}}$.

$$\vec{\mathbf{E}} = \frac{\rho r}{3\epsilon_0}\hat{\mathbf{r}}$$

The answer is different outside of the charged sphere. We will explore that next.

Region II: $r > a$.
Outside of the charged sphere, 100% of the charge is enclosed by the Gaussian surface. This changes the upper limit of the r-integration to a (since all of the charge lies inside a sphere of radius a).

$$q_{enc} = Q = \rho \int dV = \rho \int_{r=0}^{a} \int_{\theta=0}^{\pi} \int_{\varphi=0}^{2\pi} r^2 \sin\theta \, dr \, d\theta \, d\varphi$$

We don't need to work out the entire triple integral again. We'll get the same expression as before, but with a in place of r.

$$q_{enc} = Q = \frac{4\pi\rho a^3}{3}$$

Substitute this into the equation for electric field that we obtained from Gauss's law.

$$E = \frac{q_{enc}}{4\pi\epsilon_0 r^2} = \frac{4\pi\rho a^3}{3} \div 4\pi\epsilon_0 r^2 = \frac{4\pi\rho a^3}{3} \times \frac{1}{4\pi\epsilon_0 r^2} = \frac{\rho a^3}{3\epsilon_0 r^2}$$

We can turn this into a vector by including the appropriate unit vector.

$$\vec{\mathbf{E}} = \frac{\rho a^3}{3\epsilon_0 r^2}\hat{\mathbf{r}}$$

Alternate forms of the answers in regions I and II.
There are multiple ways to express our answers for regions I and II. For example, we could use the equation $\epsilon_0 = \frac{1}{4\pi k}$ to work with Coulomb's constant (k) instead of the permittivity of free space (ϵ_0). Since the total charge of the sphere is $Q = \frac{4\pi\rho a^3}{3}$ (we found this equation for region II above), we can express the electric field in terms of the total charge (Q) of the sphere instead of the charge density (ρ).

Region I: $r < a$.

$$\vec{\mathbf{E}} = \frac{\rho r}{3\epsilon_0}\hat{\mathbf{r}} = \frac{4\pi k\rho r}{3}\hat{\mathbf{r}} = \frac{Qr}{4\pi\epsilon_0 a^3}\hat{\mathbf{r}} = \frac{kQr}{a^3}\hat{\mathbf{r}}$$

Region II: $r > a$.

$$\vec{\mathbf{E}} = \frac{\rho a^3}{3\epsilon_0 r^2}\hat{\mathbf{r}} = \frac{4\pi k\rho a^3}{3r^2}\hat{\mathbf{r}} = \frac{Q}{4\pi\epsilon_0 r^2}\hat{\mathbf{r}} = \frac{kQ}{r^2}\hat{\mathbf{r}}$$

Note that the electric field in region II is identical to the electric field created by a pointlike charge (see Chapter 2). Note also that the expressions for the electric field in the two different regions both agree at the boundary: That is, in the limit that r approaches a, both expressions approach $\frac{kQ}{a^2}\hat{\mathbf{r}}$.

Example: A solid spherical insulator centered about the origin with positive charge $5Q$ has radius a and uniform charge density ρ. Concentric with the spherical insulator is a thick spherical conducting shell of inner radius b, outer radius c, and total charge $3Q$. Derive an expression for the electric field in each region.

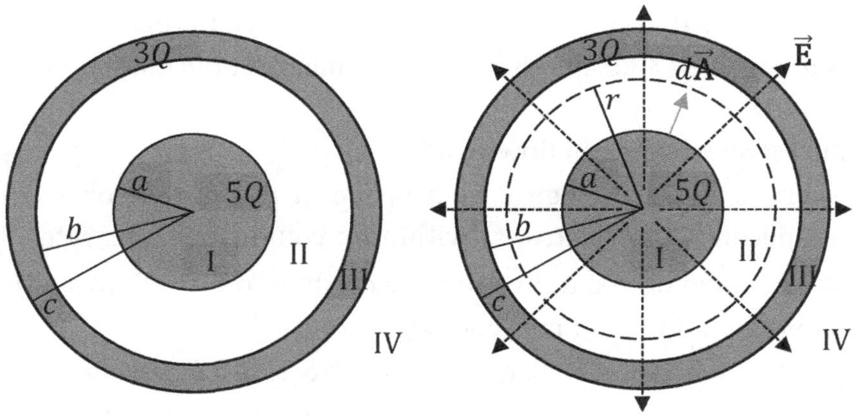

This problem is very similar to the previous example, except that there are four regions, and there is a little "trick" to working with the conducting shell for region III. The same Gaussian sphere from the previous example applies here, and the math for Gauss's law works much the same way here. The real difference is in figuring out the charge enclosed in each region.

<u>**Region I**</u>: $r < a$.
The conducting shell does <u>**not**</u> matter for region I, since none of its charge will reside in a Gaussian sphere with $r < a$. Thus, we will obtain the same result as for region I of the previous example, except that the total charge of the sphere is now $5Q$ (this was stated in the problem). We'll use one of the alternate forms of the equation for electric field that involves the total charge of the inner sphere (see the bottom of page 128), except that where we previously had Q for the total charge, we'll change that to $5Q$ for this problem.

$$\vec{\mathbf{E}} = \frac{5kQr}{a^3}\hat{\mathbf{r}}$$

<u>**Region II**</u>: $a < r < b$.
The conducting shell also does <u>**not**</u> matter for region II, since again none of its charge will reside in a Gaussian sphere with $r < b$. (Recall that Gauss's law involves the charge enclosed, q_{enc}, by the Gaussian sphere.) We obtain the same result as for region II of the previous example (see the bottom of page 128), with Q replaced by $5Q$ for this problem.

$$\vec{\mathbf{E}} = \frac{5kQ}{r^2}\hat{\mathbf{r}}$$

Region III: $b < r < c$.

Now the conducting shell matters. The electric field inside of the conducting shell equals zero once electrostatic equilibrium is attained (which just takes a fraction of a second).

$$\vec{\mathbf{E}} = 0$$

Following are the reasons that the electric field must be zero in region III:

- The conducting shell, like all other forms of macroscopic matter, consists of protons, neutrons, and electrons.
- In a **conductor**, electrons can flow readily.
- If there were a nonzero electric field inside of the conductor, it would cause the charges (especially, the electrons) within its volume to accelerate. This is because the electric field would result in a force according to $\vec{\mathbf{F}} = q\vec{\mathbf{E}}$, while Newton's second law ($\sum \vec{\mathbf{F}} = m\vec{\mathbf{a}}$) would result in acceleration.
- In a conductor, electrons redistribute in a fraction of a second until **electrostatic equilibrium** is attained (unless you connect a power supply to the conductor to create a constant flow of charge, for example, but there is no power supply involved in this problem).
- Once electrostatic equilibrium is attained, the charges won't be moving, and therefore the electric field within the conducting shell must be zero.

Because the electric field is zero in region III, we can reason how the total charge of the conducting shell (which equals $3Q$ according to the problem) is distributed. According to Gauss's law, $\oint_S \vec{\mathbf{E}} \cdot d\vec{\mathbf{A}} = \frac{q_{enc}}{\epsilon_0}$. Since $\vec{\mathbf{E}} = 0$ in region III, the right-hand side of Gauss's law must also equal zero, meaning that the charge enclosed by a Gaussian surface in region III must equal zero: $q_{enc}^{III} = 0$.

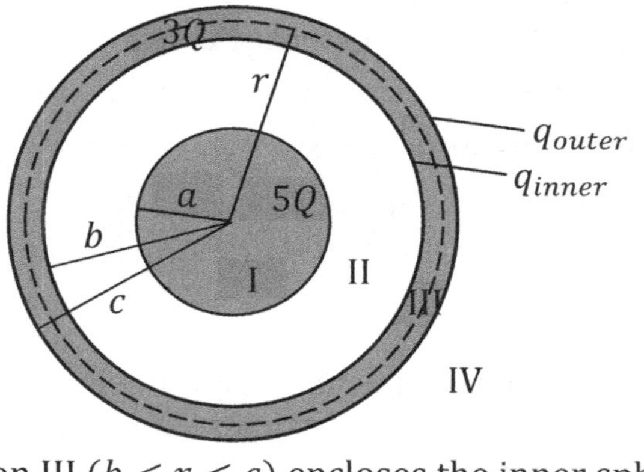

A Gaussian sphere in region III ($b < r < c$) encloses the inner sphere plus the inner surface of the conducting shell (see the dashed curve in the diagram above).

$$q_{enc}^{III} = 5Q + q_{inner} = 0$$

Since the charge enclosed in region III is zero, we can conclude that $q_{inner} = -5Q$. The total charge of the conducting shell equals $3Q$ according to the problem. Therefore, if we add the charge of the inner surface and outer surface of the conducting shell together, we must get $3Q$.

$$q_{inner} + q_{outer} = 3Q$$
$$-5Q + q_{outer} = 3Q$$
$$q_{outer} = 3Q + 5Q = 8Q$$

The conducting shell has a charge of $-5Q$ on its inner surface and a charge of $8Q$ on its outer surface, for a total charge of $3Q$. (It's important to note that "inner sphere" and "inner surface of the conducting shell" are two different things. In our notation, q_{inner} represents the charge on the "inner surface of the conducting shell." Note that q_{inner} does **not** refer to the "inner sphere.")

Region IV: $r < c$.
A Gaussian sphere in region IV encloses a total charge of $8Q$ (the $5Q$ from the inner sphere plus the $3Q$ from the conducting shell). The formula for the electric field is the same as for region II, except for changing the total charge enclosed to $8Q$.

$$\vec{\mathbf{E}} = \frac{8kQ}{r^2}\hat{\mathbf{r}}$$

Example: A solid spherical insulator centered about the origin with positive charge Q has radius a and non-uniform charge density $\rho = \beta r$, where β is a positive constant. Derive an expression for the electric field both inside and outside of the sphere.

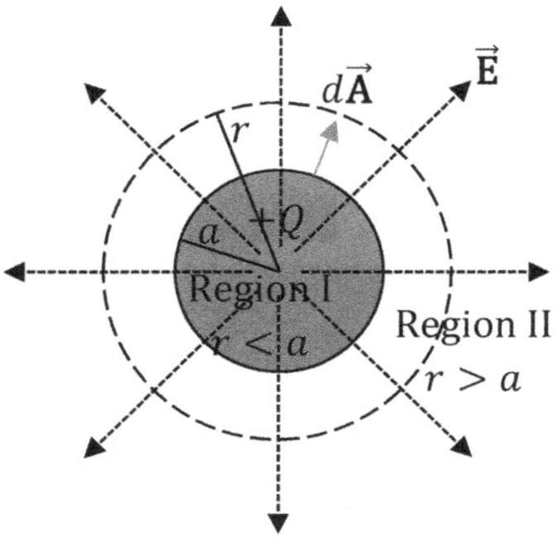

This problem is just like the problem from two examples back, except that the charge is non-uniform. The electric field lines and Gaussian surface are the same as before. To save time, we'll simply repeat the steps that are identical to the similar example, and pick up from where this solution deviates from the earlier one. It would be a good exercise to see if you can understand each step (if not, review the earlier example for the explanation).

$$\oint_S \vec{\mathbf{E}} \cdot d\vec{\mathbf{A}} = \frac{q_{enc}}{\epsilon_0}$$

$$\oint_S E \cos 0° \, dA = \frac{q_{enc}}{\epsilon_0}$$

$$\oint_S E \, dA = \frac{q_{enc}}{\epsilon_0}$$

$$E \oint_S dA = \frac{q_{enc}}{\epsilon_0}$$

$$A = \oint_S dA = \int_{\theta=0}^{\pi} \int_{\varphi=0}^{2\pi} r^2 \sin\theta \, d\theta \, d\varphi = r^2 \int_{\theta=0}^{\pi} \int_{\varphi=0}^{2\pi} \sin\theta \, d\theta \, d\varphi$$

$$A = r^2 \int_{\theta=0}^{\pi} \sin\theta \left(\int_{\varphi=0}^{2\pi} d\varphi \right) d\theta = r^2 \int_{\theta=0}^{\pi} \sin\theta \, [\varphi]_{\varphi=0}^{2\pi} \, d\theta = r^2 \int_{\theta=0}^{\pi} \sin\theta \, (2\pi) \, d\theta$$

$$A = 2\pi r^2 \int_{\theta=0}^{\pi} \sin\theta \, d\theta = 2\pi r^2 [-\cos\theta]_{\theta=0}^{\pi} = 2\pi r^2 (-\cos\pi + \cos 0) = 2\pi r^2 (1 + 1) = 4\pi r^2$$

$$E \oint_S dA = EA = E 4\pi r^2 = \frac{q_{enc}}{\epsilon_0}$$

$$E = \frac{q_{enc}}{4\pi\epsilon_0 r^2}$$

Now we have reached the point where it will matter that the charge density is non-uniform. This time, we can't pull the charge density (ρ) out of the integral. Instead, we must apply the equation $\rho = \beta r$ that was given in the problem. The symbol β, however, is constant and may come out of the integral.

$$q_{enc} = \int dq = \int \rho \, dV = \int \beta r \, dV = \beta \int r \, dV$$

<u>**Region I**</u>: $r < a$.

Inside of the charged sphere, only a fraction of the sphere's charge is enclosed by the Gaussian sphere, so the upper limit of the r-integration is the variable r. In spherical coordinates, we write the differential volume element as $dV = r^2 \sin\theta \, dr d\theta d\varphi$.

$$q_{enc} = \beta \int r \, dV = \beta \int_{r=0}^{r} \int_{\theta=0}^{\pi} \int_{\varphi=0}^{2\pi} r(r^2 \sin\theta \, dr \, d\theta \, d\varphi) = \beta \int_{r=0}^{r} \int_{\theta=0}^{\pi} \int_{\varphi=0}^{2\pi} r^3 \sin\theta \, dr \, d\theta \, d\varphi$$

This integral is separable:

$$q_{enc} = \beta \int_{r=0}^{r} r^3 \, dr \int_{\theta=0}^{\pi} \sin\theta \, d\theta \int_{\varphi=0}^{2\pi} d\varphi = \beta \left[\frac{r^4}{4}\right]_{r=0}^{r} [-\cos\theta]_{\theta=0}^{\pi} [\varphi]_{\varphi=0}^{2\pi}$$

$$q_{enc} = \beta \left(\frac{r^4}{4}\right)(-\cos\pi + \cos 0)(2\pi) = \beta \left(\frac{r^4}{4}\right)(1+1)(2\pi) = \pi\beta r^4$$

As usual with problems that feature non-uniform densities, we will eliminate the constant of proportionality (in this case, β). Note that if we integrate all of the way to $r = a$, we would obtain the total charge of the sphere, Q. We would also get the same expression as above, except with r replaced by a:

$$Q = \pi\beta a^4$$

Isolate β by dividing both sides of the equation by πa^4.

$$\beta = \frac{Q}{\pi a^4}$$

Substitute this expression into the previous equation for the charge enclosed.

$$q_{enc} = \pi\beta r^4 = \pi \left(\frac{Q}{\pi a^4}\right) r^4 = \frac{Q r^4}{a^4}$$

Now substitute the charge enclosed into the previous equation for electric field.

$$E = \frac{q_{enc}}{4\pi\epsilon_0 r^2} = \frac{Q r^4}{a^4} \div 4\pi\epsilon_0 r^2 = \frac{Q r^4}{a^4} \times \frac{1}{4\pi\epsilon_0 r^2} = \frac{Q r^2}{4\pi\epsilon_0 a^4}$$

Recall that $\frac{1}{4\pi\epsilon_0} = k$.

$$E = \frac{kQr^2}{a^4}$$

We can add a unit vector to indicate the direction of the electric field vector.

$$\vec{\mathbf{E}} = \frac{kQr^2}{a^4}\hat{\mathbf{r}}$$

<u>**Region II**</u>: $r > a$.

Outside of the charged sphere, 100% of the charge is enclosed by the Gaussian surface. This changes the upper limit of the r-integration to a (since all of the charge lies inside a sphere of radius a).

$$q_{enc} = Q$$

Substitute this into the equation for electric field which we found from Gauss's law.

$$E = \frac{q_{enc}}{4\pi\epsilon_0 r^2} = \frac{Q}{4\pi\epsilon_0 r^2}$$

Apply the equation $\frac{1}{4\pi\epsilon_0} = k$ and add a unit vector to turn the magnitude (E) of the electric field into a vector ($\vec{\mathbf{E}}$).

$$\vec{\mathbf{E}} = \frac{kQ}{r^2}\hat{\mathbf{r}}$$

Example: An infinite line of positive charge lies on the z-axis and has uniform charge density λ. Derive an expression for the electric field created by the infinite line charge.

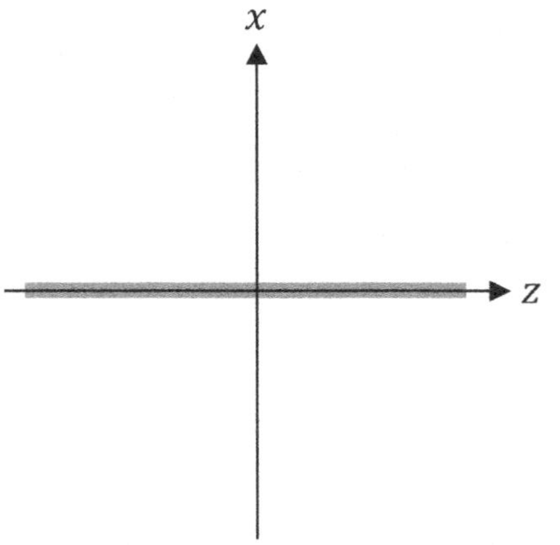

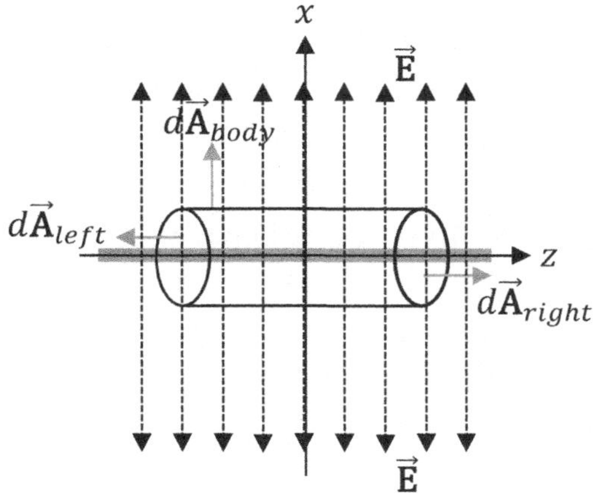

First sketch the electric field lines for the infinite line charge. Wherever a positive test charge might be placed, it would be repelled directly away from the z-axis. Therefore, the electric field lines radiate outward, as shown above on the right (but realize that the field lines really radiate outward in three dimensions). We choose our Gaussian surface to be a cylinder (see the right diagram above) coaxial with the line charge such that $\vec{\mathbf{E}}$ and $d\vec{\mathbf{A}}$ will be parallel along the body of the cylinder (and $\vec{\mathbf{E}}$ and $d\vec{\mathbf{A}}$ will be perpendicular along the ends). Write the formula for Gauss's law.

$$\oint_S \vec{\mathbf{E}} \cdot d\vec{\mathbf{A}} = \frac{q_{enc}}{\epsilon_0}$$

The scalar product is $\vec{\mathbf{E}} \cdot d\vec{\mathbf{A}} = E \cos\theta \, dA$. Study the direction of $\vec{\mathbf{E}}$ and $d\vec{\mathbf{A}}$ at each end and the body of the cylinder in the previous diagram.

- For the ends, $\theta = 90°$ because $\vec{\mathbf{E}}$ and $d\vec{\mathbf{A}}$ are perpendicular.
- For the body, $\theta = 0°$ because $\vec{\mathbf{E}}$ and $d\vec{\mathbf{A}}$ are parallel.

$$\int_{left} E \cos 90° \, dA + \int_{body} E \cos 0° \, dA + \int_{right} E \cos 90° \, dA = \frac{q_{enc}}{\epsilon_0}$$

Recall from trig that $\cos 0° = 1$ and $\cos 90° = 0$.

$$0 + \int_{body} E \, dA + 0 = \frac{q_{enc}}{\epsilon_0}$$

The magnitude of the electric field is constant over the body of the Gaussian cylinder, since every point on the body of the cylinder is equidistant from the axis of the line charge. Therefore, we may pull E out of the integral.

$$E \oint_{body} dA = \frac{q_{enc}}{\epsilon_0}$$

$$EA = \frac{q_{enc}}{\epsilon_0}$$

The area is the surface area of the Gaussian cylinder of radius r_c (which is a variable, since the electric field depends upon the distance from the line charge). Note that r_c is the distance from the z-axis (from cylindrical coordinates), and **not** the distance to the origin (which we use in spherical coordinates). The area of the body of a cylinder is $A = 2\pi r_c L$, where L is the length of the Gaussian cylinder. (We choose the Gaussian cylinder to be finite, unlike the infinite line of charge.)

$$E 2\pi r_c L = \frac{q_{enc}}{\epsilon_0}$$

Isolate the magnitude of the electric field by dividing both sides of the equation by $2\pi r_c L$.

$$E = \frac{q_{enc}}{2\pi\epsilon_0 r_c L}$$

Now we need to determine how much charge is enclosed by the Gaussian cylinder. For a line charge, we write $dq = \lambda ds$ (Step 6 on page 94). Since the line charge has uniform charge density, we may pull λ out of the integral.

$$q_{enc} = \int dq = \int \lambda\, ds = \lambda \int ds = \lambda \int dz = \lambda L$$

The charge enclosed by the Gaussian cylinder is $q_{enc} = \lambda L$, where L is the length of the Gaussian cylinder. Substitute this into the previous equation for electric field.

$$E = \frac{q_{enc}}{2\pi\epsilon_0 r_c L} = \frac{\lambda L}{2\pi\epsilon_0 r_c L} = \frac{\lambda}{2\pi\epsilon_0 r_c}$$

We can turn this into a vector by including the appropriate unit vector.

$$\vec{\mathbf{E}} = \frac{\lambda}{2\pi\epsilon_0 r_c}\hat{\mathbf{r}}_c$$

Note that $\hat{\mathbf{r}}_c$ is directed away from the z-axis.

Example: An infinite solid cylindrical conductor coaxial with the z-axis has radius a and uniform charge density σ. Derive an expression for the electric field both inside and outside of the conductor.

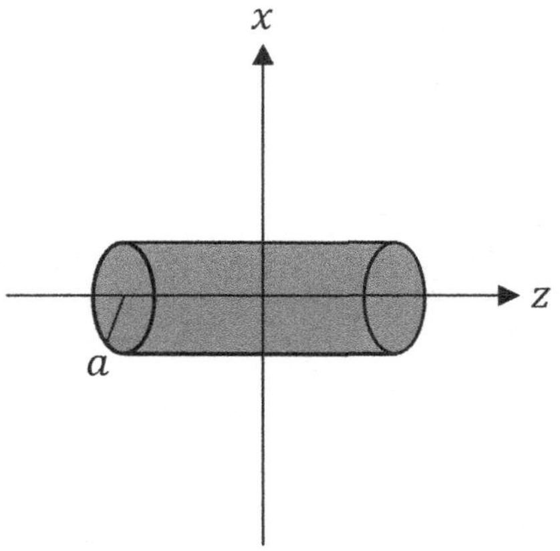

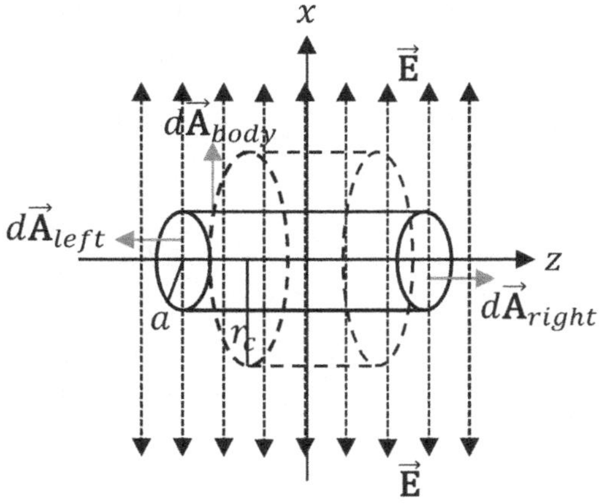

The electric field lines for this infinite charged cylinder radiate away from the z-axis just like the electric field lines for the infinite line charge in the previous example. As with the previous example, we will apply a Gaussian cylinder coaxial with the z-axis, and the math will be virtually the same as in the previous example. One difference is that this problem involves two different regions, and another difference is that this problem involves a surface charge density (σ) instead of a linear charge density (λ).

<u>**Region I**</u>: $r_c < a$.
Since the cylinder is a conductor, the electric field will be zero in region I.

$$\vec{\mathbf{E}} = 0$$

The reasoning is the same as it was a few examples back when we had a problem with a conducting shell. If there were a nonzero electric field inside the conducting cylinder, it would cause the charged particles inside of it (namely the valence electrons) to accelerate. Within a fraction of a second, the electrons in the conducting cylinder will attain electrostatic equilibrium, after which point the electric field will be zero inside the volume of the conducting cylinder. Since $\vec{\mathbf{E}} = 0$ inside the conducting cylinder and since $\oint_S \vec{\mathbf{E}} \cdot d\vec{\mathbf{A}} = \frac{q_{enc}}{\epsilon_0}$ according to Gauss's law, the net charge within the conducting cylinder must also be zero. Therefore, the net charge that resides on the conducting cylinder must reside on its surface.

<u>**Region II**</u>: $r_c > a$.
Since the electric field lines for this problem closely resemble the electric field lines of the previous example, the math for Gauss's law will start out the same. We will repeat those equations to save time, and then pick up our discussion where this solution deviates from the previous one. Again, it would be wise to try to follow along, and review the previous example if necessary.

$$\oint_S \vec{\mathbf{E}} \cdot d\vec{\mathbf{A}} = \frac{q_{enc}}{\epsilon_0}$$

$$\int_{left} E\cos 90°\, dA + \int_{body} E\cos 0°\, dA + \int_{right} E\cos 90°\, dA = \frac{q_{enc}}{\epsilon_0}$$

$$0 + \int_{body} E\, dA + 0 = \frac{q_{enc}}{\epsilon_0}$$

$$E \oint_{body} dA = \frac{q_{enc}}{\epsilon_0}$$

$$EA = \frac{q_{enc}}{\epsilon_0}$$

$$E2\pi r_c L = \frac{q_{enc}}{\epsilon_0}$$

$$E = \frac{q_{enc}}{2\pi\epsilon_0 r_c L}$$

This is the point where the solution is different for the infinite conducting cylinder than for the infinite line charge. When we determine how much charge is enclosed by the Gaussian cylinder, we will work with $dq = \sigma dA$ (Step 6 on page 94). For a solid cylinder, we would normally use ρdV, but since this is a **conducting** cylinder, as we reasoned in region I, all of the charge resides on its surface, not within its volume. (If this had been an insulator instead of a conductor, then we would use ρdV for a solid cylinder.) Since the cylinder has uniform charge density, we may pull σ out of the integral.

$$q_{enc} = \int dq = \int \sigma\, dA = \sigma \int dA = \sigma A$$

The surface area of the body of a cylinder is $A = 2\pi a L$. Here, we use the radius of the conducting cylinder (a), not the radius of the Gaussian cylinder (r_c), because the charge lies on a cylinder of radius a (we're finding the charge enclosed). Plug this expression into the equation for the charge enclosed.

$$q_{enc} = \sigma A = \sigma 2\pi a L$$

Substitute this into the previous equation for electric field.

$$E = \frac{q_{enc}}{2\pi\epsilon_0 r_c L} = \frac{\sigma 2\pi a L}{2\pi\epsilon_0 r_c L} = \frac{\sigma a}{\epsilon_0 r_c}$$

We can turn this into a vector by including the appropriate unit vector.

$$\vec{\mathbf{E}} = \frac{\sigma a}{\epsilon_0 r_c}\hat{\mathbf{r}}_c$$

It's customary to write this in terms of **charge per unit length,** $\lambda = \frac{Q}{L}$, rather than charge per unit area, $\sigma = \frac{Q}{A} = \frac{Q}{2\pi a L}$, in which case the answer is identical to the previous example:

$$\vec{\mathbf{E}} = \frac{\lambda}{2\pi\epsilon_0 r_c}\hat{\mathbf{r}}_c$$

Example: A solid spherical insulator centered about the origin has radius a and uniform positive charge density ρ. The solid sphere has a spherical cavity with radius b. The distance between the center of the charged sphere and the center of the cavity is d, for which $d + b < a$. Derive an expression for the electric field in the region inside the cavity.

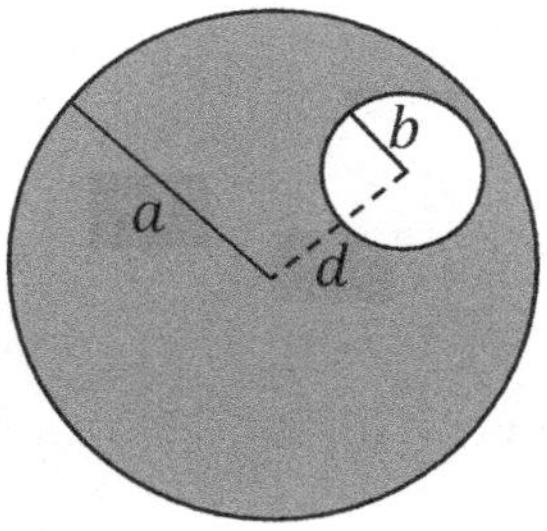

The "trick" to cavity problems is to apply the principle of **superposition** from Chapter 3. Although these aren't pointlike charges, the same principle of adding electric field vectors still applies. Geometrically, we can visualize the complete sphere as the sum of the object shown above and the small sphere that is missing from the object shown above.

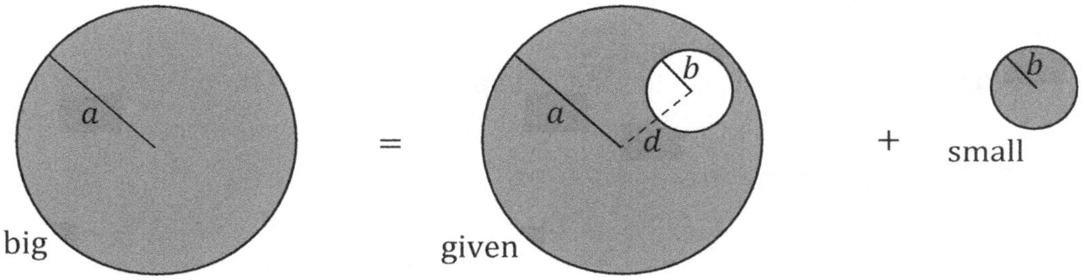

Our goal is to calculate the electric field at a point inside the cavity of the given shape. We marked such a point in the diagram below with a star (★). The electric field at the field point (★) due to the given shape (the original diagram that includes the cavity) **plus** the electric field at the field point (★) due to the small sphere (which is the same size as the cavity) **equals** the electric field at the field point (★) due to the big sphere (which doesn't have a cavity). We just wrote an equation in words, and now we will write the same equation with symbols:

$$\vec{\mathbf{E}}_{big} = \vec{\mathbf{E}}_{given} + \vec{\mathbf{E}}_{small}$$

Now we will draw the same equation with a picture:

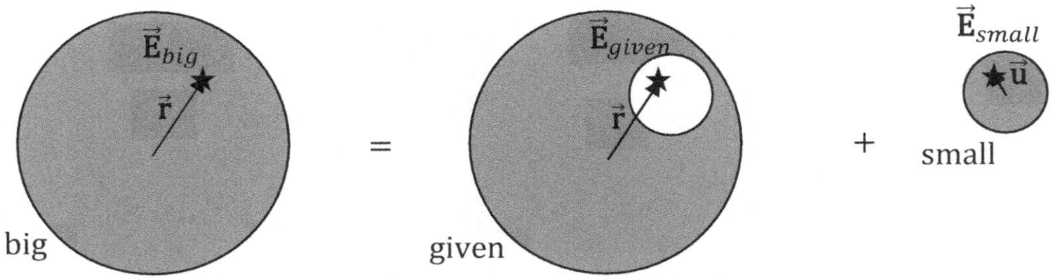

Consider the three vectors illustrated below:

- The vector $\vec{\mathbf{r}}$ extends from the center of the **big** sphere to the field point (★).
- The vector $\vec{\mathbf{u}}$ extends from the center of the **small** cavity to the field point (★).
- The vector $\vec{\mathbf{d}}$ extends from the center of the **big** sphere to the center of the **small** cavity.

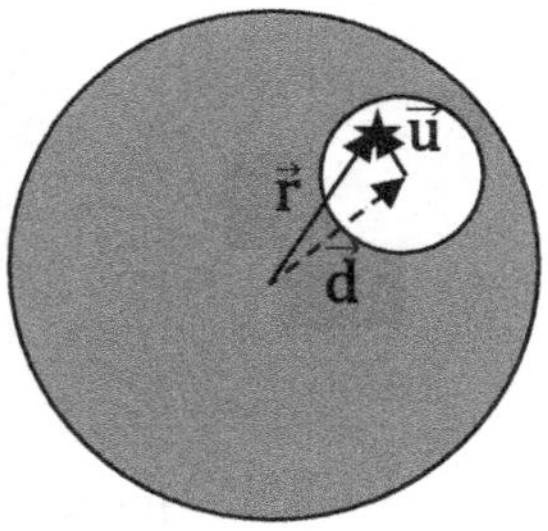

Since $\vec{\mathbf{d}}$ and $\vec{\mathbf{u}}$ join tip-to-tail to form $\vec{\mathbf{r}}$, recall from Volume 1 (vector addition) that $\vec{\mathbf{r}}$ is the resultant vector:

$$\vec{\mathbf{r}} = \vec{\mathbf{d}} + \vec{\mathbf{u}}$$

Here is our plan:

1. Find the electric field ($\vec{\mathbf{E}}_{big}$) for the big sphere (without the cavity) at the field point (★). Note that the field point (★) is inside of the big sphere. Also note from the diagrams that this involves the vector $\vec{\mathbf{r}}$.
2. Find the electric field ($\vec{\mathbf{E}}_{small}$) for the small sphere (which is the same size as the cavity) at the field point (★). Note that the field point (★) is inside of the small sphere. Also note from the diagrams that this involves the vector $\vec{\mathbf{u}}$.
3. Subtract our answers for the first two steps to obtain the electric field ($\vec{\mathbf{E}}_{given}$) for the given shape at the field point (★). Recall the equation $\vec{\mathbf{E}}_{big} = \vec{\mathbf{E}}_{given} + \vec{\mathbf{E}}_{small}$ from the previous page. We're subtracting $\vec{\mathbf{E}}_{small}$ from both sides in order to solve for $\vec{\mathbf{E}}_{given}$.

$$\vec{\mathbf{E}}_{given} = \vec{\mathbf{E}}_{big} - \vec{\mathbf{E}}_{small}$$

Step 1:

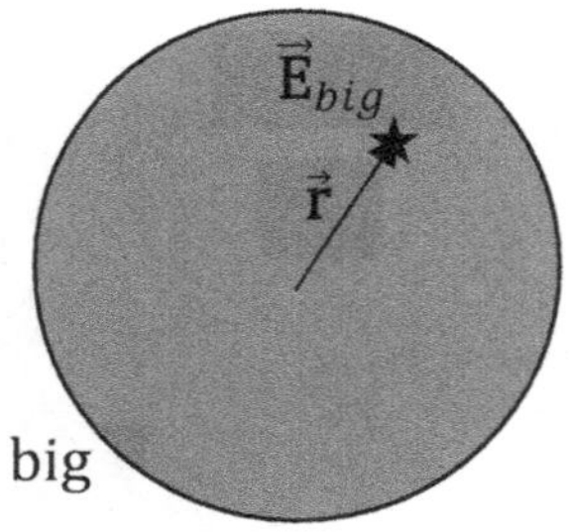

In this step, we have a single uniformly charged sphere (the big sphere) and wish to find the electric field at a point (★) inside of the sphere. We did exactly this in one of the previous examples. Rather than repeat that example, let us borrow an expression from region I ($r < a$) of that example (see page 128).

$$\vec{\mathbf{E}}_{big} = \frac{\rho r}{3\epsilon_0}\hat{\mathbf{r}}$$

<u>**Step 2**</u>:

$\vec{\mathbf{E}}_{small}$

small

In this step, we have a single uniformly charged sphere (the small sphere, which is the same size as the cavity) and wish to find the electric field at a point (★) inside of the sphere. This is identical to what we did in Step 1, except that this step involves the vector $\vec{\mathbf{u}}$ instead of the vector $\vec{\mathbf{r}}$. We will obtain the same answer with r replaced by u and $\hat{\mathbf{r}}$ replaced by $\hat{\mathbf{u}}$.

$$\vec{\mathbf{E}}_{small} = \frac{\rho u}{3\epsilon_0}\hat{\mathbf{u}}$$

<u>**Step 3**</u>:
As discussed earlier, we will simply subtract electric field vectors to obtain our answer for the electric field at the field point (★) due to the given shape.

$$\vec{\mathbf{E}}_{given} = \vec{\mathbf{E}}_{big} - \vec{\mathbf{E}}_{small}$$
$$\vec{\mathbf{E}}_{given} = \frac{\rho r}{3\epsilon_0}\hat{\mathbf{r}} - \frac{\rho u}{3\epsilon_0}\hat{\mathbf{u}}$$

We can **factor** out the $\frac{\rho}{3\epsilon_0}$.

$$\vec{\mathbf{E}}_{given} = \frac{\rho}{3\epsilon_0}(r\hat{\mathbf{r}} - u\hat{\mathbf{u}})$$

Recall that any vector can be expressed as its magnitude times its direction. For example, $\vec{\mathbf{A}} = A\hat{\mathbf{A}}$. Therefore, $\vec{\mathbf{r}} = r\hat{\mathbf{r}}$ and $\vec{\mathbf{u}} = u\hat{\mathbf{u}}$.

$$\vec{\mathbf{E}}_{given} = \frac{\rho}{3\epsilon_0}(\vec{\mathbf{r}} - \vec{\mathbf{u}})$$

Earlier, we showed that $\vec{\mathbf{r}} = \vec{\mathbf{d}} + \vec{\mathbf{u}}$. Subtracting $\vec{\mathbf{u}}$ from both sides, we get $\vec{\mathbf{r}} - \vec{\mathbf{u}} = \vec{\mathbf{d}}$.

$$\vec{\mathbf{E}}_{given} = \frac{\rho}{3\epsilon_0}\vec{\mathbf{d}}$$

Recall that $\vec{\mathbf{d}}$ is a vector joining the center of the big sphere to the center of the cavity: It is a constant. Therefore, the electric field inside of the cavity equals $\vec{\mathbf{E}}_{given} = \frac{\rho}{3\epsilon_0}\vec{\mathbf{d}}$ and is <u>**uniform**</u> in this region.

44. A solid spherical insulator centered about the origin with positive charge Q has radius a and non-uniform charge density $\rho = \beta r^2$, where β is a positive constant. Derive an expression for the electric field both inside and outside of the sphere.

Want help? Check the hints section at the back of the book.

Answers: $\vec{\mathbf{E}}_I = \frac{kQr^3}{a^5}\hat{\mathbf{r}}$, $\vec{\mathbf{E}}_{II} = \frac{kQ}{r^2}\hat{\mathbf{r}}$

45. A solid spherical insulator centered about the origin with negative charge $-6Q$ has radius a and non-uniform charge density $\rho = -\beta r$, where β is a positive constant. Concentric with the spherical insulator is a thick spherical conducting shell of inner radius b, outer radius c, and total charge $+4Q$. Derive an expression for the electric field in each region.

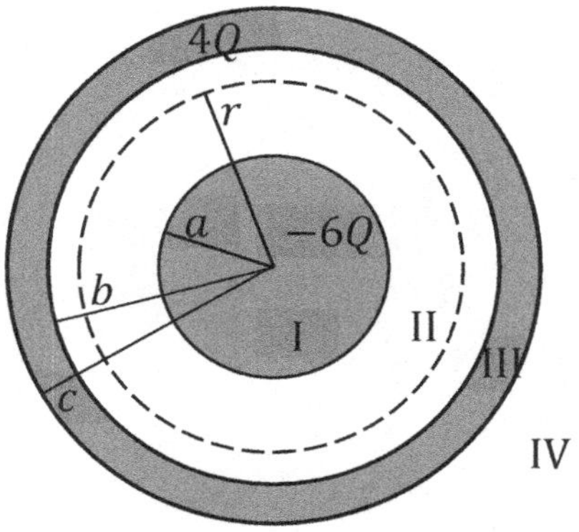

Want help? Check the hints section at the back of the book.

Answers: $\vec{\mathbf{E}}_I = -\frac{6kQr^2}{a^4}\hat{\mathbf{r}}$, $\vec{\mathbf{E}}_{II} = -\frac{6kQ}{r^2}\hat{\mathbf{r}}$, $\vec{\mathbf{E}}_{III} = 0$, $\vec{\mathbf{E}}_{IV} = -\frac{2kQ}{r^2}\hat{\mathbf{r}}$

46. An infinite solid cylindrical insulator coaxial with the z-axis has radius a and uniform positive charge density ρ. Derive an expression for the electric field both inside and outside of the charged insulator.

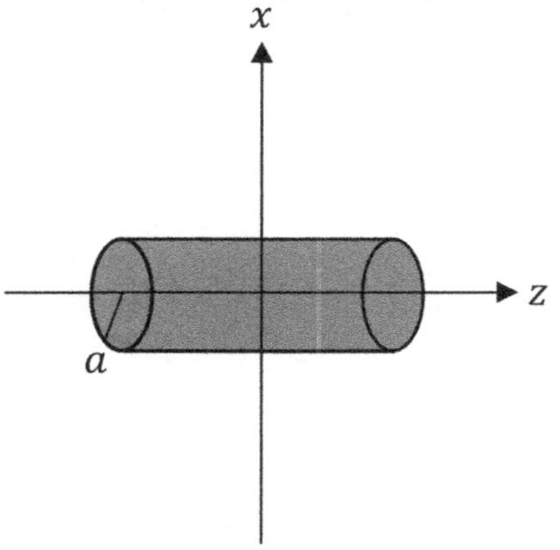

Want help? Check the hints section at the back of the book.

Answers: $\vec{\mathbf{E}}_I = \frac{\rho r_c}{2\epsilon_0}\hat{\mathbf{r}}_c$, $\vec{\mathbf{E}}_{II} = \frac{\rho a^2}{2\epsilon_0 r_c}\hat{\mathbf{r}}_c$

47. Two very thin, infinitely large sheets of charge have equal and opposite uniform charge densities $+\sigma$ and $-\sigma$. The positive sheet lies in the xy plane at $z = 0$ while the negative sheet is parallel to the first at $z = d$. Derive an expression for the electric field in each of the three regions.

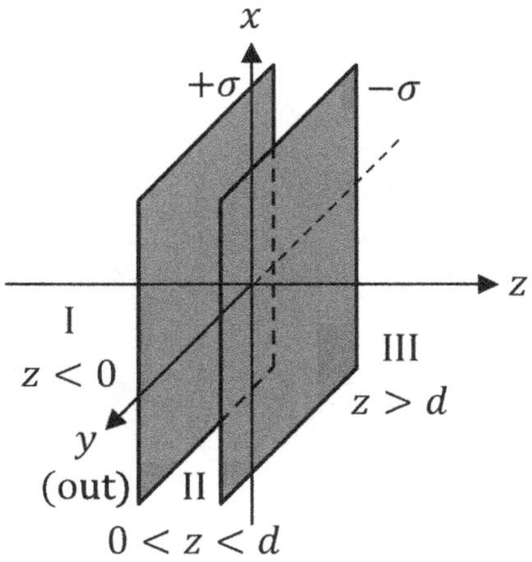

Want help? Check the hints section at the back of the book.

Answers: $\vec{\mathbf{E}}_I = 0$, $\vec{\mathbf{E}}_{II} = \frac{\sigma}{\epsilon_0}\hat{\mathbf{z}}$, $\vec{\mathbf{E}}_{III} = 0$

48. Two very thin, infinitely large sheets of charge have equal and opposite uniform charge densities $+\sigma$ and $-\sigma$. The two sheets are perpendicular to one another, with the positive sheet lying in the xy plane and with the negative sheet lying in the yz plane. Derive an expression for the electric field in the octant where x, y, and z are all positive.

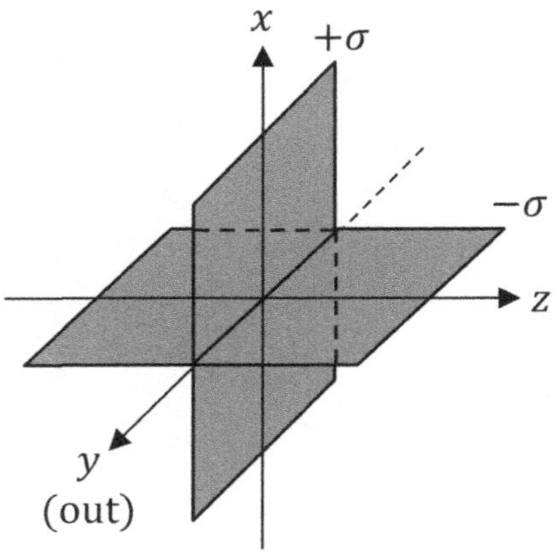

Want help? Check the hints section at the back of the book.

Answer: $\vec{\mathbf{E}} = \frac{\sigma}{2\epsilon_0}(\hat{\mathbf{z}} - \hat{\mathbf{x}})$

49. An infinite nonconducting slab with thickness T is parallel to the xy plane, centered about $z = 0$, and has non-uniform charge density $\rho = \beta|z|$, where β is a positive constant. Derive an expression for the electric field in each region.

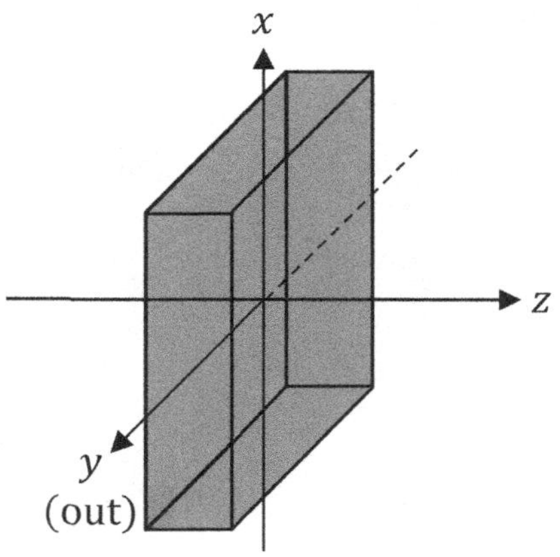

Want help? Check the hints section at the back of the book.

Answers: $\vec{\mathbf{E}}_I = -\frac{\beta T^2}{8\epsilon_0}\hat{\mathbf{z}}$, $\vec{\mathbf{E}}_{II} = \pm\frac{\beta z^2}{2\epsilon_0}\hat{\mathbf{z}}$, $\vec{\mathbf{E}}_{III} = \frac{\beta T^2}{8\epsilon_0}\hat{\mathbf{z}}$

50. An infinite solid cylindrical insulator coaxial with the z-axis has radius a and uniform positive charge density ρ. The solid cylinder has a cylindrical cavity with radius b. The distance between the axis of the charged insulator and the axis of the cavity is d, for which $d + b < a$. Derive an expression for the electric field in the region inside the cavity.

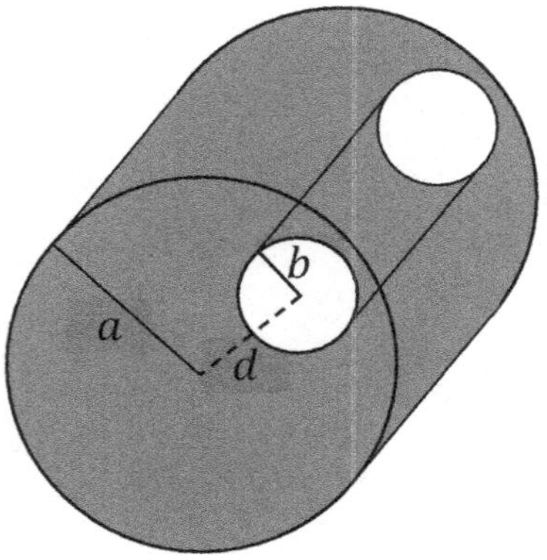

Want help? Check the hints section at the back of the book.

Answer: $\vec{\mathbf{E}} = \frac{\rho}{2\epsilon_0}\vec{\mathbf{d}}$

9 ELECTRIC POTENTIAL

Relevant Terminology

Electric charge – a fundamental property of a particle that causes the particle to experience a force in the presence of an electric field. (An electrically neutral particle has no charge and thus experiences no force in the presence of an electric field.)
Electric force – the push or pull that one charged particle exerts on another. Oppositely charged particles attract, whereas like charges (both positive or both negative) repel.
Electric field – force per unit charge.
Electric potential energy – a measure of how much electrical work a charged particle can do by changing position.
Electric potential – electric potential energy per unit charge. Specifically, the electric potential difference between two points equals the electric work per unit charge that would be involved in moving a charged particle from one point to the other.

Electric Potential Equations

If you want to find the **electric potential** created by a single **pointlike** charge, use the following equation, where R is the distance from the pointlike charge. Unlike previous equations for electric force and electric field, R is <u>not</u> squared in the equation for electric potential and there are <u>no</u> absolute values (sign matters here).

$$V = \frac{kq}{R}$$

To find the **electric potential** created by a system of pointlike charges, simply add up the electric potentials for each charge. Since electric potential is a scalar (not a vector), it's much easier to add electric potentials than it is to add electric fields (Chapter 3).

$$V_{net} = \frac{kq_1}{R_1} + \frac{kq_2}{R_2} + \cdots + \frac{kq_N}{R_N}$$

Electric potential energy (PE_e) equals charge (q) times electric potential (V).

$$PE_e = qV$$

To find the **electric potential energy** of a system of charges, first find the electric potential energy (PE_e) for each pair of charges.

$$PE_e = k\frac{q_1 q_2}{R}$$

Potential difference (ΔV) is the difference in electric potential between two points.

$$\Delta V = V_f - V_i$$

Electrical **work** (W_e) equals charge (q) times potential difference (ΔV).

$$W_e = q\Delta V$$

Symbols and SI Units

Symbol	Name	SI Units
V	electric potential	V
ΔV	potential difference	V
PE_e	electric potential energy	J
W_e	electrical work	J
q	charge	C
R	distance from the charge	m
k	Coulomb's constant	$\frac{\text{N}\cdot\text{m}^2}{\text{C}^2}$ or $\frac{\text{kg}\cdot\text{m}^3}{\text{C}^2\cdot\text{s}^2}$

Notes Regarding Units

The SI unit of electric potential (V) is the Volt (V). From the equation $PE_e = qV$, a Volt can be related to the SI unit of energy by solving for electric potential:

$$V = \frac{PE_e}{q}$$

According to this equation, one Volt (V) equals a Joule (J) per Coulomb (C), 1 V = 1 J/C, since the SI unit of energy is the Joule and the SI unit of charge is the Coulomb.

Important Distinctions

It's important to be able to distinguish between several similar terms. Units can help.

- The SI unit of electric **charge** (q) is the Coulomb (C).
- The SI units of electric **field** (E) can be expressed as $\frac{\text{N}}{\text{C}}$ or $\frac{\text{V}}{\text{m}}$.
- The SI unit of electric **force** (F_e) is the Newton (N).
- The SI unit of electric **potential** (V) is the Volt (V).
- The SI unit of electric **potential energy** (PE_e) is the Joule (J).

It's also important not to confuse similar equations.

- Coulomb's law for electric **force** has a pair of charges: $F_e = k\frac{|q_1||q_2|}{R^2}$.
- The electric **field** for a pointlike charge has one charge: $E = \frac{k|q|}{R^2}$.
- The electric **potential energy** for a system involves a pair of charges: $PE_e = k\frac{q_1 q_2}{R}$.
- The electric **potential** for a pointlike charge has one charge: $V = \frac{kq}{R}$.

Note that R is squared in the top two equations, but not the bottom equations.

Electric Potential Strategy

How you solve a problem involving electric potential depends on the type of problem:

- If you want to find the electric potential created by a single pointlike charge, use the following equation.
$$V = \frac{kq}{R}$$
R is the distance from the pointlike charge to the point where the problem asks you to find the electric potential. If the problem gives you the coordinates (x_1, y_1) of the charge and the coordinates (x_2, y_2) of the point where you need to find the electric field, apply the distance formula to find R.
$$R = \sqrt{(x_2 - x_1)^2 + (y_2 - y_1)^2}$$
- For a system of pointlike charges, add the electric potentials for each charge.
$$V_{net} = \frac{kq_1}{R_1} + \frac{kq_2}{R_2} + \cdots + \frac{kq_N}{R_N}$$
Be sure to use the signs of the charges. There are **no** absolute values.
- If you need to relate the electric potential (V) to electric potential energy (PE_e), use the following equation.
$$PE_e = qV$$
- If a problem involves a moving charge and involves electric potential, see the conservation of energy strategy of Chapter 10.
- If a problem involves electric field and potential difference, see Chapter 10.
- If a problem involves finding the electric potential created by a continuous distribution of charge (such as a line charge or charged disc), see pages 155-164.

Example: A strand of monkey fur with a charge of 6.0 μC lies at the point ($\sqrt{3}$ m, 1.0 m). A strand of gorilla fur with a charge of -36.0 μC lies at the point ($-\sqrt{3}$ m, 1.0 m). What is the net electric potential at the point ($\sqrt{3}$ m, -1.0 m)?

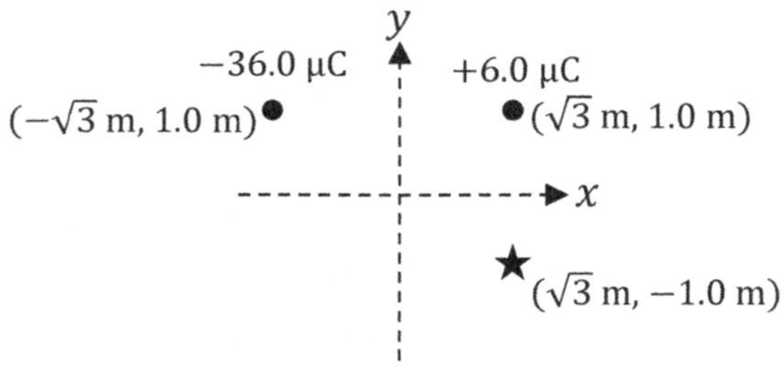

We need to determine how far each charge is from the field point at ($\sqrt{3}$ m, -1.0 m), which is marked with a star (★). The distances R_1 and R_2 are illustrated below.

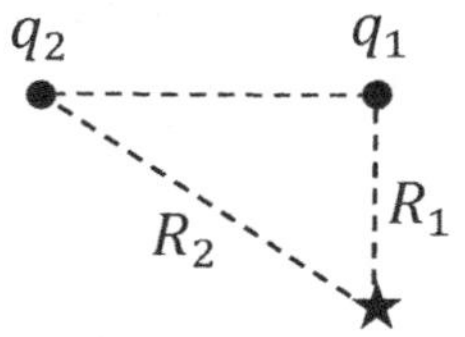

Apply the distance formula to determine these distances.

$$R_1 = \sqrt{\Delta x_1^2 + \Delta y_1^2} = \sqrt{\left(\sqrt{3} - \sqrt{3}\right)^2 + (-1 - 1)^2} = \sqrt{0^2 + (-2)^2} = \sqrt{4} = 2.0 \text{ m}$$

$$R_2 = \sqrt{\Delta x_2^2 + \Delta y_2^2} = \sqrt{\left[\sqrt{3} - \left(-\sqrt{3}\right)\right]^2 + (-1 - 1)^2} = \sqrt{\left(2\sqrt{3}\right)^2 + (-2)^2} = \sqrt{16} = 4.0 \text{ m}$$

The net electric potential equals the sum of the electric potentials for each pointlike charge.

$$V_{net} = \frac{kq_1}{R_1} + \frac{kq_2}{R_2}$$

Convert the charges to SI units: $q_1 = 6.0\ \mu\text{C} = 6.0 \times 10^{-6}$ C and $q_2 = -36.0\ \mu\text{C} = -36.0 \times 10^{-6}$ C. Recall that the metric prefix micro (μ) stands for one millionth: $\mu = 10^{-6}$. Note that we can factor out Coulomb's constant.

$$V_{net} = k\left(\frac{q_1}{R_1} + \frac{q_2}{R_2}\right) = (9 \times 10^9)\left(\frac{6 \times 10^{-6}}{2} + \frac{-36 \times 10^{-6}}{4}\right)$$

$$V_{net} = (9 \times 10^9)(3 \times 10^{-6} - 9 \times 10^{-6})$$

We can also factor out the 10^{-6}.

$$V_{net} = (9 \times 10^9)(10^{-6})(3 - 9) = (9 \times 10^9)(10^{-6})(-6) = -54 \times 10^3 \text{ V} = -5.4 \times 10^4 \text{ V}$$

Note that $10^9 10^{-6} = 10^{9-6} = 10^3$ according to the rule $x^m x^{-n} = x^{m-n}$. Also note that $54 \times 10^3 = 5.4 \times 10^4$. The answer is $V_{net} = -5.4 \times 10^4$ V, which is the same as -54×10^3 V, and can also be expressed as $V_{net} = -54$ kV, since the metric prefix kilo (k) stands for $10^3 = 1000$.

51. A monkey-shaped earring with a charge of $7\sqrt{2}$ µC lies at the point (0, 1.0 m). A banana-shaped earring with a charge of $-3\sqrt{2}$ µC lies at the point (0, −1.0 m). Your goal is to find the net electric potential at the point (1.0 m, 0).

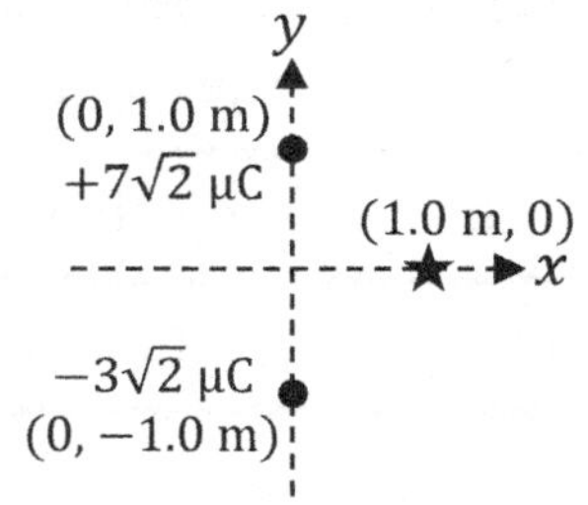

(A) Apply the distance formula to determine R_1 and R_2.

(B) Determine the net electric potential at the point (1.0 m, 0).

Want help or intermediate answers? Check the hints section at the back of the book.

Answer: 36 kV

Symbols and SI Units

Symbol	Name	SI Units
dq	differential charge element	C
Q	total charge of the object	C
V	electric potential	V
$\vec{\mathbf{E}}$	electric field	N/C or V/m
k	Coulomb's constant	$\frac{\text{N}\cdot\text{m}^2}{\text{C}^2}$ or $\frac{\text{kg}\cdot\text{m}^3}{\text{C}^2\cdot\text{s}^2}$
$\vec{\mathbf{R}}$	a vector from each dq to the field point	m
R	the distance from each dq to the field point	m
$d\vec{\mathbf{s}}$	differential displacement vector (see Chapter 21 of Volume 1)	m
x, y, z	Cartesian coordinates of dq	m, m, m
r, θ	2D polar coordinates of dq	m, rad
r_c, θ, z	cylindrical coordinates of dq	m, rad, m
r, θ, φ	spherical coordinates of dq	m, rad, rad
λ	linear charge density (for an arc)	C/m
σ	surface charge density (for an area)	C/m^2
ρ	volume charge density	C/m^3
ds	differential arc length	m
dA	differential area element	m^2
dV	differential volume element	m^3

Note: The symbols λ, σ, and ρ are the lowercase Greek letters lambda, sigma, and rho.

Strategy for Performing the Electric Potential Integral

To find the electric potential created by a continuous distribution of charge (called the **source**) at a specified point (called the **field point**), follow these steps:

1. Draw the charged object. We will call Q the total charge of the object. The total charge Q is composed of an infinite number of infinitesimal charges dq. Visualize integrating over every dq that makes up the charged object.
2. Draw and label a representative dq somewhere within the object. **Don't** draw dq at the origin or on an axis (unless the object is a rod and every point of the rod lies on an axis, in which case you have no choice). Draw an arrow from dq (the **source**) to the **field point** (the point where the problem asks you to find the electric field). Label this displacement as $\vec{\mathbf{R}}$. See the diagrams in the examples that follow.
3. Begin with the electric potential integral.
$$V = k \int \frac{dq}{R}$$
Interpret these symbols as follows:
   - V is the electric potential at the field point.
   - k is Coulomb's constant: $k = 9.0 \times 10^9 \ \frac{\text{N}\cdot\text{m}^2}{\text{C}^2}$.
   - R is the distance from each dq to the field point. In most of the problems, R is a variable and may **not** come out of the integral.
   - dq represents each infinitesimal charge element that makes up the object.

   Note: If you already have an equation for $\vec{\mathbf{E}}$, or if you can find $\vec{\mathbf{E}}$ easily through Gauss's law, you can find potential difference using the following formula. The formula below is similar to the work integral (discussed in Chapter 21 of Volume 1).
$$\Delta V = -\int \vec{\mathbf{E}} \cdot d\vec{\mathbf{s}}$$
We will apply the above formula in Chapter 12.
4. Study your diagram to express R in terms of suitable coordinates.
   - For a line or an object with all straight edges (like a polygon), work with Cartesian coordinates (x, y, z).
   - For a circular shape, work with 2D polar coordinates (r, θ).
   - For a spherical shape, work with spherical coordinates (r, θ, φ).
   - For a cylinder or cone, work with cylindrical coordinates (r_c, θ, φ). Note that r_c is the distance from the z-axis (whereas r is from the origin).
5. Make one of the following substitutions for dq, depending on the geometry:
   - $dq = \lambda ds$ for an arc length (like a rod or circular arc).
   - $dq = \sigma dA$ for a surface area (like a triangle, disc, or thin spherical shell).
   - $dq = \rho dV$ for a 3D solid (like a solid cube or a solid hemisphere).

6. Choose the appropriate coordinate system and make a substitution for ds, dA, or dV from Step 5 using the strategy from Chapter 6 (on page 86):
   - For a straight line parallel to the x-axis, $ds = dx$.
   - For a straight line parallel to the y-axis, $ds = dy$.
   - For a straight line parallel to the z-axis, $ds = dz$.
   - For a circular arc of radius a, $ds = ad\theta$.
   - For a solid polygon like a rectangle or triangle, $dA = dxdy$.
   - For a solid semicircle (**not** a circular arc) or pie slice, $dA = rdrd\theta$.
   - For a very thin spherical shell of radius a, $dA = a^2 \sin\theta \, d\theta d\varphi$.
   - For a solid polyhedron like a cube, $dV = dxdydz$.
   - For a solid cylinder or cone, $dV = r_c dr_c d\theta dz$.
   - For a portion of a solid sphere like a hemisphere, $dV = r^2 \sin\theta \, drd\theta d\varphi$.
7. Is the density uniform or non-uniform?
   - If the density is uniform, you can pull λ, σ, or ρ out of the integral.
   - If the density is non-uniform, leave λ, σ, or ρ in the integral.
8. If you have Cartesian coordinates in your integrand, but are integrating over polar, cylindrical, or spherical coordinates (or vice-versa), use the following substitutions, as needed, to put all of your coordinates in the same system.
$$x = r\cos\theta \quad , \quad y = r\sin\theta \quad \text{(2D polar)}$$
$$x = r_c\cos\theta \quad , \quad y = r_c\sin\theta \quad \text{(cylindrical)}$$
$$x = r\sin\theta\cos\varphi \quad , \quad y = r\sin\theta\sin\varphi \quad , \quad z = r\cos\theta \quad \text{(spherical)}$$
9. An integral over ds is a single integral, an integral over dA is a double integral, and an integral over dV is a triple integral. Set the limits of each integration variable that map out the region of integration, as illustrated in Chapter 6. Perform the integral using techniques from Chapter 6.
10. Perform the following integral to determine the total charge (Q) of the object:
$$Q = \int dq$$
Make the same substitutions as you made in Steps 5-9. Use the same limits of integration as you used in Step 9.
11. When you finish with Step 10, substitute your expression for Q into your original electric potential integral. Simplify the resulting expression.

Example: A rod lies on the y-axis with endpoints at the origin and $(0, L)$, as illustrated below. The positively charged rod has uniform charge density λ and total charge Q. Derive an equation for the electric potential at the point $(L, 0)$ in terms of k, Q, and L.

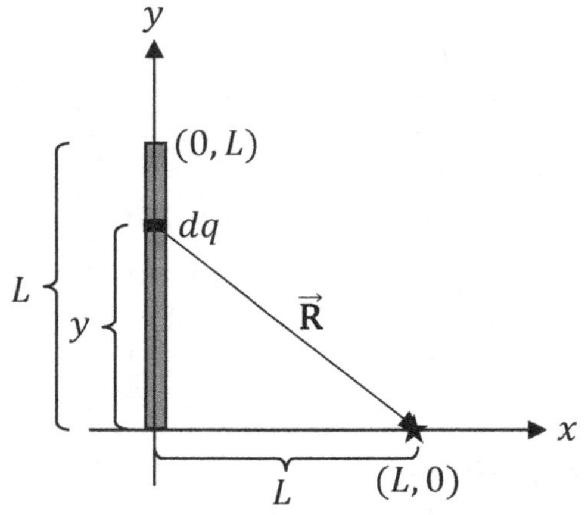

Begin with a labeled diagram. Draw a representative dq. Draw $\vec{\mathbf{R}}$ from the source, dq, to the field point $(L, 0)$. When we perform the integration, we effectively integrate over every dq that makes up the rod. Start the math with the electric potential integral.

$$V = k\int \frac{dq}{R}$$

Examine the picture above. The vector $\vec{\mathbf{R}}$ has a different length for each dq that makes up the rod. Therefore, its magnitude, R, can't come out of the integral. What we can do is express R in terms of the constant L and the variable y using the Pythagorean theorem: $R = \sqrt{L^2 + y^2}$. Substitute the expression for R into the electric potential integral.

$$V = k\int \frac{dq}{R} = k\int \frac{dq}{\sqrt{L^2 + y^2}}$$

For a thin rod, we write $dq = \lambda ds$ (Step 5 on page 155).

$$V = k\int \frac{\lambda ds}{\sqrt{L^2 + y^2}}$$

Since the rod has uniform charge density, we may pull λ out of the integral.

$$V = k\lambda\int \frac{ds}{\sqrt{L^2 + y^2}}$$

For a rod lying along the y-axis, we work with Cartesian coordinates and write the differential arc length as $ds = dy$. The limits of integration correspond to the length of the rod: $0 \leq y \leq L$.

$$V = k\lambda \int_{y=0}^{L} \frac{dy}{\sqrt{L^2 + y^2}}$$

This integral can be performed via the following trigonometric substitution:

$$y = L\tan\theta$$
$$dy = L\sec^2\theta\, d\theta$$

Solving for θ, we get $\theta = \tan^{-1}\left(\frac{y}{L}\right)$, which shows that the new limits of integration are from $\theta = \tan^{-1}(0) = 0°$ to $\theta = \tan^{-1}\left(\frac{L}{L}\right) = \tan^{-1}(1) = 45°$. Note that the denominator of the integral simplifies as follows, using the trig identity $1 + \tan^2\theta = \sec^2\theta$:

$$\sqrt{L^2 + y^2} = \sqrt{L^2 + (L\tan\theta)^2} = \sqrt{L^2(1 + \tan^2\theta)} = \sqrt{L^2\sec^2\theta} = L\sec\theta$$

In the last step, we applied the rule from algebra that $\sqrt{a^2x^2} = \sqrt{a^2}\sqrt{x^2} = ax$. Substitute the above expressions for dy and $\sqrt{L^2 + y^2}$ into the previous integral.

$$V = k\lambda \int_{y=0}^{L} \frac{dy}{\sqrt{L^2 + y^2}} = k\lambda \int_{\theta=0°}^{45°} \frac{L\sec^2\theta\, d\theta}{L\sec\theta} = k\lambda \int_{\theta=0°}^{45°} \sec\theta\, d\theta$$

Find the anti-derivative on page 70.

$$V = k\lambda[\ln|\sec\theta + \tan\theta|]_{\theta=0°}^{45°}$$

Evaluate the anti-derivative over the limits.

$$V = k\lambda(\ln|\sec 45° + \tan 45°| - \ln|\sec 0° + \tan 0°|)$$
$$V = k\lambda\left(\ln\left|\sqrt{2} + 1\right| - \ln|1 + 0|\right) = k\lambda\left(\ln\left|\sqrt{2} + 1\right| - \ln|1|\right) = k\lambda\left(\ln\left|\sqrt{2} + 1\right|\right)$$

Note that $\ln(1) = 0$. We're not finished yet. We must eliminate the constant λ from our answer. The way to do this is to integrate over dq to find the total charge of the rod, Q. This integral is convenient since we use the same substitutions from before.

$$Q = \int dq = \int \lambda\, ds = \lambda \int ds = \lambda \int_{y=0}^{L} dy = \lambda L$$

Solve for λ in terms of Q: Divide both sides of the equation by L.

$$\lambda = \frac{Q}{L}$$

Substitute this expression for λ into our previous expression for V.

$$V = k\lambda\left(\ln\left|\sqrt{2} + 1\right|\right) = \frac{kQ}{L}\left(\ln\left|\sqrt{2} + 1\right|\right)$$

The electric potential at the point $(L, 0)$ is $V = \frac{kQ}{L}\left(\ln\left|\sqrt{2} + 1\right|\right)$.

Example: A uniformly charged solid disc lies in the xy plane, centered about the origin as illustrated below. The radius of the disc is denoted by the symbol a and the positively charged disc has total charge Q. Derive an equation for the electric potential at the point $(0, 0, p)$ in terms of k, Q, a, and p.

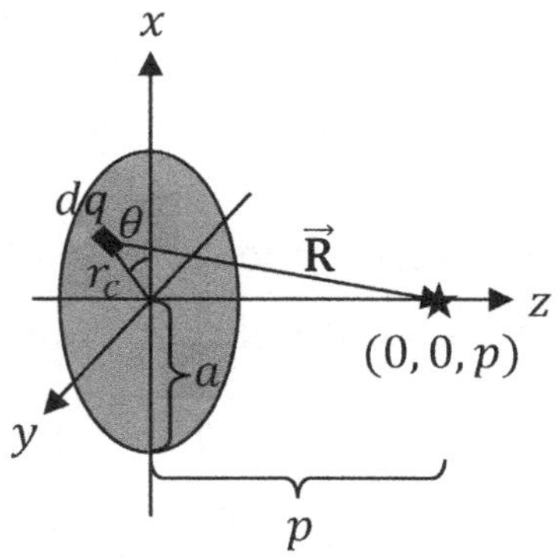

Begin with a labeled diagram. Draw a representative dq: Since this is a solid disc (unlike the previous example), most of the dq's lie within the area of the disc, so we drew dq inside the disc (not on its circumference). Draw $\vec{\mathbf{R}}$ from the source, dq, to the field point $(0, 0, p)$. When we perform the integration, we effectively integrate over every dq that makes up the disc. Start the math with the electric potential integral.

$$V = k \int \frac{dq}{R}$$

Examine the picture above. The vector $\vec{\mathbf{R}}$ has a different length for each dq that makes up the solid disc. Therefore, its magnitude, R, can't come out of the integral. What we can do is express R in terms of the variable r_c and the constant p using the Pythagorean theorem: $R = \sqrt{r_c^2 + p^2}$. Substitute the expression for R into the electric potential integral.

$$V = k \int \frac{dq}{\sqrt{r_c^2 + p^2}}$$

For a solid disc, we write $dq = \sigma dA$ (Step 5 on page 155).

$$V = k \int \frac{\sigma dA}{\sqrt{r_c^2 + p^2}}$$

Since the solid disc has uniform charge density, we may pull σ out of the integral.

$$V = k\sigma \int \frac{dA}{\sqrt{r_c^2 + p^2}}$$

For a solid disc, we write the differential area element as $dA = r_c dr_c d\theta$ (Step 6 on page 156), and since the field point at $(0, 0, p)$ does not lie in the xy plane, we work with cylindrical coordinates (which simply adds the z-coordinate to the 2D polar coordinates). The limits of integration are from $r_c = 0$ to $r_c = a$ and $\theta = 0$ to $\theta = 2\pi$ **radians**.

$$V = k\sigma \int_{r_c=0}^{a} \int_{\theta=0}^{2\pi} \frac{r_c dr_c d\theta}{\sqrt{r_c^2 + p^2}}$$

When we integrate over θ, we treat the independent variable r_c as a constant. In fact, nothing in the integrand depends on the variable θ, so we can factor everything out of the θ-integration.

$$V = k\sigma \int_{r_c=0}^{a} \frac{r_c dr_c}{\sqrt{r_c^2 + p^2}} \int_{\theta=0}^{2\pi} d\theta = k\sigma \int_{r_c=0}^{a} \frac{r_c dr_c}{\sqrt{r_c^2 + p^2}} [\theta]_{\theta=0}^{2\pi} = 2\pi k\sigma \int_{r_c=0}^{a} \frac{r_c dr_c}{\sqrt{r_c^2 + p^2}}$$

This integral can be performed via the following trigonometric substitution:

$$r_c = p \tan \psi$$
$$dr_c = p \sec^2 \psi \, d\psi$$

Solving for ψ, we get $\theta = \tan^{-1}\left(\frac{r_c}{p}\right)$, which shows that the new limits of integration are from $\psi = \tan^{-1}(0) = 0°$ to $\psi = \tan^{-1}\left(\frac{a}{p}\right)$, which for now we will simply call ψ_{max}. The denominator of the integral simplifies through the trig identity $\tan^2 \psi + 1 = \sec^2 \psi$:

$$\sqrt{r_c^2 + p^2} = \sqrt{(p \tan \psi)^2 + p^2} = \sqrt{p^2(\tan^2 \psi + 1)} = \sqrt{p^2 \sec^2 \psi} = p \sec \psi$$

In the last step, we applied the rule from algebra that $\sqrt{a^2 x^2} = \sqrt{a^2}\sqrt{x^2} = ax$. Substitute the above expressions for r_c, dr_c, and $\sqrt{r_c^2 + p^2}$ into the previous integral.

$$V = 2\pi k\sigma \int_{r_c=0}^{a} \frac{r_c dr_c}{\sqrt{r_c^2 + p^2}} = 2\pi k\sigma \int_{\psi=0°}^{\psi_{max}} \frac{(p \tan \psi)(p \sec^2 \psi \, d\psi)}{p \sec \psi}$$

Note that $\frac{p^2}{p} = p$ and $\frac{\sec^2 \psi}{\sec \psi} = \sec \psi$.

$$V = 2\pi k\sigma \int_{\psi=0°}^{\psi_{max}} p \tan \psi \sec \psi \, d\psi = 2\pi k\sigma p \int_{\psi=0°}^{\psi_{max}} \tan \psi \sec \psi \, d\psi$$

Recall from trig that $\sec \psi = \frac{1}{\cos \psi}$ and $\tan \psi = \frac{\sin \psi}{\cos \psi}$. Therefore, it follows that:

$$\tan \psi \sec \psi = \frac{\sin \psi}{\cos \psi} \frac{1}{\cos \psi} = \frac{\sin \psi}{\cos^2 \psi}$$

The electric potential integral becomes:

$$V = 2\pi k\sigma p \int_{\psi=0°}^{\psi_{max}} \frac{\sin \psi}{\cos^2 \psi} d\psi$$

One way to perform this integral is through the following substitution:

$$u = \cos \psi$$
$$du = -\sin \psi \, d\psi$$

Since $u = \cos \psi$, the new limits of integration are from $u(0) = \cos 0° = 1$ to $u(\psi_{max}) = \cos \psi_{max}$. For now, we will just call the upper limit u_{new} and worry about it later.

$$V = 2\pi k\sigma p \int_{u=1}^{u_{new}} \frac{-du}{u^2}$$

Apply the rule from algebra that $\frac{1}{u^2} = u^{-2}$.

$$V = -2\pi k\sigma p \int_{u=1}^{u_{new}} u^{-2}\, du = -2\pi k\sigma p \left[\frac{u^{-1}}{-1}\right]_{u=1}^{u_{new}} = 2\pi k\sigma p[u^{-1}]_{u=1}^{u_{new}} = 2\pi k\sigma p\left(\frac{1}{u_{u_{new}}} - 1\right)$$

Recall that what we called u_{new} is equal to $u_{new} = \cos\psi_{max}$ and that ψ_{max} is given by $\psi_{max} = \tan^{-1}\left(\frac{a}{p}\right)$, such that u_{new} is the complicated looking expression $u_{new} = \cos\left(\tan^{-1}\left(\frac{a}{p}\right)\right)$. When you find yourself taking the cosine of an inverse tangent, you can find a simpler way to write it by drawing a right triangle and applying the Pythagorean theorem. Since $\psi_{max} = \tan^{-1}\left(\frac{a}{p}\right)$, it follows that $\tan\psi_{max} = \frac{a}{p}$. We can make a right triangle from this: Since the tangent of ψ_{max} equals the opposite over the adjacent, we draw a right triangle with a opposite and p adjacent to ψ_{max}.

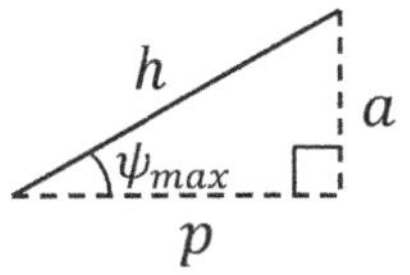

Find the hypotenuse, h, of the right triangle from the Pythagorean theorem.

$$h = \sqrt{p^2 + a^2}$$

Now we can write an expression for the cosine of ψ_{max}. It equals the adjacent (p) over the hypotenuse ($h = \sqrt{p^2 + a^2}$).

$$u_{new} = \cos\psi_{max} = \frac{p}{h} = \frac{p}{\sqrt{p^2 + a^2}}$$

The reciprocal of u_{new} is:

$$\frac{1}{u_{new}} = \frac{\sqrt{p^2 + a^2}}{p}$$

Substitute this expression into the previous equation for electric potential.

$$V = 2\pi k\sigma p\left(\frac{1}{u_{new}} - 1\right) = 2\pi k\sigma p\left(\frac{\sqrt{p^2 + a^2}}{p} - 1\right)$$

To add fractions, make a **common denominator**.

$$V = 2\pi k\sigma p\left(\frac{\sqrt{p^2 + a^2}}{p} - \frac{p}{p}\right) = 2\pi k\sigma p\left(\frac{\sqrt{p^2 + a^2} - p}{p}\right)$$

Note that the p from out front cancels with the $\frac{1}{p}$.

$$V = 2\pi k\sigma\left(\sqrt{p^2 + a^2} - p\right)$$

We're not finished yet because we need to eliminate the charge density σ from our answer. The way to do this is to integrate over dq to find the total charge of the disc, Q. This integral involves the same substitutions from before.

$$Q = \int dq = \int \sigma \, dA = \sigma \int dA = \sigma \int_{r_c=0}^{a} \int_{\theta=0}^{2\pi} r_c dr_c d\theta = \sigma \int_{r_c=0}^{a} r_c dr_c \int_{\theta=0}^{2\pi} d\theta$$

$$Q = \sigma \left[\frac{r_c^2}{2}\right]_{r_c=0}^{a} [\theta]_{\theta=0}^{2\pi} = \sigma \left(\frac{a^2}{2} - \frac{0^2}{2}\right)(2\pi - 0) = \sigma \left(\frac{a^2}{2}\right)(2\pi) = \pi\sigma a^2$$

Solve for σ in terms of Q: Divide both sides of the equation by πa^2.

$$\sigma = \frac{Q}{\pi a^2}$$

Substitute this expression for σ into our previous expression for electric potential.

$$V = 2\pi k\sigma \left(\sqrt{p^2 + a^2} - p\right) = 2\pi k \left(\frac{Q}{\pi a^2}\right)\left(\sqrt{p^2 + a^2} - p\right)$$

$$V = \frac{2kQ}{a^2}\left(\sqrt{p^2 + a^2} - p\right)$$

52. A thin wire is bent into the shape of the semicircle illustrated below. The radius of the semicircle is denoted by the symbol a and the positively charged wire has total charge Q and uniform charge density. Derive an equation for the electric potential at the origin, in terms of k, Q, and a.

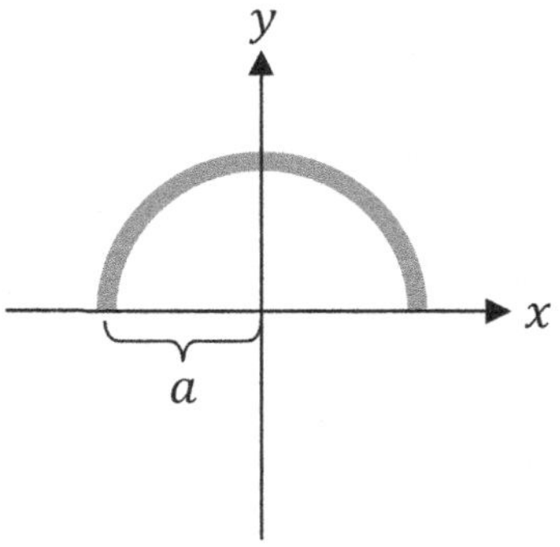

Want help? Check the hints section at the back of the book.

Answer: $V = \frac{kQ}{a}$

53. A very thin uniformly charged ring lies in the xy plane, centered about the origin as illustrated below. The radius of the ring is denoted by the symbol a and the positively charged ring has total charge Q. Derive an equation for the electric potential at the point $(0, 0, p)$ in terms of k, Q, a, and p.

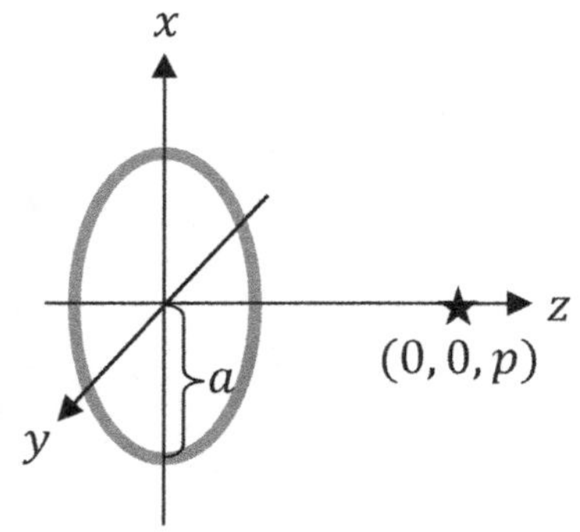

Want help? Check the hints section at the back of the book.

Answer: $V = \frac{kQ}{\sqrt{p^2+a^2}}$

10 MOTION OF A CHARGED PARTICLE IN A UNIFORM ELECTRIC FIELD

Relevant Terminology

Electric charge – a fundamental property of a particle that causes the particle to experience a force in the presence of an electric field. (An electrically neutral particle has no charge and thus experiences no force in the presence of an electric field.)
Electric force – the push or pull that one charged particle exerts on another. Oppositely charged particles attract, whereas like charges (both positive or both negative) repel.
Electric field – force per unit charge.
Electric potential energy – a measure of how much electrical work a charged particle can do by changing position.
Electric potential – electric potential energy per unit charge. Specifically, the electric potential difference between two points equals the electric work per unit charge that would be involved in moving a charged particle from one point to the other.
Potential difference – the difference in electric potential between two points. Potential difference is the work per unit charge needed to move a test charge between two points.

Uniform Electric Field Equations

A uniform electric field is one that is constant throughout a region of space. A charged particle in an **electric field** ($\vec{\mathbf{E}}$) experiences an **electric force** ($\vec{\mathbf{F}}_e$) given by:

$$\vec{\mathbf{F}}_e = q\vec{\mathbf{E}}$$

Recall from Volume 1 that the force of gravity ($\vec{\mathbf{F}}_g$), which is also called **weight**, equals mass (m) times gravitational acceleration ($\vec{\mathbf{g}}$).

$$\vec{\mathbf{F}}_g = m\vec{\mathbf{g}}$$

According to Newton's second law, net force ($\sum\vec{\mathbf{F}}$) equals mass (m) times acceleration ($\vec{\mathbf{a}}$). Recall that Newton's second law was discussed extensively in Volume 1.

$$\sum\vec{\mathbf{F}} = m\vec{\mathbf{a}}$$

Potential difference (ΔV) is the difference in electric potential between two points.

$$\Delta V = V_f - V_i$$

In a uniform electric field between two parallel plates, the magnitude of the **electric field** equals the potential difference divided by the separation between the plates (d):

$$E = \frac{\Delta V}{d}$$

Electrical **work** (W_e) equals charge (q) times potential difference (ΔV), and work also equals the negative of the change in electric potential energy.

$$W_e = q\Delta V = -\Delta PE_e$$

A charged particle traveling parallel to a uniform electric field has uniform acceleration. Recall the equations of uniform acceleration from the first volume of this book.

$$\Delta y = v_{y0}t + \frac{1}{2}a_y t^2$$

$$v_y = v_{y0} + a_y t$$

$$v_y^2 = v_{y0}^2 + 2a_y \Delta y$$

Symbols and SI Units

Symbol	Name	SI Units
E	electric field	N/C or V/m
F_e	electric force	N
m	mass	kg
g	gravitational acceleration	m/s^2
a_y	y-component of acceleration	m/s^2
v_{y0}	y-component of initial velocity	m/s
v_y	y-component of final velocity	m/s
Δy	y-component of net displacement	m
t	time	s
d	separation between parallel plates	m
V	electric potential	V
ΔV	potential difference	V
PE_e	electric potential energy	J
W_e	electrical work	J
q	charge	C

Uniform Electric Field Strategy

How you solve a problem involving a **uniform electric field** depends on the type of problem:

- For a problem where a **uniform electric field** is created by two parallel plates, you may need to relate the magnitude of the electric field to the potential difference between the plates.
$$E = \frac{\Delta V}{d}$$
- If you need to work with **acceleration** (either because it's given in the problem as a number or because the problems asks you to find it), ordinarily you will need to apply Newton's second law at some stage of the solution. We discussed Newton's second law at length in Volume 1, but here are the essentials for applying Newton's second law in the context of a uniform electric field:
    - First draw a free-body diagram (FBD) to show the forces acting on the charged object. Label the electric force as $q\vec{\mathbf{E}}$ and draw it along the electric field lines if the charge is positive or opposite to the electric field lines if the charge is negative. (If the electric field is created by two parallel plates, the electric field lines run from the positive plate to the negative plate.) Unless gravity is negligible (which may be the case), label the weight of the object as $m\vec{\mathbf{g}}$ and draw it toward the center of the planet (which is straight down in most diagrams).
    - Apply Newton's second law by stating that the net force acting on the object equals the object's mass times its acceleration.
    $$\sum \vec{\mathbf{F}} = m\vec{\mathbf{a}}$$
    In most uniform electric field problems, this becomes $\pm|q|E \pm mg = ma_y$, where you have to determine the signs based on your FBD and choice of coordinates (that is, which way $+y$ points).

  When working with acceleration, you may need additional equations:
    - If the object travels parallel to the electric field lines, you may use the equations of one-dimensional acceleration:
    $$\Delta y = v_{y0}t + \frac{1}{2}a_y t^2 \quad , \quad v_y = v_{y0} + a_y t \quad , \quad v_y^2 = v_{y0}^2 + 2a_y \Delta y$$
    - Otherwise, you need the equations of projectile motion (see Volume 1).
- If you don't know acceleration and aren't looking for it, you probably need to use work or energy instead (for example, by applying conservation of energy).
$$W_e = q\Delta V = -\Delta PE_e$$
  If a charged particle accelerates through a potential difference (and no other fields):
    - $|q|\Delta V = \frac{1}{2}m(v^2 - v_0^2)$ if it is gaining speed. Note the absolute values.
    - $|q|\Delta V = \frac{1}{2}m(v_0^2 - v^2)$ if it is losing speed. Note the absolute values.

Example: As illustrated below, two large parallel charged plates are separated by a distance of 25 cm. The potential difference between the plates is 110 V. A tiny object with a charge of $+500$ μC and mass of 10 g begins from rest at the positive plate.

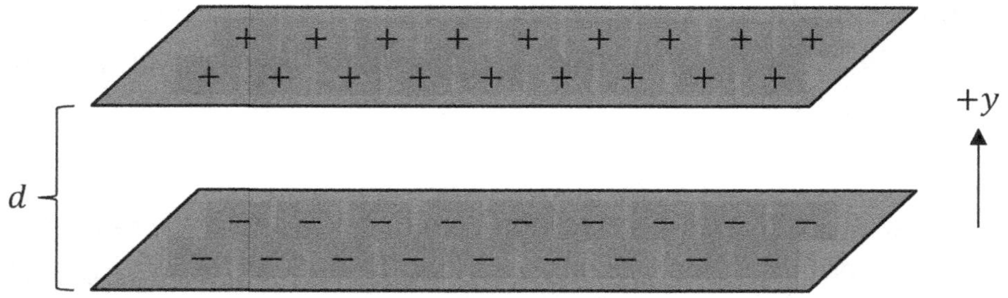

(A) Determine the acceleration of the tiny charged object.

We will eventually need to know the magnitude of the electric field (E). We can get that from the potential difference (ΔV) and the separation between the plates (d). Convert d from cm to m: $d = 25 \text{ cm} = 0.25 \text{ m}$. Note that $\frac{1}{0.25} = 4$ (there are 4 quarters in a dollar).

$$E = \frac{\Delta V}{d} = \frac{110}{0.25} = (110)(4) = 440 \text{ N/C}$$

Draw a free-body diagram (FBD) for the tiny charged object.

- The electric force ($q\vec{\mathbf{E}}$) pulls straight down. Since the charged object is positive, the electric force ($q\vec{\mathbf{E}}$) is parallel to the electric field ($\vec{\mathbf{E}}$), and since the electric field lines travel from the positive plate to the negative plate, $\vec{\mathbf{E}}$ points downward.
- The weight ($m\vec{\mathbf{g}}$) of the object pulls straight down.
- We choose $+y$ to point upward.

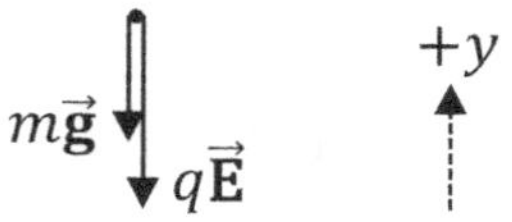

Apply Newton's second law to the tiny charged object. Since we chose $+y$ to point upward and since $q\vec{\mathbf{E}}$ and $m\vec{\mathbf{g}}$ both point downward, $|q|E$ and mg will both be negative in the sum.

$$\sum F_y = ma_y$$

$$-|q|E - mg = ma_y$$

Divide both sides of the equation by mass. Convert the charge from μC to C and the mass from g to kg: $q = 500 \text{ μC} = 5.00 \times 10^{-4}$ C and $m = 10 \text{ g} = 0.010$ kg.

$$a_y = \frac{-|q|E - mg}{m} = \frac{-|5 \times 10^{-4}|(440) - (0.01)(9.81)}{0.01}$$

In this book, we will round gravity from 9.81 to 10 m/s^2 in order to work without a calculator. (This approximation is good to 19 parts in 1000.)

$$a_y \approx \frac{-|5 \times 10^{-4}|(440) - (0.01)(10)}{0.01} = \frac{-0.22 - 0.1}{0.01} = \frac{-0.32}{0.01} = -32 \text{ m/s}^2$$

The symbol $\approx$ means "is approximately equal to." The answer is $a_y \approx -32$ m/s^2.

(If you don't round gravity, you get $a_y = -31.8 \text{ m/s}^2$, which still equals -32 m/s^2 to 2 significant figures.)

(B) How fast is the charged object moving just before it reaches the negative plate?
Now that we know the acceleration, we can use the equations of one-dimensional uniform acceleration. Begin by listing the knowns (out of Δy, v_{y0}, v_y, a_y, and t).

- $a_y \approx -32 \text{ m/s}^2$. We know this from part (A). It's **negative** because the object accelerates downward (and because we chose $+y$ to point upward).
- $\Delta y = -0.25$ m. The net displacement of the tiny charged object equals the separation between the plates. It's **negative** because it finishes below where it started (and because we chose $+y$ to point upward).
- $v_{y0} = 0$. The initial velocity is zero because it starts from rest.
- We're solving for the final velocity (v_y).

We know a_y, Δy, and v_{y0}. We're looking for v_y. Choose the equation with these symbols.

$$v_y^2 = v_{y0}^2 + 2a_y\Delta y$$
$$v_y^2 = 0^2 + 2(-32)(-.25) = 16$$
$$v_y = \sqrt{16} = \pm 4 = -4.0 \text{ m/s}$$

We chose the **negative** root because the object is heading downward in the final position (and because we chose $+y$ to point upward). The object moves 4.0 m/s just before impact.

Note: We couldn't use the equations from the bottom of page 167 in part (B) of the last example because there is a significant gravitational field in the problem. That is, $mg = 0.1$ N compared to $|q|E = 0.22$ N. Compare part (B) of the previous example with the next example, where we can use the equations from the bottom of page 167.

Example: A charged particle with a charge of +200 μC and a mass of 5.0 g accelerates from rest through a potential difference of 5000 V. The effects of gravity are negligible during this motion compared to the effects of electricity. What is the final speed of the particle?
Since the only significant field is the electric field, we may use the equation from page 167. In this case, the particle is speeding up (it started from rest). Convert the charge from μC to C and the mass from g to kg: $q = 200 \text{ μC} = 2.00 \times 10^{-4}$ C and $m = 5.0 \text{ g} = 0.0050$ kg.

$$|q|\Delta V = \frac{1}{2}m(v^2 - v_0^2)$$

$$v = \sqrt{\frac{2|q|\Delta V}{m} + v_0^2} = \sqrt{\frac{2|2 \times 10^{-4}|(5000)}{0.005} + 0^2} = \sqrt{400} = 20 \text{ m/s}$$

The final speed of the particle is $v = 20$ m/s..

54. As illustrated below, two large parallel charged plates are separated by a distance of 20 cm. The potential difference between the plates is 120 V. A tiny object with a charge of +1500 μC and mass of 18 g begins from rest at the positive plate.

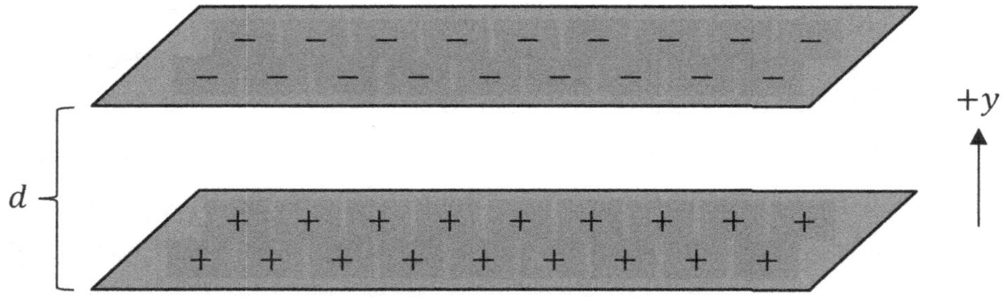

(A) Determine the acceleration of the tiny charged object.

(B) How fast is the charged object moving just before it reaches the negative plate?

Want help or intermediate answers? Check the hints section at the back of the book.

Answer: 40 m/s^2, 4.0 m/s

11 EQUIVALENT CAPACITANCE

Relevant Terminology

Conductor – a material through which electrons are able to flow readily. Metals tend to be good conductors of electricity.
Capacitor – a device that can store charge, which consists of two separated conductors (such as two parallel conducting plates).
Capacitance – a measure of how much charge a capacitor can store for a given voltage.
Equivalent capacitance – a single capacitor that is equivalent (based on how much charge it can store for a given voltage) to a given configuration of capacitors.
Charge – the amount of electric charge stored on the positive plate of a capacitor.
Potential difference – the electric work per unit charge needed to move a test charge between two points in a circuit. Potential difference is also called the **voltage**.
Voltage – the same thing as **potential difference**. However, the term "potential difference" better emphasizes what it means conceptually, whereas "voltage" only conveys the units.
Electric potential energy – a measure of how much electrical work a capacitor could do based on the charge stored on its plates and the potential difference across its plates.
DC – direct current. The direction of the current doesn't change in time.

Equivalent Capacitance Equations

The **capacitance** (C) of a capacitor equals the ratio of the **charge** (Q) stored on the positive plate to the **potential difference** (ΔV) across the plates. The same equation is written three ways below, depending upon what you need to solve for. **Tip**: Q is never downstairs.

$$C = \frac{Q}{\Delta V} \quad , \quad Q = C\Delta V \quad , \quad \Delta V = \frac{Q}{C}$$

The **energy** (U) stored by a capacitor can be expressed three ways, depending upon what you know. Plug $Q = C\Delta V$ into the first equation to obtain the second equation, and plug $\Delta V = \frac{Q}{C}$ into the first equation to obtain the third equation.

$$U = \frac{1}{2}Q\Delta V \quad , \quad U = \frac{1}{2}C\Delta V^2 \quad , \quad U = \frac{Q^2}{2C}$$

For N capacitors connected in **series**, the equivalent series capacitance is given by:

$$\frac{1}{C_s} = \frac{1}{C_1} + \frac{1}{C_2} + \cdots + \frac{1}{C_N}$$

For N capacitors connected in **parallel**, the equivalent parallel capacitance is given by:

$$C_p = C_1 + C_2 + \cdots + C_N$$

Note that the formulas for capacitors in series and parallel are **backwards** compared to the formulas for **resistors** (Chapter 13).

Symbols and SI Units

Symbol	Name	SI Units
C	capacitance	F
Q	the charge stored on the positive plate of a capacitor	C
ΔV	the potential difference between two points in a circuit	V
U	the energy stored by a capacitor	J

Notes Regarding Units

The SI unit of capacitance (C) is the Farad (F). From the equation $C = \frac{Q}{\Delta V}$, the Farad (F) can be related to the Coulomb (C) and Volt (V): 1 F = 1 C/V. The SI unit of energy (U) is the Joule (J). From the equation $U = \frac{1}{2}Q\Delta V$, we get: 1 J = 1 C·V (it's times a Volt, **not** per Volt).

Note that the C's don't match. The SI unit of capacitance (C) is the Farad (F), whereas the SI unit of charge (Q) is the Coulomb (C).

Schematic Symbols Used in Capacitor Circuits

Schematic Representation	Symbol	Name
	C	capacitor
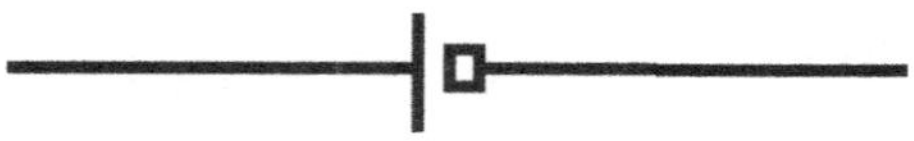	ΔV	battery or DC power supply

Many textbooks draw a capacitor as 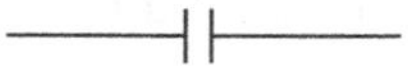 and a battery as 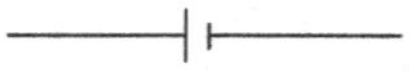. One problem with this is that capacitors and batteries look alike with sloppy handwriting. We're making one plate curve like a C (for capacitor) and making the short line a box for the battery so that students can tell them apart even if their drawing is a bit sloppy. In our notation, the symbol 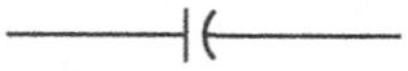 represents an **ordinary capacitor** (it does **not** represent a polarized or electrolytic capacitor in this book), and so it **won't** matter which side has the line and which side has the C (since in actuality they could both be straight, parallel plates).

Essential Concepts

The figures below show examples of what **parallel** (on the left) and **series** (on the right) combinations of capacitors look like. Some students can apply this easily, while others struggle with this visual skill. If you find yourself getting series and parallel wrong when you check your solutions, you need to make a concerted effort to memorize the following **tips** and study how to apply them. Those who learn these tips well have a big advantage.

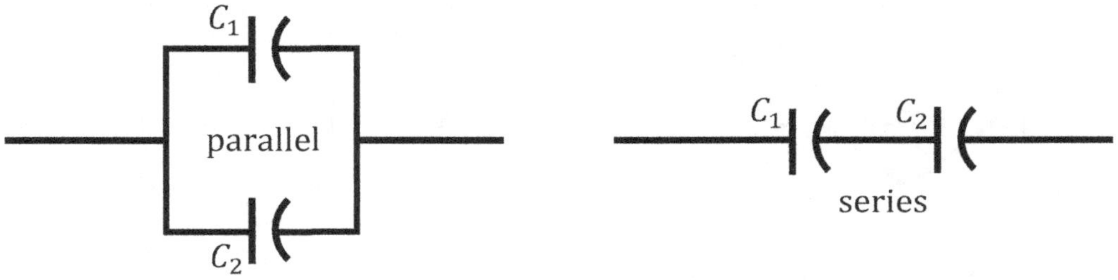

Tips

- Two capacitors are in series if there exists at least one way to reach one capacitor from the other **without crossing a junction**. (See below for an explanation of what a junction is.) For series, it's okay to cross other circuit elements (like a battery). Ask yourself if it's possible to reach one capacitor from the other without passing a junction. If **yes**, they're in **series**. If no, they're not in series. (Just because they're not in series doesn't mean they will be in parallel. They might not be in either.)
- Two capacitors are in parallel if you can place two forefingers – one from each hand – across one capacitor and move **both** forefingers across another capacitor **without crossing another circuit element** (capacitor, battery, resistor, etc.) For parallel, it's okay to cross a junction. You must be able to do this with both forefingers (just one is insufficient). If you can do it with **both** fingers, the two capacitors are in **parallel**. If not, they aren't in parallel. (Just because they're not in parallel doesn't mean they will be in series. They might not be in either series or parallel.)

A **junction** is a place where two (or more) different wires join together, or a place where one wire branches off into two (or more) wires. Imagine an electron traveling along the wire. If you want to know if a particular point is a junction, ask yourself if the electron would have a **choice of paths** (to head in two different possible directions) when it reaches that point, or if it would be forced to continue along a single path. If it has a choice, it's a junction. If it's forced to keep going along a single path, it's **not** a junction.

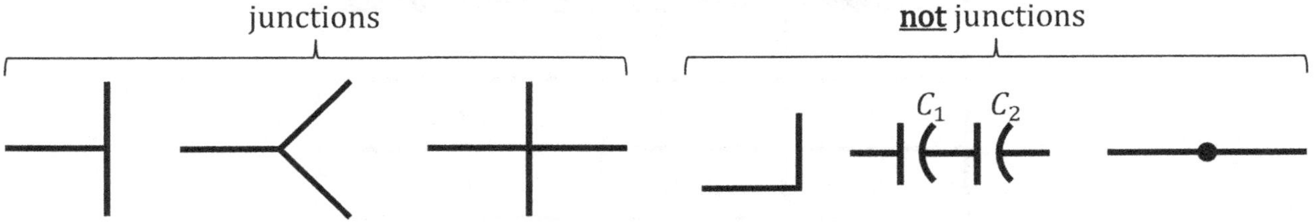

Following is an example of how to apply the rule for **series**. In the left diagram below, C_1 and C_2 are in series because an electron could travel from C_1 to C_2 without crossing a junction. In the right diagram below, none of the capacitors are in series because it would be impossible for an electron to travel from one capacitor to any other capacitor without crossing a junction. See if you can find the two junctions in the right diagram below. Note that C_3 and C_4 are in parallel. Study the parallel rule from the previous page to see why.

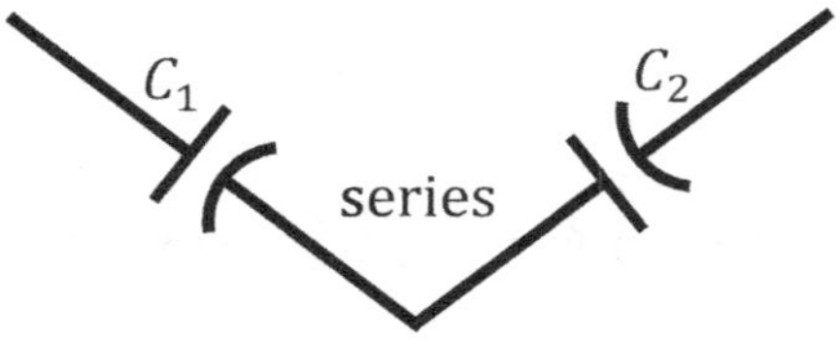

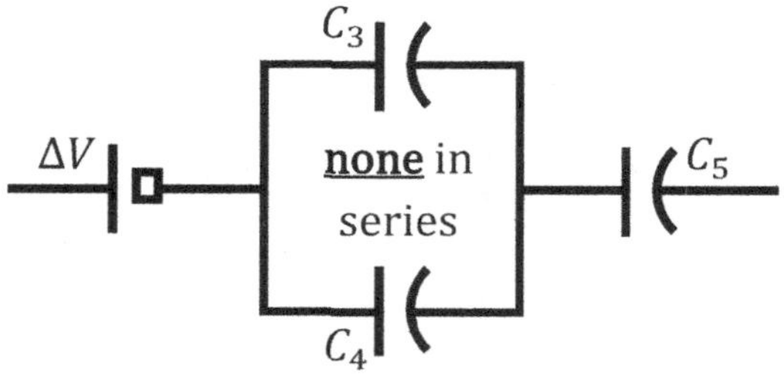

Following is an example of how to apply the rule for **parallel**. In the left diagram below, C_6 and C_7 are in parallel because you can move both fingers from one capacitor (starting with one finger on each side) to the other capacitor without crossing other capacitors or batteries (it's okay to cross junctions for the parallel rule). In the right diagram below, none of the capacitors are in parallel because you can only get one finger from one capacitor to another (the other finger would have to cross a capacitor, which isn't allowed). Note that C_9 and C_{10} are in series. Study the series rule to see why.

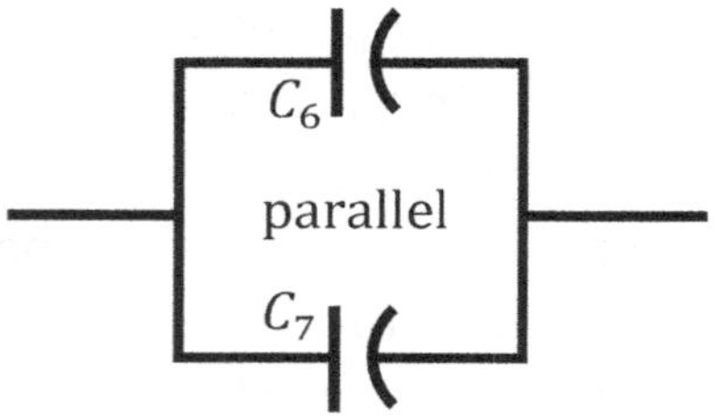

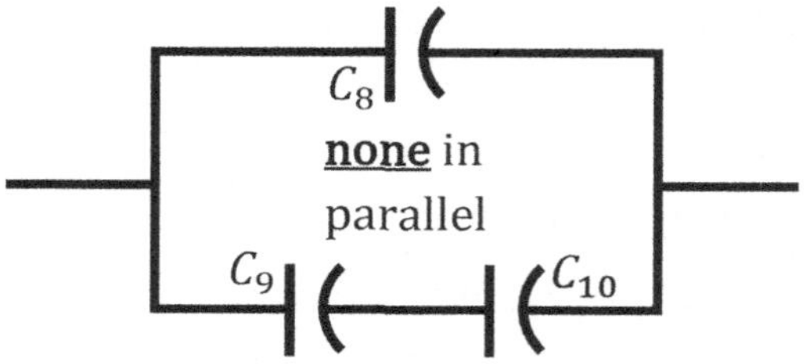

When analyzing a circuit, it's important to be able to tell which quantities are the same in series or parallel:

- Two capacitors in **series** always have the same **charge** (Q).
- Two capacitors in **parallel** always have the same **potential difference** (ΔV).

Metric Prefixes

Prefix	Name	Power of 10
μ	micro	10^{-6}
n	nano	10^{-9}
p	pico	10^{-12}

Strategy for Analyzing Capacitor Circuits

Most standard physics textbook problems involving circuits with capacitors can be solved following these steps:

1. Visually reduce the circuit one step at a time by identifying series and parallel combinations. If you pick any two capacitors at random, you **can't** force them to be in series or parallel (they might be neither). Instead, you need to look at several different pairs until you find a pair that are definitely in series or parallel. Apply the rules discussed on pages 173-174 to help identify series or parallel combinations.
   - If two (or more) capacitors are in **series**, you may remove all the capacitors that were in series and replace them with a new capacitor. Keep the wire that connected the capacitors together when you redraw the circuit. Give the new capacitor a unique name (like C_{s1}). In the math, you will solve for C_{s1} using the series formula.

   - If two (or more) capacitors are in **parallel**, when you redraw the circuit, keep one of the capacitors and its connecting wires, but remove the other capacitors that were in parallel with it and also remove their connecting wires. Give the new capacitor a unique name (like C_{p1}). In the math, you will solve for C_{p1} using the parallel formula.

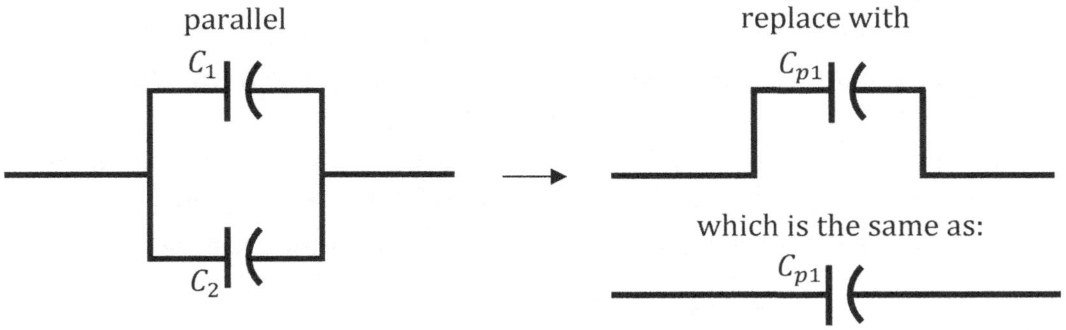

2. Continue redrawing the circuit one step at a time by identifying series and parallel combinations until there is just one capacitor left in the circuit. Label this capacitor the **equivalent capacitance** (C_{eq}).
3. Compute each series and parallel capacitance by applying the formulas below.

$$\frac{1}{C_s} = \frac{1}{C_1} + \frac{1}{C_2} + \cdots + \frac{1}{C_N}$$

$$C_p = C_1 + C_2 + \cdots + C_N$$

4. Continue applying Step 3 until you solve for the last capacitance in the circuit, C_{eq}.
5. If a problem asks you to find charge, potential difference, or energy stored, begin working **backwards** through the circuit one step at a time, as follows:
   A. Start at the very last circuit, which has the equivalent capacitance (C_{eq}). Solve for the charge (Q_{eq}) stored on the capacitor using the formula below, where ΔV_{batt} is the potential difference of the battery or DC power supply.
   $$Q_{eq} = C_{eq}\Delta V_{batt}$$
   B. Go backwards one step in your diagrams: Which diagram did the equivalent capacitance come from? In that diagram, are the capacitors connected in series or parallel?
      - If they were in **series**, write a formula like $Q_{eq} = Q_a = Q_b$ (but use their labels from your diagram, which are probably not Q_a and Q_b). Capacitors in **series** have the same **charge**.
      - If they were in **parallel**, write a formula like $\Delta V_{batt} = \Delta V_a = \Delta V_b$ (but use their labels from your diagram, which are probably not ΔV_a and ΔV_b). Capacitors in **parallel** have the same **potential difference.**
   C. In Step B, did you set charges or potential differences equal?
      - If you set charges equal (like $Q_{eq} = Q_a = Q_b$), find the **potential difference** of the desired capacitor, with a formula like $\Delta V_a = \frac{Q_a}{C_a}$.
      - If you set potential differences equal (like $\Delta V_{batt} = \Delta V_a = \Delta V_b$), find the **charge** of the desired capacitor, with a formula like $Q_a = C_a \Delta V_a$.
   D. Continue going backward one step at a time, applying Steps B and C above each time, until you solve for the desired unknown. Note that the second time you go backwards, you won't write Q_{eq} or ΔV_{batt}, but will use the label for the appropriate charge or potential difference from your diagram. (This is illustrated in the example that follows.)
   E. If a problem asks you to find the **energy** (U) stored in a capacitor, first find the charge (Q) or potential difference (ΔV) for the specified capacitor by applying Steps A-D above, and then use one of the formulas below.
   $$U = \frac{1}{2}Q\Delta V \quad , \quad U = \frac{1}{2}C\Delta V^2 \quad , \quad U = \frac{Q^2}{2C}$$

Example: Consider the circuit shown below.

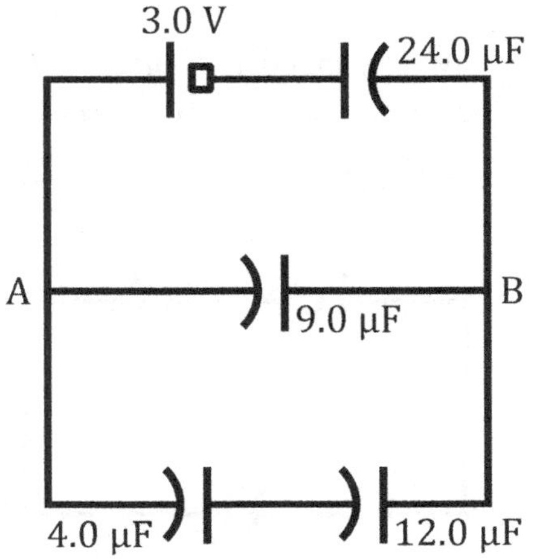

(A) Determine the equivalent capacitance of the circuit.*

Study the four capacitors in the diagram above. Can you find two capacitors that are either in series or parallel? There is only one pair to find presently: The 4.0-μF and 12.0-μF capacitors are in **series** because an electron could travel from one to the other without crossing a junction. Redraw the circuit, replacing the 4.0-μF and 12.0-μF capacitors with a single capacitor called C_{s1}. Also label the other circuit elements. Calculate C_{s1} using the formula for capacitors in series. To add fractions, make a **common denominator**.

$$\frac{1}{C_{s1}} = \frac{1}{C_4} + \frac{1}{C_{12}} = \frac{1}{4} + \frac{1}{12} = \frac{3}{12} + \frac{1}{12} = \frac{3+1}{12} = \frac{4}{12} = \frac{1}{3}$$

$$C_{s1} = 3.0\ \mu\text{F}$$

(Note that C_{s1} does **not** equal C_4 plus C_{12}. Try it: The correct answer is 3, **not** 16.) **Tip**: For series capacitors, remember to find the **reciprocal** at the end of the calculation.

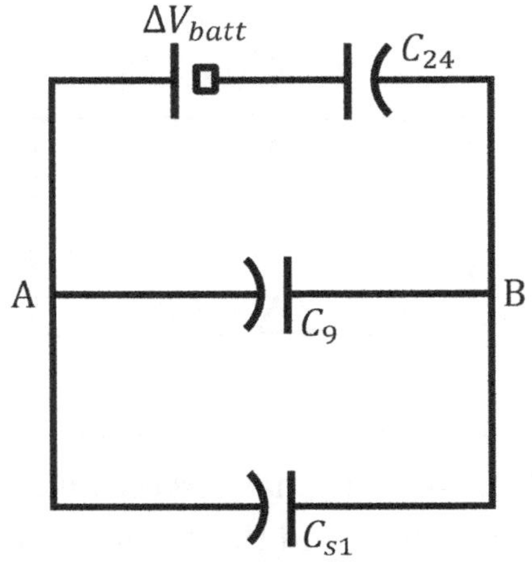

Study the three capacitors in the diagram above. Can you find two capacitors that are either in series or parallel? There is only one pair to find presently: C_{s1} and C_9 are in **parallel** because both fingers can reach C_{s1} from C_9 (starting with one finger on each side of

* That is, from one terminal of the battery to the other.

C_9) without crossing a battery or capacitor. (Remember, it's okay to cross junctions in parallel. Note that C_{24} is **not** part of the parallel combination because one finger would have to cross the battery.) Redraw the circuit, replacing C_{s1} and C_9 with a single capacitor called C_{p1}. Calculate C_{p1} using the formula for capacitors in parallel.

$$C_{p1} = C_{s1} + C_9 = 3 + 9 = 12.0 \text{ μF}$$

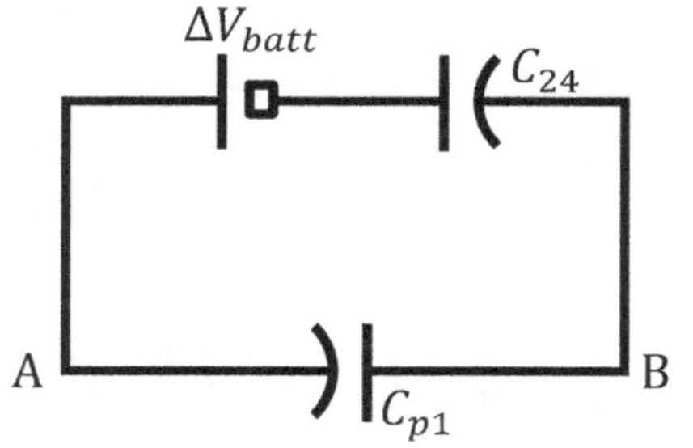

There are just two capacitors in the diagram above. Are they in series or parallel? C_{24} and C_{p1} are in **series** because it's possible for an electron to travel from C_{24} to C_{p1} without crossing a junction. (They're **not** in parallel because one finger would have to cross the battery.) Redraw the circuit, replacing C_{24} and C_{p1} with a single capacitor called C_{eq} (since this is the last capacitor remaining). Calculate C_{eq} using the formula for capacitors in series. Note that points A and B disappear in the following diagram: Since point B lies between C_{24} and C_{p1}, it's impossible to find this point on the next diagram. (It's also worth noting that, in the diagram above, points A and B are **no longer** junctions. Study this.)

$$\frac{1}{C_{eq}} = \frac{1}{C_{24}} + \frac{1}{C_{p1}} = \frac{1}{24} + \frac{1}{12} = \frac{1}{24} + \frac{2}{24} = \frac{1+2}{24} = \frac{3}{24} = \frac{1}{8}$$

$$C_{eq} = 8.0 \text{ μF}$$

The equivalent capacitance is $C_{eq} = 8.0$ μF.

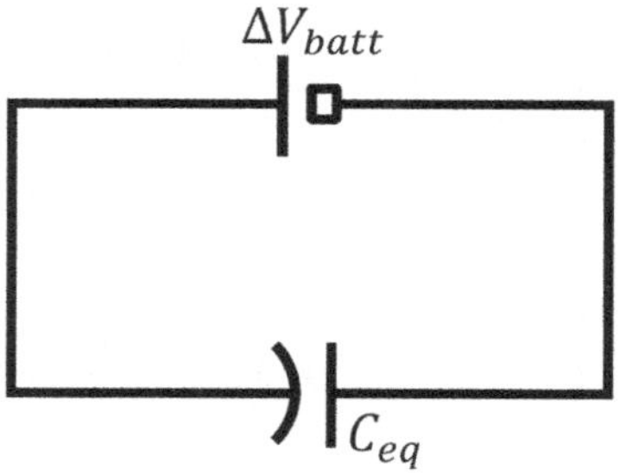

(B) What is the potential difference between points A and B?

You can find points A and B on the original circuit and on all of the simplified circuits except for the very last one. The way to find the potential difference (ΔV_{AB}) between points A and B is to identify a single capacitor that lies between these two points in any of the diagrams. Since C_{p1} is a single capacitor lying between points A and B, $\Delta V_{AB} = \Delta V_{p1}$. That is, we just need to find the potential difference across C_{p1} in order to solve this part of the problem. Note that ΔV_{p1} is **less than** ΔV_{batt} (the potential difference of the battery). The potential difference supplied by the battery was given as $\Delta V_{batt} = 3.0$ V on the original circuit. This is the work per unit charge needed to move a "test" charge from one terminal of the battery to the other. It would take less work to go from point A to point B, so ΔV_{AB} (which equals ΔV_{p1}) is **less than** 3.0 V. You can see this in the second-to-last figure: A positive "test" charge would do some work going across C_{p1} to reach point B and then it would do more work going across C_{24} to reach the negative terminal of the battery: $\Delta V_{p1} + \Delta V_{24} = 3.0$ V. We won't need this equation in the math: We're just explaining why ΔV_{p1} **isn't** 3.0 V.

We must work "backwards" through our simplified circuits, beginning with the circuit that just has C_{eq}, in order to solve for potential difference (or charge or energy stored). The math begins with the following equation, which applies to the last circuit.

$$Q_{eq} = C_{eq}\Delta V_{batt} = (8)(3) = 24.0\ \mu\text{C}$$

Note that 1 μF × V = 1 μC since 1 F × V = 1 C.

Now we will go one step backwards from the simplest circuit (with just C_{eq}) to the second-to-last circuit (which has C_{24} and C_{p1}). Are C_{24} and C_{p1} in series or parallel? We already answered this in part (A): C_{24} and C_{p1} are in **series**. What's the same in series: charge (Q) or potential difference (ΔV)? **Charge** is the same in **series**. Therefore, we set the charges of C_{24} and C_{p1} equal to one another and also set them equal to the charge of the capacitor that replaced them (C_{eq}). This is expressed in the following equation. This is Step 5B of the strategy (on pages 175-176).

$$Q_{24} = Q_{p1} = Q_{eq} = 24.0\ \mu\text{C}$$

According to Step 5C of the strategy, if we set the charges equal to one another, we must calculate potential difference. Based on the question for part (B), do we need the potential difference across C_{24} or C_{p1}? We need ΔV_{p1} since we already reasoned that $\Delta V_{AB} = \Delta V_{p1}$.

$$\Delta V_{p1} = \frac{Q_{p1}}{C_{p1}} = \frac{24\ \mu\text{C}}{12\ \mu\text{F}} = 2.0\ \text{V}$$

Note that all of the subscripts match in the above equation. Note that the μ's cancel. Your solutions to circuit problems need to be well-organized so that you can easily find previous information. In the above equation, you need to hunt for Q_{p1} and C_{p1} in earlier parts of the solution to see that $Q_{p1} = 24.0$ μC and $C_{p1} = 12.0$ μF. The potential difference between points A and B is $\Delta V_{AB} = 2.0$ V since $\Delta V_{AB} = \Delta V_{p1}$.

Does this seem like a long solution to you? It's actually very short: If you look at the math, it really consists of only a few lines. Most of the space was taken up by explanations to try to help you understand the process. The next two parts of this example will try to help you further. After you read the problem, draw the original diagram on a blank sheet of paper, and see if you can rework the solution on your own. Reread this example if you get stuck.

(C) How much charge is stored by the 4.0-μF capacitor?
We don't need to start over: Just continue working backwards from where we left off in part (B). First, let's see where we're going. Find the 4.0-μF capacitor on the original circuit and follow the path that it took as the circuit was simplified: The 4.0-μF capacitor was part of the series that became C_{s1}, and C_{s1} was part of the parallel combination that became C_{p1}. We found information about C_{p1} in part (B): $Q_{p1} = 24.0$ μC and $\Delta V_{p1} = 2.0$ V. In part (C), we will begin from there.

Note that C_{p1} came about by connecting C_9 and C_{s1} in **parallel**. What's the same in parallel: charge (Q) or potential difference (ΔV)? **Potential difference** is the same in **parallel**. Therefore, we set the potential differences of C_9 and C_{s1} equal to one another and also set them equal to the potential difference of the capacitor that replaced them (C_{p1}). This is expressed in the following equation. This is Step 5B of the strategy (see page 176).

$$\Delta V_9 = \Delta V_{s1} = \Delta V_{p1} = 2.0 \text{ V}$$

According to Step 5C of the strategy, if we set the potential differences equal to one another, we must calculate charge. Based on the question for part (C), do we need the charge across C_9 or C_{s1}? We need Q_{s1} since the 4.0-μF capacitor is part of C_{s1}.

$$Q_{s1} = C_{s1}\Delta V_{s1} = (3)(2) = 6.0 \text{ μC}$$

You have to go all the way back to part (A) to find the value for C_{s1}, whereas ΔV_{s1} appears just one equation back.

We're not quite done yet: We need to go one more step back to reach the 4.0-μF capacitor. Note that C_{s1} came about by connecting C_4 and C_{12} in **series**. What's the same in series: charge (Q) or potential difference (ΔV)? **Charge** is the same in **series**. Therefore, we set the charges of C_4 and C_{12} equal to one another and also set them equal to the charge of the capacitor that replaced them (C_{s1}). This is expressed in the following equation.

$$Q_4 = Q_{12} = Q_{s1} = 6.0 \text{ μC}$$

The charge stored by the 4.0-μF capacitor is $Q_4 = 6.0$ μC.

(D) How much energy is stored by the 4.0-μF capacitor?
Compare the questions for parts (C) and (D). They're almost identical, except that now we're looking for **energy** (U) instead of charge (Q). In part (C), we found that $Q_4 = 6.0$ μC. Look at the three equations for the energy stored by a capacitor (Step 5E on page 176). Given $C_4 = 4.0$ μF and $Q_4 = 6.0$ μC, choose the appropriate equation.

$$U_4 = \frac{Q_4^2}{2C_4} = \frac{(6 \text{ μC})^2}{2(4 \text{ μF})} = \frac{36}{8} = \frac{9}{2} \text{ μJ} = 4.5 \text{ μJ}$$

Note that $\frac{\mu^2}{\mu} = \mu$. The energy stored by the 4.0-μF capacitor is $U_4 = 4.5$ μJ (microJoules).

Note: You can find another example of series and parallel in Chapter 13.

55. Consider the circuit shown below.

(A) Redraw the circuit step by step until only a single equivalent capacitor remains. Label each reduced capacitor using a symbol with subscripts.

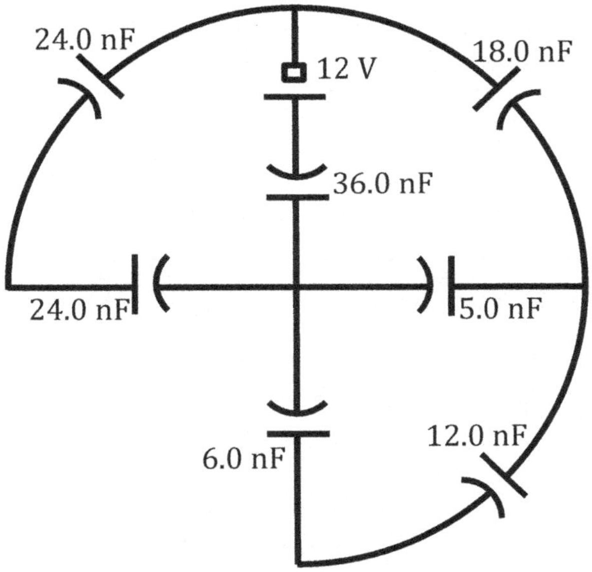

Note: You're not finished yet. This problem is **continued** on the next page.

(B) Determine the equivalent capacitance of the circuit.

(C) Determine the charge stored on each 24.0-nF capacitor.

(D) Determine the energy stored by the 18-nF capacitor.

Want help? Check the hints section at the back of the book.

Answers: 12 nF, 96 nC, 64 nJ

56. Consider the circuit shown below.

(A) Redraw the circuit step by step until only a single equivalent capacitor remains. Label each reduced capacitor using a symbol with subscripts.

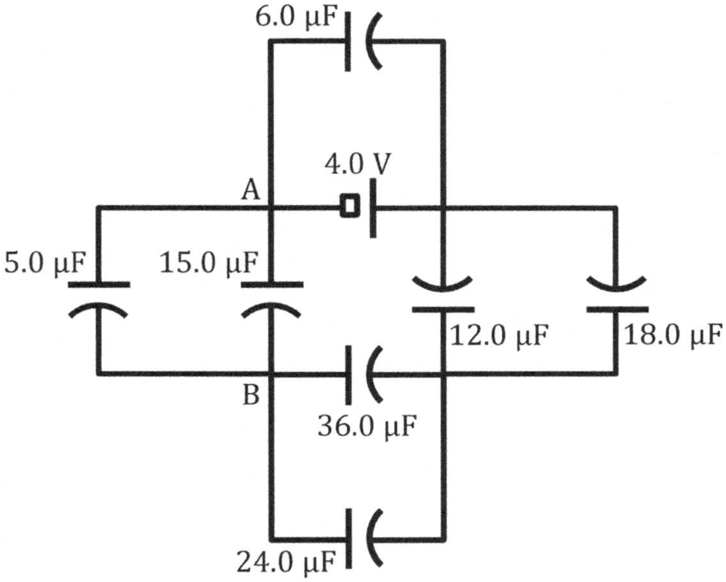

Note: You're not finished yet. This problem is **continued** on the next page.

(B) Determine the equivalent capacitance of the circuit.

(C) Determine the potential difference between points A and B.

(D) Determine the energy stored on the 5.0-μF capacitor.

Want help? Check the hints section at the back of the book.

Answers: 16 μF, 2.0 V, 10 μJ

12 PARALLEL-PLATE AND OTHER CAPACITORS

Relevant Terminology

Conductor – a material through which electrons are able to flow readily. Metals tend to be good conductors of electricity.
Capacitor – a device that can store charge, which consists of two separated conductors (such as two parallel conducting plates).
Capacitance – a measure of how much charge a capacitor can store for a given voltage.
Charge – the amount of electric charge stored on the positive plate of a capacitor.
Potential difference – the electric work per unit charge needed to move a test charge between two points in a circuit. Potential difference is also called the **voltage**.
Electric potential energy – a measure of how much electrical work a capacitor could do based on the charge stored on its plates and the potential difference across its plates.
Electric field – electric force per unit charge.
Dielectric – a nonconducting material that can sustain an electric field. The presence of a dielectric between the plates of a capacitor enhances its ability to store charge.
Dielectric constant – the enhancement factor by which a capacitor can store more charge (for a given potential difference) by including a dielectric between its plates.
Dielectric strength – the maximum electric field that a dielectric can sustain before it breaks down (at which point charge could transfer across the dielectric in the form of a spark).
Permittivity – a measure of how a dielectric material affects an electric field.

Capacitance Equations

The following equation applies only to a **parallel-plate** capacitor. Its capacitance (C) is proportional to the dielectric constant (κ), permittivity of free space (ϵ_0), and the area of one plate (A), and is inversely proportional to the separation between the plates (d).

$$C = \frac{\kappa \epsilon_0 A}{d}$$

The dielectric strength (E_{max}) is the maximum electric field that a dielectric can sustain.

$$E_{max} = \frac{\Delta V_{max}}{d}$$

The equations from Chapter 11 apply to all capacitors (not just parallel-plate capacitors):

$$C = \frac{Q}{\Delta V} \quad , \quad Q = C\Delta V \quad , \quad \Delta V = \frac{Q}{C}$$

$$U = \frac{1}{2}Q\Delta V \quad , \quad U = \frac{1}{2}C\Delta V^2 \quad , \quad U = \frac{Q^2}{2C}$$

$$\frac{1}{C_s} = \frac{1}{C_1} + \frac{1}{C_2} + \cdots + \frac{1}{C_N} \quad , \quad C_p = C_1 + C_2 + \cdots + C_N$$

Symbols and SI Units

Symbol	Name	SI Units
C	capacitance	F
ϵ_0	permittivity of free space	$\frac{\text{C}^2}{\text{N}\cdot\text{m}^2}$ or $\frac{\text{C}^2\cdot\text{s}^2}{\text{kg}\cdot\text{m}^3}$
κ	dielectric constant	unitless
A	area of one plate	m^2
d	separation between the plates	m
E_{max}	dielectric strength	N/C or V/m
ΔV_{max}	maximum potential difference	V
Q	the charge stored on the positive plate of a capacitor	C
ΔV	the potential difference between two points in a circuit	V
U	the energy stored by a capacitor	J

Note: The symbol κ is the lowercase Greek letter kappa and ϵ is epsilon.

Numerical Values

Recall (from Chapter 8) that the permittivity of free space (ϵ_0) is related to Coulomb's constant (k) via $\epsilon_0 = \frac{1}{4\pi k}$. Also recall that $k = 9.0 \times 10^9 \ \frac{\text{N}\cdot\text{m}^2}{\text{C}^2}$ and $\epsilon_0 = 8.8 \times 10^{-12} \ \frac{\text{C}^2}{\text{N}\cdot\text{m}^2} \approx \frac{10^{-9}}{36\pi} \ \frac{\text{C}^2}{\text{N}\cdot\text{m}^2}$, where the value $\frac{10^{-9}}{36\pi} \ \frac{\text{C}^2}{\text{N}\cdot\text{m}^2}$ is friendlier if you don't use a calculator.

In order to solve some textbook problems, you need to look up values of the dielectric constant (κ) or dielectric strength (E_{max}) in a table (most textbooks include such a table).

Important Distinctions

Be careful not to confuse lowercase kay (k) for Coulomb's constant $\left(k = 9.0 \times 10^9 \ \frac{\text{N}\cdot\text{m}^2}{\text{C}^2}\right)$ with the Greek letter kappa (κ), which represents the dielectric constant. You need to familiarize yourself with the equations enough to recognize which is which. Also be careful not to confuse the permittivity (ϵ) with electric field (E).

Parallel-plate Capacitor Strategy

How you solve a problem involving a **parallel-plate capacitor** depends on which quantities you are trying to relate:

1. Make a list of the symbols that you know (see the chart on page 186). The units can help you figure out which symbols you know. For a textbook problem, you may need to look up a value for the dielectric constant (κ) or dielectric strength (E_{max}). You should already know $k = 9.0 \times 10^9 \frac{\text{N}\cdot\text{m}^2}{\text{C}^2}$ and $\epsilon_0 = 8.8 \times 10^{-12} \frac{\text{C}^2}{\text{N}\cdot\text{m}^2} \approx \frac{10^{-9}}{36\pi} \frac{\text{C}^2}{\text{N}\cdot\text{m}^2}$.
2. Choose equations based on which symbols you know and which symbol you are trying to solve for:
   - The following equation applies only to parallel-plate capacitors:
   $$C = \frac{\kappa \epsilon_0 A}{d}$$
   If the plates are circular, $A = \pi a^2$. If they are square, $A = L^2$. If they are rectangular, $A = LW$.
   - The following equations apply to all capacitors:
   $$C = \frac{Q}{\Delta V} \quad , \quad Q = C\Delta V \quad , \quad \Delta V = \frac{Q}{C}$$
   - The equation for dielectric strength is:
   $$E_{max} = \frac{\Delta V_{max}}{d}$$
   - The energy stored in a capacitor can be expressed three different ways:
   $$U = \frac{1}{2} Q\Delta V \quad , \quad U = \frac{1}{2} C\Delta V^2 \quad , \quad U = \frac{Q^2}{2C}$$
3. Carry out any algebra needed to solve for the unknown.
4. If any capacitors are connected in **series** or **parallel**, you will need to apply the strategy from Chapter 11.
   - The following diagram shows two dielectrics in **series**.

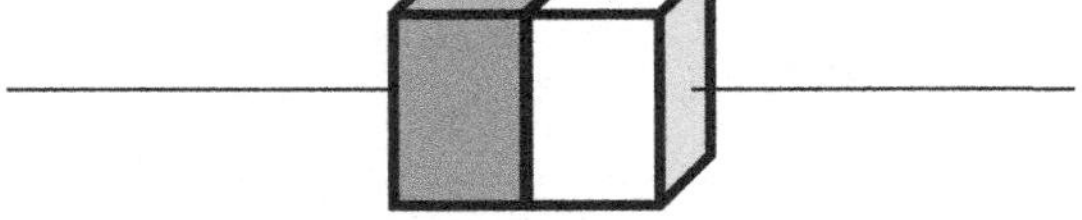

   - The following diagram shows two dielectrics in **parallel**.

5. If there is a charged particle moving between the plates, see Chapter 10.

Example: A parallel-plate capacitor has rectangular plates with a length of 25 mm and width of 50 mm. The separation between the plates is of 5.0 mm. A dielectric is inserted between the plates. The dielectric constant is 72π and the dielectric strength is $3.0 \times 10^6 \frac{\text{V}}{\text{m}}$.

(A) Determine the capacitance.

Make a list of the known quantities and identify the desired unknown symbol:

- The capacitor plates have a length $L = 25$ mm and width $W = 50$ mm.
- The separation between the plates is $d = 5.0$ mm.
- The dielectric constant is $\kappa = 72\pi$ and the dielectric strength is $E_{max} = 3.0 \times 10^6 \frac{\text{V}}{\text{m}}$.
- We also know $\epsilon_0 = 8.8 \times 10^{-12} \frac{\text{C}^2}{\text{N}\cdot\text{m}^2}$, which we will approximate as $\epsilon_0 \approx \frac{10^{-9}}{36\pi} \frac{\text{C}^2}{\text{N}\cdot\text{m}^2}$.
- The unknown we are looking for is capacitance (C).

The equation for the capacitance of a parallel-plate capacitor involves area, so we need to find area first. Before we do that, let's convert the length, width, and separation to SI units.

$$L = 25 \text{ mm} = 0.025 \text{ m} = \frac{1}{40} \text{ m} \quad , \quad W = 50 \text{ mm} = 0.050 \text{ m} = \frac{1}{20} \text{ m}$$

$$d = 5.0 \text{ mm} = 0.0050 \text{ m} = \frac{1}{200} \text{ m}$$

We converted to fractions, but you may work through the math with decimals if you prefer. Decimals are more calculator friendly, whereas fractions are sometimes simpler for working by hand (for example, fractions sometimes make it easier to spot cancellations). The area of a rectangular plate is:

$$A = LW = \left(\frac{1}{40}\right)\left(\frac{1}{20}\right) = \frac{1}{800} \text{ m}^2 = 0.00125 \text{ m}^2$$

Now we are ready to use the equation for capacitance. To divide by a fraction, multiply by its **reciprocal**. Note that the reciprocal of $\frac{1}{200}$ is 200.

$$C = \frac{\kappa\epsilon_0 A}{d} = \frac{(72\pi)\left(\frac{10^{-9}}{36\pi}\right)\left(\frac{1}{800}\right)}{\left(\frac{1}{200}\right)} = (72\pi)\left(\frac{10^{-9}}{36\pi}\right)\left(\frac{1}{800}\right)(200) = \left(\frac{72\pi}{36\pi}\right)(10^{-9})\left(\frac{200}{800}\right)$$

$$C = (2)(10^{-9})\left(\frac{1}{4}\right) = \frac{1}{2} \times 10^{-9} \text{ F} = \frac{1}{2} \text{ nF} = 0.50 \text{ nF}$$

(B) What is the maximum charge that this capacitor can store on its plates?

The key to this solution is to note the word "maximum." Let's find the maximum potential difference (ΔV_{max}) across the plates.

$$E_{max} = \frac{\Delta V_{max}}{d} \quad \Rightarrow \quad \Delta V_{max} = E_{max} d = (3 \times 10^6)\left(\frac{1}{200}\right) = 15{,}000 \text{ V}$$

For any capacitor, $Q = C\Delta V$. Therefore, the maximum charge is:

$$Q_{max} = C\Delta V_{max} = \left(\frac{1}{2} \times 10^{-9}\right)(15{,}000) = 7.5 \times 10^{-6} \text{ C} = 7.5 \text{ }\mu\text{C}$$

57. A parallel-plate capacitor has circular plates with a radius of 30 mm. The separation between the plates is of 2.0 mm. A dielectric is inserted between the plates. The dielectric constant is 8.0 and the dielectric strength is $6.0 \times 10^6 \frac{\text{V}}{\text{m}}$.

(A) Determine the capacitance.

(B) What is the maximum charge that this capacitor can store on its plates?

Want help? Check the hints section at the back of the book.

Answers: 0.10 nF, 1.2 μC

Strategy to Derive an Equation for Capacitance

Following is how to apply calculus to derive an equation for capacitance:

1. First derive an equation for the electric field between the conductors.
   - For an infinite plane, sphere, or cylinder, apply Gauss's law (Chapter 8).
   - Otherwise, perform the electric field integral (Chapter 7).
2. Use the equation you derived in Step 1 in the following definite integral for the potential difference between the plates.

   $$\Delta V = V_f - V_i = -\int_i^f \vec{\mathbf{E}} \cdot d\vec{\mathbf{s}}$$

   The differential displacement vector ($d\vec{\mathbf{s}}$) is along the path from the initial position (i) to the final position (f). Choose a path going along the electric field lines from the initial point to the final point. Note that $\vec{\mathbf{E}} \cdot d\vec{\mathbf{s}} = E \cos \psi \, ds$, where ψ is the angle between $\vec{\mathbf{E}}$ and $d\vec{\mathbf{s}}$. The magnitude of $d\vec{\mathbf{s}}$ is:
   - $ds = dx, dy$, or dz for a path parallel to the x-, y-, or z-axis.
   - $ds = dr$ for a radial path going outward from the origin (useful for a sphere).
   - $ds = dr_c$ for a path going outward from the z-axis (useful for a cylinder).
   - $ds = rd\theta$ for a path going along the arc of a circle of radius r.
3. Plug your answer from Step 2 into the following equation.

   $$C = \frac{Q}{|\Delta V|}$$

4. If necessary, perform the integral $Q = \int dq$ as discussed in Chapter 7.
5. If necessary, simplify your answer.

Example: Derive an equation for the capacitance of a parallel-plate capacitor with vacuum* or air between its plates.

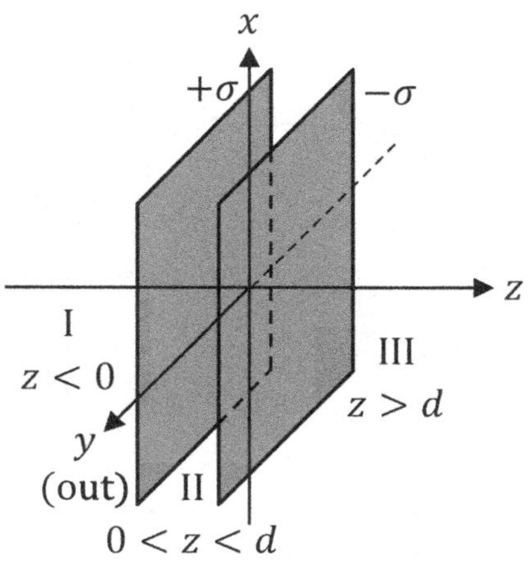

It's common for the distance between the plates (d) to be small compared to the length and width of the plates, in which case we may approximate the parallel-plate capacitor as consisting of two infinite charged planes. We found the electric field between two infinitely large, parallel, oppositely charged plates in Problem 47 of Chapter 8. Applying Gauss's law, the electric field in the region between the plates is $\vec{\mathbf{E}} = \frac{\sigma}{\epsilon_0}\hat{\mathbf{z}}$. (See Chapter 8, Problem 47, and the hints to that problem.) We'll perform the potential difference integral from the positive plate (i) to the negative plate (f), such that $d\vec{\mathbf{s}}$ points to the right (along $\hat{\mathbf{z}}$). Therefore, $\vec{\mathbf{E}} \cdot d\vec{\mathbf{s}} = E\cos 0°\, ds = Eds$. Since $d\vec{\mathbf{s}}$ is along the z-axis, $ds = dz$. The limits of integration are from $z = 0$ to $z = d$, where d is the separation between the plates.

$$\Delta V = V_f - V_i = -\int_i^f \vec{\mathbf{E}} \cdot d\vec{\mathbf{s}} = -\int_i^f E\, ds = -\int_{z=0}^{d} \frac{\sigma}{\epsilon_0} dz = -\frac{\sigma}{\epsilon_0}\int_{z=0}^{d} dz = -\frac{\sigma}{\epsilon_0}[z]_{z=0}^{d} = -\frac{\sigma d}{\epsilon_0}$$

For uniformly charged plates, σ is constant, and σ and ϵ_0 may both come out of the integral. Substitute this expression into the general equation for capacitance.

$$C = \frac{Q}{|\Delta V|} = \frac{Q}{\left|-\frac{\sigma d}{\epsilon_0}\right|} = \frac{Q}{\frac{\sigma d}{\epsilon_0}} = Q\frac{\epsilon_0}{\sigma d} = \frac{\epsilon_0 Q}{\sigma d}$$

Perform the following integral to eliminate the charge density (σ) from the answer. For a plane of charge, $dq = \sigma dA$ (Step 6 on page 94). For uniform plates, σ is constant.

$$Q = \int dq = \int \sigma\, dA = \sigma \int dA = \sigma A$$

Substitute this expression into the previous equation for capacitance.

$$C = \frac{\epsilon_0 Q}{\sigma d} = \frac{\epsilon_0(\sigma A)}{\sigma d} = \frac{\epsilon_0 A}{d}$$

* For vacuum $\kappa = 1$, and for air $\kappa \approx 1$, such that we won't need to worry about the dielectric constant. (If there is a dielectric, it simply introduces a factor of κ into the final expression.)

Example: A cylindrical capacitor consists of two very long coaxial cylindrical conductors: One is a solid cylinder with radius a, while the other is a thin cylindrical shell of radius b. There is vacuum between the two cylinders. Derive an equation for the capacitance of this cylindrical capacitor.

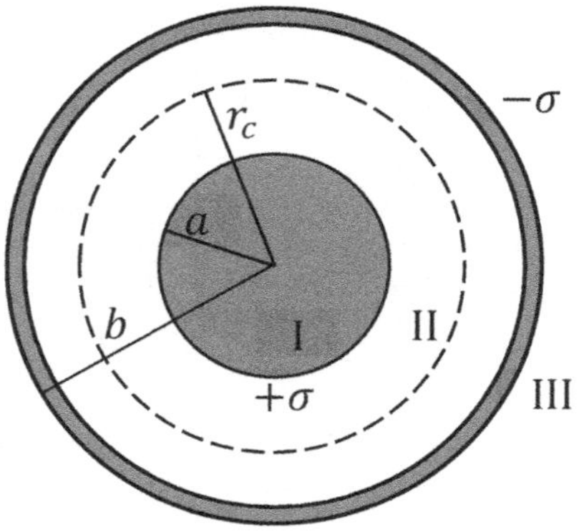

In the diagram above, we are looking at a cross section of the coaxial cylinders. The first step is to apply Gauss's law to region II ($a < r_c < b$). As discussed in Chapter 8, the charge resides on the surface of a conductor in electrostatic equilibrium. Thus, we will work with σ (and not ρ) for the charge density. In Chapter 8, we learned that the conducting shell doesn't matter in region II (since the Gaussian cylinder drawn for region II won't enclose any charge from the outer cylinder). Hence, we just need to find the electric field in region II for a solid conducting cylinder. We did that in an example in Chapter 8 (see page 138).

$$\vec{\mathbf{E}} = \frac{\sigma a}{\epsilon_0 r_c}\hat{\mathbf{r}}_c$$

We will integrate from the positive cylinder (i) to the negative shell (f), such that $d\vec{\mathbf{s}}$ points outward (along $\hat{\mathbf{r}}_c$) like $\vec{\mathbf{E}}$ does. Therefore, $\vec{\mathbf{E}} \cdot d\vec{\mathbf{s}} = E \cos 0° \, ds = E ds$. For a conductor in electrostatic equilibrium, the charge density (σ) is uniform (see Chapter 8), which means that we may pull σ out of the integral.

$$\Delta V = V_f - V_i = -\int_i^f \vec{\mathbf{E}} \cdot d\vec{\mathbf{s}} = -\int_i^f E\, ds = -\int_{r_c=a}^{b} \frac{\sigma a}{\epsilon_0 r_c} dr_c = -\frac{\sigma a}{\epsilon_0}\int_{r_c=a}^{b} \frac{dr_c}{r_c}$$

The anti-derivative for $\int \frac{dx}{x}$ is a natural logarithm (see Chapter 17).

$$\Delta V = -\frac{\sigma a}{\epsilon_0}[\ln(r_c)]_{r_c=a}^{b} = -\frac{\sigma a}{\epsilon_0}[\ln(b) - \ln(a)] = -\frac{\sigma a}{\epsilon_0}\ln\left(\frac{b}{a}\right)$$

We applied the rule $\ln\left(\frac{b}{a}\right) = \ln(b) - \ln(a)$. Plug this into the capacitance formula.

$$C = \frac{Q}{|\Delta V|} = \frac{Q}{\left|-\frac{\sigma a}{\epsilon_0}\ln\left(\frac{b}{a}\right)\right|} = \frac{Q}{\frac{\sigma a}{\epsilon_0}\ln\left(\frac{b}{a}\right)} = \frac{Q\epsilon_0}{\sigma a \ln\left(\frac{b}{a}\right)}$$

Perform the following integral to eliminate the charge density (σ) from the answer. Treat the charge density (σ) as a constant. To integrate over the surface area of the cylinder, we integrate over z (along the length of the cylinder) and θ (around the body). We write $dA = ad\theta dz$, which is like $r_c d\theta$ times dz with $r_c = a$ (since the charge that we are integrating over resides on the inner conductor which has radius a). Note that $r_c = a$ is constant over the surface of the cylinder (that is, we're not integrating over r_c - we instead integrate over θ and z to get the surface area of a cylinder).

$$Q = \int dq = \int \sigma\, dA = \sigma \int dA = \sigma \int_{z=0}^{L} \int_{\theta=0}^{2\pi} a\, d\theta\, dz = 2\pi a \sigma L$$

Substitute this expression into the previous equation for capacitance. Recall that $\epsilon_0 = \frac{1}{4\pi k}$.

$$C = \frac{Q\epsilon_0}{\sigma a \ln\left(\frac{b}{a}\right)} = \frac{2\pi a \sigma L \epsilon_0}{\sigma a \ln\left(\frac{b}{a}\right)} = \frac{2\pi \epsilon_0 L}{\ln\left(\frac{b}{a}\right)} = \frac{2\pi \left(\frac{1}{4\pi k}\right) L}{\ln\left(\frac{b}{a}\right)} = \frac{L}{2k \ln\left(\frac{b}{a}\right)}$$

For a cylindrical capacitor, it's customary to divide both sides of the equation by L to get **capacitance per unit length** (since capacitance per unit length is finite).

$$\frac{C}{L} = \frac{1}{2k \ln\left(\frac{b}{a}\right)}$$

58. A spherical capacitor consists of two spherical conductors: One is a solid sphere with radius a, while the other is a thin spherical shell of radius b. There is vacuum between the two spheres. Derive an equation for the capacitance of this spherical capacitor.

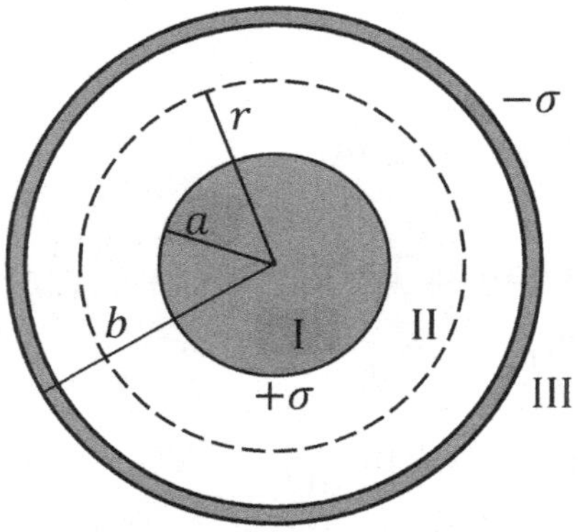

Want help? Check the hints section at the back of the book.

Answer: $C = \frac{ab}{k(b-a)}$

13 EQUIVALENT RESISTANCE

Relevant Terminology

Conductor – a material through which electrons are able to flow readily. Metals tend to be good conductors of electricity.
Resistance – a measure of how well a component in a circuit resists the flow of current.
Resistor – a component in a circuit which has a significant amount of resistance.
Equivalent resistance – a single resistor that is equivalent (based on how much current it draws for a given voltage) to a given configuration of resistors.
Current – the instantaneous rate of flow of charge through a wire.
Potential difference – the electric work per unit charge needed to move a test charge between two points in a circuit. Potential difference is also called the **voltage**.
Electric power – the instantaneous rate at which electrical work is done.
DC – direct current. The direction of the current doesn't change in time.

Equivalent Resistance Equations

According to **Ohm's law**, the **potential difference** (ΔV) across a resistor is directly proportion to the **current** (I) through the resistor by a factor of its **resistance** (R). The same equation is written three ways below, depending upon what you need to solve for.

$$\Delta V = IR \quad , \quad I = \frac{\Delta V}{R} \quad , \quad R = \frac{\Delta V}{I}$$

The **power** (P) dissipated in a resistor can be expressed three ways, depending upon what you know. Plug $\Delta V = IR$ into the first equation to obtain the second equation, and plug $I = \frac{\Delta V}{R}$ into the first equation to obtain the third equation.

$$P = I\Delta V \quad , \quad P = I^2 R \quad , \quad P = \frac{\Delta V^2}{R}$$

For N resistors connected in **series**, the equivalent series resistance is given by:

$$R_s = R_1 + R_2 + \cdots + R_N$$

For N resistors connected in **parallel**, the equivalent parallel resistance is given by:

$$\frac{1}{R_p} = \frac{1}{R_1} + \frac{1}{R_2} + \cdots + \frac{1}{R_N}$$

Note that the formulas for resistors in series and parallel are **backwards** compared to the formulas for **capacitors** (Chapter 11). You can get a feeling for why that is by comparing Ohm's law ($\Delta V = IR$) to the equation for capacitance with ΔV solved for $\left(\Delta V = \frac{Q}{C}\right)$. Since current ($I$) is the instantaneous rate of flow of charge (Q), the structural difference between the equations for resistance and capacitance is that ΔV is directly proportional to R, but inversely proportional to C.

Symbols and SI Units

Symbol	Name	SI Units
R	resistance	Ω
I	electric current	A
ΔV	the potential difference between two points in a circuit	V
P	electric power	W

Notes Regarding Units

The SI unit of current (I) is the Ampère (A) and the SI unit of resistance (R) is the Ohm (Ω). From the equation $R = \frac{\Delta V}{I}$, the Ohm (Ω) can be related to the Ampère (A) and Volt (V): 1 Ω = 1 V/A. The SI unit of power (P) is the Watt (W). From the equation $P = I\Delta V$, we get: 1 W = 1 A·V (it's an Ampère times a Volt, **<u>not</u>** per Volt).

Schematic Symbols Used in Resistor Circuits

Schematic Representation	Symbol	Name
	R	resistor
	ΔV	battery or DC power supply
	measures ΔV	voltmeter
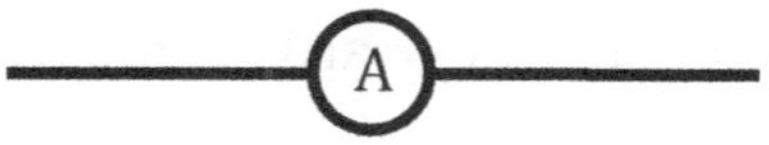	measures I	ammeter

Essential Concepts

The figures below show examples of what **parallel** (on the left) and **series** (on the right) combinations of resistors look like. Recall that we discussed how to determine whether or not two circuit elements are in series or parallel in Chapter 11. It may help to review Chapter 11 before proceeding.

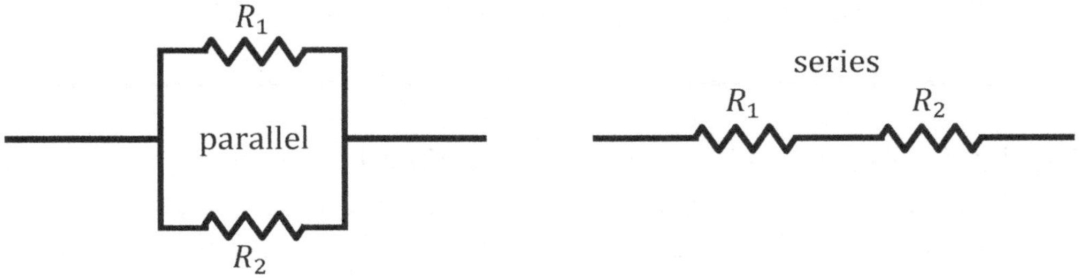

A **voltmeter** is a device that measures **potential difference**. A voltmeter has a very large internal resistance and is connected in **parallel** with a circuit element (left diagram below). A voltmeter has a large internal resistance so that it draws only a negligible amount of current.

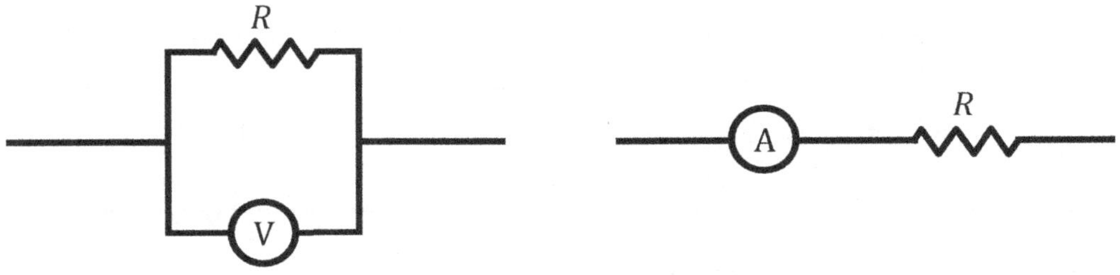

An **ammeter** is a device that measures **current**. An ammeter has a very small internal resistance and is connected in **series** with a circuit element (right diagram above). When a student accidentally connects an ammeter in parallel with a resistor, the ammeter draws a tremendous amount of current (a greater proportion of electrons take the path of less resistance, and since an ammeter has very little internal resistance most of the electrons travel through the ammeter if it's connected in parallel with a resistor), which **blows the ammeter's fuse**. Thus, students must be careful to connect an ammeter in series with a circuit element so that the current must go through a resistance in addition to the ammeter. The reason that an ammeter has very little resistance is so that it will have very little influence on the current that it measures.

Metric Prefix

Prefix	Name	Power of 10
k	kilo	10^3

Strategy for Analyzing Resistor Circuits

If a resistor circuit has **just one battery** and if it can be solved via series and parallel combinations, follow these steps (if a circuit has two or more batteries, see Chapter 15):

1. Is there a voltmeter or ammeter in the circuit? If so, redraw the circuit as follows:
   - Remove the voltmeter and also remove its connecting wires.
   - Remove the ammeter, patching it up with a line (see below).

2. Visually reduce the circuit one step at a time by identifying series and parallel combinations. If you pick any two resistors at random, you **can't** force them to be in series or parallel (they might be neither). Instead, you need to look at several different pairs until you find a pair that are definitely in series or parallel. Apply the rules discussed on pages 173-174 to help identify series or parallel combinations.
   - If two (or more) resistors are in **series**, you may remove all the resistors that were in series and replace them with a new resistor. Keep the wire that connected the resistors together when you redraw the circuit. Give the new resistor a unique name (like R_{s1}). In the math, you will solve for R_{s1} using the series formula.

   - If two (or more) resistors are in **parallel**, when you redraw the circuit, keep one of the resistors and its connecting wires, but remove the other resistors that were in parallel with it and also remove their connecting wires. Give the new resistor a unique name (like R_{p1}). In the math, you will solve for R_{p1} using the parallel formula.

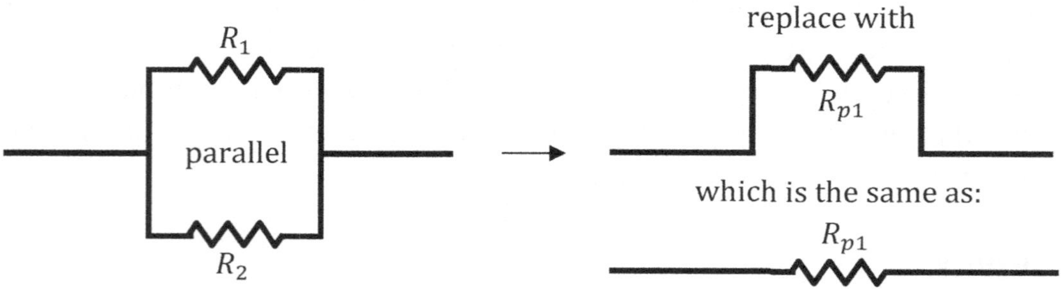

3. Continue redrawing the circuit one step at a time by identifying series and parallel combinations until there is just one resistor left in the circuit. Label this resistor the **equivalent resistance** (R_{eq}).

4. Compute each series and parallel resistance by applying the formulas below.

$$R_s = R_1 + R_2 + \cdots + R_N$$

$$\frac{1}{R_p} = \frac{1}{R_1} + \frac{1}{R_2} + \cdots + \frac{1}{R_N}$$

5. Continue applying Step 4 until you solve for the last resistance in the circuit, R_{eq}.
6. If a problem asks you to find current, potential difference, or power, begin working **backwards** through the circuit one step at a time, as follows:
   A. Start at the very last circuit, which has the equivalent resistance (R_{eq}). Solve for the current (I_{batt}) which passes through both the battery and the equivalent resistance using the formula below, where ΔV_{batt} is the potential difference of the battery or DC power supply.

   $$I_{batt} = \frac{\Delta V_{batt}}{R_{eq}}$$

   B. Go backwards one step in your diagrams: Which diagram did the equivalent resistance come from? In that diagram, are the resistors connected in series or parallel?
      - If they were in **series**, write a formula like $I_{batt} = I_a = I_b$ (but use their labels from your diagram, which are probably not I_a and I_b). Resistors in **series** have the same **current**.
      - If they were in **parallel**, write a formula like $\Delta V_{batt} = \Delta V_a = \Delta V_b$ (but use their labels from your diagram, which are probably not ΔV_a and ΔV_b). Resistors in **parallel** have the same **potential difference**.

   C. In Step B, did you set currents or potential differences equal?
      - If you set currents equal (like $I_{batt} = I_a = I_b$), find the **potential difference** of the desired resistor, with a formula like $\Delta V_a = I_a R_a$.
      - If you set potential differences equal (like $\Delta V_{batt} = \Delta V_a = \Delta V_b$), find the **current** of the desired resistor, with a formula like $I_a = \frac{\Delta V_a}{R_a}$.

   D. Continue going backward one step at a time, applying Steps B and C above each time, until you solve for the desired unknown. Note that the second time you go backwards, you won't write I_{batt} or ΔV_{batt}, but will use the label for the appropriate current or potential difference from your diagram. (This is illustrated in the example that follows.)

   E. If a problem asks you to find the **power** (P) dissipated in a resistor, first find the current (I) or potential difference (ΔV) for the specified resistor by applying Steps A-D above, and then use one of the formulas below.

   $$P = I\Delta V \quad , \quad P = I^2 R \quad , \quad P = \frac{\Delta V^2}{R}$$

Example: Consider the circuit shown below.

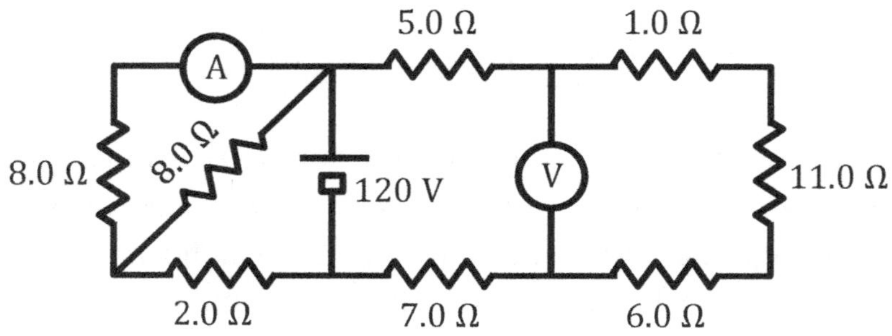

(A) Determine the equivalent resistance of the circuit.*

The first step is to redraw the circuit, treating the meters as follows:

- Remove the voltmeter and also remove its connecting wires.
- Remove the ammeter, patching it up with a line.

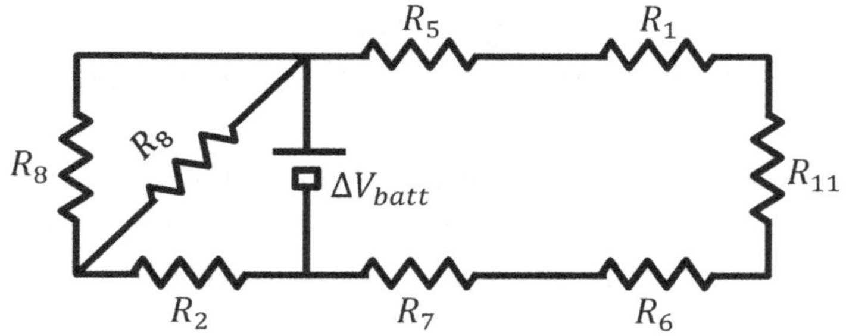

Study the eight resistors in the diagram above. Can you find two resistors that are either in series or parallel? There are two combinations to find. See if you can find them.

- R_5, R_1, R_{11}, R_6, and R_7 are all in **series** because an electron could travel through all five resistors without crossing a junction. When we redraw the circuit, we will replace these five resistors with a single resistor called R_{s1}. Calculate R_{s1} using the formula for resistors in series.

$$R_{s1} = R_5 + R_1 + R_{11} + R_6 + R_7 = 5 + 1 + 11 + 6 + 7 = 30\ \Omega$$

- The two 8.0-Ω resistors are in **parallel**. You can see this using the "finger" rule from Chapter 11 (page 173): If you put both forefingers across one R_8, you can get both forefingers across the other R_8 without crossing other circuit elements (like a battery or resistor). (Remember, it's okay to cross a junction in parallel. Note that R_2 is **not** part of the parallel combination because one finger would have to cross the battery.) When we redraw the circuit, we will replace these two resistors with a single resistor called R_{p1}. Calculate R_{p1} using the formula for resistors in parallel.

$$\frac{1}{R_{p1}} = \frac{1}{R_8} + \frac{1}{R_8} = \frac{1}{8} + \frac{1}{8} = \frac{1+1}{8} = \frac{2}{8} = \frac{1}{4}$$

$$R_{p1} = 4.0\ \Omega$$

* That is, from one terminal of the battery to the other.

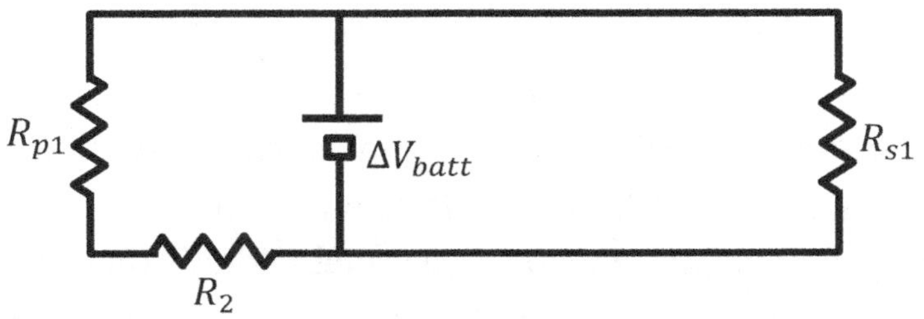

Study the three resistors in the diagram above. Can you find two resistors that are either in series or parallel? There is only one pair to find presently: R_{p1} and R_2 are in **series** because an electron could travel from one to the other without crossing a junction. Redraw the circuit, replacing R_{p1} and R_2 with a single resistor called R_{s2}. Calculate R_{s2} using the formula for resistors in series.

$$R_{s2} = R_{p1} + R_2 = 4 + 2 = 6.0\ \Omega$$

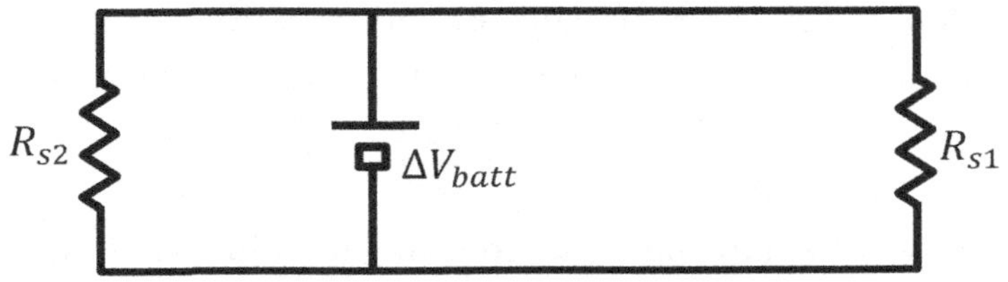

There are just two resistors in the diagram above. Are they in series or parallel? R_{s2} and R_{s1} are in **parallel** (use both forefingers to verify this). Redraw the circuit, replacing R_{s2} and R_{s1} with a single resistor called R_{eq} (since this is the last resistor remaining). Calculate R_{eq} using the formula for resistors in parallel. Add fractions with a **common denominator**.

$$\frac{1}{R_{eq}} = \frac{1}{R_{s2}} + \frac{1}{R_{s1}} = \frac{1}{6} + \frac{1}{30} = \frac{5}{30} + \frac{1}{30} = \frac{1+5}{30} = \frac{6}{30} = \frac{1}{5}$$

$$R_{eq} = 5.0\ \Omega$$

The equivalent resistance is $R_{eq} = 4.0\ \Omega$. **Tip**: For parallel resistors, remember to find the **reciprocal** at the end of the calculation.

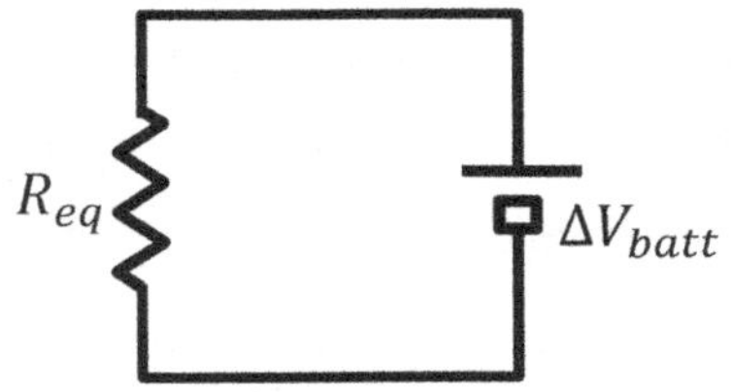

(B) What numerical value with units does the ammeter read?

An **ammeter** measures **current**. Find the ammeter in the original circuit: The ammeter is connected in series with the 8.0-Ω resistor. We need to find the current through the 8.0-Ω resistor. We must work "backwards" through our simplified circuits, beginning with the circuit that just has R_{eq}, in order to solve for current (or potential difference or power). The math begins with the following equation, which applies to the last circuit.

$$I_{batt} = \frac{\Delta V_{batt}}{R_{eq}} = \frac{120}{5} = 24 \text{ A}$$

Now we will go one step backwards from the simplest circuit (with just R_{eq}) to the second-to-last circuit (which has R_{s2} and R_{s1}). Are R_{s2} and R_{s1} in series or parallel? We already answered this in part (A): R_{s2} and R_{s1} are in **parallel**. What's the same in parallel: current (I) or potential difference (ΔV)? **Potential difference** is the same in **parallel**. Therefore, we set the potential differences of R_{s2} and R_{s1} equal to one another and also set them equal to the potential difference of the resistor that replaced them (R_{eq}). This is expressed in the following equation. This is Step 6B of the strategy. Note that the potential difference across R_{eq} is ΔV_{batt}.

$$\Delta V_{s2} = \Delta V_{s1} = \Delta V_{batt} = 120 \text{ V}$$

According to Step 6C of the strategy, if we set the potential differences equal to one another, we must calculate current. Based on the question for part (B), do we need the current through R_{s2} or R_{s1}? We need I_{s2} since the ammeter is part of R_{s2}.

$$I_{s2} = \frac{\Delta V_{s2}}{R_{s2}} = \frac{120}{6} = 20 \text{ A}$$

You have to go all the way back to part (A) to find the value for R_{s2}, whereas ΔV_{s2} appears just one equation back. Note that all of the subscripts match in the above equation.

We haven't reached the ammeter yet, so we must go back (at least) one more step. R_{s2} replaced R_{p1} and R_2. Are R_{p1} and R_2 in series or parallel? They are in **series**. What's the same in series: current (I) or potential difference (ΔV)? **Current** is the same in **series**. Therefore, we set the currents through R_{p1} and R_2 equal to one another and also set them equal to the current of the resistor that replaced them (R_{s2}). This is expressed in the following equation. This is Step 6B of the strategy (again).

$$I_{p1} = I_2 = I_{s2} = 20 \text{ A}$$

According to Step 6C of the strategy, if we set the currents equal to one another, we must calculate potential difference. Based on the question for part (B), do we need the potential difference across R_{p1} or R_2? We need ΔV_{p1} since the ammeter is part of R_{p1}.

$$\Delta V_{p1} = I_{p1} R_{p1} = (20)(4) = 80 \text{ V}$$

We must go back (at least) one more step because we still haven't reached the ammeter. R_{p1} replaced the two R_8's. Are the two R_8's in series or parallel? They are in **parallel**. What's the same in parallel: current (I) or potential difference (ΔV)? **Potential difference** is the same in **parallel**. Therefore, we set the potential differences of the two R_8's equal to one another and also set them equal to the potential difference of the resistor that replaced them (R_{p1}). This is expressed in the following equation. This is Step 6B of the strategy.

$$\Delta V_8 = \Delta V_8 = \Delta V_{p1} = 80 \text{ V}$$

According to Step 6C of the strategy, if we set the potential differences equal to one another, we must calculate current. The current through each R_8 will be the same:

$$I_8 = \frac{\Delta V_8}{R_8} = \frac{80}{8} = 10 \text{ A}$$

The ammeter reading is $I_A = I_8 = 10$ A.

(C) What numerical value with units does the voltmeter read?
A **voltmeter** measures **potential difference**. Find the voltmeter in the original circuit: The voltmeter is connected across the 1.0-Ω, 11.0-Ω, and 6.0-Ω resistors. We need to find the potential difference across these resistors, working backwards. We don't need to start over: Just continue working backwards from where we left off in part (B).

We found information about R_{s1} in part (B): $\Delta V_{s1} = \Delta V_{batt} = 120$ V. In part (C), we will begin from there. Since we already know the potential difference across R_{s1}, we'll calculate the current through R_{s1}.

$$I_{s1} = \frac{\Delta V_{s1}}{R_{s1}} = \frac{120}{30} = 4.0 \text{ A}$$

R_{s1} replaced R_5, R_1, R_{11}, R_6, and R_7. Are these five resistors in series or parallel? They are in **<u>series</u>**. What's the same in series: current (I) or potential difference (ΔV)? **Current** is the same in **series**. Therefore, we set the currents through these five resistors equal to one another and also set them equal to the current of the resistor that replaced them (R_{s1}). This is expressed in the following equation. This is Step 6B of the strategy.

$$I_5 = I_1 = I_{11} = I_6 = I_7 = I_{s1} = 4.0 \text{ A}$$

According to Step 6C of the strategy, if we set the currents equal to one another, we must calculate potential difference. Based on the question for part (B), which resistor(s) do we need the potential difference across? We need ΔV_1, ΔV_{11}, and ΔV_6 since the voltmeter is connected across those three resistors.

$$\Delta V_1 = I_1 R_1 = (4)(1) = 4 \text{ V}$$
$$\Delta V_{11} = I_{11} R_{11} = (4)(11) = 44 \text{ V}$$
$$\Delta V_6 = I_6 R_6 = (4)(6) = 24 \text{ V}$$

Since the same current passes through all three of these resistors, we add the ΔV's together to see what the voltmeter reads (Chapter 15 will show how to do this more properly).

$$\Delta V_V = \Delta V_1 + \Delta V_{11} + \Delta V_6 = 4 + 44 + 24 = 72 \text{ V}$$

The voltmeter reading is $\Delta V_V = 72$ V.

(D) How much power is dissipated in the 5.0-Ω resistor?
We found the current through R_5 in part (C): $I_5 = 4.0$ A. Choose the appropriate equation.

$$P_5 = I_5^2 R_5 = (4)^2(5) = (16)(5) = 80 \text{ W}$$

The power dissipated in the 5.0-Ω resistor is $P_5 = 80$ W. (Note that I_5 is **squared** in the equation above.)

59. Consider the circuit shown below.

(A) Redraw the circuit step by step until only a single equivalent resistor remains. Label each reduced resistor using a symbol with subscripts.

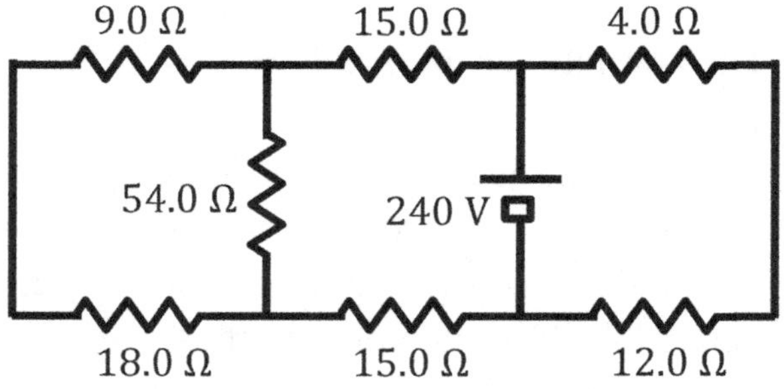

Note: You're not finished yet. This problem is **continued** on the next page.

(B) Determine the equivalent resistance of the circuit.

(C) Determine the power dissipated in either 15-Ω resistor.

Want help? Check the hints section at the back of the book.

Answers: 12 Ω, 375 W

60. Consider the circuit shown below.

(A) Redraw the circuit step by step until only a single equivalent resistor remains. Label each reduced resistor using a symbol with subscripts.

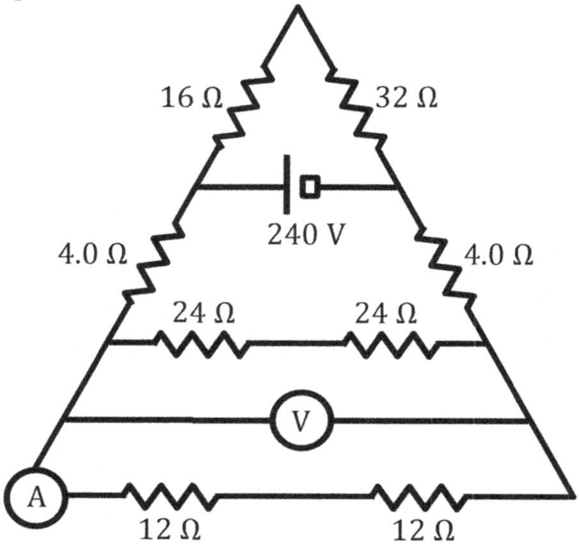

Note: You're not finished yet. This problem is **continued** on the next page.

(B) Determine the equivalent resistance of the circuit.

(C) What numerical value with units does the ammeter read?

(D) What numerical value with units does the voltmeter read?

Want help? Check the hints section at the back of the book.

Answers: $16\ \Omega, \frac{20}{3}$ A, 160 V

61. Determine the equivalent resistance of the circuit shown below.

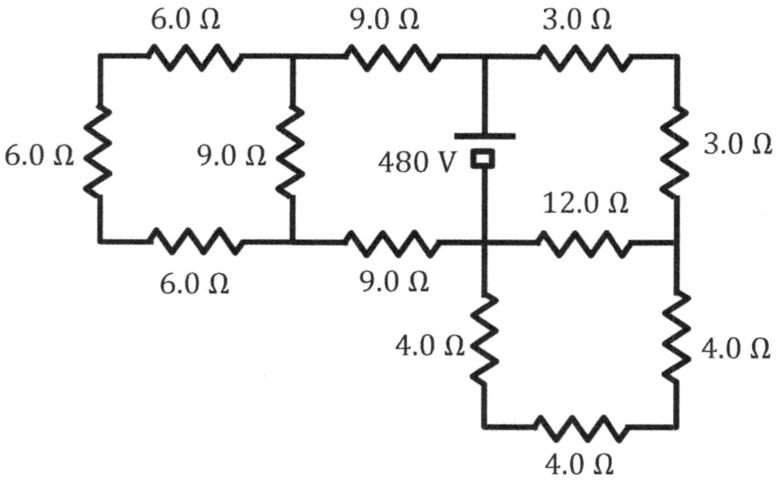

Want help? Check the hints section at the back of the book.

Answer: 8.0 Ω

14 CIRCUITS WITH SYMMETRY

Relevant Terminology

Conductor – a material through which electrons are able to flow readily. Metals tend to be good conductors of electricity.
Resistance – a measure of how well a component in a circuit resists the flow of current.
Resistor – a component in a circuit which has a significant amount of resistance.
Equivalent resistance – a single resistor that is equivalent (based on how much current it draws for a given voltage) to a given configuration of resistors.
Current – the instantaneous rate of flow of charge through a wire.
Electric potential – electric potential energy per unit charge.
Potential difference – the electric work per unit charge needed to move a test charge between two points in a circuit. Potential difference is also called the **voltage**.

Symbols and SI Units

Symbol	Name	SI Units
R	resistance	Ω
I	electric current	A
V	electric potential	V
ΔV	the potential difference between two points in a circuit	V

Schematic Symbols for Resistors and Batteries

Schematic Representation	Symbol	Name
	R	resistor
	ΔV	battery or DC power supply

Essential Concepts

A wire only impacts the equivalent resistance of a circuit if current flows through it.

- If there is **no current** flowing through a wire, the wire may be removed from the circuit without affecting the equivalent resistance.
- If a wire is added to a circuit in such a way that **no current** flows through the wire, the wire's presence won't affect the equivalent resistance.

Let's apply Ohm's law to such a wire. The potential difference across the length of the wire equals the current through the wire times the resistance of the wire. (Although the wire's resistance may be small compared to other resistances in the circuit, every wire does have some resistance.)

$$\Delta V_{wire} = I_{wire} R_{wire}$$

If the **potential difference** across the wire is **zero**, Ohm's law tells us that **there won't be any current in the wire**.

With a symmetric circuit, we consider the electric potential (V) at points between resistors, and try to find two such points that definitely have the **same electric potential**. If two points have the same electric potential, then the potential difference between those two points will be zero. We may then add or remove a wire between those two points without disturbing the equivalent resistance of the circuit.

For a circuit with a single power supply, electric potential is highest at the positive terminal and lowest at the negative terminal. Our goal is to find two points between resistors that are equal percentages – in terms of **electric potential, <u>not</u>** in terms of distance – between the two terminals of the battery.

For example, in the circuit on the left below, points B and C are each exactly halfway (in terms of electric potential) from the negative terminal to the positive terminal. In the circuit in the middle, points F and G are each one-third of the way from the negative to the positive, since 4 Ω is one-third of 12 Ω (note that 4 Ω + 8 Ω = 12 Ω). In the circuit on the right, J and K are also each one-third of the way from the negative to the positive, since 2 Ω is one-third of 6 Ω (note that 2 Ω + 4 Ω = 6 Ω) and 4 Ω is one-third of 12 Ω.

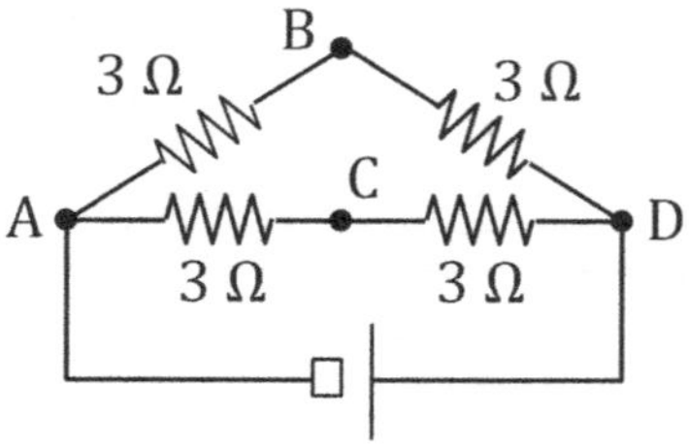

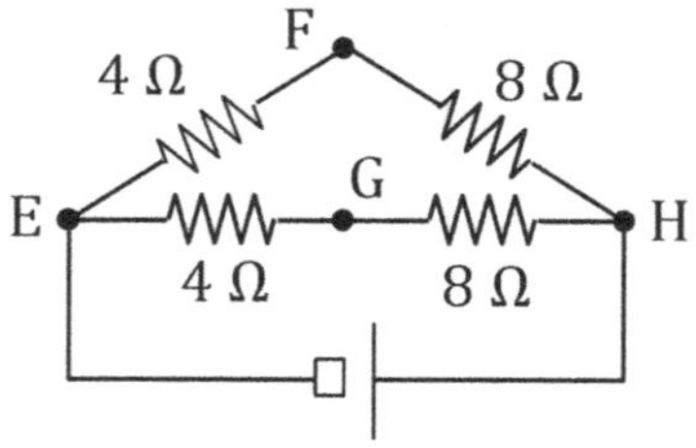

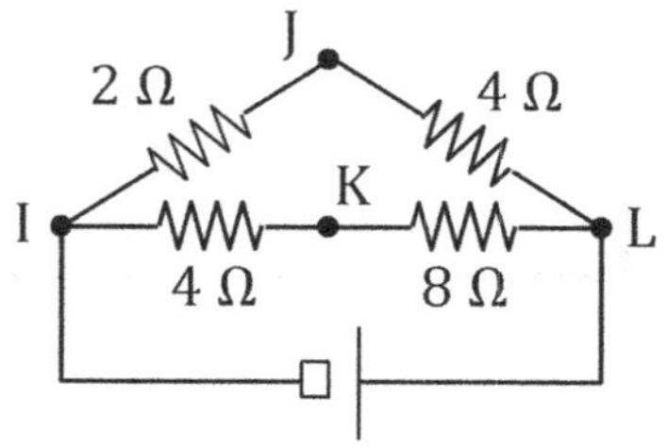

Strategy for Analyzing a Symmetric Circuit

Some resistor (or capacitor) circuits that have a single battery don't have any series or parallel combinations when you first look at them, but if there is enough symmetry in the circuit, you may still be able to apply series and parallel methods. If so, follow these steps:

1. Label the points between the resistors A, B, C, etc.
2. "Unfold" the circuit as follows.
   - Think of the negative terminal of the power supply as the "ground." Draw it at the bottom of the picture.
   - Think of the positive terminal of the power supply as the "roof." Draw it at the top of the picture.
   - All other points lie above the ground and below the roof. If a point in the original circuit is closer to the roof than it is to the ground (in terms of **electric potential**, which you can gauge percentage-wise by looking at resistance – **don't** think in terms of distance), draw it closer to the roof than to the ground in your "unfolded" circuit.
   - Look for (at least one) pair of points that are definitely the same percentage of electric potential between the ground and the roof. If either point is "closer" (in terms of electric potential, not in terms of distance) to the negative or to the positive terminal, then those points are **not** the same percentage from the ground to the roof. Once you identify two points that are the same percentage from the ground to the roof, draw them the same "height" in your unfolded circuit.
   - Study the examples that follow. They will help you learn how to visually unfold a circuit.
3. When you're 100% sure that two points have the same **electric potential**, do one of the following:
   - If there is already a wire connecting those two points, remove the wire. Since the potential difference across the wire is zero, there isn't any current in the wire, so it's safe to remove it.
   - If there isn't already a wire connecting those two points, add a wire between the points. Make it a very short wire. Make the wire so short that the two points merge together into a single point (see the examples).
4. You should now see **series** and **parallel** combinations that weren't present before. If so, now you can analyze the circuit using the strategy from Chapter 13. (When it's not possible to use series and parallel combinations, you may still apply Kirchhoff's rules, which we will learn in Chapter 15).

Example: Twelve identical 12-Ω resistors are joined together to form a cube, as illustrated below. If a battery is connected by joining its negative terminal to point F and its positive terminal to point C, what will be the equivalent resistance of the cube? (This connection is across a **body diagonal** between opposite corners of the cube.)

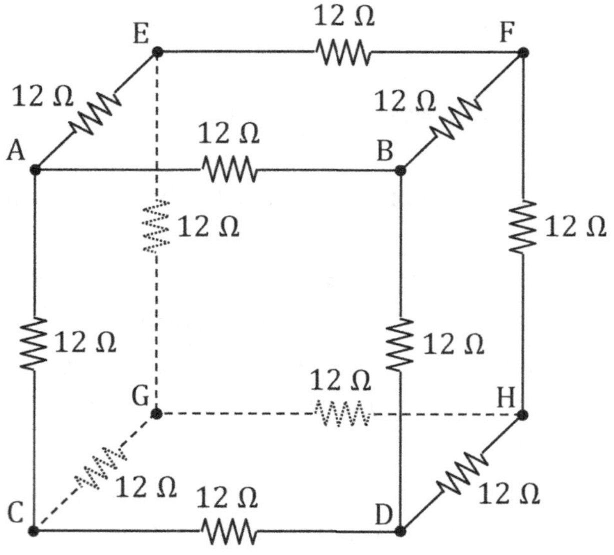

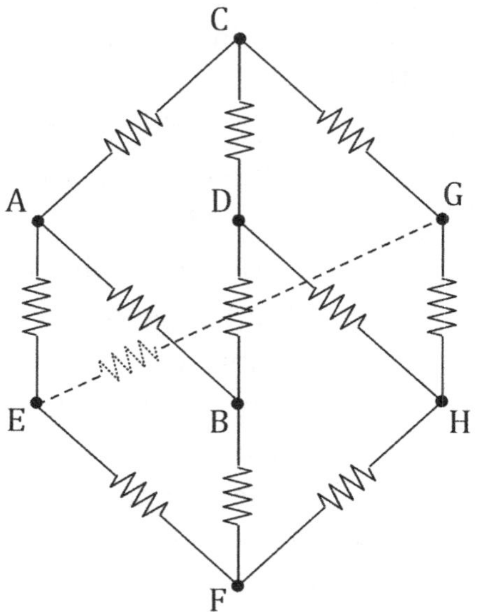

Note that no two resistors presently appear to be in series or parallel.

- Since there is a junction between any two resistors, none are in series.
- If you try the parallel rule with two forefingers (page 173), you should find that another resistor always gets in the way of one of your fingers. Thus, no two resistors are in parallel.

Fortunately, there is enough symmetry in the circuit to find points with the same potential.

We will "unfold" the circuit with point F at the bottom (call it the "ground") and point C at the top (call it the "roof"). This is based on how the battery is connected.

- Points E, B, and H are each one step from point F (the "ground") and two steps from point C (the "roof"). Neither of these points is closer to the "ground" or the "roof." Therefore, points B, E, and H have the same electric potential.
- Points A, D, and G are each two steps from point F (the "ground") and one step from point C (the "roof"). Neither of these points is closer to the "ground" or the "roof." Therefore, points A, D, and G have the same electric potential.
- Draw points E, B, and H at the same "height" in the "unfolded" circuit. Draw points A, D, and G at the same "height," too. Draw points E, B, and H closer to point F (the "ground") and points A, D, and G closer to point C (the "roof").
- Study the "unfolded" circuit at the top right and compare it to the original circuit at the top left. Try to understand the reasoning behind how it was drawn.

Consider points E, B, and H, which have the same electric potential:

- There are presently no wires connecting these three points.
- Therefore, we will add wires to connect points E, B, and H.
- Make these new wires so short that you have to move points E, B, and H toward one another. Make it so extreme that points E, B, and H merge into a single point, which we will call EBH.

Do the same thing with points A, D, and G, merging them into point ADG. With these changes, the diagram from the top right of the previous page turns into the diagram at the left below. Study the two diagrams to try to understand the reasoning behind it.

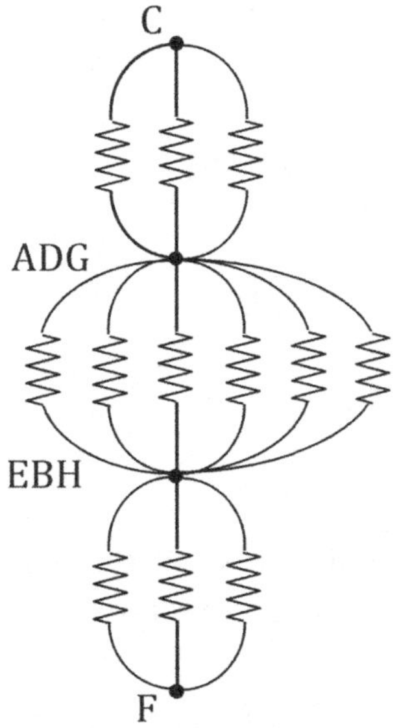

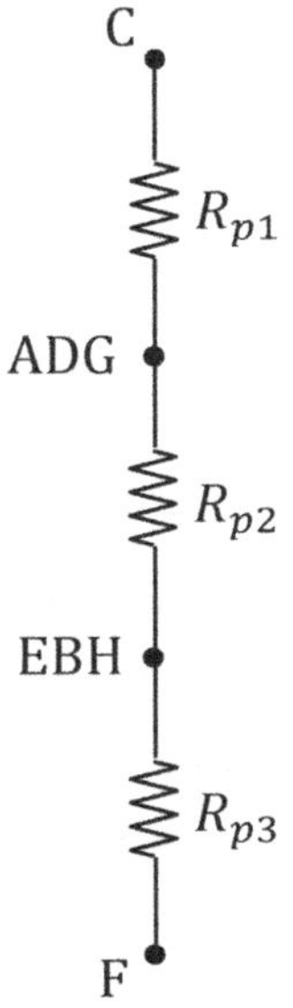

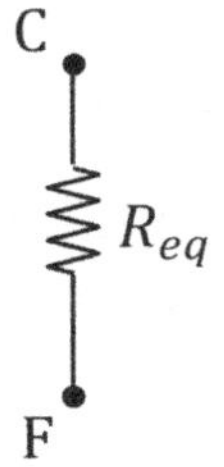

It's a good idea to count corners and resistors to make sure you don't forget one:

- The cube has 8 corners: A, B, C, D, E, F, G, and H.
- The cube has 12 resistors: one along each edge.

If you count, you should find 12 resistors in the left diagram above. The top 3 are in parallel (R_{p1}), the middle 6 are in parallel (R_{p2}), and the bottom 3 are in parallel (R_{p3}).

$$\frac{1}{R_{p1}} = \frac{1}{12} + \frac{1}{12} + \frac{1}{12} = \frac{3}{12} = \frac{1}{4} \quad \Rightarrow \quad R_{p1} = 4.0\ \Omega$$

$$\frac{1}{R_{p2}} = \frac{1}{12} + \frac{1}{12} + \frac{1}{12} + \frac{1}{12} + \frac{1}{12} + \frac{1}{12} = \frac{6}{12} = \frac{1}{2} \quad \Rightarrow \quad R_{p2} = 2.0\ \Omega$$

$$\frac{1}{R_{p3}} = \frac{1}{12} + \frac{1}{12} + \frac{1}{12} = \frac{3}{12} = \frac{1}{4} \quad \Rightarrow \quad R_{p3} = 4.0\ \Omega$$

After drawing the reduced circuit, R_{p1}, R_{p2}, and R_{p3} are in series, forming R_{eq}.

$$R_{eq} = R_{p1} + R_{p2} + R_{p3} = 4 + 2 + 4 = 10.0\ \Omega$$

The equivalent resistance of the cube along a **body diagonal** is $R_{eq} = 10.0\ \Omega$.

Example: Twelve identical 12-Ω resistors are joined together to form a cube, as illustrated below. If a battery is connected by joining its negative terminal to point H and its positive terminal to point C, what will be the equivalent resistance of the cube? (This connection is across a **face diagonal** between opposite corners of one square face of the cube. Contrast this with the previous example. It will make a huge difference in the solution.)

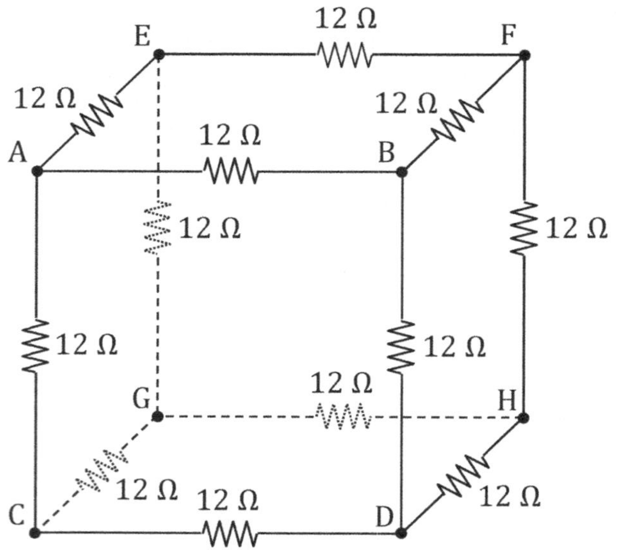

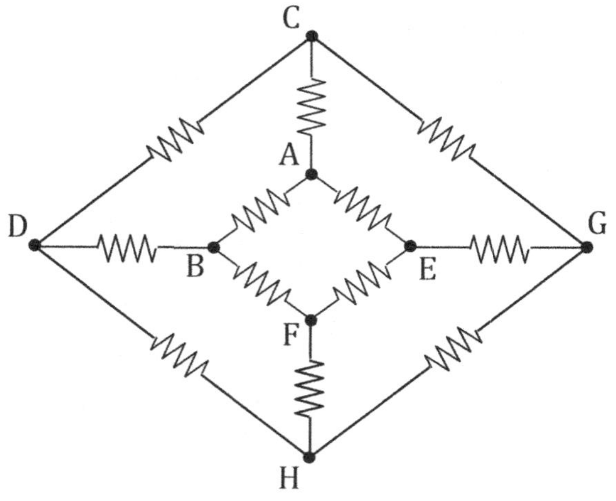

We will "unfold" the circuit with point H at the bottom (call it the "ground") and point C at the top (call it the "roof"). This is based on how the battery is connected.

- Points G and D are each one step from point H (the "ground") and also one step from point C (the "roof"). Each of these points is exactly **halfway** between the "ground" and the "roof." Therefore, points D and G have the same electric potential.
- Points B and E are each two steps from point H (the "ground") and also two steps from point C (the "roof"). Each of these points is exactly **halfway** between the "ground" and the "roof." Therefore, points B and E have the same electric potential.
- Furthermore, points G, D, B, and E all have the same electric potential, since we have reasoned that all 4 points are exactly **halfway** between the "ground" and the "roof."
- Draw points G, D, B, and E at the same height in the "unfolded" circuit.
- Draw point F closer to point H (the "ground") and point A closer to point C (the "roof").

Study the "unfolded" circuit at the top right and compare it to the original circuit at the top left. Try to understand the reasoning behind how it was drawn.

Count corners and resistors to make sure you don't forget one:

- See if you can find all 8 corners (A, B, C, D, E, F, G, and H) in the "unfolded" circuit.
- See if you can find all 12 resistors in the "unfolded" circuit.

Consider points G, D, B, and E, which have the same electric potential:

- There are presently resistors between points D and B and also between E and G.
- Therefore, we will remove these two wires.

With these changes, the diagram from the top right of the previous page turns into the diagram at the left below.

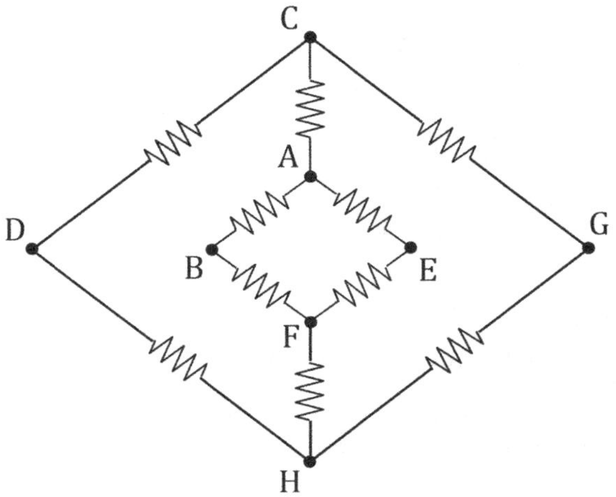

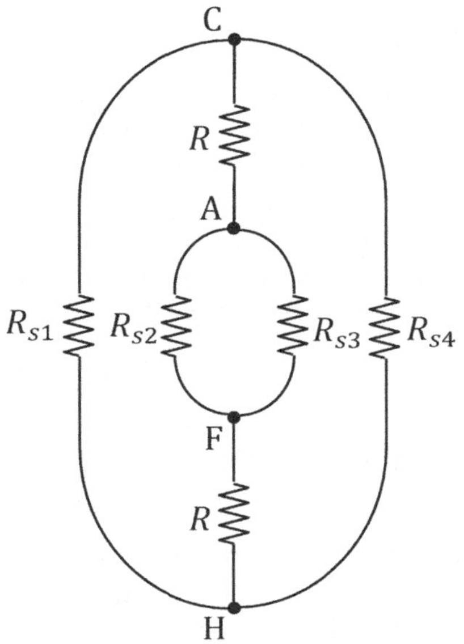

In the diagram above on the left:

- The 2 resistors from H to D and D to C are in series (R_{s1}).
- The 2 resistors from F to B and B to A are in series (R_{s2}).
- The 2 resistors from F to E and E to A are in series (R_{s3}).
- The 2 resistors from H to G and G to C are in series (R_{s4}).
- Note that points D, B, E and G are **not** junctions now that the wires connecting D to B and E to G have been removed.

$$R_{s1} = 12 + 12 = 24.0\ \Omega$$
$$R_{s2} = 12 + 12 = 24.0\ \Omega$$
$$R_{s3} = 12 + 12 = 24.0\ \Omega$$
$$R_{s4} = 12 + 12 = 24.0\ \Omega$$

In the diagram above on the right:

- R_{s1} and R_{s4} are in parallel. They form R_{p1}.
- R_{s2} and R_{s3} are in parallel. They form R_{p2}.

$$\frac{1}{R_{p1}} = \frac{1}{R_{s1}} + \frac{1}{R_{s4}} = \frac{1}{24} + \frac{1}{24} = \frac{2}{24} = \frac{1}{12} \quad \Rightarrow \quad R_{p1} = 12.0\ \Omega$$
$$\frac{1}{R_{p2}} = \frac{1}{R_{s2}} + \frac{1}{R_{s3}} = \frac{1}{24} + \frac{1}{24} = \frac{2}{24} = \frac{1}{12} \quad \Rightarrow \quad R_{p2} = 12.0\ \Omega$$

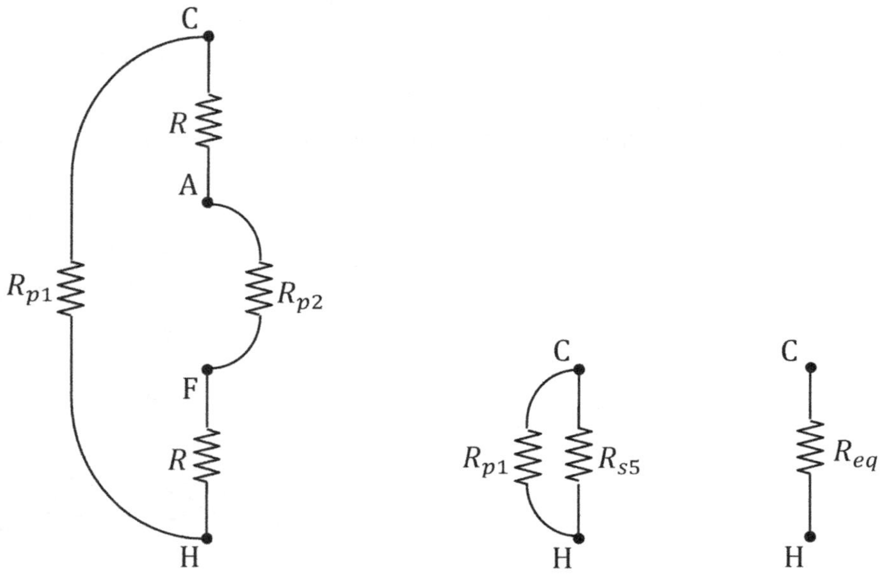

In the diagram above on the left, R, R_{p2}, and R are in series. They form R_{s5}.

$$R_{s5} = R + R_{p2} + R = 12 + 12 + 12 = 36.0\ \Omega$$

In the diagram above in the middle, R_{p1} and R_{s5} are in parallel. They form R_{eq}.

$$\frac{1}{R_{eq}} = \frac{1}{R_{p1}} + \frac{1}{R_{s5}} = \frac{1}{12} + \frac{1}{36} = \frac{3}{36} + \frac{1}{36} = \frac{4}{36} = \frac{1}{9}$$

$$R_{eq} = 9.0\ \Omega$$

The equivalent resistance of the cube along a **face diagonal** is $R_{eq} = 9.0\ \Omega$.

62. Twelve identical 12-Ω resistors are joined together to form a cube, as illustrated below. If a battery is connected by joining its negative terminal to point D and its positive terminal to point C, what will be the equivalent resistance of the cube? (This connection is across an **edge**. Contrast this with the examples. It will make a significant difference in the solution.)

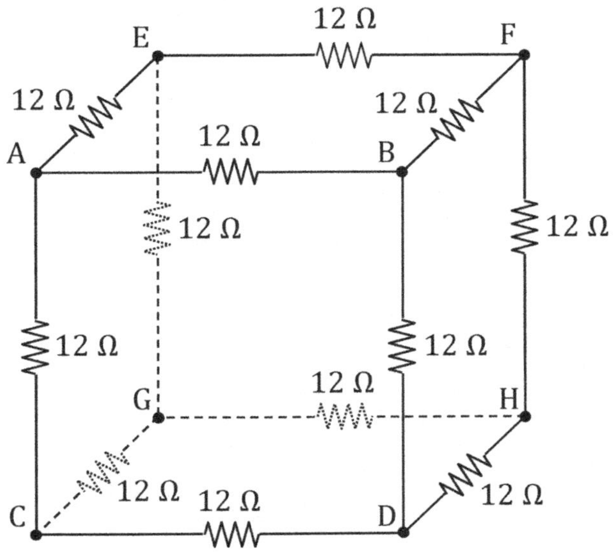

Want help? Check the hints section at the back of the book.

Answer: 7.0 Ω

63. Determine the equivalent resistance of the circuit shown below.

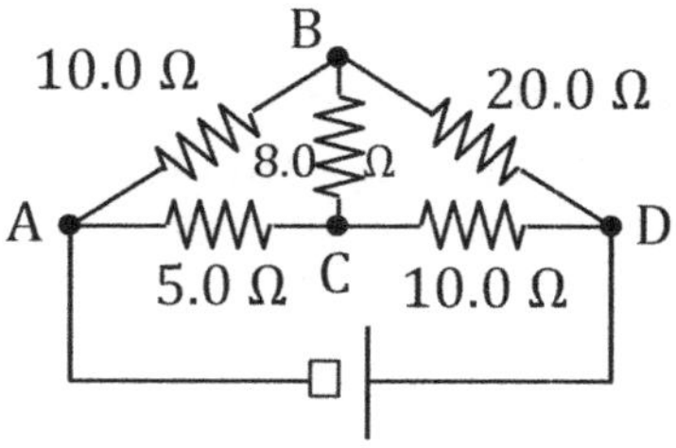

Want help? Check the hints section at the back of the book.

Answer: 10.0 Ω

15 KIRCHHOFF'S RULES

Relevant Terminology

Conductor – a material through which electrons are able to flow readily. Metals tend to be good conductors of electricity.
Resistance – a measure of how well a component in a circuit resists the flow of current.
Resistor – a component in a circuit which has a significant amount of resistance.
Current – the instantaneous rate of flow of charge through a wire.
Potential difference – the electric work per unit charge needed to move a test charge between two points in a circuit. Potential difference is also called the **voltage**.
Electric power – the instantaneous rate at which electrical work is done.
Battery – a device that supplies a potential difference between positive and negative terminals. It is a source of electric energy in a circuit.

Symbols and SI Units

Symbol	Name	SI Units
R	resistance	Ω
I	electric current	A
ΔV	the potential difference between two points in a circuit	V
P	electric power	W

Schematic Symbols

Schematic Representation	Symbol	Name
	R	resistor
	ΔV	battery or DC power supply

Note: The long line of the battery symbol represents the **positive** terminal.

Schematic Representation	Symbol	Name
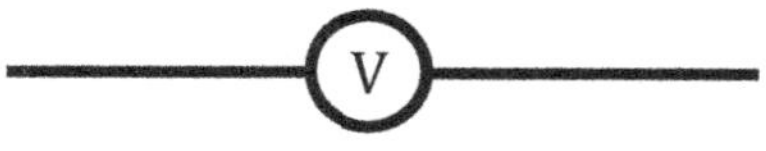	measures ΔV	voltmeter
	measures I	ammeter

Essential Concepts

Every circuit consists of loops and junctions.

Charge must be conserved at every **junction** in a circuit. Since current is the instantaneous rate of flow of charge, the sum of the currents entering a junction must equal the sum of the currents exiting the junction (such that the total charge is conserved every moment). This is Kirchhoff's (pronounced Kir-cough, **<u>not</u>** Kirch-off) **junction** rule.

$$\sum_{entering} I_i = \sum_{exiting} I_j$$

For an example of Kirchhoff's junction rule, see the junction below. Currents I_1 and I_2 go into the junction, whereas I_3 comes out of the junction. Therefore, for this particular junction, Kirchhoff's junction rule states that $I_1 + I_2 = I_3$.

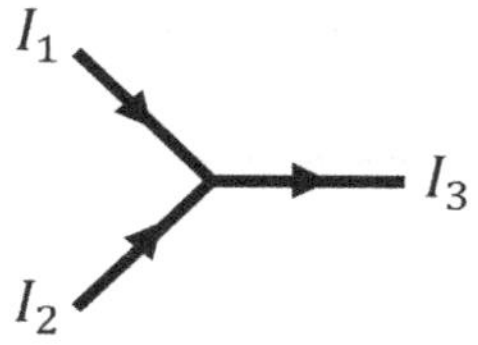

Energy must be conserved going around every **loop** in a circuit. Since energy is the ability to do work and since electrical work equals charge times potential difference, if you calculate all of the ΔV's going around a loop exactly once, they must add up to zero (such that the total energy of the system plus the surroundings is conserved). This is Kirchhoff's **loop** rule.

$$\sum_{loop} \Delta V_i = 0$$

The potential difference across a **battery** equals its value in volts, and is positive when going from the negative terminal to the positive terminal and negative when going from the positive terminal to the negative terminal. The potential difference across a **resistor** is IR (according to Ohm's law), and is positive when going against the current and negative when going with the current. A recipe for the signs is included on page 223.

Each distinct current in a circuit begins and ends at a junction. Identify the junctions in the circuit to help draw and label the currents correctly. (Recall that the term "**junction**" was defined on page 173, which includes visual examples.) The circuit shown below has two junctions (B and D) and three distinct currents shown as **solid** arrows (→):

- I_L runs from B to D.
- I_M runs from D to B.
- I_R runs from B to D.

Since I_L and I_R enter junction D, while I_M exits junction D, according to Kirchhoff's junction rule, $I_L + I_R = I_M$. (If instead you apply the junction rule to junction B, you get $I_M = I_L + I_R$, which is the same equation.) **Note**: The **dashed** arrows (⇢) below show the sense of traversal of the test charge. The **dashed** arrows (⇢) are **not** currents. See the Important Distinction note at the bottom of this page.

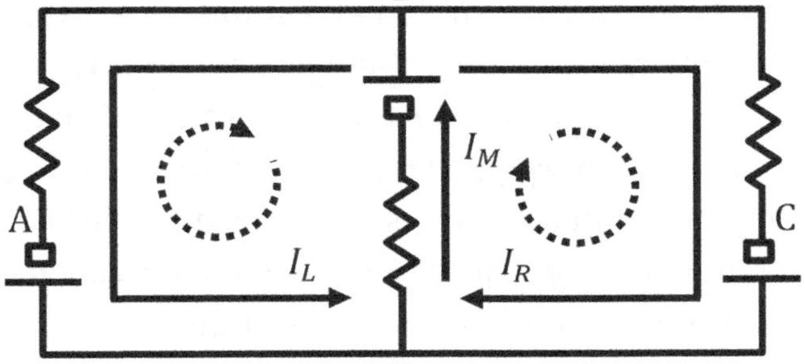

In the strategy that follows, we will count the number of "smallest loops" in a circuit. Consider the circuit shown above. There are actually 3 loops all together: There is a left loop (ABDA), a right loop (BCDB), and a big loop going all the way around (ABCDA). However, there are only 2 "smallest loops" – the left loop and the right loop. We **won't** count loops like the big one (ABCDA).

Important Distinction

There are two types of arrows used in Kirchhoff's rules problems:

- We will use solid arrows (→) for currents.
- We will use dashed arrows (⇢) to show the sense of traversal of our "test" charge.

Our "test" charge either travels around a loop clockwise or counterclockwise (it's our choice). As our "test" charge travels around the loop, we add up the ΔV's for Kirchhoff's loop rule. The "test" charge sometimes travels opposite to a current, and sometimes travels along a current. It's important to realize that **currents** and the sense of **traversal** (which way the "test" charge is traveling) are two different things. Study the circuit illustrated above to see the difference between arrows that represent currents and arrows that represent the sense of traversal of the "test" charge.

Strategy for Applying Kirchhoff's Rules

When there are **two or more batteries**, you generally need to apply Kirchhoff's rules. When there is a single battery and the circuit can be analyzed via series and parallel techniques, the strategies from Chapters 13 and 14 are usually more efficient. To apply Kirchhoff's rules to a circuit involving resistors and batteries, follow these steps:

1. Is there a voltmeter or ammeter in the circuit? If so, redraw the circuit as follows:
   - Remove the voltmeter and also remove its connecting wires.
   - Remove the ammeter, patching it up with a line (see below).

2. Draw and label a **current** in each distinction branch. (Do this **after** removing any voltmeters.) Draw an arrow to represent each current: Don't worry if you guess the direction incorrectly (you will simply get a minus sign later on if you guess wrong now – it doesn't really matter). Note that each current begins and ends at a junction.
3. Draw and label a sense of **traversal** for each of the "smallest loops." This is the direction that your "test" charge will travel as it traverses each loop. This arrow will either be clockwise or counterclockwise, and it's different from the currents.
4. Figure out how many junction equations are needed to solve the problem.
   - Count the number of distinct currents in the circuit. Call this N_C.
   - Count the number of "smallest" loops in the circuit. Call this N_L.
   - Call the number of junction equations that you need N_J.

   You need to write down $N_J = N_C - N_L$ junction equations. Here is an example. If there are $N_C = 3$ currents and $N_L = 2$ "smallest loops," you need to write down $N_J = 3 - 2 = 1$ junction equation.
5. Choose N_J junctions and apply Kirchhoff's **junction** rule to each of those junctions. For each junction, set the sum of the currents entering the junction equal to the sum of the currents exiting the junction. For example, for the junction illustrated below, the junction equation is $I_a + I_c = I_b$ (since I_a and I_c go into the junction, while I_b comes out of the junction).

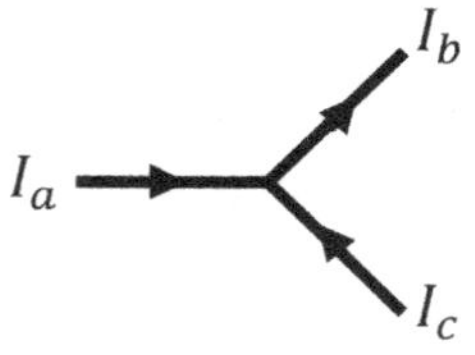

6. Label the positive and negative terminals of each battery (or DC power supply). The long line of the schematic symbol represents the positive terminal.

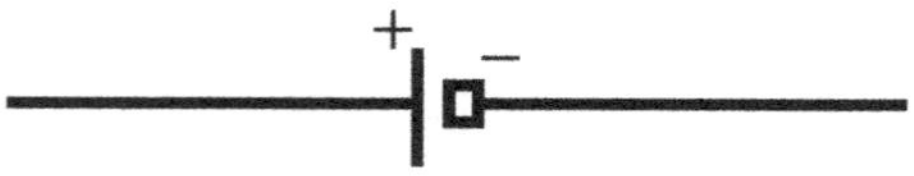

7. Apply Kirchhoff's **loop** rule to each of the "smallest loops" in the circuit. Pick a point in each loop to start and finish at. Travel around the loop in the direction that you drew in Step 3 (the sense of **traversal** of your "test" charge, which is either clockwise or counterclockwise). Imagine your "test" charge "swimming" around the loop as indicated. As your "test" charge "swims" around the loop, it will encounter resistors and batteries.

   When your "test" charge comes to a battery:
   - If the "test" charge comes to the **negative** terminal first, write a **positive** potential difference (since it rises in potential going from the negative terminal to the positive terminal).
   - If the "test" charge comes to the **positive** terminal first, write a **negative** potential difference (since it drops in potential going from the positive terminal to the negative terminal).

   When your "test" charge comes to a resistor:
   - If the "test" charge is "swimming" opposite to the current (compare the arrow for the sense of traversal to the arrow for the current), write $+IR$ (since it rises in potential when it swims "upstream" against the current).
   - If the "test" charge is "swimming" in the same direction as the current (compare the arrow for the sense of traversal to the arrow for the current), write $-IR$ (since it drops in potential when it swims "downstream" with the current).

   When the "test" charge returns to its starting point, add all of the terms up and set the sum equal to zero. Then go onto the next "smallest loop" until there are no "smallest loops" left. Study the following example, which illustrates how to apply Kirchhoff's rules.
8. You have N_C unknowns (the number of distinct currents). You also have N_C equations: N_J junctions and N_L loops. Solve these N_C equations for the unknown **currents**. The example will show you how to perform the algebra efficiently.
9. If you need to find **power**, look at your diagram to see which current is involved. Use the equation $P = I^2R$ for a resistor and $P = I\Delta V$ for a battery.
10. If you need to find the **potential difference** between two points in a circuit (or if you need to find what a **voltmeter** reads), apply the loop rule (Step 7), but **<u>don't</u>** go all of the way around the loop: Start at the first point and stop when you reach the second point. **<u>Don't</u>** set the sum equal to zero: Instead, plug in the unknowns and add up the values. The value you get equals the potential difference between the points.
11. If you need to **rank electric potential** at two or more points, set the electric potential equal to zero at one point and apply Step 10 to find the electric potential at the other points. Study the example, which shows how to rank electric potential.

Example: Consider the circuit illustrated below.

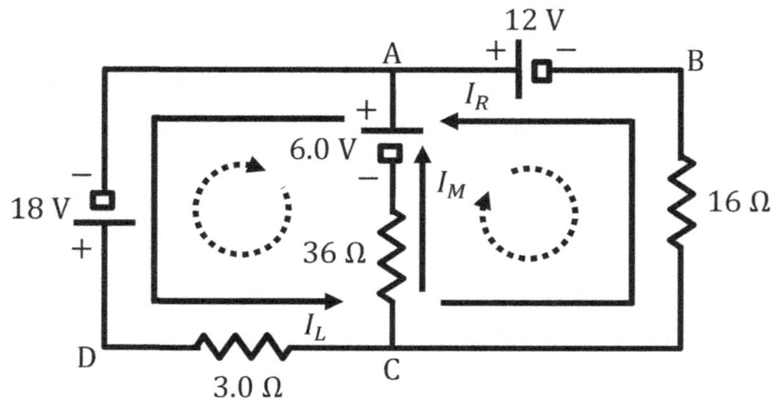

(A) Find each of the currents.

There are three distinct **currents**. See the solid arrows (→) in the diagram above:

- I_L exits junction A and enters junction C. (Note that neither B nor D is a junction.)
- I_M exits junction C and enters junction A.
- I_R exits junction C and enters junction A.

Draw the sense of **traversal** in each loop. The sense of traversal is different from current. The sense of traversal shows how your "test" charge will "swim" around each loop (which will sometimes be opposite to an actual current). See the dashed arrows (⋯→) in the diagram above: We chose our sense of traversal to be **clockwise** in each loop.

Count the number of distinct currents and "smallest" loops:

- There are $N_C = 3$ distinct currents: I_L, I_M, and I_R. These are the unknowns.
- There are $N_L = 2$ "smallest" loops: the left loop (ACDA) and the right loop (ABCA).

Therefore, we need $N_J = N_C - N_L = 3 - 2 = 1$ junction equation. We choose junction A. (We'd get the same answer for junction C in this case.)

- Which currents enter junction A? I_M and I_R go into junction A.
- Which currents exit junction A? I_L leaves junction A.

According to Kirchhoff's junction rule, for this circuit we get:

$$I_M + I_R = I_L$$

(This is different from the example on page 221. It is instructive to compare these cases.)

Next we will apply Kirchhoff's loop rule to the left loop and right loop.

- In each case, our "test" charge will start at point A. (This is our choice.)
- In each case, our "test" charge will "swim" around the loop with a **clockwise** sense of **traversal**. This is shown by the dashed arrows (⋯→).
- Study the **sign conventions** in Step 7 of strategy. If you want to solve Kirchhoff's rules problems correctly, you need to be able to get the signs correct.

Let's apply Kirchhoff's loop rule to the left loop, starting at point A and heading clockwise:

- Our "test" charge first comes to a 6.0-V battery. The "test" charge comes to the **positive terminal** of the battery **first**. According to Step 7, we write −6.0 V (since the "test" charge drops electric potential, going from positive to negative). **Note**: The direction of the current does **not** matter for a **battery**.
- Our "test" charge next comes to a 36-Ω resistor. The "test" charge is heading down from A to C presently, whereas the current I_M is drawn upward. Since the sense of **traversal** is **opposite** to the **current**, according to Step 7 we write $+36\,I_M$, multiplying current (I_M) times resistance (36 Ω) according to Ohm's law, $\Delta V = IR$. (The potential difference is positive when the test charge goes against the current when passing through a resistor.)
- Our "test" charge next comes to a 3.0-Ω resistor. The "test" charge is heading left, whereas the current I_L is heading right here. Since the **traversal** is **opposite** to the **current**, we write $+3\,I_L$.
- Finally, our "test" charge comes to an 18-V battery. The "test" charge comes to the **positive terminal** of the battery **first**. Therefore, we write −18.0 V.

Add these four terms together and set the sum equal to zero. As usual, we'll suppress the units (V for Volts and Ω for Ohms) to avoid clutter until the calculation is complete. Also as usual, we'll not worry about significant figures (like 6.0) until the end of the calculation. (When using a calculator, it's proper technique to keep extra digits and not to round until the end of the calculation. This reduces round-off error. In our case, however, we're not losing anything to rounding by writing 6.0 as 6.)

$$-6 + 36\,I_M + 3\,I_L - 18 = 0$$

Now apply Kirchhoff's loop rule to the right loop, starting at point A and heading clockwise. We're not going to include such elaborate explanations with each term this time. Study the diagram and the rules (and the explanations for the left loop), and try to follow along.

- Write −12 V for the 12.0-V battery since the "test" charge comes to the **positive terminal first**.
- Write $+16\,I_R$ for the 16-Ω resistor since the "test" charge swims **opposite** to the current I_R.
- Write $-36\,I_M$ for the 36-Ω resistor since the "test" charge swims in the **same** direction as the current I_M.
- Write +6 V for the 6.0-V battery since the "test" charge comes to the **negative terminal first**.

Add these four terms together and set the sum equal to zero.

$$-12 + 16\,I_R - 36\,I_M + 6 = 0$$

We now have three equations in three unknowns.

$$I_M + I_R = I_L$$
$$-6 + 36\,I_M + 3\,I_L - 18 = 0$$
$$-12 + 16\,I_R - 36\,I_M + 6 = 0$$

First, we will substitute the junction equation into the loop equations. According to the junction equation, I_L is the same as $I_M + I_R$. Just the first loop equation has I_L. We will rewrite the first loop equation with $I_M + I_R$ in place of I_L.

$$-6 + 36\,I_M + 3\,(I_M + I_R) - 18 = 0$$

Distribute the 3 to both terms.

$$-6 + 36\,I_M + 3\,I_M + 3\,I_R - 18 = 0$$

Combine **like terms**. The -6 and -18 are like terms: They make $-6 - 18 = -24$. The $36\,I_M$ and $3\,I_M$ are like terms: They make $36\,I_M + 3\,I_M = 39\,I_M$.

$$-24 + 39\,I_M + 3\,I_R = 0$$

Add 24 to both sides of the equation.

$$39\,I_M + 3\,I_R = 24$$

We can simplify this equation if we divide both sides of the equation by 3.

$$13\,I_M + I_R = 8$$

We'll return to this equation in a moment. Let's work with the other loop equation now.

$$-12 + 16\,I_R - 36\,I_M + 6 = 0$$

Combine **like terms**. The -12 and $+6$ are like terms: They make $-12 + 6 = -6$.

$$-6 + 16\,I_R - 36\,I_M = 0$$

Add 6 to both sides of the equation.

$$16\,I_R - 36\,I_M = 6$$

We can simplify this equation if we divide both sides of the equation by 2.

$$8\,I_R - 18\,I_M = 3$$

Let's put our two simplified equations together.

$$13\,I_M + I_R = 8$$
$$8\,I_R - 18\,I_M = 3$$

It helps to write them in the same order. Note that $8\,I_R - 18\,I_M = -18\,I_M + 8\,I_R$.

$$13\,I_M + I_R = 8$$
$$-18\,I_M + 8\,I_R = 3$$

The "trick" is to make equal and opposite coefficients for one of the currents. If we multiply the top equation by -8, we will have $-8\,I_R$ in the top equation and $+8\,I_R$ in the bottom.

$$-104\,I_M - 8\,I_R = -64$$
$$-18\,I_M + 8\,I_R = 3$$

Now I_R cancel out if we add the two equations together. The sum of the left-hand sides equals the sum of the right-hand sides.

$$-104\,I_M - 8\,I_R - 18\,I_M + 8\,I_R = -64 + 3$$
$$-104\,I_M - 18\,I_M = -61$$
$$-122\,I_M = -61$$

Divide both sides of the equation by -122.

$$I_M = -\frac{61}{-122} = +\frac{1}{2}\text{ A} = 0.50\text{ A}$$

Once you get a numerical value for one of your unknowns, you may plug this value into any of the previous equations. Look for one that will make the algebra simple. We choose:

$$13\, I_M + I_R = 8$$

$$\frac{13}{2} + I_R = 8$$

$$I_R = 8 - \frac{13}{2} = \frac{16}{2} - \frac{13}{2} = \frac{16-13}{2} = \frac{3}{2}\text{ A} = 1.50\text{ A}$$

Once you have two currents, plug them into the junction equation.

$$I_L = I_M + I_R = \frac{1}{2} + \frac{3}{2} = 0.5 + 1.5 = 2.0\text{ A}$$

The currents are $I_M = 0.50$ A, $I_R = 1.50$ A, and $I_L = 2.0$ A.

Tip: If you get a minus sign when you solve for a current:

- Keep the minus sign.
- Don't go back and alter your diagram.
- Don't rework the solution. Don't change any equations.
- If you need to plug the current into an equation, keep the minus sign.
- The minus sign simply means that the current's actual direction is opposite to the arrow that you drew in the beginning of the problem. It's not a big deal.

(B) How much power is dissipated in the 3.0-Ω resistor?

Look at the original circuit. Which current passes through the 3.0-Ω resistor? The current I_L passes through the 3.0-Ω resistor. Use the equation for **power**. Note that the current is squared in the following equation.

$$P_3 = I_L^2 R_3 = (2)^2(3) = (4)(3) = 12\text{ W}$$

(C) Find the potential difference between points A and C.

Apply Kirchhoff's loop rule, beginning at A and ending C (or vice-versa: the only difference will be a minus sign in the answer). Don't set the sum to zero (since we're not going around a complete loop). Instead, plug in the currents and add up the values. It doesn't matter which path we take (we'll get the same answer either way), but we'll choose to go down the middle branch. We're finding $V_C - V_A$: It's **final** minus **initial** (from A to C).

$$V_C - V_A = -6 + 36\, I_M = -6 + 36\left(\frac{1}{2}\right) = -6 + 18 = 12.0\text{ V}$$

The first term is -6.0 V because we came to the positive terminal first, while the second term is $+36\, I_M$ because we went opposite to the current I_M: We went down from A to C, whereas I_M is drawn up. The **potential difference** going from point A to point C is $V_C - V_A = 12.0$ V.

(D) Rank the electric potential at points A, B, C, and D.

To **rank** electric potential at two or more points, set the electric potential at one point equal to zero and then apply Kirchhoff's loop rule the same way that we did in part (C). It won't matter which point we choose to have zero electric potential, as the relative values will still come out in the same order. We choose point A to have zero electric potential.

$$V_A = 0$$

Traverse from A to B to find the potential difference $V_B - V_A$ (final minus initial). There is just a 12.0-V battery between A and B. Going from A to B, we come to the **positive terminal** first, so we write -12.0 V (review the sign conventions in Step 7, if necessary).

$$V_B - V_A = -12$$

Plug in the value for V_A.

$$V_B - 0 = -12$$
$$V_B = -12.0 \text{ V}$$

Now traverse from B to C to find the potential difference $V_C - V_B$. There is just a 16-Ω resistor between B and C. Going from B to C, we are traversing **opposite** to the current I_R, so we write $+16\ I_R$.

$$V_C - V_B = +16\ I_R$$

Plug in the values for V_B and I_R.

$$V_C - (-12) = 16(1.5)$$

Note that subtracting a negative number equates to addition.

$$V_C + 12 = 24$$
$$V_C = 24 - 12 = 12.0 \text{ V}$$

(Note that V_C and V_B are <u>**not**</u> the same: That minus sign makes a big difference.) Now traverse from C to D to find the potential difference $V_D - V_C$. There is just a 3.0-Ω resistor between C and D. Going from C to D, we are traversing **opposite** to the current I_L, so we write $+3\ I_L$.

$$V_D - V_C = +3\ I_L$$

Plug in the values for V_C and I_L.

$$V_D - 12 = 3(2)$$
$$V_D - 12 = 6$$
$$V_D = 6 + 12 = 18.0 \text{ V}$$

(You can check for consistency by now going from D to A. If you get $V_A = 0$, everything checks out). Let's tabulate the electric potentials:

- The electric potential at point A is $V_A = 0$.
- The electric potential at point B is $V_B = -12.0$ V.
- The electric potential at point C is $V_C = 12.0$ V.
- The electric potential at point D is $V_D = 18.0$ V.

Now it should be easy to rank them: D is highest, then C, then A, and B is lowest.

$$V_D > V_C > V_A > V_B$$

64. Consider the circuit illustrated below.
(A) Find each of the currents.

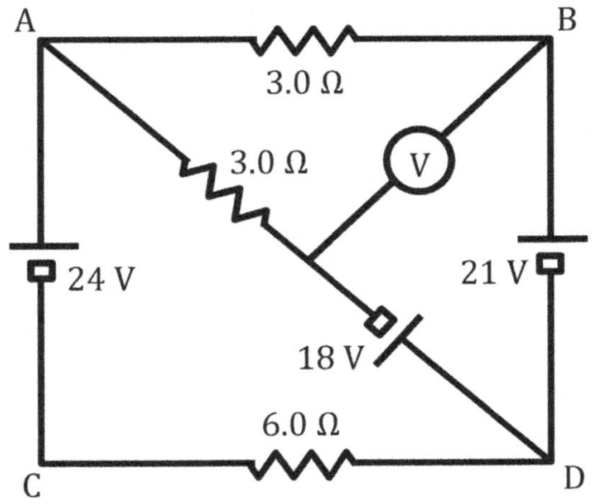

Note: You're not finished yet. This problem is **continued** on the next page.

(B) What numerical value, with units, does the voltmeter read?

(C) Rank the electric potential at points A, B, C, and D.

Want help? Check the hints section at the back of the book.

Answers: 3.0 A, 8.0 A, 5.0 A, 39 V, $V_B > V_A > V_D > V_C$

16 MORE RESISTANCE EQUATIONS

Relevant Terminology

Conductor – a material through which electrons are able to flow readily. Metals tend to be good conductors of electricity.
Conductivity – a measure of how well a given material conducts electricity.
Resistivity – a measure of how well a given material resists the flow of current. The resistivity is the reciprocal of the conductivity.
Resistance – a measure of how well a component in a circuit resists the flow of current.
Internal resistance – the resistance that is internal to a battery or power supply.
Resistor – a component in a circuit which has a significant amount of resistance.
Temperature coefficient of resistivity – a measure of how much the resistivity of a given material changes when its temperature changes.
Current – the instantaneous rate of flow of charge through a wire.
Current density – electric current per unit of cross-sectional area.
Potential difference – the electric work per unit charge needed to move a test charge between two points in a circuit. Potential difference is also called the **voltage**.
Emf – the potential difference that a battery or DC power supply would supply to a circuit neglecting its internal resistance.
Electric power – the instantaneous rate at which electrical work is done.
Battery – a device that supplies a potential difference between positive and negative terminals. It is a source of electric energy in a circuit.
Electric field – force per unit charge.

Schematic Symbols

Schematic Representation	Symbol	Name
	R	resistor
	ΔV	battery or DC power supply

Resistance Equations

Recall **Ohm's law**, which relates the potential difference (ΔV) across a resistor to the current (I) through the resistor and the resistance (R) of the resistor.

$$\Delta V = IR$$

Every battery or power supply has **internal resistance** (r). When a battery or power supply is connected in a circuit, the potential difference (ΔV) measured between its terminals is less than the actual **emf** (ε) of the battery or power supply.

$$\varepsilon = \Delta V + Ir$$

If we combine the above equations, we get the following equation, which shows that the external resistance (R) and internal resistance (r) both affect the current.

$$\varepsilon = I(R + r)$$

The conductivity is a measure of how well a given material conducts electricity, whereas the resistivity is a measure of how well a given material resists the flow of current. The **resistivity** (ρ) is the reciprocal of the **conductivity** (σ).

$$\rho = \frac{1}{\sigma}$$

The **current density** ($\vec{\mathbf{J}}$) is proportional to the electric field ($\vec{\mathbf{E}}$) and the conductivity (σ).

$$\vec{\mathbf{J}} = \sigma\vec{\mathbf{E}}$$

For a long, straight wire shaped like a right-circular cylinder, the resistance (R) of the wire is directly proportional to the **resistivity** (ρ) of the material and the length of the wire (L), and is inversely proportional to the cross-sectional area (A) of the wire.

$$R = \frac{\rho L}{A}$$

Resistivity (ρ) and resistance (R) depend on temperature, and depend on the **temperature coefficient of resistivity** (α). In the following equations, $\Delta T = T - T_0$ equals the change in temperature, and ρ_0 and R_0 are values of resistivity and resistance at some reference temperature (T_0).

$$\rho = \rho_0(1 + \alpha\Delta T)$$
$$R \approx R_0(1 + \alpha\Delta T)$$

Recall the equation for electric **power** (P), which can be expressed three ways.

$$P = I\Delta V = I^2 R = \frac{\Delta V^2}{R}$$

Also recall the formulas for **series** and **parallel** resistors (Chapter 13):

$$R_s = R_1 + R_2 + \cdots + R_N$$
$$\frac{1}{R_p} = \frac{1}{R_1} + \frac{1}{R_2} + \cdots + \frac{1}{R_N}$$

The current (I) and **current density** ($\vec{\mathbf{J}}$) are related to one another by:

$$I = \int \vec{\mathbf{J}} \cdot d\vec{\mathbf{A}}$$

In general, **resistance** can be found via calculus:

$$R = \frac{|\Delta V|}{I} = \frac{\left|\int \vec{\mathbf{E}} \cdot d\vec{\mathbf{s}}\right|}{\int \vec{\mathbf{J}} \cdot d\vec{\mathbf{A}}}$$

For the common case where $\vec{\mathbf{E}}$, $\vec{\mathbf{J}}$, $d\vec{\mathbf{s}}$, and $d\vec{\mathbf{A}}$ are all parallel, the above formula simplifies to:

$$R = \frac{\int E\, ds}{\int J\, dA}$$

Symbols and SI Units

Symbol	Name	SI Units
R	resistance	Ω
R_0	resistance at a reference temperature	Ω
r	internal resistance of a battery or power supply	Ω
ρ	resistivity	$\Omega{\cdot}\text{m}$
ρ_0	resistivity at a reference temperature	$\Omega{\cdot}\text{m}$
σ	conductivity	$\frac{1}{\Omega \cdot \text{m}}$
L	the length of the wire	m
A	cross-sectional area	m^2
α	temperature coefficient of resistivity	$\frac{1}{\text{K}}$ or $\frac{1}{°\text{C}}$
T	temperature	K
T_0	reference temperature	K
I	electric current	A
J	current density	A/m^2
ΔV	the potential difference between two points in a circuit	V
ε	emf	V
P	electric power	W

Note: The symbols ρ, σ, and α are the lowercase Greek letters rho, sigma, and alpha, while ε is a variation of the Greek letter epsilon.

Notes Regarding Units

- The SI units of resistivity (ρ) follow by solving for ρ in the equation $R = \frac{\rho L}{A}$: We get $\rho = \frac{RA}{L}$. If you plug in Ω for R, m^2 for A, and m for L, you find that the SI units for ρ are $\Omega\cdot$m. (It's Ohms times meters, **not** per meter.) Note that $\frac{\mathrm{m}^2}{\mathrm{m}} = \mathrm{m}$. Since conductivity ($\sigma$) is the reciprocal of resistivity (ρ), that is $\sigma = \frac{1}{\rho}$, it follows that the SI units of conductivity are $\frac{1}{\Omega\cdot\mathrm{m}}$.
- Since the SI unit of temperature is the Kelvin (K), it follows that the SI units of the temperature coefficient of resistivity (α) are $\frac{1}{\mathrm{K}}$. Note that α has the same numerical value in both $\frac{1}{\mathrm{K}}$ and $\frac{1}{°\mathrm{C}}$* since the change in temperature, $\Delta T = T - T_0$, works out the same in both Kelvin and Celsius (because $T_K = T_c + 273.15$).
- The SI units for current density (J) are $\mathrm{A/m^2}$, since current density is a measure of current per unit area and since the SI unit for current (I) is the Ampère (A).

Important Distinctions

Pay attention to the differences between similar quantities:

- **Resistance** (R) is measured in Ohms (Ω) and depends on the material, the length of the wire, and the thickness of the wire, whereas **resistivity** (ρ) is measured in Ωm and is just a property of the material itself. Look at the units given in a problem to help tell whether the value is resistance or resistivity.
- Emf (ε) and potential difference (ΔV) are both measured in Volts (V). The emf (ε) represents the voltage that you could get if the battery or power supply didn't have internal resistance, whereas the potential difference (ΔV) across the terminals is less than the actual emf when the battery or power supply is connected in a circuit. If you connect a voltmeter across the terminals of a battery or power supply that is connected in a circuit, it measures potential difference (ΔV).
- Emf stands for "electromotive" force, but it's not really a force and doesn't have the units of force. However, a battery creates an electric field which accelerates the charges according to $\vec{\mathbf{F}} = q\vec{\mathbf{E}}$. Emf is also not quite an "electromagnetic field," but potential difference is related to electric field through $\Delta V = -\int \vec{\mathbf{E}} \cdot d\vec{\mathbf{s}}$.
- For temperature-dependent problems, note that ρ_0 and R_0 correspond to some reference temperature T_0, while ρ and R correspond to a different temperature T.
- The SI unit of current (I) is the Ampère (A), while for current density (J) it is $\mathrm{A/m^2}$.

* Note that we only use the degree symbol (°) for Celsius (°C) and Fahrenheit (°F), but not for Kelvin (K). The reason behind this is that Kelvin is special since it is the SI unit for **absolute** temperature.

Resistor Strategy

How you solve a problem involving a **resistor** depends on the context:

1. Make a list of the symbols that you know (see the chart on page 233). The units can help you figure out which symbols you know. For a textbook problem, you may need to look up a value for the **conductivity** (σ), **resistivity** (ρ), or the **temperature coefficient of resistivity** (α). Note that you can find ρ by looking up σ, or vice-versa.
2. If a problem tells you the **gauge** of a wire (like "gauge-22 nichrome"), try looking up the gauge in your textbook. The gauge is another way of specifying thickness.
3. If a problem gives you the **color codes** of a resistor (if you are performing an experiment in a lab, many resistors are color-coded), you can determine its resistance by looking up a chart of resistor color codes.
4. Choose equations based on which symbols you know and which symbol you are trying to solve for:
   - Resistance (R) in Ω can be related to **resistivity** (ρ) in Ωm for a typical wire.
   $$R = \frac{\rho L}{A}$$
   For a wire with circular cross section, $A = \pi a^2$, where lowercase a is the radius of the wire (not to be confused with area, A). If the problem gives you the **thickness** (T) of the wire, it's the same as diameter (D): $a = \frac{D}{2} = \frac{T}{2}$. If a problem gives you conductivity (σ) instead of resistivity (ρ), note that $\rho = \frac{1}{\sigma}$.
   - If the problem involves both **temperature** and resistance (or resistivity), use the appropriate equation from below, where $\Delta T = T - T_0$.
   $$\rho = \rho_0(1 + \alpha \Delta T)$$
   $$R \approx R_0(1 + \alpha \Delta T)$$
   - The **emf** (ε) and **internal resistance** (r) of a battery or power supply are related to the potential difference (ΔV) measured across the terminals and the current (I). Ohm's law ($\Delta V = IR$) also applies to emf problems, where R is the external (or load) resistance.
   $$\varepsilon = \Delta V + Ir \quad , \quad \varepsilon = I(R + r) \quad , \quad \Delta V = IR$$
   - **Ohm's law** applies to problems that involve resistance.
   $$\Delta V = IR$$
   - Electric **power** can be expressed three different ways:
   $$P = I\Delta V = I^2 R = \frac{\Delta V^2}{R}$$
   - **Current density** ($\vec{\mathbf{J}}$) is related to electric field ($\vec{\mathbf{E}}$) and current (I) through the following equations. For **uniform** current density, $J = \frac{I}{A}$.
   $$\vec{\mathbf{J}} = \sigma \vec{\mathbf{E}} \quad , \quad I = \int \vec{\mathbf{J}} \cdot d\vec{\mathbf{A}}$$

5. Carry out any algebra needed to solve for the unknown.
6. If any resistors are connected in **series** or **parallel**, you will need to apply the strategy from Chapter 13.

Strategy to Derive an Equation for Resistance

Following is how to apply calculus to derive an equation for resistance:

1. First derive an equation for the electric field inside the resistor. Base this on how the battery or power supply is connected across the resistor: For example, if you connect a battery across the ends of a long cylinder, the electric field lines run across the length of the cylinder, whereas if you connect the terminals of a battery to the inner and outer conductors of coaxial cylinders, the electric field lines radiate outward from the axis.
   - For an infinite plane, sphere, or cylinder, apply Gauss's law (Chapter 8).
   - Otherwise, perform the electric field integral (Chapter 7).
2. Use the equation you derived in Step 1 in the following definite integral for the potential difference across the resistor.

$$\Delta V = V_f - V_i = -\int_i^f \vec{\mathbf{E}} \cdot d\vec{\mathbf{s}}$$

   The differential displacement vector ($d\vec{\mathbf{s}}$) is along the path from initial (i) to final (f). Choose a path along the electric field lines. The magnitude of $d\vec{\mathbf{s}}$ is:
   - $ds = dx$, dy, or dz for a path parallel to the x-, y-, or z-axis.
   - $ds = dr$ for a radial path going outward from the origin (useful for a sphere).
   - $ds = dr_c$ for a path going outward from the z-axis (useful for a cylinder).
   - $ds = r d\theta$ for a path going along the arc of a circle of radius r.

   Note that $\vec{\mathbf{E}} \cdot d\vec{\mathbf{s}} = E \cos\psi \, ds$, where ψ is the angle between $\vec{\mathbf{E}}$ and $d\vec{\mathbf{s}}$.
3. Use the equation you derived in Step 1 in the following definite integral for the current through the resistor, where the current density is $\vec{\mathbf{J}} = \sigma\vec{\mathbf{E}}$.

$$I = \int_S \vec{\mathbf{J}} \cdot d\vec{\mathbf{A}}$$

   The differential area vector ($d\vec{\mathbf{A}}$) is over the cross-sectional area (S). The direction of $d\vec{\mathbf{A}}$ is perpendicular to the cross-sectional area. The magnitude of dA is:
   - $dA = dxdy$ for a flat surface with straight sides lying in the xy plane.
   - $dA = rdrd\theta$ or $r_c dr_c d\theta$ for a pie slice or circle (in 2D polar or cylindrical).
   - $dA = a^2 \sin\theta \, d\theta d\varphi$ for the surface of a sphere of radius a.

   Note that $\vec{\mathbf{E}} \cdot d\vec{\mathbf{A}} = E \cos\gamma \, dA$, where γ is the angle between $\vec{\mathbf{E}}$ and $d\vec{\mathbf{A}}$.
4. Plug your answers from Steps 2-3 into the following equation and simplify: $R = \frac{|\Delta V|}{I}$.

Example: A long, straight wire has the shape of a right-circular cylinder. The length of the wire is 8.0 m and the thickness of the wire is 0.40 mm. The resistivity of the wire is $\pi \times 10^{-8}$ Ω·m. What is the resistance of the wire?

Make a list of the known quantities and identify the desired unknown symbol:

- The length of the wire is $L = 8.0$ m and the thickness of the wire is $T = 0.40$ mm.
- The resistivity is $\rho = \pi \times 10^{-8}$ Ω·m.
- The unknown we are looking for is resistance (R).

The equation relating resistance to resistivity involves area, so we need to find area first. Before we do that, let's convert the thickness to SI units.

$$T = 0.40 \text{ mm} = 0.00040 \text{ m} = 4 \times 10^{-4} \text{ m}$$

The radius of the circular cross section is one-half the diameter, and the diameter is the same as the thickness.

$$a = \frac{D}{2} = \frac{T}{2} = \frac{4 \times 10^{-4}}{2} = 2 \times 10^{-4} \text{ m}$$

The cross-sectional area of the wire is the area of a circle:

$$A = \pi a^2 = \pi(2 \times 10^{-4})^2 = \pi(4 \times 10^{-8}) \text{ m}^2 = 4\pi \times 10^{-8} \text{ m}^2$$

Now we are ready to use the equation for resistance.

$$R = \frac{\rho L}{A} = \frac{(\pi \times 10^{-8})(8)}{(4\pi \times 10^{-8})} = 2.0 \ \Omega$$

The resistance is $R = 2.0$ Ω. (Note that the π's and 10^{-8}'s both cancel out.)

Example: A resistor is made from a material that has a temperature coefficient of resistivity of 5.0×10^{-3} /°C. Its resistance is 40 Ω at 20 °C. What is its temperature at 80 °C?

Make a list of the known quantities and identify the desired unknown symbol:

- The reference temperature is $T_0 = 20$ °C and the specified temperature is $T = 80$ °C.
- The reference resistance is $R_0 = 40$ Ω.
- The temperature coefficient of resistivity is $\alpha = 5.0 \times 10^{-3}$ /°C.
- The unknown we are looking for is the resistance (R) corresponding to $T = 80$ °C.

Apply the equation for resistance that depends on temperature.

$$R \approx R_0(1 + \alpha \Delta T) = R_0[1 + \alpha(T - T_0)] = (40)[1 + (5 \times 10^{-3})(80 - 20)]$$

$$R \approx (40)[1 + (5.0 \times 10^{-3})(60)] = (40)[1 + 300 \times 10^{-3}]$$

$$R \approx (40)(1 + 0.3) = (40)(1.3) = 52 \ \Omega$$

The resistance of the resistor at $T = 80$ °C is $R \approx 52$ Ω.†

† The similar equation for resistivity involves an equal sign, whereas with resistance it is approximately equal. The reason this is approximate is that we're neglecting the effect that thermal expansion has on the length and thickness. The effect that temperature has on resistivity is generally more significant to the change in resistance than the effect that temperature has on the length and thickness of the wire, so the approximation is usually very good.

Example: A battery has an emf of 12 V and an internal resistance of 2.0 Ω. The battery is connected across a 4.0-Ω resistor.
(A) What current would an ammeter measure through the 4.0-Ω resistor?
Make a list of the known quantities and identify the desired unknown symbol:

- The emf of the battery is $\varepsilon = 12$ V and its internal resistance is $r = 2.0\ \Omega$.
- The load resistance (or external resistance) is $R = 4.0\ \Omega$.
- The unknown we are looking for is the current (I).

Choose the emf equation that involves the above symbols.

$$\varepsilon = I(R + r)$$

Divide both sides of the equation by $(R + r)$.

$$I = \frac{\varepsilon}{R + r} = \frac{12}{4 + 2} = \frac{12}{6} = 2.0 \text{ A}$$

The ammeter would measure $I = 2.0$ A. (Note that if the battery had negligible internal resistance, the ammeter would have measured $\frac{12}{4} = 3.0$ A. In this example, the internal resistance is quite significant, but in most good batteries, the internal resistance would be much smaller than 2.0 Ω, and thus would not be significant unless the load resistance were relatively small.)
(B) What potential difference would a voltmeter measure across the 4.0-Ω resistor?
Now we are solving for ΔV. Apply Ohm's law.

$$\Delta V = IR = (2)(4) = 8.0 \text{ V}$$

The voltmeter would measure $\Delta V = 8.0$ V. (Again, this is a significant deviation from the 12-V emf of the battery because this example has a dramatically large internal resistance.)

Example: Derive an equation for the resistance of a right-circular cylinder of length L, radius a, uniform cross-section, and uniform resistivity, where the terminals of a battery are connected across the circular ends of the cylinder to create an approximately uniform electric field along the length of the cylinder.

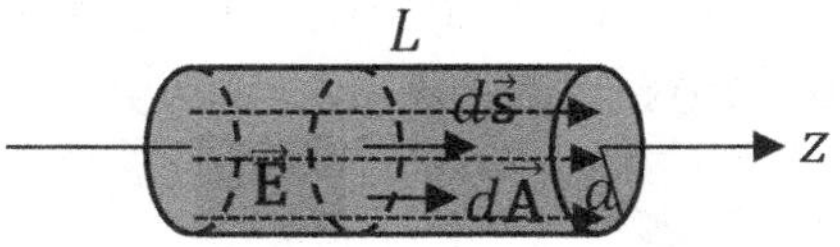

It is convenient to combine Steps 2 and 3 of the strategy for deriving an equation for resistance (see page 236) into a ratio.

$$R = \frac{|\Delta V|}{I} = \frac{\left|\int \vec{\mathbf{E}} \cdot d\vec{\mathbf{s}}\right|}{\int \vec{\mathbf{J}} \cdot d\vec{\mathbf{A}}}$$

The calculus of this example will be trivial because the electric field ($\vec{\mathbf{E}}$) running along the length of the cylinder is uniform. Since $\vec{\mathbf{E}}$ is uniform (constant), we may pull it out of the integrals. (In this example, unlike the following example and most problems that involve deriving an equation for resistance, we won't need to integrate or use Gauss's law to find $\vec{\mathbf{E}}$. The only reason is that in this problem, $\vec{\mathbf{E}}$ happens to be constant.) In this example, we may simply write $\vec{\mathbf{E}} = E\hat{\mathbf{z}}$ since the electric field is along the $+z$-axis. $\vec{\mathbf{J}}$ is along the $+z$-axis since the current density is $\vec{\mathbf{J}} = \sigma\vec{\mathbf{E}}$. Note that $d\vec{\mathbf{s}}$ and $d\vec{\mathbf{A}}$ are also along the $+z$-axis, such that $\vec{\mathbf{E}} \cdot d\vec{\mathbf{s}} = E \cos 0° \, ds = Eds$ and $\vec{\mathbf{J}} \cdot d\vec{\mathbf{A}} = J \cos 0° \, dA = JdA$. (In most problems, you can't pull E out of the integrals, but since the electric field is uniform in this example, we may now.)

$$R = \frac{\int E\,ds}{\int J\,dA} = \frac{\int E\,ds}{\int \sigma E\,dA} = \frac{E\int ds}{\sigma E \int dA} = \frac{EL}{\sigma EA} = \frac{L}{\sigma A} = \frac{\rho L}{A}$$

The integral over ds corresponds to potential difference: Since the battery is connected to the ends of the cylinder, the integral over ds is along the length (L) of the cylinder. In the last step, we used the equation $\sigma = \frac{1}{\rho}$, which can also be written $\rho = \frac{1}{\sigma}$ (the conductivity and resistivity have a reciprocal relationship, as each is the reciprocal of the other).

Example: A cylindrical resistor consists of two very long thin coaxial cylindrical conducting shells: The inner cylindrical shell has radius a, while the outer shell has radius b. The region between the two shells is filled with a semiconductor (such as silicon). Derive an equation for the resistance of this cylindrical resistor when one terminal of a battery is connected to the inner conductor while the other terminal is connected to the outer conductor, creating electric field lines that radiate outward from the axis of the cylinders.

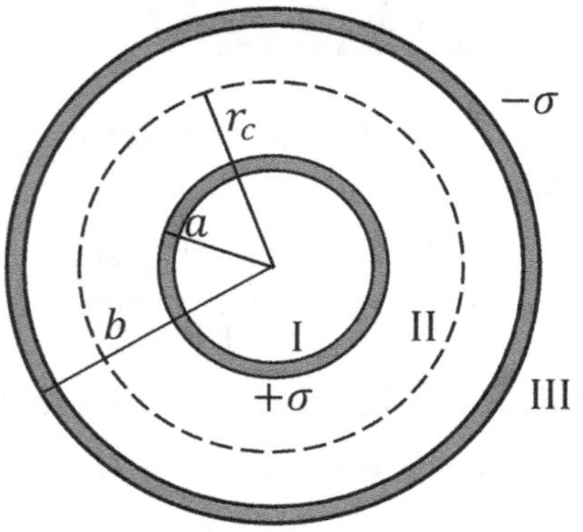

In the diagram above, we are looking at a cross section of the coaxial cylinders. The first step is to apply Gauss's law to region II ($a < r_c < b$). In Chapter 8, we learned that the outer conducting shell doesn't matter in region II (since the Gaussian cylinder drawn for region II won't enclose any charge from the outer cylinder). Hence, we just need to find the electric field in region II for a thin conducting cylindrical shell. We will get the same result as in an example from Chapter 8 that featured a long conducting cylinder (see page 138).

$$\vec{\mathbf{E}} = \frac{\sigma_{cd} a}{\epsilon_0 r_c} \hat{\mathbf{r}}_c$$

Note the symbol σ_{cd}, where the cd stands for "charge density." Unfortunately, we use the same symbol (σ) for conductivity and for surface charge density. So to help avoid confusion, we will add subscripts to the symbol for surface charge density, σ_{cd}, to distinguish it from the symbol for conductivity, σ.

$\vec{\mathbf{E}}$, $\vec{\mathbf{J}}$, $d\vec{\mathbf{s}}$, and $d\vec{\mathbf{A}}$ all radiate outward along $\hat{\mathbf{r}}_c$. (It's common for these four vectors to be parallel in problems where you're deriving an equation for resistance.) When we integrate over $\vec{\mathbf{E}} \cdot d\vec{\mathbf{s}}$ to get ΔV, we integrate outward along dr_c from $r_c = a$ to $r_c = b$. When we integrate over $\vec{\mathbf{J}} \cdot d\vec{\mathbf{A}}$ to get I, the area is over the surface area of a cylinder of radius r_c, where $dA = r_c d\theta dz$ (note that θ and z vary with r_c fixed to make a cylinder). We combine Steps 2 and 3 of the strategy for deriving an equation for resistance (page 236) into a ratio.

$$R = \frac{|\Delta V|}{I} = \frac{\left|\int \vec{\boldsymbol{E}} \cdot d\vec{\mathbf{s}}\right|}{\int \vec{\mathbf{J}} \cdot d\vec{\mathbf{A}}} = \frac{\int E\, ds}{\int J\, dA} = \frac{\int E\, ds}{\int \sigma E\, dA} = \frac{\int E\, ds}{\sigma \int E\, dA}$$

$$R = \frac{\int_{r_c=a}^{r_c=b} \frac{\sigma_{cd} a}{\epsilon_0 r_c} dr_c}{\sigma \int_{z=0}^{L} \int_{\theta=0}^{2\pi} \frac{\sigma_{cd} a}{\epsilon_0 r_c} r_c \, d\theta dz} = \frac{\frac{\sigma_{cd} a}{\epsilon_0} \int_{r_c=a}^{r_c=b} \frac{dr_c}{r_c}}{\sigma \frac{\sigma_{cd} a}{\epsilon_0} \int_{z=0}^{L} \int_{\theta=0}^{2\pi} d\theta \, dz}$$

In the denominator, note that the r_c from $dA = r_c d\theta dz$ cancels the r_c from $E = \frac{\sigma_{cd} a}{\epsilon_0 r_c}$. Also, the constants $\frac{\sigma_{cd} a}{\epsilon_0}$ from the numerator and denominator all cancel.

$$R = \frac{\int_{r_c=a}^{r_c=b} \frac{dr_c}{r_c}}{\sigma \int_{z=0}^{L} \int_{\theta=0}^{2\pi} d\theta \, dz} = \frac{[\ln(r_c)]_{r_c=a}^{b}}{2\pi\sigma L}$$

The anti-derivative for $\int \frac{dx}{x}$ is a natural logarithm (see Chapter 17).

$$R = \frac{\ln(b) - \ln(a)}{2\pi\sigma L}$$

The conductivity can be written in terms of the resistivity according to $\sigma = \frac{1}{\rho}$. Substitute this into the previous equation. Note that $\frac{1}{1/\rho} = \rho$.

$$R = \frac{\rho}{2\pi L}[\ln(b) - \ln(a)]$$

Apply the rule for logarithms that $\ln\left(\frac{b}{a}\right) = \ln(b) - \ln(a)$. (See Chapter 17.)

$$R = \frac{\rho}{2\pi L} \ln\left(\frac{b}{a}\right)$$

65. A long, straight wire has the shape of a right-circular cylinder. The length of the wire is π m and the thickness of the wire is 0.80 mm. The resistance of the wire is 5.0 Ω. What is the resistivity of the wire?

66. A resistor is made from a material that has a temperature coefficient of resistivity of 2.0×10^{-3} /°C. Its resistance is 30 Ω at 20 °C. At what temperature is its resistance equal to 33 Ω?

Want help? Check the hints section at the back of the book.

Answers: 8.0×10^{-7} Ω·m, 70 °C

67. A battery has an emf of 36 V. When the battery is connected across an 8.0-Ω resistor, a voltmeter measures the potential difference across the resistor to be 32 V.

(A) What is the internal resistance of the battery?

(B) How much power is dissipated in the 8.0-Ω resistor?

Want help? Check the hints section at the back of the book.

Answers: 1.0 Ω, 128 W

68. A spherical resistor consists of two thin spherical conducting shells: The inner shell has radius a while the outer shell has radius b. There is a semiconductor between the two spheres. Derive an equation for the resistance of this spherical resistor.

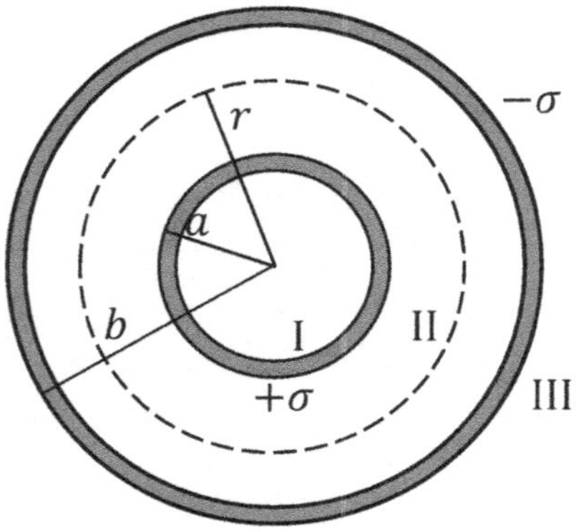

Want help? Check the hints section at the back of the book.

Answer: $R = \frac{\rho(b-a)}{4\pi ab}$

17 LOGARITHMS AND EXPONENTIALS

Relevant Terminology

In the expression x^n, the quantity n can be referred to either as the **power** or as the **exponent** of the quantity x, while the quantity x is called the **base**.

Essential Concepts

Consider the expression $y = b^x$. Solving for y is easy. For example, when $y = 4^3$, this means to multiply 4 by itself 3 times: $y = 4^3 = (4)(4)(4) = 64$. As another example, you could calculate $y = 9^{1.5}$ on your calculator by entering 9^1.5. The answer is 27.

What if you want to solve for x in the equation $y = b^x$? Some cases you can reason out. For example, when $81 = 3^x$, you might figure out that $x = 4$ since $3^4 = 81$. You're asking yourself, "What power do you need to raise 3 to in order to make 81?" But what if the problem is $40 = 10^x$? What would you enter on your calculator to figure out x?

The answer has to do with logarithms. A **logarithm** is a function that helps you solve for the exponent in an equation like $y = b^x$.

The logarithm $\log_b y = x$ is exactly the same equation as $y = b^x$, except that the logarithm gives you the answer for the exponent (x). Following are a few examples.

- $\log_{10} 1000 = x$ is the same thing as $1000 = 10^x$. Read it as, "Log base 10 of 1000." It means the same thing as, "10 to the power of what makes 1000?" The answer is $x = 3$ because $10^3 = (10)(10)(10) = 1000$.
- The answer to $40 = 10^x$ can be found by writing the logarithm $\log_{10} 40 = x$. This says that x is the log base 10 of 40, meaning 10 to what power makes 40? You can't figure this one out in your head. If you have a calculator that has both the "log" function and the "ln" function, the "log" button is usually for base 10. In that case, enter log(40) on your calculator, as it means the same thing as $\log_{10} 40$. Try it. The correct answer is 1.602 (and the digits go on indefinitely). You can check that this is the correct answer by entering $10^{1.602}$ on your calculator. You'll get 39.99, which is very close to 40 (the slight difference comes from rounding to 1.602).
- $\log_2 32$ means, "What power of 2 makes 32?" It's the same problem as $2^x = 32$, where $x = \log_2 32$. The answer is $\log_2 32 = 5$ because $2^5 = 2 \times 2 \times 2 \times 2 \times 2 = 32$. This problem isn't as easy to do on your calculator,* since most calculators don't have a built-in log-base-2 button (it's usually log for base 10 and ln for base e). However, it is easy to check that $2^5 = 32$.

* You could do it with the change of base formula: $\log_2 32 = \log_{10} 32 / \log_{10} 2$. Try it. You should get 5.

Euler's number (e) comes up frequently in physics. Numerically, Euler's number is:

$$e = 2.718281828 \ldots$$

The ... represents that the digits continue forever without repeating (similar to how the digits of π continue forever without repeating).

The function e^x is called an **exponential**. When we write e^x, it's approximately the same thing as writing 2.718^x (just like 3.14 is approximately the same number as π).

A logarithm of base e is called a natural log. We use ln to represent a natural log. For example, $\ln x$ means the natural log of x, which also means $\log_e x$ (log base e of x).

Logarithm and Exponential Equations

Following are some handy logarithm and exponential identities:

$$\ln(xy) = \ln(x) + \ln(y)$$

$$\ln\left(\frac{x}{y}\right) = \ln(x) - \ln(y)$$

$$\ln\left(\frac{1}{x}\right) = \ln(x^{-1}) = -\ln(x)$$

$$\ln(x^a) = a\ln(x)$$

$$\ln(1) = 0$$

$$e^{x+y} = e^x e^y$$

$$e^{x-y} = e^x e^{-y}$$

$$e^{-x} = \frac{1}{e^x}$$

$$(e^x)^a = e^{ax}$$

$$e^0 = 1 \quad , \quad \ln(e) = 1$$

$$\ln(e^x) = x \quad , \quad e^{\ln(x)} = x$$

The change of base formula is handy when you need to take a logarithm of a base other than 10 or e on a calculator. This formula lets you use log base 10 instead of log base b.

$$\log_b y = \frac{\log_{10} y}{\log_{10} b}$$

Calculus with Logarithms and Exponentials

We will encounter the following derivatives and anti-derivatives in this book. These come up frequently in physics, and so are worth memorizing.

$$\frac{de^x}{dx} = e^x \quad , \quad \frac{de^{ax}}{dx} = ae^{ax} \quad , \quad \int e^x\,dx = e^x \quad , \quad \int e^{ax}\,dx = \frac{e^{ax}}{a}$$

$$\frac{d}{dx}\ln(x) = \frac{1}{x} \quad , \quad \int \frac{dx}{x} = \int x^{-1}\,dx = \ln(x)$$

Graphs of Exponentials and Logarithms

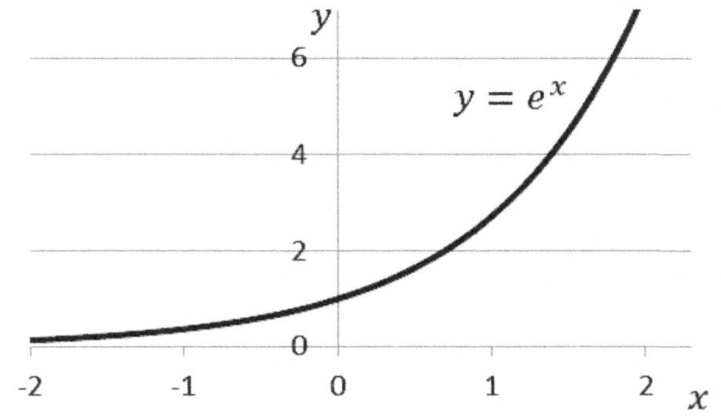

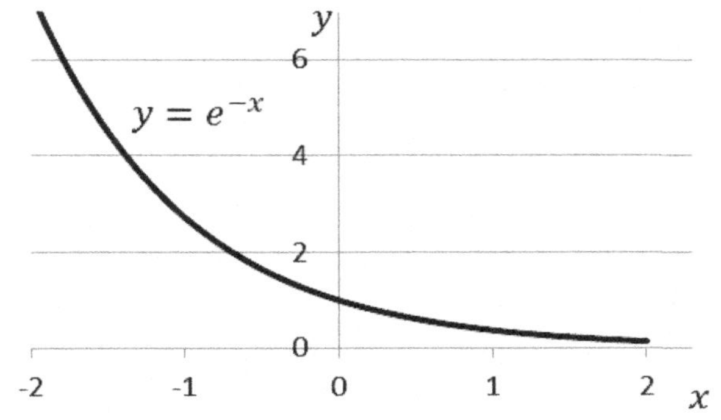

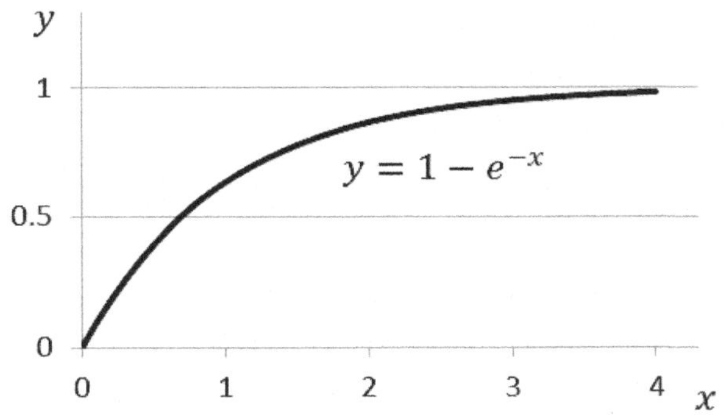

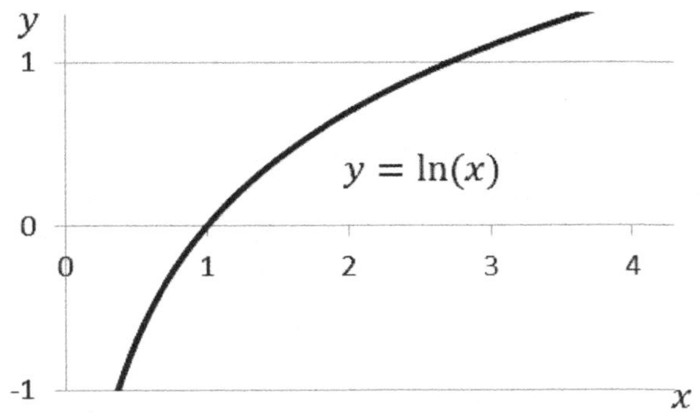

Example: Simplify $\ln\left(\frac{1}{e}\right)$.
Apply the rule $\ln\left(\frac{x}{y}\right) = \ln(x) - \ln(y)$.

$$\ln\left(\frac{1}{e}\right) = \ln(1) - \ln(e) = 0 - 1 = -1$$

Note that $\ln(1) = 0$ and $\ln(e) = 1$. These values are worth memorizing.

Example: What is $\ln(2) + \ln(0.5)$?
Apply the rule $\ln(xy) = \ln(x) + \ln(y)$.

$$\ln(2) + \ln(0.5) = \ln[(2)(0.5)] = \ln(1) = 0$$

Once again we need to know that $\ln(1) = 0$.

Example: Evaluate $\log_4 64$.
This means: "4 raised to what power makes 64?"
The answer is 3 because $4^3 = 64$.

Example: Simplify $2\ln(\sqrt{3})$.
Apply the rule $\ln(x^a) = a\ln(x)$.

$$2\ln(\sqrt{3}) = \ln\left[(\sqrt{3})^2\right] = \ln(3)$$

Example: Simplify $\ln(8e) - \ln(4e)$.
Apply the rule $\ln\left(\frac{x}{y}\right) = \ln(x) - \ln(y)$.

$$\ln(8e) - \ln(4e) = \ln\left(\frac{8e}{4e}\right) = \ln\left(\frac{8}{4}\right) = \ln(2)$$

Example: Solve for x in the equation $e^x = 2$.
Take the natural logarithm of both sides of the equation.

$$\ln(e^x) = \ln(2)$$

Apply the rule $\ln(e^x) = x$.

$$x = \ln(2)$$

Example: Solve for x in the equation $\ln(x) = 3$.
Exponentiate both sides of the equation.

$$e^{\ln(x)} = e^3$$

Apply the rule $e^{\ln(x)} = x$.

$$x = e^3$$

Example: Evaluate $\frac{d}{dx}(3e^{2x})$ at $x = 0.5$.

Take the derivative with respect to x before plugging in the numerical value. Factor out the coefficient (3). Apply the rule $\frac{de^{ax}}{dx} = ae^{ax}$ with $a = 2$.

$$\frac{d}{dx}(3e^{2x}) = 3\frac{d}{dx}(e^{2x}) = 3(2)e^{2x} = 6e^{2x}$$

Now evaluate the derivative at $x = 0.5$.

$$\frac{d}{dx}(3e^{2x})\bigg|_{x=0.5} = 6e^{2x}|_{x=0.5} = 6e^{2(0.5)} = 6e^1 = 6e$$

Note that $e^1 = e$.

Example: Prove that the positive constant a has no effect on the result of $\frac{d}{dx}\ln(ax)$.

There are a couple of ways to do this. One way is to apply the rule $\ln(xy) = \ln(x) + \ln(y)$.

$$\frac{d}{dx}\ln(ax) = \frac{d}{dx}[\ln(a) + \ln(x)] = \frac{d}{dx}\ln(a) + \frac{d}{dx}\ln(x) = 0 + \frac{d}{dx}\ln(x) = \frac{1}{x}$$

Note that $\frac{d}{dx}\ln(a) = 0$ because the derivative of any constant with respect to x equals zero, and note that $\frac{d}{dx}\ln(x) = \frac{1}{x}$. The left-hand side of the above equation is $\frac{d}{dx}\ln(ax)$ and the right-hand side is $\frac{1}{x}$, meaning that $\frac{d}{dx}\ln(ax) = \frac{1}{x}$. The answer, $\frac{1}{x}$, is independent of a.

This property of logarithms doesn't apply to most functions. For example, $\frac{d}{dx}\sin(ax) = a\cos(ax)$. The constant a very much matters in the derivative of the sine function.

An alternative way to show that $\frac{d}{dx}\ln(ax) = \frac{1}{x}$ is to make the substitution $u = ax$ and apply the chain rule:

$$\frac{d}{dx}\ln(ax) = \frac{d}{dx}\ln(u) = \frac{d}{du}\frac{du}{dx}\ln(u) = \frac{du}{dx}\frac{d}{du}\ln(u) = \left(\frac{d}{dx}ax\right)\left(\frac{d}{du}\ln u\right)$$

$$\frac{d}{dx}\ln(ax) = (a)\left(\frac{1}{u}\right) = \frac{a}{u} = \frac{a}{ax} = \frac{1}{x}$$

The constant a cancels out.

Example: Perform the following definite integral.

$$\int_{x=0}^{\infty} e^{-x}\,dx = [-e^{-x}]_{x=0}^{\infty} = \left[-\frac{1}{e^x}\right]_{x=0}^{\infty} = -\lim_{x\to\infty}\left(\frac{1}{e^x}\right) + \frac{1}{e^0} = 0 + \frac{1}{1} = 1$$

Note that $e^{-x} = \frac{1}{e^x}$ and $e^0 = 1$. Also note that $\frac{1}{e^x}$ approaches zero as x gets larger and larger (approaching infinity). That is, 1 divided by a very large number is very tiny, such that 1 divided by infinity equals zero.

69. Evaluate $\log_5 625$.

70. Solve for x in the equation $6e^{-x/2} = 2$.

71. Perform the following definite integral.

$$\int_{x=0}^{2} \left(1 - e^{-x/2}\right) dx =$$

72. Perform the following definite integral.

$$\int_{x=1}^{2} \frac{dx}{x} =$$

Want help? Check the hints section at the back of the book.

Answers: 4, $2\ln(3)$, $2/e$, $\ln(2)$

18 RC CIRCUITS

Relevant Terminology

Half-life – the time it takes to decay to one-half of the initial value, or to grow to one-half of the final value.

Time constant – the time it takes to decay to $\frac{1}{e}$ or $\approx 37\%$ of the initial value, or to grow to $\left(1 - \frac{1}{e}\right)$ or $\approx 63\%$ of the final value.

Capacitor – a device that can store charge, which consists of two separated conductors (such as two parallel conducting plates).

Capacitance – a measure of how much charge a capacitor can store for a given voltage.

Charge – the amount of electric charge stored on the positive plate of a capacitor.

Potential difference – the electric work per unit charge needed to move a test charge between two points in a circuit. Potential difference is also called the **voltage**.

Electric potential energy – a measure of how much electrical work a component of an electric circuit can do.

Resistance – a measure of how well a component in a circuit resists the flow of current.

Resistor – a component in a circuit which has a significant amount of resistance.

Current – the instantaneous rate of flow of charge through a wire.

Electric power – the instantaneous rate at which electrical work is done.

Schematic Symbols Used in RC Circuits

Schematic Representation	Symbol	Name
	R	resistor
	ΔV	battery or DC power supply
	C	capacitor

Discharging a Capacitor in an RC Circuit

In a simple RC circuit like the one shown below, a capacitor with initial charge Q_m (the subscript m stands for "maximum") loses its charge with exponential decay. The current that transfers the charge between the two plates also decays exponentially from an initial value of I_m.

$$Q = Q_m e^{-t/\tau} \quad , \quad I = I_m e^{-t/\tau}$$

The **time constant** (τ) equals the product of the resistance (R) and capacitance (C).

$$\tau = RC$$

The **half-life** ($t_{1/2}$) is related to the time constant (τ) by:

$$t_{1/2} = \tau \ln(2)$$

The resistor and capacitor equations still apply:

$$Q = C\Delta V \quad , \quad \Delta V = IR \quad , \quad Q_m = C\Delta V_m \quad , \quad \Delta V_m = I_m R$$

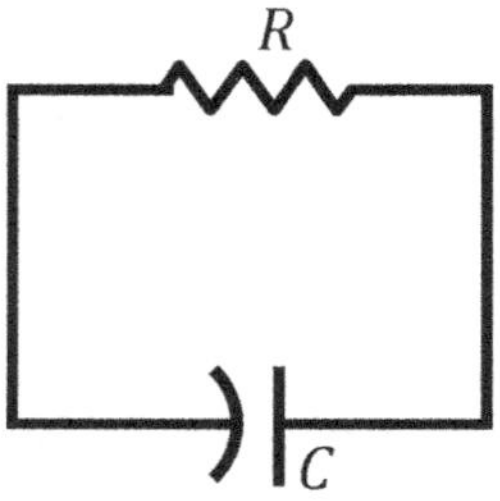

Charging a Capacitor in an RC Circuit

When a battery is connected to a resistor and capacitor in series as shown below, an initially uncharged capacitor has the charge on its plates grow according to $1 - e^{-t/\tau}$. The current that transfers the charge between the two plates instead decays exponentially from an initial value of I_m.

$$Q = Q_m(1 - e^{-t/\tau}) \quad , \quad I = I_m e^{-t/\tau}$$

The **time constant** (τ) and **half-life** ($t_{1/2}$) obey the same equations:

$$\tau = RC \quad , \quad t_{1/2} = \tau \ln(2)$$

The resistor and capacitor equations still apply. In the case of charging, the symbol ΔV_{batt} represents the potential difference supplied by the battery, while ΔV_C and ΔV_R represent the potential differences across the capacitor and resistor, respectively.

$$Q = C\Delta V_C \quad , \quad \Delta V_R = IR \quad , \quad Q_m = C\Delta V_{batt} \quad , \quad \Delta V_{batt} = I_m R$$

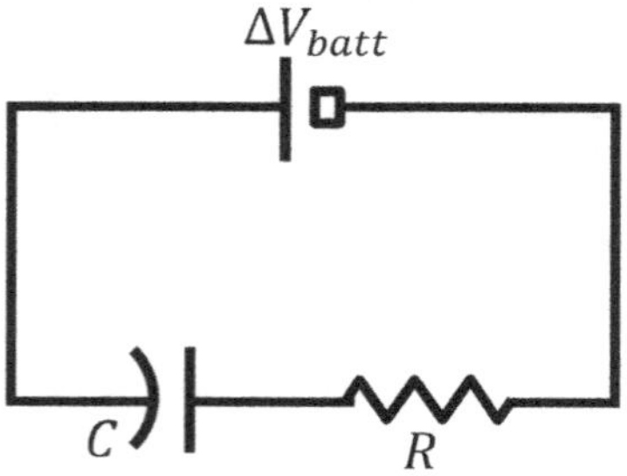

Where Do The RC Circuit Equations Come From?

The equations for discharging and charging capacitors in RC circuits come from Kirchhoff's rules (Chapter 15). Consider the case of a **discharging** capacitor. If you apply Kirchhoff's loop rule to the corresponding circuit on the previous page, you will find that the potential difference across the capacitor and resistor must be equal.

$$|\Delta V_C| = |\Delta V_R|$$

Use the equations $Q = C\Delta V_C$ and $\Delta V_R = IR$.

$$\frac{Q}{C} = IR$$

Divide both sides of the equation by R.

$$\frac{Q}{RC} = I$$

Current (I) is the instantaneous rate of flow of charge (Q), meaning that current is a derivative of charge with respect to time: $I = -\frac{dQ}{dt}$. It's negative for a discharging capacitor because charge is a **decreasing** function of time (so its slope, or derivative, is negative).

$$\frac{Q}{RC} = -\frac{dQ}{dt}$$

This simple differential equation is separable, meaning that we can put one variable (time) on one side and the other variable (charge) on the other side via algebra.

$$-\frac{dt}{RC} = \frac{dQ}{Q}$$

Now we integrate both sides of the equation, pulling out the constants. As time grows from 0 to t, the charge on the plates diminishes from its initial maximum value Q_m to Q.

$$-\frac{1}{RC}\int_{t=0}^{t} dt = \int_{Q=Q_m}^{Q} \frac{dQ}{Q}$$

In Chapter 17, we learned that $\int \frac{dx}{x} = \ln(x)$ and $\ln\left(\frac{x}{y}\right) = \ln(x) - \ln(y)$.

$$-\frac{t}{RC} = [\ln(Q)]_{Q=Q_m}^{Q} = \ln(Q) - \ln(Q_m) = \ln\left(\frac{Q}{Q_m}\right)$$

Exponentiate both sides of the equation. Recall that $e^{\ln(x)} = x$.

$$\frac{Q}{Q_m} = e^{-t/RC}$$

Multiply both sides of the equation by Q_m. Recall that $\tau = RC$.

$$Q = Q_m e^{-t/RC} = Q_m e^{-t/\tau}$$

To find the current, take a derivative of charge with respect to time. Recall the minus sign.

$$I = -\frac{dQ}{dt} = \frac{Q_m}{RC}e^{-t/RC} = \frac{\Delta V_m}{R}e^{-t/RC}$$

$$I = I_m e^{-t/RC} = I_m e^{-t/\tau}$$

The equations for a charging capacitor can be found via a similar method.

Symbols and SI Units

Symbol	Name	SI Units
t	time	s
$t_{1/2}$	half-life	s
τ	time constant	s
C	capacitance	F
Q	the charge stored on the positive plate of a capacitor	C
ΔV	the potential difference between two points in a circuit	V
U	energy stored	J
R	resistance	Ω
I	electric current	A
P	electric power	W

Note: The symbol τ is the lowercase Greek letter tau.

Note Regarding Units

From the equation $\tau = RC$, the SI units for time constant (τ) must equal an Ohm (Ω) times a Farad (F) – the SI units of resistance (R) and capacitance (C). From the equation $C = \frac{Q}{\Delta V}$, we can write a Farad (F) as $\frac{\text{C}}{\text{V}}$ since charge (Q) is measured in Coulombs (C). Similarly, from $R = \frac{\Delta V}{I}$, we can write an Ohm (Ω) as $\frac{\text{V}}{\text{A}}$ since current (I) is measured in Ampères (A). Thus, the units of τ are $\frac{\text{V}}{\text{A}} \cdot \frac{\text{C}}{\text{V}} = \frac{\text{C}}{\text{A}}$. Current is the rate of flow of charge: $I = \frac{dQ}{dt}$. Thus, an Ampère equals a Coulomb per second: $1 \text{ A} = 1\frac{\text{C}}{\text{s}}$, which can be written $\frac{\text{C}}{\text{A}} = \text{s}$ with a little algebra. Therefore, the SI unit of the time constant (τ) is the second (s).

Important Distinctions

Note the three different times involved in RC circuits:

- t (without a subscript) represents the elapsed **time**.
- $t_{1/2}$ is the **half-life**. It's the time it takes to reach one-half the maximum value.
- τ is the **time constant**: $\tau = RC$. Note that $t_{1/2} = \tau \ln(2)$.

Strategy for RC Circuits

Apply the equations that relate to RC circuits:

- For a **discharging** capacitor:
$$Q = Q_m e^{-t/\tau} \quad , \quad I = I_m e^{-t/\tau}$$
$$Q_m = C\Delta V_m \quad , \quad \Delta V_m = I_m R$$
- For a **charging** capacitor:
$$Q = Q_m(1 - e^{-t/\tau}) \quad , \quad I = I_m e^{-t/\tau}$$
$$Q_m = C\Delta V_{batt} \quad , \quad \Delta V_{batt} = I_m R$$
- In either case, the **time constant** is:
$$\tau = RC$$
- You can find the **half-life** from the time constant:
$$t_{1/2} = \tau \ln(2)$$
- The equations for capacitance and resistance apply:
$$Q = C\Delta V_C \quad , \quad \Delta V_R = IR$$
- To find energy, see Chapter 11. To find power, see Chapter 13.
- If necessary, apply rules from Chapter 17 regarding logarithms and exponentials.

Example: A 3.0-μF capacitor with an initial charge of 24 μC discharges while connected in series with a 4.0-Ω resistor.

(A) What is the initial potential difference across the capacitor?

Use the equation for capacitance: $Q_m = C\Delta V_m$. Divide both sides of the equation by C.

$$\Delta V_m = \frac{Q_m}{C} = \frac{24 \times 10^{-6}}{3 \times 10^{-6}} = 8.0 \text{ V}$$

Recall that the metric prefix micro (μ) stands for 10^{-6}. It cancels out here. The initial potential difference across the capacitor is $\Delta V_m = 8.0$ V.

(B) What is the initial current?

The current begins at its maximum value. Apply Ohm's law: $\Delta V_m = I_m R$. Divide by R.

$$I_m = \frac{\Delta V_m}{R} = \frac{8}{4} = 2.0 \text{ A}$$

The initial current is $I_m = 2.0$ A.

(C) What is the time constant for the circuit?

Use the equation for time constant involving resistance and capacitance.

$$\tau = RC = (4)(3 \times 10^{-6}) = 12 \times 10^{-6} \text{ s} = 12 \text{ μs}$$

The time constant is $\tau = 12$ μs, which can be expressed as 12×10^{-6} s or 1.2×10^{-5} s.

(D) How long will it take for the capacitor to have one-half of its initial charge?

Solve for the half-life.

$$t_{1/2} = \tau \ln(2) = 12 \ln(2) \text{ μs}$$

The half-life is $t_{1/2} = 12 \ln(2)$ μs. If you use a calculator, it's $t_{1/2} = 8.3$ μs.

Example: Prove that $t_{1/2} = \tau \ln(2)$ for a discharging capacitor.
Start with the equation for charge.

$$Q = Q_m e^{-t/\tau}$$

The above equation is true for any time t. When the time happens to equal one-half life, the charge will equal $\frac{Q_m}{2}$, one-half of its maximum value. (That's what the definition of **half-life** states.) Plug in $\frac{Q_m}{2}$ for Q and $t_{1/2}$ for t.

$$\frac{Q_m}{2} = Q_m e^{-t_{1/2}/\tau}$$

Divide both sides of the equation by Q_m. It will cancel.

$$\frac{1}{2} = e^{-t_{1/2}/\tau}$$

Take the natural log of both sides of the equation.

$$\ln\left(\frac{1}{2}\right) = \ln\left(e^{-t_{1/2}/\tau}\right)$$

Recall from Chapter 17 that $\ln(e^x) = x$.

$$\ln\left(\frac{1}{2}\right) = -t_{1/2}/\tau$$

Recall from Chapter 17 that $\ln\left(\frac{1}{x}\right) = -\ln(x)$.

$$-\ln(2) = -t_{1/2}/\tau$$

Multiply both sides of the equation by $-\tau$

$$\tau \ln(2) = t_{1/2}$$

Example: A capacitor discharges while connected in series with a 50-kΩ resistor. The current drops from its initial value of 4.0 A down to 2.0 A after 500 ms. What is the capacitance of the capacitor?

The "trick" to this problem is to realize that 500 ms is the half-life because the current has dropped to one-half of its initial value. Convert ms to s and convert kΩ to Ω.

$$t_{1/2} = 500 \text{ ms} = 0.500 \text{ s} = \frac{1}{2} \text{ s}$$

$$R = 50 \text{ k}\Omega = 50{,}000\ \Omega = 5.0 \times 10^4\ \Omega$$

Find the time constant from the half-life. Divide by $\ln(2)$ in the equation $\tau \ln(2) = t_{1/2}$.

$$\tau = \frac{t_{1/2}}{\ln(2)} = \frac{1}{2\ln(2)}$$

Now use the other equation for time constant: $\tau = RC$. Divide both sides by R.

$$C = \frac{\tau}{R} = \left[\frac{1}{2\ln(2)}\right]\left(\frac{1}{5.0 \times 10^4}\right) = \frac{1}{10 \times 10^4 \ln(2)} = \frac{1}{10^5 \ln(2)} = \frac{10^{-5}}{\ln(2)} \text{ F}$$

The capacitance is $C = \frac{10^{-5}}{\ln(2)}$ F. If you use a calculator, it is $C = 14$ μF, which is the same as 14×10^{-6} F or 1.4×10^{-5} F.

73. A 5.0-μF capacitor with an initial charge of 60 μC discharges while connected in series with a 20-kΩ resistor.

(A) What is the initial potential difference across the capacitor?

(B) What is the initial current?

(C) What is the half-life?

(D) How much charge is stored on the capacitor after 0.20 s?

Want help? Check the hints section at the back of the book.

Answers: 12 V, 0.60 mA, $\frac{\ln(2)}{10}$ s (or 0.069 s), $\frac{60}{e^2}$ μC (or 8.1 μC)

74. A 4.0-μF capacitor discharges while connected in series with a resistor. The current drops from its initial value of 6.0 A down to 3.0 A after 200 ms.

(A) What is the time constant?

(B) What is the resistance of the resistor?

Want help? Check the hints section at the back of the book.

Answers: $\frac{1}{5\ln(2)}$ s (or 0.29 s), $\frac{50}{\ln(2)}$ kΩ (or 72 kΩ)

19 SCALAR AND VECTOR PRODUCTS

Note: The first volume of this series covered the scalar product and vector product. This chapter provides a quick review.

Vector Equations

Any vector can be expressed in terms of its components (A_x, A_y, and A_z) and Cartesian unit vectors ($\hat{\mathbf{x}}$, $\hat{\mathbf{y}}$, and $\hat{\mathbf{z}}$) in the following form:

$$\vec{\mathbf{A}} = A_x\hat{\mathbf{x}} + A_y\hat{\mathbf{y}} + A_z\hat{\mathbf{z}}$$

There are two equivalent ways of expressing the **scalar** product (or **dot** product):

- Use the following equation to compute the scalar product when you know the **components** of the given vectors.
$$\vec{\mathbf{A}} \cdot \vec{\mathbf{B}} = A_x B_x + A_y B_y + A_z B_z$$
- Use the following equation to compute the scalar product when you know the magnitudes of $\vec{\mathbf{A}}$ and $\vec{\mathbf{B}}$ along with the angle between the two vectors.
$$\vec{\mathbf{A}} \cdot \vec{\mathbf{B}} = AB\cos\theta$$
- You can use both equations together to **find the angle** between two vectors, as shown in one of the examples that follow.

The **vector** product (or **cross** product) $\vec{\mathbf{A}} \times \vec{\mathbf{B}}$ is defined according to a 3×3 determinant:

$$\vec{\mathbf{A}} \times \vec{\mathbf{B}} = \begin{vmatrix} \hat{\mathbf{x}} & \hat{\mathbf{y}} & \hat{\mathbf{z}} \\ A_x & A_y & A_z \\ B_x & B_y & B_z \end{vmatrix} = \hat{\mathbf{x}}\begin{vmatrix} A_y & A_z \\ B_y & B_z \end{vmatrix} - \hat{\mathbf{y}}\begin{vmatrix} A_x & A_z \\ B_x & B_z \end{vmatrix} + \hat{\mathbf{z}}\begin{vmatrix} A_x & A_y \\ B_x & B_y \end{vmatrix}$$

$$\vec{\mathbf{A}} \times \vec{\mathbf{B}} = \left(A_y B_z - A_z B_y\right)\hat{\mathbf{x}} - \left(A_x B_z - A_z B_x\right)\hat{\mathbf{y}} + \left(A_x B_y - A_y B_x\right)\hat{\mathbf{z}}$$

$$\vec{\mathbf{A}} \times \vec{\mathbf{B}} = A_y B_z\hat{\mathbf{x}} - A_z B_y\hat{\mathbf{x}} + A_z B_x\hat{\mathbf{y}} - A_x B_z\hat{\mathbf{y}} + A_x B_y\hat{\mathbf{z}} - A_y B_x\hat{\mathbf{z}}$$

The magnitude of the vector product, $\|\vec{\mathbf{A}} \times \vec{\mathbf{B}}\|$, has a form similar to the scalar product, except that it involves a sine instead of a cosine:

$$\|\vec{\mathbf{A}} \times \vec{\mathbf{B}}\| = AB\sin\theta$$

The angle θ is the angle between the two vectors $\vec{\mathbf{A}}$ and $\vec{\mathbf{B}}$.

Following are the scalar and vector products between Cartesian unit vectors.

$$\hat{\mathbf{x}} \cdot \hat{\mathbf{x}} = 1 \quad , \quad \hat{\mathbf{y}} \cdot \hat{\mathbf{y}} = 1 \quad , \quad \hat{\mathbf{z}} \cdot \hat{\mathbf{z}} = 1$$
$$\hat{\mathbf{x}} \cdot \hat{\mathbf{y}} = 0 \quad , \quad \hat{\mathbf{y}} \cdot \hat{\mathbf{z}} = 0 \quad , \quad \hat{\mathbf{z}} \cdot \hat{\mathbf{x}} = 0$$
$$\hat{\mathbf{y}} \cdot \hat{\mathbf{x}} = 0 \quad , \quad \hat{\mathbf{z}} \cdot \hat{\mathbf{y}} = 0 \quad , \quad \hat{\mathbf{x}} \cdot \hat{\mathbf{z}} = 0$$

$$\hat{\mathbf{x}} \times \hat{\mathbf{x}} = 0 \quad , \quad \hat{\mathbf{y}} \times \hat{\mathbf{y}} = 0 \quad , \quad \hat{\mathbf{z}} \times \hat{\mathbf{z}} = 0$$
$$\hat{\mathbf{x}} \times \hat{\mathbf{y}} = \hat{\mathbf{z}} \quad , \quad \hat{\mathbf{y}} \times \hat{\mathbf{z}} = \hat{\mathbf{x}} \quad , \quad \hat{\mathbf{z}} \times \hat{\mathbf{x}} = \hat{\mathbf{y}}$$
$$\hat{\mathbf{y}} \times \hat{\mathbf{x}} = -\hat{\mathbf{z}} \quad , \quad \hat{\mathbf{z}} \times \hat{\mathbf{y}} = -\hat{\mathbf{x}} \quad , \quad \hat{\mathbf{x}} \times \hat{\mathbf{z}} = -\hat{\mathbf{y}}$$

Notation

The cross ($\times$) is used to designate the vector (or cross) product, as in $\vec{\mathbf{A}} \times \vec{\mathbf{B}}$, which results in a vector, whereas the dot ($\cdot$) is used to designate the scalar (or dot) product, as in $\vec{\mathbf{A}} \cdot \vec{\mathbf{B}}$, which results in a scalar. When multiplying ordinary numbers, we use the cross and dot interchangeably. For example, $3 \times 2 = 3 \cdot 2 = (3)(2) = 6$. However, you may **not** swap the cross and dot in the context of vector multiplication: $\vec{\mathbf{A}} \times \vec{\mathbf{B}}$ is much different from $\vec{\mathbf{A}} \cdot \vec{\mathbf{B}}$.

Strategy for Finding or Applying the Scalar Product

To compute the scalar product between two vectors, or to determine the angle between two vectors, follow these steps:

- If you know the components of the given vectors, use $\vec{\mathbf{A}} \cdot \vec{\mathbf{B}} = A_x B_x + A_y B_y + A_z B_z$.
- If you know the magnitudes of the given vectors along with the angle between the two vectors, use $\vec{\mathbf{A}} \cdot \vec{\mathbf{B}} = AB \cos\theta$.
- To find the angle between two vectors, first compute the scalar product as in Step 1 and then solve for θ using the equation from Step 2. Find the magnitudes of the given vectors using the Pythagorean theorem. For example, $A = \sqrt{A_x^2 + A_y^2 + A_z^2}$.

Strategy for Finding the Vector Product

To compute the vector product $\vec{\mathbf{C}} = \vec{\mathbf{A}} \times \vec{\mathbf{B}}$, follow these steps:

- If you know the components of the given vectors, find the following determinant.

$$\vec{\mathbf{C}} = \vec{\mathbf{A}} \times \vec{\mathbf{B}} = \begin{vmatrix} \hat{\mathbf{x}} & \hat{\mathbf{y}} & \hat{\mathbf{z}} \\ A_x & A_y & A_z \\ B_x & B_y & B_z \end{vmatrix} = \hat{\mathbf{x}} \begin{vmatrix} A_y & A_z \\ B_y & B_z \end{vmatrix} - \hat{\mathbf{y}} \begin{vmatrix} A_x & A_z \\ B_x & B_z \end{vmatrix} + \hat{\mathbf{z}} \begin{vmatrix} A_x & A_y \\ B_x & B_y \end{vmatrix}$$

  Once you find the vector product, if you would like to determine the magnitude of the vector product, you can use the three-dimensional generalization of the Pythagorean theorem:

$$C = \left\|\vec{\mathbf{A}} \times \vec{\mathbf{B}}\right\| = \sqrt{C_x^2 + C_y^2 + C_z^2}$$

  (C_x is the coefficient of $\hat{\mathbf{x}}$, C_y is the coefficient of $\hat{\mathbf{y}}$, and C_z is the coefficient of $\hat{\mathbf{z}}$ in the expression for $\vec{\mathbf{C}}$ resulting from the determinant.)
- If you know the magnitudes of the given vectors along with the angle between the two vectors, you can find the magnitude of the vector product using trig:

$$C = \left\|\vec{\mathbf{A}} \times \vec{\mathbf{B}}\right\| = AB \sin\theta$$

Example: (A) Find the scalar product between the vectors $\vec{\mathbf{A}} = 3\,\hat{\mathbf{x}} - 2\,\hat{\mathbf{y}} + 6\,\hat{\mathbf{z}}$ and $\vec{\mathbf{B}} = -6\,\hat{\mathbf{x}} + 4\,\hat{\mathbf{y}} - 12\,\hat{\mathbf{z}}$. (B) What is the angle between these two vectors?

(A) Compare $\vec{\mathbf{A}} = 3\,\hat{\mathbf{x}} - 2\,\hat{\mathbf{y}} + 6\,\hat{\mathbf{z}}$ to $\vec{\mathbf{A}} = A_x\hat{\mathbf{x}} + A_y\hat{\mathbf{y}} + A_z\hat{\mathbf{z}}$ to see that $A_x = 3$, $A_y = -2$, and $A_z = 6$. Similarly, $B_x = -6$, $B_y = 4$, and $B_z = -12$. Use the appropriate equation:

$$\vec{\mathbf{A}} \cdot \vec{\mathbf{B}} = A_xB_x + A_yB_y + A_zB_z = (3)(-6) + (-2)(4) + (6)(-12) = -18 - 8 - 72 = -98$$

(B) Find the magnitudes of $\vec{\mathbf{A}}$ and $\vec{\mathbf{B}}$ using the Pythagorean theorem.

$$A = \sqrt{A_x^2 + A_y^2 + A_z^2} = \sqrt{3^2 + (-2)^2 + 6^2} = \sqrt{9 + 4 + 36} = \sqrt{49} = 7$$

$$B = \sqrt{B_x^2 + B_y^2 + B_z^2} = \sqrt{(-6)^2 + 4^2 + (-12)^2} = \sqrt{36 + 16 + 144} = \sqrt{196} = 14$$

Now apply the other equation for the scalar product.

$$\vec{\mathbf{A}} \cdot \vec{\mathbf{B}} = AB\cos\theta$$

$$-98 = (7)(14)\cos\theta$$

$$-98 = 98\cos\theta$$

$$-1 = \cos\theta$$

$$\theta = \cos^{-1}(-1) = 180°$$

Example: Find the vector product between $\vec{\mathbf{A}} = \hat{\mathbf{x}} - 2\,\hat{\mathbf{y}} + 3\,\hat{\mathbf{z}}$ and $\vec{\mathbf{B}} = 2\,\hat{\mathbf{x}} + 3\,\hat{\mathbf{y}} - \hat{\mathbf{z}}$.

Compare $\vec{\mathbf{A}} = \hat{\mathbf{x}} - 2\,\hat{\mathbf{y}} + 3\,\hat{\mathbf{z}}$ to $\vec{\mathbf{A}} = A_x\hat{\mathbf{x}} + A_y\hat{\mathbf{y}} + A_z\hat{\mathbf{z}}$ to see that $A_x = 1$, $A_y = -2$, and $A_z = 3$. Similarly, $B_x = 2$, $B_y = 3$, and $B_z = -1$. Plug these values into the determinant form of the vector product:

$$\vec{\mathbf{A}} \times \vec{\mathbf{B}} = \begin{vmatrix} \hat{\mathbf{x}} & \hat{\mathbf{y}} & \hat{\mathbf{z}} \\ A_x & A_y & A_z \\ B_x & B_y & B_z \end{vmatrix} = \begin{vmatrix} \hat{\mathbf{x}} & \hat{\mathbf{y}} & \hat{\mathbf{z}} \\ 1 & -2 & 3 \\ 2 & 3 & -1 \end{vmatrix} = \hat{\mathbf{x}}\begin{vmatrix} -2 & 3 \\ 3 & -1 \end{vmatrix} - \hat{\mathbf{y}}\begin{vmatrix} 1 & 3 \\ 2 & -1 \end{vmatrix} + \hat{\mathbf{z}}\begin{vmatrix} 1 & -2 \\ 2 & 3 \end{vmatrix}$$

$$\vec{\mathbf{A}} \times \vec{\mathbf{B}} = \hat{\mathbf{x}}[(-2)(-1) - (3)(3)] - \hat{\mathbf{y}}[(1)(-1) - (3)(2)] + \hat{\mathbf{z}}[(1)(3) - (-2)(2)]$$

$$\vec{\mathbf{A}} \times \vec{\mathbf{B}} = \hat{\mathbf{x}}(2 - 9) - \hat{\mathbf{y}}(-1 - 6) + \hat{\mathbf{z}}(3 + 4)$$

$$\vec{\mathbf{A}} \times \vec{\mathbf{B}} = -7\hat{\mathbf{x}} - (-7)\hat{\mathbf{y}} + 7\hat{\mathbf{z}}$$

$$\vec{\mathbf{A}} \times \vec{\mathbf{B}} = -7\hat{\mathbf{x}} + 7\hat{\mathbf{y}} + 7\hat{\mathbf{z}}$$

Example: $\vec{\mathbf{A}}$ has a magnitude of 7 and $\vec{\mathbf{B}}$ has a magnitude of 4. The angle between $\vec{\mathbf{A}}$ and $\vec{\mathbf{B}}$ is 30°. Find the magnitude of $\vec{\mathbf{C}}$, where $\vec{\mathbf{C}} = \vec{\mathbf{A}} \times \vec{\mathbf{B}}$.

Use the trigonometric form of the vector product.

$$C = \|\vec{\mathbf{A}} \times \vec{\mathbf{B}}\| = AB\sin\theta$$

$$C = (7)(4)\sin 30° = 28\left(\frac{1}{2}\right) = 14$$

75. Consider the two vectors $\vec{\mathbf{A}} = 5\,\hat{\mathbf{x}} + 2\,\hat{\mathbf{y}} + 3\,\hat{\mathbf{z}}$ and $\vec{\mathbf{B}} = 4\,\hat{\mathbf{x}} - 6\,\hat{\mathbf{y}} - 2\,\hat{\mathbf{z}}$.

(A) Find the scalar product between $\vec{\mathbf{A}}$ and $\vec{\mathbf{B}}$.

(B) Find the vector product between $\vec{\mathbf{A}}$ and $\vec{\mathbf{B}}$.

Want help? Check the hints section at the back of the book.

Answers: 2, $14\,\hat{\mathbf{x}} + 22\,\hat{\mathbf{y}} - 38\,\hat{\mathbf{z}}$

76. Consider the two vectors $\vec{\mathbf{A}} = 3\,\hat{\mathbf{x}} - \hat{\mathbf{y}} - 4\,\hat{\mathbf{z}}$ and $\vec{\mathbf{B}} = 2\,\hat{\mathbf{x}} - \hat{\mathbf{z}}$.

(A) Find the scalar product between $\vec{\mathbf{A}}$ and $\vec{\mathbf{B}}$.

(B) Find the vector product between $\vec{\mathbf{A}}$ and $\vec{\mathbf{B}}$.

Want help? Check the hints section at the back of the book.

Answers: 10, $\hat{\mathbf{x}} - 5\,\hat{\mathbf{y}} + 2\,\hat{\mathbf{z}}$

77. The magnitude of $\vec{\mathbf{A}}$ is 8 and the magnitude of $\vec{\mathbf{B}}$ is 5. The angle between $\vec{\mathbf{A}}$ and $\vec{\mathbf{B}}$ is 60°.

(A) Find the scalar product between $\vec{\mathbf{A}}$ and $\vec{\mathbf{B}}$.

(B) Find the magnitude of the vector product between $\vec{\mathbf{A}}$ and $\vec{\mathbf{B}}$.

Want help? Check the hints section at the back of the book.

Answers: $20, 20\sqrt{3}$

20 BAR MAGNETS

Relevant Terminology

Magnetic field – a magnetic effect created by a moving charge (or current).

Essential Concepts

A bar magnet creates magnetic field lines and interacts with other magnets in such a way that it appears to have well-defined **north** and **south** poles.

- **Like magnetic poles repel.** For example, the north pole of one magnet repels the north pole of another magnet.
- **Opposite magnetic poles attract.** The north pole of one magnet attracts the south pole of another magnet.

The magnetic field lines outside of a bar magnet closely resemble the electric field lines of the electric dipole (Chapter 4). (Inside the magnet is much different: See Chapter 32.)

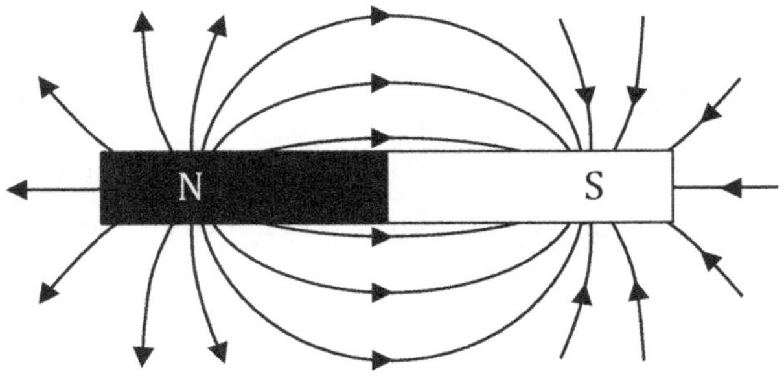

It's worth studying the magnetic field lines illustrated above.

- The magnetic field lines exit the north end of the magnet and enter the south end of the magnet.
- In between the north and south poles, the magnetic field lines run from north to south. However, note that in the above diagram the magnetic field lines run the opposite direction in the other two regions.
    - Left of the north pole, the magnetic field lines are headed to the left.
    - Right of the south pole, the magnetic field lines are also headed to the left.
    - However, between the two poles, the magnetic field lines run to the right.

Important Distinction

Magnets have **north** and **south poles**, whereas positive and negative charges create electric fields. It's improper to use the term "charge" to refer to a magnetic pole, or to use the adjectives "positive" or "negative" to describe a magnetic pole.

How Does a Magnet Work?

Moving charges create magnetic fields. Atoms consist of protons, neutrons, and electrons. Protons and electrons have electric charge.* Each atom has its own tiny magnetic field due to the motions† of its charges. Most macroscopic materials are nonmagnetic because their atomic magnetic fields are randomly aligned, such that on average their magnetic fields cancel out. A magnet is a material where the atomic magnetic fields are at least partially aligned, creating a significant net magnetic field.

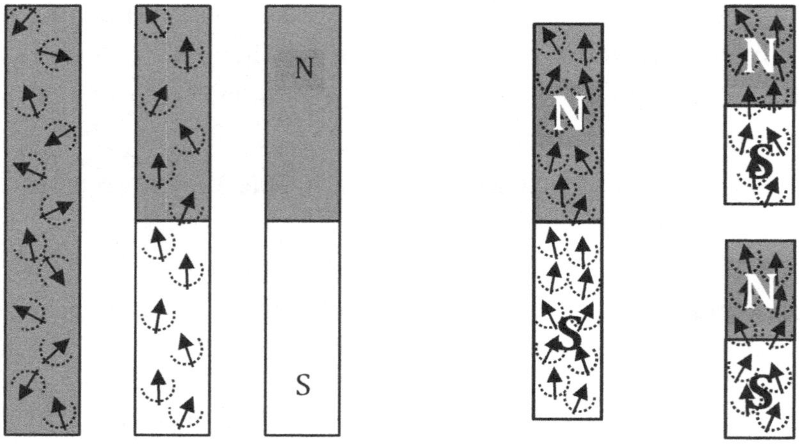

If you break a magnet in half, you don't get two chunks that each have a single magnetic pole. Instead, you get two smaller magnets, each with their own north and south poles.

Symbols and SI Units

Symbol	Name	SI Units
$\vec{\mathbf{B}}$	magnetic field	T

* Although neutrons are electrically neutral, they are composed of fractionally charged particles called quarks. Those quarks actually create magnetic fields which give the neutron a magnetic field. So even though the neutron is electrically neutral, it still creates a magnetic field. However, we'll focus on protons and electrons. If you see a question in a physics course asking about the magnetic field made by a neutron, the answer is probably "zero" (if your class ignores quarks). The main idea is that a moving **charge** creates a magnetic field.

† Technically, even a stationary electron or proton creates a magnetic field due to an intrinsic property called spin. Electrons have both orbital and spin angular momentum. (The earth has orbital angular momentum from its annual revolution around the sun and the earth also has spin angular momentum from its daily rotation about its axis. We can measure two similar kinds of angular momentum for an electron, except that unlike the earth, electrons are evidently pointlike. How can a "point" spin? Good question. That's why we say that the spin angular momentum of an electron is an **intrinsic** property.) Both kinds of angular momentum give charged particles magnetic fields. If you see a question in a physics course asking you which kinds of particles create magnetic fields, the correct answer is probably "moving charges." The main idea is that **moving** charges create magnetic fields.

How Does a Compass Work?

A compass needle is a tiny bar magnet. The earth behaves like a giant magnet with its magnetic south pole near geographic north. The north end of a compass needle is attracted to the magnetic south pole of the earth (since opposite magnetic poles attract).

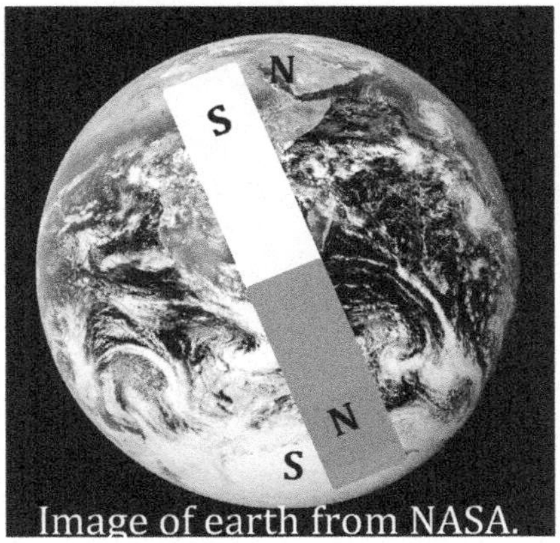

Image of earth from NASA.

Strategy to Determine the Direction of the Magnetic Field of a Bar Magnet

If you're given a bar magnet and want to determine the direction of the magnetic field at a specified point:

- First sketch the magnetic field lines that the bar magnet creates. Study the diagram on page 265 to aid with this. Remember that magnetic field lines leave the north end of the magnet and enter the south end of the magnet.
- One of the magnetic field lines will pass through the specified point. (If you draw enough magnetic field lines in your diagram, at least one will come near the specified point.) What is the direction of the magnetic field line when it passes through the specified point? That's the answer to the question.

Example: Sketch the magnetic field at points A, B, and C for the magnet shown below.

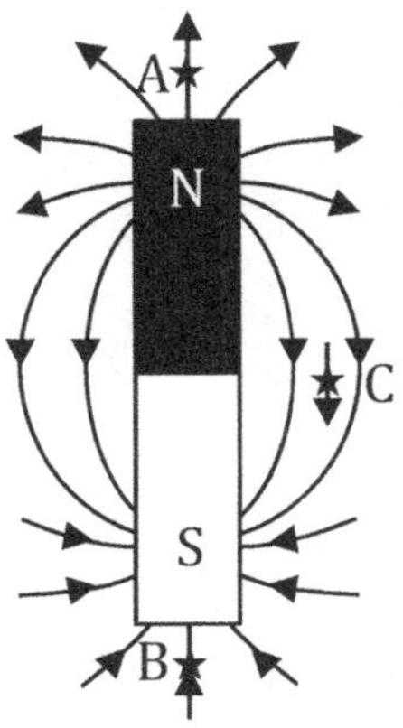

Sketch the magnetic field lines by rotating the diagram on page 265. The magnetic field points up (↑) at point A, up (↑) at point B, and down (↓) at point C.

78. Sketch the magnetic field at the points indicated in each diagram below.

(A)

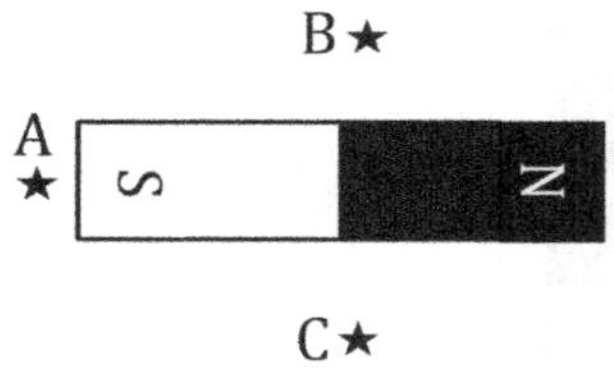

(B)

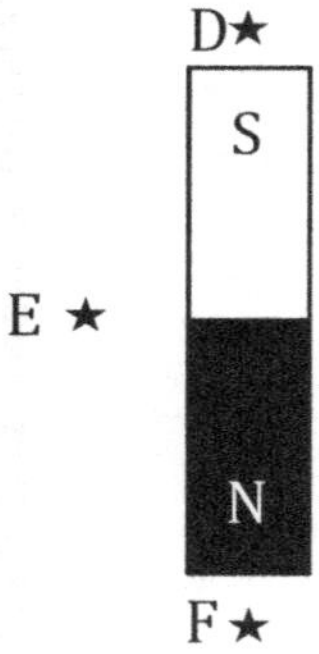

(C)

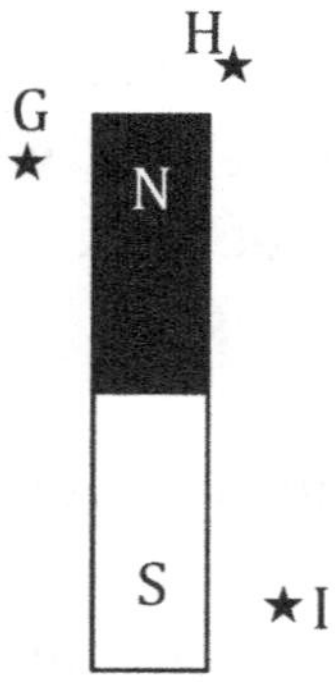

(D)

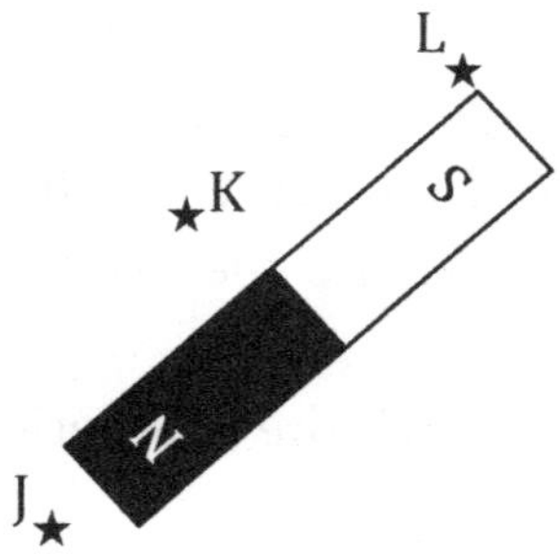

Want help? This problem is fully sketched in the back of the book.

21 RIGHT-HAND RULE FOR MAGNETIC FORCE

Relevant Terminology

Electric charge – a fundamental property of a particle that causes the particle to experience a force in the presence of an electric field. (An electrically neutral particle has no charge and thus experiences no force in the presence of an electric field.)
Velocity – a combination of speed and direction.
Current – the instantaneous rate of flow of charge through a wire.
Magnetic field – a magnetic effect created by a moving charge (or current).
Magnetic force – the push or pull that a moving charge (or current) experiences in the presence of a magnetic field.

Essential Concepts

A moving charge (or current-carrying wire) in the presence of an external magnetic field ($\vec{\mathbf{B}}$) experiences a magnetic force ($\vec{\mathbf{F}}_m$). The direction of the magnetic force is non-obvious: It's <u>**not**</u> along the magnetic field, and it's <u>**not**</u> along the velocity (or current). Fortunately, the following right-hand rule correctly provides the direction of the magnetic force.

To find the direction of the magnetic force ($\vec{\mathbf{F}}_m$) exerted on a moving charge (or current-carrying wire) in the presence of an external magnetic field ($\vec{\mathbf{B}}$):

- Point the extended fingers of your right hand along the velocity ($\vec{v}$) or current (I).
- Rotate your forearm until your palm faces the magnetic field ($\vec{\mathbf{B}}$), meaning that your palm will be perpendicular to the magnetic field.
- When your right-hand is simultaneously doing both of the first two steps, your extended thumb will point along the magnetic force ($\vec{\mathbf{F}}_m$).
- As a check, if your fingers are pointing toward the velocity ($\vec{v}$) or current (I) while your thumb points along the magnetic force ($\vec{\mathbf{F}}_m$), if you then bend your fingers as shown below, your fingers should now point along the magnetic field ($\vec{\mathbf{B}}$).

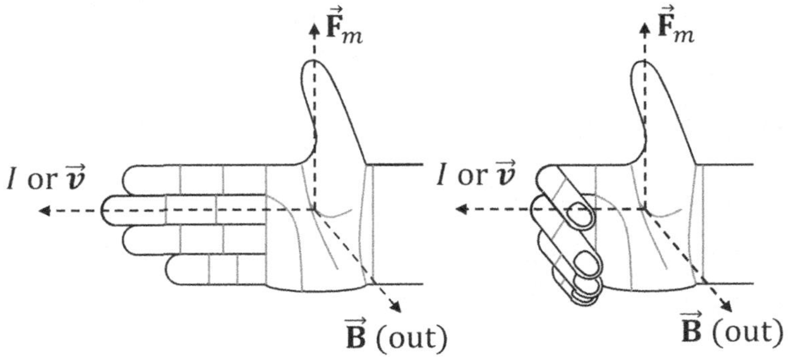

Important Exception

If the velocity ($\vec{v}$) or current (I) is **parallel** or **anti-parallel** to the magnetic field ($\vec{\mathbf{B}}$), the magnetic force ($\vec{\mathbf{F}}_m$) is **zero**. In Chapter 24, we'll learn that in these two extreme cases, $\theta = 0°$ or $180°$ such that $\sin\theta = 0$.

Symbols and SI Units

Symbol	Name	SI Units
$\vec{\mathbf{B}}$	magnetic field	T
$\vec{\mathbf{F}}_m$	magnetic force	N
$\vec{v}$	velocity	m/s
I	current	A

Special Symbols

Symbol	Name
$\otimes$	into the page
$\odot$	out of the page
p	proton
n	neutron
e^-	electron
N	north pole
S	south pole

Protons, Neutrons, and Electrons

- **Protons** have **positive** electric charge.
- Neutrons are electrically neutral.
- **Electrons** have **negative** electric charge.

Strategy to Apply the Right-hand Rule for Magnetic Force

Note the meaning of the following symbols:

- The symbol ⊗ represents an arrow going **into** the page.
- The symbol ⊙ represents an arrow coming **out** of the page.

There are a variety of problems that involve right-hand rules:

- If there is a **magnet** in the problem, first sketch the magnetic field ($\vec{\mathbf{B}}$) lines for the magnet as described in Chapter 20. This will help you find the direction of $\vec{\mathbf{B}}$.
- To find the direction of the magnetic force ($\vec{\mathbf{F}}_m$) exerted on a moving charge (or current-carrying wire) in the presence of an external magnetic field ($\vec{\mathbf{B}}$):*
    - **Note**: If the velocity ($\vec{v}$) or current (I) happens to be parallel or anti-parallel to the magnetic field ($\vec{\mathbf{B}}$), the magnetic force ($\vec{\mathbf{F}}_m$) is **zero**. Stop here.
    - Point the fingers of your right hand along the velocity ($\vec{v}$) or current (I).
    - Rotate your forearm until your palm faces the magnetic field ($\vec{\mathbf{B}}$).
    - Make sure that you are doing both of the first two steps simultaneously.
    - Your extended thumb is pointing along the magnetic force ($\vec{\mathbf{F}}_m$).
    - **Note**: For a **negative** charge (like an **electron**), the answer is **backwards**.
- If you already know the direction of $\vec{\mathbf{F}}_m$ and either $\vec{v}$ or I, and are instead looking for magnetic field ($\vec{\mathbf{B}}$), point your fingers along $\vec{v}$ or I, point your thumb along $\vec{\mathbf{F}}_m$, and your palm will face $\vec{\mathbf{B}}$.
- If you already know the direction of $\vec{\mathbf{F}}_m$ and $\vec{\mathbf{B}}$, and are instead looking for $\vec{v}$ or I, face your palm toward $\vec{\mathbf{B}}$, point your thumb along $\vec{\mathbf{F}}_m$, and your fingers will point toward $\vec{v}$ or I.
- If a **loop** of wire is in the presence of a magnetic field ($\vec{\mathbf{B}}$) and a question asks you about the loop **rotating** (or expanding or contracting), first apply the right-hand rule to find the direction of the magnetic force ($\vec{\mathbf{F}}_m$) exerted on each part of the wire.
- If a problem involves a charge traveling in a circle, apply the right-hand rule at a point in the circle. The magnetic force ($\vec{\mathbf{F}}_m$) will point toward the center of the circle because a **centripetal** force causes an object to travel in a circle.
- If you don't know the direction of $\vec{\mathbf{F}}_m$ and you're not looking for the direction of $\vec{\mathbf{F}}_m$, use the right-hand rule from Chapter 22 instead.
- To find the direction of the magnetic field ($\vec{\mathbf{B}}$) created by a moving charge or current-carrying wire, apply the right-hand rule from Chapter 22.
- To find the force that one wire (or moving charge) exerts on another wire (or moving charge), apply the strategy from Chapter 23.

---

* Unfortunately, not all textbooks apply this right-hand rule the same way, though the other versions of this right-hand rule are equivalent to this one. Thus, if you read another book, it may teach this rule differently.

Example: What is the direction of the magnetic force exerted on the current shown below?

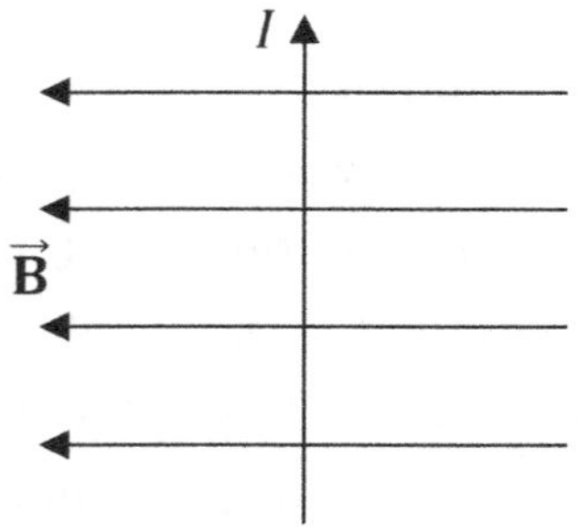

Apply the right-hand rule for magnetic force:

- Point your fingers up (↑), along the current (I).
- At the same time, face your palm to the left (←), along the magnetic field ($\vec{\mathbf{B}}$).
- If your fingers point up (↑) at the same time as your palm faces left (←), your thumb will be pointing out of the page (⊙). **Tip**: Make sure you're using your **right** hand.†
- Your thumb is the answer: The magnetic force ($\vec{\mathbf{F}}_m$) is out of the page (⊙).
- Note: The symbol ⊙ represents an arrow pointing out of the page.

Example: What is the direction of the magnetic force exerted on the proton shown below?

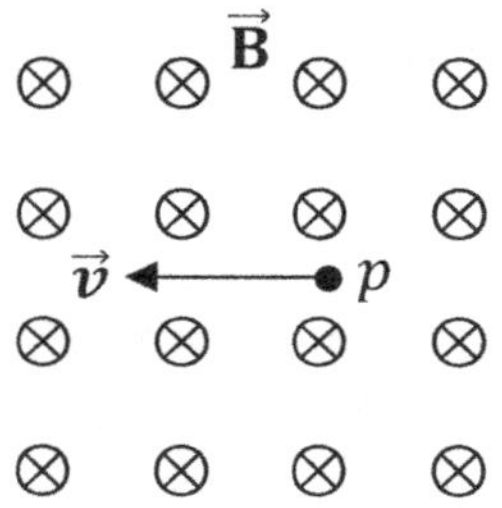

Apply the right-hand rule for magnetic force:

- Point your fingers left (←), along the velocity ($\vec{v}$) of the proton (p).
- At the same time, face your palm into the page (⊗), along the magnetic field ($\vec{\mathbf{B}}$).
- Your thumb points down: The magnetic force ($\vec{\mathbf{F}}_m$) is down (↓).

Example: What is the direction of the magnetic force exerted on the current shown below?

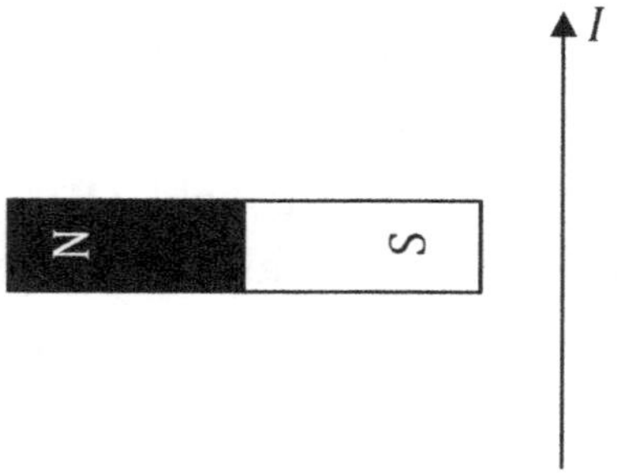

† This should seem obvious, right? But guess what: If you're right-handed, when you're taking a test, your right hand is busy writing, so it's instinctive to want to use your free hand, which is the wrong one.

First, sketch the magnetic field lines for the bar magnet (recall Chapter 20).

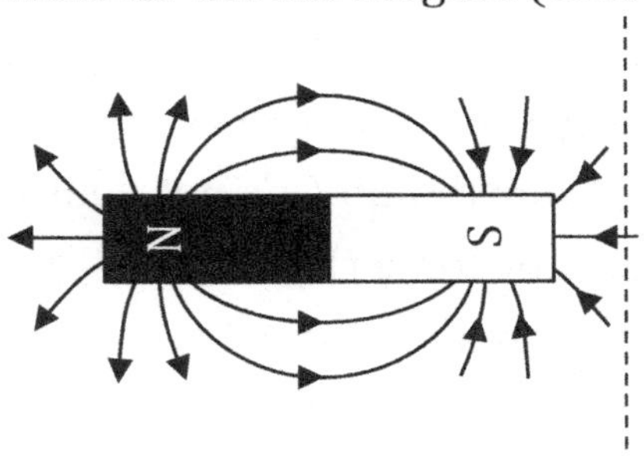

The dashed line in the diagram above shows the location of the current. On average, what is the direction of the magnetic field ($\vec{\mathbf{B}}$) lines where the dashed line (current) is? Where the current is located, the magnetic field lines point to the left on average. Now we are prepared to apply the right-hand rule for magnetic force:

- Point your fingers up (↑), along the current (I).
- At the same time, face your palm to the left (←), along the magnetic field ($\vec{\mathbf{B}}$).
- Your thumb points out of the page: The magnetic force ($\vec{\mathbf{F}}_m$) is out of the page (⊙).

Example: What is the direction of the magnetic force exerted on the electron shown below?

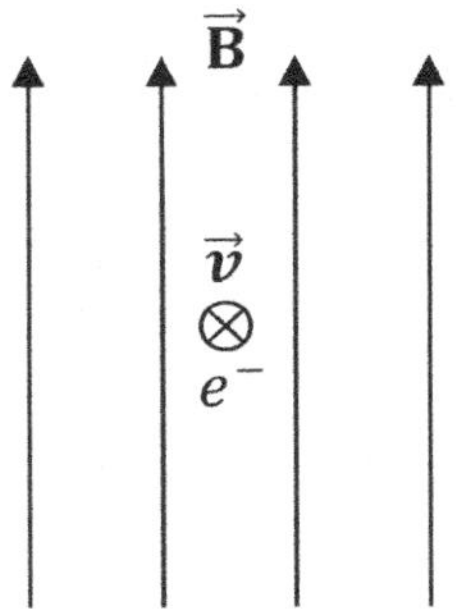

Apply the right-hand rule for magnetic force:

- Point your fingers into the page (⊗), along the velocity ($\vec{v}$) of the electron (e^-).
- At the same time, face your palm up (↑), along the magnetic field ($\vec{\mathbf{B}}$).
- Your thumb points to the right (→). However, electrons are negatively charged, so the answer is **backwards** for electrons.
- Therefore, the magnetic force ($\vec{\mathbf{F}}_m$) is to the left (←).

Example: What is the direction of the magnetic field for the situation shown below, assuming that the magnetic field is perpendicular‡ to the current?

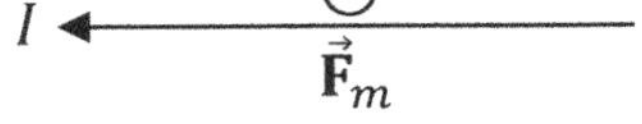

‡ The magnetic force is always perpendicular to both the current and the magnetic field, but the current and magnetic field need not be perpendicular. We will work with the angle between I and $\vec{\mathbf{B}}$ in Chapter 24.

Invert the right-hand rule for magnetic force. In this example, we already know the direction of $\vec{\mathbf{F}}_m$, and are instead solving for the direction of $\vec{\mathbf{B}}$.

- Point your fingers to the left (←), along the current (I).
- At the same time, point your **thumb** (it's **not** your palm in this example) out of the page (⊙), along the magnetic force ($\vec{\mathbf{F}}_m$).
- Your palm faces down: The magnetic field ($\vec{\mathbf{B}}$) is down (↓).

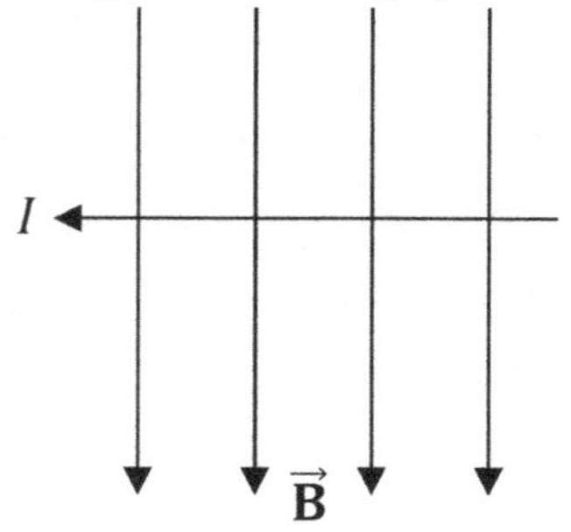

Check your answer by applying the right-hand rule the usual way. Look at the diagram above. Point your fingers to the left (←), along I, and face your palm down (↓), along $\vec{\mathbf{B}}$. Your thumb will point out of the page (⊙), along $\vec{\mathbf{F}}_m$. It all checks out.

Example: What is the direction of the magnetic force exerted on the current shown below?

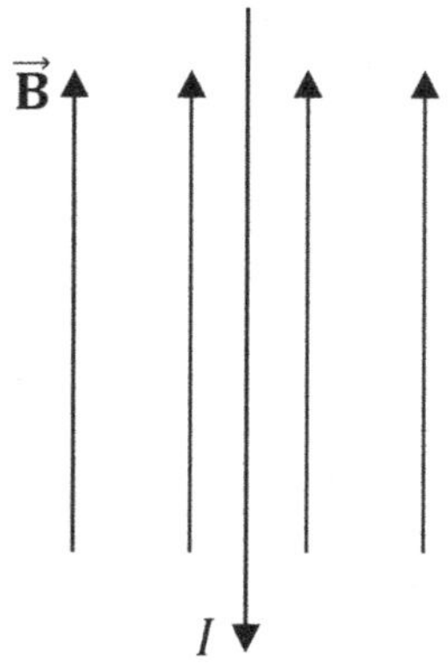

This is a "trick" question. Well, it's an "easy" question when you remember the trick. See the note in the strategy on page 271: In this example, the current (I) is anti-parallel to the magnetic field ($\vec{\mathbf{B}}$). Therefore, the magnetic force ($\vec{\mathbf{F}}_m$) is **zero** (and thus has no direction).

Example: What must be the direction of the magnetic field in order for the proton to travel in the circle shown below?

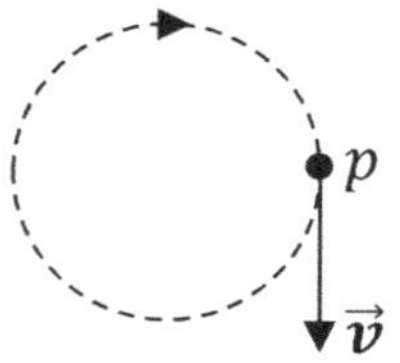

In Volume 1 of this series, we learned that an object traveling in a circle experiences a **centripetal** force. This means that the magnetic force ($\vec{\mathbf{F}}_m$) must be pushing on the proton towards the center of the circle. See the diagram on the left below: For the position indicated, the velocity ($\vec{v}$) is down (along a tangent) and the magnetic force ($\vec{\mathbf{F}}_m$) is to the left (toward the center). Invert the right-hand rule to find the magnetic field.

- Point your fingers down (↓), along the velocity ($\vec{v}$).
- At the same time, point your <u>**thumb**</u> (it's <u>**not**</u> your palm in this example) to the left (←), along the magnetic force ($\vec{\mathbf{F}}_m$).
- Your palm faces out of the page: The magnetic field ($\vec{\mathbf{B}}$) is out of the page (⊙).

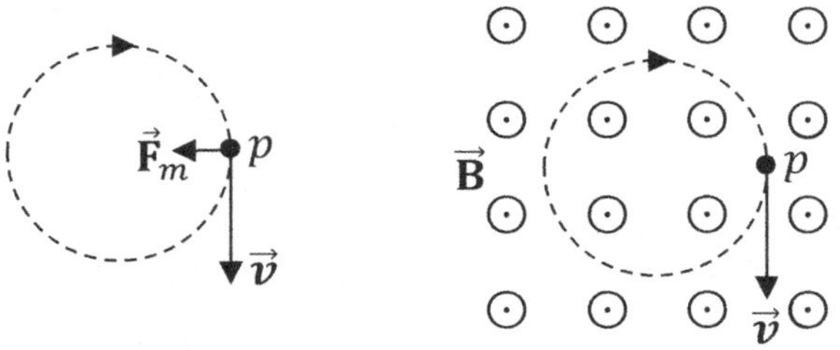

Example: Would the rectangular loop shown below tend to rotate, contract, or expand?

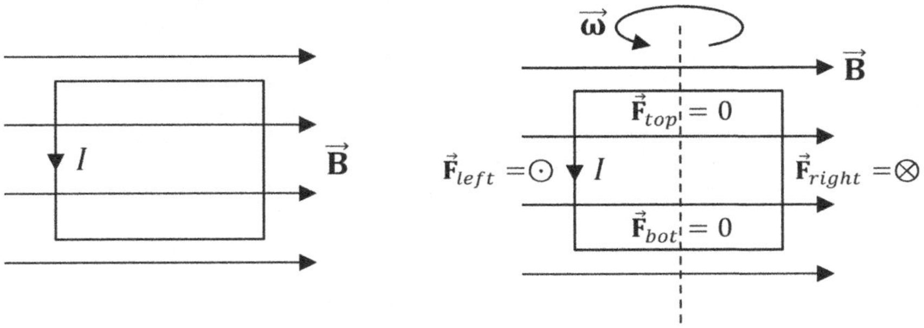

First, apply the right-hand rule for magnetic force to each side of the rectangular loop:

- <u>**Left side**</u>: Point your fingers down (↓) along the current (I) and your palm to the right (→) along the magnetic field ($\vec{\mathbf{B}}$). The magnetic force ($\vec{\mathbf{F}}_m$) is out of the page (⊙).
- <u>**Bottom side**</u>: The current (I) points right (→) and the magnetic field ($\vec{\mathbf{B}}$) also points right (→). Since I and $\vec{\mathbf{B}}$ are parallel, the magnetic force ($\vec{\mathbf{F}}_m$) is <u>**zero**</u>.
- <u>**Right side**</u>: Point your fingers up (↑) along the current (I) and your palm to the right (→) along the magnetic field ($\vec{\mathbf{B}}$). The magnetic force ($\vec{\mathbf{F}}_m$) is into the page (⊗).
- <u>**Top side**</u>: The current (I) points left (←) and the magnetic field ($\vec{\mathbf{B}}$) points right (→). Since I and $\vec{\mathbf{B}}$ are anti-parallel, the magnetic force ($\vec{\mathbf{F}}_m$) is <u>**zero**</u>.

We drew these forces on the diagram on the right above. The left side of the loop is pulled out of the page, while the right side of the loop is pushed into the page. What will happen? The loop will rotate about the dashed axis.

79. Apply the right-hand rule for magnetic force to answer each question below.

(A) What is the direction of the magnetic force exerted on the current?

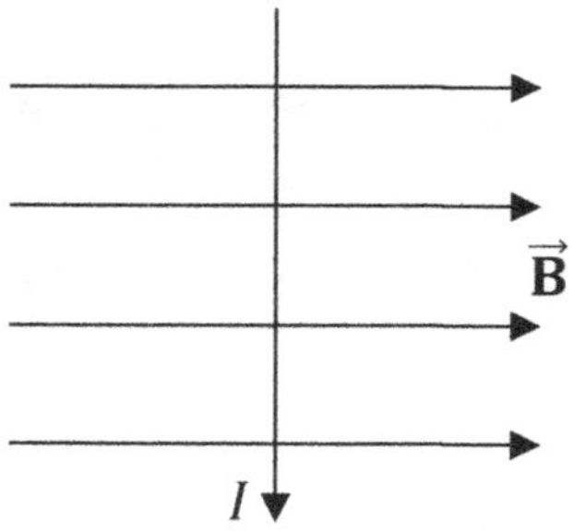

(B) What is the direction of the magnetic force exerted on the proton?

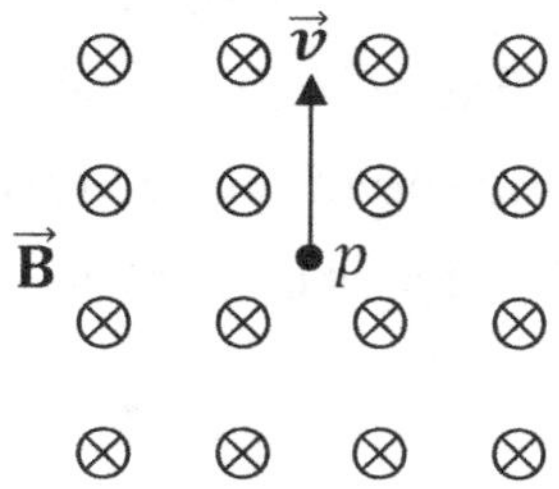

(C) What is the direction of the magnetic force exerted on the current?

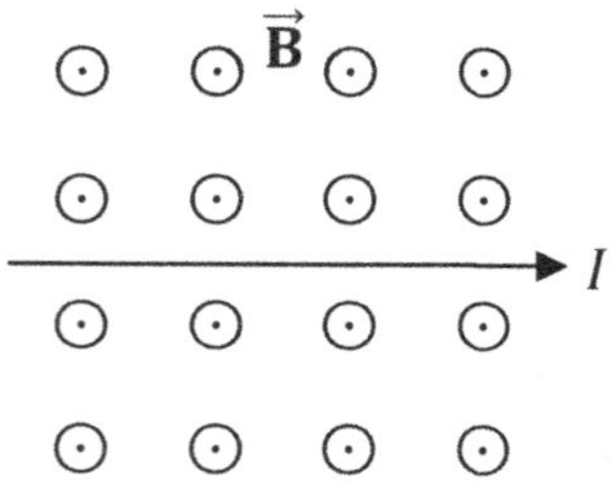

(D) What is the direction of the magnetic force exerted on the electron?

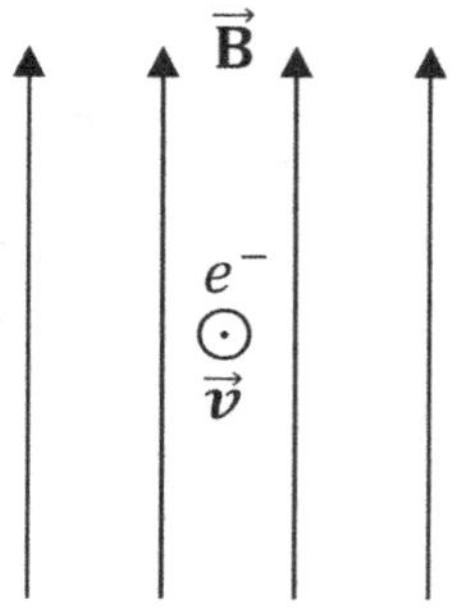

(E) What is the direction of the magnetic force exerted on the proton?

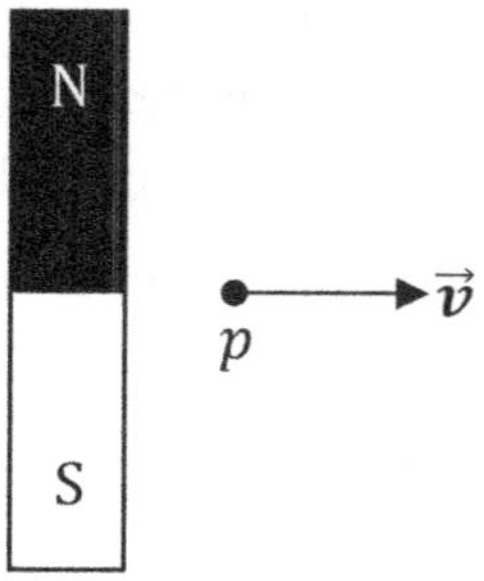

(F) What is the direction of the magnetic field, assuming that the magnetic field is perpendicular to the current?

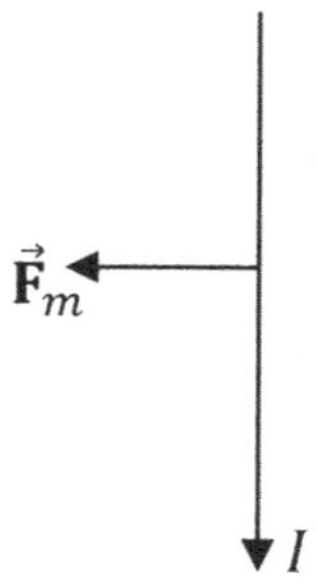

Want help? The problems from Chapter 21 are fully solved in the back of the book.

80. Apply the right-hand rule for magnetic force to answer each question below.

(A) What is the direction of the magnetic force exerted on the proton?

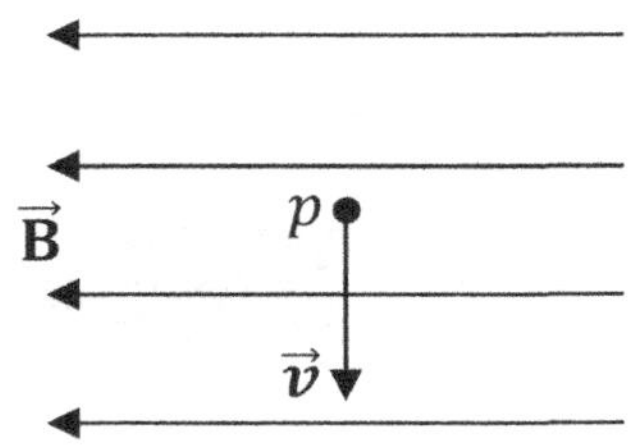

(B) What is the direction of the magnetic force exerted on the current?

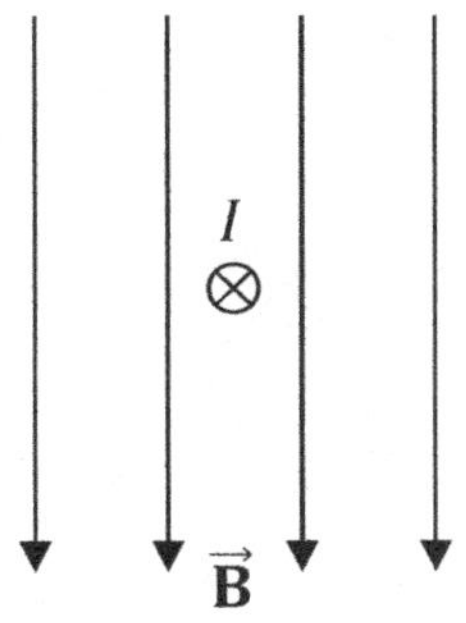

(C) What is the direction of the magnetic force exerted on the current?

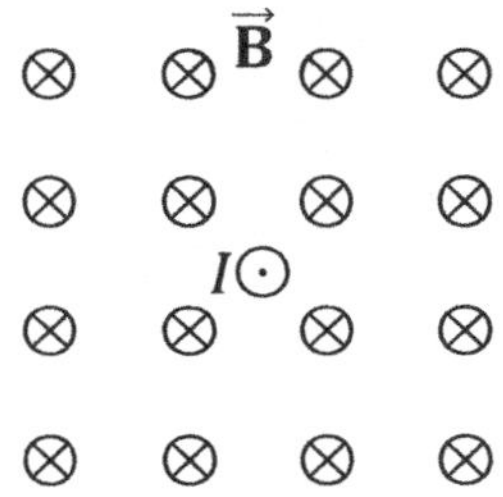

(D) What is the direction of the magnetic force exerted on the proton?

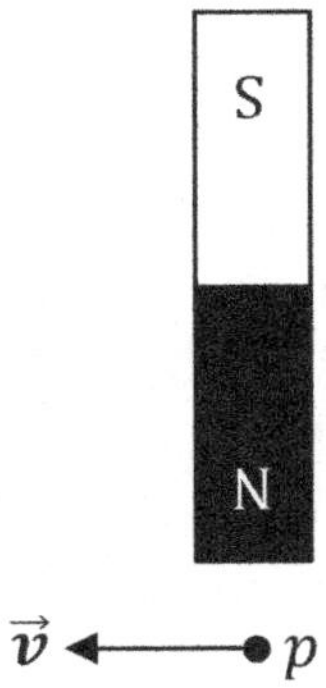

(E) What is the direction of the magnetic force exerted on the electron?

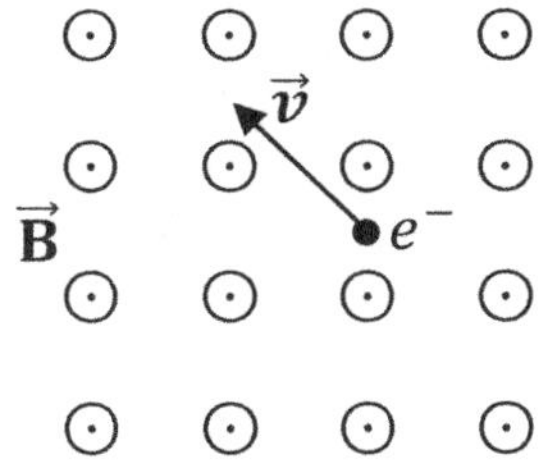

(F) What is the direction of the magnetic field, assuming that the magnetic field is perpendicular to the current?

Want help? The problems from Chapter 21 are fully solved in the back of the book.

81. Apply the right-hand rule for magnetic force to answer each question below.

(A) What is the direction of the magnetic force exerted on the current?

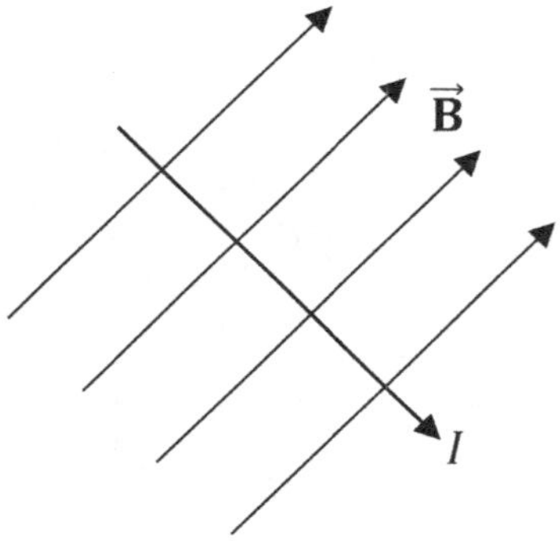

(B) What is the direction of the magnetic force exerted on the electron?

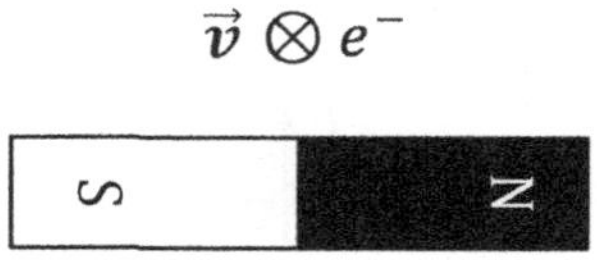

(C) What must be the direction of the magnetic field in order for the proton to travel in the circle shown below?

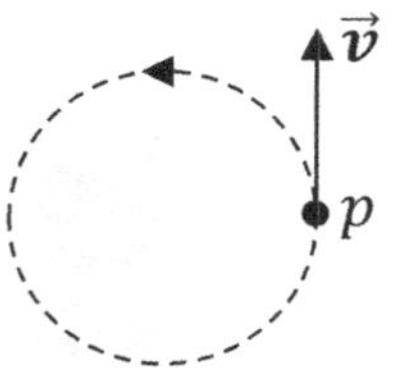

(D) What must be the direction of the magnetic field in order for the electron to travel in the circle shown below?

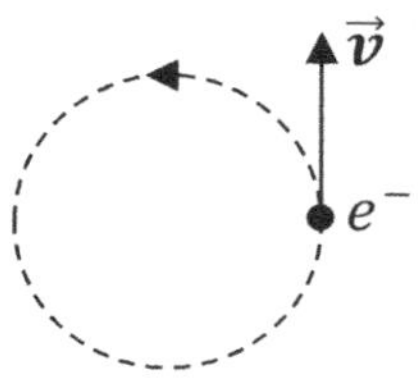

(E) Would the rectangular loop tend to rotate, expand, or contract?

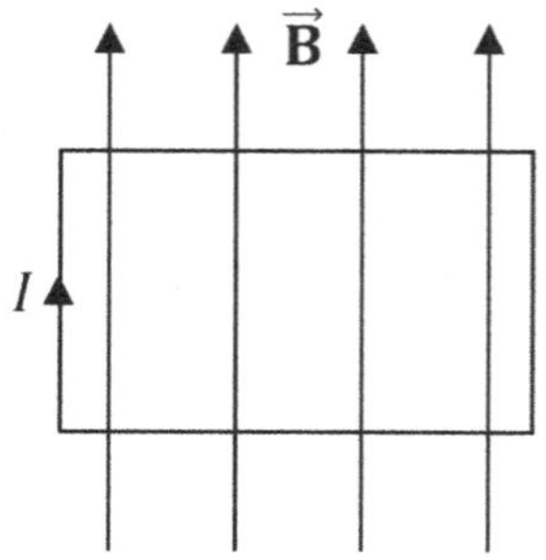

(F) Would the rectangular loop tend to rotate, expand, or contract?

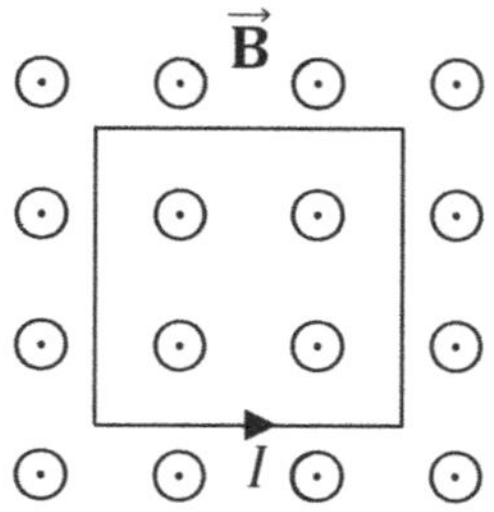

Want help? The problems from Chapter 21 are fully solved in the back of the book.

22 RIGHT-HAND RULE FOR MAGNETIC FIELD

Relevant Terminology

Current – the instantaneous rate of flow of charge through a wire.
Magnetic field – a magnetic effect created by a moving charge (or current).
Solenoid – a coil of wire in the shape of a right-circular cylinder.

Essential Concepts

A long straight wire creates magnetic field lines that circulate around the current, as shown below.

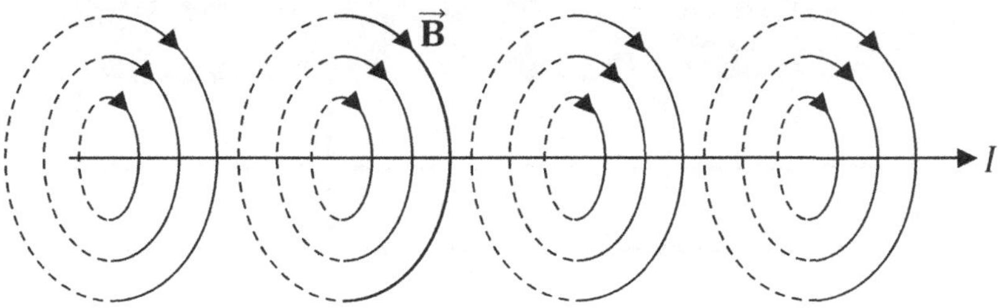

We use a right-hand rule (different from the right-hand rule that we learned in Chapter 21) to determine which way the magnetic field lines circulate. This right-hand rule gives you the direction of the magnetic field ($\vec{\mathbf{B}}$) created by a current (I) or moving charge:

- Imagine grabbing the wire with your right hand. **Tip**: You can use a pencil to represent the wire and actually grab the pencil.
- Grab the wire with your thumb pointing along the current (I).
- Your fingers represent **circular** magnetic field lines traveling around the wire **toward your fingertips**. At a given point, the direction of the magnetic field is **tangent** to these circles (your fingers).

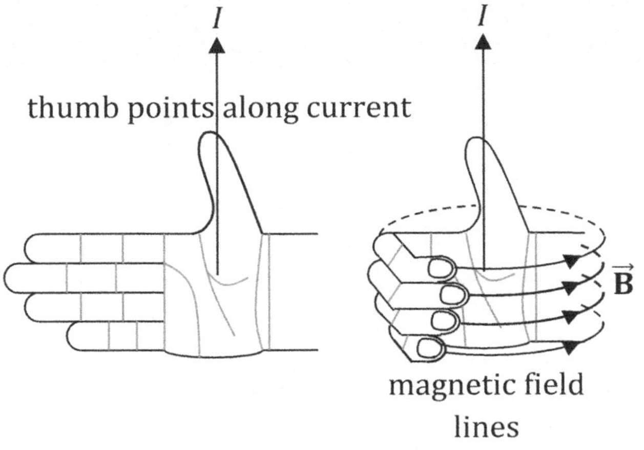

Important Distinction

There are two different right-hand rules used in magnetism:

- One right-hand rule is used to find the direction of the **magnetic force** ($\vec{\mathbf{F}}_m$) exerted on a current or moving charge in the presence of an external magnetic field ($\vec{\mathbf{B}}$). We discussed that right-hand rule in Chapter 21.
- A second right-hand rule is used to find the direction of the **magnetic field** ($\vec{\mathbf{B}}$) created by a current or moving charge. This is the subject of the current chapter. Note that this right-hand rule does **not** involve magnetic force ($\vec{\mathbf{F}}_m$).

Symbols and SI Units

Symbol	Name	SI Units
$\vec{\mathbf{B}}$	magnetic field	T
I	current	A

Special Symbols

Symbol	Name
⊗	into the page
⊙	out of the page

Schematic Symbols

Schematic Representation	Symbol	Name
	R	resistor
	ΔV	battery or DC power supply

Recall that the long line represents the positive terminal, while the small rectangle represents the negative terminal. Current runs from positive to negative.

Strategy to Apply the Right-hand Rule for Magnetic Field

Note the meaning of the following symbols:

- The symbol ⊗ represents an arrow going <u>**into**</u> the page.
- The symbol ⊙ represents an arrow coming <u>**out**</u> of the page.

If there is a battery in the diagram, label the positive and negative terminals of each battery (or DC power supply). The long line of the schematic symbol represents the positive terminal, as shown below. Draw the conventional* current the way that positive charges would flow: from the positive terminal to the negative terminal.

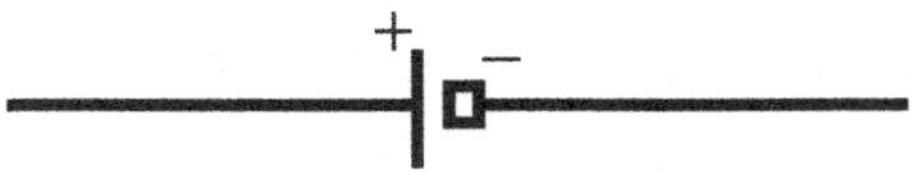

There are a variety of problems that involve right-hand rules:

- To find the direction of the magnetic field ($\vec{\mathbf{B}}$) at a specified point that is created by a current (I) or moving charge:
  - Grab the wire with your thumb pointing along the current (I), with your fingers wrapped in circles around the wire.
  - Your fingers represent **circular** magnetic field lines traveling around the wire **toward your fingertips**. At a given point, the direction of the magnetic field is **tangent** to these circles (your fingers).
  - **Note**: For a <u>**negative**</u> charge (like an <u>**electron**</u>), the answer is <u>**backwards**</u>.
- If you already know the direction of the magnetic field ($\vec{\mathbf{B}}$) and are instead looking for the current (I), invert the right-hand rule for magnetic field as follows:
  - Grab the wire with your fingers wrapped in circles around the wire such that your fingers match the magnetic field at the specified point. Your fingers represent **circular** magnetic field lines traveling around the wire **toward your fingertips**. The direction of the magnetic field is **tangent** to these circles.
  - Your thumb represents the direction of the current along the wire.
  - **Note**: For a <u>**negative**</u> charge (like an <u>**electron**</u>), the answer is <u>**backwards**</u>.
- If the problem involves magnetic force ($\vec{\mathbf{F}}_m$), use the strategy from Chapter 21 or 23:
  - To find the direction of the magnetic force exerted on a current-carrying wire (or moving charge) in the presence of an external magnetic field, apply the strategy from Chapter 21.
  - To find the force that one wire (or moving charge) exerts on another wire (or moving charge), apply the strategy from Chapter 23. The strategy from Chapter 23 combines the two right-hand rules from Chapters 21 and 22.

---

* Of course, it's really the electrons moving through the wire, not positive charges. However, all of the signs in physics are based on positive charges, so it's "conventional" to draw the current based on what a positive charge would do. Remember, all the rules are backwards for electrons, due to the negative charge.

Example: What is the direction of the magnetic field at points A and C for the current shown below?

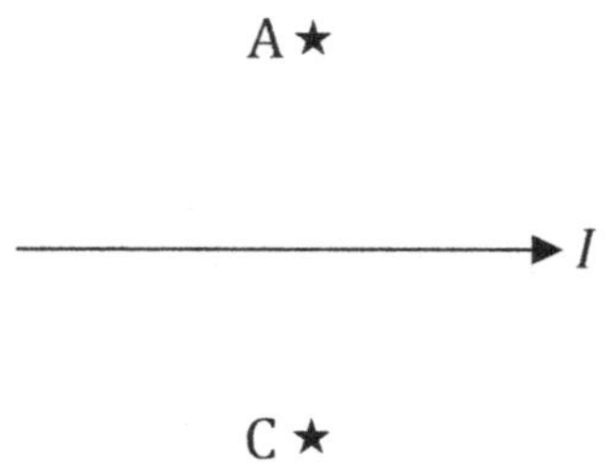

Apply the right-hand rule for magnetic field:

- Grab the current with your thumb pointing to the right (→), along the current (I).
- Your fingers make circles around the wire (toward your fingertips), as shown in the diagram below on the left.
- The magnetic field ($\vec{\mathbf{B}}$) at a specified point is tangent to these circles, as shown in the diagram below on the right. Try to visualize the circles that your fingers make: Above the wire, your fingers are coming out of the page, while below the wire, your fingers are going back into the page. (Note that in order to truly "grab" the wire, your fingers would actually go "through" the page, with part of your fingers on each side of the page. Your fingers would intersect the paper above the wire, where they are headed out of the page, and also intersect the paper below the wire, where they are headed back into the page.) At point A the magnetic field ($\vec{\mathbf{B}}_A$) points out of the page (⊙), while at point C the magnetic field ($\vec{\mathbf{B}}_C$) points into the page (⊗).
- It may help to study the diagram on the top of page 279 and compare it with the diagrams shown below. These are actually all the same pictures.

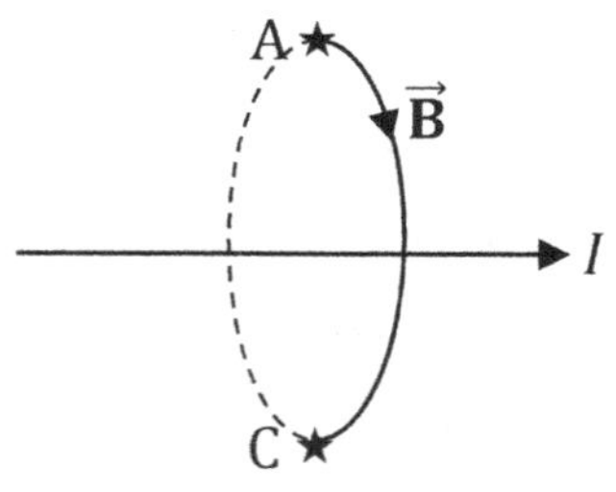

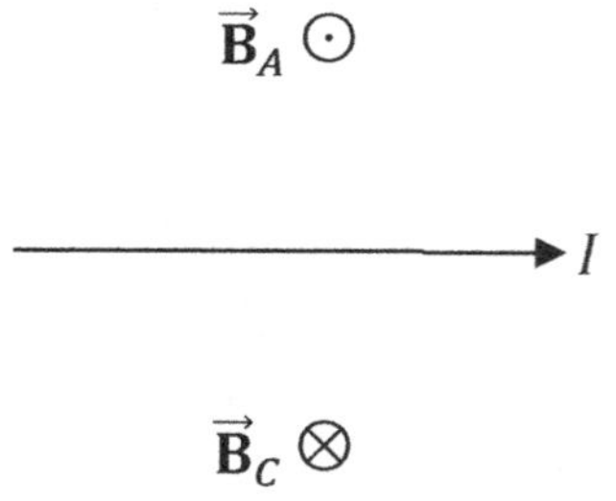

Example: What is the direction of the magnetic field at points D and E for the current shown below?

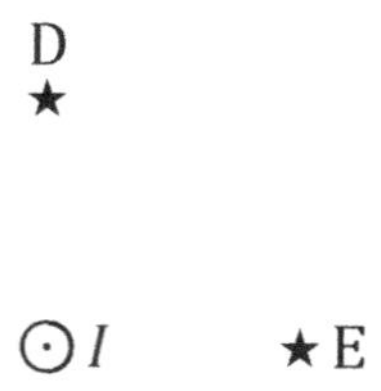

Apply the right-hand rule for magnetic field:

- Grab the current with your thumb pointing out of the page ($\odot$), along the current (I).
- Your fingers make **counterclockwise** (use the right-hand rule to see this) circles around the wire (toward your fingertips), as shown in the diagram below on the left.
- The magnetic field ($\vec{\mathbf{B}}$) at a specified point is **tangent** to these circles, as shown in the diagram below on the right. Draw tangent lines at points D and E with the arrows headed **counterclockwise**. See the diagram below on the right. At point D the magnetic field ($\vec{\mathbf{B}}_D$) points to the left ($\leftarrow$), while at point E the magnetic field ($\vec{\mathbf{B}}_E$) points up ($\uparrow$).

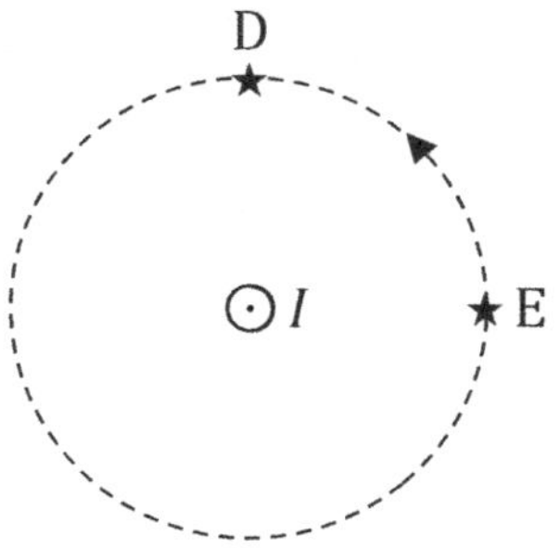

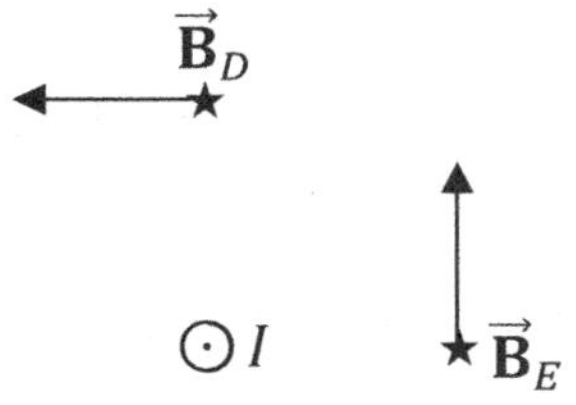

Example: What is the direction of the magnetic field at points F and G for the current shown below?

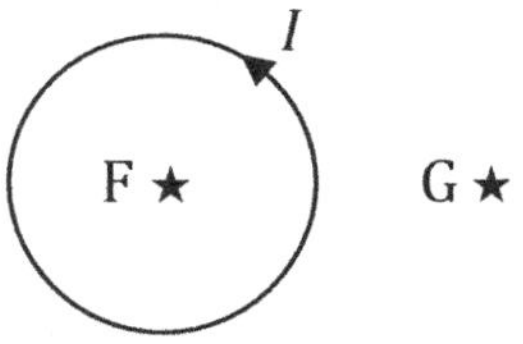

Apply the right-hand rule for magnetic field:

- Imagine grabbing the steering wheel of a car with your right hand, such that your thumb points **counterclockwise** (since that's how the current is drawn above). No matter where you grab the steering wheel, your fingers are coming out of the page ($\odot$) at point F. The magnetic field ($\vec{\mathbf{B}}_F$) points out of the page ($\odot$) at point F.
- For point G, grab the steering wheel at the rightmost point (that point is nearest to point G, so it will have the dominant effect). Your fingers are going into of the page ($\otimes$) at point G. The magnetic field ($\vec{\mathbf{B}}_G$) points into the page ($\otimes$) at point G.

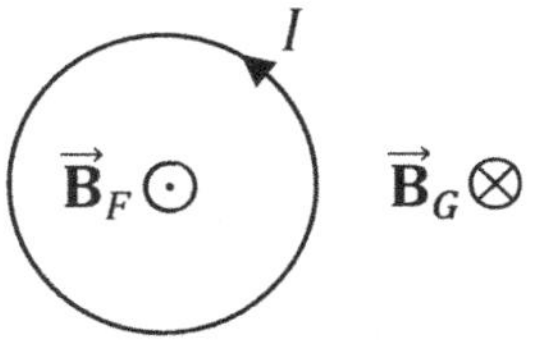

Tip: The magnetic field **outside** of the loop is **opposite** to its direction **inside** the loop.

Example: What is the direction of the magnetic field at point H for the loop shown below?

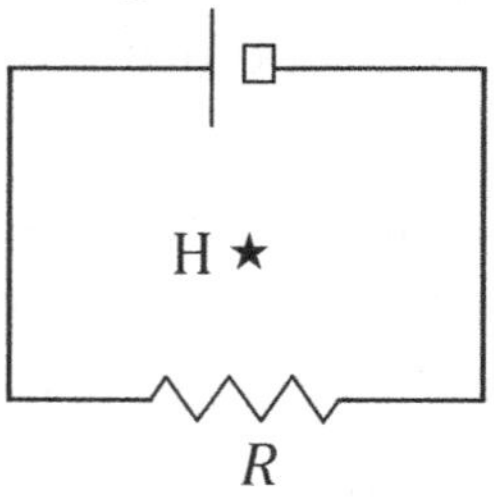

First label the positive (+) and negative (−) terminals of the battery (the long line is the positive terminal) and draw the current from the positive terminal to the negative terminal. See the diagram below on the left. Then apply the right-hand rule for magnetic field. It turns out to be identical to point F in the previous example, since again the current is traveling through the loop in a **counterclockwise** path. The magnetic field ($\vec{\mathbf{B}}_F$) points out of the page (⊙) at point H.

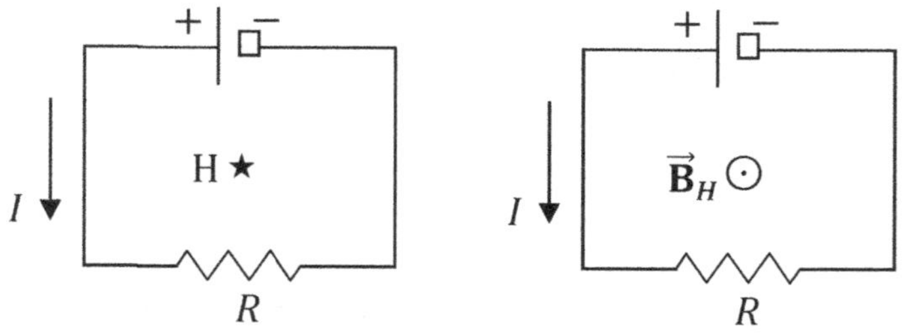

Example: What is the direction of the magnetic field at point J for the loop shown below?

Note that this loop (unlike the two previous examples) does __**not**__ lie in the plane of the paper. Rather, this loop is a **horizontal** circle with the solid (—) semicircle in front of the paper and the dashed (---) semicircle behind the paper. It's like the rim of a basketball hoop. Apply the right-hand rule for magnetic field:

- Imagine grabbing the front of the rim of a basketball hoop with your right hand, such that your thumb points to your right (since in the diagram above, the current is heading to the right in the front of the loop).
- Your fingers are going up (↑) at point J inside of the loop. The magnetic field ($\vec{\mathbf{B}}_J$) points up (↑) at point J.

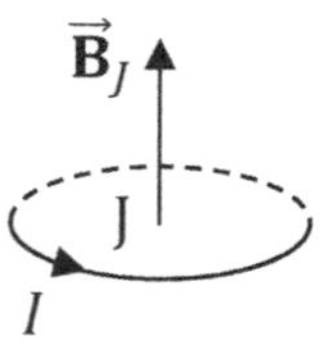

Example: What is the direction of the magnetic field inside the solenoid shown below?

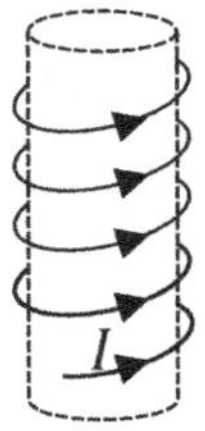

A solenoid is a coil of wire wrapped in the shape of a right-circular cylinder. The solenoid above essentially consists of several (approximately) horizontal loops. Each horizontal loop is just like the previous example. Note that the current (I) is heading the same way (it is pointing to the right in the front of each loop). Therefore, just as in the previous example, the magnetic field ($\vec{\mathbf{B}}$) points up ($\uparrow$) inside of the solenoid.

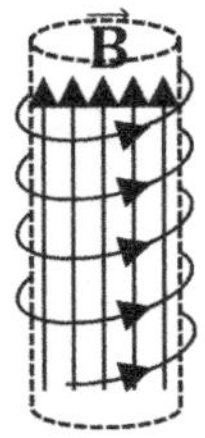

82. Determine the direction of the magnetic field at each point indicated below.

(A)

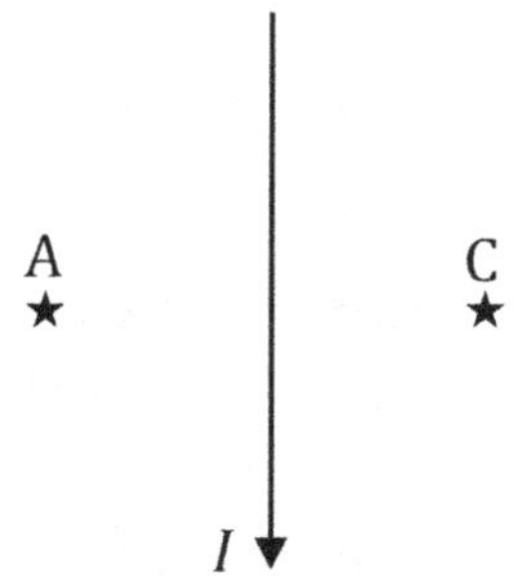

(B)

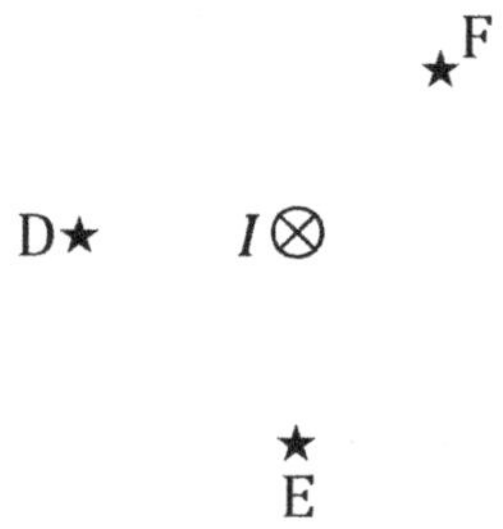

(C)

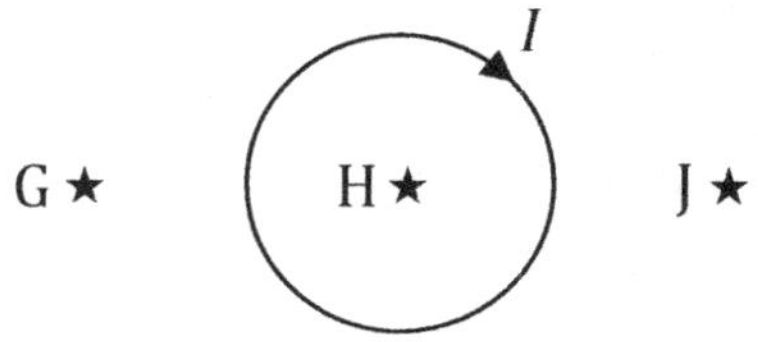

(D)

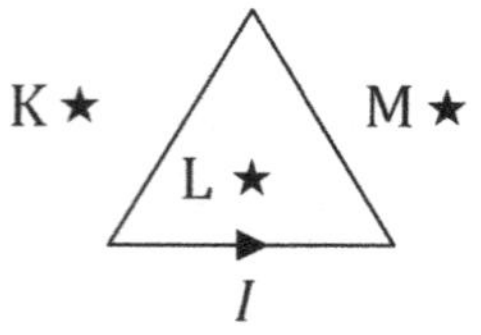

(E)

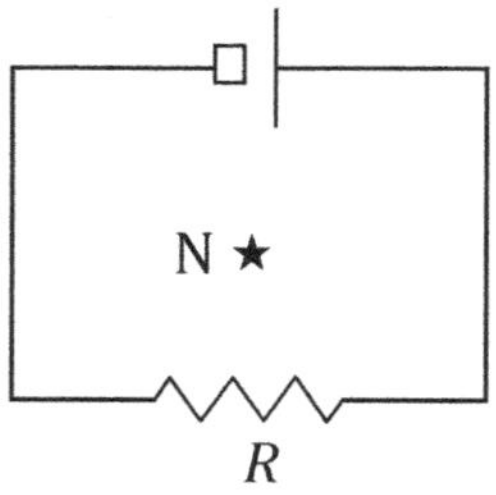

(F)

(G)

(H)

(I)

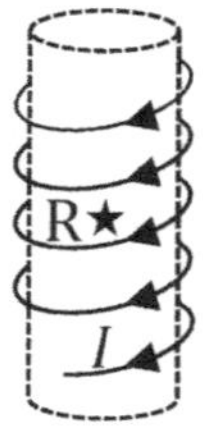

Want help? This problem is fully solved in the back of the book.

23 COMBINING THE TWO RIGHT-HAND RULES

Relevant Terminology

Electric charge – a fundamental property of a particle that causes the particle to experience a force in the presence of an electric field. (An electrically neutral particle has no charge and thus experiences no force in the presence of an electric field.)
Velocity – a combination of speed and direction.
Current – the instantaneous rate of flow of charge through a wire.
Magnetic field – a magnetic effect created by a moving charge (or current).
Magnetic force – the push or pull that a moving charge (or current) experiences in the presence of a magnetic field.
Solenoid – a coil of wire in the shape of a right-circular cylinder.

Essential Concepts

To find the magnetic force that one current-carrying wire (call it I_a) exerts on another current-carrying wire (call it I_b), apply both right-hand rules in combination. In this example, we're thinking of I_a as exerting the force and I_b as being pushed or pulled by the force. (Of course, it's mutual: I_a exerts a force on I_b, and I_b also exerts a force on I_a. However, we will calculate just one force at a time. So for the purposes of the calculation, let's consider the force that I_a exerts on I_b.)

- First find the direction of the magnetic field ($\vec{\mathbf{B}}_a$) created by the current (I_a), which we're thinking of as exerting the force, **at the location of the second current** (I_b). Put the field point (★) on I_b and ask yourself, "What is the direction of $\vec{\mathbf{B}}_a$ at the ★?" Apply the right-hand rule for magnetic **field** (Chapter 22) to find $\vec{\mathbf{B}}_a$ from I_a.
- Now find the direction of the magnetic force ($\vec{\mathbf{F}}_a$) that the first current (I_a) exerts on the second current (I_b). Apply the right-hand rule for magnetic **force** (Chapter 21) to find $\vec{\mathbf{F}}_a$ from I_b and $\vec{\mathbf{B}}_a$. In this step, we use the second current (I_b) because that is the current we're thinking of as being pushed or pulled by the force.
- Note that both currents get used: The first current (I_a), which we're thinking of as exerting the force, is used in the first step to find the magnetic **field** ($\vec{\mathbf{B}}_a$), and the second current (I_b), which we're thinking of as being pushed or pulled by the force, is used in the second step to find the magnetic **force** ($\vec{\mathbf{F}}_a$).

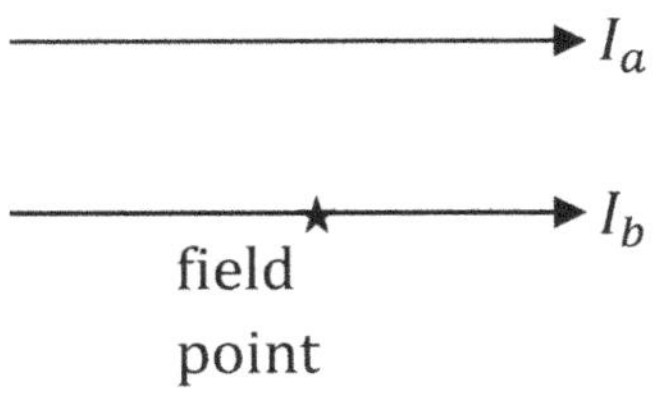

Symbols and SI Units

Symbol	Name	SI Units
$\vec{\mathbf{B}}$	magnetic field	T
$\vec{\mathbf{F}}_m$	magnetic force	N
$\vec{\mathbf{v}}$	velocity	m/s
I	current	A

Special Symbols

Symbol	Name
$\otimes$	into the page
$\odot$	out of the page
p	proton
n	neutron
e^-	electron

Schematic Symbols

Schematic Representation	Symbol	Name
	R	resistor
	ΔV	battery or DC power supply

Recall that the long line represents the positive terminal.

Strategy to Find the Direction of the Magnetic Force that Is Exerted by One Current on Another Current

Note the meaning of the following symbols:

- The symbol ⊗ represents an arrow going **into** the page.
- The symbol ⊙ represents an arrow coming **out** of the page.

If there is a battery in the diagram, label the positive and negative terminals of each battery (or DC power supply). The long line of the schematic symbol represents the positive terminal, as shown below. Draw the conventional current the way that positive charges would flow: from the positive terminal to the negative terminal.

To find the magnetic force that one current-carrying wire exerts on another current-carrying wire, follow these steps:

1. Determine which current the problem is thinking of as exerting the force. We call that I_a below. Determine which current the problem is thinking of as being pushed or pulled by the force. We call that I_b below. (**Note**: A current does not exert a net force on itself. We will find the force that one current exerts on another.)
2. Draw a field point (★) on I_b. Apply the right-hand rule for magnetic **field** (Chapter 22) to find the direction of the magnetic field ($\vec{\mathbf{B}}_a$) that I_a creates at the field point.
3. Apply the right-hand rule for magnetic **force** (Chapter 21) to find the direction of the magnetic force ($\vec{\mathbf{F}}_a$) exerted on I_b in the presence of the magnetic field ($\vec{\mathbf{B}}_a$).

Example: What is the direction of the magnetic force that the top current (I_1) exerts on the bottom current (I_2) in the diagram below?

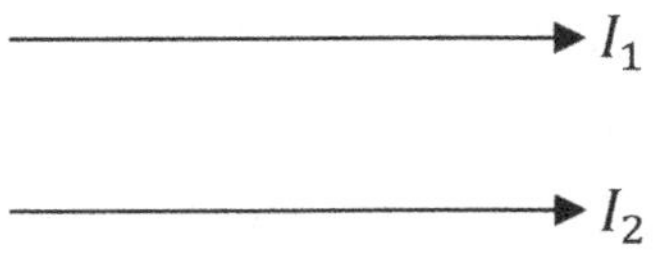

This problem is thinking of the magnetic force that is exerted by I_1 and which is pushing or pulling I_2. We therefore draw a field point on I_2, as shown below.

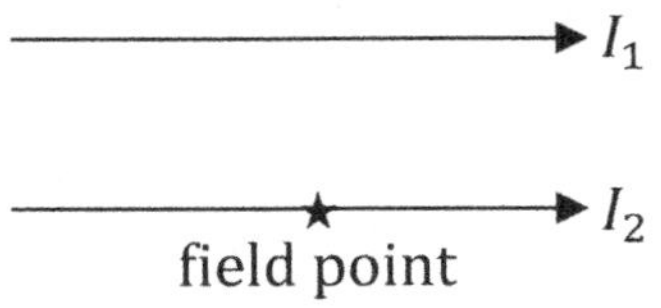

Apply the right-hand rule for magnetic **field** (Chapter 22) to I_1. Grab I_1 with your thumb along I_1 and your fingers wrapped around I_1. What are your fingers doing at the field point (★)? They are going into the page (⊗) at the field point (★). The magnetic field ($\vec{\mathbf{B}}_1$) that I_1 makes at the field point (★) is into the page (⊗).

Now apply the right-hand rule for magnetic force (Chapter 21) to I_2. Point your fingers to the right (→), along I_2. At the same time, face your palm into the page (⊗), along $\vec{\mathbf{B}}_1$. Your thumb points up (↑), along the magnetic force ($\vec{\mathbf{F}}_1$). The top current (I_1) pulls the bottom current (I_2) upward (↑).

Example: What is the direction of the magnetic force that the bottom current (I_2) exerts on the top current (I_1) in the diagram below?

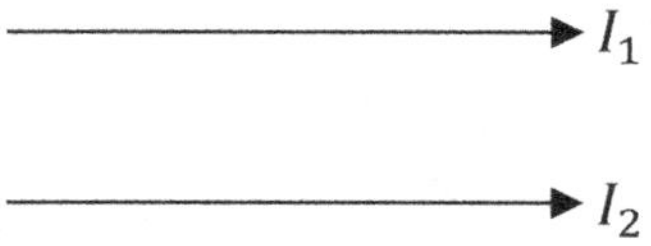

This time, the problem is thinking of the magnetic force that is exerted by I_2 and which is pushing or pulling I_1. We therefore draw a field point on I_1, as shown below.

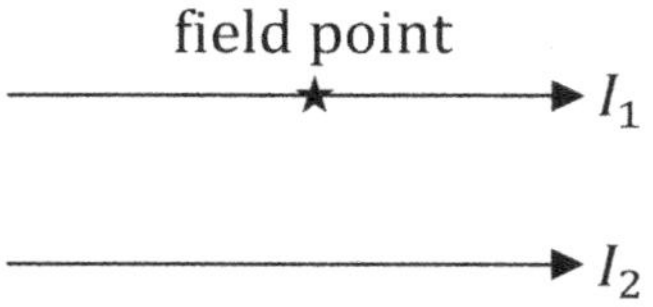

Apply the right-hand rule for magnetic **field** (Chapter 22) to I_2. Grab I_2 with your thumb along I_2 and your fingers wrapped around I_2. What are your fingers doing at the field point (★)? They are coming out of the page (⊙) at the field point (★). The magnetic field ($\vec{\mathbf{B}}_2$) that I_2 makes at the field point (★) is out of the page (⊙).

Now apply the right-hand rule for magnetic force (Chapter 21) to I_1. Point your fingers to the right (→), along I_1. At the same time, face your palm out of the page (⊙), along $\vec{\mathbf{B}}_2$. Your thumb points down (↓), along the magnetic force ($\vec{\mathbf{F}}_2$). The bottom current (I_2) pulls the top current (I_1) downward (↓).

It is instructive to compare these two examples. If you put them together, what you see is that two **parallel** currents **attract**. If you apply the right-hand rules to anti-parallel currents, you will discover that two **anti-parallel** currents **repel**.

Example: What is the direction of the magnetic force that the top current (I_1) exerts on the bottom current (I_2) in the diagram below?

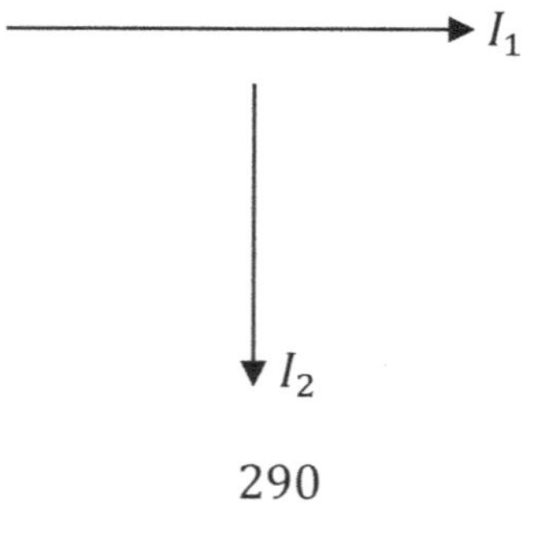

This problem is thinking of the magnetic force that is exerted by I_1 and which is pushing or pulling I_2. We therefore draw a field point on I_2, as shown below.

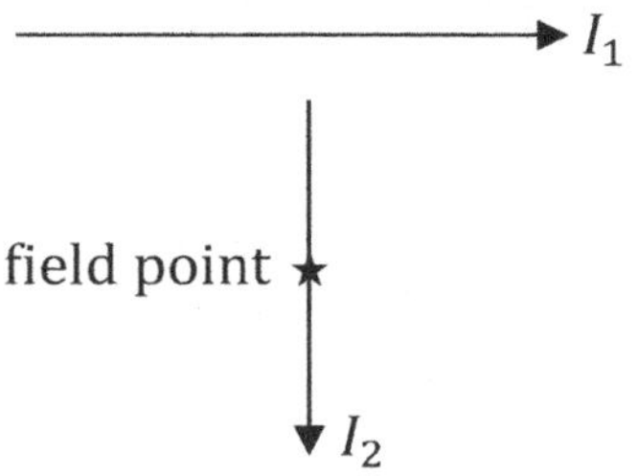

Apply the right-hand rule for magnetic **field** (Chapter 22) to I_1. Grab I_1 with your thumb along I_1 and your fingers wrapped around I_1. What are your fingers doing at the field point (★)? They are going into the page (⊗) at the field point (★). The magnetic field ($\vec{\mathbf{B}}_1$) that I_1 makes at the field point (★) is into the page (⊗).

Now apply the right-hand rule for magnetic force (Chapter 21) to I_2. Point your fingers down (↓), along I_2. At the same time, face your palm into the page (⊗), along $\vec{\mathbf{B}}_1$. Your thumb points to the right (→), along the magnetic force ($\vec{\mathbf{F}}_1$). The top current (I_1) pushes the bottom current (I_2) to the right (→).

Example: What is the direction of the magnetic force that the outer current (I_1) exerts on the inner current (I_2) at point A in the diagram below?

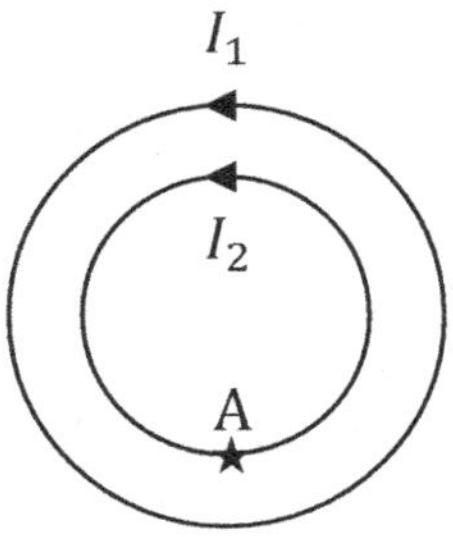

This problem is thinking of the magnetic force that is exerted by I_1 and which is pushing or pulling I_2. In this example, the field point (★) is already marked as point A. Apply the right-hand rule for magnetic **field** (Chapter 22) to I_1. Grab I_1 with your thumb along I_1 and your fingers wrapped around I_1. What are your fingers doing at point A (★)? They are coming out of the page (⊙) at point A (★). The magnetic field ($\vec{\mathbf{B}}_1$) that I_1 makes at the point A (★) is out of the page (⊙).

Now apply the right-hand rule for magnetic force (Chapter 21) to I_2. Do this at point A. Point your fingers to the right (→) at point A (since I_2 is heading to the right when it passes through point A). At the same time, face your palm out of the page (⊙), along $\vec{\mathbf{B}}_1$. **Tip**: Turn the book to make this more comfortable. Your thumb points down (↓), along the magnetic force ($\vec{\mathbf{F}}_1$). The top current (I_1) pulls the bottom current (I_2) down (↓) at point A.

83. Determine the direction of the magnetic force at the indicated point.

(A) that I_1 exerts on I_2

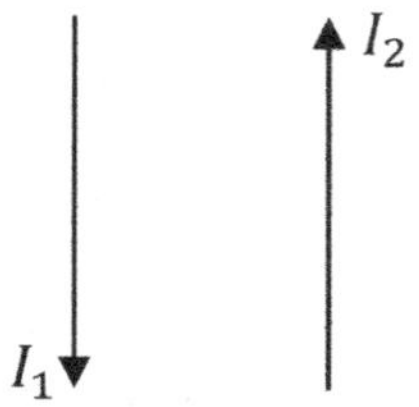

(B) that I_1 exerts on I_2

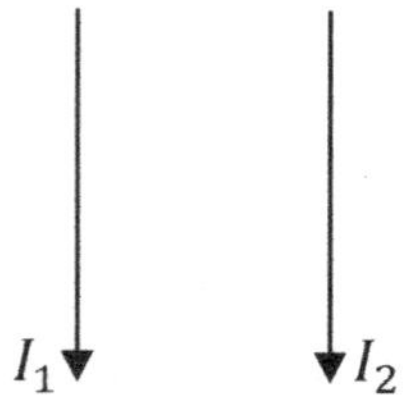

(C) that I_2 exerts on I_1

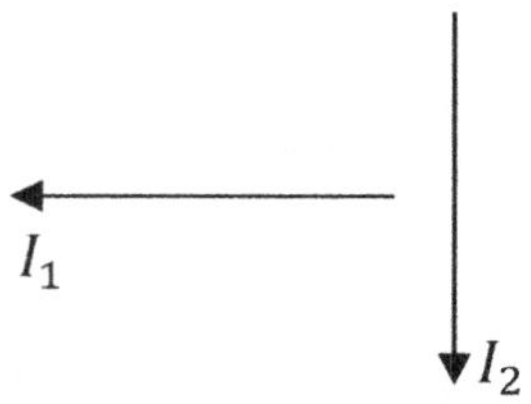

(D) that I_1 exerts on I_2 at A

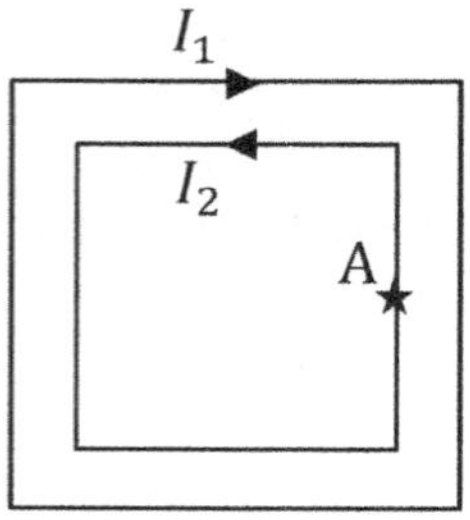

(E) that is exerted on I_2

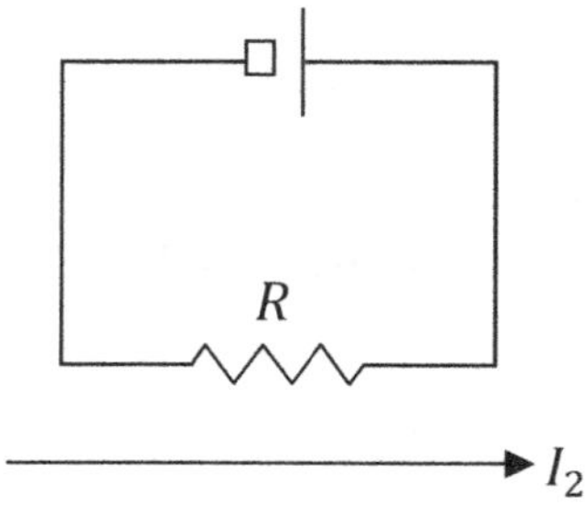

(F) that I_1 exerts on I_2

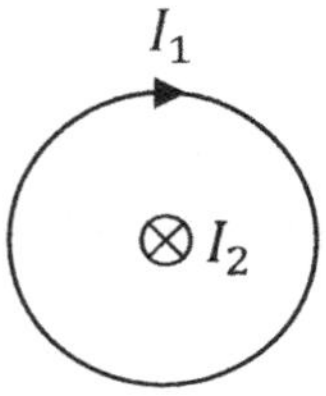

(G) that I_1 exerts on the p

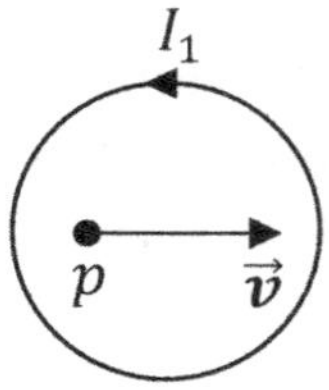

(H) that I_1 exerts on I_2

(I) that I_1 exerts on I_2 at C

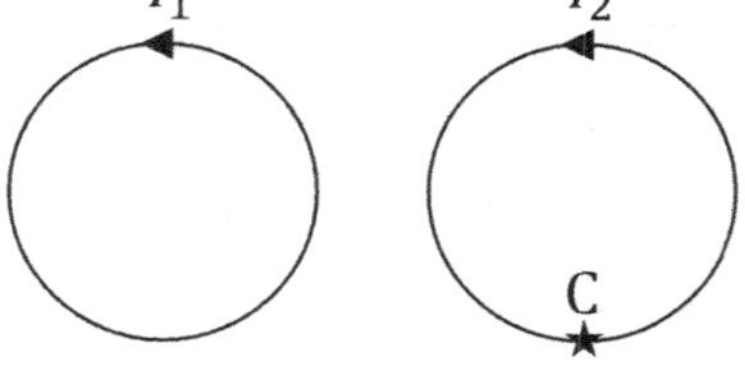

Want help? This problem is fully solved in the back of the book.

24 MAGNETIC FORCE

Relevant Terminology

Electric charge – a fundamental property of a particle that causes the particle to experience a force in the presence of an electric field. (An electrically neutral particle has no charge and thus experiences no force in the presence of an electric field.)
Velocity – a combination of speed and direction.
Current – the instantaneous rate of flow of charge through a wire.
Magnetic field – a magnetic effect created by a moving charge (or current).
Magnetic force – the push or pull that a moving charge (or current) experiences in the presence of a magnetic field.

Magnetic Force Equations

A charge moving in the presence of an external magnetic field ($\vec{\mathbf{B}}$) experiences a **magnetic force** ($\vec{\mathbf{F}}_m$) that involves the **vector product** between velocity ($\vec{\boldsymbol{v}}$) and magnetic field.

$$\vec{\mathbf{F}}_m = q\vec{\boldsymbol{v}} \times \vec{\mathbf{B}}$$

A current consists of a stream of moving charges. Thus, a current-carrying conductor similarly experiences a magnetic force in the presence of a magnetic field. In the equation below, $\vec{\mathbf{L}}$ is a displacement vector along the wire in the direction of the current.

$$\vec{\mathbf{F}}_m = I\vec{\mathbf{L}} \times \vec{\mathbf{B}}$$

Recall the equations for the **vector product** from Chapter 19:

$$\vec{\mathbf{A}} \times \vec{\mathbf{B}} = \begin{vmatrix} \hat{\mathbf{x}} & \hat{\mathbf{y}} & \hat{\mathbf{z}} \\ A_x & A_y & A_z \\ B_x & B_y & B_z \end{vmatrix} = \hat{\mathbf{x}} \begin{vmatrix} A_y & A_z \\ B_y & B_z \end{vmatrix} - \hat{\mathbf{y}} \begin{vmatrix} A_x & A_z \\ B_x & B_z \end{vmatrix} + \hat{\mathbf{z}} \begin{vmatrix} A_x & A_y \\ B_x & B_y \end{vmatrix}$$

$$\vec{\mathbf{A}} \times \vec{\mathbf{B}} = (A_y B_z - A_z B_y)\hat{\mathbf{x}} - (A_x B_z - A_z B_x)\hat{\mathbf{y}} + (A_x B_y - A_y B_x)\hat{\mathbf{z}}$$

$$\vec{\mathbf{A}} \times \vec{\mathbf{B}} = A_y B_z \hat{\mathbf{x}} - A_z B_y \hat{\mathbf{x}} + A_z B_x \hat{\mathbf{y}} - A_x B_z \hat{\mathbf{y}} + A_x B_y \hat{\mathbf{z}} - A_y B_x \hat{\mathbf{z}}$$

If you just need the magnitude of the magnetic force, use the following equations, where θ is the angle between $\vec{\boldsymbol{v}}$ and $\vec{\mathbf{B}}$ or between I and $\vec{\mathbf{B}}$.

$$F_m = |q|vB\sin\theta \quad \text{or} \quad F_m = ILB\sin\theta$$

If a charged particle moves in a **circle** in a uniform magnetic field, apply Newton's second law in the context of uniform circular motion, where $\sum F_{in}$ represents the sum of the **inward** components of the forces and a_c is the **centripetal** acceleration (discussed in Volume 1):

$$\sum F_{in} = ma_c \quad , \quad a_c = \frac{v^2}{R}$$

If you need to find the net **torque** exerted on a current loop with area A, use the following equation. Recall that torque (τ) was discussed in Volume 1 of this series.

$$\tau_{net} = IAB\sin\theta$$

Essential Concepts

Magnetic field is one of three kinds of fields that we have learned about:

- **Current** (I) or moving charge is the source of a **magnetic field** ($\vec{\mathbf{B}}$).
- **Charge** (q) is the source of an **electric field** ($\vec{\mathbf{E}}$).
- **Mass** (m) is the source of a **gravitational field** ($\vec{\mathbf{g}}$).

Source	Field	Force
mass (m)	gravitational field ($\vec{\mathbf{g}}$)	$\vec{\mathbf{F}}_g = m\vec{\mathbf{g}}$
charge (q)	electric field ($\vec{\mathbf{E}}$)	$\vec{\mathbf{F}}_e = q\vec{\mathbf{E}}$
current (I)	magnetic field ($\vec{\mathbf{B}}$)	$\vec{\mathbf{F}}_m = I\vec{\mathbf{L}} \times \vec{\mathbf{B}}$

Consider a few special cases of the equations $F_m = |q|vB \sin\theta$ and $F_m = ILB \sin\theta$:

- When the velocity ($\vec{v}$) or current (I) is **parallel** to the magnetic field ($\vec{\mathbf{B}}$), the angle is $\theta = 0°$ and the magnetic force ($\vec{\mathbf{F}}_m$) is **zero** because $\sin 0° = 0$.
- When the velocity ($\vec{v}$) or current (I) is **anti-parallel** to the magnetic field ($\vec{\mathbf{B}}$), the angle is $\theta = 180°$ and the magnetic force ($\vec{\mathbf{F}}_m$) is **zero** because $\sin 180° = 0$.
- When the velocity ($\vec{v}$) or current (I) is **perpendicular** to the magnetic field ($\vec{\mathbf{B}}$), the angle is $\theta = 90°$ and the magnetic force ($\vec{\mathbf{F}}_m$) is **maximum** because $\sin 90° = 1$.

Magnetic field supplies a **centripetal** force: Magnetic fields tend to cause moving charges to travel in circles. A charged particle traveling in a uniform magnetic field for which no other forces are significant will experience one of the following types of motion:

- The charge will travel in a **straight line** if $\theta = 0°$ or $\theta = 180°$. This is the case when the velocity ($\vec{v}$) is **parallel** or **anti-parallel** to the magnetic field ($\vec{\mathbf{B}}$).
- The charge will travel in a **circle** if $\theta = 90°$. This is the case when the velocity ($\vec{v}$) is **perpendicular** to the magnetic field ($\vec{\mathbf{B}}$).
- Otherwise, the charge will travel along a **helix**.

The two equations $\vec{\mathbf{F}}_m = q\vec{v} \times \vec{\mathbf{B}}$ and $\vec{\mathbf{F}}_m = I\vec{\mathbf{L}} \times \vec{\mathbf{B}}$ correspond to the right-hand rule for magnetic force that we learned in Chapter 21. The right-hand rule for magnetic force helps to visualize the direction of the magnetic force ($\vec{\mathbf{F}}_m$).

Recall the equation for torque (τ) from Volume 1 of this series, where θ in this equation is the angle between $\vec{\mathbf{r}}$ and $\vec{\mathbf{F}}$:

$$\tau = rF \sin\theta$$

When a magnetic field exerts a net torque on a rectangular current loop, the force is $F_m = ILB \sin\theta$. There are two torques (one on each half of the loop), where r is half the width of the loop: $r = \frac{W}{2}$. The net torque is then $\tau_{net} = \tau_1 + \tau_2 = \left(\frac{W}{2}\right)(ILB \sin\theta) + \left(\frac{W}{2}\right)(ILB \sin\theta) = ILWB \sin\theta$. Note that $A = LW$ is the area of the loop, such that:

$$\tau_{net} = IAB \sin\theta$$

This θ is the angle between the magnetic field ($\vec{\mathbf{B}}$) and the axis of the loop.

Protons, Neutrons, and Electrons

- **Protons** have **positive** electric charge.
- Neutrons are electrically neutral (although they are made up of fractionally charged particles called quarks).
- **Electrons** have **negative** electric charge.

Protons have a charge equal to 1.60×10^{-19} C (to three significant figures). We call this elementary charge and give it the symbol e. Electrons have the same charge, except for being negative. Thus, protons have charge $+e$, while electrons have charge $-e$. If you need to use the charge of a proton or electron to solve a problem, use the value of e below.

Elementary Charge
$e = 1.60 \times 10^{-19}$ C

Special Symbols

Symbol	Name
$\otimes$	into the page
$\odot$	out of the page
p	proton
n	neutron
e^-	electron

Symbols and Units

Symbol	Name	Units
$\vec{\mathbf{B}}$	magnetic field	T
B	magnitude of the magnetic field	T
$\vec{\mathbf{F}}_m$	magnetic force	N
F_m	magnitude of the magnetic force	N
q	electric charge	C
$\vec{\boldsymbol{v}}$	velocity	m/s
v	speed	m/s
I	current	A
$\vec{\mathbf{L}}$	displacement vector along the current	m
L	length of the wire	m
θ	angle between $\vec{\boldsymbol{v}}$ and $\vec{\mathbf{B}}$ or between I and $\vec{\mathbf{B}}$	° or rad

Notes Regarding Units

The SI unit of magnetic field (B) is the Tesla (T). A Tesla can be related to other SI units by solving for B in the equation $F_m = ILB \sin\theta$:

$$B = \frac{F_m}{IL \sin\theta}$$

Recall that the SI unit of magnetic force (F_m) is the Newton (N), the SI unit of current (I) is the Ampère (A), the SI unit of length (L) is the meter (m), and $\sin\theta$ is unitless (sine equals the ratio of two sides of a triangle, and the units cancel out in the ratio). Plugging these units into the above equation, a Tesla (T) equals:

$$1 \text{ T} = 1 \frac{\text{N}}{\text{A·m}}$$

It's better to write A·m than to write the m first because mA could be confused with milliAmps (mA). Recall from first-semester physics that a Newton is equivalent to:

$$1 \text{ N} = 1 \frac{\text{kg·m}}{\text{s}^2}$$

Plugging this into $\frac{\text{N}}{\text{A·m}}$, a Tesla can alternatively be expressed as $\frac{\text{kg}}{\text{A·s}^2}$. Magnetic field is sometimes expressed in Gauss (G), where $1 \text{ T} = 10^4 \text{ G}$, but the Tesla (T) is the SI unit.

Strategy to Find Magnetic Force

Note: If a problem gives you the magnetic field in Gauss (G), convert to Tesla (T):

$$1 \text{ G} = 10^{-4} \text{ T}$$

(Previously, we wrote this as $1 \text{ T} = 10^4$ G, which is equivalent. We write $1 \text{ G} = 10^{-4}$ T to convert from Gauss to Tesla, and $1 \text{ T} = 10^4$ G to convert back from Tesla to Gauss.)

How you solve a problem involving magnetic force depends on the context:

- You can relate the **magnitude** of the magnetic field (B) to the magnitude of the magnetic force (F_m) using trig:
$$F_m = |q|vB\sin\theta \quad \text{or} \quad F_m = ILB\sin\theta$$
Here, θ is the angle between $\vec{\boldsymbol{v}}$ and $\vec{\mathbf{B}}$ or between I and $\vec{\mathbf{B}}$.
- If a problem gives you vector expressions of the form $\vec{\mathbf{B}} = B_x\hat{\mathbf{x}} + B_y\hat{\mathbf{y}} + B_z\hat{\mathbf{z}}$, apply the **vector product** equations:
$$\vec{\mathbf{F}}_m = q\vec{\boldsymbol{v}} \times \vec{\mathbf{B}} \quad \text{or} \quad \vec{\mathbf{F}}_m = I\vec{\mathbf{L}} \times \vec{\mathbf{B}}$$
The vector product can be expressed as follows:
$$\vec{\mathbf{A}} \times \vec{\mathbf{B}} = \begin{vmatrix} \hat{\mathbf{x}} & \hat{\mathbf{y}} & \hat{\mathbf{z}} \\ A_x & A_y & A_z \\ B_x & B_y & B_z \end{vmatrix} = \hat{\mathbf{x}}\begin{vmatrix} A_y & A_z \\ B_y & B_z \end{vmatrix} - \hat{\mathbf{y}}\begin{vmatrix} A_x & A_z \\ B_x & B_z \end{vmatrix} + \hat{\mathbf{z}}\begin{vmatrix} A_x & A_y \\ B_x & B_y \end{vmatrix}$$
$$\vec{\mathbf{A}} \times \vec{\mathbf{B}} = (A_yB_z - A_zB_y)\hat{\mathbf{x}} - (A_xB_z - A_zB_x)\hat{\mathbf{y}} + (A_xB_y - A_yB_x)\hat{\mathbf{z}}$$
$$\vec{\mathbf{A}} \times \vec{\mathbf{B}} = A_yB_z\hat{\mathbf{x}} - A_zB_y\hat{\mathbf{x}} + A_zB_x\hat{\mathbf{y}} - A_xB_z\hat{\mathbf{y}} + A_xB_y\hat{\mathbf{z}} - A_yB_x\hat{\mathbf{z}}$$
- If a charged particle is traveling in a circle in an uniform magnetic field, apply Newton's second law in the context of uniform circular motion, where $\sum F_{in}$ is the sum of the **inward** components of the forces and a_c is the **centripetal** acceleration:
$$\sum F_{in} = ma_c$$
In most of the problems encountered in first-year physics courses, the only force acting on the charged particle is the magnetic force, such that:
$$|q|vB\sin\theta = ma_c$$
If the magnetic field is perpendicular to the velocity, $\theta = 90°$ and $\sin\theta = 1$. You may need to recall UCM equations, such as $a_c = \frac{v^2}{R}$, $v = R\omega$, and $\omega = \frac{2\pi}{T}$.
- To find the net torque exerted on a current loop in a magnetic field, apply the following equation:
$$\tau_{net} = IAB\sin\theta$$
The symbols in this equation are:
  - I is the current running through the loop.
  - A is the area of the loop. For example, for a rectangle, $A = LW$.
  - B is the magnitude of the magnetic field.
  - θ is the angle between the magnetic field ($\vec{\mathbf{B}}$) and the axis of the loop.
- To find the force that one current exerts on another, see Chapter 25.

Example: A particle with a charge of 200 μC is traveling vertically straight upward with a speed of 4.0 km/s in a region where there is a uniform magnetic field. The magnetic field has a magnitude of 500,000 G and is oriented horizontally to the east. What are the magnitude and direction of the magnetic force exerted on the charged particle?

Make a list of the known quantities:

- The charge is $q = 200$ μC. Convert this to SI units: $q = 2.00 \times 10^{-4}$ C. Recall that the metric prefix μ stands for 10^{-6}.
- The speed is $v = 4.0$ km/s. Convert this to SI units: $v = 4000$ m/s. Recall that the metric prefix k stands for 1000.
- The magnetic field has a magnitude of $B = 500{,}000$ G. Convert this to SI units using the conversion factor 1 G $= 10^{-4}$ T. The magnetic field is $B = 50$ T.
- The angle between $\vec{v}$ and $\vec{\mathbf{B}}$ is $\theta = 90°$ because the velocity and magnetic field are perpendicular: $\vec{v}$ is vertical, whereas $\vec{\mathbf{B}}$ is horizontal.

Use the appropriate trig equation to find the magnitude of the magnetic force.

$$F_m = |q|vB \sin\theta = (2 \times 10^{-4})(4000)(50) \sin 90° = 40 \text{ N}$$

One way to find the direction of the magnetic force is to apply the right-hand rule. Let's establish a coordinate system:*

- On a map, we usually choose $+x$ to point east.
- On a map, we usually choose $+y$ to point north.
- Then $+z$ is vertically upward. Note that in our picture, "up" means out of the page.

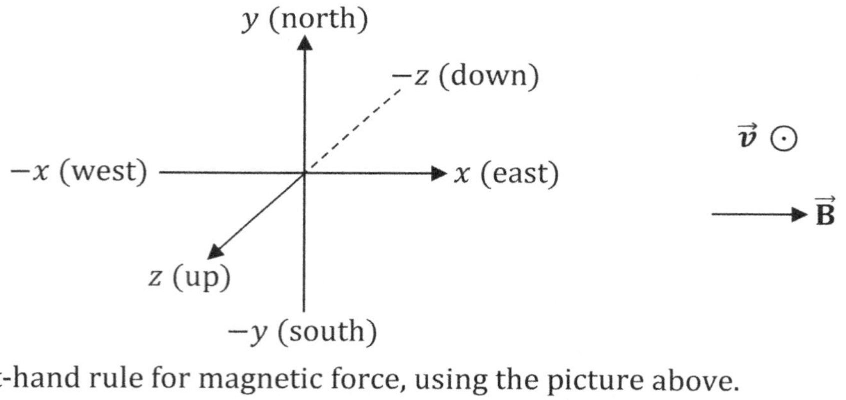

Apply the right-hand rule for magnetic force, using the picture above.

- Point your fingers out of the page (⊙), along the velocity ($\vec{v}$). Note that vertically "upward" is out of the page (<u>**not**</u> "up") in the map above.
- At the same time, face your palm to the right (→), along the magnetic field ($\vec{\mathbf{B}}$).
- **Tip**: Rotate your book to make it more comfortable to get your hand in this position.
- Your thumb points north (↑). This direction is north (<u>**not**</u> "up") in this context.

The direction ($\vec{\mathbf{F}}_m$) of the magnetic force is to the north.

---

* Your coordinate system needs to be right-handed: If you point the fingers of your right hand toward $+x$, curl your fingers toward $+y$, and your thumb points along $+z$, the coordinate system is right-handed.

There is an alternate way to solve the previous example. We could use our coordinate system to express the velocity and magnetic field as vectors:

- The velocity is $\vec{\boldsymbol{v}} = 4000\ \hat{\mathbf{z}}$ because vertically "upward" is along $+z$, which is out of the page ($\odot$) in our map.
- The magnetic field is $\vec{\mathbf{B}} = 50\ \hat{\mathbf{x}}$ because east is to the right ($\rightarrow$) along $+x$ in our map.

The magnetic force can then be found via the cross product:

$$\vec{\mathbf{F}}_m = q\vec{\boldsymbol{v}} \times \vec{\mathbf{B}} = q\begin{vmatrix} \hat{\mathbf{x}} & \hat{\mathbf{y}} & \hat{\mathbf{z}} \\ v_x & v_y & v_z \\ B_x & B_y & B_z \end{vmatrix} = (2 \times 10^{-4})\begin{vmatrix} \hat{\mathbf{x}} & \hat{\mathbf{y}} & \hat{\mathbf{z}} \\ 0 & 0 & 4000 \\ 50 & 0 & 0 \end{vmatrix}$$

You can work out this determinant the same way that we will in a couple of examples, but in this example, the problem is simpler:

$$\vec{\mathbf{F}}_m = q\vec{\boldsymbol{v}} \times \vec{\mathbf{B}} = (2 \times 10^{-4})(4000\ \hat{\mathbf{z}}) \times (50\ \hat{\mathbf{x}}) = 40\ \hat{\mathbf{z}} \times \hat{\mathbf{x}}$$

If you review Chapter 19, you will see that $\hat{\mathbf{z}} \times \hat{\mathbf{x}} = \hat{\mathbf{y}}$.

$$\vec{\mathbf{F}}_m = 40\ \hat{\mathbf{y}}$$

The magnetic force has a magnitude of $F_m = 40$ N and a direction of $\hat{\mathbf{y}}$. If you look at our map, you will see that $\hat{\mathbf{y}}$ points to the north.

Example: A current of 3.0 A runs along a 2.0-m long wire in a region where there is a magnetic field of 5.0 T. The current makes a 30° with the magnetic field. What is the magnitude of the magnetic force exerted on the wire?

Make a list of the known quantities:

- The current is $I = 3.0$ A.
- The length of the wire is $L = 2.0$ m.
- The magnetic field has a magnitude of $B = 5.0$ T.
- The angle between I and $\vec{\mathbf{B}}$ is $\theta = 30°$.

Use the appropriate trig equation to find the magnitude of the magnetic force.

$$F_m = ILB \sin\theta = (3)(2)(5) \sin 30° = 30\left(\frac{1}{2}\right) = 15 \text{ N}$$

The magnetic force exerted on the wire is $F_m = 15$ N.

Example: A current of 3.0 A runs along a wire with a displacement of $\vec{\mathbf{L}} = 4\,\hat{\mathbf{x}} + \hat{\mathbf{y}} - 5\,\hat{\mathbf{z}}$ in a region where the magnetic field is $\vec{\mathbf{B}} = \hat{\mathbf{x}} + 2\,\hat{\mathbf{y}} + 3\,\hat{\mathbf{z}}$, where SI units have been suppressed. Find the magnetic force exerted on the wire.

Compare $\vec{\mathbf{L}} = 4\,\hat{\mathbf{x}} + \hat{\mathbf{y}} - 5\,\hat{\mathbf{z}}$ to $\vec{\mathbf{L}} = L_x\hat{\mathbf{x}} + L_y\hat{\mathbf{y}} + L_z\hat{\mathbf{z}}$ to see that $L_x = 4$, $L_y = 1$, and $L_z = -5$. Similarly, compare $\vec{\mathbf{B}} = \hat{\mathbf{x}} + 2\,\hat{\mathbf{y}} + 3\,\hat{\mathbf{z}}$ to $\vec{\mathbf{B}} = B_x\hat{\mathbf{x}} + B_y\hat{\mathbf{y}} + B_z\hat{\mathbf{z}}$ to see that $B_x = 1$, $B_y = 2$, and $B_z = 3$. Plug these values into the determinant form of the vector product:

$$\vec{\mathbf{F}}_m = I\vec{\mathbf{L}} \times \vec{\mathbf{B}} = I\begin{vmatrix} \hat{\mathbf{x}} & \hat{\mathbf{y}} & \hat{\mathbf{z}} \\ L_x & L_y & L_z \\ B_x & B_y & B_z \end{vmatrix} = (3)\begin{vmatrix} \hat{\mathbf{x}} & \hat{\mathbf{y}} & \hat{\mathbf{z}} \\ 4 & 1 & -5 \\ 1 & 2 & 3 \end{vmatrix}$$

It may be helpful to review the vector product from Chapter 19. Note that the current ($I = 3.0$ A) is multiplying the determinant.

$$\vec{\mathbf{F}}_m = 3\,\hat{\mathbf{x}}\begin{vmatrix} 1 & -5 \\ 2 & 3 \end{vmatrix} - 3\,\hat{\mathbf{y}}\begin{vmatrix} 4 & -5 \\ 1 & 3 \end{vmatrix} + 3\,\hat{\mathbf{z}}\begin{vmatrix} 4 & 1 \\ 1 & 2 \end{vmatrix}$$

$$\vec{\mathbf{F}}_m = 3\,\hat{\mathbf{x}}[(1)(3) - (-5)(2)] - 3\,\hat{\mathbf{y}}[(4)(3) - (-5)(1)] + 3\,\hat{\mathbf{z}}[(4)(2) - (1)(1)]$$

$$\vec{\mathbf{F}}_m = 3\,\hat{\mathbf{x}}(3 + 10) - 3\,\hat{\mathbf{y}}(12 + 5) + 3\,\hat{\mathbf{z}}(8 - 1) = 3\,\hat{\mathbf{x}}(13) - 3\,\hat{\mathbf{y}}(17) + 3\,\hat{\mathbf{z}}(7)$$

$$\vec{\mathbf{F}}_m = 39\,\hat{\mathbf{x}} - 51\,\hat{\mathbf{y}} + 21\,\hat{\mathbf{z}}$$

Example: A particle with a positive charge of 500 μC and a mass of 4.0 g travels in a circle with a radius of 2.0 m with constant speed in a uniform magnetic field of 60 T. How fast is the particle traveling?

Apply Newton's second law to the particle. Since the particle travels with uniform circular motion (meaning constant speed in a circle), the acceleration is **centripetal** (inward):

$$\sum F_{in} = ma_c$$

The magnetic force supplies the needed centripetal force. Apply the equation $F_m = |q|vB\sin\theta$. Since the particle travels in a circle (and not a helix), we know that $\theta = 90°$.

$$|q|vB\sin 90° = ma_c$$

In Volume 1 of this series, we learned that $a_c = \frac{v^2}{R}$. Note that $\sin 90° = 1$.

$$|q|vB = m\frac{v^2}{R}$$

Divide both sides of the equation by the speed. Note that $\frac{v^2}{v} = v$.

$$|q|B = m\frac{v}{R}$$

Multiply both sides by R and divide by m. Convert the charge and mass to SI units: $q = 5.00 \times 10^{-4}$ C and $m = 4.0 \times 10^{-3}$ kg. Note that $\frac{10^{-4}}{10^{-3}} = 10^{-4} \times 10^{3} = 10^{-4+3} = 10^{-1}$.

$$v = \frac{|q|BR}{m} = \frac{(5 \times 10^{-4})(60)(2)}{4 \times 10^{-3}} = \frac{(5)(60)(2)}{4}\frac{10^{-4}}{10^{-3}} = 150 \times 10^{-1} = 15\text{ m/s}$$

Example: The rectangular loop illustrated below carries a current of 2.0 A. The magnetic field has a magnitude of 30 T. Determine the net torque that is exerted on the loop.

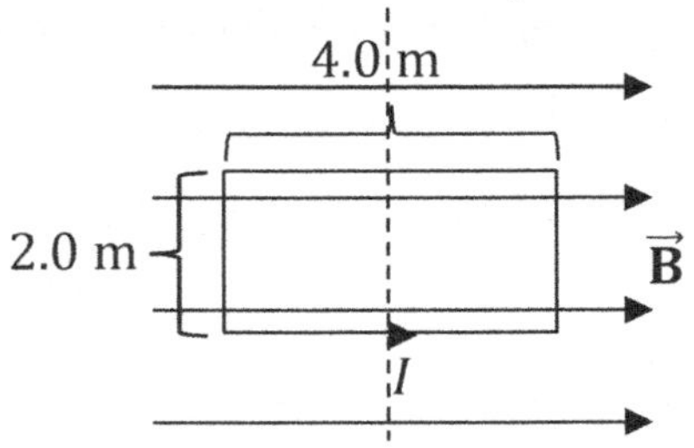

We saw conceptual problems like this in Chapter 21. If you apply the right-hand rule for magnetic force to each side of the loop, you will see that the loop will rotate about the dashed axis.

- The magnetic force ($\vec{\mathbf{F}}_m$) is zero for the top and bottom wires because the current (I) is parallel or anti-parallel to the magnetic field ($\vec{\mathbf{B}}$): $\theta = 0°$ or $\theta = 180°$.
- The magnetic force ($\vec{\mathbf{F}}_m$) pushes the right wire into the page (⊗) because your fingers point up (↑) along I and your palm faces right (→) along $\vec{\mathbf{B}}$, such that your thumb points into the page (⊗).
- The magnetic force ($\vec{\mathbf{F}}_m$) pushes the left wire out of the page (⊙) because your fingers point down (↓) along I and your palm faces right (→) along $\vec{\mathbf{B}}$, such that your thumb points out of the page (⊙).

Apply the equation for the net torque exerted on a current loop. The area of the loop is $A = LW$. The angle in the torque equation is $\theta = 90°$ because the magnetic field is perpendicular to the axis of the loop (which is different from the axis of rotation: the axis of rotation is the vertical dashed line about which the loop rotates, whereas the axis of the loop is perpendicular to the loop and passing through its center – this may be easier to visualize if you look at a picture of a solenoid in Chapter 22 and think about the axis of the solenoid, which is the same as the axis of each of its loops). The magnetic field is horizontal, while the axis of the loop (**<u>not</u>** the axis of rotation) is perpendicular to the page.

$$\tau_{net} = IAB\sin\theta = ILWB\sin\theta = (2)(4)(2)(30)\sin 90° = 480 \text{ Nm}$$

84. There is a uniform magnetic field of 0.50 T directed along the positive y-axis.

(A) Determine the force exerted on a 4.0-A current in a 3.0-m long wire heading along the negative z-axis.

(B) Determine the force on a 200-μC charge moving 60 km/s at an angle of 30° below the $+x$-axis.

(C) In which direction(s) could a proton travel and experience <u>zero</u> magnetic force?

Want help? Check the hints section at the back of the book.

Answers: 6.0 N along $\hat{\mathbf{x}}$, $3\sqrt{3}$ N along $\hat{\mathbf{z}}$, $\pm\hat{\mathbf{y}}$

85. A 200-μC charge has a velocity of $\vec{\boldsymbol{v}} = 3\,\hat{\mathbf{x}} - 2\,\hat{\mathbf{y}} - \hat{\mathbf{z}}$ in a region where the magnetic field is $\vec{\mathbf{B}} = 5\,\hat{\mathbf{x}} + \hat{\mathbf{y}} - 4\,\hat{\mathbf{z}}$, where SI units have been suppressed. Find the magnetic force exerted on the charge.

86. As illustrated below, a 0.25-g object with a charge of −400 μC travels in a circle with a constant speed of 4000 m/s in an approximately zero-gravity region where there is a uniform magnetic field of 200,000 G perpendicular to the page.

(A) What is the direction of the magnetic field?

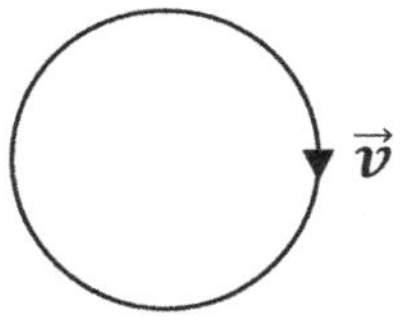

(B) What is the radius of the circle?

Want help? Check the hints section at the back of the book.

Answers: $(1.8\,\hat{\mathbf{x}} + 1.4\,\hat{\mathbf{y}} + 2.6\,\hat{\mathbf{z}}) \times 10^{-3}$ N, $\otimes$, 125 m

87. A 30-A current runs through the rectangular loop of wire illustrated below. There is a uniform magnetic field of 8000 G directed downward. The width (which is horizontal) of the rectangle is 50 cm and the height (which is vertical) of the rectangle is 25 cm.

(A) Find the magnitude of the net force exerted on the loop.

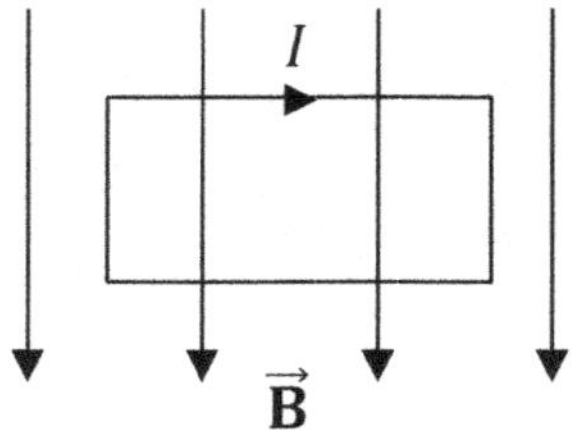

(B) Find the magnitude of the net torque exerted on the loop.

Want help? Check the hints section at the back of the book.

Answers: 0, 3.0 Nm

25 MAGNETIC FIELD

Relevant Terminology

Electric charge – a fundamental property of a particle that causes the particle to experience a force in the presence of an electric field. (An electrically neutral particle has no charge and thus experiences no force in the presence of an electric field.)
Current – the instantaneous rate of flow of charge through a wire.
Magnetic field – a magnetic effect created by a moving charge (or current).
Magnetic force – the push or pull that a moving charge (or current) experiences in the presence of a magnetic field.
Solenoid – a coil of wire in the shape of a right-circular cylinder.
Turns – the loops of a solenoid.
Permeability – a measure of how a substance affects a magnetic field.

Magnetic Force Equations

To find the magnetic field (B) created by a **long straight wire** a distance r_c from the axis of the wire (assuming that r_c is small compared to the length of the wire), use the following equation, where the constant μ_0 is the **permeability of free space.**

$$B = \frac{\mu_0 I}{2\pi r_c}$$

To find the magnetic field created by a **circular loop** of wire at the center of the loop, use the following equation, where a is the radius of the loop.

$$B = \frac{\mu_0 I}{2a}$$

To find the magnetic field created by a long, tightly wound **solenoid** near the center of the solenoid, use the following equation, where N is the number of loops (called **turns**), L is the length of the solenoid, and $n = \frac{N}{L}$ is the number of turns per unit length.

$$B = \frac{\mu_0 N I}{L} = \mu_0 n I$$

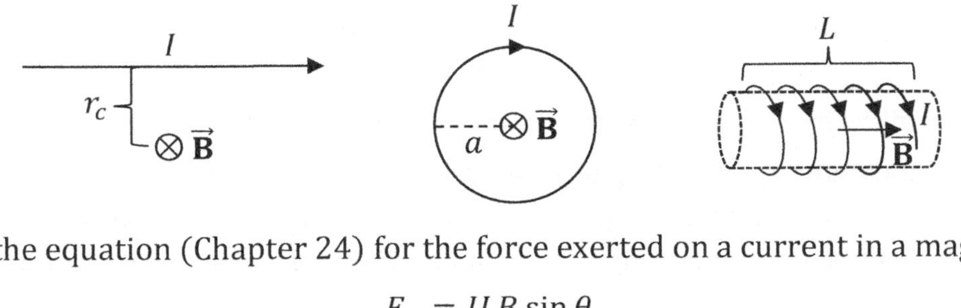

Also recall the equation (Chapter 24) for the force exerted on a current in a magnetic field.

$$F_m = ILB \sin\theta$$

Special Symbols

Symbol	Name
⊗	into the page
⊙	out of the page

Symbols and Units

Symbol	Name	Units
B	magnitude of the magnetic field	T
I	current	A
μ_0	permeability of free space	$\frac{\text{T·m}}{\text{A}}$
r_c	distance from a long, straight wire	m
a	radius of a loop	m
N	number of loops (or turns)	unitless
n	number of turns per unit length	$\frac{1}{\text{m}}$
L	length of a wire or length of a solenoid	m
F_m	magnitude of the magnetic force	N
θ	angle between $\vec{v}$ and $\vec{\mathbf{B}}$ or between I and $\vec{\mathbf{B}}$	° or rad

Important Distinction

Note that permittivity and permeability are two different quantities from two different contexts:

- The **permittivity** (ϵ) is an **electric** quantity (Chapter 12) relating to **electric** field ($\vec{\mathbf{E}}$).
- The **permeability** (μ) is a **magnetic** quantity relating to **magnetic** field ($\vec{\mathbf{B}}$).

In Chapter 32, we will see that the permittivity of free space (ϵ_0) and the permeability of free space (μ_0) can be combined together to make the speed of light in vacuum.

Strategy to Find Magnetic Field

Note: If a problem gives you the magnetic field in Gauss (G), convert to Tesla (T):

$$1 \text{ G} = 10^{-4} \text{ T}$$

How you solve a problem involving magnetic field depends on the context:

- To find the magnetic field created by a long straight wire, at the center of a circular loop, or at the center of a solenoid, use the appropriate equation:
  - At a distance r_c from the axis of a long **straight wire** (left figure on page 305):
  $$B = \frac{\mu_0 I}{2\pi r_c}$$
  - At the center of a **circular loop** of radius a:
  $$B = \frac{\mu_0 I}{2a}$$
  - Near the center of a long, tightly wound **solenoid** with N loops and length L:
  $$B = \frac{\mu_0 N I}{L} = \mu_0 n I$$
  Here, $n = \frac{N}{L}$ is the number of turns (or loops) per unit length.

  Note that the **permeability of free space** is $\mu_0 = 4\pi \times 10^{-7} \frac{\text{T·m}}{\text{A}}$.
- If you also need to find the **direction** of the magnetic field ($\vec{\mathbf{B}}$), apply the right-hand rule for magnetic field (Chapter 22).
- If there are two or more currents, to find the magnitude of the <u>net</u> **magnetic field** (B_{net}) at a specified point (called the **field point**), first find the magnetic field ($B_1, B_2, \cdots$) at the field point due to each current using the equations above, and then find the net magnetic field using the principle of **superposition.** This means to find the direction of $\vec{\mathbf{B}}_1, \vec{\mathbf{B}}_2, \cdots$ using the right-hand rule for magnetic field (Chapter 22), and then find B_{net} depending on the situation, as noted below:
  - If $\vec{\mathbf{B}}_1$ and $\vec{\mathbf{B}}_2$ are parallel, simply add their magnitudes: $B_{net} = B_1 + B_2$.
  - If $\vec{\mathbf{B}}_1$ and $\vec{\mathbf{B}}_2$ are anti-parallel, subtract their magnitudes: $B_{net} = |B_1 - B_2|$.
  - If $\vec{\mathbf{B}}_1$ is perpendicular to $\vec{\mathbf{B}}_2$, use the Pythagorean theorem: $B_{net} = \sqrt{B_1^2 + B_2^2}$.
  - Otherwise, add $\vec{\mathbf{B}}_1$ and $\vec{\mathbf{B}}_2$ according to vector addition (as in Chapter 3).
- If you need to find the magnitude of the magnetic force (F_m) exerted on a current (I) in an external magnetic field (B), apply the equation $F = ILB \sin\theta$ from Chapter 24. If you also need to find the direction of the magnetic force, apply the right-hand rule for magnetic force (Chapter 21).
- If there are two parallel or anti-parallel currents and you need to find the magnetic force that one current (call it I_a) exerts on the other current (call it I_b), first find the magnetic field that I_a creates at the location of I_b using $B_a = \frac{\mu_0 I_a}{2\pi d}$, where d is the distance between the currents. Next, find the force exerted on I_b using the equation

$F_a = I_b L_b B_a \sin\theta$. Note that both currents (I_a and I_b) get used in the math, but in different steps. Also note that L_b is the length of the wire for I_b, what we labeled as F_a is the force exerted on I_b (by I_a), and θ is the angle between I_b and $\vec{\mathbf{B}}_a$. If you also need to find the direction of the magnetic force that one current exerts on another, apply the technique discussed in Chapter 23.

- If there are three or more parallel or anti-parallel currents and you need to find the magnitude of the **net magnetic force** (F_{net}) exerted on one of the currents, first find the magnetic force ($F_1, F_2, \cdots$) exerted on the specified current due to each of the other currents using the technique from the previous step (regarding how to find the magnetic force that one current exerts on another current), and then find the net magnetic force using the principle of **superposition**. This means to find the direction of $\vec{\mathbf{F}}_1, \vec{\mathbf{F}}_2, \cdots$ using the technique from Chapter 23, and then find F_{net} depending on the situation, as noted below:
  - If $\vec{\mathbf{F}}_1$ and $\vec{\mathbf{F}}_2$ are parallel, simply add their magnitudes: $F_{net} = F_1 + F_2$.
  - If $\vec{\mathbf{F}}_1$ and $\vec{\mathbf{F}}_2$ are anti-parallel, subtract their magnitudes: $F_{net} = |F_1 - F_2|$.
  - If $\vec{\mathbf{F}}_1$ is perpendicular to $\vec{\mathbf{F}}_2$, use the Pythagorean theorem: $F_{net} = \sqrt{F_1^2 + F_2^2}$.
  - Otherwise, add $\vec{\mathbf{F}}_1$ and $\vec{\mathbf{F}}_2$ according to vector addition (as in Chapter 3).
- To find the **net magnetic force** (F_{net}) that one current exerts on a rectangular loop of wire, first apply the technique from Chapter 23 to find the direction of the magnetic force exerted on each side of the rectangular loop. If the long straight wire is perpendicular to two sides of the rectangular loop, two of these forces will cancel out, and then you can apply the technique from the previous step twice to solve the problem. This is illustrated in the last example of this chapter.
- If you need to **derive** an equation for magnetic field, see Chapters 26-27.

The Permeability of Free Space

The constant μ_0 is called the permeability of **free space** (meaning **vacuum**). The permeability of free space (μ_0) is $\mu_0 = 4\pi \times 10^{-7}\ \frac{\text{T}\cdot\text{m}}{\text{A}}$. The units of the permeability can be found by solving for μ_0 in the equation $B = \frac{\mu_0 I}{2\pi r_c}$ to get $\mu_0 = \frac{2\pi r_c B}{I}$. Since the SI unit of magnetic field (B) is the Tesla (T), the SI unit of current (I) is the Ampère (A), and the SI unit of distance (r_c) is the meter (m), it follows that the SI units of μ_0 are $\frac{\text{T}\cdot\text{m}}{\text{A}}$. It's better to write T·m than to write the m first because mT could be confused with milliTesla (mT). Recall from Chapter 24 that a Tesla (T) equals $1\ \text{T} = 1\ \frac{\text{N}}{\text{A}\cdot\text{m}}$. Plugging this in for a Tesla in the units of μ_0, we find that the units of μ_0 can alternatively be expressed as $\frac{\text{N}}{\text{A}^2}$. If you recall that a Newton is equivalent to $1\ \text{N} = 1\ \frac{\text{kg}\cdot\text{m}}{\text{s}^2}$, yet another way to write the units of μ_0 is $\frac{\text{kg}\cdot\text{m}}{\text{A}^2\text{s}^2}$.

Example: The long straight wire shown below carries a current of 5.0 A. What are the magnitude and direction of the magnetic field at the point marked with a star (★), which is 25 cm from the wire?

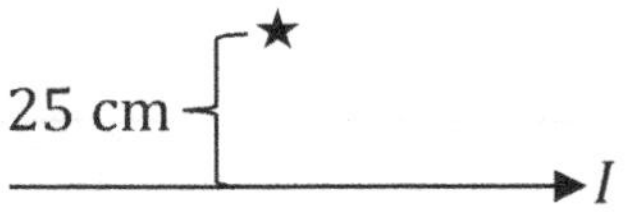

Convert the distance r_c from cm to m: $r_c = 25 \text{ cm} = 0.25 \text{ m} = \frac{1}{4}$ m. Apply the equation for the magnetic field created by a long straight wire.

$$B = \frac{\mu_0 I}{2\pi r_c} = \frac{(4\pi \times 10^{-7})(5)}{2\pi \left(\frac{1}{4}\right)}$$

To divide by a fraction, multiply by its **reciprocal**. Note that the reciprocal of $\frac{1}{4}$ is 4.

$$B = \frac{(4\pi \times 10^{-7})(5)}{2\pi}(4) = 40 \times 10^{-7} \text{ T} = 4.0 \times 10^{-6} \text{ T}$$

Apply the right-hand rule for magnetic field (Chapter 22) to find the direction of the magnetic field at the field point:

- Grab the current with your thumb pointing to the right (→), along the current (I).
- Your fingers make circles around the wire (toward your fingertips).
- The magnetic field ($\vec{\mathbf{B}}$) at a specified point is tangent to these circles. At the field point (★), the magnetic field ($\vec{\mathbf{B}}$) points out of the page (⊙).

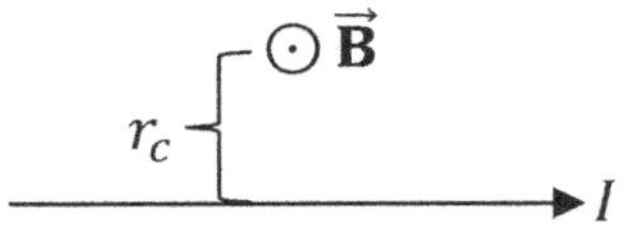

The magnitude of the magnetic field at the field point (★) is $B = 4.0 \times 10^{-6}$ T and its direction is out of the page (⊙).

Example: A tightly wound solenoid has a length of 50 cm, has 200 loops, and carries a current of 3.0 A. What is the magnitude of the magnetic field at the center of the solenoid?

Convert the distance L from cm to m: $L = 50 \text{ cm} = 0.50 \text{ m} = \frac{1}{2}$ m. Apply the equation for the magnetic field at the center of a solenoid.

$$B = \frac{\mu_0 N I}{L} = \frac{(4\pi \times 10^{-7})(200)(3)}{\frac{1}{2}}$$

To divide by a fraction, multiply by its **reciprocal**. Note that the reciprocal of $\frac{1}{2}$ is 2.

$$B = (4\pi \times 10^{-7})(200)(3)(2) = 4800\pi \times 10^{-7} \text{ T} = 4.8\pi \times 10^{-4} \text{ T}$$

The magnitude of the magnetic field near the center of the solenoid is $B = 4.8\pi \times 10^{-4}$ T, which could also be expressed in milliTesla (mT) as $B = 0.48\pi \text{ mT} = 1.5 \text{ mT}$.

Example: In the diagram below, the top wire carries a current of 3.0 A, the bottom wire carries a current of 4.0 A, each wire is 9.0 m long, and the distance between the wires is 0.50 m. What are the magnitude and direction of the net magnetic field at a point (★) that is midway between the two wires?

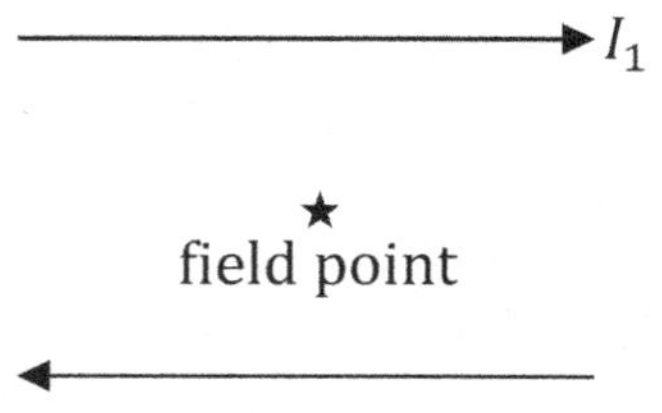

First find the magnitude of the magnetic fields created at the field point by each of the currents. In each case, $r_c = \frac{d}{2} = \frac{0.5}{2} = 0.25$ m because the field point (★) is halfway between the two wires.

$$B_1 = \frac{\mu_0 I_1}{2\pi r_c} = \frac{(4\pi \times 10^{-7})(3)}{2\pi(0.25)} = 24 \times 10^{-7}\text{ T} = 2.4 \times 10^{-6}\text{ T}$$

$$B_2 = \frac{\mu_0 I_2}{2\pi r_c} = \frac{(4\pi \times 10^{-7})(4)}{2\pi(0.25)} = 32 \times 10^{-7}\text{ T} = 3.2 \times 10^{-6}\text{ T}$$

Before we can determine how to combine these magnetic fields, we must apply the right-hand rule for magnetic field (Chapter 22) in order to determine the direction of each of these magnetic fields.

- Grab the current with your thumb pointing along the current. When you do this for I_1, your thumb will point right (→), and when you do this for I_2, your thumb will point left (←).
- Your fingers make circles around the wire (toward your fingertips).
- The magnetic field ($\vec{\mathbf{B}}$) at a specified point is tangent to these circles. At the field point (★), the magnetic fields ($\vec{\mathbf{B}}_1$ and $\vec{\mathbf{B}}_2$) both point into the page (⊗).
- Here's why: When you grab I_1 with your thumb pointing right, your fingers are below the wire and going into the page. When you grab I_2 with your thumb pointing left, your fingers are above that wire and are also going into the page.

Since $\vec{\mathbf{B}}_1$ and $\vec{\mathbf{B}}_2$ both point in the same direction, which is into the page (⊗), we add their magnitudes in order to find the magnitude of the net magnetic field.

$$B_{net} = B_1 + B_2 = 2.4 \times 10^{-6} + 3.2 \times 10^{-6} = 5.6 \times 10^{-6}\text{ T}$$

The net magnetic field at the field point has a magnitude of $B_{net} = 5.6 \times 10^{-6}$ T and a direction that is into the page (⊗).

Example: In the diagram below, the top wire carries a current of 2.0 A, the bottom wire carries a current of 3.0 A, each wire is 6.0 m long, and the distance between the wires is 0.20 m. What are the magnitude and direction of the magnetic force that the top current (I_1) exerts on the bottom current (I_2)?

First imagine a field point (★) at the location of I_2 (since the force specified in the problem is exerted on I_2), and find the magnetic field at the field point (★) created by I_1. When we do this, we use $I_1 = 2.0$ A and $r_c = d = 0.20$ m (since the field point is 0.20 m from I_1).

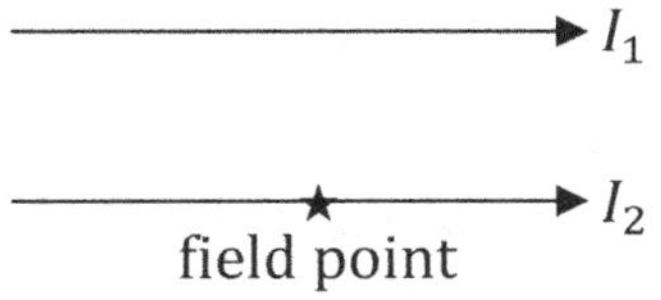

$$B_1 = \frac{\mu_0 I_1}{2\pi d} = \frac{(4\pi \times 10^{-7})(2)}{2\pi(0.2)} = 20 \times 10^{-7}\text{ T} = 2.0 \times 10^{-6}\text{ T}$$

Now we can find the force exerted on I_2. When we do this, we use $I_2 = 3.0$ A (since I_2 is experiencing the force specified in the problem) and $\theta = 90°$ (since I_2 is to the right and $\vec{\mathbf{B}}_1$ is into the page – as discussed below). What we're calling F_1 is the magnitude of the force that I_1 exerts on I_2.

$$F_1 = I_2 L_2 B_1 \sin\theta = (3)(6)(2 \times 10^{-6}) = 36 \times 10^{-6}\text{ N} = 3.6 \times 10^{-5}\text{ N}$$

To find the direction of this force, apply the technique from Chapter 23. First apply the right-hand rule for magnetic **field** (Chapter 22) to I_1. Grab I_1 with your thumb along I_1 and your fingers wrapped around I_1. What are your fingers doing at the field point (★)? They are going into the page (⊗) at the field point (★). The magnetic field ($\vec{\mathbf{B}}_1$) that I_1 makes at the field point (★) is into the page (⊗).

Now apply the right-hand rule for magnetic **force** (Chapter 21) to I_2. Point your fingers to the right (→), along I_2. At the same time, face your palm into the page (⊗), along $\vec{\mathbf{B}}_1$. Your thumb points up (↑), along the magnetic force ($\vec{\mathbf{F}}_1$). The top current (I_1) pulls the bottom current (I_2) upward (↑).

The magnetic force that I_1 exerts on I_2 has a magnitude of $F_1 = 3.6 \times 10^{-5}$ N and a direction that is straight upward (↑). (You might recall from Chapter 23 that **parallel** currents **attract** one another.)

Example: In the diagram below, the left wire carries a current of 4.0 A, the middle wire carries a current of 5.0 A, the right wire carries a current of 8.0 A, each wire is 3.0 m long, and the distance between neighboring wires is 0.10 m. What are the magnitude and direction of the net magnetic force exerted on the right current (I_3)?

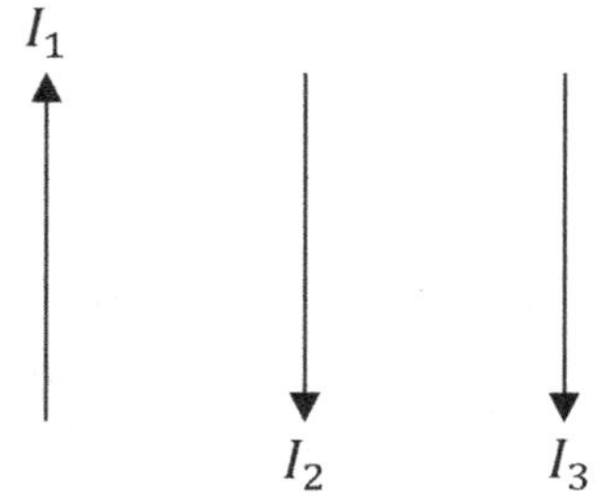

Here is how we will solve this problem:

- We'll find the force that I_1 exerts on I_3 the way that we solved the previous example.
- We'll similarly find the force that I_2 exerts on I_3.
- Once we know the magnitudes and directions of both forces, we will know how to combine them (see the top bullet point on page 308).

First imagine a field point (★) at the location of I_3 (since the force specified in the problem is exerted on I_3), and find the magnetic fields at the field point (★) created by I_1 and I_2. When we do this, note that $d_1 = 0.1 + 0.1 = 0.20$ m and $d_2 = 0.10$ m (since these are the distances from I_1 to I_3 and from I_2 to I_3, respectively).

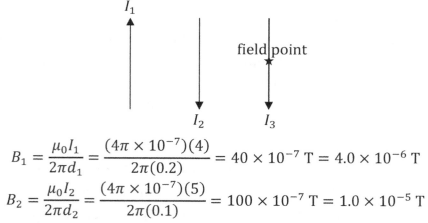

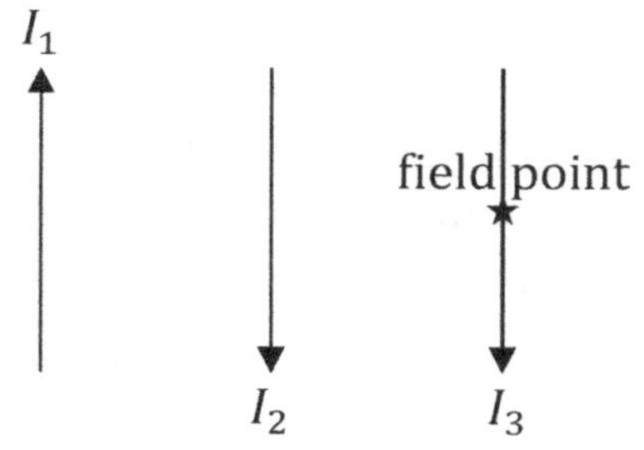

$$B_1 = \frac{\mu_0 I_1}{2\pi d_1} = \frac{(4\pi \times 10^{-7})(4)}{2\pi(0.2)} = 40 \times 10^{-7} \text{ T} = 4.0 \times 10^{-6} \text{ T}$$

$$B_2 = \frac{\mu_0 I_2}{2\pi d_2} = \frac{(4\pi \times 10^{-7})(5)}{2\pi(0.1)} = 100 \times 10^{-7} \text{ T} = 1.0 \times 10^{-5} \text{ T}$$

Now we can find the forces that I_1 and I_2 exert on I_3. When we do this, we use $I_3 = 8.0$ A (since I_3 is experiencing the force specified in the problem) and $\theta = 90°$ (since I_3 is down and since $\vec{\mathbf{B}}_1$ and $\vec{\mathbf{B}}_2$ are perpendicular to the page – as discussed on the following page). What we're calling F_1 is the magnitude of the force that I_1 exerts on I_3, and what we're calling F_2 is the magnitude of the force that I_2 exerts on I_3.

$$F_1 = I_3 L_3 B_1 \sin\theta = (8)(3)(4 \times 10^{-6}) \sin 90° = 96 \times 10^{-6} \text{ N} = 9.6 \times 10^{-5} \text{ N}$$

$$F_2 = I_3 L_3 B_2 \sin\theta = (8)(3)(1 \times 10^{-5}) \sin 90° = 24 \times 10^{-5} \text{ N} = 2.4 \times 10^{-4} \text{ N}$$

Before we can determine how to combine these magnetic forces, we must apply the technique from Chapter 23 in order to determine the direction of each of these forces.

First apply the right-hand rule for magnetic **field** (Chapter 22) to I_1 and I_2. When you grab I_1 with your thumb along I_1 and your fingers wrapped around I_1, your fingers are going into the page (⊗) at the field point (★). The magnetic field ($\vec{\mathbf{B}}_1$) that I_1 makes at the field point (★) is into the page (⊗). When you grab I_2 with your thumb along I_2 and your fingers wrapped around I_2, your fingers are coming out of the page (⊙) at the field point (★). The magnetic field ($\vec{\mathbf{B}}_2$) that I_2 makes at the field point (★) is out of the page (⊙).

Now apply the right-hand rule for magnetic **force** (Chapter 21) to I_3. Point your fingers down (↓), along I_3. At the same time, face your palm into the page (⊗), along $\vec{\mathbf{B}}_1$. Your thumb points to the right (→), along the magnetic force ($\vec{\mathbf{F}}_1$) that I_1 exerts on I_3. The current I_1 pushes I_3 to the right (→). Once again, point your fingers down (↓), along I_3. At the same time, face your palm out of the page (⊙), along $\vec{\mathbf{B}}_2$. Now your thumb points to the left (←), along the magnetic force ($\vec{\mathbf{F}}_2$) that I_2 exerts on I_3. (You might recall from Chapter 23 that parallel currents, like I_2 and I_3, attract one another and **anti-parallel** currents, like I_1 and I_3, **repel** one another.)

Since $\vec{\mathbf{F}}_1$ and $\vec{\mathbf{F}}_2$ point in opposite directions, as $\vec{\mathbf{F}}_1$ points right (→) and $\vec{\mathbf{F}}_2$ points left (←), we subtract their magnitudes in order to find the magnitude of the net magnetic force. We use absolute values because the magnitude of the net magnetic force can't be negative.

$$F_{net} = |F_1 - F_2| = |9.6 \times 10^{-5} - 2.4 \times 10^{-4}|$$

You need to express both numbers in the same power of 10 before you subtract them. Note that $2.4 \times 10^{-4} = 24 \times 10^{-5}$. (Enter both numbers on your calculator and compare them, if necessary.)

$$F_{net} = |9.6 \times 10^{-5} - 24 \times 10^{-5}| = |-14.4 \times 10^{-5} \text{ N}| = 14.4 \times 10^{-5} \text{ N} = 1.44 \times 10^{-4} \text{ N}$$

The net magnetic force that I_1 and I_2 exert on I_3 has a magnitude of $F_{net} = 1.4 \times 10^{-4}$ N (to two significant figures) and a direction that is to the left (←). The reason that it's to the left is that F_2 (which equals 2.4×10^{-4} N, which is the same as 24×10^{-5} N) is greater than F_1 (which equals 9.6×10^{-5} N), and the dominant force $\vec{\mathbf{F}}_2$ points to the left. When you put both numbers in the same power of 10, it's easier to see that 24×10^{-5} N is greater than 9.6×10^{-5} N. (If the two forces aren't parallel or anti-parallel, you would need to apply trig, as in Chapter 3, in order to find the direction of the net magnetic force.)

Example: In the diagram below, the long straight wire is 14.0 m long and carries a current of 2.0 A, while the square loop carries a current of 3.0 A. What are the magnitude and direction of the net magnetic force that the top current (I_1) exerts on the square loop (I_2)?

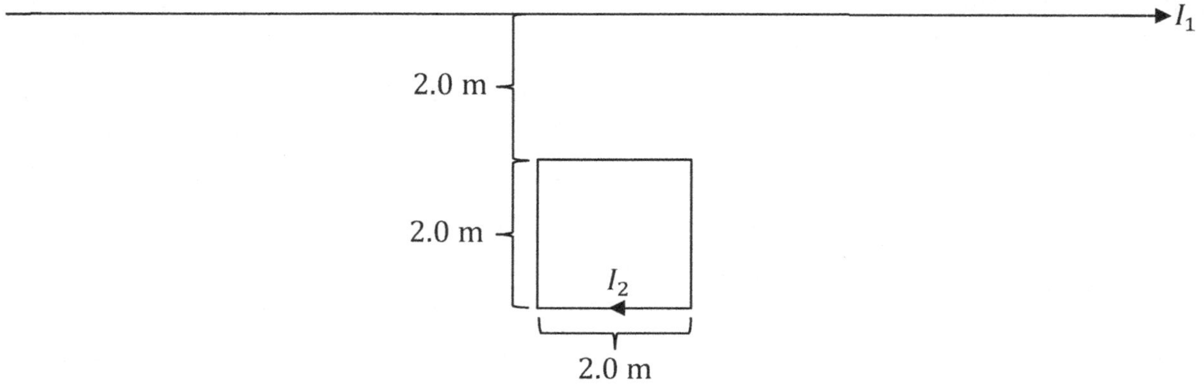

The first step is to determine the direction of the magnetic force that I_1 exerts on each side of the square loop. Apply the right-hand rule for magnetic **field** (Chapter 22) to I_1. Grab I_1 with your thumb along I_1 and your fingers wrapped around I_1. What are your fingers doing below I_1, where the square loop is? They are going into the page (⊗) where the square loop is. The magnetic field ($\vec{\mathbf{B}}_1$) that I_1 makes at the square loop is into the page (⊗).

Now apply the right-hand rule for magnetic **force** (Chapter 21) to each side of the loop.

- <u>**Bottom side**</u>: Point your fingers to the left (←) along the current (I_2) and your palm into the page (⊗) along the magnetic field ($\vec{\mathbf{B}}_1$). The magnetic force ($\vec{\mathbf{F}}_{bot}$) is down (↓).
- <u>**Left side**</u>: Point your fingers up (↑) along the current (I_2) and your palm into the page (⊗) along the magnetic field ($\vec{\mathbf{B}}_1$). The magnetic force ($\vec{\mathbf{F}}_{left}$) is left (←).
- <u>**Top side**</u>: Point your fingers to the right (→) along the current (I_2) and your palm into the page (⊗) along the magnetic field ($\vec{\mathbf{B}}_1$). The magnetic force ($\vec{\mathbf{F}}_{top}$) is up (↑).
- <u>**Right side**</u>: Point your fingers down (↓) along the current (I_2) and your palm into the page (⊗) along the magnetic field ($\vec{\mathbf{B}}_1$). The magnetic force ($\vec{\mathbf{F}}_{right}$) is right (→).

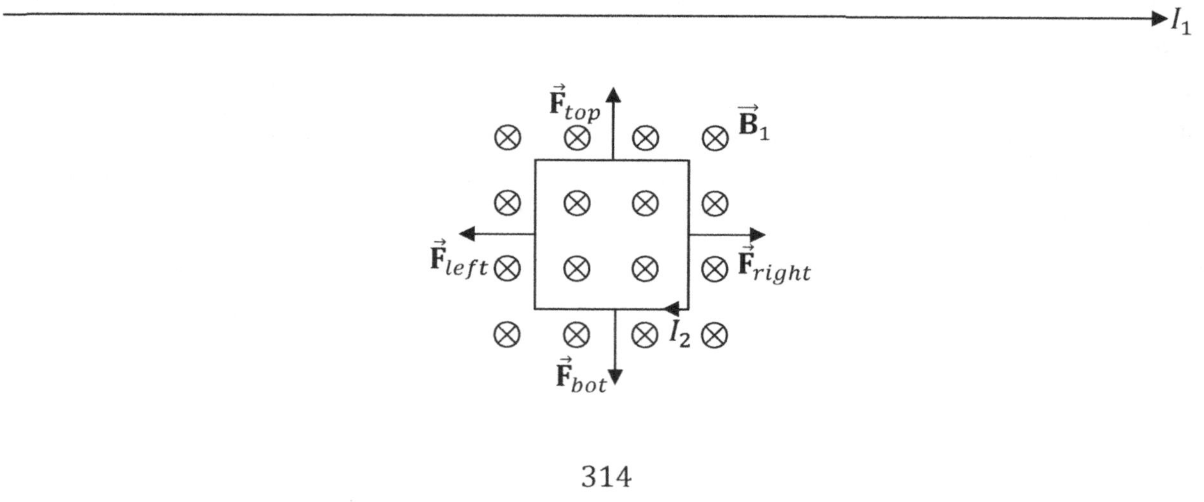

Study the diagram at the bottom of the previous page. You should see that $\vec{\mathbf{F}}_{left}$ and $\vec{\mathbf{F}}_{right}$ cancel out:

- $\vec{\mathbf{F}}_{left}$ and $\vec{\mathbf{F}}_{right}$ point in **opposite** directions: One points right, the other points left.
- $\vec{\mathbf{F}}_{left}$ and $\vec{\mathbf{F}}_{right}$ have **equal** magnitudes: They are the same distance from I_1.

We don't need to calculate $\vec{\mathbf{F}}_{left}$ and $\vec{\mathbf{F}}_{right}$ because they will cancel out later when we find the magnitude of the net force. (These would also be a challenge to calculate since they are perpendicular to I_1: That problem involves calculus.)

You might note that $\vec{\mathbf{F}}_{top}$ and $\vec{\mathbf{F}}_{bot}$ also have opposite directions (one points up, the other points down). However, $\vec{\mathbf{F}}_{top}$ and $\vec{\mathbf{F}}_{bot}$ do **<u>not</u>** cancel because $\vec{\mathbf{F}}_{top}$ is closer to I_1 and $\vec{\mathbf{F}}_{bot}$ is further from I_1.

To begin the math, find the magnetic fields created by I_1 at the top and bottom of the square loop. When we do this, note that $d_{top} = 2.0$ m and $d_{bot} = 2 + 2 = 4.0$ m (since these are the distances from I_1 to the top and bottom of the square loop, respectively). We use $I_1 = 2.0$ A in each case because I_1 is creating these two magnetic fields.

$$B_{top} = \frac{\mu_0 I_1}{2\pi d_{top}} = \frac{(4\pi \times 10^{-7})(2)}{2\pi(2)} = 2.0 \times 10^{-7} \text{ T}$$

$$B_{bot} = \frac{\mu_0 I_1}{2\pi d_{bot}} = \frac{(4\pi \times 10^{-7})(2)}{2\pi(4)} = 1.0 \times 10^{-7} \text{ T}$$

Now we can find the forces that I_1 exerts on the top and bottom sides of the square loop. When we do this, we use $I_2 = 3.0$ A (since I_2 is experiencing the force specified in the problem) and $\theta = 90°$ (since we already determined that $\vec{\mathbf{B}}_1$, which points into the page, is perpendicular to the loop). We also use the width of the square, $L_2 = 2.0$ m, since that is the distance that I_2 travels in the top and bottom sides of the square loop.

$$F_{top} = I_2 L_2 B_{top} \sin\theta = (3)(2)(2.0 \times 10^{-7}) \sin 90° = 12.0 \times 10^{-7} \text{ N}$$

$$F_{bot} = I_2 L_2 B_{bot} \sin\theta = (3)(2)(1.0 \times 10^{-7}) \sin 90° = 6.0 \times 10^{-7} \text{ N}$$

Since $\vec{\mathbf{F}}_{top}$ and $\vec{\mathbf{F}}_{bot}$ point in opposite directions, as $\vec{\mathbf{F}}_{top}$ points up (↑) and $\vec{\mathbf{F}}_{bot}$ points down (↓), we subtract their magnitudes in order to find the magnitude of the net magnetic force. We use absolute values because the magnitude of the net magnetic force can't be negative.

$$F_{net} = |F_{top} - F_{bot}| = |12.0 \times 10^{-7} - 6.0 \times 10^{-7}| = 6.0 \times 10^{-7} \text{ N}$$

The net magnetic force that I_1 exerts on I_2 has a magnitude of $F_{net} = 6.0 \times 10^{-7}$ N and a direction that is straight upward (↑). The reason that it's upward is that F_{top} (which equals 12.0×10^{-7} N) is greater than F_{bot} (which equals 6.0×10^{-7} N), and the dominant force $\vec{\mathbf{F}}_{top}$ points upward. (If the two forces aren't parallel or anti-parallel, you would need to apply trig, as in Chapter 3, in order to find the direction of the net magnetic force.)

88. Three currents are shown below. The currents run perpendicular to the page in the directions indicated. The triangle, which is **not** equilateral, has a base of 4.0 m and a height of 4.0 m. Find the magnitude and direction of the net magnetic field at the midpoint of the base.

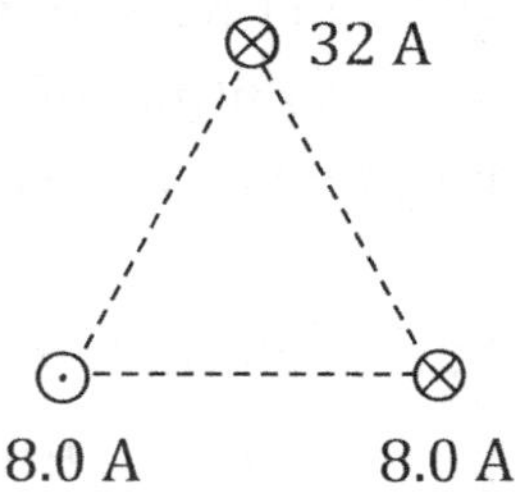

Want help? Check the hints section at the back of the book.

Answers: $16\sqrt{2} \times 10^{-7}$ T, 135°

89. In the diagram below, the top wire carries a current of 8.0 A, the bottom wire carries a current of 5.0 A, each wire is 3.0 m long, and the distance between the wires is 0.050 m. What are the magnitude and direction of the magnetic force that the top current (I_1) exerts on the bottom current (I_2)?

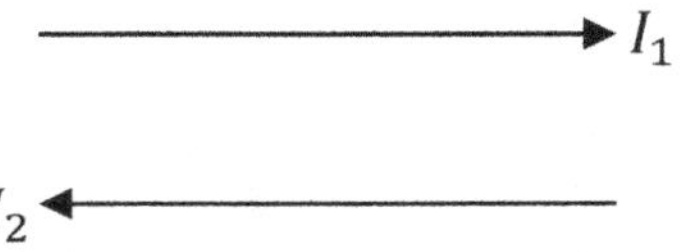

Want help? Check the hints section at the back of the book.

Answers: 4.8×10^{-4} N, down

90. In the diagram below, the left wire carries a current of 3.0 A, the middle wire carries a current of 4.0 A, the right wire carries a current of 6.0 A, each wire is 5.0 m long, and the distance between neighboring wires is 0.25 m. What are the magnitude and direction of the net magnetic force exerted on the right current (I_3)?

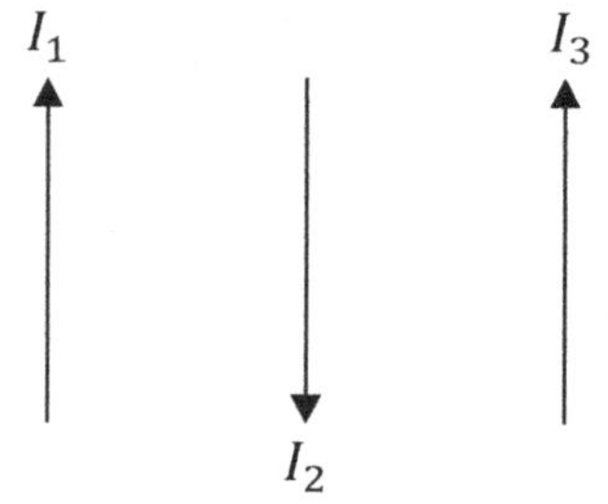

Want help? Check the hints section at the back of the book.

Answers: 6.0×10^{-5} N, right

91. The three currents below lie at the three corners of a square. The currents run perpendicular to the page in the directions indicated. The currents run through 3.0-m long wires, while the square has 25-cm long edges. Find the magnitude and direction of the net magnetic force exerted on the 2.0-A current.

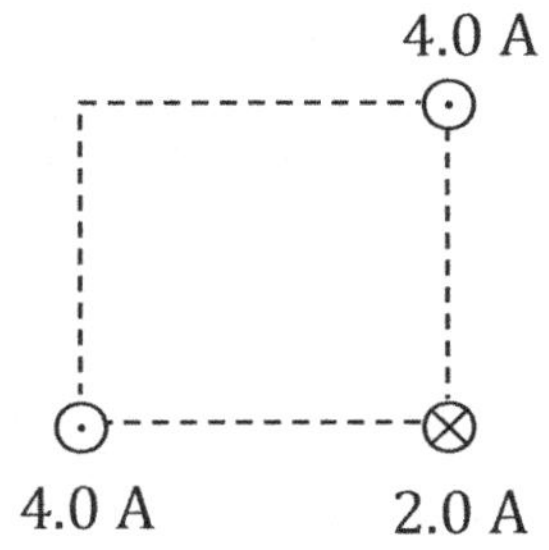

Want help? Check the hints section at the back of the book.

Answers: $192\sqrt{2} \times 10^{-7}$ N, 315°

92. In the diagram below, the long straight wire is 5.0 m long and carries a current of 6.0 A, while the rectangular loop carries a current of 8.0 A. What are the magnitude and direction of the net magnetic force that the top current (I_1) exerts on the rectangular loop (I_2)?

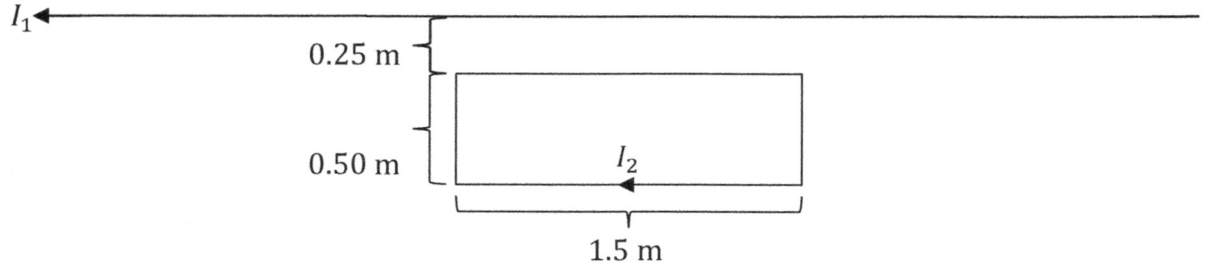

Want help? Check the hints section at the back of the book.

Answers: 3.84×10^{-5} N, down

26 THE LAW OF BIOT-SAVART

Relevant Terminology

Electric charge – a fundamental property of a particle that causes the particle to experience a force in the presence of an electric field. (An electrically neutral particle has no charge and thus experiences no force in the presence of an electric field.)
Current – the instantaneous rate of flow of charge through a conductor.
Filamentary current – a current that runs through a very thin wire.
Magnetic field – a magnetic effect created by a moving charge (or current).
Magnetic force – the push or pull that a moving charge (or current) experiences in the presence of a magnetic field.
Field point – the point where you are trying to calculate the magnetic field. The field point is ordinarily specified in the problem. There often is **not** a current at the field point.
Source – the current (or moving charge) that is creating the magnetic field.

Current Densities

We use different kinds of current densities when performing magnetic field integrals, depending upon the geometry.

- A very thin current is called a filamentary current. We don't use a current density for a filamentary current. We just work with the current, I, in this case.
- A surface current density $\vec{\mathbf{K}}$ applies to current that runs along a conducting surface, such as a gold plate or a strip of copper.
- Current density $\vec{\mathbf{J}}$ is distributed throughout a volume, such as current that runs along the length of a solid right-circular cylinder.
- For a moving charge, such as a rotating charged disc, we work with velocity ($\vec{\boldsymbol{v}}$) rather than current densities.

The current densities $\vec{\mathbf{K}}$ and $\vec{\mathbf{J}}$ are analogous to the charge densities λ, σ, and ρ, which we encountered in Chapters 7-8. It may help to review Chapter 7 before you proceed, since the law of Biot-Savart is similar in many ways to the electric field integrals in Chapter 7.

Notation

Unfortunately, not all electricity and magnetism books adopt the same notation. If you're also reading another book, be sure to compare notation carefully (this workbook has a handy chart in each chapter defining each symbol). For example, another book might use $d\ell$ for arc length (instead of ds), dS (uppercase) for surface area (instead of dA), or $d\tau$ (instead of dV) for volume.

The Law of Biot-Savart

The law of Biot-Savart is a magnetic field integral. How we express the law of Biot-Savart depends on the context:

- For a filamentary current (which runs through a very thin wire), we write:
$$\vec{\mathbf{B}} = \frac{\mu_0}{4\pi}\int \frac{I\, d\vec{\mathbf{s}} \times \hat{\mathbf{R}}}{R^2}$$
Here the integral is over the length of the wire.
- For a surface current (running along a surface, like a conducting strip), we write:
$$\vec{\mathbf{B}} = \frac{\mu_0}{4\pi}\int \frac{\vec{\mathbf{K}}\, dA \times \hat{\mathbf{R}}}{R^2}$$
Here, $\vec{\mathbf{K}}$ represents surface current density, the integral is over the surface area of the strip, and $I = \int \vec{\mathbf{K}} \cdot d\vec{\boldsymbol{\ell}}$, where $\vec{\boldsymbol{\ell}}$ is a distance across the width of the strip. The vector $d\vec{\boldsymbol{\ell}}$ has a direction that is perpendicular to the width of the strip (the direction of $d\vec{\boldsymbol{\ell}}$ is generally along the current density $\vec{\mathbf{K}}$).
- For a current distributed through a volume (like a thick wire), we write:
$$\vec{\mathbf{B}} = \frac{\mu_0}{4\pi}\int \frac{\vec{\mathbf{J}}\, dV \times \hat{\mathbf{R}}}{R^2}$$
Here, $\vec{\mathbf{J}}$ represents current density, the integral is over the volume of the conductor, and $I = \int \vec{\mathbf{J}} \cdot d\vec{\mathbf{A}}$, where $\vec{\mathbf{A}}$ represents cross-sectional area. The vector $d\vec{\mathbf{A}}$ has a direction which is perpendicular to the cross-sectional area (the direction of $d\vec{\mathbf{A}}$ is generally along the current density $\vec{\mathbf{J}}$).
- For moving charges (such as a rotating charged disc), we write:
$$\vec{\mathbf{B}} = \frac{\mu_0}{4\pi}\int \frac{\vec{\boldsymbol{v}}\, dq \times \hat{\mathbf{R}}}{R^2}$$
Here, $\vec{\boldsymbol{v}}$ represents velocity. If the charged object is rotating, note that $v = r_{rot}\omega$, where r_{rot} is the radius of the circle that each dq makes during the rotation.

Note: In all of these integrals, it is to be understood that $\hat{\mathbf{R}}$ is inside the integrand. Perform the vector product (according to Chapter 19) **before** integrating, and include the result of the vector product as part of what you're integrating over.

Special Symbols

Symbol	Name
$\otimes$	into the page
$\odot$	out of the page

Symbols and SI Units

Symbol	Name	SI Units
I	current	A
$\vec{\mathbf{B}}$	magnetic field	T
μ_0	the permeability of free space	$\frac{\text{T}\cdot\text{m}}{\text{A}}$
$\vec{\mathbf{R}}$	a vector from each differential element to the field point	m
$\hat{\mathbf{R}}$	a unit vector along $\vec{\mathbf{R}}$	unitless
R	the distance corresponding to $\vec{\mathbf{R}}$	m
x, y, z	Cartesian coordinates	m, m, m
$\hat{\mathbf{x}}, \hat{\mathbf{y}}, \hat{\mathbf{z}}$	unit vectors along the $+x$-, $+y$-, $+z$-axes	unitless
r, θ	2D polar coordinates	m, rad
r_c, θ, z	cylindrical coordinates	m, rad, m
r, θ, φ	spherical coordinates	m, rad, rad
$\hat{\mathbf{r}}, \hat{\boldsymbol{\theta}}, \hat{\boldsymbol{\varphi}}$	unit vectors along spherical coordinate axes	unitless
$\hat{\mathbf{r}}_c$	a unit vector pointing away from the $+z$-axis	unitless
$\vec{\mathbf{K}}$	surface current density (distributed over a surface)	A/m
$\vec{\mathbf{J}}$	current density (distributed throughout a volume)	A/m^2
ds	differential arc length	m
dA	differential area element	m^2
dV	differential volume element	m^3
$\vec{\boldsymbol{v}}$	velocity	m/s
ω	angular speed	rad/s
dq	differential charge element	C
Q	total charge of the object	C

Strategy for Performing the Law of Biot-Savart Magnetic Field Integral

To derive an equation for the magnetic field created by a current (called the **source**) at a specified point (called the **field point**), follow these steps:

1. Draw the conductor. (A rare problem may instead involve a rotating charged object.) The conductor is composed of an infinite number of differential elements. Visualize integrating over every differential element that makes up the conductor.
2. Draw and label a representative differential element somewhere within the object. Depending on the context (see below), label the differential element as $d\vec{\mathbf{s}}$, dA, dV, or dq. <u>**Don't**</u> draw the differential element at the origin or on an axis (unless the object is a filamentary current lying on a coordinate axis, in which case you have no choice). Draw an arrow from the differential element (the **source**) to the **field point** (the point where the problem asks you to find the magnetic field). Label this displacement as $\vec{\mathbf{R}}$. See the diagrams in the examples that follow.
3. Begin with the magnetic field integral for the law of Biot-Savart.
   - For a **filamentary** current (which runs through a very thin wire):
   $$\vec{\mathbf{B}} = \frac{\mu_0}{4\pi}\int \frac{I\, d\vec{\mathbf{s}} \times \hat{\mathbf{R}}}{R^2}$$
   - For a **surface** current (running along a surface, like a conducting strip):
   $$\vec{\mathbf{B}} = \frac{\mu_0}{4\pi}\int \frac{\vec{\mathbf{K}}\, dA \times \hat{\mathbf{R}}}{R^2}$$
   - For a current distributed through a **volume** (like a thick wire):
   $$\vec{\mathbf{B}} = \frac{\mu_0}{4\pi}\int \frac{\vec{\mathbf{J}}\, dV \times \hat{\mathbf{R}}}{R^2}$$
   - For **moving charges** (such as a rotating charged disc):
   $$\vec{\mathbf{B}} = \frac{\mu_0}{4\pi}\int \frac{\vec{v}\, dq \times \hat{\mathbf{R}}}{R^2}$$

   Interpret these symbols as follows:
   - $\vec{\mathbf{B}}$ is the **magnetic field** at the field point.
   - μ_0 is the **permeability of free space**: $\mu_0 = 4\pi \times 10^{-7} \frac{\text{T}\cdot\text{m}}{\text{A}}$.
   - $\hat{\mathbf{R}}$ is a unit vector pointing from each differential element to the field point.
   - R is the distance from each differential element to the field point.
   - I is the total current running along the conductor.
   - $\vec{\mathbf{K}}$ and $\vec{\mathbf{J}}$ are **current densities**.
   - $d\vec{\mathbf{s}}$ is a differential displacement vector along a filamentary current.
   - dA is a differential area element along a conducting surface.
   - dV is a differential volume element in a three-dimensional solid conductor.
   - dq is an infinitesimal charge element in a charged object.
   - $\vec{v}$ is the velocity of a moving charge.

4. Study your diagram to express R in terms of suitable coordinates and to express $\hat{\mathbf{R}}$ in terms of suitable unit vectors.
   - For a line or an object with all straight edges (like a polygon), work with Cartesian coordinates (x, y, z) and unit vectors $(\hat{\mathbf{x}}, \hat{\mathbf{y}}, \hat{\mathbf{z}})$. Recall that $\hat{\mathbf{x}}$ points along $+x$, $\hat{\mathbf{y}}$ points along $+y$, and $\hat{\mathbf{z}}$ points along $+z$.
   - For a circular shape, work with 2D polar coordinates (r, θ) and unit vectors $(\hat{\mathbf{r}}, \hat{\boldsymbol{\theta}})$. Recall that $\hat{\mathbf{r}}$ points outward from the origin and $\hat{\boldsymbol{\theta}}$ is tangential.
   $$\hat{\mathbf{r}} = \hat{\mathbf{x}} \cos\theta + \hat{\mathbf{y}} \sin\theta \quad , \quad \hat{\boldsymbol{\theta}} = -\hat{\mathbf{x}} \sin\theta + \hat{\mathbf{y}} \cos\theta$$
   - For a spherical shape, work with spherical coordinates (r, θ, φ) and unit vectors $(\hat{\mathbf{r}}, \hat{\boldsymbol{\theta}}, \hat{\boldsymbol{\varphi}})$. Recall that $\hat{\mathbf{r}}$ points outward from the origin.
   $$\hat{\mathbf{r}} = \hat{\mathbf{x}} \cos\varphi \sin\theta + \hat{\mathbf{y}} \sin\varphi \sin\theta + \hat{\mathbf{z}} \cos\theta$$
   - For a cylinder or cone, work with cylindrical coordinates (r_c, θ, φ) and unit vectors $(\hat{\mathbf{r}}_c, \hat{\boldsymbol{\theta}}, \hat{\mathbf{z}})$. Recall that $\hat{\mathbf{r}}_c$ and $\hat{\boldsymbol{\theta}}$ are the same as the 2D polar unit vectors and $\hat{\mathbf{z}}$ points along the $+z$-axis.
   $$\hat{\mathbf{r}}_c = \hat{\mathbf{x}} \cos\theta + \hat{\mathbf{y}} \sin\theta$$
5. For a problem where the unit vector $\hat{\mathbf{R}}$ (which is the direction of $\vec{\mathbf{R}}$) doesn't seem easy to express in terms of other unit vectors, it may be easier to express $\vec{\mathbf{R}}$ in terms of unit vectors and then divide by its magnitude: $\hat{\mathbf{R}} = \frac{\vec{\mathbf{R}}}{R}$. (We did this in Chapter 7.)
6. For a problem that involves working with moving charges rather than working with current or current density (such as a rotating charged disc), make one of the following substitutions for dq, depending on the geometry:
   - $dq = \lambda ds$ for an arc length (like a rod or circular arc).
   - $dq = \sigma dA$ for a surface area (like a triangle, disc, or thin spherical shell).
   - $dq = \rho dV$ for a 3D solid (like a solid cube or a solid hemisphere).
7. Choose the appropriate coordinate system and make a substitution for ds, dA, or dV using the strategy from Chapter 6 (on page 86):
   - For a filamentary current parallel to the x-axis, $d\vec{\mathbf{s}} = \hat{\mathbf{x}}\, dx$.
   - For a filamentary current parallel to the y-axis, $d\vec{\mathbf{s}} = \hat{\mathbf{y}}\, dy$.
   - For a filamentary current parallel to the z-axis, $d\vec{\mathbf{s}} = \hat{\mathbf{z}}\, dz$.
   - For a filamentary current that is a circular arc of radius a lying in the xy plane with center at the origin, $d\vec{\mathbf{s}} = \pm\hat{\boldsymbol{\theta}}\, a d\theta$ (negative if clockwise).
   - For a solid polygon like a rectangle or triangle, $dA = dxdy$.
   - For a solid semicircle (**<u>not</u>** a circular arc) or pie slice, $dA = r dr d\theta$.
   - For a very thin spherical shell of radius a, $dA = a^2 \sin\theta\, d\theta d\varphi$.
   - For a solid polyhedron like a cube, $dV = dxdydz$.
   - For a solid cylinder or cone, $dV = r_c dr_c d\theta dz$.
   - For a portion of a solid sphere like a hemisphere, $dV = r^2 \sin\theta\, dr d\theta d\varphi$.

8. Is the current density (or charge density) uniform or non-uniform?
   - If the density is uniform, you can pull $\vec{\mathbf{J}}$, $\vec{\mathbf{K}}$, λ, σ, or ρ out of the integral.
   - If the density is non-uniform, leave $\vec{\mathbf{J}}$, $\vec{\mathbf{K}}$, λ, σ, or ρ in the integral. In this case, the problem generally gives you an equation to substitute for the density.
9. Perform the vector product $d\vec{s} \times \hat{\mathbf{R}}$, $\vec{\mathbf{K}} \times \hat{\mathbf{R}}$, $\vec{\mathbf{J}} \times \hat{\mathbf{R}}$, or $\vec{v} \times \hat{\mathbf{R}}$ following the technique from Chapter 19. (Be sure to apply the **vector** product, involving a **determinant**, and <u>**not**</u> the scalar product.) The answer will involve unit vectors, and is part of the integral. **Note**: For a moving charged object, you will need to write the velocity ($\vec{v}$) as speed (v) times an appropriate unit vector in order to perform the vector product. Then use $v = r_{rot}\omega$ and express r_{rot} in terms of suitable coordinates (see the last example in this chapter).
10. If you have Cartesian coordinates in your integrand, but are integrating over polar, cylindrical, or spherical coordinates (or vice-versa), use the following substitutions, as needed, to put all of your coordinates in the same system.

$$x = r\cos\theta \quad , \quad y = r\sin\theta \quad \text{(2D polar)}$$
$$x = r_c\cos\theta \quad , \quad y = r_c\sin\theta \quad \text{(cylindrical)}$$
$$x = r\sin\theta\cos\varphi \quad , \quad y = r\sin\theta\sin\varphi \quad , \quad z = r\cos\theta \quad \text{(spherical)}$$

11. An integral over ds is a single integral, an integral over dA is a double integral, and an integral over dV is a triple integral. Set the limits of each integration variable that map out the region of integration, as illustrated in Chapter 6. Perform the integral using techniques from Chapter 6.
12. Perform the appropriate integral below in order to determine the total current (I) or total charge (Q) of the object:

$$I = \int \vec{\mathbf{K}} \cdot d\vec{\boldsymbol{\ell}}$$
$$I = \int \vec{\mathbf{J}} \cdot d\vec{\mathbf{A}}$$
$$Q = \int dq$$

   Make the same substitutions as you made in Steps 6-10. Use the same limits of integration as you used in Step 11. The first two integrals are over the intersection of the Ampèrian loop and the conducting surface. For a surface current density ($\vec{\mathbf{K}}$), the intersection is the length ($\vec{\boldsymbol{\ell}}$) across the width of the surface (which is perpendicular to the current). For a 3D current density ($\vec{\mathbf{J}}$), the intersection is the cross-sectional area ($\vec{\mathbf{A}}$). The third integral is just like those from Chapter 7.
13. When you finish with Step 12, substitute your expression for I or Q into the result from your original magnetic field integral (Step 11). Simplify your expression.

Example: A filamentary current travels along the circular wire illustrated below. The radius of the circle is denoted by the symbol a. Derive an equation for the magnetic field at the center of the circle in terms of μ_0, I, a, and appropriate unit vectors.

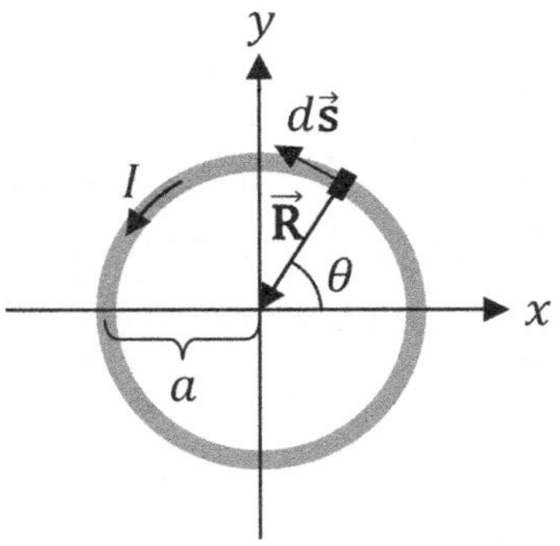

Begin with a labeled diagram. For a filamentary current, draw a representative $d\vec{\mathbf{s}}$: Note that $d\vec{\mathbf{s}}$ must lie on the circular wire and has a direction that is tangent to the current (I). Draw $\vec{\mathbf{R}}$ from the source, $d\vec{\mathbf{s}}$, to the field point (in this problem, it's at the origin). When we perform the integration, we effectively integrate over every $d\vec{\mathbf{s}}$ that makes up the circle. Apply the law of Biot-Savart to the filamentary current.

$$\vec{\mathbf{B}} = \frac{\mu_0}{4\pi}\int \frac{I\, d\vec{\mathbf{s}} \times \hat{\mathbf{R}}}{R^2}$$

Examine the picture above:

- The vector $\vec{\mathbf{R}}$ has the same length for each dq that makes up the circle. Its magnitude, R, equals the radius of the circle: $R = a$.
- $\vec{\mathbf{R}}$ points **inward**, toward the center of the circle. Since the unit vector $\hat{\mathbf{r}}$ of 2D polar coordinates points **outward**, we can write $\hat{\mathbf{R}} = -\hat{\mathbf{r}}$.

For a circular filamentary current, write $d\vec{\mathbf{s}} = \hat{\boldsymbol{\theta}}\, a d\theta$ (it's positive because I is counter-clockwise in this example). This is Step 7 on page 325. Substitute the expressions $R = a$, $\hat{\mathbf{R}} = -\hat{\mathbf{r}}$, and $d\vec{\mathbf{s}} = \hat{\boldsymbol{\theta}}\, a d\theta$ into the magnetic field integral. The current I and radius a are constants and may come out of the integral. Note that $\frac{a}{a^2} = \frac{1}{a}$. Also note that the $(-\hat{\mathbf{r}})$ is actually part of the integration. The limits of integration are from $\theta = 0$ to $\theta = 2\pi$ **radians** for a full circle.

$$\vec{\mathbf{B}} = \frac{\mu_0}{4\pi}\int \frac{I\, d\vec{\mathbf{s}} \times \hat{\mathbf{R}}}{R^2} = \frac{\mu_0}{4\pi}\int_{\theta=0}^{2\pi} \frac{I(\hat{\boldsymbol{\theta}}\, a d\theta) \times (-\hat{\mathbf{r}})}{a^2} = -\frac{\mu_0 I}{4\pi a}\int_{\theta=0}^{2\pi} \hat{\boldsymbol{\theta}} \times \hat{\mathbf{r}}\, d\theta$$

Now we perform the vector product $\hat{\boldsymbol{\theta}} \times \hat{\mathbf{r}}$. There are two ways to do this. One way is to insert the expressions $\hat{\mathbf{r}} = \hat{\mathbf{x}}\cos\theta + \hat{\mathbf{y}}\sin\theta$ and $\hat{\boldsymbol{\theta}} = -\hat{\mathbf{x}}\sin\theta + \hat{\mathbf{y}}\cos\theta$ (see Step 4 on page 325) into the determinant form of the vector product (Chapter 19):

$$\hat{\boldsymbol{\theta}} \times \hat{\mathbf{r}} = \begin{vmatrix} \hat{\mathbf{x}} & \hat{\mathbf{y}} & \hat{\mathbf{z}} \\ -\sin\theta & \cos\theta & 0 \\ \cos\theta & \sin\theta & 0 \end{vmatrix} = \hat{\mathbf{z}}(-\sin^2\theta - \cos^2\theta) = -\hat{\mathbf{z}}(\sin^2\theta + \cos^2\theta) = -\hat{\mathbf{z}}$$

We used the trigonometric identity $\sin^2\theta + \cos^2\theta = 1$.

There is an alternative method to determine that $\hat{\boldsymbol{\theta}} \times \hat{\mathbf{r}} = -\hat{\mathbf{z}}$: Apply the right-hand rule as follows.

Point your fingers along $\hat{\boldsymbol{\theta}}$ (a counterclockwise tangent along $d\vec{\mathbf{s}}$) and face your palm along $\hat{\mathbf{r}}$ (opposite to $\hat{\mathbf{R}}$, since $\hat{\mathbf{r}}$ points outward). Your thumb points into the page ($\otimes$), along $-\hat{\mathbf{z}}$ (since $+z$ comes out of the page). Since $\hat{\boldsymbol{\theta}}$ is perpendicular to $\hat{\mathbf{r}}$, and since these are unit vectors, the vector product $\hat{\boldsymbol{\theta}} \times \hat{\mathbf{r}}$ has a magnitude of $\|\hat{\boldsymbol{\theta}} \times \hat{\mathbf{r}}\| = (1)(1)\sin 90° = 1$ according to the formula $\|\vec{\mathbf{A}} \times \vec{\mathbf{B}}\| = AB \sin\theta$ from Chapter 19. Thus, $\hat{\boldsymbol{\theta}} \times \hat{\mathbf{r}} = (1)(-\hat{\mathbf{z}}) = -\hat{\mathbf{z}}$.

Substitute $\hat{\boldsymbol{\theta}} \times \hat{\mathbf{r}} = -\hat{\mathbf{z}}$ into our previous expression for the magnetic field integral. Note that the two minus signs make a plus sign.

$$\vec{\mathbf{B}} = -\frac{\mu_0 I}{4\pi a} \int_{\theta=0}^{2\pi} \hat{\boldsymbol{\theta}} \times \hat{\mathbf{r}}\, d\theta = \frac{\mu_0 I}{4\pi a} \int_{\theta=0}^{2\pi} \hat{\mathbf{z}}\, d\theta$$

The unit vector $\hat{\mathbf{z}}$ is a constant because it always points one unit along the $+z$-axis. (Note that the unit vectors $\hat{\boldsymbol{\theta}}$ and $\hat{\mathbf{r}}$ are **not** constants, since their directions are different for each point on the circle. It may help to study the diagram on the previous page and draw these unit vectors for a few different points on the circle.) Since $\hat{\mathbf{z}}$ is constant, we may pull it out of the integral. The remaining integral is trivial.

$$\vec{\mathbf{B}} = \frac{\mu_0 I}{4\pi a}\hat{\mathbf{z}} \int_{\theta=0}^{2\pi} d\theta = \frac{\mu_0 I}{4\pi a}\hat{\mathbf{z}}[\theta]_{\theta=0}^{2\pi} = \frac{\mu_0 I}{4\pi a}\hat{\mathbf{z}}(2\pi - 0) = \frac{\mu_0 I}{2a}\hat{\mathbf{z}}$$

The magnetic field at the origin is $\vec{\mathbf{B}} = \frac{\mu_0 I}{2a}\hat{\mathbf{z}}$. It has a magnitude of $B = \frac{\mu_0 I}{2a}$ and a direction of $\hat{\mathbf{B}} = \hat{\mathbf{z}}$. If you apply the right-hand rule for magnetic **field** (Chapter 22) to the diagram on the previous page, you should see that your fingers are coming out of the page ($\odot$) along $\hat{\mathbf{z}}$ at the origin, which agrees with the direction of our answer above.

Tips: It's a good habit to check that the direction of your answer from applying the law of Biot-Savart agrees with the right-hand rule for magnetic field. It's easy to make a mistake with the vector product or to make a sign mistake in the algebra, so it's good to have a quick way to help check your answer for consistency.

Another thing that you can check is the units. Your final expression for magnetic field should include the units of μ_0 times the unit of current divided by the unit of length. Our answer, $B = \frac{\mu_0 I}{2a}$, meets this criteria.

Example: A filamentary current travels along the path illustrated below, which includes a semicircular arc with radius a. Derive an equation for the magnetic field at the origin in terms of μ_0, I, a, and appropriate unit vectors.

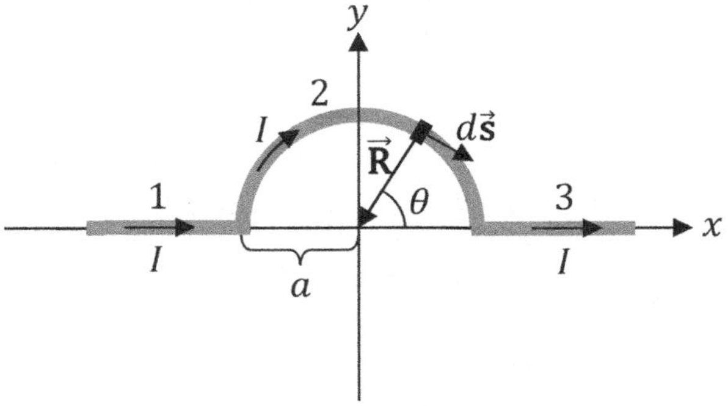

One "trick" to problems like this, which involve a circular arc and straight sections, is to divide the shape up into pieces:

- Piece 1 is the straight current on the left.
- Piece 2 is the semicircular current in the middle.
- Piece 3 is the straight current on the right.

A second "trick" is to realize that $\vec{\mathbf{B}}_1 = 0$ and $\vec{\mathbf{B}}_3 = 0$. For these two straight sections, $\hat{\mathbf{R}}$ points along $\hat{\mathbf{x}}$ or $-\hat{\mathbf{x}}$, while $d\vec{\mathbf{s}}$ points along $\hat{\mathbf{x}}$. For these two sections, $d\vec{\mathbf{s}} \times \hat{\mathbf{R}} = 0$ because $\hat{\mathbf{x}} \times \hat{\mathbf{x}} = 0$ (see Chapter 19). Whenever two vectors are parallel or anti-parallel, their vector product equals zero (since $\theta = 0°$ or $180°$, so that $\|\vec{\mathbf{A}} \times \vec{\mathbf{B}}\| = AB\sin\theta = AB\sin 0° = 0$).

Therefore, we only need to find the magnetic field ($\vec{\mathbf{B}}_2$) created by the semicircular section at the origin. The math for this is nearly identical to the previous example. The only differences between these two examples are:

- The limits of integration will now be from $\theta = 0$ to $\theta = \pi$ **radians** for the semicircle (instead of from $\theta = 0$ to $\theta = 2\pi$ for a full circle).
- The current in this example runs clockwise instead of counterclockwise, such that $(-\hat{\boldsymbol{\theta}}) \times \hat{\mathbf{r}} = \hat{\mathbf{z}}$ (instead of $\hat{\boldsymbol{\theta}} \times \hat{\mathbf{r}} = -\hat{\mathbf{z}}$). The different direction of the current in this example will simply change the overall sign of the answer.

The solution to this example will be the same as for the previous problem, except that the magnetic field integral will become:

$$\vec{\mathbf{B}}_2 = -\frac{\mu_0 I}{4\pi a}\hat{\mathbf{z}} \int_{\theta=0}^{\pi} d\theta = -\frac{\mu_0 I}{4a}\hat{\mathbf{z}}$$

The magnetic field at the origin is $\vec{\mathbf{B}} = \vec{\mathbf{B}}_2 = -\frac{\mu_0 I}{4a}\hat{\mathbf{z}}$. It has a magnitude of $B = \frac{\mu_0 I}{4a}$ and a direction of $\hat{\mathbf{B}} = -\hat{\mathbf{z}}$. If you apply the right-hand rule for magnetic **field** (Chapter 22) to the diagram above, you should see that your fingers are going into the page (⊗) along $-\hat{\mathbf{z}}$, which agrees with the direction of our answer above.

Example: A filamentary current travels along the circular wire illustrated below. The circular loop lies in the xy plane, centered about the origin. The radius of the circle is denoted by the symbol a. Derive an equation for the magnetic field at the point $(0,0,p)$ in terms of μ_0, I, a, p, and appropriate unit vectors.

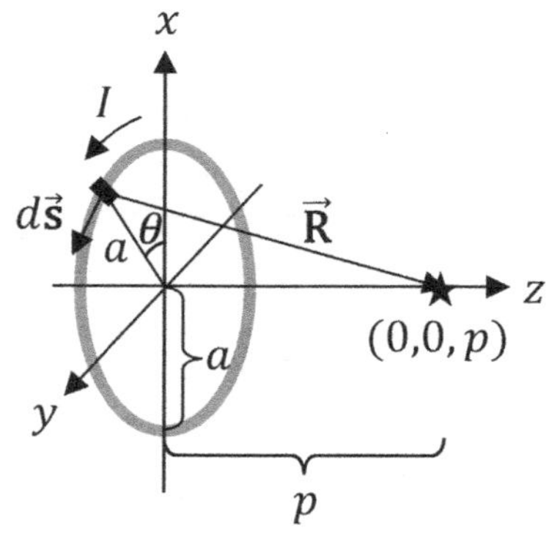

Begin with a labeled diagram. For a filamentary current, draw a representative $d\vec{\mathbf{s}}$ that is tangent to the current (I). Draw $\vec{\mathbf{R}}$ from the source, $d\vec{\mathbf{s}}$, to the field point $(0,0,p)$. When we perform the integration, we effectively integrate over every $d\vec{\mathbf{s}}$ that makes up the current loop. Apply the law of Biot-Savart to the filamentary current.

$$\vec{\mathbf{B}} = \frac{\mu_0}{4\pi}\int \frac{I\, d\vec{\mathbf{s}} \times \hat{\mathbf{R}}}{R^2}$$

Examine the picture above:

- The vector $\vec{\mathbf{R}}$ extends a units inward, towards the z-axis (along $-\hat{\mathbf{r}}_c$), and p units along the z-axis (along $\hat{\mathbf{z}}$). Therefore, $\vec{\mathbf{R}} = -a\hat{\mathbf{r}}_c + p\hat{\mathbf{z}}$. Note that this is the $\hat{\mathbf{r}}_c$ of **cylindrical** coordinates.
- Apply the Pythagorean theorem to find the magnitude of $\vec{\mathbf{R}}$.

$$R = \sqrt{a^2 + p^2}$$

- Divide $\vec{\mathbf{R}}$ by R to find the direction of $\vec{\mathbf{R}}$.

$$\hat{\mathbf{R}} = \frac{\vec{\mathbf{R}}}{R} = \frac{-a\hat{\mathbf{r}}_c + p\hat{\mathbf{z}}}{\sqrt{a^2 + p^2}}$$

For a circular filamentary current, write $d\vec{\mathbf{s}} = \hat{\boldsymbol{\theta}}\, a d\theta$ (it's positive because I is counter-clockwise, looking from the $+z$-axis, in this example). This is Step 7 on page 325. The limits of integration are from $\theta = 0$ to $\theta = 2\pi$ **radians** for a full circle. Watch how we rewrite the magnetic field integral in order to manage the substitutions:

$$\vec{\mathbf{B}} = \frac{\mu_0}{4\pi}\int \frac{I\, d\vec{\mathbf{s}} \times \hat{\mathbf{R}}}{R^2} = \frac{\mu_0}{4\pi}\int I\, d\vec{\mathbf{s}}\left(\frac{1}{R^2}\right) \times \hat{\mathbf{R}}$$

Now substitute the expressions for R, $\hat{\mathbf{R}}$, and $d\vec{\mathbf{s}}$ into the integral. Compare these two lines.

$$\vec{\mathbf{B}} = \frac{\mu_0}{4\pi}\int_{\theta=0}^{2\pi} I(\hat{\boldsymbol{\theta}}\, a d\theta)\frac{1}{a^2 + p^2} \times \frac{-a\hat{\mathbf{r}}_c + p\hat{\mathbf{z}}}{\sqrt{a^2 + p^2}}$$

Note that $R^2 = \left(\sqrt{a^2 + p^2}\right)^2 = a^2 + p^2$. In the next step, we will apply the rule from algebra that $(a^2 + p^2)\sqrt{a^2 + p^2} = (a^2 + p^2)^1(a^2 + p^2)^{1/2} = (a^2 + p^2)^{3/2}$. We will also pull the constants I and a out of the integral.

$$\vec{\mathbf{B}} = \frac{\mu_0 I a}{4\pi} \int_{\theta=0}^{2\pi} \frac{\hat{\boldsymbol{\theta}} \times (-a\hat{\mathbf{r}}_c + p\hat{\mathbf{z}})}{(a^2 + p^2)^{3/2}} d\theta$$

(This power of $3/2$ is very common in electric and magnetic field integrals. If you get a different power, you should check your work very carefully for a possible mistake.)

Work out the vector product $\hat{\boldsymbol{\theta}} \times (-a\hat{\mathbf{r}}_c + p\hat{\mathbf{z}})$ according to Chapter 19. It's convenient to work with cylindrical coordinates for the determinant. It works the same way as in Cartesian coordinates: The first row has the unit vectors and the subsequent rows have the components of the two vectors corresponding to those unit vectors. The cylindrical components correspond to the unit vectors $\hat{\mathbf{r}}_c$, $\hat{\boldsymbol{\theta}}$, and $\hat{\mathbf{z}}$. For the first vector ($\hat{\boldsymbol{\theta}}$), the cylindrical components are simply 0, 1, and 0 (since $\hat{\boldsymbol{\theta}}$ is one unit long), while for the second vector $(-a\hat{\mathbf{r}}_c + p\hat{\mathbf{z}})$ they are $-a$, 0, and p. (In either case, the components are the coefficients of the unit vectors.)

$$\hat{\boldsymbol{\theta}} \times (-a\hat{\mathbf{r}}_c + p\hat{\mathbf{z}}) = \begin{vmatrix} \hat{\mathbf{r}}_c & \hat{\boldsymbol{\theta}} & \hat{\mathbf{z}} \\ 0 & 1 & 0 \\ -a & 0 & p \end{vmatrix} = \hat{\mathbf{r}}_c \begin{vmatrix} 1 & 0 \\ 0 & p \end{vmatrix} - \hat{\boldsymbol{\theta}} \begin{vmatrix} 0 & 0 \\ -a & p \end{vmatrix} + \hat{\mathbf{z}} \begin{vmatrix} 0 & 1 \\ -a & 0 \end{vmatrix}$$

$$\hat{\boldsymbol{\theta}} \times (-a\hat{\mathbf{r}}_c + p\hat{\mathbf{z}}) = \hat{\mathbf{r}}_c(p - 0) - \hat{\boldsymbol{\theta}}(0 - 0) + \hat{\mathbf{z}}(0 + a) = p\,\hat{\mathbf{r}}_c + a\,\hat{\mathbf{z}}$$

Substitute this result into the magnetic field integral.

$$\vec{\mathbf{B}} = \frac{\mu_0 I a}{4\pi} \int_{\theta=0}^{2\pi} \frac{p\,\hat{\mathbf{r}}_c + a\,\hat{\mathbf{z}}}{(a^2 + p^2)^{3/2}} d\theta = \frac{\mu_0 I a}{4\pi(a^2 + p^2)^{3/2}} \int_{\theta=0}^{2\pi} (p\,\hat{\mathbf{r}}_c + a\,\hat{\mathbf{z}})\, d\theta$$

Note that $(a^2 + p^2)^{3/2}$ is a constant and may come out of the integral. Separate the integral into two terms.

$$\vec{\mathbf{B}} = \frac{\mu_0 I a}{4\pi(a^2 + p^2)^{3/2}} \int_{\theta=0}^{2\pi} p\,\hat{\mathbf{r}}_c\, d\theta + \frac{\mu_0 I a}{4\pi(a^2 + p^2)^{3/2}} \int_{\theta=0}^{2\pi} a\,\hat{\mathbf{z}}\, d\theta$$

Note that p, a, and $\hat{\mathbf{z}}$ are constants, and thus may come out of their integrals, but $\hat{\mathbf{r}}_c$ is **not** constant: Since $\hat{\mathbf{r}}_c$ points away from the z-axis, its direction is different for each $d\vec{\mathbf{s}}$ that makes up the current loop. We may **not** pull $\hat{\mathbf{r}}_c$ out of the integral. Instead, we use the handy equation from Step 4 on page 325.

$$\hat{\mathbf{r}}_c = \hat{\mathbf{x}} \cos\theta + \hat{\mathbf{y}} \sin\theta$$

Substitute this equation into the previous expression for magnetic field.

$$\vec{\mathbf{B}} = \frac{\mu_0 I a p}{4\pi(a^2 + p^2)^{3/2}} \int_{\theta=0}^{2\pi} (\hat{\mathbf{x}} \cos\theta + \hat{\mathbf{y}} \sin\theta)\, d\theta + \frac{\mu_0 I a^2}{4\pi(a^2 + p^2)^{3/2}} \hat{\mathbf{z}} \int_{\theta=0}^{2\pi} d\theta$$

The first integral we again separate into two terms. The last integral is trivial.

$$\vec{\mathbf{B}} = \frac{\mu_0 I a p}{4\pi(a^2+p^2)^{3/2}} \int_{\theta=0}^{2\pi} \hat{\mathbf{x}} \cos\theta \, d\theta + \frac{\mu_0 I a p}{4\pi(a^2+p^2)^{3/2}} \int_{\theta=0}^{2\pi} \hat{\mathbf{y}} \sin\theta \, d\theta + \frac{\mu_0 I a^2}{4\pi(a^2+p^2)^{3/2}} \hat{\mathbf{z}}(2\pi)$$

The unit vectors $\hat{\mathbf{x}}$ and $\hat{\mathbf{y}}$ are constants, so they may come out of their integrals.

$$\vec{\mathbf{B}} = \frac{\mu_0 I a p}{4\pi(a^2+p^2)^{3/2}} \hat{\mathbf{x}} \int_{\theta=0}^{2\pi} \cos\theta \, d\theta + \frac{\mu_0 I a p}{4\pi(a^2+p^2)^{3/2}} \hat{\mathbf{y}} \int_{\theta=0}^{2\pi} \sin\theta \, d\theta + \frac{\mu_0 I a^2}{2(a^2+p^2)^{3/2}} \hat{\mathbf{z}}$$

$$\vec{\mathbf{B}} = \frac{\mu_0 I a p}{4\pi(a^2+p^2)^{3/2}} \hat{\mathbf{x}} \left[\sin\theta\right]_{\theta=0}^{2\pi} + \frac{\mu_0 I a p}{4\pi(a^2+p^2)^{3/2}} \hat{\mathbf{y}} \left[-\cos\theta\right]_{\theta=0}^{2\pi} + \frac{\mu_0 I a^2}{2(a^2+p^2)^{3/2}} \hat{\mathbf{z}}$$

$$\vec{\mathbf{B}} = \frac{\mu_0 I a p}{4\pi(a^2+p^2)^{3/2}} \hat{\mathbf{x}} \left(\sin 2\pi - \sin 0\right) + \frac{\mu_0 I a p}{4\pi(a^2+p^2)^{3/2}} \hat{\mathbf{y}} \left(-\cos 2\pi + \cos 0\right) + \frac{\mu_0 I a^2}{2(a^2+p^2)^{3/2}} \hat{\mathbf{z}}$$

$$\vec{\mathbf{B}} = \frac{\mu_0 I a p}{4\pi(a^2+p^2)^{3/2}} \hat{\mathbf{x}} \,(0-0) + \frac{\mu_0 I a p}{4\pi(a^2+p^2)^{3/2}} \hat{\mathbf{y}} \,(-1+1) + \frac{\mu_0 I a^2}{2(a^2+p^2)^{3/2}} \hat{\mathbf{z}}$$

$$\vec{\mathbf{B}} = \frac{\mu_0 I a p}{4\pi(a^2+p^2)^{3/2}} \hat{\mathbf{x}} \,(0) + \frac{\mu_0 I a p}{4\pi(a^2+p^2)^{3/2}} \hat{\mathbf{y}} \,(0) + \frac{\mu_0 I a^2}{2(a^2+p^2)^{3/2}} \hat{\mathbf{z}}$$

$$\vec{\mathbf{B}} = \frac{\mu_0 I a^2}{2(a^2+p^2)^{3/2}} \hat{\mathbf{z}}$$

Look at what happened: The integral $\int_{\theta=0}^{2\pi} p \, \hat{\mathbf{r}}_c \, d\theta$ turned out to be exactly zero. If you really understand the magnetic field lines from Chapter 22, you could have deduced this earlier and saved the trouble of doing that extra math. The magnetic field at the point $(0,0,p)$ will point along $\hat{\mathbf{z}}$, so only the integral $\int_{\theta=0}^{2\pi} a \, \hat{\mathbf{z}} \, d\theta$ mattered.

The magnetic field at the point $(0,0,p)$ is $\vec{\mathbf{B}} = \frac{\mu_0 I a^2}{2(a^2+p^2)^{3/2}} \hat{\mathbf{z}}$. It has a magnitude of $B = \frac{\mu_0 I a^2}{2(a^2+p^2)^{3/2}}$ and a direction of $\hat{\mathbf{B}} = \hat{\mathbf{z}}$.

Example: An infinite filamentary current travels along the z-axis as illustrated below. Derive an equation for the magnetic field at the point $(a, 0, 0)$, where a is a constant, in terms of μ_0, I, a, and appropriate unit vectors.

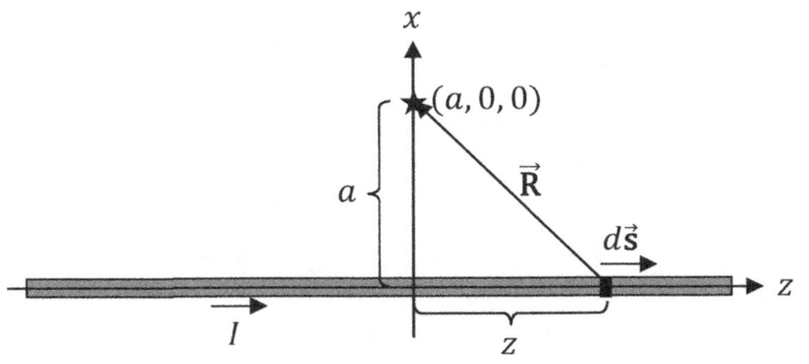

Begin with a labeled diagram. For a filamentary current, draw a representative $d\vec{\mathbf{s}}$ along the current (I). Draw $\vec{\mathbf{R}}$ from the source, $d\vec{\mathbf{s}}$, to the field point $(a, 0, 0)$. When we perform the integration, we effectively integrate over every $d\vec{\mathbf{s}}$ that makes up the current. Apply the law of Biot-Savart to the filamentary current.

$$\vec{\mathbf{B}} = \frac{\mu_0}{4\pi} \int \frac{I\, d\vec{\mathbf{s}} \times \hat{\mathbf{R}}}{R^2}$$

Examine the picture above:

- The vector $\vec{\mathbf{R}}$ extends a units up, along the x-axis (along $\hat{\mathbf{x}}$), and z units to the left, along the negative z-axis (along $-\hat{\mathbf{z}}$). Therefore, $\vec{\mathbf{R}} = a\hat{\mathbf{x}} - z\hat{\mathbf{z}}$.
- Apply the Pythagorean theorem to find the magnitude of $\vec{\mathbf{R}}$.

$$R = \sqrt{a^2 + z^2}$$

- Divide $\vec{\mathbf{R}}$ by R to find the direction of $\vec{\mathbf{R}}$.

$$\hat{\mathbf{R}} = \frac{\vec{\mathbf{R}}}{R} = \frac{a\hat{\mathbf{x}} - z\hat{\mathbf{z}}}{\sqrt{a^2 + z^2}}$$

For a straight filamentary current along the z-axis, write $d\vec{\mathbf{s}} = \hat{\mathbf{z}}\, dz$. This is Step 7 on page 325. The limits of integration are from $z = -\infty$ to $z = \infty$ (for the infinite wire). Substitute the expressions for R, $\hat{\mathbf{R}}$, and $d\vec{\mathbf{s}}$ into the integral.

$$\vec{\mathbf{B}} = \frac{\mu_0}{4\pi} \int \frac{I\, d\vec{\mathbf{s}} \times \hat{\mathbf{R}}}{R^2} = \frac{\mu_0}{4\pi} \int I\, d\vec{\mathbf{s}} \left(\frac{1}{R^2}\right) \times \hat{\mathbf{R}} = \frac{\mu_0}{4\pi} \int_{z=-\infty}^{\infty} I(\hat{\mathbf{z}}\, dz) \frac{1}{a^2 + z^2} \times \frac{a\hat{\mathbf{x}} - z\hat{\mathbf{z}}}{\sqrt{a^2 + z^2}}$$

Note that $R^2 = \left(\sqrt{a^2 + z^2}\right)^2 = a^2 + z^2$. In the next step, we will apply the rule from algebra that $(a^2 + z^2)\sqrt{a^2 + z^2} = (a^2 + z^2)^1 (a^2 + z^2)^{1/2} = (a^2 + z^2)^{3/2}$.

$$\vec{\mathbf{B}} = \frac{\mu_0}{4\pi} \int_{z=-\infty}^{\infty} \frac{I\, \hat{\mathbf{z}} \times (a\hat{\mathbf{x}} - z\hat{\mathbf{z}})}{(a^2 + z^2)^{3/2}} dz = \frac{\mu_0 I}{4\pi} \int_{z=-\infty}^{\infty} \frac{\hat{\mathbf{z}} \times (a\hat{\mathbf{x}} - z\hat{\mathbf{z}})}{(a^2 + z^2)^{3/2}} dz$$

Note that I is a constant, which may come out of the integral.

Work out the vector product $\hat{\mathbf{z}} \times (a\hat{\mathbf{x}} - z\hat{\mathbf{z}})$ according to Chapter 19, where the first vector just has a z-component (equal to 1) and the second vector has components a, 0, and z. (The components of the vectors are the coefficients of the unit vectors.)

$$\hat{\mathbf{z}} \times (a\hat{\mathbf{x}} - z\hat{\mathbf{z}}) = \begin{vmatrix} \hat{\mathbf{x}} & \hat{\mathbf{y}} & \hat{\mathbf{z}} \\ 0 & 0 & 1 \\ a & 0 & -z \end{vmatrix} = \hat{\mathbf{x}} \begin{vmatrix} 0 & 1 \\ 0 & -z \end{vmatrix} - \hat{\mathbf{y}} \begin{vmatrix} 0 & 1 \\ a & -z \end{vmatrix} + \hat{\mathbf{z}} \begin{vmatrix} 0 & 0 \\ a & 0 \end{vmatrix}$$

$$\hat{\mathbf{z}} \times (a\hat{\mathbf{x}} - z\hat{\mathbf{z}}) = \hat{\mathbf{x}}(0-0) - \hat{\mathbf{y}}(0-a) + \hat{\mathbf{z}}(0-0) = a\,\hat{\mathbf{y}}$$

Substitute this result into the magnetic field integral.

$$\vec{\mathbf{B}} = \frac{\mu_0 I}{4\pi} \int_{z=-\infty}^{\infty} \frac{a\,\hat{\mathbf{y}}}{(a^2+z^2)^{3/2}} dz$$

The constants a and $\hat{\mathbf{y}}$ may come out of the integral.

$$\vec{\mathbf{B}} = \frac{\mu_0 I a}{4\pi} \hat{\mathbf{y}} \int_{z=-\infty}^{\infty} \frac{dz}{(a^2+z^2)^{3/2}}$$

This integral can be performed via the following trigonometric substitution:

$$z = a \tan\theta$$

$$dz = a \sec^2\theta\, d\theta$$

Solving for θ, we get $\theta = \tan^{-1}\left(\frac{z}{a}\right)$, which shows that the new limits of integration are from $\theta = \tan^{-1}(-\infty) = -90°$ to $\theta = \tan^{-1}(\infty) = 90°$ (since $\tan\theta$ approaches infinity in the limit that θ approaches 90°). Note that the denominator of the integral simplifies as follows, using the trig identity $1 + \tan^2\theta = \sec^2\theta$:

$$(a^2+z^2)^{3/2} = [a^2 + (a\tan\theta)^2]^{3/2} = [a^2(1+\tan^2\theta)]^{3/2} = (a^2\sec^2\theta)^{3/2} = a^3\sec^3\theta$$

In the last step, we applied the rule from algebra that $(a^2x^2)^{3/2} = (a^2)^{3/2}(x^2)^{3/2} = a^3x^3$. Substitute the above expressions for dz and $(a^2+z^2)^{3/2}$ into the previous integral.

$$\vec{\mathbf{B}} = \frac{\mu_0 I a}{4\pi} \hat{\mathbf{y}} \int_{\theta=-90°}^{90°} \frac{a\sec^2\theta\, d\theta}{a^3\sec^3\theta} = \frac{\mu_0 I a}{4\pi} \hat{\mathbf{y}} \int_{\theta=-90°}^{90°} \frac{d\theta}{a^2\sec\theta}$$

Recall that $\sec\theta = \frac{1}{\cos\theta}$. Pull the constant $\frac{1}{a^2}$ out of the integral. Note that $a\left(\frac{1}{a^2}\right) = \frac{1}{a}$.

$$\vec{\mathbf{B}} = \frac{\mu_0 I}{4\pi a} \hat{\mathbf{y}} \int_{\theta=-90°}^{90°} \cos\theta\, d\theta = \frac{\mu_0 I}{4\pi a} \hat{\mathbf{y}} [\sin\theta]_{\theta=-90°}^{90°} = \frac{\mu_0 I}{4\pi a} \hat{\mathbf{y}} [\sin(90°) - \sin(-90°)]$$

$$\vec{\mathbf{B}} = \frac{\mu_0 I}{4\pi a} \hat{\mathbf{y}} [(1) - (-1)] = \frac{\mu_0 I}{4\pi a} \hat{\mathbf{y}} (1+1) = \frac{\mu_0 I}{4\pi a} \hat{\mathbf{y}} (2) = \frac{\mu_0 I}{2\pi a} \hat{\mathbf{y}}$$

The magnetic field at the point $(a, 0, 0)$ is $\vec{\mathbf{B}} = \frac{\mu_0 I}{2\pi a} \hat{\mathbf{y}}$. It has a magnitude of $B = \frac{\mu_0 I}{2\pi a}$ and a direction of $\hat{\mathbf{B}} = \hat{\mathbf{y}}$. If you apply the right-hand rule for magnetic **field** (Chapter 22) to the previous diagram, you should see that your fingers are coming out of the page (⊙) along $\hat{\mathbf{y}}$, at the point $(a, 0, 0)$, which agrees with the direction of our answer above.

Example: A solid disc lying in the xy plane and centered about the origin has positive charge Q, radius a, and uniform charge density σ. The solid disc rotates about the z-axis with constant angular speed ω. Derive an equation for the magnetic field at the point $(0,0,p)$, where p is a constant, in terms of μ_0, Q, a, p, ω, and appropriate unit vectors.

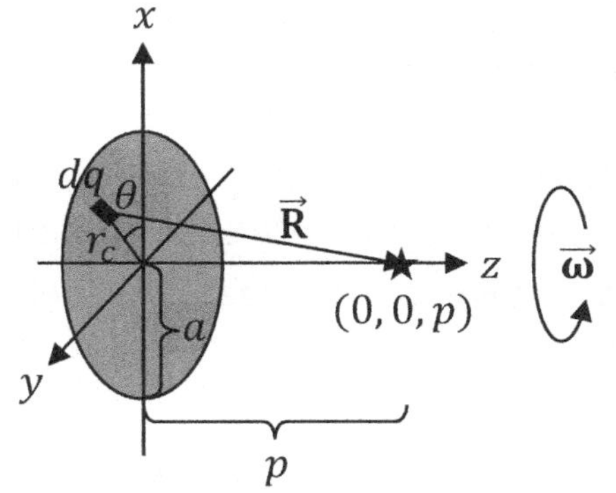

Begin with a labeled diagram. For a rotating charged disc, draw a representative dq within the area of the disc. Draw $\vec{\mathbf{R}}$ from the source, dq, to the field point $(0,0,p)$. When we perform the integration, we effectively integrate over every dq that makes up the solid disc. Apply the law of Biot-Savart to the rotating charged disc (Step 3 on page 324).

$$\vec{\mathbf{B}} = \frac{\mu_0}{4\pi}\int \frac{\vec{v}\,dq \times \hat{\mathbf{R}}}{R^2}$$

Examine the picture above:

- The vector $\vec{\mathbf{R}}$ extends r_c units inward, towards the z-axis (along $-\hat{\mathbf{r}}_c$), and p units along the z-axis (along $\hat{\mathbf{z}}$). Therefore, $\vec{\mathbf{R}} = -r_c\hat{\mathbf{r}}_c + p\hat{\mathbf{z}}$. Note that these are the r_c and $\hat{\mathbf{r}}_c$ of **cylindrical** coordinates.
- Apply the Pythagorean theorem to find the magnitude of $\vec{\mathbf{R}}$.
$$R = \sqrt{r_c^2 + p^2}$$
- Divide $\vec{\mathbf{R}}$ by R to find the direction of $\vec{\mathbf{R}}$.
$$\hat{\mathbf{R}} = \frac{\vec{\mathbf{R}}}{R} = \frac{-r_c\hat{\mathbf{r}}_c + p\hat{\mathbf{z}}}{\sqrt{r_c^2 + p^2}}$$

For a solid disc, $dq = \sigma dA$ and $dA = r_c dr_c d\theta$ (Steps 6-7 on page 325, except that we are using the cylindrical coordinate r_c for this three-dimensional problem), such that $dq = \sigma r_c dr_c d\theta$. The limits of integration are from $r_c = 0$ to $r_c = a$ and $\theta = 0$ to $\theta = 2\pi$ **radians**. Now substitute the expressions for R, $\hat{\mathbf{R}}$, and dq into the integral.

$$\vec{\mathbf{B}} = \frac{\mu_0}{4\pi}\int \frac{\vec{v}\,dq \times \hat{\mathbf{R}}}{R^2} = \frac{\mu_0}{4\pi}\int \vec{v}\,dq\left(\frac{1}{R^2}\right) \times \hat{\mathbf{R}} = \frac{\mu_0}{4\pi}\int_{r_c=0}^{a}\int_{\theta=0}^{2\pi} \vec{v}\,(\sigma r_c dr_c d\theta)\frac{1}{r_c^2+p^2} \times \frac{-r_c\hat{\mathbf{r}}_c + p\hat{\mathbf{z}}}{\sqrt{r_c^2+p^2}}$$

Note that $R^2 = \left(\sqrt{r_c^2+p^2}\right)^2 = r_c^2 + p^2$. In the next step, we will apply the rule from algebra that $(r_c^2+p^2)\sqrt{r_c^2+p^2} = (r_c^2+p^2)^1(r_c^2+p^2)^{1/2} = (r_c^2+p^2)^{3/2}$. The charge density σ is constant since the disc is **uniform**, so it may come out of the integral.

$$\vec{\mathbf{B}} = \frac{\mu_0 \sigma}{4\pi} \int_{r_c=0}^{a} \int_{\theta=0}^{2\pi} \frac{\vec{\boldsymbol{v}} \times (-r_c \hat{\mathbf{r}}_c + p\hat{\mathbf{z}})}{(r_c^2 + p^2)^{3/2}} r_c dr_c d\theta$$

Since the velocity ($\vec{\boldsymbol{v}}$) of each dq is tangential to the circle of rotation, and since the cylindrical unit vector $\hat{\boldsymbol{\theta}}$ is tangential to a circle lying in the xy plane, we may write $\vec{\boldsymbol{v}} = v\,\hat{\boldsymbol{\theta}}$. See the note in Step 9 on page 326. (Note that $\vec{\boldsymbol{v}}$ is along positive $\hat{\boldsymbol{\theta}}$ if the solid disc rotates counterclockwise as viewed from the $+z$-axis.) For a rotating charged object, we write $v = r_{rot}\omega$, where r_{rot} is the radius of rotation of each dq. Based on how this solid disc is rotating, $r_{rot} = r_c$. (However, if the disc were instead "flipping" like a coin, it would be different.) Substitute $\vec{\boldsymbol{v}} = v\,\hat{\boldsymbol{\theta}} = r_c\omega\hat{\boldsymbol{\theta}}$ into the magnetic field integral. The constant ω may come out of the integral.

$$\vec{\mathbf{B}} = \frac{\mu_0 \sigma}{4\pi} \int_{r_c=0}^{a} \int_{\theta=0}^{2\pi} \frac{(r_c\omega\hat{\boldsymbol{\theta}}) \times (-r_c \hat{\mathbf{r}}_c + p\hat{\mathbf{z}})}{(r_c^2 + p^2)^{3/2}} r_c dr_c d\theta$$

$$\vec{\mathbf{B}} = \frac{\mu_0 \sigma \omega}{4\pi} \int_{r_c=0}^{a} \int_{\theta=0}^{2\pi} \frac{(r_c\hat{\boldsymbol{\theta}}) \times (-r_c \hat{\mathbf{r}}_c + p\hat{\mathbf{z}})}{(r_c^2 + p^2)^{3/2}} r_c dr_c d\theta$$

We worked out virtually the same vector product as $(r_c\hat{\boldsymbol{\theta}}) \times (-r_c\hat{\mathbf{r}}_c + p\hat{\mathbf{z}})$ a couple of examples back (compare with page 331). This time, we'll skip the discussion and proceed straight to the math:

$$(r_c\hat{\boldsymbol{\theta}}) \times (-r_c\hat{\mathbf{r}}_c + p\hat{\mathbf{z}}) = \begin{vmatrix} \hat{\mathbf{r}}_c & \hat{\boldsymbol{\theta}} & \hat{\mathbf{z}} \\ 0 & r_c & 0 \\ -r_c & 0 & p \end{vmatrix} = \hat{\mathbf{r}}_c \begin{vmatrix} r_c & 0 \\ 0 & p \end{vmatrix} - \hat{\boldsymbol{\theta}} \begin{vmatrix} 0 & 0 \\ -r_c & p \end{vmatrix} + \hat{\mathbf{z}} \begin{vmatrix} 0 & r_c \\ -r_c & 0 \end{vmatrix}$$

$$(r_c\hat{\boldsymbol{\theta}}) \times (-r_c\hat{\mathbf{r}}_c + p\hat{\mathbf{z}}) = \left[\hat{\mathbf{r}}_c(r_c p - 0) - \hat{\boldsymbol{\theta}}(0 - 0) + \hat{\mathbf{z}}(0 + r_c^2)\right] = r_c p\,\hat{\mathbf{r}}_c + r_c^2\,\hat{\mathbf{z}}$$

Substitute this result into the magnetic field integral.

$$\vec{\mathbf{B}} = \frac{\mu_0 \sigma \omega}{4\pi} \int_{r_c=0}^{a} \int_{\theta=0}^{2\pi} \frac{(r_c p\,\hat{\mathbf{r}}_c + r_c^2\,\hat{\mathbf{z}})}{(r_c^2 + p^2)^{3/2}} r_c dr_c d\theta$$

Separate the integral into two terms.

$$\vec{\mathbf{B}} = \frac{\mu_0 \sigma \omega}{4\pi} \int_{r_c=0}^{a} \int_{\theta=0}^{2\pi} \frac{r_c p\,\hat{\mathbf{r}}_c}{(r_c^2 + p^2)^{3/2}} r_c dr_c d\theta + \frac{\mu_0 \sigma \omega}{4\pi} \int_{r_c=0}^{a} \int_{\theta=0}^{2\pi} \frac{r_c^2\,\hat{\mathbf{z}}}{(r_c^2 + p^2)^{3/2}} r_c dr_c d\theta$$

Note that p and $\hat{\mathbf{z}}$ are constants, and thus may come out of their integrals, but $\hat{\mathbf{r}}_c$ is **<u>not</u>** constant. Also note that $r_c r_c = r_c^2$ and that $r_c^2 r_c = r_c^3$.

$$\vec{\mathbf{B}} = \frac{\mu_0 \sigma \omega p}{4\pi} \int_{r_c=0}^{a} \int_{\theta=0}^{2\pi} \frac{r_c^2\,\hat{\mathbf{r}}_c}{(r_c^2 + p^2)^{3/2}} dr_c d\theta + \frac{\mu_0 \sigma \omega}{4\pi}\hat{\mathbf{z}} \int_{r_c=0}^{a} \int_{\theta=0}^{2\pi} \frac{r_c^3}{(r_c^2 + p^2)^{3/2}} dr_c d\theta$$

We encountered a very similar situation a couple of examples back. We found that the integral $\int_{\theta=0}^{2\pi} \hat{\mathbf{r}}_c\, d\theta$ equals zero. Although this is now part of a double integral, we will still get the integral $\int_{\theta=0}^{2\pi} \hat{\mathbf{r}}_c\, d\theta$ as part of the angular integration, and so once again the first term will be zero. If you're adept at visualizing magnetic field lines like those from Chapter 23, you might be able to see that the magnetic field at the field point will only have a $\hat{\mathbf{z}}$ contribution. Whether you do the first integral or not, it will still be zero. Only the second integral yields a nonzero result.

$$\vec{\mathbf{B}} = \frac{\mu_0 \sigma \omega}{4\pi} \hat{\mathbf{z}} \int_{r_c=0}^{a} \int_{\theta=0}^{2\pi} \frac{r_c^3}{(r_c^2 + p^2)^{3/2}} dr_c d\theta$$

This double integral is separable. Treat the variable r_c as if it were a constant when integrating over the independent variable θ.

$$\vec{\mathbf{B}} = \frac{\mu_0 \sigma \omega}{4\pi} \hat{\mathbf{z}} \int_{r_c=0}^{a} \frac{r_c^3}{(r_c^2 + p^2)^{3/2}} dr_c \int_{\theta=0}^{2\pi} d\theta = \frac{\mu_0 \sigma \omega}{4\pi} \hat{\mathbf{z}} \int_{r_c=0}^{a} \frac{r_c^3}{(r_c^2 + p^2)^{3/2}} dr_c \,(2\pi)$$

$$\vec{\mathbf{B}} = \frac{\mu_0 \sigma \omega}{2} \hat{\mathbf{z}} \int_{r_c=0}^{a} \frac{r_c^3}{(r_c^2 + p^2)^{3/2}} dr_c$$

This integral can be performed via the following trigonometric substitution:

$$r_c = p \tan \psi$$
$$dr_c = p \sec^2 \psi \, d\psi$$

Solving for ψ, we get $\psi = \tan^{-1}\left(\frac{r_c}{p}\right)$, which shows that the new limits of integration are from $\psi = \tan^{-1}(0) =$ to $\psi_{max} = \tan^{-1}\left(\frac{a}{p}\right)$. Note that the denominator of the integral simplifies as follows, using the trig identity $1 + \tan^2 \psi = \tan^2 \psi + 1 = \sec^2 \psi$:

$$(r_c^2 + p^2)^{3/2} = [(p \tan \psi)^2 + p^2]^{3/2} = [p^2(\tan^2 \psi + 1)]^{3/2} = (p^2 \sec^2 \psi)^{3/2} = p^3 \sec^3 \psi$$

In the last step, we applied the rule from algebra that $(a^2 x^2)^{3/2} = (a^2)^{3/2}(x^2)^{3/2} = a^3 x^3$. Substitute the above expressions for dr_c, r_c, and $(r_c^2 + p^2)^{3/2}$ into the previous integral.

$$\vec{\mathbf{B}} = \frac{\mu_0 \sigma \omega}{2} \hat{\mathbf{z}} \int_{r_c=0}^{a} \frac{r_c^3}{(r_c^2 + p^2)^{3/2}} dr_c = \frac{\mu_0 \sigma \omega}{2} \hat{\mathbf{z}} \int_{\psi=0}^{\psi_{max}} \frac{p^3 \tan^3 \psi}{p^3 \sec^3 \psi} p \sec^2 \psi \, d\psi$$

$$\vec{\mathbf{B}} = \frac{\mu_0 \sigma \omega}{2} \hat{\mathbf{z}} \int_{\psi=0}^{\psi_{max}} \frac{\tan^3 \psi}{\sec \psi} p \, d\psi = \frac{\mu_0 \sigma \omega p}{2} \hat{\mathbf{z}} \int_{\psi=0}^{\psi_{max}} \frac{\tan^3 \psi}{\sec \psi} d\psi$$

In the last step, we pulled the constant p out of the integral. Recall from trig that $\sec \psi = \frac{1}{\cos \psi}$ and $\tan \psi = \frac{\sin \psi}{\cos \psi}$. Therefore, it follows that:

$$\frac{\tan^3 \psi}{\sec \psi} = \frac{\sin^3 \psi}{\cos^3 \psi} \div \frac{1}{\cos \psi} = \frac{\sin^3 \psi}{\cos^3 \psi} \times \frac{\cos \psi}{1} = \frac{\sin^3 \psi}{\cos^2 \psi}$$

To divide by a fraction, multiply by its **reciprocal**. The magnetic field integral becomes:

$$\vec{\mathbf{B}} = \frac{\mu_0 \sigma \omega}{2} \hat{\mathbf{z}} \int_{\psi=0}^{\psi_{max}} \frac{\tan^3 \psi}{\sec \psi} p \, d\psi = \frac{\mu_0 \sigma \omega p}{2} \hat{\mathbf{z}} \int_{\psi=0}^{\psi_{max}} \frac{\sin^3 \psi}{\cos^2 \psi} d\psi$$

Now we will use the trig identity $\sin^2\psi + \cos^2\psi = 1$, which can be written in the form $\sin^2\psi = 1 - \cos^2\psi$. This allows us to write the numerator as:

$$\sin^3\psi = \sin\psi\sin^2\psi = \sin\psi\,(1 - \cos^2\psi) = \sin\psi - \sin\psi\cos^2\psi$$

Substitute this expression into the previous integral, and separate the integral into terms.

$$\vec{\mathbf{B}} = \frac{\mu_0\sigma\omega p}{2}\hat{\mathbf{z}}\int_{\psi=0}^{\psi_{max}} \frac{\sin\psi - \sin\psi\cos^2\psi}{\cos^2\psi}d\psi$$

$$\vec{\mathbf{B}} = \frac{\mu_0\sigma\omega p}{2}\hat{\mathbf{z}}\int_{\psi=0}^{\psi_{max}} \frac{\sin\psi}{\cos^2\psi}d\psi - \frac{\mu_0\sigma\omega p}{2}\hat{\mathbf{z}}\int_{\psi=0}^{\psi_{max}} \frac{\sin\psi\cos^2\psi}{\cos^2\psi}d\psi$$

$$\vec{\mathbf{B}} = \frac{\mu_0\sigma\omega p}{2}\hat{\mathbf{z}}\int_{\psi=0}^{\psi_{max}} \frac{\sin\psi}{\cos^2\psi}d\psi - \frac{\mu_0\sigma\omega p}{2}\hat{\mathbf{z}}\int_{\psi=0}^{\psi_{max}} \sin\psi\,d\psi$$

We will perform the first integral using the following substitution:

$$u = \cos\psi$$

$$du = -\sin\psi\,d\psi$$

We'll deal with the new limits later. In the meantime, the magnetic field becomes:

$$\vec{\mathbf{B}} = \frac{\mu_0\sigma\omega p}{2}\hat{\mathbf{z}}\int_{u=u_{lower}}^{u_{upper}} \frac{-du}{u^2} - \frac{\mu_0\sigma\omega p}{2}\hat{\mathbf{z}}\int_{\psi=0}^{\psi_{max}} \sin\psi\,d\psi$$

$$\vec{\mathbf{B}} = \frac{\mu_0\sigma\omega p}{2}\hat{\mathbf{z}}\left[\frac{1}{u}\right]_{u=u_{lower}}^{u_{upper}} - \frac{\mu_0\sigma\omega p}{2}\hat{\mathbf{z}}[-\cos\psi]_{\psi=0}^{\psi_{max}}$$

Instead of worrying about the new limits over u, let's just express u back in terms of ψ. Recall that $u = \cos\psi$. Note that the two minus signs combine to make a plus sign.

$$\vec{\mathbf{B}} = \frac{\mu_0\sigma\omega p}{2}\hat{\mathbf{z}}\left[\frac{1}{\cos\psi}\right]_{\psi=0}^{\psi_{max}} + \frac{\mu_0\sigma\omega p}{2}\hat{\mathbf{z}}[\cos\psi]_{\psi=0}^{\psi_{max}}$$

$$\vec{\mathbf{B}} = \frac{\mu_0\sigma\omega p}{2}\hat{\mathbf{z}}\left(\frac{1}{\cos\psi_{max}} - \frac{1}{\cos 0}\right) + \frac{\mu_0\sigma\omega p}{2}\hat{\mathbf{z}}(\cos\psi_{max} - \cos 0)$$

$$\vec{\mathbf{B}} = \frac{\mu_0\sigma\omega p}{2}\hat{\mathbf{z}}\left(\frac{1}{\cos\psi_{max}} - 1\right) + \frac{\mu_0\sigma\omega p}{2}\hat{\mathbf{z}}(\cos\psi_{max} - 1)$$

Factor the $\frac{\mu_0\sigma\omega p}{2}$ out of all of the terms. Note that $-1 - 1 = -2$.

$$\vec{\mathbf{B}} = \frac{\mu_0\sigma\omega p}{2}\hat{\mathbf{z}}\left(\frac{1}{\cos\psi_{max}} - 1 + \cos\psi_{max} - 1\right) = \frac{\mu_0\sigma\omega p}{2}\hat{\mathbf{z}}\left(\frac{1}{\cos\psi_{max}} + \cos\psi_{max} - 2\right)$$

Recall that $\psi_{max} = \tan^{-1}\left(\frac{a}{p}\right)$, such that $\cos\psi_{max}$ is the complicated looking expression $\cos\left(\tan^{-1}\left(\frac{a}{p}\right)\right)$. Instead of taking the cosine of an inverse tangent, it is simpler to draw a right triangle and apply the Pythagorean theorem. Note that $\tan\psi_{max} = \frac{a}{p}$. We can make a right triangle from this: Since the tangent of ψ_{max} equals the opposite over the adjacent, we draw a right triangle with a opposite and p adjacent to ψ_{max}.

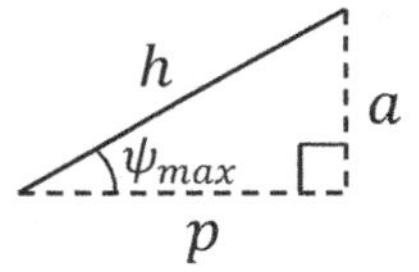

Find the hypotenuse, h, of the right triangle from the Pythagorean theorem.

$$h = \sqrt{p^2 + a^2}$$

Now we can write an expression for the cosine of ψ_{max}. It equals the adjacent (p) over the hypotenuse ($h = \sqrt{p^2 + a^2}$).

$$\cos\psi_{max} = \frac{p}{h} = \frac{p}{\sqrt{p^2 + a^2}}$$

Substitute this expression into the previous equation for magnetic field.

$$\vec{\mathbf{B}} = \frac{\mu_0 \sigma \omega p}{2}\hat{\mathbf{z}}\left(\frac{1}{\cos\psi_{max}} + \cos\psi_{max} - 2\right)$$

$$\vec{\mathbf{B}} = \frac{\mu_0 \sigma \omega p}{2}\hat{\mathbf{z}}\left(\frac{\sqrt{p^2 + a^2}}{p} + \frac{p}{\sqrt{p^2 + a^2}} - 2\right)$$

Add fractions using a **common denominator**. Make a common denominator of $p\sqrt{p^2 + a^2}$ by multiplying the first term by $\frac{\sqrt{p^2+a^2}}{\sqrt{p^2+a^2}}$, the second term by $\frac{p}{p}$, and the third term by $\frac{p\sqrt{p^2+a^2}}{p\sqrt{p^2+a^2}}$. Recall from algebra that $\sqrt{p^2 + a^2}\sqrt{p^2 + a^2} = p^2 + a^2$.

$$\vec{\mathbf{B}} = \frac{\mu_0 \sigma \omega p}{2}\hat{\mathbf{z}}\left(\frac{\sqrt{p^2 + a^2}\sqrt{p^2 + a^2}}{p\sqrt{p^2 + a^2}} + \frac{p^2}{p\sqrt{p^2 + a^2}} - \frac{2p\sqrt{p^2 + a^2}}{p\sqrt{p^2 + a^2}}\right)$$

$$\vec{\mathbf{B}} = \frac{\mu_0 \sigma \omega p}{2}\hat{\mathbf{z}}\left(\frac{p^2 + a^2}{p\sqrt{p^2 + a^2}} + \frac{p^2}{p\sqrt{p^2 + a^2}} - \frac{2p\sqrt{p^2 + a^2}}{p\sqrt{p^2 + a^2}}\right)$$

$$\vec{\mathbf{B}} = \frac{\mu_0 \sigma \omega p}{2}\hat{\mathbf{z}}\left(\frac{p^2 + a^2 + p^2 - 2p\sqrt{p^2 + a^2}}{p\sqrt{p^2 + a^2}}\right)$$

Note that $p^2 + p^2 = 2p^2$.

$$\vec{\mathbf{B}} = \frac{\mu_0 \sigma \omega p}{2}\hat{\mathbf{z}}\left(\frac{2p^2 + a^2 - 2p\sqrt{p^2 + a^2}}{p\sqrt{p^2 + a^2}}\right)$$

The p from out front cancels the p from the denominator.

$$\vec{\mathbf{B}} = \frac{\mu_0 \sigma \omega}{2}\hat{\mathbf{z}}\left(\frac{2p^2 + a^2 - 2p\sqrt{p^2 + a^2}}{\sqrt{p^2 + a^2}}\right)$$

Note: Some physics instructors allow their students to use a handy **table of integrals**. In that case, you could have **skipped** several steps of integral calculus, trig, and algebra.

We're **not finished yet** because we need to eliminate the charge density σ from our answer. The way to do this is to integrate over dq to find the total charge of the disc, Q. This integral involves the same substitutions from before.

$$Q = \int dq = \int \sigma \, dA = \sigma \int dA = \sigma \int_{r_c=0}^{a} \int_{\theta=0}^{2\pi} r_c dr_c d\theta = \sigma \int_{r_c=0}^{a} r_c dr_c \int_{\theta=0}^{2\pi} d\theta$$

$$Q = \sigma \left[\frac{r_c^2}{2}\right]_{r_c=0}^{a} [\theta]_{\theta=0}^{2\pi} = \sigma\left(\frac{a^2}{2} - \frac{0^2}{2}\right)(2\pi - 0) = \sigma\left(\frac{a^2}{2}\right)(2\pi) = \pi\sigma a^2$$

Solve for σ in terms of Q: Divide both sides of the equation by πa^2.

$$\sigma = \frac{Q}{\pi a^2}$$

Substitute this expression for σ into our previous expression for $\vec{\mathbf{B}}$.

$$\vec{\mathbf{B}} = \frac{\mu_0 \sigma \omega}{2} \hat{\mathbf{z}} \left(\frac{2p^2 + a^2 - 2p\sqrt{p^2 + a^2}}{\sqrt{p^2 + a^2}}\right)$$

$$\vec{\mathbf{B}} = \frac{\mu_0 \omega}{2} \left(\frac{Q}{\pi a^2}\right) \hat{\mathbf{z}} \left(\frac{2p^2 + a^2 - 2p\sqrt{p^2 + a^2}}{\sqrt{p^2 + a^2}}\right)$$

$$\vec{\mathbf{B}} = \frac{\mu_0 Q \omega}{2\pi a^2} \hat{\mathbf{z}} \left(\frac{2p^2 + a^2 - 2p\sqrt{p^2 + a^2}}{\sqrt{p^2 + a^2}}\right)$$

The magnetic field at the point $(0, 0, p)$ is $\vec{\mathbf{B}} = \frac{\mu_0 Q \omega}{2\pi a^2} \hat{\mathbf{z}} \left(\frac{2p^2+a^2-2p\sqrt{p^2+a^2}}{\sqrt{p^2+a^2}}\right)$. It has a magnitude of $B = \frac{\mu_0 Q \omega}{2\pi a^2} \left(\frac{2p^2+a^2-2p\sqrt{p^2+a^2}}{\sqrt{p^2+a^2}}\right)$ and a direction of $\hat{\mathbf{B}} = \hat{\mathbf{z}}$ (assuming that the charged disc rotates counterclockwise as viewed from the $+z$-axis).

93. A filamentary current travels along the path illustrated below, which includes two circular arcs with radii of $2a$ and $3a$ (where the "complete" circle would be centered about the origin) and three straight line segments. The angles are in radians, measured counterclockwise from the $+x$-axis. Derive an equation for the magnetic field at the origin in terms of μ_0, I, a, and appropriate unit vectors.

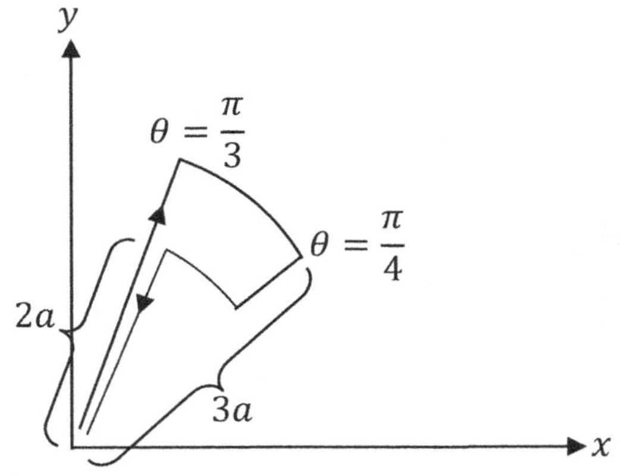

Want help? Check the hints section at the back of the book.

Answer: $\vec{\mathbf{B}} = \frac{\mu_0 I}{288a}\hat{\mathbf{z}}$

94. A filamentary current travels along a finite length of wire on the z-axis with endpoints at $\left(0,0,-\frac{L}{2}\right)$ and $\left(0,0,\frac{L}{2}\right)$, as illustrated below. Derive an equation for the magnetic field at the point $(a,0,0)$, where a is a constant, in terms of μ_0, I, L, a, and appropriate unit vectors.

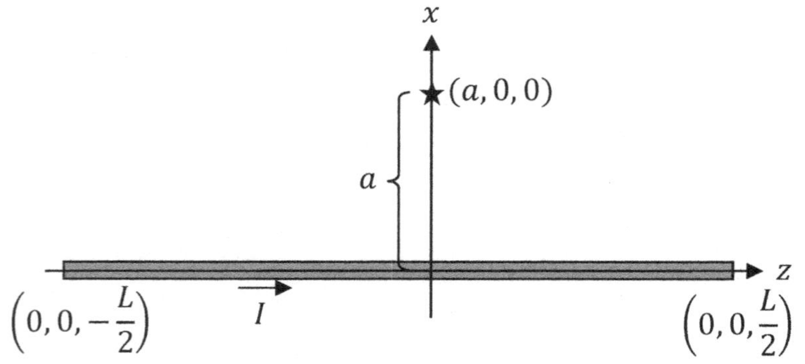

Want help? Check the hints section at the back of the book.

Answer: $\vec{\mathbf{B}} = \frac{\mu_0 I L}{4\pi a\sqrt{\frac{L^2}{4}+a^2}}\hat{\mathbf{y}}$

95. In the Helmholtz coils illustrated below, equal filamentary currents travel in the same direction along two parallel loops of wire. The circular loops are parallel to the xy plane, with one at $z = -\frac{a}{2}$ and the other at $z = \frac{a}{2}$ (such that the loops are separated by a distance a, which also equals the radius of each loop). Derive an equation for the magnetic field at the origin in terms of μ_0, I, a, and appropriate unit vectors.

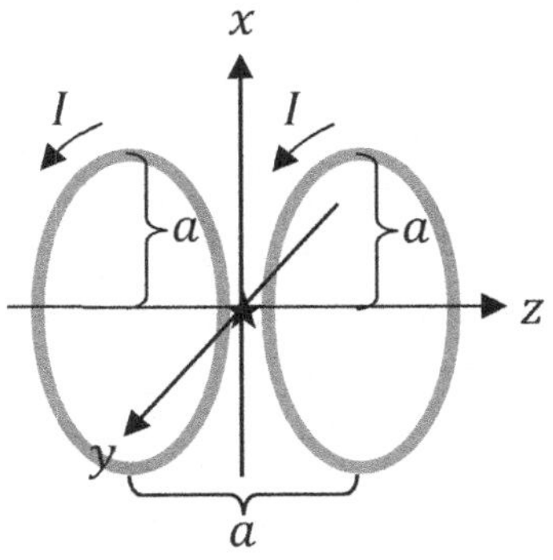

Want help? Check the hints section at the back of the book.

Answer: $\vec{\mathbf{B}} = \frac{8}{(5)^{3/2}} \left(\frac{\mu_0 I}{a}\right) \hat{\mathbf{z}}$

96. A solid disc lying in the xy plane and centered about the origin has positive charge Q, radius a, and uniform charge density σ. The solid disc rotates about the z-axis counter-clockwise with constant angular speed ω. Derive an equation for the magnetic field at the origin in terms of μ_0, Q, a, ω, and appropriate unit vectors. (Do this using calculus, **not** by simply setting p to zero with the result from the similar example.) **Note**: Because the field point lies at the origin, the calculus involved in this problem is **much simpler** than the similar example.

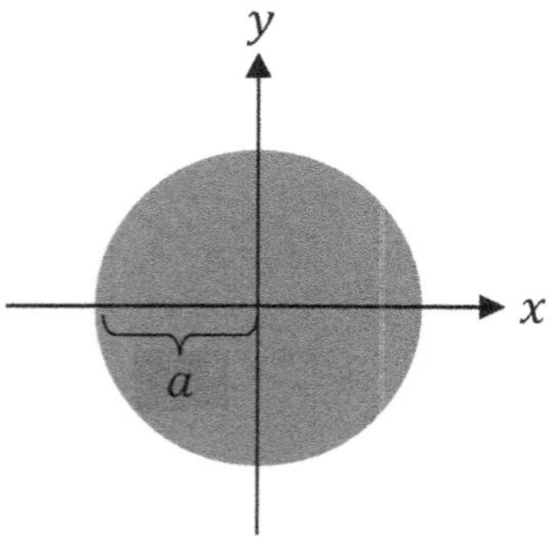

Want help? Check the hints section at the back of the book.

Answer: $\vec{\mathbf{B}} = \frac{\mu_0 Q \omega}{2\pi a}\hat{\mathbf{z}}$

27 AMPÈRE'S LAW

Relevant Terminology

Electric charge – a fundamental property of a particle that causes the particle to experience a force in the presence of an electric field. (An electrically neutral particle has no charge and thus experiences no force in the presence of an electric field.)
Current – the instantaneous rate of flow of charge through a conductor.
Filamentary current – a current that runs through a very thin wire.
Magnetic field – a magnetic effect created by a moving charge (or current).
Magnetic force – the push or pull that a moving charge (or current) experiences in the presence of a magnetic field.
Field point – the point where you are trying to calculate the magnetic field. The field point is ordinarily specified in the problem. There often is **not** a current at the field point.
Source – the current (or moving charge) that is creating the magnetic field.
Open path – a curve that doesn't bound an area. A U-shaped curve (∪) is an example of an open path because it doesn't separate the region inside of it from the region outside of it.
Closed path – a curve that bounds an area. A triangle (△) is an example of a closed path because the region inside of the triangle is completely isolated from the region outside of it.
Solenoid – a coil of wire in the shape of a right-circular cylinder.
Toroidal coil – a coil of wire in the shape of a single-holed ring torus (which is a donut).

Ampère's Law

According to **Ampère's law**, the integral of $\vec{\mathbf{B}} \cdot d\vec{\mathbf{s}}$ (called a line integral) over a **closed** path (often called an Ampèrian loop) is proportional to the current enclosed (I_{enc}) by the path.

$$\oint_C \vec{\mathbf{B}} \cdot d\vec{\mathbf{s}} = \mu_0 I_{enc}$$

The symbol $\oint$ is called a **closed** integral: It represents that the integral is over a **closed** path. The constant μ_0 is called the **permeability of free space**.

Current Densities

We use different kinds of current densities (analogous to the charge densities λ, σ, and ρ of Chapter 7) when performing magnetic field integrals, depending upon the geometry.

- A surface current density $\vec{\mathbf{K}}$ applies to current that runs along a conducting surface, such as a strip of copper.
- Current density $\vec{\mathbf{J}}$ is distributed throughout a volume, such as current that runs along the length of a solid right-circular cylinder.

Symbols and SI Units

Symbol	Name	SI Units
I_{enc}	the current enclosed by the Ampèrian loop	A
I	the total current	A
$\vec{\mathbf{B}}$	magnetic field	T
μ_0	the permeability of free space	$\frac{\text{T}\cdot\text{m}}{\text{A}}$
x, y, z	Cartesian coordinates	m, m, m
$\hat{\mathbf{x}}, \hat{\mathbf{y}}, \hat{\mathbf{z}}$	unit vectors along the $+x$-, $+y$-, $+z$-axes	unitless
r, θ	2D polar coordinates	m, rad
r_c, θ, z	cylindrical coordinates	m, rad, m
r, θ, φ	spherical coordinates	m, rad, rad
$\hat{\mathbf{r}}, \hat{\boldsymbol{\theta}}, \hat{\boldsymbol{\varphi}}$	unit vectors along spherical coordinate axes	unitless
$\hat{\mathbf{r}}_c$	a unit vector pointing away from the $+z$-axis	unitless
$\vec{\mathbf{K}}$	surface current density (distributed over a surface)	A/m
$\vec{\mathbf{J}}$	current density (distributed throughout a volume)	A/m^2
$d\vec{\mathbf{s}}$	differential displacement along a filamentary current	m
dA	differential area element	m^2
dV	differential volume element	m^3
N	number of loops (or turns)	unitless

Special Symbols

Symbol	Name
$\otimes$	into the page
$\odot$	out of the page

Strategy for Applying Ampère's Law

If there is enough symmetry in a magnetic field problem for it to be practical to take advantage of Ampère's law, follow these steps:

1. Sketch the magnetic field lines by applying the right-hand rule for magnetic field. It may be helpful to review Chapter 22 before proceeding.
2. Look at these magnetic field lines. Try to visualize a closed path (like a circle or rectangle) called an **Ampèrian loop** for which the magnetic field lines would always be tangential or perpendicular to (or a combination of these) the path no matter which part of the path they pass through. When the magnetic field lines are tangential to the path, you want the magnitude of the magnetic field to be **constant** over that part of the path. These features make the left-hand side of Ampère's law very easy to compute. The closed path must also enclose some of the current.
3. Study the examples that follow. Most Ampère's law problems have a geometry that is very similar to one of these examples (an infinitely long cylinder, an infinite plane, a solenoid, or a toroidal coil). The Ampèrian loops that we draw in the examples are basically the same as the Ampèrian loops encountered in most of the problems. This makes Steps 1-2 very easy.
4. Write down the formula for Ampère's law.
$$\oint_C \vec{\mathbf{B}} \cdot d\vec{\mathbf{s}} = \mu_0 I_{enc}$$
If the path is a circle, there will just be one integral. If the path is a rectangle, break the closed integral up into four open integrals: one for each side (see the examples).
5. Simplify the left-hand side of Ampère's law. Note that $d\vec{\mathbf{s}}$ is **tangential** to the path.
   - If $\vec{\mathbf{B}}$ is parallel to $d\vec{\mathbf{s}}$, then $\vec{\mathbf{B}} \cdot d\vec{\mathbf{s}} = Bds$.
   - If $\vec{\mathbf{B}}$ is anti-parallel to $d\vec{\mathbf{s}}$, then $\vec{\mathbf{B}} \cdot d\vec{\mathbf{s}} = -Bds$.
   - If $\vec{\mathbf{B}}$ is perpendicular to $d\vec{\mathbf{s}}$, then $\vec{\mathbf{B}} \cdot d\vec{\mathbf{s}} = 0$.
6. If by symmetry the magnitude of the magnetic field (B) is constant over any part of the path, $\int_C B\, ds = B \int_C ds = Bs$ for that part of the path. If you choose your Ampèrian loop wisely in Step 2, you will either get Bs or zero for each part of the closed path, such that Ampère's law simplifies to:
$$B_1 s_1 + B_2 s_2 + \cdots + B_N s_N = \mu_0 I_{enc}$$
For a circle, there is just one term. For a rectangle, there are four terms - one for each side (though some terms may be zero).
7. Replace each length with the appropriate expression, depending upon the geometry.
   - The circumference of a circle is $s = 2\pi r$.
   - For a rectangle, $s = L$ or $s = W$ for each side.

   Ampère's law problems with cylinders generally involve two different radii: The radius of an Ampèrian circle is r_c, whereas the radius of a conducting cylinder is a different symbol (which we will usually call a in this book, but may be called R in other textbooks – we are using a to avoid possible confusion between lowercase and uppercase r's).

8. Isolate the magnetic field in your simplified equation from Ampère's law. (This should be a simple algebra exercise.)
9. Consider each region in the problem. There will ordinarily be at least two regions. One region may be inside of a conductor and another region may be outside of the conductor, or the two regions might be on opposite sides of a conducting sheet, for example. We will label the regions with Roman numerals (I, II, III, IV, V, etc.).
10. Determine the net current enclosed in each region. In a given region, if the Ampèrian loop encloses the entire current, the current enclosed (I_{enc}) will equal the total current (I). However, if the Ampèrian loop encloses only a fraction of the current, you will need to integrate in order to find the current enclosed, applying the technique from Chapter 26 (Step 12 on page 326).

$$I = \int \vec{\mathbf{K}} \cdot d\vec{\boldsymbol{\ell}}$$

$$I = \int \vec{\mathbf{J}} \cdot d\vec{\mathbf{A}}$$

These integrals are over the intersection of the Ampèrian loop and the conducting surface. For a surface current density ($\vec{\mathbf{K}}$), the intersection is the length ($\vec{\boldsymbol{\ell}}$) across the width of the surface (which is perpendicular to the current). For a 3D current density ($\vec{\mathbf{J}}$), the intersection is the cross-sectional area ($\vec{\mathbf{A}}$) of the solid conductor.

11. For each region, substitute the current enclosed (from Step 10) into the simplified expression for the magnetic field (from Step 8).

Important Distinctions

Although Ampère's law is similar to Gauss's law (Chapter 8), the differences are important:

- Ampère's law involves **magnetic** field ($\vec{\mathbf{B}}$), not electric field ($\vec{\mathbf{E}}$).
- Ampère's law is over a closed **path** (C), not a closed surface (S).
- Ampère's law involves differential displacement ($d\vec{\mathbf{s}}$), not differential area ($d\vec{\mathbf{A}}$).
- Ampère's law involves the **current** enclosed (I_{enc}), not the charge enclosed (q_{enc}).
- Ampère's law involves the **permeability** of free space (μ_0), not permittivity (ϵ_0).

Ampère's law for magnetism	$\oint_C \vec{\mathbf{B}} \cdot d\vec{\mathbf{s}} = \mu_0 I_{enc}$
Gauss's law for electrostatics	$\oint_S \vec{\mathbf{E}} \cdot d\vec{\mathbf{A}} = \frac{q_{enc}}{\epsilon_0}$

Example: An infinite solid cylindrical conductor coaxial with the z-axis has radius a, uniform current density $\vec{\mathbf{J}}$ along $\hat{\mathbf{z}}$, and carries total current I. Derive an expression for the magnetic field both inside and outside of the cylinder.

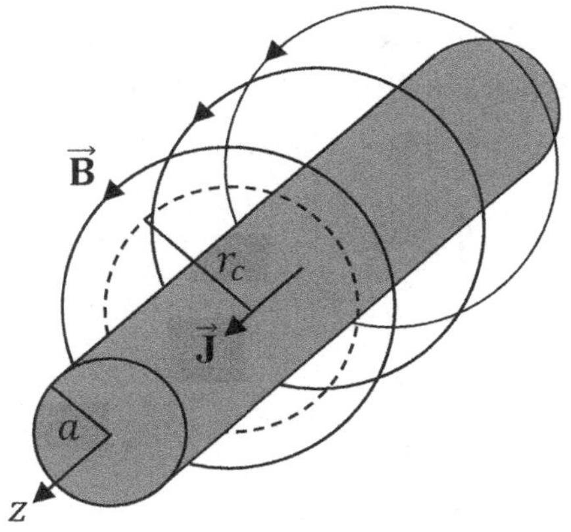

First sketch the magnetic field lines for the conducting cylinder. It's hard to draw, but the cylinder is perpendicular to the page with the current (I) coming out of the page ($\odot$). Apply the right-hand rule for magnetic field (Chapter 22): Grab the current with your thumb pointing out of the page ($\odot$), along the current (I). Your fingers make counter-clockwise (use the right-hand rule to see this) circles around the wire (toward your fingertips), as shown in the diagram above.

We choose our Ampèrian loop to be a circle (dashed line above) coaxial with the conducting cylinder such that $\vec{\mathbf{B}}$ and $d\vec{\mathbf{s}}$ (which is tangent to the Ampèrian loop) will be parallel and the magnitude of $\vec{\mathbf{B}}$ will be constant over the Ampèrian loop (since every point on the Ampèrian loop is equidistant from the axis of the conducting cylinder).

Note: In Chapter 26, when we applied the law of Biot-Savart, $d\vec{\mathbf{s}}$ was tangent to the current. However, in Chapter 27, when we apply Ampère's law, $d\vec{\mathbf{s}}$ is instead tangent to the Ampèrian loop, **not** tangent to the current.

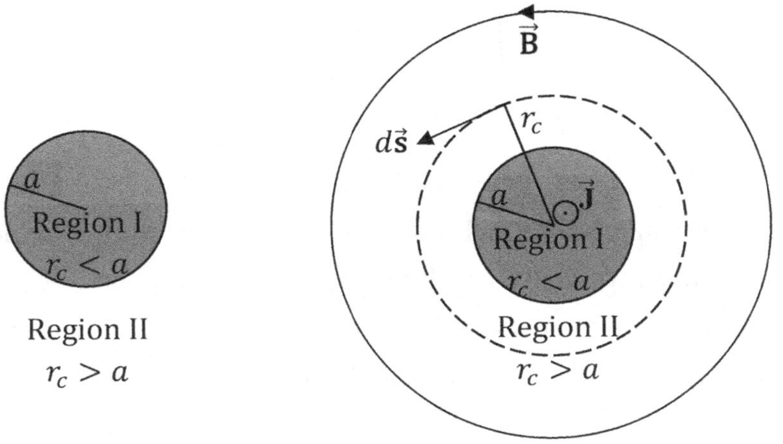

Write the formula for Ampère's law.

$$\oint_C \vec{\mathbf{B}} \cdot d\vec{\mathbf{s}} = \mu_0 I_{enc}$$

The scalar product is $\vec{\mathbf{B}} \cdot d\vec{\mathbf{s}} = B\cos\theta\, ds$, and $\theta = 0°$ since the magnetic field lines make circles coaxial with our Ampèrian circle (and are therefore parallel to $d\vec{\mathbf{s}}$, which is tangent to the Ampèrian circle). See the previous diagram.

$$\oint_C B\cos 0°\, ds = \mu_0 I_{enc}$$

Recall from trig that $\cos 0° = 1$.

$$\oint_C B\, ds = \mu_0 I_{enc}$$

The magnitude of the magnetic field is constant over the Ampèrian circle, since every point on the circle is equidistant from the axis of the conducting cylinder. Therefore, we may pull B out of the integral.

$$B\oint_C ds = \mu_0 I_{enc}$$

This integral is over the path of the Ampèrian circle of radius r_c, where r_c is a variable because the magnetic field depends on how close or far the field point is away from the conducting cylinder. We work with cylindrical coordinates (Chapter 6) and write $ds = r_c d\theta$. This is a purely angular integration (over the angle θ), and we treat r_c as a constant in this integral because every point on the path of the Ampèrian circle has the same value of r_c.

$$\oint_C ds = \int_{\theta=0}^{2\pi} r_c\, d\theta = r_c \int_{\theta=0}^{2\pi} d\theta = 2\pi r_c$$

This is the circumference of the Ampèrian circle. Substitute this expression for circumference into the previous equation for magnetic field.

$$B\oint_C ds = B 2\pi r_c = \mu_0 I_{enc}$$

Isolate the magnitude of the magnetic field by dividing both sides of the equation by $2\pi r_c$.

$$B = \frac{\mu_0 I_{enc}}{2\pi r_c}$$

Now we need to determine how much current is enclosed by the Ampèrian circle. For a solid conducting cylinder, we write $I = \int \vec{\mathbf{J}} \cdot d\vec{\mathbf{A}}$ (Step 10 on page 348). Since the cylinder has uniform current density, we may pull $\vec{\mathbf{J}}$ out of the integral. This integral is over the area of the Ampèrian circle. Since $d\vec{\mathbf{A}}$ is perpendicular to the cross-sectional area, $\vec{\mathbf{J}}$ is parallel to $d\vec{\mathbf{A}}$ (both come out of the page, along the axis of the cylinder). Therefore, the scalar product is $\vec{\mathbf{J}} \cdot d\vec{\mathbf{A}} = J \cos 0° \, dA = JdA$.

$$I_{enc} = \int \vec{\mathbf{J}} \cdot d\vec{\mathbf{A}} = \int J \cos 0° \, dA = \int J \, dA = J \int dA$$

We work with cylindrical coordinates and write the differential area element as $dA = r_c dr_c d\theta$ (see Chapter 6). Since we are now integrating over area (not arc length like before), r_c is a variable and we will have a double integral.

We must consider two different regions:

- The Ampèrian circle could be smaller than the conducting cylinder. This will help us find the magnetic field in region I.
- The Ampèrian circle could be larger than the conducting cylinder. This will help us find the magnetic field in region II.

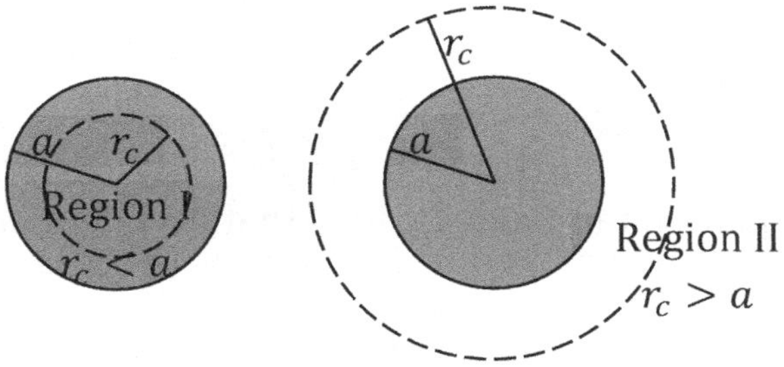

<u>**Region I**</u>: $r_c < a$.

Inside of the conducting cylinder, only a fraction of the cylinder's current is enclosed by the Ampèrian circle. In this region, the upper limit of the r_c-integration is the variable r_c: The larger the Ampèrian circle, the more current it encloses, up to a maximum radius of a (the radius of the conducting cylinder).

$$I_{enc} = J \int dA = J \int_{r_c=0}^{r_c} \int_{\theta=0}^{2\pi} r_c \, dr_c \, d\theta$$

This integral is separable:

$$I_{enc} = J \int_{r_c=0}^{r_c} r_c \, dr_c \int_{\theta=0}^{2\pi} d\theta = J \left[\frac{r_c^2}{2} \right]_{r_c=0}^{r_c} [\theta]_{\theta=0}^{2\pi} = J \left(\frac{r_c^2}{2} \right) (2\pi) = \pi r_c^2 J$$

You should recognize that πr_c^2 is the area of a circle. For a conducting cylinder with uniform current density, current equals J times area. (For a non-uniform current density, you would need to integrate: Then you couldn't just multiply J by area. Although we could have skipped the calculus in this example, we performed the integral so that it would be easier for you to adapt the solution to a non-uniform current density.) Substitute this expression for the current enclosed into the previous equation for magnetic field.

$$B = \frac{\mu_0 I_{enc}}{2\pi r_c} = \frac{\mu_0}{2\pi r_c} (\pi r_c^2 J) = \frac{\mu_0 J r_c}{2}$$

Since the magnetic field lines circulate around the conducting cylinder, we can include a direction with the magnetic field by adding on the cylindrical unit vector $\hat{\boldsymbol{\theta}}$.

$$\vec{\mathbf{B}} = \frac{\mu_0 J r_c}{2} \hat{\boldsymbol{\theta}}$$

The answer is different outside of the conducting cylinder. We will explore that next.

<u>Region II</u>: $r_c > a$.

Outside of the conducting cylinder, 100% of the current is enclosed by the Ampèrian circle. This changes the upper limit of the r_c-integration to a (since all of the current lies inside a cylinder of radius a).

$$I_{enc} = I = J \int dA = J \int_{r_c=0}^{r_c} \int_{\theta=0}^{2\pi} r_c \, dr_c \, d\theta$$

We don't need to work out the entire double integral again. We'll get the same expression as before, but with a in place of r_c.

$$I_{enc} = I = \pi a^2 J$$

Substitute this into the equation for magnetic field that we obtained from Ampère's law.

$$B = \frac{\mu_0 I_{enc}}{2\pi r_c} = \frac{\mu_0}{2\pi r_c}(\pi a^2 J) = \frac{\mu_0 J a^2}{2 r_c}$$

We can turn this into a vector by including the appropriate unit vector.

$$\vec{\mathbf{B}} = \frac{\mu_0 J a^2}{2 r_c} \hat{\boldsymbol{\theta}}$$

<u>Alternate forms of the answers in regions I and II</u>.

Since the total current is $I = \pi a^2 J$ (we found this equation for region II above), we can alternatively express the magnetic field in terms of the total current (I) of the conducting cylinder instead of the current density (J).

<u>Region I</u>: $r_c < a$.

$$\vec{\mathbf{B}} = \frac{\mu_0 J r_c}{2} \hat{\boldsymbol{\theta}} = \frac{\mu_0 I r_c}{2\pi a^2} \hat{\boldsymbol{\theta}}$$

<u>Region II</u>: $r_c > a$.

$$\vec{\mathbf{B}} = \frac{\mu_0 J a^2}{2 r_c} \hat{\boldsymbol{\theta}} = \frac{\mu_0 I}{2\pi r_c} \hat{\boldsymbol{\theta}}$$

Note that the magnetic field in region II is identical to the magnetic field created by an infinite filamentary current (see Chapter 26, but note that a has a different meaning there; you can also find this formula in Chapter 25 for a long straight wire). Note also that the expressions for the magnetic field in the two different regions both agree at the boundary: That is, in the limit that r_c approaches a, both expressions approach $\frac{\mu_0 I}{2\pi a} \hat{\boldsymbol{\theta}}$.

Example: The infinite current sheet illustrated below is a very thin infinite conducting plane with uniform current density* $\vec{\mathbf{K}}$ coming out of the page. The infinite current sheet lies in the xy plane at $z = 0$. Derive an expression for the magnetic field on either side of the infinite current sheet.

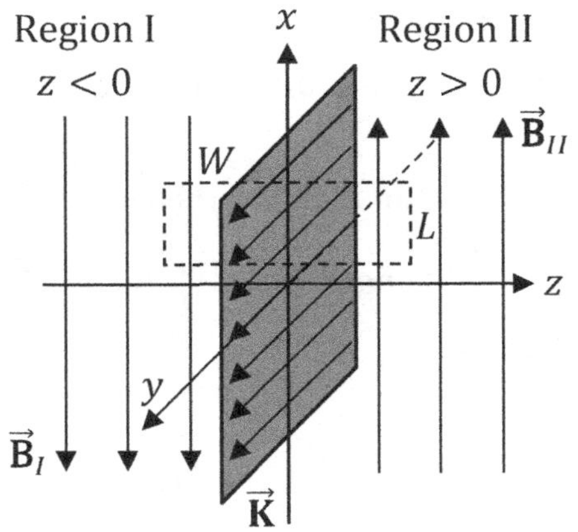

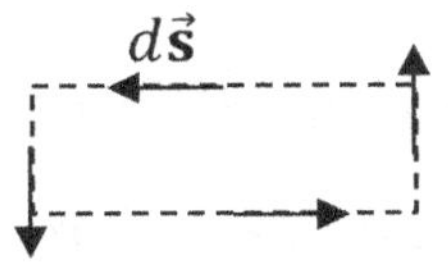

First sketch the magnetic field lines for the current sheet. Note that the current (I) is coming out of the page (⊙) along $\vec{\mathbf{K}}$. Apply the right-hand rule for magnetic field (Chapter 22): Grab the current with your thumb pointing out of the page (⊙), along the current (I). As shown above, the magnetic field ($\vec{\mathbf{B}}$) lines are straight up (↑) to the right of the sheet and straight down (↓) to the left of the sheet.

We choose our Ampèrian loop to be a rectangle lying in the zx plane (the same as the plane of this page) such that $\vec{\mathbf{B}}$ and $d\vec{\mathbf{s}}$ (which is tangent to the Ampèrian loop) will be parallel or perpendicular at each side of the rectangle. The Ampèrian rectangle is centered about the current sheet, as shown above on the left (we redrew the Ampèrian rectangle again on the right in order to make it easier to visualize $d\vec{\mathbf{s}}$).

- Along the top and bottom sides, $\vec{\mathbf{B}}$ is vertical and $d\vec{\mathbf{s}}$ is horizontal, such that $\vec{\mathbf{B}}$ and $d\vec{\mathbf{s}}$ are perpendicular.
- Along the right and left sides, $\vec{\mathbf{B}}$ and $d\vec{\mathbf{s}}$ are both parallel (they either both point up or both point down).

Write the formula for Ampère's law.

$$\oint_C \vec{\mathbf{B}} \cdot d\vec{\mathbf{s}} = \mu_0 I_{enc}$$

The closed integral on the left-hand side of the equation involves integrating over the complete path of the Ampèrian rectangle. The rectangle includes four sides: the right, top, left, and bottom edges.

* Note that a few textbooks may use different symbols for the current densities, $\vec{\mathbf{K}}$ and $\vec{\mathbf{J}}$.

$$\int_{right} \vec{\mathbf{B}} \cdot d\vec{\mathbf{s}} + \int_{top} \vec{\mathbf{B}} \cdot d\vec{\mathbf{s}} + \int_{left} \vec{\mathbf{B}} \cdot d\vec{\mathbf{s}} + \int_{bot} \vec{\mathbf{B}} \cdot d\vec{\mathbf{s}} = \mu_0 I_{enc}$$

The scalar product is $\vec{\mathbf{B}} \cdot d\vec{\mathbf{s}} = B\cos\theta\, ds$, where θ is the angle between $\vec{\mathbf{B}}$ and $d\vec{\mathbf{s}}$. Also recall that the direction of $d\vec{\mathbf{s}}$ is tangential to the Ampèrian loop. Study the direction of $\vec{\mathbf{B}}$ and $d\vec{\mathbf{s}}$ at each side of the rectangle in the previous diagram.

- For the right and left sides, $\theta = 0°$ because $\vec{\mathbf{B}}$ and $d\vec{\mathbf{s}}$ either both point up or both point down.
- For the top and bottom sides, $\theta = 90°$ because $\vec{\mathbf{B}}$ and $d\vec{\mathbf{s}}$ are perpendicular.

$$\int_{right} B\cos 0°\, ds + \int_{top} B\cos 90°\, ds + \int_{left} B\cos 0°\, ds + \int_{bot} B\cos 90°\, ds = \mu_0 I_{enc}$$

Recall from trig that $\cos 0° = 1$ and $\cos 90° = 0$.

$$\int_{right} Bds + 0 + \int_{left} Bds + 0 = \mu_0 I_{enc}$$

Over the right or left side of the Ampèrian rectangle, the magnitude of the magnetic field is constant, since every point on either side of the rectangle is equidistant from the infinite sheet. Therefore, we may pull B out of the integrals. (We choose our Ampèrian rectangle to be centered about the infinite sheet such that the value of B is the same† at both ends.)

$$B \int_{right} ds + B \int_{left} ds = \mu_0 I_{enc}$$

The remaining integrals are trivial: $\int ds = L$ for the left and right sides of the rectangle. Note that L is the height of the Ampèrian rectangle, **not** the length of the current sheet (which is infinite).

$$BL + BL = 2BL = \mu_0 I_{enc}$$

Isolate the magnitude of the magnetic field by dividing both sides of the equation by $2L$.

$$B = \frac{\mu_0 I_{enc}}{2L}$$

Now we need to determine how much current is enclosed by the Ampèrian rectangle. For a current sheet, we write $I = \int \vec{\mathbf{K}} \cdot d\vec{\boldsymbol{\ell}}$ (Step 10 on page 348). Since the current sheet has uniform current density, we may pull $\vec{\mathbf{K}}$ out of the integral. This integral is over the length L of the Ampèrian rectangle (since that's the direction that encompasses some current). Note that the direction of $d\vec{\boldsymbol{\ell}}$ is perpendicular to the length L (consistent with how $d\vec{\mathbf{A}}$ is perpendicular to the cross-sectional area when working with $\vec{\mathbf{J}} \cdot d\vec{\mathbf{A}}$), which is parallel to $\vec{\mathbf{K}}$. Therefore, the scalar product is $\vec{\mathbf{K}} \cdot d\vec{\boldsymbol{\ell}} = K\cos 0°\, d\ell = Kd\ell$.

† Once we reach our final answer, we will see that this doesn't matter: It turns out that the magnetic field is independent of the distance from the infinite current sheet.

$$I_{enc} = \int \vec{\mathbf{K}} \cdot d\vec{\boldsymbol{\ell}} = \int K \cos 0° \, d\ell = \int K \, d\ell = K \int d\ell = KL$$

Substitute this expression for the current enclosed into the previous equation for magnetic field.

$$B = \frac{\mu_0 I_{enc}}{2L} = \frac{\mu_0}{2L}(KL) = \frac{\mu_0 K}{2}$$

The magnitude of the magnetic field is $B = \frac{\mu_0 K}{2}$, which is a constant. Thus, the magnetic field created by an infinite current sheet is uniform. We can use a unit vector to include the direction of the magnetic field with our answer: $\vec{\mathbf{B}} = \frac{\mu_0 K}{2}\hat{\mathbf{x}}$ for $z > 0$ (to the right of the sheet) and $\vec{\mathbf{B}} = -\frac{\mu_0 K}{2}\hat{\mathbf{x}}$ for $z < 0$ (to the left of the sheet), since $\hat{\mathbf{x}}$ points one unit along the $+x$-axis (which is upward in our original diagram). There is a clever way to combine both results into a single equation: We can simply write $\vec{\mathbf{B}} = \frac{\mu_0 K}{2}\frac{|z|}{z}\hat{\mathbf{x}}$, since $\frac{|z|}{z} = +1$ if $z > 0$ and $\frac{|z|}{z} = -1$ if $z < 0$. (Note carefully the combination of z and $\hat{\mathbf{x}}$: The z-coordinate tells us whether we're on the right or left side of the sheet, while the unit vector $\hat{\mathbf{x}}$ points upward.)

Example: The infinite tightly wound solenoid illustrated below is coaxial with the z-axis and carries an insulated filamentary current I. Derive an expression for the magnetic field inside of the solenoid.

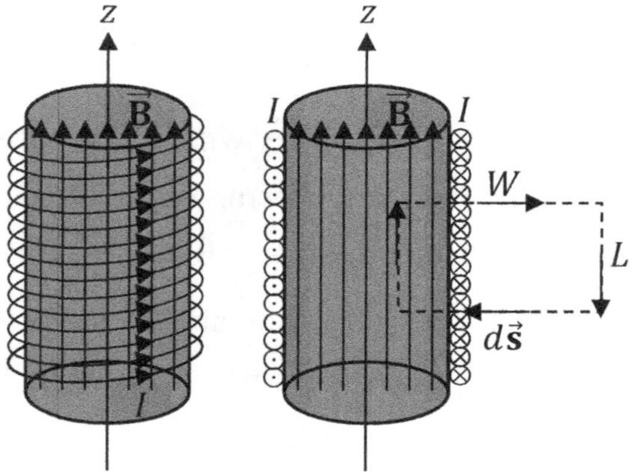

First sketch the magnetic field lines for the solenoid. Recall that we have already done this in Chapter 22: As shown above, the magnetic field ($\vec{\mathbf{B}}$) lines are straight up (↑) inside of the solenoid.

We choose our Ampèrian loop to be a rectangle with one edge perpendicular to the axis of the solenoid such that $\vec{\mathbf{B}}$ and $d\vec{\mathbf{s}}$ (which is tangent to the Ampèrian loop) will be parallel or perpendicular at each side of the rectangle.

- Along the top and bottom sides, $\vec{\mathbf{B}}$ is vertical and $d\vec{\mathbf{s}}$ is horizontal, such that $\vec{\mathbf{B}}$ and $d\vec{\mathbf{s}}$ are perpendicular.
- Along the right and left sides, $\vec{\mathbf{B}}$ and $d\vec{\mathbf{s}}$ are both parallel (they either both point up or both point down).

Write the formula for Ampère's law.

$$\oint_C \vec{\mathbf{B}} \cdot d\vec{\mathbf{s}} = \mu_0 I_{enc}$$

The closed integral on the left-hand side of the equation involves integrating over the complete path of the Ampèrian rectangle. The rectangle includes four sides.

$$\int_{right} \vec{\mathbf{B}} \cdot d\vec{\mathbf{s}} + \int_{top} \vec{\mathbf{B}} \cdot d\vec{\mathbf{s}} + \int_{left} \vec{\mathbf{B}} \cdot d\vec{\mathbf{s}} + \int_{bot} \vec{\mathbf{B}} \cdot d\vec{\mathbf{s}} = \mu_0 I_{enc}$$

The scalar product is $\vec{\mathbf{B}} \cdot d\vec{\mathbf{s}} = B \cos\theta \, ds$, where θ is the angle between $\vec{\mathbf{B}}$ and $d\vec{\mathbf{s}}$. Also recall that the direction of $d\vec{\mathbf{s}}$ is tangential to the Ampèrian loop. Study the direction of $\vec{\mathbf{B}}$ and $d\vec{\mathbf{s}}$ at each side of the rectangle in the previous diagram.

- For the right and left sides, $\theta = 0°$ because $\vec{\mathbf{B}}$ and $d\vec{\mathbf{s}}$ either both point up or both point down.
- For the top and bottom sides, $\theta = 90°$ because $\vec{\mathbf{B}}$ and $d\vec{\mathbf{s}}$ are perpendicular.

$$\int_{right} B\cos 0°\, ds + \int_{top} B\cos 90°\, ds + \int_{left} B\cos 0°\, ds + \int_{bot} B\cos 90°\, ds = \mu_0 I_{enc}$$

Recall from trig that $\cos 0° = 1$ and $\cos 90° = 0$.

$$\int_{right} Bds + 0 + \int_{left} Bds + 0 = \mu_0 I_{enc}$$

It turns out that the magnetic field outside of the solenoid is very weak compared to the magnetic field inside of the solenoid. With this in mind, we will approximate $B \approx 0$ outside of the solenoid, such that $\int_{right} Bds \approx 0$.

$$\int_{left} Bds = \mu_0 I_{enc}$$

Over the left side of the Ampèrian rectangle, the magnitude of the magnetic field is constant, since every point on the left side of the rectangle is equidistant from the axis of the infinite solenoid. Therefore, we may pull B out of the integral.

$$B\int_{left} ds = \mu_0 I_{enc}$$

The remaining integral is trivial: $\int ds = L$ for the left side of the rectangle.

$$BL = \mu_0 I_{enc}$$

Isolate the magnitude of the magnetic field by dividing both sides of the equation by L.

$$B = \frac{\mu_0 I_{enc}}{L}$$

Now we need to determine how much current is enclosed by the Ampèrian rectangle. The answer is simple: The current passes through the Ampèrian rectangle N times. Therefore, the current enclosed by the Ampèrian rectangle is $I_{enc} = NI$. Substitute this expression for the current enclosed into the previous equation for magnetic field.

$$B = \frac{\mu_0 I_{enc}}{L} = \frac{\mu_0 NI}{L} = \mu_0 nI$$

Note that lowercase n is the number of turns (or loops) per unit length: $n = \frac{N}{L}$. For a truly infinite solenoid, N and L would each be infinite, yet n is finite. The magnitude of the magnetic field inside of the solenoid is $B = \mu_0 nI$, which is a constant. Thus, the magnetic field inside of an infinitely long, tightly wound solenoid is uniform. We can use a unit vector to include the direction of the magnetic field with our answer: $\vec{\mathbf{B}} = \mu_0 nI\hat{\mathbf{z}}$.

Example: The tightly wound toroidal coil illustrated below carries an insulated filamentary current I. Derive an expression for the magnetic field in the region that is shaded gray in the diagram below.

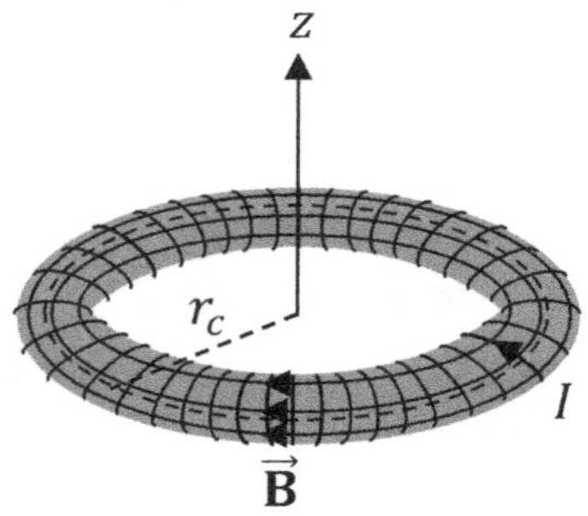

Although some students struggle to visualize or draw the toroidal coil, you shouldn't be afraid of the math: As we will see, the mathematics involved in this application of Ampère's law is very simple.

First sketch the magnetic field lines for the toroidal coil. Apply the right-hand rule for magnetic field (Chapter 22): As shown above, the magnetic field ($\vec{\mathbf{B}}$) lines are circles running along the (circular) axis of the toroidal coil. We choose our Ampèrian loop to be a circle inside of the toroid (dashed circle above) such that $\vec{\mathbf{B}}$ and $d\vec{\mathbf{s}}$ (which is tangent to the Ampèrian loop) will be parallel. Write the formula for Ampère's law.

$$\oint_C \vec{\mathbf{B}} \cdot d\vec{\mathbf{s}} = \mu_0 I_{enc}$$

The scalar product is $\vec{\mathbf{B}} \cdot d\vec{\mathbf{s}} = B \cos\theta \, ds$, where $\theta = 0°$, and the magnitude of the magnetic field is constant over the Ampèrian circle. Ampère's law thus reduces to:

$$B \oint_C ds = \mu_0 I_{enc}$$

This integral is over the path of the Ampèrian circle. Note that r_c is constant for a circle.

$$B \oint_C ds = B \oint_{\theta=0}^{2\pi} r_c \, d\theta = Br_c \oint_{\theta=0}^{2\pi} d\theta = B2\pi r_c = \mu_0 I_{enc}$$

Isolate the magnitude of the magnetic field by dividing both sides of the equation by $2\pi r_c$.

$$B = \frac{\mu_0 I_{enc}}{2\pi r_c}$$

Similar to the previous example, the current enclosed by the Ampèrian loop is $I_{enc} = NI$.

$$B = \frac{\mu_0 NI}{2\pi r_c}$$

The magnitude of the magnetic field inside of the toroidal coil is $B = \frac{\mu_0 NI}{2\pi r_c}$. We can use a unit vector to include the direction of the magnetic field with our answer: $\vec{\mathbf{B}} = \frac{\mu_0 NI}{2\pi r_c}\hat{\boldsymbol{\theta}}$.

97. An infinite solid cylindrical conductor coaxial with the z-axis has radius a and carries total current I. The current density is non-uniform: $\vec{\mathbf{J}} = \beta r_c \hat{\mathbf{z}}$, where β is a positive constant, Derive an expression for the magnetic field both inside and outside of the cylinder.

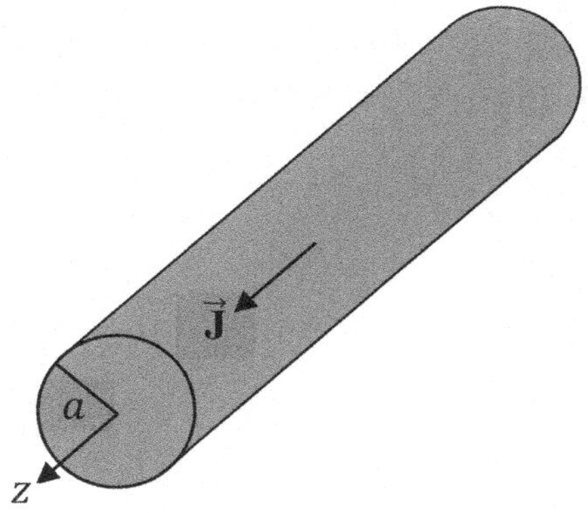

Want help? Check the hints section at the back of the book.

Answers: $\vec{\mathbf{B}}_I = \frac{\mu_0 I r_c^2}{2\pi a^3}\hat{\boldsymbol{\theta}}$, $\vec{\mathbf{B}}_{II} = \frac{\mu_0 I}{2\pi r_c}\hat{\boldsymbol{\theta}}$

98. An infinite solid cylindrical conductor coaxial with the z-axis has radius a and uniform current density $\vec{\mathbf{J}}$. Coaxial with the solid cylindrical conductor is a thick infinite cylindrical conducting shell of inner radius b, outer radius c, and uniform current density $\vec{\mathbf{J}}$. The conducting cylinder carries total current I coming out of the page, while the conducting cylindrical shell carries the same total current I, except that its current is going into the page. Derive an expression for the magnetic field in each region. (This is a **coaxial cable**.)

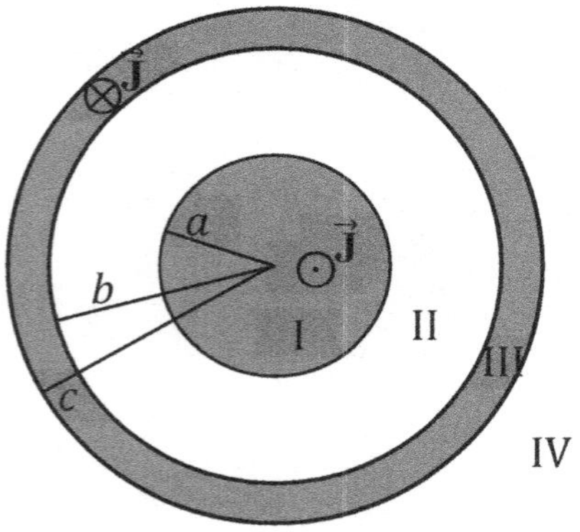

Want help? Check the hints section at the back of the book.

Answers: $\vec{\mathbf{B}}_I = \frac{\mu_0 I r_c}{2\pi a^2}\hat{\boldsymbol{\theta}}$, $\vec{\mathbf{B}}_{II} = \frac{\mu_0 I}{2\pi r_c}\hat{\boldsymbol{\theta}}$, $\vec{\mathbf{B}}_{III} = \frac{\mu_0 I (c^2 - r_c^2)}{2\pi r_c (c^2 - b^2)}\hat{\boldsymbol{\theta}}$ (see the note in the hints), $\vec{\mathbf{B}}_{IV} = 0$

99. An infinite conducting slab with thickness T is parallel to the xy plane, centered about $z = 0$, and has uniform current density $\vec{\mathbf{J}}$ coming out of the page. Derive an expression for the magnetic field in each region.

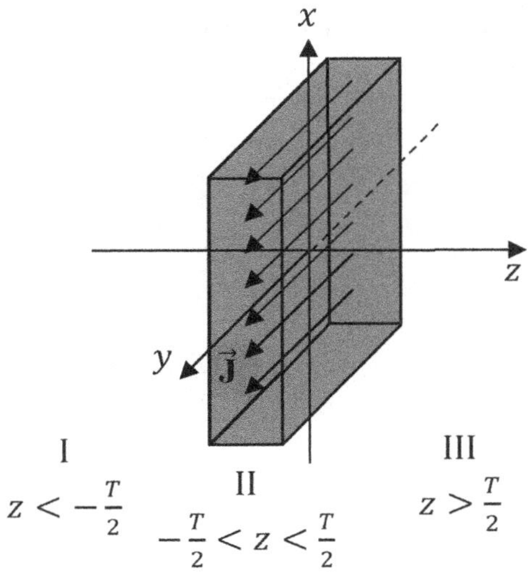

Want help? Check the hints section at the back of the book.

Answers: $\vec{\mathbf{B}}_I = -\frac{\mu_0 JT}{2}\hat{\mathbf{x}}$, $\vec{\mathbf{B}}_{II} = \mu_0 Jz\,\hat{\mathbf{x}}$, $\vec{\mathbf{B}}_{III} = \frac{\mu_0 JT}{2}\hat{\mathbf{x}}$

100. An infinite solid cylindrical conductor coaxial with the z-axis has radius a, uniform current density $\vec{\mathbf{J}}$ along $\hat{\mathbf{z}}$,, and carries total current I. As illustrated below, the solid cylinder has a cylindrical cavity with radius $\frac{a}{2}$ centered about a line parallel to the z-axis and passing through the point $\left(\frac{a}{2}, 0, 0\right)$. Determine the magnitude and direction of the magnetic field at the point $\left(\frac{3a}{2}, 0, 0\right)$.

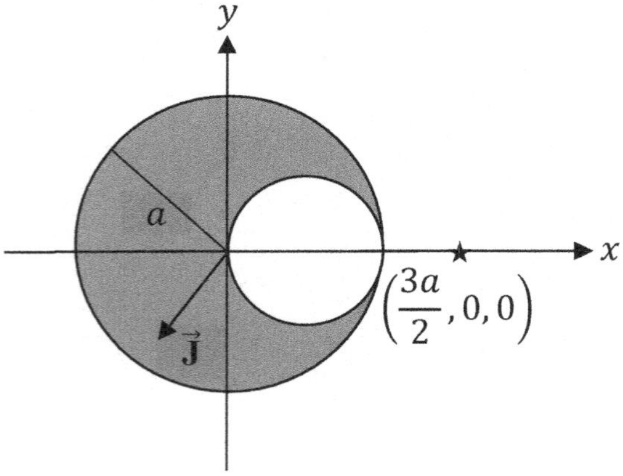

Want help? Check the hints section at the back of the book.

Answer: $\vec{\mathbf{B}} = \frac{5\mu_0 I}{18\pi a}\hat{\mathbf{y}}$

28 LENZ'S LAW

Relevant Terminology

Current – the instantaneous rate of flow of charge through a wire.
Magnetic field – a magnetic effect created by a moving charge (or current).
Magnetic flux –a measure of the relative number of magnetic field lines that pass through a surface.
Solenoid – a coil of wire in the shape of a right-circular cylinder.

Essential Concepts

According to **Faraday's law**, a **current** (I_{ind}) is induced in a loop of wire when there is a **changing magnetic flux** through the loop. The **magnetic flux** (Φ_m) through the loop is a measure of the relative number of magnetic field lines passing through the loop. There are three ways for the **magnetic flux** (Φ_m) to change:

- The average value of the **magnetic field** ($\vec{\mathbf{B}}_{ext}$) in the area of the loop may change. One way for this to happen is if a magnet (or current-carrying wire) is getting closer to or further from the loop.
- The **area** of the loop may change. This is possible if the shape of the loop changes.
- The **orientation** of the loop or magnetic field lines may change. This can happen, for example, if the loop or a magnet **rotates**.

Note: If the magnetic flux through a loop isn't changing, **no** current is induced in the loop.

In Chapter 29, we'll learn how to apply Faraday's law to calculate the induced emf or the induced current. In Chapter 28, we'll focus on how to determine the **direction** of the induced current. Lenz's law gives the direction of the **induced current** (I_{ind}).

According to **Lenz's law**, the induced current runs through the loop in a direction such that the induced magnetic field created by the induced current opposes the change in the magnetic flux. That's the "fancy" way of explaining Lenz's law. Following is what the "fancy" definition really means:

- If the magnetic flux (Φ_m) through the area of the loop is **increasing**, the induced magnetic field ($\vec{\mathbf{B}}_{ind}$) created by the induced current (I_{ind}) will be **opposite** to the external magnetic field ($\vec{\mathbf{B}}_{ext}$).
- If the magnetic flux (Φ_m) through the area of the loop is **decreasing**, the induced magnetic field ($\vec{\mathbf{B}}_{ind}$) created by the induced current (I_{ind}) will be **parallel** to the external magnetic field ($\vec{\mathbf{B}}_{ext}$).

In the strategy, we'll break Lenz's law down into four precise steps.

Special Symbols

Symbol	Name
⊗	into the page
⊙	out of the page
N	north pole
S	south pole

Symbols and SI Units

Symbol	Name	SI Units
$\vec{\mathbf{B}}_{ext}$	external magnetic field (that is, external to the loop)	T
$\vec{\mathbf{B}}_{ind}$	induced magnetic field (created by the induced current)	T
Φ_m	magnetic flux through the area of the loop	$T{\cdot}m^2$ or Wb
I_{ind}	induced current (that is, induced in the loop)	A
$\vec{\boldsymbol{v}}$	velocity	m/s

Schematic Symbols

Schematic Representation	Symbol	Name
	R	resistor
	ΔV	battery or DC power supply

Recall that the long line represents the positive terminal, while the small rectangle represents the negative terminal. Current runs from positive to negative.

Strategy to Apply Lenz's Law

Note the meaning of the following symbols:

- The symbol ⊗ represents an arrow going **into** the page.
- The symbol ⊙ represents an arrow coming **out** of the page.

If there is a battery in the diagram, label the positive and negative terminals of each battery (or DC power supply). The long line of the schematic symbol represents the positive terminal, as shown below. Draw the conventional* current the way that positive charges would flow: from the positive terminal to the negative terminal.

To determine the direction of the induced current in a loop of wire, apply Lenz's law in four steps as follows:

1. What is the direction of the **external magnetic field** ($\vec{\mathbf{B}}_{ext}$) in the area of the loop? Draw an arrow to show the direction of $\vec{\mathbf{B}}_{ext}$.
   - What is creating a magnetic field in the area of the loop to begin with?
   - If it is a bar magnet, you'll need to know what the magnetic field lines of a bar magnet look like (see Chapter 20).
   - You want to know the direction of $\vec{\mathbf{B}}_{ext}$ specifically in the area of the loop.
   - If $\vec{\mathbf{B}}_{ext}$ points in multiple directions within the area of the loop, ask yourself which way $\vec{\mathbf{B}}_{ext}$ points on average.
2. Is the **magnetic flux** (Φ_m) through the loop increasing or decreasing? Write one of the following words: increasing, decreasing, or constant.
   - Is the relative number of magnetic field lines passing through the loop increasing or decreasing?
   - Is a magnet or external current getting closer or further from the loop?
   - Is something rotating? If so, ask yourself if more or fewer magnetic field lines will pass through the loop while it rotates.
3. What is the direction of the **induced magnetic field** ($\vec{\mathbf{B}}_{ind}$) created by the induced current? Determine this from your answer for Φ_m in Step 2 and your answer for $\vec{\mathbf{B}}_{ext}$ in Step 1 as follows. Draw an arrow to show the direction of $\vec{\mathbf{B}}_{ind}$.
   - If Φ_m is **increasing**, draw $\vec{\mathbf{B}}_{ind}$ **opposite** to your answer for Step 1.
   - If Φ_m is **decreasing**, draw $\vec{\mathbf{B}}_{ind}$ **parallel** to your answer for Step 1.
   - If Φ_m is **constant**, then $\vec{\mathbf{B}}_{ind} = 0$ and there won't be any induced current.

---

* Of course, it's really the electrons moving through the wire, not positive charges. However, all of the signs in physics are based on positive charges, so it's "conventional" to draw the current based on what a positive charge would do. Remember, all the rules are backwards for electrons, due to the negative charge.

4. Apply the **right-hand rule for magnetic field** (Chapter 22) to determine the direction of the **induced current** (I_{ind}) from your answer to Step 3 for the induced magnetic field ($\vec{\mathbf{B}}_{ind}$). Draw and label your answer for I_{ind} on the diagram.
   - Be sure to use your answer to Step 3 and <u>**not**</u> your answer to Step 1.
   - You're actually inverting the right-hand rule. In Chapter 22, we knew the direction of the current and applied the right-hand rule to determine the direction of the magnetic field. With Lenz's law, we know the direction of the magnetic field (from Step 3), and we're applying the right-hand rule to determine the direction of the induced current.
   - The right-hand rule for magnetic field still works the same way. The difference is that now you'll need to grab the loop both ways (try it both with your thumb clockwise and with your thumb counterclockwise to see which way points your fingers correctly inside of the loop).
   - What matters is which way your fingers are pointing inside of the loop (they must match your answer to Step 3) and which way your thumb points (this is your answer for the direction of the induced current).

Important Distinctions

Magnetic field ($\vec{\mathbf{B}}$) and **magnetic flux** (Φ_m) are two entirely different quantities. The magnetic flux (Φ_m) through a loop of wire depends on three things:

- the average value of magnetic field over the loop ($\vec{\mathbf{B}}_{ext}$)
- the area of the loop (A)
- the direction (θ) of the magnetic field relative to the axis of the loop.

Two different magnetic fields are involved in Lenz's law:

- There is **external** magnetic field ($\vec{\mathbf{B}}_{ext}$) in the area of the loop, which is produced by some source other than the loop in question. The source could be a magnet or it could be another current.
- An **induced** magnetic field ($\vec{\mathbf{B}}_{ind}$) is created by the induced current (I_{ind}) in response to a changing magnetic flux (Φ_m).

Example: If the magnetic field is increasing in the diagram below, what is the direction of the current induced in the loop?

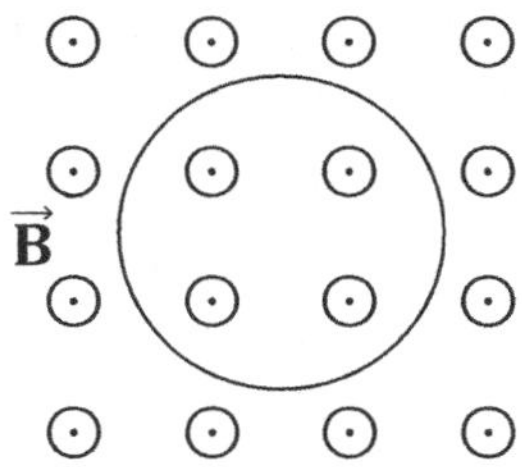

Apply the four steps of Lenz's law:

1. The **external** magnetic field ($\vec{\mathbf{B}}_{ext}$) is out of the page (⊙). It happens to already be drawn in the problem.
2. The magnetic flux (Φ_m) is **increasing** because the problem states that the external magnetic field is increasing.
3. The **induced** magnetic field ($\vec{\mathbf{B}}_{ind}$) is into the page (⊗). Since Φ_m is increasing, the direction of $\vec{\mathbf{B}}_{ind}$ is **opposite** to the direction of $\vec{\mathbf{B}}_{ext}$ from Step 1.
4. The **induced current** (I_{ind}) is clockwise, as drawn below. If you grab the loop with your thumb pointing clockwise and your fingers wrapped around the wire, inside the loop your fingers will go into the page (⊗). Remember, you want your fingers to match Step 3 inside the loop: You **<u>don't</u>** want your fingers to match Step 1 or the magnetic field lines already drawn in the beginning of the problem.

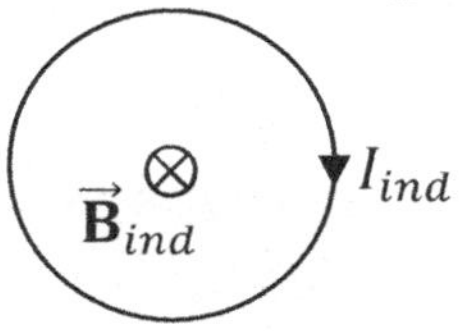

Example: If the magnetic field is decreasing in the diagram below, what is the direction of the current induced in the loop?

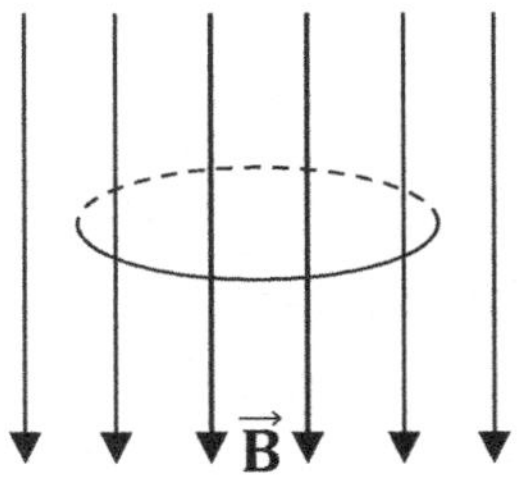

Apply the four steps of Lenz's law:

1. The **external** magnetic field ($\vec{\mathbf{B}}_{ext}$) is down (↓). It happens to already be drawn in the problem.
2. The magnetic flux (Φ_m) is **decreasing** because the problem states that the external magnetic field is decreasing.
3. The **induced** magnetic field ($\vec{\mathbf{B}}_{ind}$) is down (↓). Since Φ_m is decreasing, the direction of $\vec{\mathbf{B}}_{ind}$ is the **same** as the direction of $\vec{\mathbf{B}}_{ext}$ from Step 1.

4. The **induced current** (I_{ind}) runs to the left in the front of the loop (and therefore runs to the right in the back of the loop), as drawn below. Note that this loop is horizontal. If you grab the front of the loop with your thumb pointing to the left and your fingers wrapped around the wire, inside the loop your fingers will go down (↓).

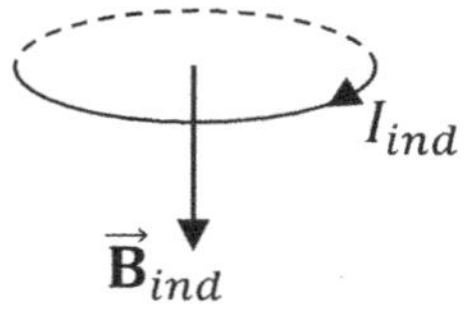

Example: As the magnet is moving away from the loop in the diagram below, what is the direction of the current induced in the loop?

Apply the four steps of Lenz's law:

1. The **external** magnetic field ($\vec{\mathbf{B}}_{ext}$) is to the right (→). This is because the magnetic field lines of the magnet are going to the right, away from the north (N) pole, in the area of the loop as illustrated below. **Tip**: When you view the diagram below, ask yourself which way, on average, the magnetic field lines would be headed if you extend the diagram to where the loop is.

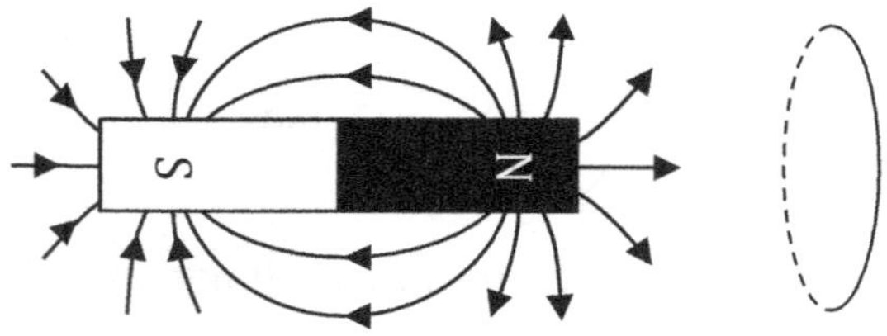

2. The magnetic flux (Φ_m) is **decreasing** because the magnet is getting further away from the loop.
3. The **induced** magnetic field ($\vec{\mathbf{B}}_{ind}$) is to the right (→). Since Φ_m is decreasing, the direction of $\vec{\mathbf{B}}_{ind}$ is the **same** as the direction of $\vec{\mathbf{B}}_{ext}$ from Step 1.
4. The **induced current** (I_{ind}) runs down the front of the loop (and therefore runs up the back of the loop), as drawn below. Note that this loop is vertical. If you grab the front of the loop with your thumb pointing down and your fingers wrapped around the wire, inside the loop your fingers will go to the right (→).

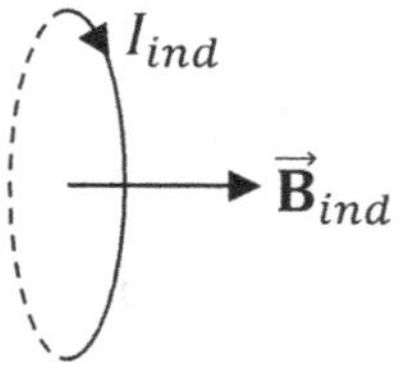

Example: As the right loop travels toward the left loop in the diagram below, what is the direction of the current induced in the right loop?

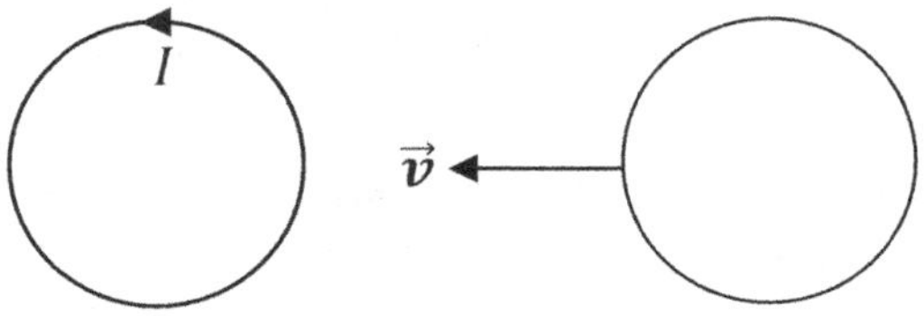

1. The **external** magnetic field ($\vec{\mathbf{B}}_{ext}$) is into the page ($\otimes$). This is because the left loop creates a magnetic field that is into the page in the region where the right loop is. Get this from the right-hand rule for magnetic field (Chapter 22). When you grab the left loop with your thumb pointed counterclockwise (along the given current) and your fingers wrapped around the wire, outside of the left loop (because the right loop is outside of the left loop) your fingers point into the page.

2. The magnetic flux (Φ_m) is **increasing** because the right loop is getting closer to the left loop so the magnetic field created by the left loop is getting stronger in the area of the right loop.
3. The **induced** magnetic field ($\vec{\mathbf{B}}_{ind}$) is out of the page ($\odot$). Since Φ_m is increasing, the direction of $\vec{\mathbf{B}}_{ind}$ is **opposite** to the direction of $\vec{\mathbf{B}}_{ext}$ from Step 1.
4. The **induced current** (I_{ind}) is counterclockwise, as drawn below. If you grab the loop with your thumb pointing counterclockwise and your fingers wrapped around the wire, inside the loop your fingers will come out of the page ($\odot$). Remember, you want your fingers to match Step 3 inside the right loop (**<u>not</u>** Step 1).

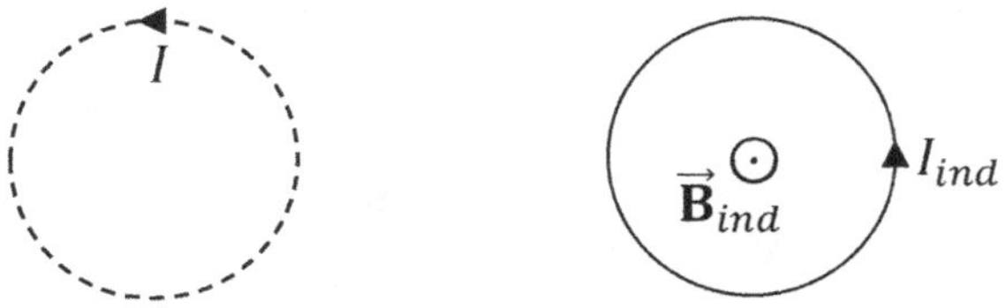

Example: In the diagram below, a conducting bar slides to the right along the rails of a bare U-channel conductor in the presence of a constant magnetic field. What is the direction of the current induced in the conducting bar?

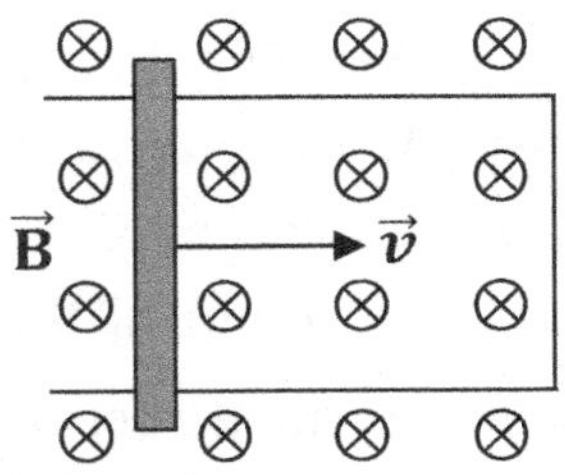

1. The **external** magnetic field ($\vec{\mathbf{B}}_{ext}$) is into the page (⊗). It happens to already be drawn in the problem.
2. The magnetic flux (Φ_m) is **decreasing** because the area of the loop is getting smaller as the conducting bar travels to the right. Note that the conducting bar makes electrical contract where it touches the bare U-channel conductor. The dashed (---) line below illustrates how the area of the loop is getting smaller as the conducting bar travels to the right.

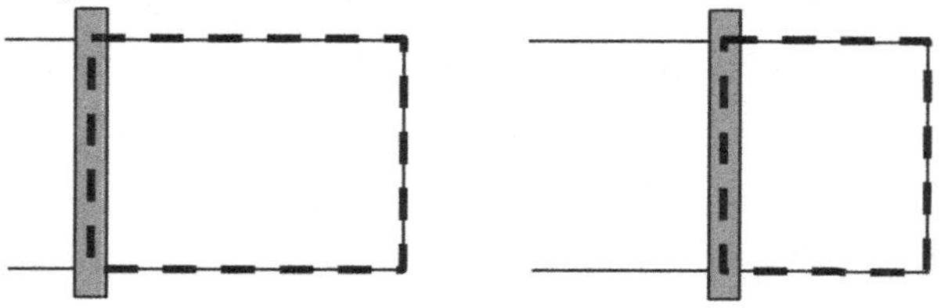

3. The **induced** magnetic field ($\vec{\mathbf{B}}_{ind}$) is into the page (⊗). Since Φ_m is decreasing, the direction of $\vec{\mathbf{B}}_{ind}$ is the **same** as the direction of $\vec{\mathbf{B}}_{ext}$ from Step 1.
4. The **induced current** (I_{ind}) is clockwise, as drawn below. If you grab the loop with your thumb pointing clockwise and your fingers wrapped around the wire, inside the loop your fingers will go into the page (⊗). Since the induced current is clockwise in the loop, the induced current runs up (↑) the conducting bar.

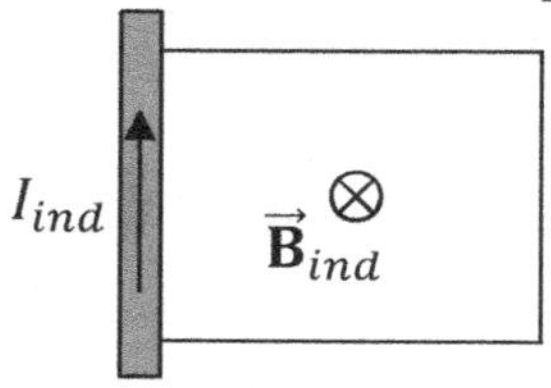

Example: In the diagrams below, the loop rotates, with the bottom of the loop coming out of the page and the top of the loop going into the page, making the loop horizontal. What is the direction of the current induced in the loop during this 90° rotation?

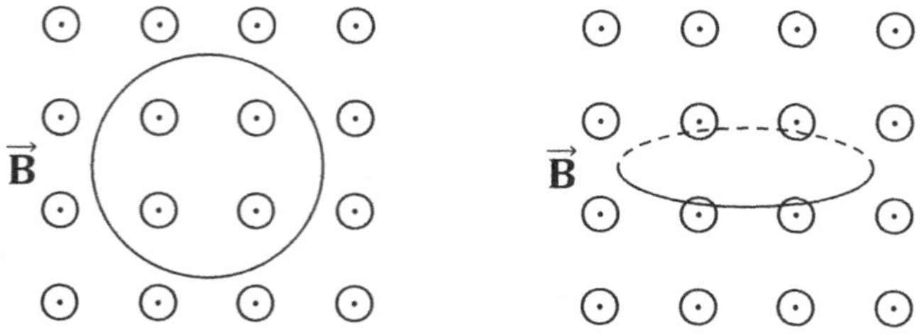

1. The **external** magnetic field ($\vec{\mathbf{B}}_{ext}$) is out of the page (⊙). It happens to already be drawn in the problem.
2. The magnetic flux (Φ_m) is **decreasing** because fewer magnetic field lines pass through the loop as it rotates. At the end of the described 90° rotation, the loop is horizontal and the final magnetic flux is zero.
3. The **induced** magnetic field ($\vec{\mathbf{B}}_{ind}$) is out of the page (⊙). Since Φ_m is decreasing, the direction of $\vec{\mathbf{B}}_{ind}$ is the **same** as the direction of $\vec{\mathbf{B}}_{ext}$ from Step 1.

4. The **induced current** (I_{ind}) is counterclockwise, as drawn below. If you grab the loop with your thumb pointing counterclockwise and your fingers wrapped around the wire, inside the loop your fingers will come out of the page (⊙).

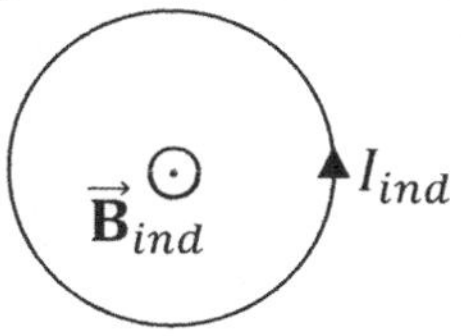

Example: In the diagrams below, the loop rotates such that the point (•) at the top of the loop moves to the right of the loop. What is the direction of the current induced in the loop during this 90° rotation?

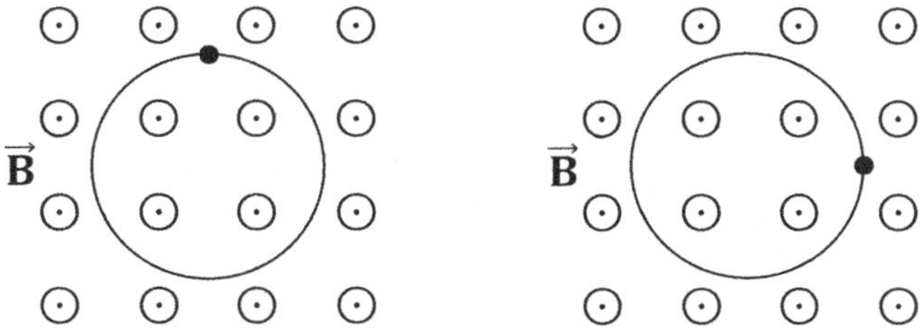

1. The **external** magnetic field ($\vec{\mathbf{B}}_{ext}$) is out of the page (⊙). It happens to already be drawn in the problem.
2. The magnetic flux (Φ_m) is **constant** because the number of magnetic field lines passing through the loop doesn't change. It is instructive to compare this example to the previous example.
3. The **induced** magnetic field ($\vec{\mathbf{B}}_{ind}$) is **zero** because the magnetic flux (Φ_m) through the loop is **constant**.
4. The **induced current** (I_{ind}) is also **zero** because the magnetic flux (Φ_m) through the loop is **constant**.

101. If the magnetic field is increasing in the diagram below, what is the direction of the current induced in the loop?

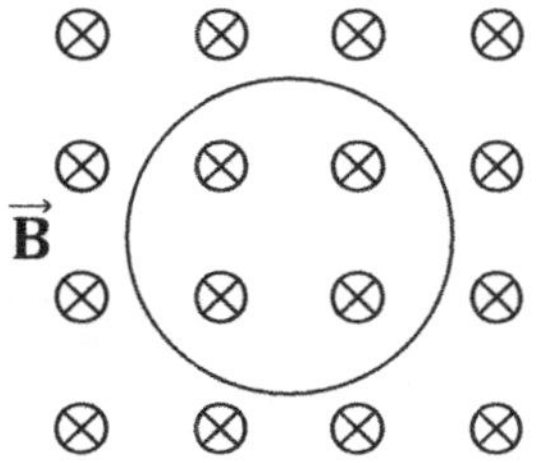

1. $\vec{\mathbf{B}}_{ext}$ is ________. (Draw an arrow.)
2. Φ_m is increasing / decreasing / constant. (Circle one.)
3. $\vec{\mathbf{B}}_{ind}$ is ________. (Draw an arrow.)
4. Draw and label I_{ind} in the diagram.

102. If the magnetic field is decreasing in the diagram below, what is the direction of the current induced in the loop?

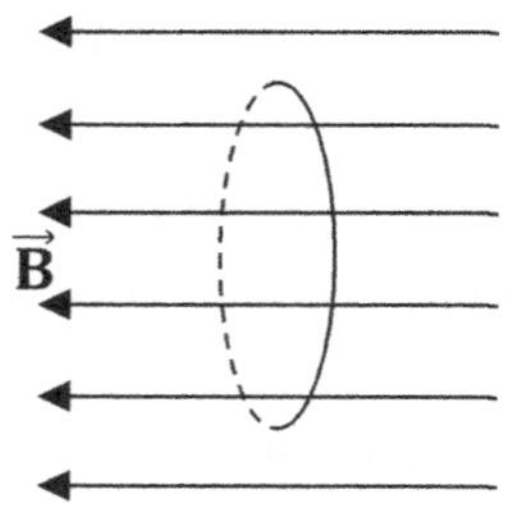

1. $\vec{\mathbf{B}}_{ext}$ is ________. (Draw an arrow.)
2. Φ_m is increasing / decreasing / constant. (Circle one.)
3. $\vec{\mathbf{B}}_{ind}$ is ________. (Draw an arrow.)
4. Draw and label I_{ind} in the diagram.

103. If the magnetic field is increasing in the diagram below, what is the direction of the current induced in the loop?

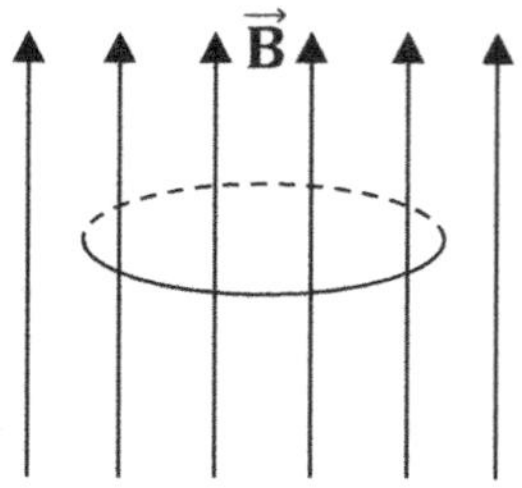

1. $\vec{\mathbf{B}}_{ext}$ is ________. (Draw an arrow.)
2. Φ_m is increasing / decreasing / constant. (Circle one.)
3. $\vec{\mathbf{B}}_{ind}$ is ________. (Draw an arrow.)
4. Draw and label I_{ind} in the diagram.

Want help? The problems from Chapter 28 are fully solved in the back of the book.

104. As the magnet is moving towards the loop in the diagram below, what is the direction of the current induced in the loop?

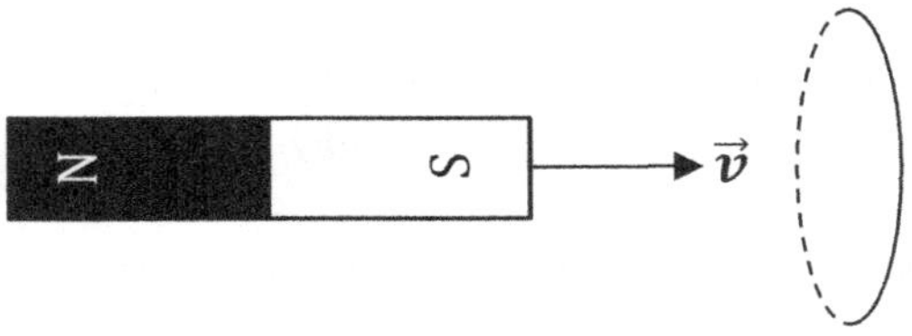

1. $\vec{\mathbf{B}}_{ext}$ is ________. (Draw an arrow.)
2. Φ_m is increasing / decreasing / constant. (Circle one.)
3. $\vec{\mathbf{B}}_{ind}$ is ________. (Draw an arrow.)
4. Draw and label I_{ind} in the diagram.

105. As the magnet is moving away from the loop in the diagram below, what is the direction of the current induced in the loop?

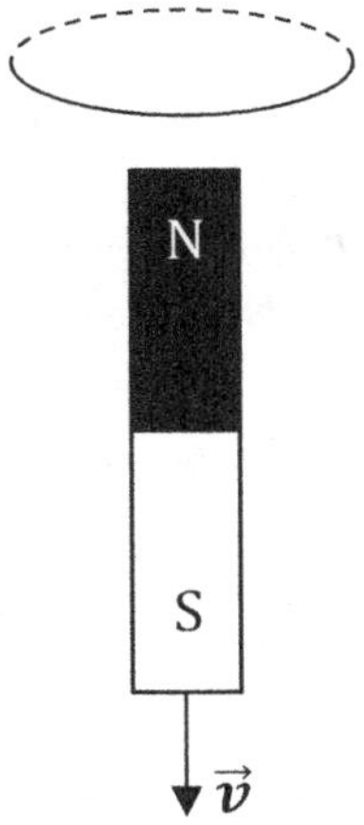

1. $\vec{\mathbf{B}}_{ext}$ is ________. (Draw an arrow.)
2. Φ_m is increasing / decreasing / constant. (Circle one.)
3. $\vec{\mathbf{B}}_{ind}$ is ________. (Draw an arrow.)
4. Draw and label I_{ind} in the diagram.

106. As the magnet (which is perpendicular to the page) is moving into the page and towards the loop in the diagram below, what is the direction of the current induced in the loop? (Unlike the magnet, the loop lies in the plane of the page.)

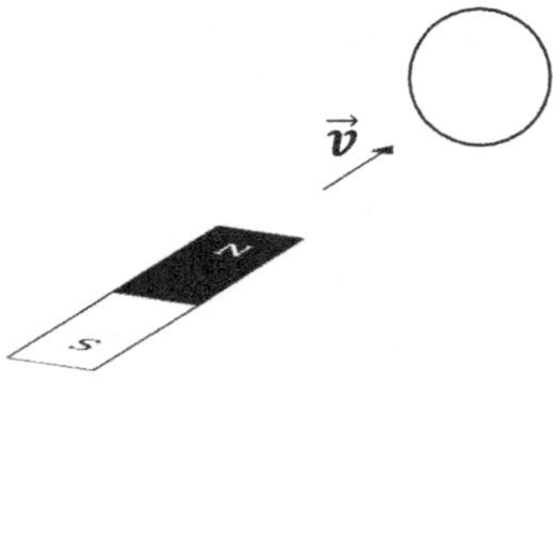

1. $\vec{\mathbf{B}}_{ext}$ is ________. (Draw an arrow.)
2. Φ_m is increasing / decreasing / constant. (Circle one.)
3. $\vec{\mathbf{B}}_{ind}$ is ________. (Draw an arrow.)
4. Draw and label I_{ind} in the diagram.

Want help? The problems from Chapter 28 are fully solved in the back of the book.

107. As the current increases in the outer loop in the diagram below, what is the direction of the current induced in the inner loop?

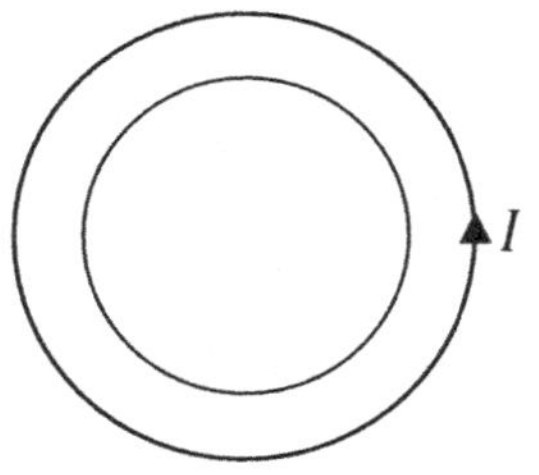

1. $\vec{\mathbf{B}}_{ext}$ is ________. (Draw an arrow.)
2. Φ_m is increasing / decreasing / constant. (Circle one.)
3. $\vec{\mathbf{B}}_{ind}$ is ________. (Draw an arrow.)
4. Draw and label I_{ind} in the diagram.

108. As the current in the straight conductor decreases in the diagram below, what is the direction of the current induced in the rectangular loop?

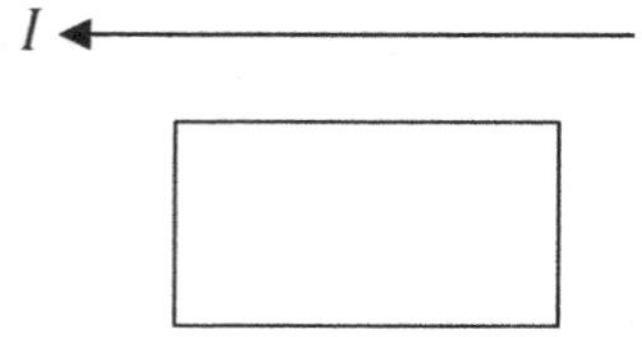

1. $\vec{\mathbf{B}}_{ext}$ is ________. (Draw an arrow.)
2. Φ_m is increasing / decreasing / constant. (Circle one.)
3. $\vec{\mathbf{B}}_{ind}$ is ________. (Draw an arrow.)
4. Draw and label I_{ind} in the diagram.

109. As the potential difference of the DC power supply increases in the diagram below, what is the direction of the current induced in the right loop?

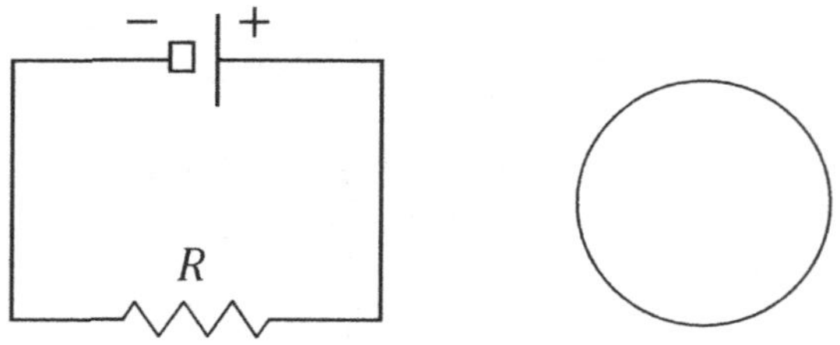

1. $\vec{\mathbf{B}}_{ext}$ is ________. (Draw an arrow.)
2. Φ_m is increasing / decreasing / constant. (Circle one.)
3. $\vec{\mathbf{B}}_{ind}$ is ________. (Draw an arrow.)
4. Draw and label I_{ind} in the diagram.

Want help? The problems from Chapter 28 are fully solved in the back of the book.

110. As the conducting bar illustrated below slides to the right along the rails of a bare U-channel conductor, what is the direction of the current induced in the conducting bar?

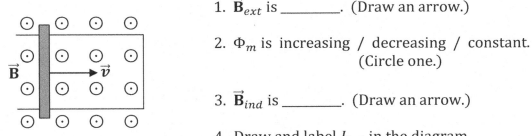

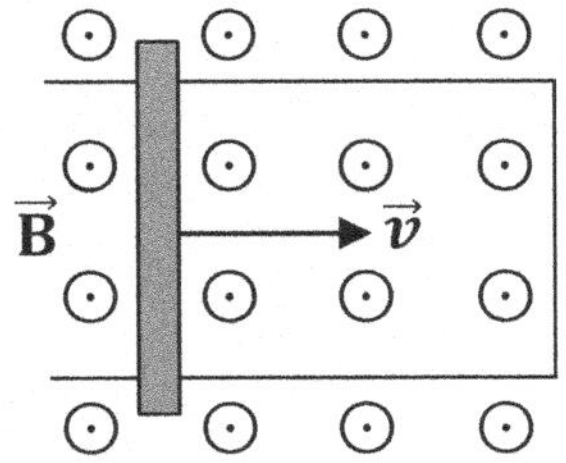

1. $\vec{\mathbf{B}}_{ext}$ is ________. (Draw an arrow.)
2. Φ_m is increasing / decreasing / constant. (Circle one.)
3. $\vec{\mathbf{B}}_{ind}$ is ________. (Draw an arrow.)
4. Draw and label I_{ind} in the diagram.

111. As the conducting bar illustrated below slides to the left along the rails of a bare U-channel conductor, what is the direction of the current induced in the conducting bar?

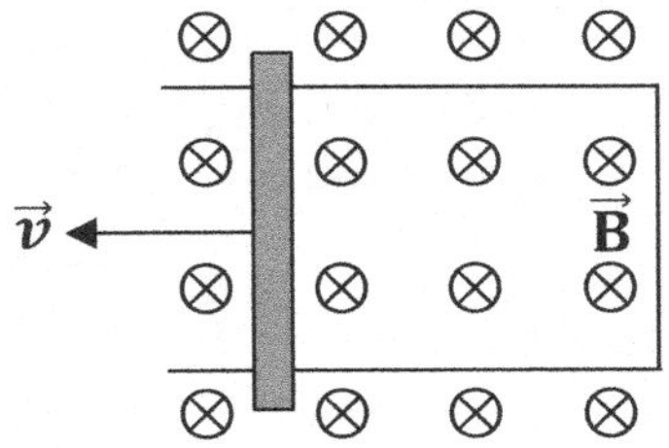

1. $\vec{\mathbf{B}}_{ext}$ is ________. (Draw an arrow.)
2. Φ_m is increasing / decreasing / constant. (Circle one.)
3. $\vec{\mathbf{B}}_{ind}$ is ________. (Draw an arrow.)
4. Draw and label I_{ind} in the diagram.

112. As the vertex of the triangle illustrated below is pushed downward from point A to point C, what is the direction of the current induced in the triangular loop?

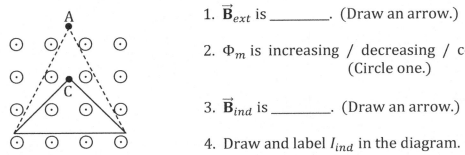

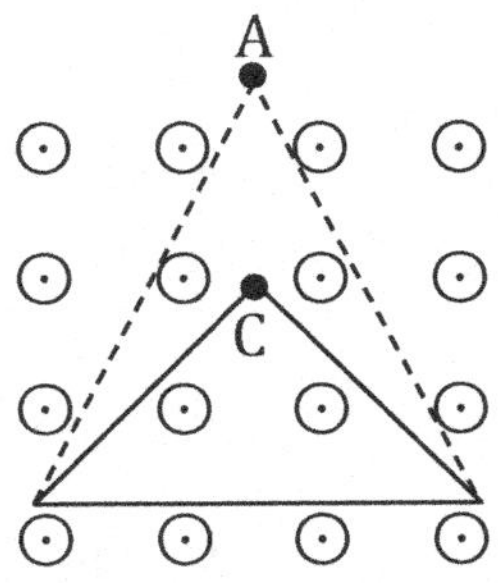

1. $\vec{\mathbf{B}}_{ext}$ is ________. (Draw an arrow.)
2. Φ_m is increasing / decreasing / constant. (Circle one.)
3. $\vec{\mathbf{B}}_{ind}$ is ________. (Draw an arrow.)
4. Draw and label I_{ind} in the diagram.

Want help? The problems from Chapter 28 are fully solved in the back of the book.

113. The loop in the diagram below rotates with the right side of the loop coming out of the page. What is the direction of the current induced in the loop during this 90° rotation?

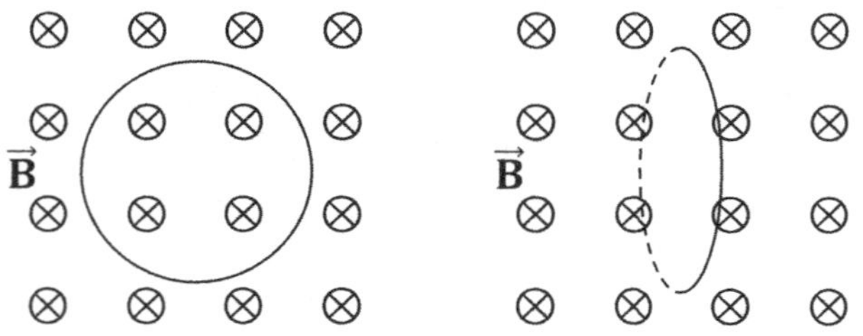

1. $\vec{\mathbf{B}}_{ext}$ is ________. (Draw an arrow.)
2. Φ_m is increasing / decreasing / constant. (Circle one.)
3. $\vec{\mathbf{B}}_{ind}$ is ________. (Draw an arrow.)
4. Draw and label I_{ind} in the diagram.

114. The magnetic field in the diagram below rotates 90° clockwise. What is the direction of the current induced in the loop during this 90° rotation?

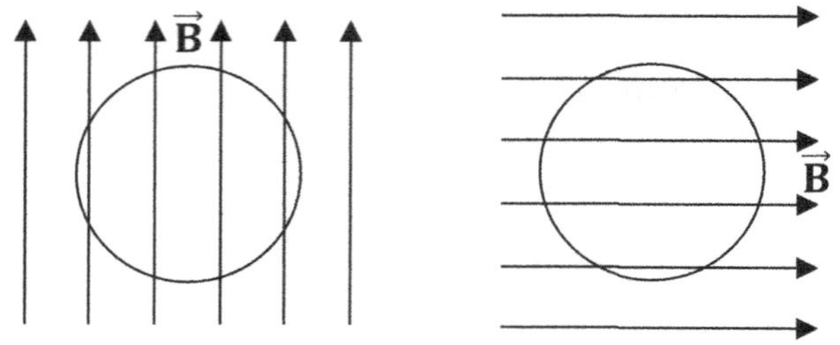

1. $\vec{\mathbf{B}}_{ext}$ is ________. (Draw an arrow.)
2. Φ_m is increasing / decreasing / constant. (Circle one.)
3. $\vec{\mathbf{B}}_{ind}$ is ________. (Draw an arrow.)
4. Draw and label I_{ind} in the diagram.

115. The loop in the diagram below rotates with the top of the loop coming out of the page. What is the direction of the current induced in the loop during this 90° rotation?

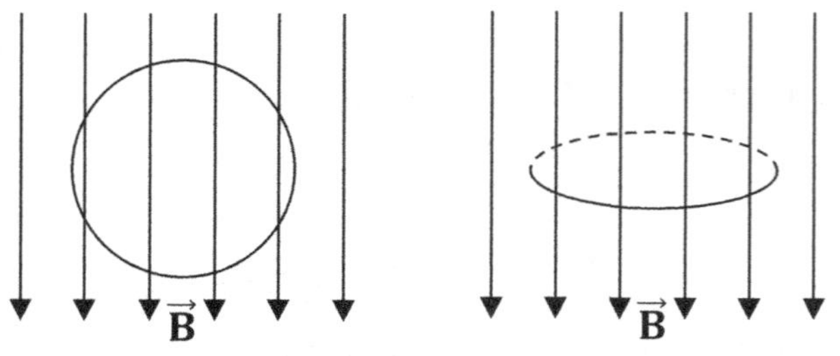

1. $\vec{\mathbf{B}}_{ext}$ is ________. (Draw an arrow.)
2. Φ_m is increasing / decreasing / constant. (Circle one.)
3. $\vec{\mathbf{B}}_{ind}$ is ________. (Draw an arrow.)
4. Draw and label I_{ind} in the diagram.

Want help? The problems from Chapter 28 are fully solved in the back of the book.

29 FARADAY'S LAW

Relevant Terminology

Electric charge – a fundamental property of a particle that causes the particle to experience a force in the presence of an electric field. (An electrically neutral particle has no charge and thus experiences no force in the presence of an electric field.)
Current – the instantaneous rate of flow of charge through a wire.
Potential difference – the electric work per unit charge needed to move a test charge between two points in a circuit. Potential difference is also called the **voltage**.
Emf – the potential difference that a battery or DC power supply would supply to a circuit neglecting its internal resistance.
Magnetic field – a magnetic effect created by a moving charge (or current).
Magnetic flux –a measure of the relative number of magnetic field lines that pass through a surface.
Solenoid – a coil of wire in the shape of a right-circular cylinder.
Turns – the loops of a coil of wire (such as a solenoid).
Permeability – a measure of how a substance affects a magnetic field.

Magnetic Flux Equations

Magnetic flux (Φ_m) is a measure of the relative number of magnetic field lines passing through a surface. The equation for magnetic flux involves the scalar product between magnetic field ($\vec{\mathbf{B}}$) and the differential area element ($d\vec{\mathbf{A}}$), where the direction of $d\vec{\mathbf{A}}$ is **perpendicular to the surface**. The symbol S below stands for the surface over which area is integrated. (Compare magnetic flux to electric flux: Review the beginning of Chapter 8.)

$$\Phi_m = \int_S \vec{\mathbf{B}} \cdot d\vec{\mathbf{A}}$$

Recall the scalar product from Volume 1, where θ is the angle between $\vec{\mathbf{B}}$ and $d\vec{\mathbf{A}}$. Note that $\vec{\mathbf{B}} \cdot d\vec{\mathbf{A}}$ is maximum when $\theta = 0°$ and zero when $\theta = 90°$.

$$\vec{\mathbf{B}} \cdot d\vec{\mathbf{A}} = B \cos\theta \, dA$$

For the case of a loop of wire in the presence of a **uniform** magnetic field, the equation for magnetic flux reduces to:

$$\Phi_m = BA\cos\theta$$

In this case, θ is the angle between the **axis of the loop** (which is perpendicular to the plane of the loop and passes through the center of the loop) and the direction of the magnetic field. Magnetic flux is maximum when magnetic field lines are perpendicular to the loop.

Faraday's Law Equations

According to **Faraday's law**, a **current** (I_{ind}) is induced in a loop of wire when there is a **changing magnetic flux** through the loop. Lenz's law provides the direction of the induced current, while Faraday's law provides the **emf** (ε_{ind}) induced in the loop (from which the induced current can be determined).

$$\varepsilon_{ind} = -N\frac{d\Phi_m}{dt}$$

Faraday's law involves a derivative of magnetic flux (Φ_m) with respect to time. The factor of N represents the total number of loops. In general, magnetic flux is given by an integral.

$$\Phi_m = \int_S \vec{\mathbf{B}} \cdot d\vec{\mathbf{A}}$$

Many physics problems relating to Faraday's law involve a magnetic field that is **uniform** throughout the area of the loop, in which case the equation for magnetic flux reduces to:

$$\Phi_m = BA\cos\theta$$

For a magnetic field that is **uniform** throughout the area of the loop, Faraday's law becomes:

$$\varepsilon_{ind} = -N\frac{d}{dt}(BA\cos\theta)$$

Unlike the emf of a battery or power supply such as we discussed in Chapter 16, the emf induced via Faraday's law involves no internal resistance. Thus, when we apply Ohm's law to solve for the induced current, the equation looks like this:

$$\varepsilon_{ind} = I_{ind}R_{loops}$$

Here, R_{loop} is the combined resistance of the N loops. If a problem gives you the resistivity (ρ) of the wire, you can find the resistance using the following equation from Chapter 16.

$$R_{loops} = \frac{\rho\ell}{A_{wire}}$$

Note that ℓ is the total length of the wire. (For a solenoid, this ℓ is **not** the same as the length of the solenoid.) Note also that A_{wire} is the cross-sectional area of the wire, involving the radius of the wire, and **not** the area of the loop. Magnetic flux involves the area of the **loop** (A), whereas the resistivity equation involves the area of the **wire** (A_{wire}).

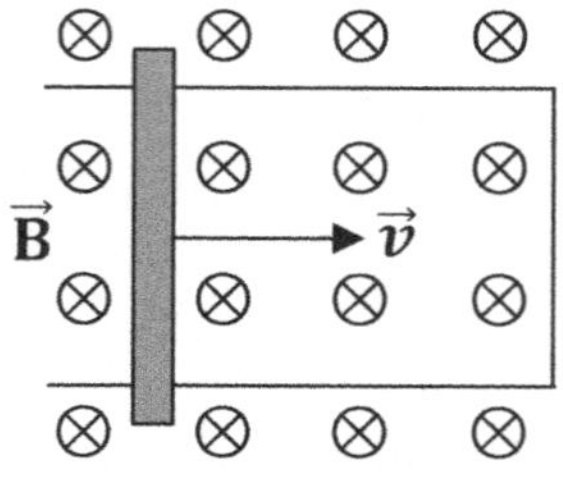

Motional emf (ε_{ind}) arises when a conducting bar of length ℓ travels with a velocity $\vec{v}$ through a uniform magnetic field ($\vec{B}$). An example of motional emf is illustrated above.

$$\varepsilon_{ind} = -B\ell v$$

Magnetic Field Equations

Recall the magnetic field equations from Chapter 25. If the magnetic field isn't given in the problem, you may need to apply one of the following equations to find the external magnetic field in the area of the loop before you can find the magnetic flux, or you may need to apply a strategy from Chapters 26-27 to derive an equation for magnetic field.

$$B = \frac{\mu_0 I}{2\pi r_c} \quad \text{(long straight wire)}$$

$$B = \frac{\mu_0 N I}{L} = \mu_0 n I \quad \text{(center of a solenoid)}$$

Recall that the permeability of free space is $\mu_0 = 4\pi \times 10^{-7}\ \frac{\text{T}\cdot\text{m}}{\text{A}}$.

Essential Concepts

The **magnetic flux** through a loop of wire must **<u>change</u>** in order for a current to be induced in the loop via Faraday's law. There are three ways for the **magnetic flux** (Φ_m) to change:

- The average value of the **magnetic field** ($\vec{\mathbf{B}}_{ext}$) in the area of the loop may change. One way for this to happen is if a magnet (or current-carrying wire) is getting closer to or further from the loop.
- The **area** of the loop may change. This is possible if the shape of the loop changes.
- The **orientation** of the loop or magnetic field lines may change. This can happen, for example, if the loop or a magnet **rotates**.

Note: If the magnetic flux through a loop isn't changing, <u>**no**</u> current is induced in the loop.

Important Distinction

Magnetic field ($\vec{\mathbf{B}}$) and **magnetic flux** (Φ_m) are two entirely different quantities. The magnetic flux (Φ_m) through a loop of wire depends on three things:

- the average value of magnetic field over the loop ($\vec{\mathbf{B}}_{ext}$)
- the area of the loop (A)
- the direction (θ) of the magnetic field relative to the axis of the loop.

Note that two different areas are involved in the equations:

- A is the area of the **loop**, involving the radius (a) of the loop.
- A_{wire} is the cross-sectional area of the **wire**, involving the radius (a_{wire}) of the wire.

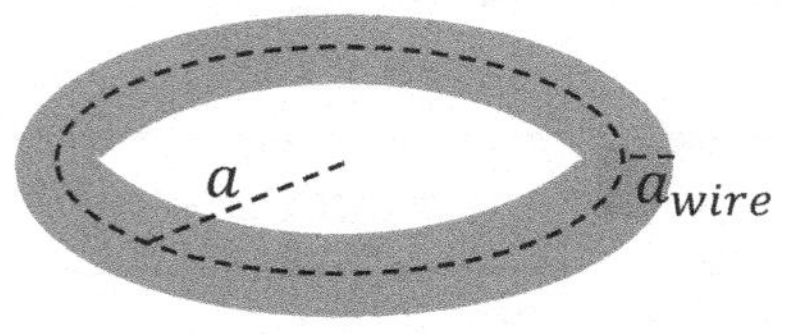

Symbols and Units

Symbol	Name	Units
Φ_m	magnetic flux	T·m^2 or Wb
B	magnitude of the external magnetic field	T
A	area of the loop	m^2
A_{wire}	area of the wire	m^2
θ	angle between $\vec{\mathbf{B}}$ and the axis of the loop, or the angle between $\vec{\mathbf{B}}$ and $d\vec{\mathbf{A}}$	° or rad
ε_{ind}	induced emf	V
I_{ind}	induced current	A
R_{loops}	resistance of the loops	Ω
ρ	resistivity	Ω·m
N	number of loops (or turns)	unitless
n	number of turns per unit length	$\frac{1}{\text{m}}$
t	time	s
μ_0	permeability of free space	$\frac{\text{T}\cdot\text{m}}{\text{A}}$
r_c	distance from a long, straight wire	m
a	radius of a loop	m
a_{wire}	radius of a wire	m
ℓ	length of the wire or length of conducting bar	m
L	length of a solenoid	m
v	speed	m/s
ω	angular speed	rad/s

Note: The symbol Φ is the uppercase Greek letter phi.

Special Symbols

Symbol	Name
⊗	into the page
⊙	out of the page

Notes Regarding Units

The SI unit of magnetic flux (Φ_m) is the Weber (Wb), which is equivalent to T·m^2. It's fairly common to express magnetic flux in T·m^2 rather than Wb, perhaps to help avoid possible confusion with the Watt (W). If you do use the Weber (Wb), be careful not to forget the b.

The units of magnetic flux (Φ_m) follow from the equation $\Phi_m = BA\cos\theta$. Since the SI unit of magnetic field (B) is the Tesla (T), the SI units of area (A) are square meters (m^2), and the cosine function is unitless (since it's a ratio of two distances), it follows that the SI units of magnetic flux (Φ_m) are T·m^2 (the SI unit of B times the SI units of A).

Strategy to Apply Faraday's Law

Note: If a problem gives you the magnetic field in Gauss (G), convert to Tesla (T):

$$1\text{ G} = 10^{-4}\text{ T}$$

Note the meaning of the following symbols:

- The symbol ⊗ represents an arrow going **<u>into</u>** the page.
- The symbol ⊙ represents an arrow coming **<u>out</u>** of the page.

To solve a problem involving Faraday's law, make a list of the known quantities and choose equations that relate them to solve for the desired unknown.

- At some stage in the problem, you will need to apply one of the two equations for Faraday's law:
  - The main Faraday's law equation relates the induced emf (ε_{ind}) to the change in the magnetic flux (Φ_m).

    $$\varepsilon_{ind} = -N\frac{d\Phi_m}{dt}$$

    In this case, you will ordinarily need to find the magnetic flux before you can use the equation. See the bullet point regarding magnetic flux.
  - The motional emf equation applies to a conducting bar of length ℓ traveling with speed v through a uniform magnetic field, where the length ℓ is perpendicular to the velocity of the conducting bar.

    $$\varepsilon_{ind} = -B\ell v$$

- If a problem asks you to find the **average** emf induced (this is also what you'll need if a problem asks for the **average** current induced), use the following equation.
$$\varepsilon_{ave} = -N\frac{\Delta\Phi_m}{\Delta t}$$
When you find the average, you divide the change in the magnetic flux by the elapsed time, where $\Delta\Phi_m = \Phi_m - \Phi_{m0}$.
- Most Faraday's law problems (except for motional emf) require finding magnetic flux (Φ_m) before you apply Faraday's law.
  - If the magnetic field ($\vec{\mathbf{B}}$) is **uniform** throughout the area (A) of a planar loop, you may use the following equation to find magnetic flux:
  $$\Phi_m = BA\cos\theta$$
  Here, θ is the angle between $\vec{\mathbf{B}}$ and the **axis** of the loop (which is perpendicular to the loop and passes through its center). In Faraday's law, you will ultimately need a **derivative** of magnetic flux with respect to time:
  $$\frac{d\Phi_m}{dt} = \frac{d}{dt}(BA\cos\theta)$$
  If A and θ are both constant, then $\frac{d\Phi_m}{dt} = A\cos\theta\frac{dB}{dt}$. If B and θ are both constant, then $\frac{d\Phi_m}{dt} = B\cos\theta\frac{dA}{dt}$. If A and B are both constant, then $\frac{d\Phi_m}{dt} = -AB\sin\theta\frac{d\theta}{dt} = -AB\omega\sin\theta$, where $\omega = \frac{d\theta}{dt}$ is the angular speed.*
  - When the magnetic field isn't uniform throughout the area of the loop, you must perform an **integral** to find magnetic flux.
  $$\Phi_m = \int_S \vec{\mathbf{B}} \cdot d\vec{\mathbf{A}}$$
  Note that $\vec{\mathbf{B}} \cdot d\vec{\mathbf{A}} = B\cos\theta\, dA$. Also note that this is an open integral over an open surface (the area of the loop is an open surface), unlike the similar closed integrals that we did for electric flux with Gauss's law (Chapter 8).
- Recall the formulas for area for some common objects. For a circle, $A = \pi a^2$. For a rectangle, $A = LW$. For a square, $A = L^2$. For a triangle, $A = \frac{1}{2}bh$.
- To find the **induced current** (I_{ind}), apply Ohm's law.
$$\varepsilon_{ind} = I_{ind}R_{loops}$$
If you know the resistivity (ρ) or can look it up, use it to find resistance as follows:
$$R_{loops} = \frac{\rho\ell}{A_{wire}}$$
- **Note**: If you're not given the value of the magnetic field in the area of the loop, you may need to use an equation at the top of page 379 (or apply Chapters 26-27).

* We applied the chain rule from calculus: $\frac{df}{dx} = \frac{df}{du}\frac{du}{dx}$. If you let $f = \cos\theta$, $u = \theta$, and $x = t$, the chain rule becomes $\frac{d}{dt}\cos\theta = \frac{d}{d\theta}\cos\theta\frac{d\theta}{dt}$. Since $\omega = \frac{d\theta}{dt}$ and $\frac{d}{d\theta}\cos\theta = -\sin\theta$, it follows that $\frac{d}{dt}\cos\theta = -\omega\sin\theta$.

Example: A rectangular loop lies in the xy plane, centered about the origin. The loop is 3.0 m long and 2.0 m wide. A uniform magnetic field of 40,000 G makes a 30° angle with the z-axis. What is the magnetic flux through the area of the loop?

First convert the magnetic field from Gauss (G) to Tesla (T) given that $1 \text{ G} = 10^{-4}$ T: $B = 40{,}000 \text{ G} = 4.0 \text{ T}$. We need to find the area of the rectangle before we can find the magnetic flux:

$$A = LW = (3)(2) = 6.0 \text{ m}^2$$

Use the equation for the magnetic flux for a **uniform** magnetic field through a planar surface. Since θ is the angle between the magnetic field and the axis of the loop, in this example $\theta = 30°$. Note that the z-axis is the axis of the rectangular loop because the axis of a loop is defined as the line that is perpendicular to the loop and passing through the center of the loop.

$$\Phi_m = BA\cos\theta = (4)(6)\cos 30° = 24\left(\frac{\sqrt{3}}{2}\right) = 12\sqrt{3} \text{ T·m}^2$$

The magnetic flux through the loop is $\Phi_m = 12\sqrt{3}$ T·m².

Example: A circular loop of wire with a radius of 25 cm and resistance of 6.0 Ω lies in the yz plane. A magnetic field is oriented along the $+x$ direction and is uniform throughout the loop at any given moment. The magnetic field increases from 4.0 T to 16.0 T in 250 ms.

(A) What is the **average** emf induced in the loop during this time?

First convert the radius and time to SI units: $a = 25 \text{ cm} = 0.25 \text{ m} = \frac{1}{4}$ m and $t = 250$ ms $= 0.250 \text{ s} = \frac{1}{4}$ s. We need to find the area of the circle before we can find the magnetic flux.

$$A = \pi a^2 = \pi\left(\frac{1}{4}\right)^2 = \frac{\pi}{16} \text{ m}^2$$

(A benefit of using the symbol a for the radius of the loop is to avoid possible confusion with resistance, but then you must still distinguish lowercase a for radius from uppercase A for area.) Next find the initial and final magnetic flux. Use the equation for the magnetic flux through a loop in a uniform magnetic field. Note that $\theta = \theta_0 = 0°$ because the x-axis is the axis of the loop (because the axis of a loop is defined as the line that is perpendicular to the loop and passing through the center of the loop) and the magnetic field lines run parallel to the x-axis. Recall from trig that $\cos 0° = 1$.

$$\Phi_{m0} = B_0 A_0 \cos\theta_0 = (4)\left(\frac{\pi}{16}\right)\cos 0° = \frac{\pi}{4} \text{ T·m}^2$$

$$\Phi_m = BA\cos\theta = (16)\left(\frac{\pi}{16}\right)\cos 0° = \pi \text{ T·m}^2$$

Subtract these values to find the change in the magnetic flux.

$$\Delta\Phi_m = \Phi_m - \Phi_m = \pi - \frac{\pi}{4} = \frac{4\pi}{4} - \frac{\pi}{4} = \frac{3\pi}{4} \text{ T·m}^2$$

Note that this problem asked for the **average** emf induced (as opposed to the instantaneous value of the emf). The distinction is that the formula for the average emf involves the change in magnetic flux over the elapsed time, whereas the formula for the instantaneous emf involves a derivative. There is just one loop in this example such that $N = 1$. Recall that $t = 250 \text{ ms} = \frac{1}{4}$ s.

$$\varepsilon_{ave} = -N\frac{\Delta\Phi_m}{\Delta t} = -(1)\frac{\frac{3\pi}{4}}{\frac{1}{4}} = -\frac{3\pi}{4}(4) = -3\pi \text{ V}$$

To divide by a fraction, multiply by its **reciprocal**. Note that the reciprocal of $\frac{1}{4}$ is 4. The average emf induced in the loop is $\varepsilon_{ave} = -3\pi$ V. If you use a calculator, this comes out to $\varepsilon_{ave} = -9.4$ V.

(B) What is the **average** current induced in the loop during this time?

Apply Ohm's law.

$$\varepsilon_{ave} = I_{ave}R_{loop}$$

$$I_{ave} = \frac{\varepsilon_{ave}}{R_{loop}} = -\frac{3\pi}{6} = -\frac{\pi}{2} \text{ A}$$

The average current induced in the loop is $I_{ave} = -\frac{\pi}{2}$ A. If you use a calculator, this comes out to $I_{ave} = -1.6$ A.

Example: The rectangular loop illustrated below is stretched at a constant rate, such that the right edge travels with a constant speed of 2.0 m/s to the right while the left edge remains fixed. The height of the rectangle is 25 cm. The magnetic field is uniform and has a magnitude of 5000 G. What is the emf induced in the loop?

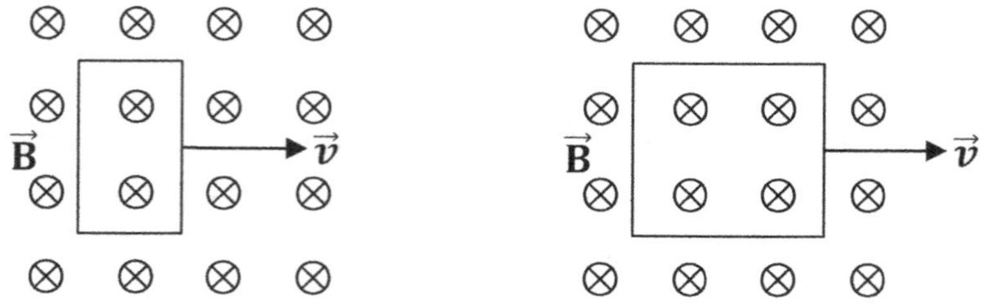

First convert the height and magnetic field to SI units. Recall that $1 \text{ G} = 10^{-4}$ T.

$$h = 25 \text{ cm} = 0.25 \text{ m} = \frac{1}{4} \text{ m}$$

$$B = 5000 \text{ G} = 0.50 \text{ T} = \frac{1}{2} \text{ T}$$

Since the length of the rectangle is changing, we can't get a fixed numerical value for the area. That's not a problem though: We'll just express the area in symbols (length times height, using the variable x to represent the length of the rectangle):

$$A = hx$$

Substitute this expression into the equation for the magnetic flux through a loop in a uniform magnetic field.

$$\Phi_m = BA\cos\theta = Bhx\cos\theta$$

Now substitute the expression for magnetic flux into Faraday's law.

$$\varepsilon_{ind} = -N\frac{d\Phi_m}{dt} = -N\frac{d}{dt}(Bhx\cos\theta)$$

The magnetic field, height, and angle are constants. Only x is changing in this problem. We may factor the constants out of the derivative.

$$\varepsilon_{ind} = -NBh\cos\theta\frac{dx}{dt}$$

You should recognize $\frac{dx}{dt}$ as the velocity of an object moving along the x-axis: $v = \frac{dx}{dt}$.

$$\varepsilon_{ind} = -NBhv\cos\theta$$

There is just one loop, so $N = 1$. Note that $\theta = 0°$ because the axis of the loop and the magnetic field lines are both perpendicular to the plane of the page in the given diagram.

$$\varepsilon_{ind} = -(1)\left(\frac{1}{2}\right)\left(\frac{1}{4}\right)(2)\cos 0° = -\frac{1}{4}\text{ V} = -0.25\text{ V}$$

The emf induced in the loop is $\varepsilon_{ind} = -\frac{1}{4}$ V $= -0.25$ V. Note that you could obtain the same answer by applying the motional emf equation to the right side (which is moving) of the loop, using the height ($h = 0.25$ m) for ℓ in the equation $\varepsilon_{ind} = -B\ell v$. Try it.

Example: The conducting bar illustrated below has a length of 5.0 cm and travels with a constant speed of 40 m/s through a uniform magnetic field of 3000 G. What emf is induced across the ends of the conducting bar?

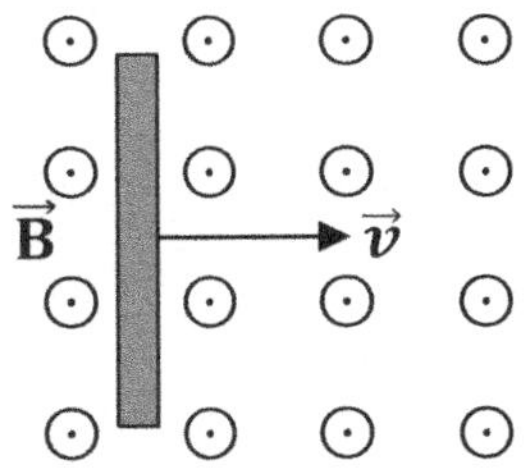

First convert the height and magnetic field to SI units. Recall that $1\text{ G} = 10^{-4}$ T.

$$\ell = 5.0\text{ cm} = 0.050\text{ m} = \frac{1}{20}\text{ m}$$

$$B = 3000\text{ G} = 0.30\text{ T} = \frac{3}{10}\text{ T}$$

Use the equation for motional emf.

$$\varepsilon_{ind} = -B\ell v = -\left(\frac{3}{10}\right)\left(\frac{1}{20}\right)(40) = -\frac{6}{10} = -\frac{3}{5}\text{ V} = -0.60\text{ V}$$

The emf induced across the conducting bar is $\varepsilon_{ind} = -\frac{3}{5}$ V $= -0.60$ V.

116. A circular loop of wire with a diameter of 8.0 cm lies in the zx plane, centered about the origin. A uniform magnetic field of 2,500 G is oriented along the $+y$ direction. What is the magnetic flux through the area of the loop?

117. The vertical loop illustrated below has the shape of a square with 2.0-m long edges. The uniform magnetic field shown makes a 30° angle with the plane of the square loop and has a magnitude of 7.0 T. What is the magnetic flux through the area of the loop?

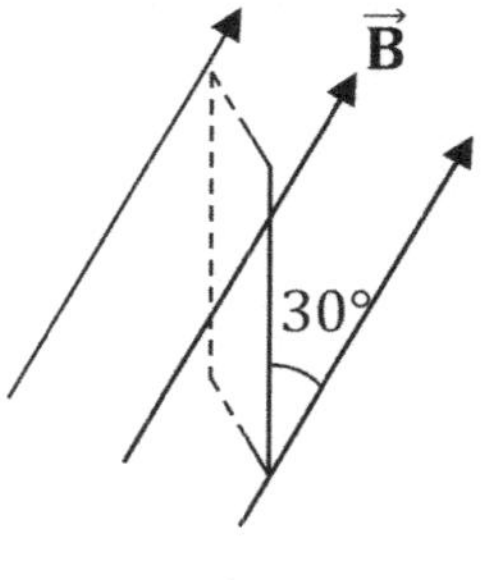

Want help? Check the hints section at the back of the book.

Answers: $4\pi \times 10^{-4}$ T·m^2, 14 T·m^2

118. A rectangular loop of wire with a length of 50 cm, a width of 30 cm, and a resistance of 4.0 Ω lies in the xy plane. A magnetic field is oriented along the $+z$ direction and is uniform throughout the loop at any given moment. The magnetic field increases from 6,000 G to 8,000 G in 500 ms.

(A) Find the initial magnetic flux through the loop.

(B) Find the final magnetic flux through the loop.

(C) What is the average emf induced in the loop during this time?

(D) What is the average current induced in the loop during this time?

Want help? Check the hints section at the back of the book.

Answers: 0.090 T·m^2, 0.120 T·m^2, −60 mV, −15 mA

119. The top vertex of the triangular loops of wire illustrated below is pushed up in 250 ms. The base of the triangle is 50 cm, the initial height is 25 cm, and the final height is 50 cm. The magnetic field has a magnitude of 4000 G. There are 200 loops.

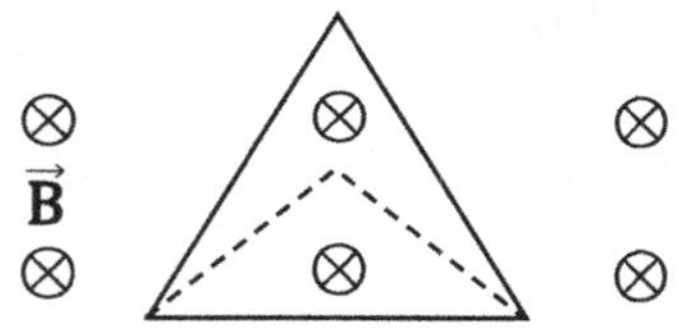

(A) Find the initial magnetic flux through each loop.

(B) Find the final magnetic flux through each loop.

(C) What is the average emf induced in the loop during this time?

(D) Find the direction of the induced current.

Want help? Check the hints section at the back of the book.

Answers: $\frac{1}{40}$ T·m^2, $\frac{1}{20}$ T·m^2, −20 V, counterclockwise

120. Consider the loop in the figure below, which initially has the shape of a square. Each side is 2.0 m long and has a resistance of 5.0 Ω. There is a uniform magnetic field of $5000\sqrt{3}$ G directed into the page. The loop is hinged at each vertex. Monkeys pull corners K and M apart until corners L and N are 2.0 m apart, changing the shape to that of a rhombus. The duration of this process is $(2 - \sqrt{3})$ s.

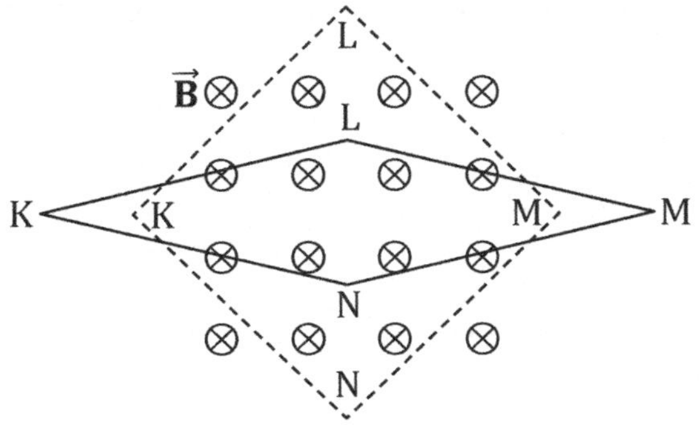

(A) Find the initial magnetic flux through the loop.

(B) Find the final magnetic flux through the loop.

(C) What is the average current induced in the loop during this time?

(D) Find the direction of the induced current.

Want help? Check the hints section at the back of the book.

Answers: $2\sqrt{3}$ T·m², 3.0 T·m², $\frac{\sqrt{3}}{20}$ A, clockwise

121. A solenoid that is not connected to any power supply has 300 turns, a length of 18 cm, and a radius of 4.0 cm. A uniform magnetic field with a magnitude of 500,000 G is initially oriented along the axis of the solenoid. The axis of the solenoid rotates through an angle of 30° relative to the magnetic field with a constant angular speed of 20 rad/s.

(A) What instantaneous emf is induced across the solenoid as the angle reaches 30°?

(B) What average emf is induced during the 30° rotation?

(C) Why should your answer for the "average" emf that you found in part (B) be about one-half of the "instantaneous" emf that you found in part (A)?

Want help? Check the hints section at the back of the book.

Answer: 240π V $\approx$ 754 V, $1440(2 - \sqrt{3})$ V $\approx$ 386 V

122. In the diagram below, a conducting bar with a length of 12 cm slides to the right with a constant speed of 3.0 m/s along the rails of a bare U-channel conductor in the presence of a uniform magnetic field of 25 T. The conducting bar has a resistance of 3.0 Ω and the U-channel conductor has negligible resistance.

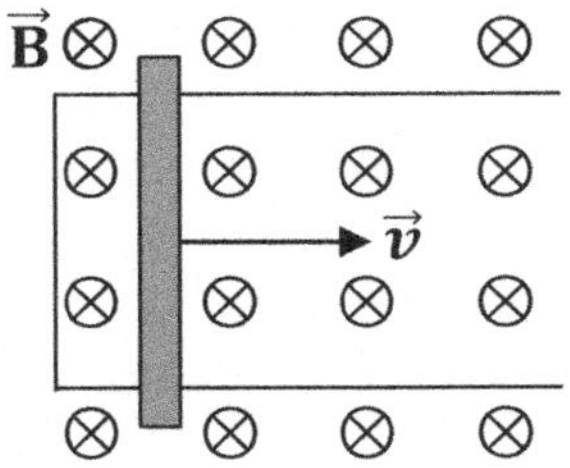

(A) What emf is induced across the ends of the conducting bar?

(B) How much current is induced in the loop?

(C) Find the direction of the induced current.

Want help? Check the hints section at the back of the book.

Answers: −9.0 V, −3.0 A, counterclockwise

123. The conducting bar illustrated below has a length of 8.0 cm and travels with a constant speed of 25 m/s through a uniform magnetic field of 40,000 G. What emf is induced across the ends of the conducting bar?

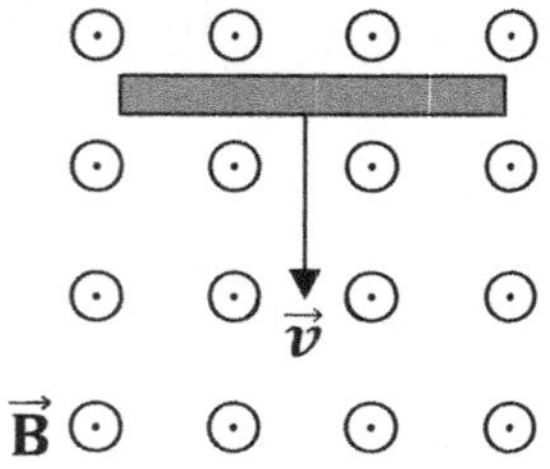

Want help? Check the hints section at the back of the book.

Answer: −8.0 V

124. A solenoid has 3,000 turns, a length of 50 cm, and a radius of 5.0 mm. The current running through the solenoid is increased from 3.0 A to 7.0 A in 16 ms. A second solenoid coaxial with the first solenoid has 2,000 turns, a length of 20 cm, a radius of 5.0 mm, and a resistance of 5.0 Ω. The second solenoid is not connected to the first solenoid or any power supply.

(A) What initial magnetic field does the first solenoid create inside of its coils?

(B) What is the initial magnetic flux through each loop of the second solenoid?

(C) What is the final magnetic flux through each loop of the second solenoid?

(D) What is the average current induced in the second solenoid during this time?

Want help? Check the hints section at the back of the book.

Answers: $72\pi \times 10^{-4}$ T ≈ 0.023 T, $18\pi^2 \times 10^{-8}$ T·m^2 $\approx 1.8 \times 10^{-6}$ T·m^2, $42\pi^2 \times 10^{-8}$ T·m^2 $\approx 4.1 \times 10^{-6}$ T·m^2, $-6\pi^2 \times 10^{-3}$ A ≈ -0.059 A

125. A rectangular loop of wire with a length of 150 cm and a width of 80 cm lies in the xy plane. A magnetic field is oriented along the $+z$ direction and its magnitude is given by the following equation, where SI units have been suppressed.

$$B = 15e^{-t/3}$$

(A) What is the instantaneous emf induced in the loop at $t = 3.0$ s?

(B) What is the average emf induced in the loop from $t = 0$ to $t = 3.0$ s?

(C) Compare the values from (A) and (B) and comment on the difference between them.

Want help? Check the hints section at the back of the book.

Answers: $\frac{6}{e}$ V $\approx$ 2.2 V, $\left(6 - \frac{6}{e}\right)$ V $\approx$ 3.8 V

30 INDUCTANCE

Relevant Terminology

Inductor – a coil of any geometry. Even a single loop of wire serves as an inductor.
Inductance – the property of an inductor for which a changing current causes an emf to be induced in the inductor (as well as in any other nearby conductors).
Solenoid – a coil of wire in the shape of a right-circular cylinder.
Toroidal coil – a coil of wire in the shape of a single-holed ring torus (which is a donut).
Turns – the loops of a coil of wire (such as a solenoid).
Half-life – the time it takes to decay to one-half of the initial value, or to grow to one-half of the final value.
Time constant – the time it takes to decay to $\frac{1}{e}$ or $\approx 37\%$ of the initial value, or to grow to $\left(1 - \frac{1}{e}\right)$ or $\approx 63\%$ of the final value.
Electric charge – a fundamental property of a particle that causes the particle to experience a force in the presence of an electric field. (An electrically neutral particle has no charge and thus experiences no force in the presence of an electric field.)
Current – the instantaneous rate of flow of charge through a wire.
DC – direct current. The direction of the current doesn't change in time.
Potential difference – the electric work per unit charge needed to move a test charge between two points in a circuit. Potential difference is also called the **voltage**.
Emf – the potential difference that a battery or DC power supply would supply to a circuit neglecting its internal resistance.
Back emf – the emf induced in an inductor via Faraday's law. The back emf acts against the applied potential difference.
Magnetic field – a magnetic effect created by a moving charge (or current).
Magnetic flux – a measure of the relative number of magnetic field lines that pass through a surface.
Permeability – a measure of how a substance affects a magnetic field.
Resistance – a measure of how well a component in a circuit resists the flow of current.
Resistor – a component in a circuit which has a significant amount of resistance.
Capacitor – a device that can store charge, which consists of two separated conductors (such as two parallel conducting plates).
Capacitance – a measure of how much charge a capacitor can store for a given voltage.
Magnetic energy – a measure of how much magnetic work an inductor could do based on the current running through its loops.
Electric potential energy – a measure of how much electrical work a capacitor could do based on the charge stored on its plates and the potential difference across its plates.

Inductance Equations

An inductor experiences a self-induced emf via Faraday's law when the current running through the inductor changes. The **self-induced emf** (ε_L) is proportional to the **inductance** (L) and the instantaneous rate of change of the current (I).

$$\varepsilon_L = -L\frac{dI}{dt}$$

Since $\varepsilon_{ind} = -N\frac{d\Phi_m}{dt}$ according to Faraday's law, it follows that $-L\frac{dI}{dt} = -N\frac{d\Phi_m}{dt}$. If you integrate both sides of this equation, you get the following handy formula:

$$L = \frac{N\Phi_m}{I}$$

Two nearby inductors experience mutual inductance (in addition to their self-inductances). The equations of mutual inductance (M) are similar to the equations for self-inductance. The symbol Φ_{12} represents the magnetic flux through inductor 1 created by inductor 2.

$$\varepsilon_1 = -M_{12}\frac{dI_2}{dt} \quad , \quad \varepsilon_2 = -M_{21}\frac{dI_1}{dt} \quad , \quad M_{12} = \frac{N_1\Phi_{12}}{I_2} \quad , \quad M_{21} = \frac{N_2\Phi_{21}}{I_1}$$

Recall the definition of **magnetic flux** (Φ_m).

$$\Phi_m = \int_S \vec{\mathbf{B}} \cdot d\vec{\mathbf{A}}$$

The **magnetic energy** stored in an inductor is:

$$U_L = \frac{1}{2}LI^2$$

Schematic Symbols Used in Circuits

Schematic Representation	Symbol	Name
	R	resistor
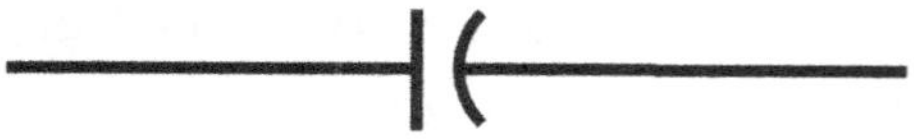	C	capacitor
	L	inductor
	ΔV	battery or DC power supply

Symbols and Units

Symbol	Name	Units
L	inductance (or self-inductance)	H
M	mutual inductance	H
ε_L	self-induced emf	V
ε_{ind}	induced emf	V
I	current	A
t	time	s
Φ_m	magnetic flux	$\text{T}\cdot\text{m}^2$ or Wb
N	number of loops (or turns)	unitless
n	number of turns per unit length	$\frac{1}{\text{m}}$
B	magnitude of the external magnetic field	T
A	area of the loop	m^2
θ	angle between $\vec{\mathbf{B}}$ and the axis of the loop, or the angle between $\vec{\mathbf{B}}$ and $d\vec{\mathbf{A}}$	° or rad
R	resistance	Ω
C	capacitance	F
U_L	magnetic energy stored in an inductor	J
Q	the charge stored on the positive plate of a capacitor	C
ΔV	the potential difference between two points in a circuit	V
$t_{1/2}$	half-life	s
τ	time constant	s
ω	angular frequency	rad/s
φ	phase angle	rad

Note: The symbols Φ and φ are the uppercase and lowercase forms of the Greek letter phi.

Inductors in RL Circuits

In a simple RL circuit like the one shown below on the left, the current drops from its maximum value I_m (the subscript m stands for "maximum") with exponential decay. When a DC power supply is added to the circuit like the diagram below in the middle, the current rises to its maximum value I_m according to $1 - e^{-t/\tau}$. **Note**: If there is an AC power supply involved, use the equations from Chapter 31 instead.

$$I = I_m e^{-t/\tau} \quad \text{(left diagram)}$$
$$I = I_m(1 - e^{-t/\tau}) \quad \text{(middle diagram)}$$

The **time constant** (τ) equals the inductance (L) divided by the resistance (R).

$$\tau = \frac{L}{R}$$

The **half-life** ($t_{½}$) is related to the time constant (τ) by:

$$t_{½} = \tau \ln(2)$$

The resistor and inductor equations still apply, where ΔV_R is the potential difference across the resistor:

$$\Delta V_R = IR \quad , \quad \varepsilon_L = -L\frac{dI}{dt}$$

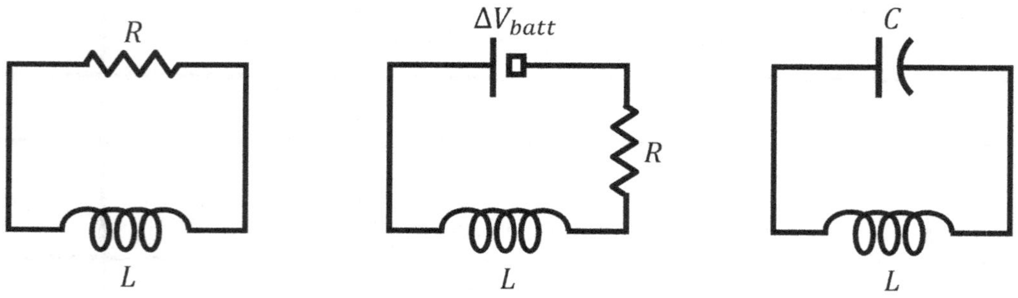

Inductors in LC Circuits

In a simple LC circuit like the one shown above on the right, the current (I) and charge (Q) both oscillate in time (provided that there was some current or charge to begin with). **Note**: If there is an AC power supply involved, use the equations from Chapter 31 instead.

$$I = -I_m \sin(\omega t + \varphi) \quad \text{(right diagram)}$$
$$Q = Q_m \cos(\omega t + \varphi) \quad \text{(right diagram)}$$

The symbol φ represents the **phase constant** and ω represents the angular frequency. The phase constant shifts the sine (or cosine) function horizontally.

$$\omega = \frac{1}{\sqrt{LC}}$$

Recall that angular frequency (ω) is related to frequency (f) by $\omega = 2\pi f$. The maximum charge stored on the capacitor is related to the maximum current through the inductor:

$$I_m = \omega Q_m$$

Inductors in RLC Circuits

In a simple RLC circuit like the one shown below, **for a small value of resistance** (R), the current (I) and charge (Q) both undergo damped harmonic motion (similar to an oscillating spring with friction present). **Note**: If there is an AC power supply involved, use the equations from Chapter 31 instead.

$$Q = Q_m e^{-Rt/2L} \cos(\omega_d t + \varphi)$$

$$I = \frac{dQ}{dt}$$

For the RLC circuit, the angular frequency (ω_d) is (the subscript d stands for "damped"):

$$\omega_d = \sqrt{\frac{1}{LC} - \left(\frac{R}{2L}\right)^2}$$

There is a critical resistance, R_c, for which no damping occurs. We find this value by setting ω_d equal to zero, which happens if $\frac{1}{LC} = \left(\frac{R_c}{2L}\right)^2$. Squareroot both sides to get $\frac{1}{\sqrt{LC}} = \frac{R_c}{2L}$. Multiply both sides by $2L$ to get $\frac{2L}{\sqrt{LC}} = R_c$. Since $2L = \sqrt{4L^2}$, we can write $\sqrt{\frac{4L^2}{LC}} = R_c$, which reduces to the equation below.

$$R_c = \sqrt{\frac{4L}{C}}$$

How damped the harmonic motion is depends upon the value of the critical resistance:

- If $R > R_c$, the motion is said to be **overdamped**.
- If $R = R_c$, the motion is said to be **critically damped**.
- If $R < R_c$, the motion is said to be **underdamped**.
- If $R << R_c$ (read this as "much less than"), the behavior is that of a simple LC circuit.

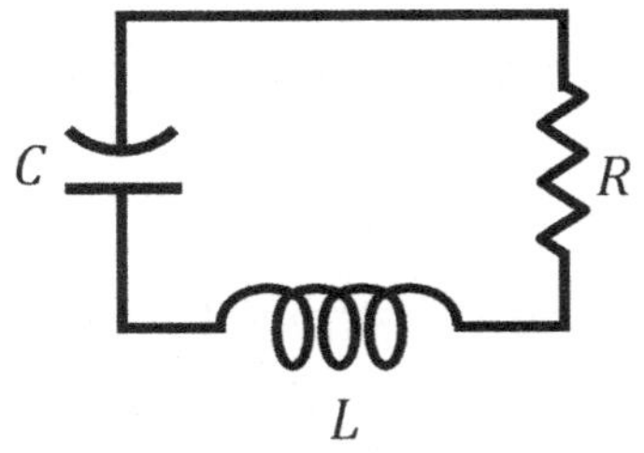

The resistor, capacitor, and inductor equations still apply to RLC circuits. (Again, if there is an AC power supply involved, use equations from Chapter 31 instead).

$$Q = C\Delta V_C \quad , \quad \Delta V_R = IR \quad , \quad \varepsilon_L = -L\frac{dI}{dt}$$

Essential Concepts

Self-inductance and mutual inductance relate to **Lenz's law** and **Faraday's law** (Chapters 28-29). When the current running through an inductor changes, the magnetic field created by the current changes, which causes the **magnetic flux** through the inductor (as well as any other inductors that happen to be nearby) to change. This changing magnetic flux induces an emf in the inductor. When the emf is induced in the same inductor (as opposed to the case of mutual inductance), we call this the **self-induced emf**.

How much emf is induced depends on how rapidly the current is changing.

Lenz's law and Faraday's law similarly cause a **back emf** to be induced in a motor.

Notes Regarding Units

The SI unit of inductance (L) is the Henry (H). A Henry can be related to other SI units according to the equation $L = \frac{N\Phi_m}{I}$. Since the number of loops (N) is unitless, the SI units of magnetic flux (Φ_m) are ($\mathrm{T{\cdot}m^2}$), and the SI unit of current (I) is the Ampère (A), it follows from the previous equation that a Henry (H) equals:

$$1\ \mathrm{H} = 1\frac{\mathrm{T{\cdot}m^2}}{\mathrm{A}}$$

From the equation $\tau = \frac{L}{R}$, the SI unit for time constant (τ) must equal a Henry (H) divided by an Ohm (Ω). That is, $1\ \mathrm{s} = 1\ \frac{\mathrm{H}}{\Omega}$. We can also get this from the equation $\varepsilon_L = -L\frac{dI}{dt}$, which shows that a Volt (V) is related to a Henry (H) by: $1\ \mathrm{V} = 1\ \mathrm{H}{\cdot}\frac{\mathrm{A}}{\mathrm{s}}$. Rearranging this equation, we get $1\ \mathrm{s} = 1\ \mathrm{H}{\cdot}\frac{\mathrm{A}}{\mathrm{V}}$. From the equation $R = \frac{\Delta V}{I}$, we can write an Ohm (Ω) as $\frac{\mathrm{V}}{\mathrm{A}}$. Thus, the $\frac{\mathrm{A}}{\mathrm{V}}$ equals $\frac{1}{\Omega}$, and we again see that $1\ \mathrm{s} = 1\ \frac{\mathrm{H}}{\Omega}$. Put another way, $1\ \mathrm{H} = 1\ \Omega{\cdot}\mathrm{s}$ (it's an Ohm times a second, **<u>not</u>** per second).

Recall from Chapter 18 that the time constant of an RC circuit was $\tau = RC$. We noted in Chapter 18 that the SI unit of capacitance (C), which is the Farad (F), can be expressed as $1\ \mathrm{F} = 1\ \frac{\mathrm{s}}{\Omega}$. When inductance and capacitance are multiplied, their units simplify as follows: $\mathrm{H{\cdot}F} = (\Omega{\cdot}\mathrm{s})\left(\frac{\mathrm{s}}{\Omega}\right) = \mathrm{s}^2$. Therefore, the SI unit of $\sqrt{LC}$ is the second and $\frac{1}{\sqrt{LC}}$ has units consistent with angular frequency (ω), which is measured in radians per second.

Strategy for Solving Inductor Problems

How you solve a problem involving an **inductor** depends on the context:

- **Inductance** (L) can be related to **magnetic flux** (Φ_m).

$$L = \frac{N\Phi_m}{I}$$

  Magnetic flux (Φ_m) is related to **magnetic field** ($\vec{\mathbf{B}}$) via an integral (Chapter 29).

$$\Phi_m = \int_S \vec{\mathbf{B}} \cdot d\vec{\mathbf{A}}$$

  If a problem asks you derive an equation for inductance, see the last strategy of this chapter (two strategies after this one).
- **Self-induced emf** (ε_L) is related to **inductance** (L) by the change in the current.

$$\varepsilon_L = -L\frac{dI}{dt}$$

- For **mutual inductance**, use the following equations, where Φ_{12} is the magnetic flux through inductor 1 created by inductor 2.

$$\varepsilon_1 = -M_{12}\frac{dI_2}{dt} \quad , \quad \varepsilon_2 = -M_{21}\frac{dI_1}{dt} \quad , \quad M_{12} = \frac{N_1\Phi_{12}}{I_2} \quad , \quad M_{21} = \frac{N_2\Phi_{21}}{I_1}$$

- The magnetic **energy** stored in an inductor is:

$$U_L = \frac{1}{2}LI^2$$

- If a problem involves a circuit, see the following strategy (unless it involves an AC power supply, then see Chapter 31 instead).

Strategy for Circuits with Inductors

Note: If the circuit involves an AC power supply, see Chapter 31 instead.
For an RL, LC, or RLC circuit that doesn't have an AC power supply, use these equations:

- For an RL circuit with <u>no</u> battery, where there is initial current I_m:

$$I = I_m e^{-t/\tau}$$

$$\tau = \frac{L}{R} \quad , \quad t_{1/2} = \tau \ln(2)$$

$$\Delta V_R = IR \quad , \quad \varepsilon_L = -L\frac{dI}{dt}$$

- For an RL circuit with a battery, where the initial current is zero:

$$I = I_m(1 - e^{-t/\tau})$$

$$\tau = \frac{L}{R} \quad , \quad t_{1/2} = \tau \ln(2)$$

$$\Delta V_R = IR \quad , \quad \varepsilon_L = -L\frac{dI}{dt}$$

- For an LC circuit with **no** battery, where the initial charge is Q_m:

$$I = -I_m \sin(\omega t + \varphi) \quad , \quad Q = Q_m \cos(\omega t + \varphi)$$

$$\omega = \frac{1}{\sqrt{LC}} = 2\pi f$$

$$I_m = \omega Q_m$$

- For an RLC circuit with **no** battery, where the maximum charge is Q_m:

$$Q = Q_m e^{-Rt/2L} \cos(\omega_d t + \varphi)$$

$$I = \frac{dQ}{dt}$$

$$\omega_d = \sqrt{\frac{1}{LC} - \left(\frac{R}{2L}\right)^2}$$

The **critical resistance** (R_c) determines if the system is underdamped ($R < R_c$), critically damped ($R = R_c$), or overdamped ($R > R_c$).

$$R_c = \sqrt{\frac{4L}{C}}$$

- For any of these circuits, to find energy or power, use the following equations:

$$U_C = \frac{Q^2}{2C} \quad , \quad P_R = I\Delta V_R \quad , \quad U_L = \frac{LI^2}{2}$$

- For a general circuit, you can derive the needed equations for charge and current by applying Kirchhoff's rules. For the loop rule, $\Delta V_R = IR$, $\Delta V_C = \frac{Q}{C}$, and $\Delta V_L = -L\frac{dI}{dt}$. Use the equations $I = \frac{dQ}{dt}$ and $\frac{dI}{dt} = \frac{d^2Q}{dt^2}$, if necessary, to write the loop rule exclusively with current or with charge (not a combination of the two). Then solve the differential equation. We did an example of this in Chapter 18 for an RC circuit.
- If necessary, apply rules from Chapter 17 regarding logarithms and exponentials.

Strategy to Derive an Equation for Inductance

Following is how to apply calculus to derive an equation for inductance:

1. First derive an equation for the **magnetic field** inside of the inductor.
   - For a long solenoid or for a toroidal coil, apply Ampère's law (Chapter 27).
   - Otherwise, apply the law of Biot-Savart (Chapter 26).
2. Use the equation you derived in Step 1 in the **magnetic flux** integral.

$$\Phi_m = \int_S \vec{\mathbf{B}} \cdot d\vec{\mathbf{A}}$$

The differential area vector ($d\vec{\mathbf{A}}$) is perpendicular to the surface, which generally means that it's along the axis of the loop. The magnitude of $d\vec{\mathbf{A}}$ is:

- $dA = dxdy$ for a flat surface with straight sides lying in the xy plane.
- $dA = rdrd\theta$ or $r_c dr_c d\theta$ for a pie slice or circle (in 2D polar or cylindrical).
- $dA = a^2 \sin\theta \, d\theta d\varphi$ for the surface of a sphere of radius a.

Note that $\vec{\mathbf{B}} \cdot d\vec{\mathbf{A}} = B \cos\psi \, dA$, where ψ is the angle between $\vec{\mathbf{B}}$ and $d\vec{\mathbf{A}}$.

3. Plug your equations from Steps 1 and 2 into the following equation for **inductance**.

$$L = \frac{N\Phi_m}{I}$$

4. For **mutual inductance**, find the magnetic field that inductor 2 creates inside of inductor 1 in Step 1, and then find the magnetic flux (Φ_{12}) through inductor 1 created by inductor 2 (or swap all of the 1's and 2's to find Φ_{21}).

$$M_{12} = \frac{N_1 \Phi_{12}}{I_2} \quad , \quad M_{21} = \frac{N_2 \Phi_{21}}{I_1}$$

Important Distinctions

Note that we use the symbol L for **inductance**. Therefore, if there is an inductor involved in a problem, you should use a different symbol, such as lowercase ℓ, for **length**.

DC stands for **direct** current, whereas AC stands for **alternating** current. This chapter involved circuits that either have a DC power supply or no power supply at all. If a problem involves an AC power supply, see Chapter 31 instead.

Note the three different times involved in RL, LC, and RLC circuits:

- t (without a subscript) represents the elapsed **time**.
- $t_{1/2}$ is the **half-life**. It's the time it takes to reach one-half the maximum value.
- τ is the **time constant**: $\tau = \frac{L}{R}$. Note that $t_{1/2} = \tau \ln(2)$.

Don't confuse the terms **inductor** and **insulator**.

- An **inductor** is a coil of wire and relates to Faraday's law.
- An **insulator** is a material through which electrons don't flow readily.

Note that the opposite of a conductor is an **insulator**. An **inductor** is <u>**not**</u> the opposite of a conductor. In fact, an inductor is constructed from a conductor that is wound in the shape of a coil.

A coil of any geometry is termed an **inductor**. A **solenoid** is a specific type of inductor where the coil has the shape of a tight helix wound around a right-circular cylinder. All solenoids are inductors, but not all inductors are solenoids. One example of an inductor that isn't a solenoid is the toroidal coil (see Chapter 27).

Example: The emf induced in an inductor is -6.0 mV when the current through the inductor increases at a rate of 3.0 A/s. What is the inductance?

First convert the self-induced emf from milliVolts (mV) to Volts (V): $\varepsilon_L = -0.0060$ V. Note that $\frac{dI}{dt} = 3.0$ A/s is the rate at which the current increases. Use the equation that relates the self-induced emf to the inductance.

$$\varepsilon_L = -L\frac{dI}{dt}$$

$$-0.006 = -L(3)$$

Divide both sides of the equation by -3. The minus signs cancel.

$$L = \frac{0.006}{3} = 0.0020 \text{ H} = 2.0 \times 10^{-3} \text{ H} = 2.0 \text{ mH}$$

The inductance is $L = 2.0$ mH. That's in milliHenry (mH): Recall that the metric prefix milli (m) stands for 10^{-3}.

Example: An inductor has an inductance of 12 mH.

(A) What is the time constant if the inductor is connected in series with a 3.0-Ω resistor?

First convert the inductance from milliHenry (mH) to Henry (H): $L = 0.012$ H. Use the equation for the time constant of an RL circuit.

$$\tau = \frac{L}{R} = \frac{0.012}{3} = 0.0040 \text{ s} = 4.0 \text{ ms}$$

The time constant is $\tau = 4.0$ ms. (That's in milliseconds.)

(B) What is the angular frequency of oscillations in the current if the inductor is connected in series with a 30-μF capacitor?

First convert the capacitance from microFarads (μF) to Farads (F): $C = 3.0 \times 10^{-5}$ F. Use the equation for the angular frequency in an LC circuit.

$$\omega = \frac{1}{\sqrt{LC}} = \frac{1}{\sqrt{(0.012)(3.0 \times 10^{-5})}} = \frac{1}{\sqrt{0.036 \times 10^{-5}}} = \frac{1}{\sqrt{36 \times 10^{-8}}}$$

Note that $3.6 \times 10^{-7} = 36 \times 10^{-8}$. It's simpler to take a square root of an even power of 10, which makes it preferable to work with 10^{-8} instead of 10^{-7}. Apply the rule from algebra that $\sqrt{ax} = \sqrt{a}\sqrt{x}$.

$$\omega = \frac{1}{\sqrt{36}}\frac{1}{\sqrt{10^{-8}}} = \frac{1}{6}\frac{1}{10^{-4}} = \frac{1}{6} \times 10^4 \text{ rad/s}$$

The angular frequency is $\omega = \frac{1}{6} \times 10^4$ rad/s. If you use a calculator, this works out to $\omega = 1.7 \times 10^3$ rad/s.

Example: The long, tightly wound solenoid illustrated below (both diagrams show the same solenoid) is coaxial with the z-axis and carries an insulated filamentary current I. Derive an expression for the self-inductance of the solenoid.

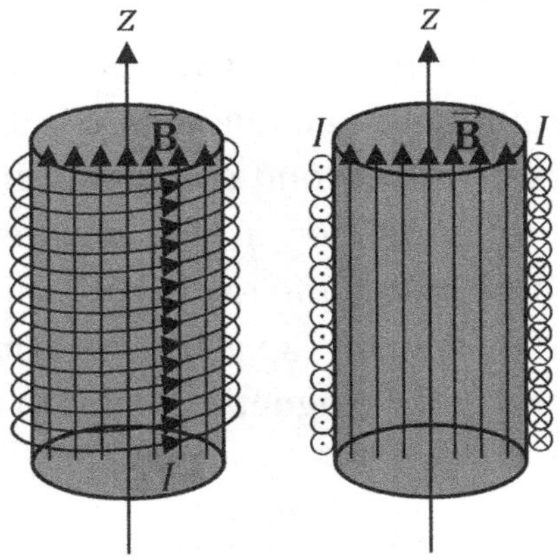

We applied Ampère's law to derive an equation for the magnetic field inside of a similar solenoid in Chapter 27. The result was:

$$\vec{\mathbf{B}} = \mu_0 n I \hat{\mathbf{z}} = \frac{\mu_0 N I \hat{\mathbf{z}}}{\ell}$$

Here, n is the number of turns per unit length, N is the number of turns, and ℓ is the length of the solenoid. Plug this equation for magnetic field into the magnetic flux integral.

$$\Phi_m = \int_S \vec{\mathbf{B}} \cdot d\vec{\mathbf{A}} = \int_S \frac{\mu_0 N I \hat{\mathbf{z}}}{\ell} \cdot d\vec{\mathbf{A}} = \frac{\mu_0 N I}{\ell} \int_S \hat{\mathbf{z}} \cdot d\vec{\mathbf{A}}$$

The surface is the area of one loop. The corresponding differential area element has a magnitude of $dA = r_c dr_c d\theta$. The direction of $d\vec{\mathbf{A}}$ is along the z-axis, since $d\vec{\mathbf{A}}$ is in general perpendicular to the surface and the z-axis is perpendicular to the loop. Combining this together, we get $d\vec{\mathbf{A}} = \hat{\mathbf{z}}\, dA = \hat{\mathbf{z}}\, r_c dr_c d\theta$. The limits are from $r_c = 0$ to $r_c = a$ (the radius of the solenoid) and from $\theta = 0$ to $\theta = 2\pi$.

$$\Phi_m = \frac{\mu_0 N I}{\ell} \int_{r_c=0}^{a} \int_{\theta=0}^{2\pi} \hat{\mathbf{z}} \cdot \hat{\mathbf{z}}\, r_c\, dr_c\, d\theta$$

The scalar product is $\hat{\mathbf{z}} \cdot \hat{\mathbf{z}} = 1$ (see Chapter 19).

$$\Phi_m = \frac{\mu_0 N I}{\ell} \int_{r_c=0}^{a} \int_{\theta=0}^{2\pi} r_c dr_c d\theta = \frac{\mu_0 N I}{\ell} \int_{r_c=0}^{a} r_c\, dr_c \int_{\theta=0}^{2\pi} d\theta$$

$$\Phi_m = \frac{\mu_0 N I}{\ell} \left[\frac{r_c^2}{2}\right]_{r_c=0}^{a} [\theta]_{\theta=0}^{2\pi} = \frac{\pi \mu_0 N I a^2}{\ell}$$

Plug this expression into the equation for self-inductance.

$$L = \frac{N\Phi_m}{I} = \frac{N}{I}\left(\frac{\pi \mu_0 N I a^2}{\ell}\right) = \frac{\pi \mu_0 N^2 a^2}{\ell}$$

Note that it's the same as $L = \frac{\mu_0 N^2 A}{\ell}$, where $A = \pi a^2$ is the area of each loop.

Example: Two long, tightly wound solenoids are coaxial with the z-axis, have the same radius (a), and carry insulated filamentary currents I_1 and I_2. Derive an expression for the mutual inductance (M_{21}) of solenoid 2 with respect to solenoid 1.

This example is very similar to the previous example, except that we will need to use subscripts to keep track of the mutual inductance. We begin with the magnetic field that the first solenoid creates inside of the second solenoid (since Φ_{21}, which corresponds to M_{21} as specified in the problem, is defined as the magnetic flux created by solenoid 1 through solenoid 2 – although note that some textbooks call this the magnetic flux through solenoid 2 created by solenoid 1, which is exactly the **same thing** worded in a different order). We use the same expression for magnetic field as in the previous example, but with subscripts.

$$\vec{\mathbf{B}}_1 = \frac{\mu_0 N_1 I_1 \hat{\mathbf{z}}}{\ell_1}$$

Next, we find the magnetic flux (Φ_{21}) through solenoid 2 created by solenoid 1.

$$\Phi_{21} = \int_{S_2} \vec{\mathbf{B}}_1 \cdot d\vec{\mathbf{A}}$$

We use the magnetic field ($\vec{\mathbf{B}}_1$) created by solenoid 1, but the integral is over the surface (S_2) of solenoid 2 (which is the area of one of solenoid 2's loops) – although in this problem it doesn't matter because the two solenoids happen to have the same radius (a). Just as in the previous example, $d\vec{\mathbf{A}} = \hat{\mathbf{z}}\, dA = \hat{\mathbf{z}}\, r_c dr_c d\theta$ and $\hat{\mathbf{z}} \cdot \hat{\mathbf{z}} = 1$. The limits are from $r_c = 0$ to $r_c = a$ (technically, the radius of solenoid 2, since it's the flux through solenoid 2, though in this problem the two solenoids have the same radius) and from $\theta = 0$ to $\theta = 2\pi$.

$$\Phi_{21} = \int_{S_2} \vec{\mathbf{B}}_1 \cdot d\vec{\mathbf{A}} = \int_{S_2} \frac{\mu_0 N_1 I_1 \hat{\mathbf{z}}}{\ell_1} \cdot d\vec{\mathbf{A}} = \frac{\mu_0 N_1 I_1}{\ell_1} \int_{r_c=0}^{a} \int_{\theta=0}^{2\pi} \hat{\mathbf{z}} \cdot \hat{\mathbf{z}}\, r_c\, dr_c\, d\theta$$

$$\Phi_{21} = \frac{\mu_0 N_1 I_1}{\ell_1} \int_{r_c=0}^{a} r_c\, dr_c \int_{\theta=0}^{2\pi} d\theta = \frac{\mu_0 N_1 I_1}{\ell_1} \left[\frac{r_c^2}{2}\right]_{r_c=0}^{a} [\theta]_{\theta=0}^{2\pi} = \frac{\pi \mu_0 N_1 I_1 a^2}{\ell_1}$$

Plug this expression into the equation for mutual inductance.

$$M = \frac{N_2 \Phi_{21}}{I_1} = \frac{N_2}{I_1}\left(\frac{\pi \mu_0 N_1 I_1 a^2}{\ell_1}\right) = \frac{\pi \mu_0 N_1 N_2 a^2}{\ell_1}$$

Note that it's the same as $M = \frac{\mu_0 N_1 N_2 A_2}{\ell_1}$, where $A_2 = \pi a_2^2 = \pi a^2$ is the area of each loop of solenoid 2.

126. The emf induced in an 80-mH inductor is -0.72 V when the current through the inductor increases at a constant rate. What is the rate at which the current increases?

127. A 50.0-Ω resistor is connected in series with a 20-mH inductor. What is the time constant?

128. A 40.0-μF capacitor that is initially charged is connected in series with a 16.0-mH inductor. What is the angular frequency of oscillations in the current?

Want help? Check the hints section at the back of the book.

Answers: 9.0 A/s, 0.40 ms, 1250 rad/s

129. The tightly wound (even though the diagram is not drawn to look that way) toroidal coil illustrated below has square cross section and carries an insulated filamentary current I. Derive an expression for the self-inductance of the toroidal coil.

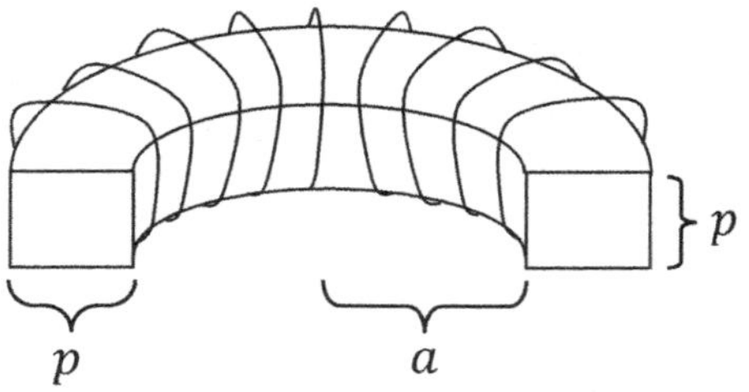

Want help? Check the hints section at the back of the book.

Answer: $L = \frac{\mu_0 N^2 p}{2\pi} \ln\left(\frac{a+p}{a}\right)$

31 AC CIRCUITS

Relevant Terminology

AC – alternating current. The direction of the current alternates in time.
DC – direct current. The direction of the current doesn't change in time.
Root-mean square (rms) value – square the values, average them, and then take the square-root. AC multimeters measure rms values of current or potential difference.
Frequency – the number of cycles completed per unit time.
Angular frequency – the number of radians completed per unit time.
Period – the time it takes to complete one cycle.
Electric charge – a fundamental property of a particle that causes the particle to experience a force in the presence of an electric field.
Current – the instantaneous rate of flow of charge through a wire.
Potential difference – the electric work per unit charge needed to move a test charge between two points in a circuit. Potential difference is also called the **voltage**.
Emf – the potential difference that a battery or DC power supply would supply to a circuit neglecting its internal resistance.
Magnetic field – a magnetic effect created by a moving charge (or current).
Resistance – a measure of how well a component in a circuit resists the flow of current.
Resistor – a component in a circuit which has a significant amount of resistance.
Capacitor – a device that can store charge, which consists of two separated conductors (such as two parallel conducting plates).
Capacitance – a measure of how much charge a capacitor can store for a given voltage.
Inductor – a coil of any geometry. Even a single loop of wire serves as an inductor.
Inductance – the property of an inductor for which a changing current causes an emf to be induced in the inductor (as well as in any other nearby conductors).
Impedance – the ratio of the maximum potential difference to the maximum current in an AC circuit. It combines the effects of resistance and reactance.
Reactance – the effect that capacitors and inductors have on the current in an AC circuit, causing the current to be out of phase with the potential difference.
Inductive reactance – the effect that an inductor has on the current in an AC circuit, causing the current to **lag** the potential difference across an inductor by 90°.
Capacitive reactance – the effect that a capacitor has on the current in an AC circuit, causing the current to **lead** the potential difference across a capacitor by 90°.
Electric power – the instantaneous rate at which electrical work is done.
Transformer – a device that steps AC potential difference up or down.
Primary – refers to the input inductor of a transformer connected to an AC power supply.
Secondary – refers to the output inductor of a transformer, yielding the adjusted voltage.

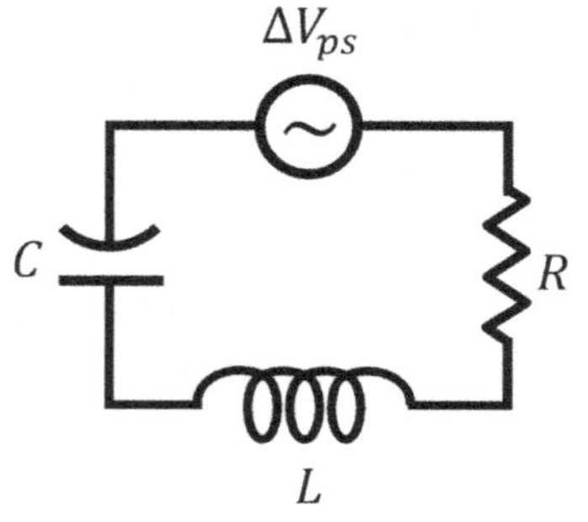

Equations for a Series RLC Circuit with an AC Power Supply

Consider the series RLC circuit illustrated above, which has an AC power supply. The following equations apply to such a circuit:

- The **potential difference** across each circuit element is (where "ps" stands for "power supply"):
$$\Delta V_{ps} = \Delta V_m \sin(\omega t) \quad , \quad \Delta V_R = \Delta V_{Rm} \sin(\omega t)$$
$$\Delta V_L = \Delta V_{Lm} \cos(\omega t) \quad , \quad \Delta V_C = -\Delta V_{Cm} \cos(\omega t)$$
The values on the left-hand sides of the above equations are instantaneous values; they vary in time. The quantities with a subscript m (which stands for "maximum") are maximum values. The **maximum** values are related to the maximum current by:
$$\Delta V_m = I_m Z \quad , \quad \Delta V_{Rm} = I_m R \quad , \quad \Delta V_{Lm} = I_m X_L \quad , \quad \Delta V_{Cm} = I_m X_C$$
The symbol Z is called the impedance and X is called the reactance. See below.
- The **current** in the circuit is:
$$I = I_m \sin(\omega t - \varphi)$$
The angle φ is called the **phase angle**. See the following page.
- Ohm's law is effectively generalized by defining a quantity called **impedance** (Z).
$$\Delta V_m = I_m Z \quad , \quad Z = \sqrt{R^2 + (X_L - X_C)^2}$$
The impedance (Z) includes both resistance (R) and reactances (X_L and X_C).
- There are two kinds of reactance:
    - **Inductive reactance** (X_L) is the ratio of the maximum potential difference across the inductor to the maximum current. Its SI unit is the Ohm (Ω). It is directly proportional to the angular frequency (ω) and the inductance (L).
    $$X_L = \omega L \quad , \quad \Delta V_{Lm} = I_m X_L$$
    - **Capacitive reactance** (X_C) is the ratio of the maximum potential difference across the capacitor to the maximum current. Its SI unit is the Ohm (Ω). It is inversely proportional to the angular frequency (ω) and the capacitance (C).
    $$X_C = \frac{1}{\omega C} \quad , \quad \Delta V_{Cm} = I_m X_C$$
- The **root-mean-square** (rms) values are related to the maximum values by:
$$\Delta V_{rms} = \frac{\Delta V_m}{\sqrt{2}} \quad , \quad I_{rms} = \frac{I_m}{\sqrt{2}}$$

- The **phase angle** (φ) describes the horizontal shift in the graph of the current compared to the graph of the potential difference.

$$\varphi = \tan^{-1}\left(\frac{X_L - X_C}{R}\right)$$

The **sign** of the phase angle has the following significance:
  - If $X_L > X_C$, the phase angle is **positive**, meaning that the current **lags** behind the potential difference of the power supply.
  - If $X_L < X_C$, the phase angle is **negative**, meaning that the current **leads** the potential difference of the power supply.
  - If $X_L = X_C$, the phase angle is **zero**, meaning that the current is in phase with the potential difference of the power supply. This corresponds to the **resonance frequency** (see below).
- The **angular frequency** (ω), **frequency** (f), and **period** (T) are related by:

$$\omega = 2\pi f = \frac{2\pi}{T} \quad , \quad f = \frac{1}{T}$$

The **resonance frequency** is the frequency which gives the greatest value of the rms current. (In an AC circuit, it turns out that the value of I_{rms} is a function of the frequency.) Resonance occurs when $X_L = X_C$ (since this minimizes the impedance, it thereby maximizes the rms current). At resonance, the angular frequency is:

$$\omega_0 = \frac{1}{\sqrt{LC}} = 2\pi f_0$$

Note that some texts refer to the "resonance frequency," but mean ω_0 rather than f_0.
- The **instantaneous power** (P) delivered by the AC power supply is:

$$P = I\Delta V_{ps} = I_m \Delta V_m \sin(\omega t)\sin(\omega t - \varphi)$$

The **average power** (P_{av}) is:

$$P_{av} = I_{rms}\Delta V_{rms}\cos\varphi = I_{rms}^2 R$$

The factor of $\cos\varphi$ is called the **power factor**.
- The **quality factor** (Q_0), where the Q_0 does **not** refer to charge, provides a measure of the sharpness of a graph of P_{av} as a function of ω.

$$Q_0 = \frac{\omega_0}{\Delta\omega} = \frac{\omega_0 L}{R}$$

Here, ω_0 is the resonance frequency (see above) and $\Delta\omega$ is the width of the curve of P_{av} as a function of ω between the two points for which P_{av} equals half of its maximum value. Hence, $\Delta\omega$ is termed the "full width at half maximum" (FWHM).
- The **current** (I) and the **charge** (Q) stored on the capacitor are related by:

$$I = \frac{dQ}{dt}$$

Symbols and SI Units

Symbol	Name	SI Units
ΔV_{ps}	instantaneous potential difference across the power supply	V
ΔV_R	instantaneous potential difference across the resistor	V
ΔV_L	instantaneous potential difference across the inductor	V
ΔV_C	instantaneous potential difference across the capacitor	V
ΔV_m	maximum potential difference across the power supply	V
ΔV_{rms}	root-mean-square value of the potential difference	V
ΔV_{Rm}	maximum potential difference across the resistor	V
ΔV_{Lm}	maximum potential difference across the inductor	V
ΔV_{Cm}	maximum potential difference across the capacitor	V
ΔV_{in}	input voltage	V
ΔV_{out}	output voltage	V
N_p	number of loops in the primary inductor	unitless
N_s	number of loops in the secondary inductor	unitless
I	instantaneous current	A
I_m	maximum value (amplitude) of the current	A
I_{rms}	root-mean-square value of the current	A
R	resistance	Ω
L	inductance	H
C	capacitance	F
Z	impedance	Ω
X_L	inductive reactance	Ω
X_C	capacitive reactance	Ω

P	instantaneous power	W
P_{av}	average power	W
t	time	s
ω	angular frequency	rad/s
ω_0	resonance (angular) frequency	rad/s
$\Delta\omega$	full-width at half maximum (FWHM)	rad/s
f	frequency	Hz
f_0	resonance frequency	Hz
T	period	s
φ	phase angle	rad
Q_0	quality factor (**not** charge)	unitless
Q	instantaneous charge stored on the capacitor	C

Schematic Symbols Used in AC Circuits

Schematic Representation	Symbol	Name
	R	resistor
	C	capacitor
	L	inductor
	ΔV	AC power supply
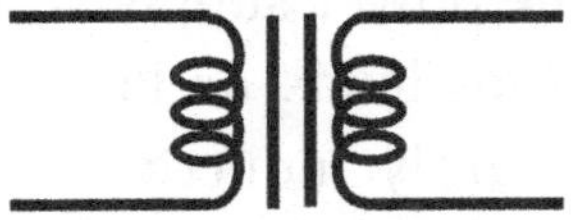		transformer

Phasors

In an AC circuit, the current is generally not in phase with the potential difference, meaning that a graph of current as a function of time is shifted horizontally compared to a graph of the power supply potential difference as a function of time. Both the current and potential difference are sine waves, but the phase angle (φ) in $I = I_m \sin(\omega t - \varphi)$ shifts the current's graph compared to $\Delta V_{ps} = \Delta V_m \sin(\omega t)$.

- When φ is **positive**, the current **lags** behind the power supply's potential difference.
- When φ is **negative**, the current **leads** the power supply's potential difference.
- When φ is **zero**, the current is **in phase** with the power supply's potential difference.

Inductors and capacitors cause the phase shift in an AC circuit:

- Because of Faraday's law, which can be expressed as $\Delta V_L = -L\frac{dI}{dt}$ (see Chapter 30), the current in an **inductor** lags the inductor's potential difference by 90°.
- Because current is the instantaneous rate of flow of charge, $I = \frac{dQ}{dt}$, the current in a capacitor **leads** the capacitor's potential difference by 90°.

Mathematically, calculus causes the phase shifts:

- $\int \sin\theta \, d\theta = -\cos\theta$. An integral of the sine function shifts it 90° to the right, turning it into a negative cosine. This is what happens for an inductor, since $\Delta V_L = -L\frac{dI}{dt}$. (Solving for the current requires an integral.)
- $\frac{d}{d\theta}\sin\theta = \cos\theta$. A derivative of the sine function **shifts** it 90° to the left, turning it into a cosine. This is what happens for a capacitor, since $I = \frac{dQ}{dt}$ and $Q = C\Delta V_C$.

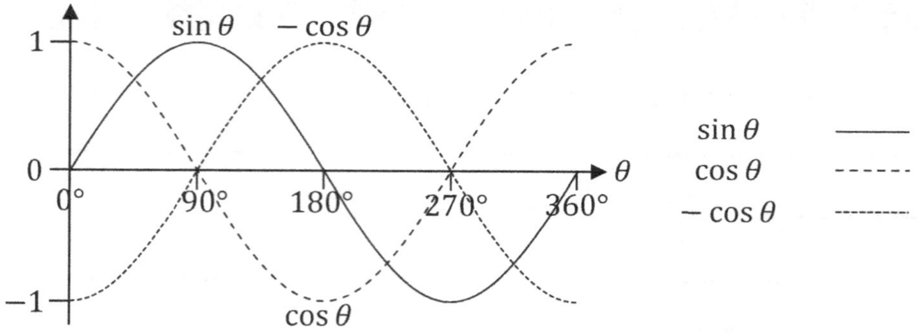

Visually, in the graph above, you can see that:

- Shifting a sine wave 90° to the left turns it into a cosine function.
- Shifting a sine wave 90° to the right turns it into a negative cosine function.

In a series RLC circuit with an AC power supply, the phase angle (φ) can be any value between −90° and 90°. The current can lead or lag the power supply voltage by any value from 0° to 90°. The extreme cases where the current leads or lags the power supply voltage by 90° only occur for purely capacitive or purely inductive circuits (which are technically impossible, as the wires and connections always have some resistance).

We use phasor diagrams in order to determine the phase angle in an AC circuit. A **phasor** is basically a vector, with a magnitude and direction as follows:

- The amplitude* (ΔV_m) of the potential difference is the magnitude of the phasor.
- The phase angle serves as the direction of the phasor.

The phase angle (φ) equals 90° for an inductor, 0° for a resistor, and −90° for a capacitor. Therefore, a phasor diagram for a series RLC circuit with an AC power supply looks the diagram below on the left.

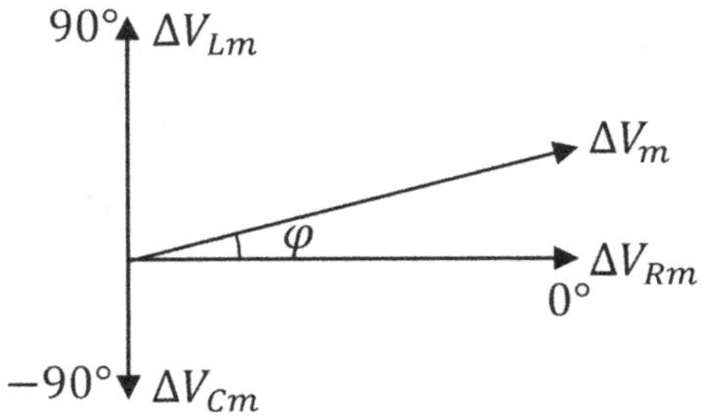

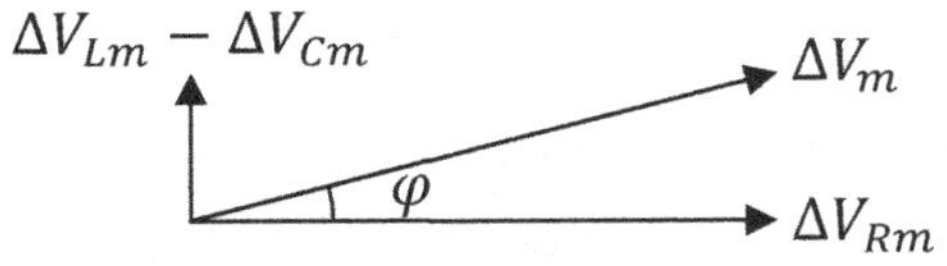

The right diagram above combines the phasors for the inductor and capacitor. Since these phasors point in opposite directions, we subtract the two magnitudes when doing the vector addition (or phasor addition) for these two phasors. From the right diagram above, we see that the amplitude (ΔV_m) of the potential difference supplied by the power supply can be found from the Pythagorean theorem, since ΔV_{Rm} is perpendicular to $\Delta V_{Lm} - \Delta V_{Cm}$.

$$\Delta V_m = \sqrt{\Delta V_{Rm}^2 + (\Delta V_{Lm} - \Delta V_{Cm})^2}$$

Since each potential difference amplitude is proportional to the amplitude of the current ($\Delta V_m = I_m Z$, $\Delta V_{Rm} = I_m R$, $\Delta V_{Lm} = I_m X_L$, and $\Delta V_{Cm} = I_m X_C$), we can divide the previous equation by I_m to get an equation for impedance (Z):

$$Z = \sqrt{R^2 + (X_L - X_C)^2}$$

Therefore, it would be just as effective to draw an **impedance triangle** like the one below.

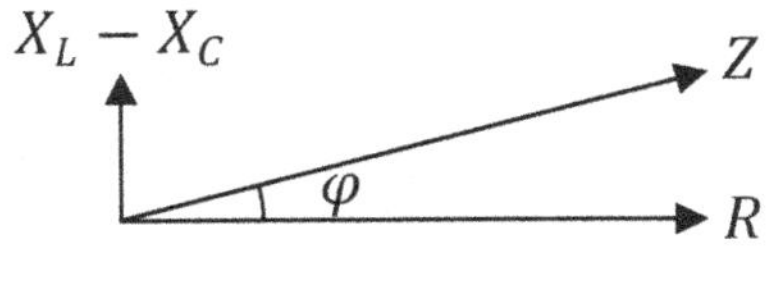

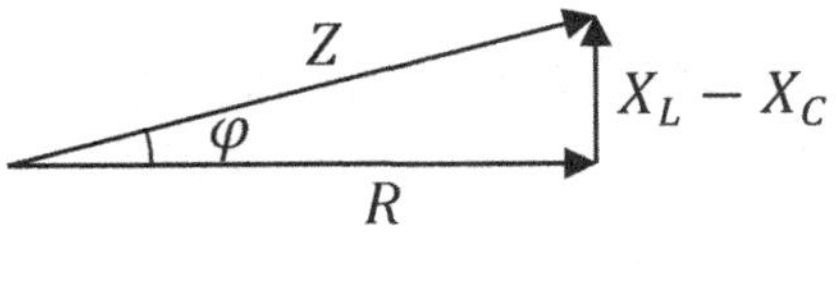

From the impedance triangle, we can find the phase angle (φ) through trig.

$$\tan\varphi = \frac{X_L - X_C}{R}$$

* It works the same if you use rms values (ΔV_{rms}) instead, since a simple $\sqrt{2}$ is involved.

Resonance in an RLC Circuit

Inductive and capacitive reactance both depend on **frequency** in an AC circuit:

- **Inductive reactance** is directly proportional to frequency according to $X_L = \omega L$. The reason for this is that at higher frequencies, the current changes rapidly, and rapid changes in current increase the induced emf according to Faraday's law: $\varepsilon_L = -L\frac{dI}{dt}$.
- **Capacitive reactance** is inversely proportional to frequency according to $X_C = \frac{1}{\omega C}$. The reason for this is that when the frequency approaches zero, the capacitive reactance must be very high, since the current ($I = \frac{\Delta V_C}{X_C}$) must be zero in the extreme case that ω is zero. (That's because when ω equals zero, you get a DC circuit, and no current passes through the capacitor in a steady-state DC circuit.)

Because inductive and capacitive reactance depend on frequency, the amplitude of the current also depends on frequency.

$$I_m = \frac{\Delta V_m}{Z} = \frac{\Delta V_m}{\sqrt{R^2 + (X_L - X_C)^2}} = \frac{\Delta V_m}{\sqrt{R^2 + \left(\omega L - \frac{1}{\omega C}\right)^2}}$$

For given values of ΔV_m, R, L, and C, the amplitude of the current is maximized when $X_L = X_C$ (since the denominator reaches its minimum possible value, R, when $X_L - X_C = 0$, and since a smaller denominator makes the current greater). If you set $X_L = X_C$, you get:

$$\omega_0 L = \frac{1}{\omega_0 C}$$

Multiply both sides of the equation by ω_0 (which does <u>**not**</u> cancel) and divide by L:

$$\omega_0^2 = \frac{1}{LC}$$

Squareroot both sides of the equation.

$$\omega_0 = \frac{1}{\sqrt{LC}}$$

The **resonance frequency** (ω_0) is the value of ω that maximizes I_m. The maximum possible current (I_{max}, which is in general different from I_m) is:

$$I_{max} = \frac{\Delta V_m}{R}$$

Note that $I_m = \frac{\Delta V_m}{Z}$, whereas $I_{max} = \frac{\Delta V_m}{R}$. The quantity I_m is the amplitude of the current: The current varies as a sine wave, and I_m is the peak of that sine wave. The peak value of the sine wave, I_m, depends on the frequency. You get the maximum possible peak, I_{max}, at the resonance frequency. That is, I_{max} is the greatest possible I_m. It corresponds to the minimum possible impedance: $Z_{min} = R$. Less impedance yields more current.

Quality Factor

The **average power** (P_{av}) is plotted below as a function of angular frequency (ω) for a series RLC circuit with an AC power supply. The average power is maximum for the resonance frequency (ω_0). The quantity $\Delta\omega$ shown below is the **full-width at half the maximum** (FWHM). To find $\Delta\omega$, read off the maximum value of the graph, $P_{av,max}$, divide this quantity by 2 to get $\frac{P_{av,max}}{2}$, find the two values of ω where the curve has a height equal to $\frac{P_{av,max}}{2}$, and subtract these two values of ω: $\Delta\omega = \omega_2 - \omega_1$.

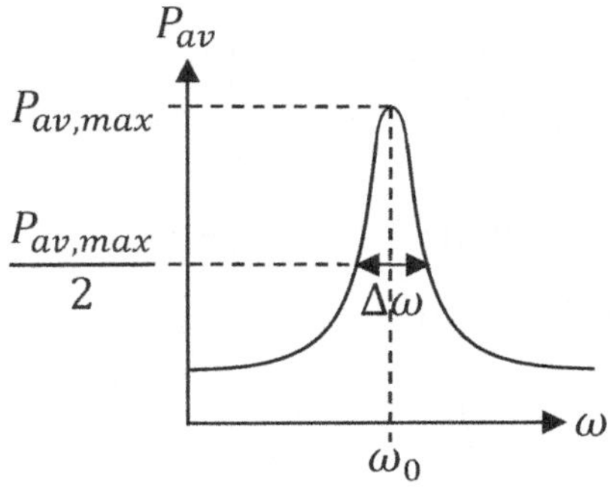

There are two ways to determine the **quality factor** (Q_0). If you are given a graph of average power as a function of angular frequency, read off the values of ω_0 and $\Delta\omega$ (as described in the previous paragraph, and as shown above).

$$Q_0 = \frac{\omega_0}{\Delta\omega}$$

If you are given L and R, you can use these values instead of ω_0 and $\Delta\omega$.

$$Q_0 = \frac{\omega_0 L}{R}$$

The quality factor provides a measure of the sharpness of the average power curve. A higher value of Q_0 corresponds to an average power curve with a narrower peak.

Root-mean-square (rms) Values

An AC multimeter used as an AC ammeter or AC voltmeter measures **root-mean-square** (rms) values. They don't measure "average" values in the usual since: A sine wave is zero on "average," since it's above the horizontal axis half the time and equally below the horizontal axis the other half of the time. Here is what root-mean-square means:

- First square all of the values.
- Then find the average of the squared values.
- Take the squareroot of the previous answer.

Although the ordinary "average" of a sine wave is zero over one cycle, the rms value of a sine wave isn't zero. That's because everything becomes nonnegative in the second step, when every value is squared.

The rms Value of a Sine Wave

The current and potential difference are sinusoidal waves. The instantaneous values are constantly varying. The peak values, I_m and ΔV_m, are the amplitudes of the sine waves. When we measure rms values, these measured values are related to the peak values by the average value of a sine wave.

$$\Delta V_{rms} = \Delta V_m \sqrt{\overline{\sin^2(\theta)}} \quad , \quad I_{rms} = I_m \sqrt{\overline{\sin^2(\theta)}}$$

The expression $\sqrt{\overline{\sin^2(\theta)}}$ represents the rms value of sine theta: First square the function, then average sine squared, and then take the squareroot of the answer. The **mean-value theorem** (MVT) from calculus allows us to find the average value of a function.

$$\overline{\sin^2(\theta)} = \frac{\int_{\theta=0}^{2\pi} \sin^2\theta \, d\theta}{2\pi - 0}$$

One way to perform this integral is to apply the following trig identity:

$$\sin^2\theta = \frac{1 - \cos 2\theta}{2}$$

Substitute this equation into the previous integral.

$$\overline{\sin^2(\theta)} = \frac{\int_{\theta=0}^{2\pi} \frac{1 - \cos 2\theta}{2} d\theta}{2\pi} = \frac{\int_{\theta=0}^{2\pi} \frac{1}{2} d\theta - \int_{\theta=0}^{2\pi} \frac{\cos 2\theta}{2} d\theta}{2\pi}$$

Note that the anti-derivative of $\cos 2\theta$ equals $\cos 2\theta = \frac{\sin 2\theta}{2}$, so that $\int \frac{\cos 2\theta}{2} d\theta = \frac{\sin 2\theta}{4}$.

$$\overline{\sin^2(\theta)} = \frac{\left[\frac{\theta}{2}\right]_{\theta=0}^{2\pi} - \left[\frac{\sin 2\theta}{4}\right]_{\theta=0}^{2\pi}}{2\pi} = \frac{\left(\frac{2\pi}{2} - 0\right) - \left(\frac{\sin 4\pi}{4} - \frac{\sin 0}{4}\right)}{2\pi} = \frac{\frac{2\pi}{2} - 0 - 0 + 0}{2\pi} = \frac{1}{2}$$

The average value of $\sin^2\theta$ over one cycle equals one-half. To find the rms value, take the squareroot of both sides.

$$\text{rms value of} \sin\theta = \sqrt{\overline{\sin^2(\theta)}} = \frac{1}{\sqrt{2}} = \frac{1}{\sqrt{2}}\frac{\sqrt{2}}{\sqrt{2}} = \frac{\sqrt{2}}{2}$$

In the last step, we rationalized the denominator by multiplying the numerator and denominator both by $\sqrt{2}$. If you enter them correctly on your calculator, you will see that $\frac{1}{\sqrt{2}}$ and $\frac{\sqrt{2}}{2}$ are both the same and approximately equal to 0.7071.

If we substitute $\sqrt{\overline{\sin^2(\theta)}} = \frac{1}{\sqrt{2}}$ (which is the same as saying $\sqrt{\overline{\sin^2(\theta)}} = \frac{\sqrt{2}}{2}$) into the equations at the top of the page, we get:

$$\Delta V_{rms} = \frac{\Delta V_m}{\sqrt{2}} \quad , \quad I_{rms} = \frac{I_m}{\sqrt{2}}$$

Therefore, what an AC multimeter measures is $\frac{1}{\sqrt{2}}$ (which is about 70.71%) of the amplitude of the AC current or potential difference (depending on the type of meter used).

Transformers

A **transformer** is a device consisting of two inductors, which utilizes Faraday's law to step AC voltage up or down. A typical transformer is illustrated below.

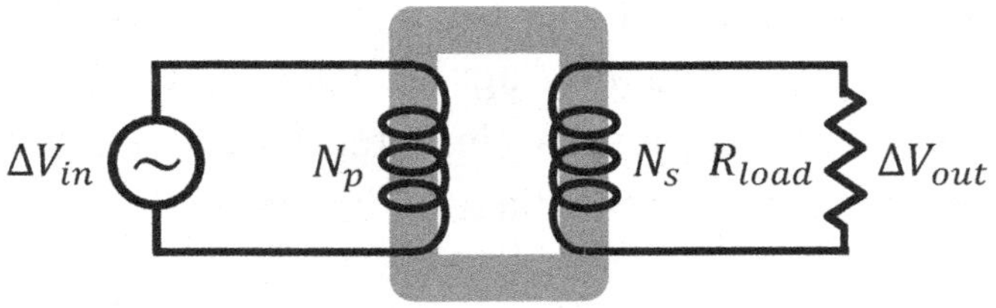

The transformer shown above consists of two inductors, called the primary and secondary. An AC power supply provides the input voltage (potential difference) to the **primary**. The alternating current in the primary creates a changing magnetic flux in the **secondary**, which induces a potential difference (and therefore also a current) in the secondary according to Faraday's law. The presence of the magnet (shown in gray above) helps to reduce losses.

The **ratio** of the potential difference in the secondary (ΔV_{out}) to the potential difference in the primary (ΔV_{in}) equals the ratio of the number of turns (or loops) in the secondary (N_s) to the number of turns in the primary (N_p).

$$\frac{\Delta V_{out}}{\Delta V_{in}} = \frac{N_s}{N_p}$$

Neglecting any losses, the **power** supplied to the primary equals the power delivered to the secondary. Recall that power (P) equals current (I) times potential difference (ΔV).

$$I_{in}\Delta V_{in} = I_{out}\Delta V_{out}$$

According to Ohm's law, $I_{in} = \frac{\Delta V_{in}}{R_{eq}}$ and $I_{out} = \frac{\Delta V_{out}}{R_{load}}$.

$$\frac{\Delta V_{in}}{R_{eq}}\Delta V_{in} = \frac{\Delta V_{out}}{R_{load}}\Delta V_{out}$$

$$\frac{\Delta V_{in}^2}{R_{eq}} = \frac{\Delta V_{out}^2}{R_{load}}$$

Since $\Delta V_{out} = \Delta V_{in}\frac{N_s}{N_p}$, this becomes:

$$\frac{\Delta V_{in}^2}{R_{eq}} = \frac{1}{R_{load}}\left(\Delta V_{in}\frac{N_s}{N_p}\right)^2 = \frac{\Delta V_{in}^2}{R_{load}}\left(\frac{N_s}{N_p}\right)^2$$

$$\frac{1}{R_{eq}} = \frac{1}{R_{load}}\left(\frac{N_s}{N_p}\right)^2$$

Reciprocate both sides of the equation.

$$R_{eq} = R_{load}\left(\frac{N_p}{N_s}\right)^2$$

This is the **equivalent resistance** of the transformer from the primary perspective.

High-pass and Low-pass Filters

The frequency dependence of capacitive reactance can be utilized to create high-pass and low-pass filters.

- A **high-pass filter** effectively blocks (or attenuates) low frequencies and lets high frequencies pass through. See the diagram below on the left.
- A **low-pass filter** effectively blocks (or attenuates) high frequencies and lets low frequencies pass through. See the diagram below on the right.

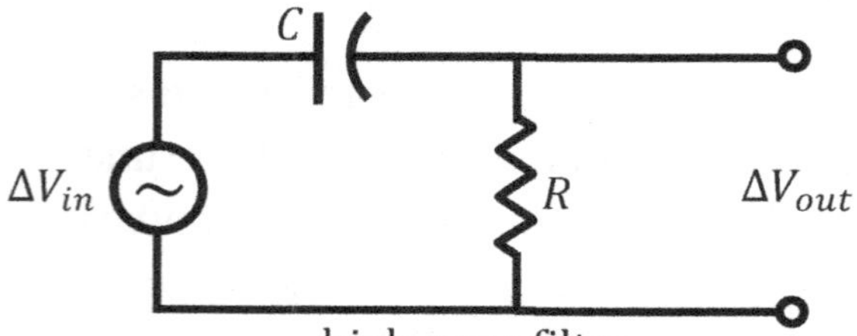

high-pass filter

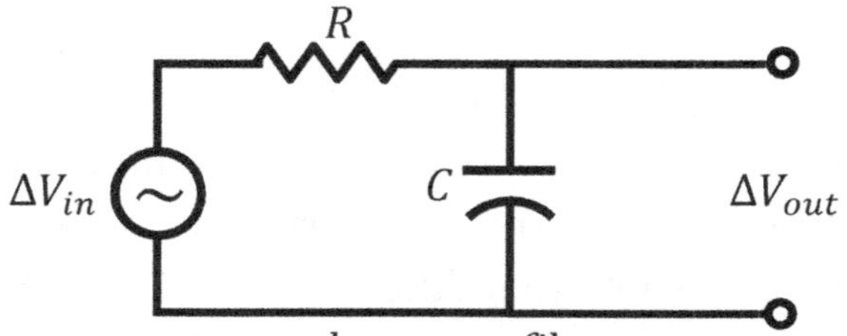

low-pass filter

In both types of filters, the input voltage equals the potential difference of the AC power supply for an RLC circuit where the inductive reactance is set equal to zero ($X_L = 0$), since there is no inductor in the circuit. Recall that capacitive reactance is $X_C = \frac{1}{\omega C}$.

$$\Delta V_{in} = I_m Z = I_m\sqrt{R^2 + (0 - X_C)^2} = I_m\sqrt{R^2 + X_C^2} = I_m\sqrt{R^2 + \left(\frac{1}{\omega C}\right)^2}$$

For a **high-pass** filter, the output voltage equals the potential difference across the resistor.

$$\Delta V_{out} = I_m R$$

$$\frac{\Delta V_{out}}{\Delta V_{in}} = \frac{I_m R}{I_m\sqrt{R^2 + \left(\frac{1}{\omega C}\right)^2}} = \frac{R}{\sqrt{R^2 + \left(\frac{1}{\omega C}\right)^2}}$$

When the angular frequency (ω) is high, the ratio $\frac{\Delta V_{in}}{\Delta V_{out}}$ is close to 100% (since $\frac{1}{\omega}$ is small for large values of ω), and when the angular frequency is (ω) low, the ratio $\frac{\Delta V_{in}}{\Delta V_{out}}$ is close to 0% (since $\frac{1}{\omega}$ is large for small values of ω, which makes the denominator of $\frac{\Delta V_{in}}{\Delta V_{out}}$ large, and a large denominator makes a fraction smaller).

For a **low-pass** filter, the output voltage equals the potential difference across the capacitor.

$$\Delta V_{out} = I_m X_C$$

$$\frac{\Delta V_{out}}{\Delta V_{in}} = \frac{I_m X_C}{I_m\sqrt{R^2 + \left(\frac{1}{\omega C}\right)^2}} = \frac{\frac{1}{\omega C}}{\sqrt{R^2 + \left(\frac{1}{\omega C}\right)^2}}$$

In this case, when the angular frequency (ω) is low, the ratio $\frac{\Delta V_{in}}{\Delta V_{out}}$ is close to 100%, and when the angular frequency is (ω) high, the ratio $\frac{\Delta V_{in}}{\Delta V_{out}}$ is close to 0%.

Important Distinctions

DC stands for **direct** current, whereas AC stands for **alternating** current. This chapter involves circuits that have an AC power supply. If a problem involves a DC power supply (or no power supply at all), see Chapter 30 instead (or Chapter 18 for an RC circuit).

The subscript m on potential difference (such as ΔV_m or ΔV_{Rm}) or on current (I_m) indicates that it's the amplitude of the corresponding sine wave (it's a peak or maximum value). (Some texts use a different notation for this, but have some way of telling which quantities are amplitudes and which are not.) Similarly, the subscripts rms (as in I_{rms}) stand for root-mean-square values. Potential differences (such as ΔV_{ps} or ΔV_R) or currents (I) which do not have a subscript m or rms are **instantaneous** values. It's very important to tell whether or not a potential difference or current is an instantaneous value or not: The reason is that some equations apply only to maximum or rms values and are not true for instantaneous values. It would be a mistake, for example, to use an equation which has subscript m's and plug in instantaneous values instead.

The "maximum current" is I_m, which means for a given value of angular frequency (ω), as the current oscillates in time according to $I = I_m \sin(\omega t - \varphi)$, the current oscillates between a "maximum" value (called the amplitude) of I_m and minimum value of $-I_m$. Since there is another sense in which current is maximized, it's important to distinguish between I_m and I_{max}. (Again, some books adopt different notation, but still have some way of telling these symbols apart.) The current I_{max} is the maximum possible value of I_m, which occurs at the resonance frequency (ω_0).

- $\Delta V_m = I_m Z$ involves the amplitude of the current and the impedance, and is true at any value of ω.
- $\Delta V_m = I_{max} R$ involves the maximum possible current and the resistance, and is true only when the angular frequency equals ω_0 (the **resonance frequency**).

Remember that we use the symbol Q_0 to represent the **quality factor** of an AC circuit. Don't confuse the quality factor with the charge (Q) stored on the capacitor.

Technically, the symbol ω is the angular frequency in rad/s, whereas the symbol f is the frequency in Hertz (Hz). The angular frequency and frequency are related by $\omega = 2\pi f$. Unfortunately, there are books which use the term "frequency," but really mean ω instead of f, or which refer to "resonance frequency," but really mean ω_0 instead of f_0. If you're using another book, read the chapter on AC circuits carefully and study the examples, as this will show you exactly what that book means by the word "frequency." (This workbook has a handy chart in each chapter to clarify exactly how we are using each symbol.)

Strategy for AC Circuits

Note: If a circuit involves a DC power supply (or no power supply), see Chapter 30 instead.

For an RL, RC, LC, or RLC circuit that has an AC power supply, the following equations apply. Choose the relevant equation based on which quantities you are given in a problem and which quantity you are solving for.

- The units following a number can help you identify the given information.
  - A value in Hertz (Hz) is **frequency** (f). If it says "**resonance**," it's f_0.
  - A value in rad/s is **angular frequency** (ω). If it says "**resonance**," it's ω_0. Another quantity measured in rad/s is the full-width at half max ($\Delta\omega$).
  - A value in seconds (s) may be the **time** (t), but could also be **period** (T).
  - A value in Henry (H) is **inductance** (L). Note that $1\text{mH} = 0.001$ H.
  - A value in Farads (F) is **capacitance** (C). Note that $\mu = 10^{-6}$ and $\text{n} = 10^{-9}$.
  - A value in Ohms (Ω) may be **resistance** (R), but could also be **impedance** (Z), **inductive reactance** (X_L), or **capacitive reactance** (X_C).
  - A value in Ampères (A) is **current**, but you must read carefully to see if it's I_m (which may be called the amplitude, maximum, or peak value), I_{rms} (the root-mean square value), I (the instantaneous value), or I_{max} (the maximum possible value of I_m, which occurs at resonance).
  - A value in Volts (V) is **potential difference** (also called **voltage**), and involves even more choices: the amplitude of the power supply voltage (ΔV_m), the amplitude for the voltage across a specific element (ΔV_{Rm}, ΔV_{Lm}, ΔV_{Cm}), an rms value (ΔV_{rms}, ΔV_{Rrms}, ΔV_{Lrms}, ΔV_{Crms}), or an instantaneous value (ΔV_{ps}, ΔV_R, ΔV_L, ΔV_C). Some books adopt different notation, such as using lowercase letters (like v_R) for instantaneous values and uppercase values (like V_R for amplitudes). When using another book, compare notation carefully. The handy chart of symbols in this chapter should aid in such a comparison.
  - A value in Watts (W) is the instantaneous **power** (P) or **average power** (P_{av}).
  - A value in radians (rad) is the **phase angle** (φ). If a value is given in degrees (°), convert it to **radians** using $180° = \pi$ rad. Check that your calculator is in radians mode (**<u>not</u>** degrees mode). That's because ω is in rad/s.
- The equations involve **angular frequency** (ω). If you're given the frequency (f) in Hertz (Hz) or period (T) in sec, first calculate the angular frequency: $\omega = 2\pi f = \frac{2\pi}{T}$.
- Note that some equations require finding the reactance before you can use them.
  - The **inductive reactance** is $X_L = \omega L$.
  - The **capacitive reactance** is $X_C = \frac{1}{\omega C}$.
- The main equation for **impedance** is:

$$Z = \sqrt{R^2 + (X_L - X_C)^2}$$

- The main equation for the **phase angle** is:

$$\varphi = \tan^{-1}\left(\frac{X_L - X_C}{R}\right)$$

- The **resonance** angular frequency (ω_0) and resonance frequency (f_0) are:

$$\omega_0 = \frac{1}{\sqrt{LC}} = 2\pi f_0$$

- If you see the word "power," use one of the following equations:
  - The **instantaneous power** (P) is: $P = I\Delta V_{ps} = I_m \Delta V_m \sin(\omega t)\sin(\omega t - \varphi)$.
  - The **average power** (P_{av}) is: $P_{av} = I_{rms}\Delta V_{rms}\cos\varphi = I_{rms}^2 R$.
  - The factor of $\cos\varphi$ is called the **power factor**.
- There are two ways to determine the **quality factor**:
  - If you know the resistance (R) and inductance (L), use the equation $Q_0 = \frac{\omega_0 L}{R}$.
  - If you're given a graph of average power (P_{av}) as a function of angular frequency (ω), read off the values of ω_0 and $\Delta\omega$ (as described and shown on page 417), and then use the equation $Q_0 = \frac{\omega_0}{\Delta\omega}$.
- The maximum (or peak) voltages (also called potential differences) can be related through the equation $\Delta V_m = \sqrt{\Delta V_{Rm}^2 + (\Delta V_{Lm} - \Delta V_{Cm})^2}$, where ΔV_m is the power supply voltage. These values are also related to the maximum current through $\Delta V_m = I_m Z$, $\Delta V_{Rm} = I_m R$, $\Delta V_{Lm} = I_m X_L$, and $\Delta V_{Cm} = I_m X_C$. These five equations can also be written with rms values instead of maximum values.
- In general, $\Delta V_m = I_m Z$, but at **resonance** this becomes $\Delta V_m = I_{max} R$, where I_m is the maximum current for a given ω and I_{max} is the maximum possible value of I_m, which occurs at the resonance frequency (ω_0).
- If you need to switch between **rms** and maximum (or peak) values, note that:

$$\Delta V_{rms} = \frac{\Delta V_m}{\sqrt{2}} \quad , \quad I_{rms} = \frac{I_m}{\sqrt{2}}$$

- The **instantaneous** voltages and currents are:

$$\Delta V_{ps} = \Delta V_m \sin(\omega t) \quad , \quad \Delta V_R = \Delta V_{Rm}\sin(\omega t)$$
$$\Delta V_L = \Delta V_{Lm}\cos(\omega t) \quad , \quad \Delta V_C = -\Delta V_{Cm}\cos(\omega t)$$
$$I = I_m \sin(\omega t - \varphi)$$

  Note that the **instantaneous** values depend on **time** (t).
- If you need to draw or work with a **phasor**, it's basically a vector for which:
  - The amplitude (ΔV_m) of the potential difference is the magnitude of the phasor. (You could use rms values instead of peak values.)
  - The phase angle serves as the direction of the phasor. Measure the angle counterclockwise from the $+x$-axis.
  - The phase angle (φ) equals 90° for an inductor, 0° for a resistor, and −90° for a capacitor.

  Use the technique of **vector addition** (like in Chapter 3) to add phasors together.

- If a problem involves a **transformer**, see the next strategy.
- If a problem involves a **high-pass** or **low-pass filter**, see the last strategy below.
- For a general circuit, you can derive the needed equations for charge and current by applying Kirchhoff's rules. For the loop rule, $\Delta V_R = IR$, $\Delta V_C = \frac{Q}{C}$, and $\Delta V_L = -L\frac{dI}{dt}$. Use the equations $I = \frac{dQ}{dt}$ and $\frac{dI}{dt} = \frac{d^2Q}{dt^2}$, if necessary, to write the loop rule exclusively with current or with charge (not a combination of the two). Then solve the differential equation. We did an example of this in Chapter 18 for an RC circuit.

Strategy for Transformers

To solve a problem involving a transformer:

- The ratio of the **secondary** voltage (ΔV_{out}) to the **primary** voltage (ΔV_{in}) equals the ratio of the number of turns (or loops) in the secondary (N_s) to the number of turns in the primary (N_p):

$$\frac{\Delta V_{out}}{\Delta V_{in}} = \frac{N_s}{N_p}$$

- The **equivalent resistance** of the transformer from the primary perspective is:

$$R_{eq} = R_{load}\left(\frac{N_p}{N_s}\right)^2$$

Strategy for High-pass or Low-pass Filters

To solve a problem with a simple high-pass or low-pass filter:

- For the high-pass filter shown on the left of page 420:

$$\frac{\Delta V_{out}}{\Delta V_{in}} = \frac{R}{\sqrt{R^2 + \left(\frac{1}{\omega C}\right)^2}}$$

- For the low-pass filter shown on the right of page 420:

$$\frac{\Delta V_{out}}{\Delta V_{in}} = \frac{\frac{1}{\omega C}}{\sqrt{R^2 + \left(\frac{1}{\omega C}\right)^2}}$$

Example: An AC power supply that provides an rms voltage of 20 V and operates at an angular frequency of 40 rad/s is connected in series with a 1.5-kΩ resistor, a 40-H inductor, and a 250-μF capacitor.

Begin by making a list of the given quantities in SI units:

- The rms voltage is $\Delta V_{rms} = 20$ V. This relates to the power supply.
- The angular frequency is $\omega = 40$ rad/s. **Note**: It's **not** ω_0 (that is, it isn't resonance).
- The resistance is $R = 1500\ \Omega$. The metric prefix kilo (k) stands for 1000.
- The inductance is $L = 40$ H.
- The capacitance is $C = 2.5 \times 10^{-4}$ F. The metric prefix micro (μ) stands for 10^{-6}.

(A) What is the amplitude of the AC power supply voltage?
We're solving for ΔV_m. Find the equation that relates ΔV_{rms} to ΔV_m.

$$\Delta V_{rms} = \frac{\Delta V_m}{\sqrt{2}}$$

Multiply both sides of the equation by $\sqrt{2}$.

$$\Delta V_m = \Delta V_{rms}\sqrt{2} = (20)(\sqrt{2}) = 20\sqrt{2} \text{ V}$$

The amplitude of the AC power supply voltage is $\Delta V_m = 20\sqrt{2}$ V. If you use a calculator, you get $\Delta V_m = 28$ V to two significant figures.
(B) What is the frequency of the AC power supply in Hertz?
Use the equation that relates angular frequency to frequency.

$$\omega = 2\pi f$$

Divide both sides of the equation by 2π.

$$f = \frac{\omega}{2\pi} = \frac{40}{2\pi} = \frac{20}{\pi} \text{ Hz}$$

The frequency is $f = \frac{20}{\pi}$ Hz. Using a calculator, this comes out to $f = 6.4$ Hz.
(C) What is the inductive reactance?
Use the equation that relates inductive reactance to inductance.

$$X_L = \omega L = (40)(40) = 1600\ \Omega$$

The inductive reactance is $X_L = 1600\ \Omega$.
(D) What is the capacitive reactance?
Use the equation that relates capacitive reactance to capacitance.

$$X_C = \frac{1}{\omega C} = \frac{1}{(40)(2.5 \times 10^{-4})} = \frac{1}{100 \times 10^{-4}} = \frac{1}{10^{-2}} = 10^2 = 100\ \Omega$$

The capacitive reactance is $X_C = 100\ \Omega$.
(E) What is the impedance of the RLC circuit?
Use the equation that relates impedance to resistance and reactance.

$$Z = \sqrt{R^2 + (X_L - X_C)^2} = \sqrt{1500^2 + (1600 - 100)^2} = \sqrt{1500^2 + 1500^2} = 1500\sqrt{2}\ \Omega$$

The impedance is $Z = 1500\sqrt{2}\ \Omega$. Using a calculator, this comes out to $Z = 2.1$ kΩ, where the metric prefix kilo (k) stands for 1000.

(F) What is the phase angle for the current with respect to the power supply voltage?
Use the equation that relates the phase angle to resistance and reactance.

$$\varphi = \tan^{-1}\left(\frac{X_L - X_C}{R}\right) = \tan^{-1}\left(\frac{1600 - 100}{1500}\right) = \tan^{-1}\left(\frac{1500}{1500}\right) = \tan^{-1}(1) = 45°$$

The phase angle is $\varphi = 45°$. Since the phase angle is **positive**, the current **lags** behind the power supply's potential difference by 45° (one-eighth of a cycle, since 360° is full-cycle).

(G) What is the rms current for the RLC circuit?
One way to do this is to first solve for the amplitude (I_m) of the current from the amplitude (ΔV_m) of the potential difference of the power supply and the impedance (Z). Recall that we found $\Delta V_m = 20\sqrt{2}$ V in part (A) and $Z = 1500\sqrt{2}\ \Omega$ in part (E).

$$\Delta V_m = I_m Z$$

Divide both sides of the equation by Z.

$$I_m = \frac{\Delta V_m}{Z} = \frac{20\sqrt{2}}{1500\sqrt{2}} = \frac{1}{75}\text{ A}$$

Now use the equation that relates I_{rms} to I_m.

$$I_{rms} = \frac{I_m}{\sqrt{2}} = \frac{1}{75\sqrt{2}} = \frac{1}{75\sqrt{2}}\frac{\sqrt{2}}{\sqrt{2}} = \frac{\sqrt{2}}{150}\text{ A}$$

Note that we multiplied by $\frac{\sqrt{2}}{\sqrt{2}}$ in order to **rationalize the denominator**. Also note that $\sqrt{2}\sqrt{2} = 2$. The rms current is $I_{rms} = \frac{\sqrt{2}}{150}$ A. If you use a calculator, this works out to $I_{rms} = 0.0094$ A, which can also be expressed as $I_{rms} = 9.4$ mA since the prefix milli (m) stands for $\text{m} = 10^{-3}$. An alternative way to solve this problem would be to use the equation $\Delta V_{rms} = I_{rms} Z$ instead of $\Delta V_m = I_m Z$. You would get the same answer either way (provided that you do the math correctly).

(H) What frequency (in Hz) should the power supply be adjusted to in order to create the maximum possible current?
Qualitatively, the answer is the **resonance frequency** (f_0). Quantitatively, first solve for ω_0 from the inductance and capacitance.

$$\omega_0 = \frac{1}{\sqrt{LC}} = \frac{1}{\sqrt{(40)(2.5 \times 10^{-4})}} = \frac{1}{\sqrt{100 \times 10^{-4}}} = \frac{1}{\sqrt{10^{-2}}} = \frac{1}{10^{-1}} = 10\text{ rad/s}$$

Technically, ω_0 is the "resonance angular frequency" (though it's not uncommon for books or scientists to say "resonance frequency" and really mean f_0 - however, this problem specifically says "in Hz," so we're definitely looking for f_0, **not** ω_0 since the units of ω_0 are rad/s). To find f_0, use the following equation.

$$\omega_0 = 2\pi f_0$$

Divide both sides of the equation by 2π.

$$f_0 = \frac{\omega_0}{2\pi} = \frac{10}{2\pi} = \frac{5}{\pi}\text{ Hz}$$

The resonance frequency in Hertz is $f_0 = \frac{5}{\pi}$ Hz. Using a calculator, it is $f_0 = 1.6$ Hz.

(I) What is the maximum possible amplitude of the current that could be obtained by adjusting the frequency of the power supply?

The answer is **not** the $I_m = \frac{1}{75}$ A (which equates to $\approx$0.013 A) that we found as an intermediate answer in part (G). The symbol I_m is the maximum value of the instantaneous current for a given frequency (the current oscillates between I_m and $-I_m$), but we can get an even larger value of I_m using a different frequency. Which frequency gives you the most current? The answer is the resonance frequency. We found the resonance angular frequency to be $\omega_0 = 10$ rad/s in part (H), but we don't actually need this number: The resonance frequency is the frequency for which $X_L = X_C$. When $X_L = X_C$, the impedance equals the resistance, as expressed in the following equation.

$$\Delta V_m = I_{max} R$$

Note that for any frequency, $\Delta V_m = I_m Z$, but for the resonance frequency, this simplifies to $\Delta V_m = I_{max} R$ (since Z equals R at resonance). The symbol I_{max} represents the maximum possible value of I_m. Divide both sides of the equation by R.

$$I_{max} = \frac{\Delta V_m}{R}$$

Recall that we found $\Delta V_m = 20\sqrt{2}$ V in part (A).

$$I_{max} = \frac{20\sqrt{2}}{1500} = \frac{\sqrt{2}}{75} \text{ A}$$

The maximum possible current that could be obtained by adjusting the frequency is $I_{max} = \frac{\sqrt{2}}{75}$ A, and this occurs when the frequency is adjusted to $f_0 = \frac{5}{\pi}$ Hz $= 1.6$ Hz (or when the angular frequency equals $\omega_0 = 10$ rad/s). If you use a calculator, the answer works out to $I_{max} = 0.019$ A $= 19$ mA. As a check, note that this value is larger than the value of $I_m = \frac{1}{75}$ A $= 13$ mA that we found in part (G). Just to be clear, our **final answer** to this part of the problem is $I_{max} = \frac{\sqrt{2}}{75}$ A $= 0.019$ A $= 19$ mA.

(J) What is the minimum possible impedance that could be obtained by adjusting the frequency of the power supply?

The maximum current and minimum impedance go hand in hand: Both occur at resonance. It should make sense: Less impedance results in more current (for a fixed amplitude of potential difference). As we discussed in part (I), at resonance, the impedance equals the resistance. There is no math to do! Just write the following (well, if you're taking a class, it would also be wise, and perhaps required, for you to explain your answer):

$$Z_{min} = R = 1500\ \Omega$$

The minimum possible impedance that could be obtained by adjusting the frequency is $Z_{min} = 1500\ \Omega$, and this occurs when the frequency equals $f_0 = \frac{5}{\pi}$ Hz $= 1.6$ Hz (or when the angular frequency equals $\omega_0 = 10$ rad/s). **Tip**: Since the minimum possible impedance equals the resistance, for any problem in which you calculate Z, check that your answer for Z is greater than R (otherwise, you know that you made a mistake).

(K) What is the average power delivered by the AC power supply?
Use one of the equations for average power.

$$P_{av} = I_{rms}\Delta V_{rms}\cos\varphi = I_{rms}^2 R$$

Either equation will work. Recall that we found $I_{rms} = \frac{\sqrt{2}}{150}$ A in part (G) and $\varphi = 45°$ in part (F), while $\Delta V_{rms} = 20$ V was given in the problem.

$$P_{av} = I_{rms}\Delta V_{rms}\cos\varphi = \left(\frac{\sqrt{2}}{150}\right)(20)\cos 45° = \left(\frac{\sqrt{2}}{150}\right)(20)\left(\frac{\sqrt{2}}{2}\right) = \frac{(2)(20)}{300} = \frac{40}{300} = \frac{2}{15}\text{ W}$$

Note that $\sqrt{2}\sqrt{2} = 2$. The average power that the AC power supply delivers to the circuit is $P_{av} = \frac{2}{15}$ W. If you use a calculator, this comes out to $P_{av} = 0.13$ W, which could also be expressed as $P_{av} = 130$ mW using the metric prefix milli (m), since m $= 10^{-3}$. Note that you would get the same answer using the other equation:

$$P_{av} = I_{rms}^2 R = \left(\frac{\sqrt{2}}{150}\right)^2 (1500) = \frac{2}{15}\text{ W}$$

(L) What is the quality factor for the RLC circuit described in the problem?
Use the equation for quality factor (Q_0) that involves the resonance angular frequency (ω_0), resistance (R), and inductance (L).

$$Q_0 = \frac{\omega_0 L}{R} = \frac{(10)(40)}{1500} = \frac{400}{1500} = \frac{4}{15}$$

The quality factor for this RLC circuit is $Q_0 = \frac{4}{15}$. (Note that the numbers in this problem are **not** typical of most RLC circuits. We used numbers that made the arithmetic simpler so that you could focus more on the strategy and follow along more easily. A typical RLC circuit has a much higher quality factor – above 10, perhaps as high as 100. One of the unusual things about this problem is that the resistance, which is 1500 Ω, is very large.)

(M) What would an AC voltmeter measure across the resistor?
The AC voltmeter would measure the rms value across the resistor, which is ΔV_{Rrms}. Note that we can rewrite the equation $\Delta V_{Rm} = I_m R$ in terms of rms values as:†

$$\Delta V_{Rrms} = I_{rms} R = \left(\frac{\sqrt{2}}{150}\right)(1500) = 10\sqrt{2}\text{ V}$$

Recall that we found $I_{rms} = \frac{\sqrt{2}}{150}$ A in part (G). An AC voltmeter would measure the voltage across the resistor to be $\Delta V_{Rrms} = 10\sqrt{2}$ V. Using a calculator, it is $\Delta V_{Rrms} = 14$ V.

(N) What would an AC voltmeter measure across the inductor?
The AC voltmeter would measure the rms value across the inductor, which is ΔV_{Lrms}. Note that we can rewrite the equation $\Delta V_{Lm} = I_m X_L$ in terms of rms values as:

$$\Delta V_{Lrms} = I_{rms} X_L = \left(\frac{\sqrt{2}}{150}\right)(1600) = \frac{32\sqrt{2}}{3}\text{ V}$$

† The reason this works is that $\Delta V_{rms} = \frac{\Delta V_m}{\sqrt{2}}$ and $I_{rms} = \frac{I_m}{\sqrt{2}}$. When you substitute these equations into $\Delta V_{Rm} = I_m R$, the $\sqrt{2}$'s cancel out and you get $\Delta V_{Rrms} = I_{rms} R$.

An AC voltmeter would measure the voltage across the inductor to be $\Delta V_{Lrms} = \frac{32\sqrt{2}}{3}$ V. Using a calculator, it is $\Delta V_{Lrms} = 15$ V.

(O) What would an AC voltmeter measure across the capacitor?

The AC voltmeter would measure the rms value across the capacitor, which is ΔV_{Crms}. Note that we can rewrite the equation $\Delta V_{Cm} = I_m X_C$ in terms of rms values as

$$\Delta V_{Crms} = I_{rms} X_C = \left(\frac{\sqrt{2}}{150}\right)(100) = \frac{2\sqrt{2}}{3} \text{ V}$$

An AC voltmeter would measure the voltage across the inductor to be $\Delta V_{Crms} = \frac{2\sqrt{2}}{3}$ V. Using a calculator, it is $\Delta V_{Crms} = 0.94$ V.

(P) What would an AC voltmeter measure if the two probes were connected across the inductor-capacitor combination?

We must do phasor addition in order to figure this out. See the discussion of phasors on pages 414-415. The phase angle for the inductor is 90°, so the phasor for the inductor points straight up. The phase angle for the capacitor is −90°, so the phasor for the capacitor points straight down. Since these two phasors are opposite, we subtract the rms potential differences across the inductor and capacitor:

$$\Delta V_{LCrms} = |\Delta V_{Lrms} - \Delta V_{Crms}| = \left|\frac{32\sqrt{2}}{3} - \frac{2\sqrt{2}}{3}\right| = \frac{30\sqrt{2}}{3} = 10\sqrt{2} \text{ V}$$

An AC voltmeter would measure the voltage across the inductor-capacitor combination to be $\Delta V_{LCrms} = 10\sqrt{2}$ V. Using a calculator, it is $\Delta V_{LCrms} = 14$ V.

(Q) Show that the answers to parts (M) through (O) are consistent with the 20-V rms voltage supplied by the AC power supply.

Use the following equation to check the rms value of the AC power supply.

$$\Delta V_{rms} = \sqrt{\Delta V_{Rrms}^2 + (\Delta V_{Lrms} - \Delta V_{Crms})^2} = \sqrt{\left(10\sqrt{2}\right)^2 + \left(\frac{32\sqrt{2}}{3} - \frac{2\sqrt{2}}{3}\right)^2}$$

$$\Delta V_{rms} = \sqrt{\left(10\sqrt{2}\right)^2 + \left(\frac{30\sqrt{2}}{3}\right)^2} = \sqrt{\left(10\sqrt{2}\right)^2 + \left(10\sqrt{2}\right)^2} = \sqrt{200 + 200} = \sqrt{400} = 20 \text{ V}$$

This checks out with the value of $\Delta V_{rms} = 20$ V given in the problem.

It's instructive to note that it would be **incorrect** to add the answers for ΔV_{Rrms}, ΔV_{Lrms}, and ΔV_{Crms} together. If you did that, you would get the incorrect answer of 30 V. The correct way to combine these values is through phasor addition, which accounts for the direction (or phase angle) of each phasor. The equation $\Delta V_{rms} = \sqrt{\Delta V_{Rrms}^2 + (\Delta V_{Lrms} - \Delta V_{Crms})^2}$ combines the values together correctly, as explained on pages 414-415.

Example: What is the quality factor for the graph of average power shown below?

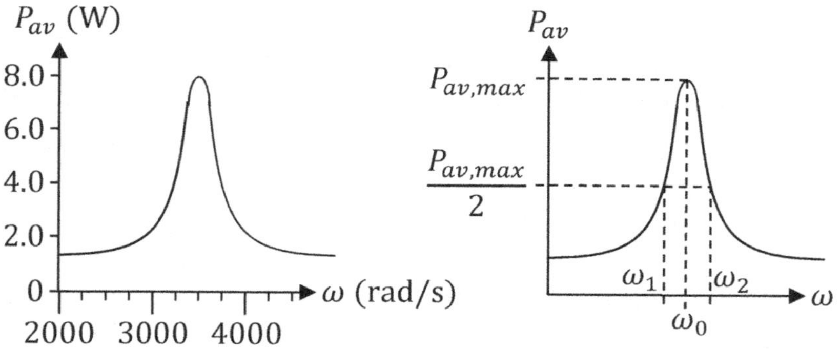

Read off the needed values from the graph:

- The resonance angular frequency is $\omega_0 = 3500$ rad/s. This is the angular frequency for which the curve reaches its peak.
- The maximum average power is $P_{av,max} = 8.0$ W. This is the maximum vertical value of the curve.
- One-half of the maximum average power is $\frac{P_{av,max}}{2} = \frac{8}{2} = 4.0$ W.
- Draw a horizontal line on the graph where the vertical value of the curve is equal to 4.0 W, which corresponds to $\frac{P_{av,max}}{2}$.
- Draw two vertical lines on the graph where the curve intersects the horizontal line that you drew in the previous step. See the right figure above. These two vertical lines correspond to ω_1 and ω_2.
- Read off ω_1 and ω_2 from the graph: $\omega_1 = 3250$ rad/s and $\omega_2 = 3750$ rad/s.
- Subtract these values: $\Delta\omega = \omega_2 - \omega_1 = 3750 - 3250 = 500$ rad/s. This is the **full-width at half max**.

Use the equation for quality factor that involves $\Delta\omega$ and ω_0.

$$Q_0 = \frac{\omega_0}{\Delta\omega} = \frac{3500}{500} = 7.0$$

Example: An AC power supply with an rms voltage of 120 V is connected across the primary coil of a transformer. The transformer has 300 turns in the primary coil and 60 turns in the secondary coil. What is the rms output voltage?

Use the ratio equation for a **transformer**.

$$\frac{\Delta V_{out}}{\Delta V_{in}} = \frac{N_s}{N_p}$$

Cross-multiply.

$$\Delta V_{out} N_p = \Delta V_{in} N_s$$

$$\Delta V_{out} = \frac{\Delta V_{in} N_s}{N_p} = \frac{(120)(60)}{300} = 24 \text{ V}$$

130. The current varies as a sine wave in an AC circuit with an amplitude of 2.0 A. What is the rms current?

131. An AC circuit where the power supply's frequency is set to 50 Hz includes a 30-mH inductor. What is the inductive reactance?

132. An AC circuit where the power supply's frequency is set to 100 Hz includes a 2.0-μF capacitor. What is the capacitive reactance?

Want help? Check the hints section at the back of the book.

Answers: $\sqrt{2}$ A, 3π Ω, $\frac{2500}{\pi}$ Ω

133. An AC power supply that operates at an angular frequency of 25 rad/s is connected in series with a $100\sqrt{3}$-Ω resistor, a 6.0-H inductor, and an 800-μF capacitor.

(A) What is the impedance for this RLC circuit?

(B) What is the phase angle for the current with respect to the power supply voltage?

Want help? Check the hints section at the back of the book.

Answers: 200 Ω, 30°

134. An AC power supply that provides an rms voltage of 200 V and operates at an angular frequency of 50 rad/s is connected in series with a $50\sqrt{3}$-Ω resistor, a 3.0-H inductor, and a 100-μF capacitor.

(A) What would an AC ammeter measure for this RLC circuit?

(B) What would an AC voltmeter measure across the resistor?

(C) What would an AC voltmeter measure across the inductor?

(D) What would an AC voltmeter measure across the capacitor?

(E) What would an AC voltmeter measure if the two probes were connected across the inductor-capacitor combination?

(F) Show that the answers to parts (B) through (D) are consistent with the 200-V rms voltage supplied by the AC power supply.

Want help? Check the hints section at the back of the book.

Answers: 2.0 A, $100\sqrt{3}$ V, 300 V, 400 V, 100 V

135. An AC power supply is connected in series with a $100\sqrt{3}$-Ω resistor, an 80-mH inductor, and a 50-μF capacitor.

(A) What angular frequency would produce resonance for this RLC circuit?

(B) What is the resonance frequency in Hertz?

Want help? Check the hints section at the back of the book.

Answers: 500 rad/s, $\frac{250}{\pi}$ Hz

136. An AC power supply that provides an rms voltage of 120 V is connected in series with a 30-Ω resistor, a 60-H inductor, and a 90-μF capacitor.

(A) What is the maximum possible rms current that the AC power supply could provide to this RLC circuit?

(B) What is the minimum possible impedance for this RLC circuit?

Want help? Check the hints section at the back of the book.

Answers: 4.0 A, 30 Ω

137. An AC power supply that provides an rms voltage of 3.0 kV and operates at an angular frequency of 25 rad/s is connected in series with a $500\sqrt{3}$-Ω resistor, an 80-H inductor, and an 80-μF capacitor.

(A) What is the rms current for this RLC circuit?

(B) What is the average power delivered by the AC power supply?

(C) What angular frequency would produce resonance for this RLC circuit?

(D) What is the maximum possible rms current that could be obtained by adjusting the frequency of the power supply?

Want help? Check the hints section at the back of the book.

Answers: $\sqrt{3}$ A, $1500\sqrt{3}$ W, 12.5 rad/s, $2\sqrt{3}$ A

138. An AC power supply is connected in series with a 2.0-Ω resistor, a 30-mH inductor, and a 12-μF capacitor. What is the quality factor for this RLC circuit?

139. What is the quality factor for the graph of average power shown below?

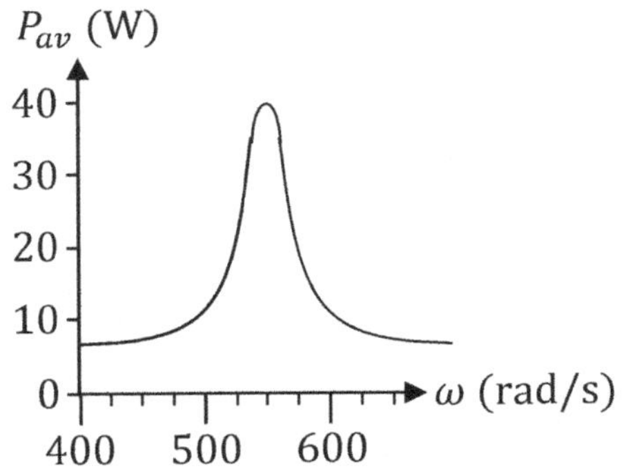

Want help? Check the hints section at the back of the book.

Answers: 25, 11

140. An AC power supply with an rms voltage of 240 V is connected across the primary coil of a transformer. The transformer has 400 turns in the primary coil and 100 turns in the secondary coil. What is the rms output voltage?

141. An AC power supply with an rms voltage of 40 V is connected across the primary coil of a transformer. The transformer has 200 turns in the primary coil. How many turns are in the secondary coil if the rms output voltage is 120 V?

Want help? Check the hints section at the back of the book.

Answers: 60 V, 600 turns

32 MAXWELL'S EQUATIONS

Relevant Terminology

Electric charge – a fundamental property of a particle that causes the particle to experience a force in the presence of an electric field. (An electrically neutral particle has no charge and thus experiences no force in the presence of an electric field.)
Velocity – a combination of speed and direction.
Current – the instantaneous rate of flow of charge through a wire.
Potential difference – the electric work per unit charge needed to move a test charge between two points in a circuit. Potential difference is also called the **voltage**.
Emf – the potential difference that a battery or DC power supply would supply to a circuit neglecting its internal resistance.
Electric field – electric force per unit charge.
Magnetic field – a magnetic effect created by a moving charge (or current).
Magnetic force – the push or pull that a moving charge (or current) experiences in the presence of a magnetic field.
Electric flux – a measure of the relative number of electric field lines that pass through a surface.
Magnetic flux –a measure of the relative number of magnetic field lines that pass through a surface.
Dielectric – a nonconducting material that can sustain an electric field.
Permittivity – a measure of how a dielectric material affects an electric field.
Permeability – a measure of how a substance affects a magnetic field.
Magnetic monopole – a hypothetical particle that could create magnetic fields similar to the way that a stationary charge creates electric fields.
Displacement current – the changing electric flux in a region of discontinuity (such as a capacitor) in a circuit produces magnetic field lines like an ordinary current. The term "displacement current" refers to the changing electric flux in such a discontinuity.
Poynting vector – the power per unit area of an electromagnetic wave.
Electric potential energy – a measure of how much electrical work a charged particle can do by changing position.
Magnetic energy – a measure of how much magnetic work an inductor could do based on the current running through its loops.
Electric potential – electric potential energy per unit charge. Specifically, the electric potential difference between two points equals the electric work per unit charge that would be involved in moving a charged particle from one point to the other.

Maxwell's Equations in Integral Form (in Vacuum)

Gauss's law in electricity (Chapter 8) states that the **net electric flux** ($\Phi_{e,net} = \oint_S \vec{\mathbf{E}} \cdot d\vec{\mathbf{A}}$) is proportional to the **charge enclosed** by the closed Gaussian surface.

$$\oint_S \vec{\mathbf{E}} \cdot d\vec{\mathbf{A}} = \frac{q_{enc}}{\epsilon_0}$$

Gauss's law in magnetism states that the **net magnetic flux** ($\Phi_{m,net} = \oint_S \vec{\mathbf{B}} \cdot d\vec{\mathbf{A}}$) always equals **zero**. That's because only moving charges create magnetic fields. Nobody has ever discovered a magnetic **monopole** – the hypothetical magnetic equivalent of electric charge.

$$\oint_S \vec{\mathbf{B}} \cdot d\vec{\mathbf{A}} = 0$$

Faraday's law (Chapter 29) states that a **changing magnetic flux** $\left(\frac{d\Phi_m}{dt}\right)$ through a loop of wire induces an emf ($\varepsilon_{ind} = \oint_C \vec{\mathbf{E}} \cdot d\vec{\mathbf{s}}$) in the loop of wire.

$$\oint_C \vec{\mathbf{E}} \cdot d\vec{\mathbf{s}} = -\frac{d\Phi_m}{dt}$$

Ampère's law (Chapter 27) states that the **line integral of magnetic field** ($\oint_C \vec{\mathbf{B}} \cdot d\vec{\mathbf{s}}$) is proportional to the **current enclosed** (I_{enc}) by the Ampèrian loop. Maxwell generalized Ampère's law to account for a **displacement current** (I_d), which is associated with a **changing electric flux** $\left(\frac{d\Phi_e}{dt}\right)$: $I_d = \epsilon_0 \frac{d\Phi_e}{dt}$.

$$\oint_C \vec{\mathbf{B}} \cdot d\vec{\mathbf{s}} = \mu_0 I_{enc} + \epsilon_0 \mu_0 \frac{d\Phi_e}{dt}$$

Maxwell's four equations, which we have written in integral form, describe electromagnetism. With Faraday's law, we see that a changing magnetic flux induces an **electric field** (which causes an emf and current to be induced in a loop of wire). Similarly, the generalized form of Ampère's law shows that a changing electric flux induces a **magnetic field**. Maxwell's equations can be applied to describe **electromagnetic waves** (or **light**).

Lorentz Force

Recall from Chapter 2 that a charged particle in the presence of an electric field ($\vec{\mathbf{E}}$) experiences an **electric** force $\vec{\mathbf{F}}_e = q\vec{\mathbf{E}}$, and recall from Chapter 24 that a moving charge in the presence of a magnetic field ($\vec{\mathbf{B}}$) experiences a **magnetic** force $\vec{\mathbf{F}}_m = q\vec{\boldsymbol{v}} \times \vec{\mathbf{B}}$. When a moving charge is in the presence of both electric and magnetic fields, it experiences a Lorentz force, which combines these two equations together.

$$\vec{\mathbf{F}}_{net} = q\vec{\mathbf{E}} + q\vec{\boldsymbol{v}} \times \vec{\mathbf{B}}$$

Symbols and SI Units

Symbol	Name	SI Units
q	charge	C
I	current	A
$\vec{\mathbf{E}}$	electric field	N/C or V/m
$\vec{\mathbf{B}}$	magnetic field	T
$\vec{\mathbf{S}}$	the Poynting vector	$\mathrm{W/m^2}$
ϵ_0	permittivity of free space	$\frac{\mathrm{C^2}}{\mathrm{N \cdot m^2}}$ or $\frac{\mathrm{C^2 \cdot s^2}}{\mathrm{kg \cdot m^3}}$
μ_0	permeability of free space	$\frac{\mathrm{T \cdot m}}{\mathrm{A}}$
$d\vec{\mathbf{s}}$	differential displacement (vector)	m
$d\vec{\mathbf{A}}$	differential area element (vector)	$\mathrm{m^2}$
t	time	s
$\vec{\mathbf{F}}_e$	electric force	N
$\vec{\mathbf{F}}_m$	magnetic force	N
Φ_e	electric flux	$\frac{\mathrm{N \cdot m^2}}{\mathrm{C}}$ or $\frac{\mathrm{kg \cdot m^3}}{\mathrm{C \cdot s^2}}$
Φ_m	magnetic flux	$\mathrm{T \cdot m^2}$ or Wb
ε_{ind}	induced emf	V
ρ	volume charge density	$\mathrm{C/m^3}$
$\vec{\mathbf{J}}$	current density (distributed throughout a volume)	$\mathrm{A/m^2}$
$\vec{v}$	velocity	m/s
c	speed of light in vacuum	m/s
U	energy	J
u	energy density	$\mathrm{J/m^3}$

Gauss's Law in Magnetism

Gauss's law applies to both electricity and magnetism, except that the right-hand side is zero in the case of magnetism.

$$\oint_S \vec{\mathbf{E}} \cdot d\vec{\mathbf{A}} = \frac{q_{enc}}{\epsilon_0}$$

$$\oint_S \vec{\mathbf{B}} \cdot d\vec{\mathbf{A}} = 0$$

The reason for this is that there evidently is no such thing as a magnetic **monopole** – a hypothetical particle that would serve as sort of a "magnetic charge," analogous to electric charge. When you study magnets macroscopically, at first they appear to consist of two separate poles (north and south), but it turns out that such north and south poles always come in pairs. If you cut a magnet in half, you **don't** get two pieces that each have only one pole: Instead, you get two new smaller magnets, where each new magnet has both a north and south pole. (If you keep cutting the magnet in half forever, eventually you will have to split a single atom.) As explained in Chapter 20, **all magnetic fields**, including those created by magnets, **are produced by moving charges**. A magnet doesn't really have a north and south pole inside of it: Rather, each atom in the magnet produces its own tiny magnetic field, and these atomic magnetic fields create the net magnetic field of the magnet.

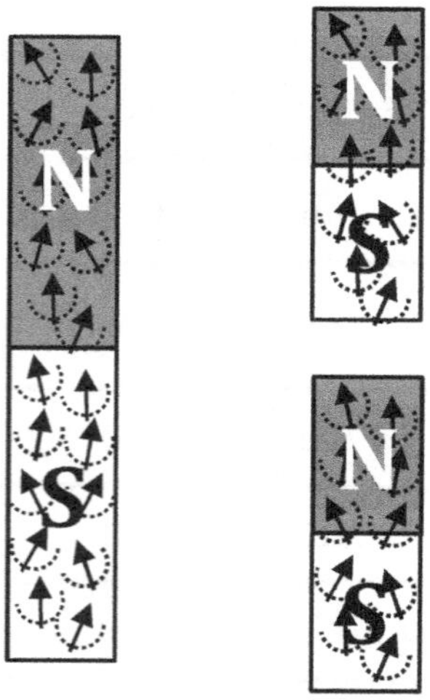

It is instructive to compare the magnetic field lines for a bar magnet (Chapter 20) to the electric field lines of an electric dipole (Chapter 4). Outside of a bar magnet, the magnetic field lines resemble the electric field lines of an electric dipole. However, inside of the magnet, the magnetic field lines look much different. According to Gauss's law in magnetism, **the net magnetic flux through any closed surface is always zero**: The same number of **magnetic** field lines always enter and exit any closed surface.

Compare the electric field map for the electric dipole shown below on the left with the magnetic field map for a bar magnet shown below on the right.

- In the electric field map, if a closed surface surrounds the negative charge (as in surface X drawn below), there is a net electric flux through the surface, but if a closed surface surrounds zero charge (or a net charge of zero), the net electric flux through the surface is zero (as in surface W drawn below).
- In the magnetic field map, the net magnetic flux through any closed surface is zero (see surfaces Y and Z below).
- Pay close attention to the distinction between surfaces X and Z in the two different diagrams: The net electric flux through surface X is negative, while the net magnetic flux through surface Z is zero. (The net flux through surfaces W and Y is also zero.)

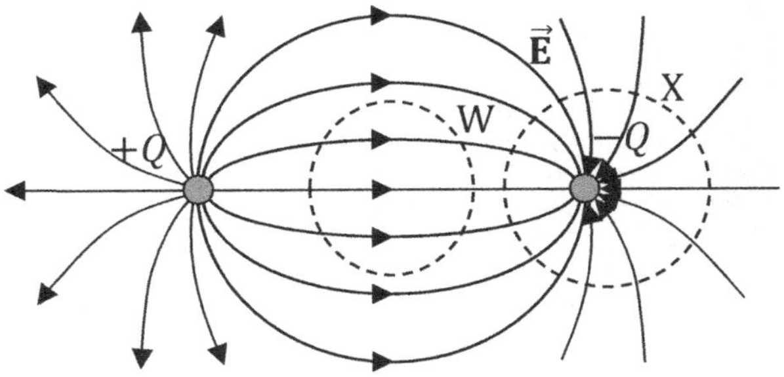

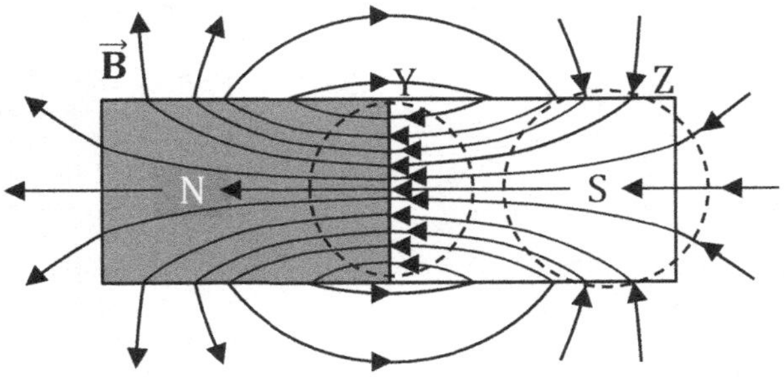

Because the right-hand side of Gauss's law is zero for a magnetic field, Gauss's law isn't as practical for deriving equations for magnetic field as it is for deriving equations for electric field (like we did in Chapter 8). However, if you want to derive an equation for magnetic field, we have Ampère's law (Chapter 27) for that, or the law of Biot-Savart (Chapter 26).

Stokes' Theorem

Faraday's law and Ampère's law in integral form are two applications of Stokes' theorem.

$$\oint_C \vec{\mathbf{E}} \cdot d\vec{\mathbf{s}} = -\frac{d\Phi_m}{dt}$$

$$\oint_C \vec{\mathbf{B}} \cdot d\vec{\mathbf{s}} = \mu_0 I_{enc} + \epsilon_0 \mu_0 \frac{d\Phi_e}{dt}$$

Stokes' theorem is a powerful geometric formula like Gauss's law (Chapter 8), except that:

- Stokes' theorem involves a line integral, $\oint_C \vec{\mathbf{E}} \cdot d\vec{\mathbf{s}}$ or $\oint_C \vec{\mathbf{B}} \cdot d\vec{\mathbf{s}}$, over a closed **path**, whereas Gauss's law involves an integral, $\oint_S \vec{\mathbf{E}} \cdot d\vec{\mathbf{A}}$ or $\oint_S \vec{\mathbf{B}} \cdot d\vec{\mathbf{A}}$, over a closed **surface**.
- The right-hand side of Stokes' theorem involves a source that creates circulating field lines (e.g. magnetic field lines **circulate** around a current, as shown in Chapter 22), whereas the right-hand side of Gauss's law is nonzero if the source creates radial field lines (e.g. electric field lines that **radiate** from a point charge).

The Generalized Form of Ampère's Law

Maxwell generalized Ampère's law to account for a gap in a circuit, such as the space between the plates of a capacitor. For example, consider the very long straight (horizontal) wire illustrated below, carrying a steady current (I), which features a parallel-plate capacitor with two large, closely spaced plates of uniform charge density. We applied Ampère's law ($\oint_C \vec{\mathbf{B}} \cdot d\vec{\mathbf{s}} = \mu_0 I_{enc}$) to a very long straight wire in Chapter 27, where we found that the left-hand side was $2\pi r_c B$ for an Ampèrian circle coaxial with the current.

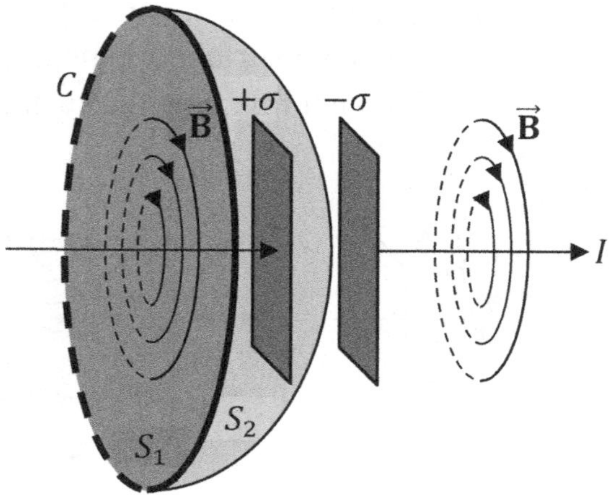

The right-hand side of Ampère's law includes $\mu_0 I_{enc}$. The current enclosed (I_{enc}) by the Ampèrian loop equals the current that passes through a surface that is bounded by the path C in the integral $\oint_C \vec{\mathbf{B}} \cdot d\vec{\mathbf{s}}$: The path C is the Ampèrian circle shown above. The need for Maxwell's displacement current (I_d) can be understood by considering the following surfaces which are both bounded by the Ampèrian circle:

- The surface S_1 is the flat solid disc bounded by C. Since the current I passes through S_1, in this case the current enclosed is $I_{enc} = I$.
- The surface S_2 is a hemisphere that is bounded by C, but which passes between the plates of the capacitor like a butterfly net. The current I does **not** pass through S_2.

However, we must get the same answer for the magnetic field when we apply Ampère's law regardless of whether we choose the surface S_1 or S_2: Our result must apply for any surface that is bounded by the path C. The **generalized form** of Ampère's law, $\oint_C \vec{\mathbf{B}} \cdot d\vec{\mathbf{s}} = \mu_0 I_{enc} + \epsilon_0 \mu_0 \frac{d\Phi_e}{dt}$, states that there is indeed a current, called **displacement current**, passing through S_2, where the displacement current equals $I_d = \epsilon_0 \frac{d\Phi_e}{dt}$.

Why does the displacement current equal $I_d = \epsilon_0 \frac{d\Phi_e}{dt}$? According to Gauss's law, the electric flux between the parallel plates is $\Phi_e = \int_S \vec{\mathbf{E}} \cdot d\vec{\mathbf{A}} = \frac{q_{enc}}{\epsilon_0}$, such that $I_d = \epsilon_0 \frac{d\Phi_e}{dt} = \epsilon_0 \frac{d}{dt}\left(\frac{q_{enc}}{\epsilon_0}\right) = \frac{dq_{enc}}{dt}$. Recall that current is the rate of flow of charge: $I_d = \frac{dq_{enc}}{dt}$.

Faraday's Law as a Line Integral

We discussed Faraday's law in Chapter 29, where we expressed Faraday's law as $\varepsilon_{ind} = -N\frac{d\Phi_m}{dt}$. Recall that a **changing magnetic flux** (Φ_m) induces an emf (ε_{ind}) in a loop of wire. In Maxwell's equations, we instead write Faraday's law as a line integral: $\oint_C \vec{\mathbf{E}} \cdot d\vec{\mathbf{s}} = -\frac{d\Phi_m}{dt}$. These are really the same equation. Recall from Chapter 9 that $\Delta V = -\int \vec{\mathbf{E}} \cdot d\vec{\mathbf{s}}$. Also recall from Chapter 16 that the distinction between emf (ε) and potential difference (ΔV) is the internal resistance of the power supply - but in the case of Faraday's law, the emf is **induced**. Thus, we can replace ε_{ind} with $\oint_C \vec{\mathbf{E}} \cdot d\vec{\mathbf{s}}$. There are two reasons this is done when writing Maxwell's equations (but the equation $\varepsilon_{ind} = -N\frac{d\Phi_m}{dt}$ is more practical for solving typical physics problems involving Faraday's law):

- It makes the left-hand side of the equation, $\oint_C \vec{\mathbf{E}} \cdot d\vec{\mathbf{s}}$, look more like the left-hand side of Ampère's law, $\oint_C \vec{\mathbf{B}} \cdot d\vec{\mathbf{s}}$. That is, it makes the equations look more symmetric.
- The electric field is induced whether or not there is a physical loop of wire present. An induced emf (ε_{ind}) only makes sense if there is a loop of wire present, but $\oint_C \vec{\mathbf{E}} \cdot d\vec{\mathbf{s}}$ is true in general, even if there is no loop of wire lying around. (When a loop of wire happens to be present, the circulating electric field created by the changing magnetic flux causes charges to accelerate in the wire according to $\vec{\mathbf{F}}_e = q\vec{\mathbf{E}}$, which gives rise to the induced current in the wire.)

Symmetry in Electromagnetism: Almost

It's tempting to try to "complete" the symmetry in Maxwell's equations by introducing a hypothetical magnetic **monopole** with "magnetic charge" q_m (to be distinguished from ordinary electric charge q_e) with SI units of Ampères times meters (Am).

$$\oint_S \vec{\mathbf{E}} \cdot d\vec{\mathbf{A}} = \frac{q_{enc}}{\epsilon_0} \quad , \quad \oint_S \vec{\mathbf{B}} \cdot d\vec{\mathbf{A}} = \mu_0 q_m$$

$$\oint_C \vec{\mathbf{E}} \cdot d\vec{\mathbf{s}} = -\mu_0\frac{dq_m}{dt} - \frac{d}{dt}\int_S \vec{\mathbf{B}} \cdot d\vec{\mathbf{A}} \quad , \quad \oint_C \vec{\mathbf{B}} \cdot d\vec{\mathbf{s}} = \mu_0\frac{dq_e}{dt} + \epsilon_0\mu_0\frac{d}{dt}\int_S \vec{\mathbf{E}} \cdot d\vec{\mathbf{A}}$$

To further emphasize the near symmetry, we have made the substitutions $I_{enc} = \frac{dq_e}{dt}$, $\Phi_e = \int_S \vec{\mathbf{E}} \cdot d\vec{\mathbf{A}}$, and $\Phi_m = \int_S \vec{\mathbf{B}} \cdot d\vec{\mathbf{A}}$. However, this is all **hypothetical**: Magnetic monopoles have not been discovered (yet). Furthermore, nature has demonstrated **broken symmetries** in other fields of science, such as the symmetry-breaking **Higgs mechanism** in particle physics. So there very well might not be such a thing as a magnetic monopole. Maybe there is, and maybe there isn't. We leave q_m out of the equations until a monopole is discovered.

The Gradient, Divergence, and Curl

We can cast Maxwell's equations in another form by defining the following mathematical operators. We begin with the **partial derivative**: If f is a function of two independent variables, x and y, when taking a partial derivative of f with respect to x, treat y as a constant, and when taking a partial derivative of f with respect to y, treat x as a constant. We use the symbol ∂, which is a rounded version of the letter d, to distinguish partial derivatives from total derivatives.

Example: Given $f = 5x^3y^2$, find a partial derivative of f with respect to x, and also find a partial derivative of f with respect to y.

- To find a partial derivative of f with respect to x, treat y as a constant.
$$\frac{\partial f}{\partial x} = \frac{\partial}{\partial x}(5x^3y^2) = 5y^2\frac{\partial}{\partial x}(x^3) = 5y^2(3x^2) = 15x^2y^2$$
- To find a partial derivative of f with respect to y, treat x as a constant.
$$\frac{\partial f}{\partial y} = \frac{\partial}{\partial y}(5x^3y^2) = 5x^3\frac{\partial}{\partial y}(y^2) = 5x^3(2y) = 10x^3y$$

The **gradient** ($\vec{\nabla}$) is a differential operator which applies partial derivatives to a scalar function to create a **vector** function. We use the symbol $\vec{\nabla}$ (called the **del** operator) to represent the gradient operator. Note that the del symbol (∇) looks like an upside down uppercase delta (Δ). The gradient operator is defined as:
$$\vec{\nabla} = \hat{\mathbf{x}}\frac{\partial}{\partial x} + \hat{\mathbf{y}}\frac{\partial}{\partial y} + \hat{\mathbf{z}}\frac{\partial}{\partial z}$$
Recall that the **unit vectors** $\hat{\mathbf{x}}$, $\hat{\mathbf{y}}$, and $\hat{\mathbf{z}}$ point one unit along the $+x$-, $+y$-, and $+z$-axes, respectively. When the gradient operator acts on a scalar function f, you get:
$$\vec{\nabla} f = \hat{\mathbf{x}}\frac{\partial f}{\partial x} + \hat{\mathbf{y}}\frac{\partial f}{\partial y} + \hat{\mathbf{z}}\frac{\partial f}{\partial z}$$
The gradient is a sort of three-dimensional, multivariable derivative. Like an ordinary derivative, the gradient represents the **slope** of a tangent. The direction of the gradient at a given point on the function is along the direction in which the function increases the most (**the direction of greatest increase**) from that point.

The **divergence** operator ($\vec{\nabla}\cdot$) is a differential operator that uses the del operator ($\vec{\nabla}$) in a **scalar product** (we discussed the scalar product in Chapter 19). We apply the divergence to a vector function and obtain a **scalar** function as a result.
$$\vec{\nabla}\cdot\vec{\mathbf{F}} = \left(\hat{\mathbf{x}}\frac{\partial}{\partial x} + \hat{\mathbf{y}}\frac{\partial}{\partial y} + \hat{\mathbf{z}}\frac{\partial}{\partial z}\right)\cdot\left(F_x\hat{\mathbf{x}} + F_y\hat{\mathbf{y}} + F_z\hat{\mathbf{z}}\right) = \frac{\partial F_x}{\partial x} + \frac{\partial F_y}{\partial y} + \frac{\partial F_z}{\partial z}$$
The divergence of a vector field provides a measure of how much the field lines of a vector **diverge** (or spread out) from a given point.

The **curl** operator ($\vec{\nabla}\times$) is a differential operator that uses the del operator ($\vec{\nabla}$) in a **vector product** (we discussed the vector product in Chapter 19). We apply the curl to a vector function and obtain a different **vector** function as a result. **Note**: When finding a determinant that includes differential operators $\left(\text{like } \frac{\partial}{\partial x}\right)$, it's important to work out the determinant from **top to bottom** along each diagonal.

$$\vec{\nabla}\times\vec{\mathbf{F}} = \begin{vmatrix} \hat{\mathbf{x}} & \hat{\mathbf{y}} & \hat{\mathbf{z}} \\ \frac{\partial}{\partial x} & \frac{\partial}{\partial y} & \frac{\partial}{\partial z} \\ F_x & F_y & F_z \end{vmatrix} = \hat{\mathbf{x}}\begin{vmatrix} \frac{\partial}{\partial y} & \frac{\partial}{\partial z} \\ F_y & F_z \end{vmatrix} - \hat{\mathbf{y}}\begin{vmatrix} \frac{\partial}{\partial x} & \frac{\partial}{\partial z} \\ F_x & F_z \end{vmatrix} + \hat{\mathbf{z}}\begin{vmatrix} \frac{\partial}{\partial x} & \frac{\partial}{\partial y} \\ F_x & F_y \end{vmatrix}$$

$$\vec{\nabla}\times\vec{\mathbf{F}} = \left(\frac{\partial}{\partial y}F_z - \frac{\partial}{\partial z}F_y\right)\hat{\mathbf{x}} - \left(\frac{\partial}{\partial x}F_z - \frac{\partial}{\partial z}F_x\right)\hat{\mathbf{y}} + \left(\frac{\partial}{\partial x}F_y - \frac{\partial}{\partial y}F_x\right)\hat{\mathbf{z}}$$

The curl of a vector field expresses the **circulation** of the field lines.

Maxwell's Equations in Differential Form (in Vacuum)

We can use the **divergence** and **curl** operators to write Maxwell's equations in differential form (whereas on page 440 we wrote them in integral form).

$$\vec{\nabla}\cdot\vec{\mathbf{E}} = \frac{\rho}{\epsilon_0} \quad \text{(Gauss's law in electricity)}$$

$$\vec{\nabla}\cdot\vec{\mathbf{B}} = 0 \quad \text{(Gauss's law in magnetism)}$$

$$\vec{\nabla}\times\vec{\mathbf{E}} = -\frac{\partial\vec{\mathbf{B}}}{\partial t} \quad \text{(Faraday's law)}$$

$$\vec{\nabla}\times\vec{\mathbf{B}} = \mu_0\vec{\mathbf{J}} + \epsilon_0\mu_0\frac{\partial\vec{\mathbf{E}}}{\partial t} \quad \text{(generalized Ampère's law)}$$

The divergence ($\vec{\nabla}\cdot$) and curl ($\vec{\nabla}\times$) operators help to interpret these equations:

- Gauss's law in electricity, $\vec{\nabla}\cdot\vec{\mathbf{E}} = \frac{\rho}{\epsilon_0}$, states that the electric field lines tend to **radiate** away from (or towards) a pointlike charge, since the divergence of a vector field is nonzero when the field lines diverge.
- Gauss's law in magnetism, $\vec{\nabla}\cdot\vec{\mathbf{B}} = 0$, states that magnetic field lines **never diverge**. Rather, magnetic field lines tend to circulate (as shown by the curl in Ampère's law).
- Faraday's law, $\vec{\nabla}\times\vec{\mathbf{E}} = -\frac{\partial\vec{\mathbf{B}}}{\partial t}$, states that a changing magnetic field creates electric field lines that **circulate** (which could accelerate charges in a loop of wire, inducing a current in the loop).
- Ampère's law, $\vec{\nabla}\times\vec{\mathbf{B}} = \mu_0\vec{\mathbf{J}}$, states that magnetic field lines tend to **circulate** around currents, since the curl of a vector field is nonzero when the field lines circulate. The generalized form of Ampère's law, $\vec{\nabla}\times\vec{\mathbf{B}} = \mu_0\vec{\mathbf{J}} + \epsilon_0\mu_0\frac{\partial\vec{\mathbf{E}}}{\partial t}$, states that a changing electric flux associated with the displacement current ($I_d = \epsilon_0\frac{d\Phi_e}{dt}$) also creates magnetic field lines that circulate.

Light Is an Electromagnetic Wave

Light is an electromagnetic wave. The figure below illustrates (in part) what this means. The graph below shows a light wave propagating to the right along the $+z$-axis. As the wave propagates to the right, the electric field ($\vec{\mathbf{E}}$) oscillates up and down and the magnetic field ($\vec{\mathbf{B}}$) oscillates into and out of the page.

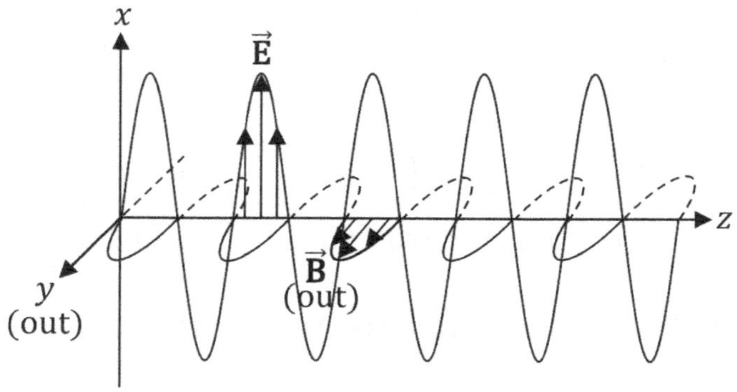

The **permittivity** of free space (ϵ_0) – which we encountered in Chapter 8 in the context of **electricity** with Gauss's law – and the **permeability** of free space (μ_0) – which we encountered in Chapters 26-27 in the context of **magnetism** with the law of Biot-Savart and Ampère's law – combine together to form the **speed of light** in vacuum (c) as follows.

$$c = \frac{1}{\sqrt{\epsilon_0 \mu_0}}$$

We can verify this by plugging in numbers. Recall that the **permittivity** of free space (ϵ_0) is related to **Coulomb's constant** (k) by $\epsilon_0 = \frac{1}{4\pi k}$. Also recall that **Coulomb's constant** is $k = 9.0 \times 10^9 \ \frac{\text{N}\cdot\text{m}^2}{\text{C}^2}$ and that the **permeability** of free space is $\mu_0 = 4\pi \times 10^{-7} \ \frac{\text{T}\cdot\text{m}}{\text{A}}$.

$$c = \frac{1}{\sqrt{\epsilon_0 \mu_0}} = \frac{1}{\sqrt{\frac{1}{4\pi k}\mu_0}} = \frac{1}{\sqrt{\frac{1}{4\pi(9 \times 10^9)}(4\pi \times 10^{-7})}} = \frac{1}{\sqrt{\frac{10^{-7}}{9 \times 10^9}}} = \frac{1}{\sqrt{\frac{10^{-7-9}}{9}}}$$

$$c = \frac{1}{\sqrt{\frac{10^{-16}}{9}}} = \sqrt{\frac{9}{10^{-16}}} = \sqrt{9 \times 10^{16}} = \sqrt{9}\sqrt{10^{16}} = 3.0 \times 10^8 \text{ m/s}$$

The speed of light in vacuum equals $c = 3.0 \times 10^8$ m/s to two (really, 3) significant figures.

The units work out as follows. Since ϵ_0's units are $\frac{\text{C}^2}{\text{N}\cdot\text{m}^2}$ (the reciprocal of k's units) and μ_0's units are $\frac{\text{T}\cdot\text{m}}{\text{A}}$, the units of $\epsilon_0\mu_0$ are $\frac{\text{C}^2\cdot\text{T}}{\text{N}\cdot\text{m}\cdot\text{A}}$. In Chapter 24, we showed that a Tesla can be expressed as $1 \text{ T} = 1 \ \frac{\text{N}}{\text{A}\cdot\text{m}}$, such that $\epsilon_0\mu_0$'s units become $\frac{\text{C}^2}{\text{m}^2\cdot\text{A}^2}$. Since $I = \frac{dq}{dt}$, an Ampère is $1 \text{ A} = 1 \ \frac{\text{C}}{\text{s}}$. With this, $\epsilon_0\mu_0$'s units are $\frac{\text{s}^2}{\text{m}^2}$, and $\frac{1}{\sqrt{\epsilon_0\mu_0}}$ has units of m/s.

Electromagnetic Spectrum

The electromagnetic spectrum consists of radio waves, microwaves, infrared, visible light, ultraviolet, x-rays, and gamma rays, as shown below. The visible spectrum is a just a very narrow slice (much narrower than it would appear by looking at the chart below – the blocks do **not** really have equal width when drawn to scale) of the full electromagnetic spectrum. The acronym Roy G. Biv stands for Red Orange Yellow Green Blue Indigo Violet, and helps to remember the order of the visible colors in increasing frequency. Wavelength (λ) and frequency (f) share a reciprocal relationship: $\lambda f = c$, such that $\lambda = \frac{c}{f}$ and $f = \frac{c}{\lambda}$. Higher frequency thus corresponds to shorter wavelength, while lower frequency corresponds to longer wavelength.

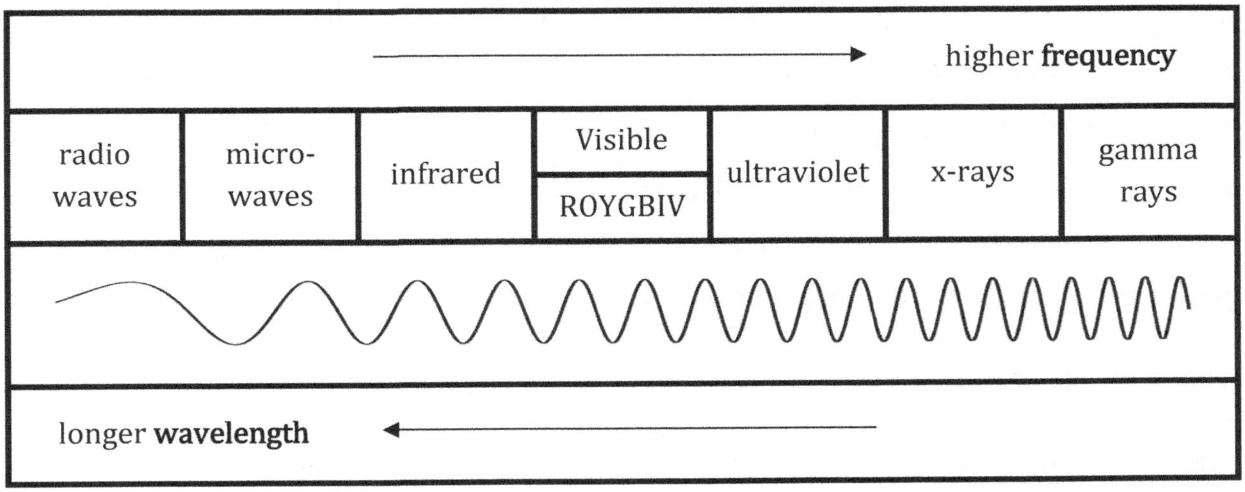

The Poynting Vector

The Poynting vector ($\vec{\mathbf{S}}$) describes the power per unit area – or the **rate of flow of energy**, since power is the rate at which work is done, while energy is the ability to do work – of an electromagnetic wave.

$$\vec{\mathbf{S}} = \frac{1}{\mu_0}\vec{\mathbf{E}} \times \vec{\mathbf{B}}$$

The wave intensity (I) equals the average value of the Poynting vector (S_{av}) over one cycle.

$$I = S_{av} = \frac{E_m B_m}{2\mu_0}$$

Here, E_m and B_m are the amplitudes (maximum values) of the electric and magnetic fields.

Notes Regarding Units

The SI units of the Poynting vector ($\vec{\mathbf{S}}$) are $\frac{\text{W}}{\text{m}^2}$. This follows from the equation $\vec{\mathbf{S}} = \frac{1}{\mu_0}\vec{\mathbf{E}} \times \vec{\mathbf{B}}$. Recall that the SI units of the permeability of free space (μ_0) are $\frac{\text{T·m}}{\text{A}}$ (see Chapter 25), the SI units of electric field ($\vec{\mathbf{E}}$) are $\frac{\text{N}}{\text{C}}$ (see Chapter 2), and the SI unit of magnetic field ($\vec{\mathbf{B}}$) is the Tesla (T). Note that $\frac{1}{\mu_0}$ has units of $\frac{\text{A}}{\text{T·m}}$. It therefore follows from the equation $\vec{\mathbf{S}} = \frac{1}{\mu_0}\vec{\mathbf{E}} \times \vec{\mathbf{B}}$ that the SI units of the Poynting vector ($\vec{\mathbf{S}}$) are $\left(\frac{\text{A}}{\text{T·m}}\right)\left(\frac{\text{N}}{\text{C}}\right)(\text{T}) = \frac{\text{N·A}}{\text{C·m}}$. Since current ($I$) is the rate of flow of charge (q), we can write $I = \frac{dq}{dt}$, which shows that an Ampère (A) equals a Coulomb (C) per second (s): $1 \text{ A} = 1 \frac{\text{C}}{\text{s}}$. With this, the SI units of the Poynting vector ($\vec{\mathbf{S}}$) become $\frac{\text{N}}{\text{m·s}}$. From the equation for work, $W = \int \vec{\mathbf{F}} \cdot d\vec{\mathbf{s}}$, we see that a Newton (N) times a meter (m) equals a Joule (J), such that $1 \text{ N} = 1 \frac{\text{J}}{\text{m}}$. This transforms the SI units of the Poynting vector ($\vec{\mathbf{S}}$) into $\frac{\text{J}}{\text{m}^2\text{·s}}$. Finally, the equation for power, $P = \frac{dW}{dt}$, shows that a Joule (J) per second (s) equals a Watt (W), or $1 \text{ W} = 1 \frac{\text{J}}{\text{s}}$. Thus, the SI units of the Poynting vector ($\vec{\mathbf{S}}$) are $\frac{\text{W}}{\text{m}^2}$. These units, Watts per square meter, show that the Poynting vector represents **power per unit area** (or the **rate of flow of energy**).

This workbook uses SI units – specifically, the MKS system of units, which uses meters, kilograms, seconds, and Coulombs (or, equivalently, Ampères) as the base units. The MKS system of units expresses charge (q) in Coulombs (C). Note that some books on electricity and magnetism (especially, more advanced texts) adopt a different system of units, in which case the constants in Maxwell's equations and related equations may look different.

For example, the Gaussian system of units works with centimeters, grams, and seconds as the base units, but the huge difference is that charge (q) is expressed in **electrostatic units** (esu), which effectively absorbs Coulomb's constant into the units of charge (so that electric force becomes $\frac{q_1 q_2}{R^2}$ without a proportionality constant – that is, the units of charge are calibrated so that Coulomb's constant equals one).

The point is that if you read two or more electricity and magnetism books, if the equations seem a little bit different, it could be that the two books are using different systems of units.

Where Are the Problems?

The goal of this chapter is to help you try to understand how the material that we learned in Chapters 1-31 fits into the bigger picture. This chapter has <u>no</u> problems.

HINTS, INTERMEDIATE ANSWERS, AND EXPLANATIONS

How to Use This Section Effectively

Think of hints and intermediate answers as training wheels. They help you proceed with your solution. When you stray from the right path, the hints help you get back on track. The answers also help to build your confidence.

However, if you want to succeed in a physics course, you must eventually learn to rely less and less on the hints and intermediate answers. Make your best effort to solve the problem on your own before checking for hints, answers, and explanations. When you need a hint, try to find just the hint that you need to get over your current hurdle. Refrain from reading additional hints until you get further into the solution.

When you make a mistake, think about what you did wrong and what you should have done differently. Try to learn from your mistake so that you don't repeat the mistake in other solutions.

It's natural for students to check hints and intermediate answers repeatedly in the early chapters. However, at some stage, you would like to be able to consult this section less frequently. When you can solve more problems without help, you know that you're really beginning to master physics.

Would You Prefer to See Full Solutions?

Full solutions are like a security blanket: Having them makes students feel better. But full solutions are also dangerous: Too many students rely too heavily on the full solutions, or simply read through the solutions instead of working through the solutions on their own. Students who struggle through their solutions and improve their solutions only as needed tend to earn better grades in physics (though comparing solutions **after** solving a problem is always helpful).

It's a challenge to get just the right amount of help. In the ideal case, you would think your way through every solution on your own, seek just the help you need to make continued progress with your solution, and wait until you've solved the problem as best you can before consulting full solutions or reading every explanation.

With this in mind, full solutions to all problems are contained in a separate book. This workbook contains hints, intermediate answers, explanations, and several directions to help walk you through the steps of every solution, which should be enough to help most students figure out how to solve all of the problems. However, if you need the security of seeing **full solutions to all problems**, look for the book 100 Instructive Calculus-based Physics Examples with ISBN 978-1-941691-13-7. The solution to every problem in this workbook can be found in that book.

How to Cover up Hints that You Don't Want to See too Soon

There is a simple and effective way to cover up hints and answers that you don't want to see too soon:

- Fold a blank sheet of paper in half and place it in the hints and answers section. This will also help you bookmark this handy section of the book.
- Place the folded sheet of paper just below your current reading position. The folded sheet of paper will block out the text below.
- When you want to see the next hint or intermediate answer, just drop the folded sheet of paper down slowly, just enough to reveal the next line.
- This way, you won't reveal more hints or answers than you need.

You learn more when you force yourself to struggle through the problem. Consult the hints and answers when you really need them, but try it yourself first. After you read a hint, try it out and think it through as best you can before consulting another hint. Similarly, when checking intermediate answers to build confidence, try not to see the next answer before you have a chance to try it on your own first.

Chapter 1: Coulomb's Law

1. First identify the given quantities.
   - The knowns are $q_1 = -8.0$ µC, $q_2 = -3.0$ µC, $R = 2.0$ m, and $k = 9.0 \times 10^9 \frac{\text{N}\cdot\text{m}^2}{\text{C}^2}$.
   - Convert the charges from microCoulombs (µC) to Coulombs (C). Note that the metric prefix micro (µ) stands for 10^{-6}, such that 1 µC = 10^{-6} C. The charges are $q_1 = -8.0 \times 10^{-6}$ C and $q_2 = -3.0 \times 10^{-6}$ C.
   - Plug these values into Coulomb's law: $F_e = k\frac{|q_1||q_2|}{R^2}$.
   - Apply the rule $x^m x^n = x^{m+n}$. Note that $10^9 10^{-6} 10^{-6} = 10^{9-12} = 10^{-3}$.
   - Note that $R^2 = 2^2 = 4.0$ m². Also note that $54 \times 10^{-3} = 5.4 \times 10^{-2} = 0.054$.
   - The answer is $F_e = 0.054$ N. It can also be expressed as 54×10^{-3} N, 5.4×10^{-2} N, or 54 mN (meaning milliNewtons, where the prefix milli, m, stands for 10^{-3}).
   - The force is **repulsive** because two negative charges repel one another.

---

2. First identify the given quantities.
   - The knowns are $q_1 = 800$ nC, $q_2 = -800$ nC, $R = 20$ cm, and $k = 9.0 \times 10^9 \frac{\text{N}\cdot\text{m}^2}{\text{C}^2}$.
   - Convert the charges from nanoCoulombs (nC) to Coulombs (C). Note that the metric prefix nano (n) stands for 10^{-9}, such that 1 nC = 10^{-9} C. The charges are $q_1 = 800 \times 10^{-9}$ C $= 8.0 \times 10^{-7}$ C and $q_2 = -800 \times 10^{-9}$ C $= -8.0 \times 10^{-7}$ C. Note that $100 \times 10^{-9} = 10^{-7}$. (It doesn't matter which charge is negative.)

- Convert the distance from centimeters (cm) to meters (m): $R = 20 \text{ cm} = 0.20 \text{ m}$.
- Plug these values into Coulomb's law: $F_e = k\frac{|q_1||q_2|}{R^2}$.
- Apply the rule $x^m x^n = x^{m+n}$. Note that $10^9 10^{-9} 10^{-9} = 10^{9-18} = 10^{-9}$.
- Note that $R^2 = 0.2^2 = 0.04 \text{ m}^2$. Also note that $144 \times 10^{-3} = 1.44 \times 10^{-1} = 0.144$.
- The answer is $F_e = 0.144$ N. It can also be expressed as 144×10^{-3} N, 1.44×10^{-1} N, or 144 mN (meaning milliNewtons, where the prefix milli, m, equals 10^{-3}).
- The force is **attractive** because opposite charges attract one another. We know that the charges are opposite because when the charge was transferred (during the process of rubbing), one object gained electrons while the other lost electrons, making one object negative and the other object positive (since both were neutral prior to rubbing).

---

3. First identify the given quantities.

(A) The knowns are $q_1 = -2.0 \text{ µC}$, $q_2 = 8.0 \text{ µC}$, $R = 3.0 \text{ m}$, and $k = 9.0 \times 10^9 \frac{\text{N·m}^2}{\text{C}^2}$.

- Convert the charges from microCoulombs (µC) to Coulombs (C). Note that the metric prefix micro (µ) stands for 10^{-6}, such that $1 \text{ µC} = 10^{-6}$ C. The charges are $q_1 = -2.0 \times 10^{-6}$ C and $q_2 = 8.0 \times 10^{-6}$ C.
- Plug these values into Coulomb's law: $F_e = k\frac{|q_1||q_2|}{R^2}$.
- Apply the rule $x^m x^n = x^{m+n}$. Note that $10^9 10^{-6} 10^{-6} = 10^{9-12} = 10^{-3}$.
- Note that $R^2 = 3^2 = 9.0 \text{ m}^2$. Also note that $16 \times 10^{-3} = 1.6 \times 10^{-2} = 0.016$.
- The answer is $F_e = 0.016$ N. It can also be expressed as 16×10^{-3} N, 1.6×10^{-2} N, or 16 mN (meaning milliNewtons, where the prefix milli, m, stands for 10^{-3}).
- The force is **attractive** because opposite charges attract one another.

(B) To the extent possible, the charges would like to **neutralize** when contact is made. The −2.0 µC isn't enough to completely neutralize the +8.0 µC, so what will happen is that the −2.0 µC will pair up with +2.0 µC from the +8.0 µC, leaving a net excess charge of +6.0 µC. **One-half** of the **net excess charge**, +6.0 µC, resides on each earring. That is, after contact, each earring will have a charge of $q = +3.0 \text{ µC}$. That's what happens conceptually. You can arrive at the same answer mathematically using the formula below.

$$q = \frac{q_1 + q_2}{2}$$

- The knowns are $q = 3.0 \text{ µC} = 3.0 \times 10^{-6} \text{ C}$, $R = 3.0 \text{ m}$, and $k = 9.0 \times 10^9 \frac{\text{N·m}^2}{\text{C}^2}$.
- Set the two charges equal in Coulomb's law: $F_e = k\frac{q^2}{R^2}$.
- The answer is $F_e = 0.0090$ N. It can also be expressed as 9.0×10^{-3} N or 9.0 mN (meaning milliNewtons, where the prefix milli, m, stands for 10^{-3}).
- The force is **repulsive** because two positive charges repel one another. (After they make contact, both earrings become **positively** charged.)

Chapter 2: Electric Field

4. First identify the given quantities and the desired unknown.
   - The knowns are $q = -300$ μC and $E = 80{,}000$ N/C. Solve for F_e.
   - Convert the charge from microCoulombs (μC) to Coulombs (C). Note that the metric prefix micro (μ) stands for 10^{-6}. The charge is $q = -3.0 \times 10^{-4}$ C.
   - Plug these values into the following equation: $F_e = |q|E$. The answer is $F_e = 24$ N.

---

5. First identify the given quantities and the desired unknown.
   - The knowns are $q = 800$ μC, $R = 2.0$ m, and $k = 9.0 \times 10^9 \frac{\text{N·m}^2}{\text{C}^2}$. Solve for E.
   - Convert the charge from microCoulombs (μC) to Coulombs (C). Note that the metric prefix micro (μ) stands for 10^{-6}. The charge is $q = 8.0 \times 10^{-4}$ C.
   - Plug these values into the following equation: $E = \frac{k|q|}{R^2}$.
   - Apply the rule $x^m x^n = x^{m+n}$. Note that $10^9 10^{-4} = 10^{9-4} = 10^5$.
   - Note that $R^2 = 2^2 = 4.0$ m^2. Also note that $18 \times 10^5 = 1.8 \times 10^6$.
   - The answer is $E = 1.8 \times 10^6$ N/C. It can also be expressed as 18×10^5 N/C.

---

6. First identify the given quantities and the desired unknown.
   - The knowns are $F_e = 12$ N and $E = 30{,}000$ N/C. Solve for $|q|$.
   - Plug these values into the following equation: $F_e = |q|E$.
   - The absolute value of the charge is $|q| = 4.0 \times 10^{-4}$ C.
   - The charge must be <u>**negative**</u> since the force ($\vec{\mathbf{F}}_e$) is opposite to the electric field ($\vec{\mathbf{E}}$).
   - The answer is $q = -4.0 \times 10^{-4}$ C, which can also be expressed as $q = -400$ μC.

---

7. First identify the given quantities and the desired unknown.
   - The knowns are $q = 80$ μC, $E = 20{,}000$ N/C, and $k = 9.0 \times 10^9 \frac{\text{N·m}^2}{\text{C}^2}$. Solve for R.
   - Convert the charge from microCoulombs (μC) to Coulombs (C). Note that the metric prefix micro (μ) stands for 10^{-6}. The charge is $q = 8.0 \times 10^{-5}$ C.
   - Plug these values into the following equation: $E = \frac{k|q|}{R^2}$.
   - Solve for R. Multiply both sides by R^2. Divide both sides by E. Squareroot both sides. You should get $R = \sqrt{\frac{k|q|}{E}}$.
   - Apply the rule $x^m x^n = x^{m+n}$. Note that $10^9 10^{-5} = 10^{9-5} = 10^4$.
   - Apply the rule $\frac{x^m}{x^n} = x^{m-n}$. Note that $10^4 10^3 = 10^{4-3} = 10^1 = 10$.
   - Note that $\sqrt{3.6 \times 10} = \sqrt{36} = 6$.
   - The answer is $R = 6.0$ m.

8. First apply the distance formula, $R = \sqrt{(x_2 - x_1)^2 + (y_2 - y_1)^2}$, to determine how far the point (−5.0 m, 12.0 m) is from (3.0 m, 6.0 m). You should get $R = \sqrt{(-8)^2 + 6^2} = 10$ m.
(A) Identify the given quantities and the desired unknown.

- The knowns are $q = 30$ µC, $R = 10$ m, and $k = 9.0 \times 10^9 \frac{\text{N·m}^2}{\text{C}^2}$. Solve for E.
- Convert the charge from microCoulombs (µC) to Coulombs (C). Note that the metric prefix micro (µ) stands for 10^{-6}. The charge is $q = 3.0 \times 10^{-5}$ C.
- Plug these values into the following equation: $E = \frac{k|q|}{R^2}$.
- Apply the rule $x^m x^n = x^{m+n}$. Note that $10^9 10^{-5} = 10^{9-5} = 10^4$.
- The answer is $E = 2.7 \times 10^3$ N/C. It can also be expressed as 2700 N/C.

(B) Identify the given quantities and the desired unknown.

- The knowns are $q = 500$ µC and $E = 2700$ N/C. Solve for F_e.
- Convert the charge from microCoulombs (µC) to Coulombs (C). Note that the metric prefix micro (µ) stands for 10^{-6}. The charge is $q = 5.0 \times 10^{-4}$ C.
- Plug these values into the following equation: $F_e = |q|E$. The answer is $F_e = 1.35$ N.

Chapter 3: Superposition of Electric Fields

9. Begin by sketching the electric field vectors created by each of the charges. We choose to call the positive charge $q_1 = 6.0$ µC and the negative charge $q_2 = -6.0$ µC.
(A) Imagine a positive "test" charge at the point (2.0 m, 0), marked by a star.

- A positive "test" charge at (2.0 m, 0) would be repelled by $q_1 = 6.0$ µC. Thus, we draw $\vec{\mathbf{E}}_1$ directly away from $q_1 = 6.0$ µC (diagonally down and to the right).
- A positive "test" charge at (2.0 m, 0) would be attracted to $q_2 = -6.0$ µC. Thus, we draw $\vec{\mathbf{E}}_2$ toward $q_2 = -6.0$ µC (diagonally down and to the left).

(We have drawn multiple diagrams below to help label the different parts clearly.)

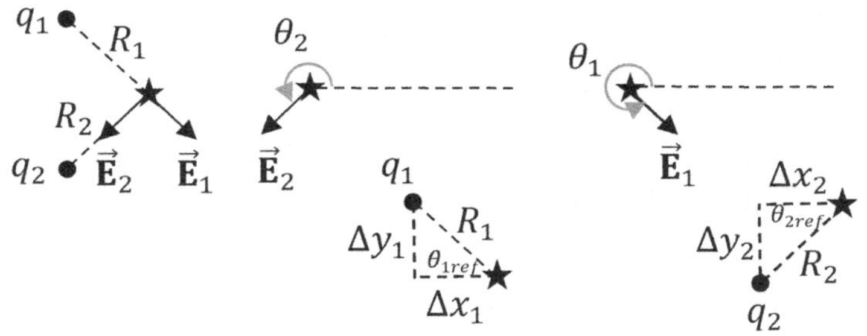

(B) Use the distance formula.

$$R_1 = \sqrt{\Delta x_1^2 + \Delta y_1^2} \quad , \quad R_2 = \sqrt{\Delta x_2^2 + \Delta y_2^2}$$

- Check your distances: $|\Delta x_1| = |\Delta x_2| = 2.0$ m and $|\Delta y_1| = |\Delta y_2| = 2.0$ m.
- You should get $R_1 = R_2 = 2\sqrt{2}$ m.

(C) Use an inverse tangent to determine each reference angle.

$$\theta_{1ref} = \tan^{-1}\left|\frac{\Delta y_1}{\Delta x_1}\right| \quad , \quad \theta_{2ref} = \tan^{-1}\left|\frac{\Delta y_2}{\Delta x_2}\right|$$

- The reference angle is the smallest angle between the vector and the horizontal.
- You should get $\theta_{1ref} = \theta_{2ref} = 45°$.

(D) Use the reference angles to determine the direction of each electric field vector counterclockwise from the $+x$-axis. Recall from trig that 0° points along $+x$, 90° points along $+y$, 180° points along $-x$, and 270° points along $-y$.

- $\vec{\mathbf{E}}_1$ lies in Quadrant IV:

$$\theta_1 = 360° - \theta_{1ref}$$

- $\vec{\mathbf{E}}_2$ lies in Quadrant III:

$$\theta_2 = 180° + \theta_{2ref}$$

- You should get $\theta_1 = 315°$ and $\theta_2 = 225°$.

(E) First convert the charges from μC to C: $q_1 = 6.0 \times 10^{-6}$ C and $q_2 = -6.0 \times 10^{-6}$ C.

- Use the following equations. Note the absolute values.

$$E_1 = \frac{k|q_1|}{R_1^2} \quad , \quad E_2 = \frac{k|q_2|}{R_2^2}$$

- Note that $\frac{27}{4} = 6.75$ and $10^9 10^{-6} = 10^3$. Also note that $(2\sqrt{2})^2 = (4)(2) = 8$.
- The magnitudes of the electric fields are $E_1 = 6{,}750$ N/C and $E_2 = 6{,}750$ N/C.

(F) Use trig to determine the components of the electric field vectors.

$$E_{1x} = E_1 \cos\theta_1 \quad , \quad E_{1y} = E_1 \sin\theta_1 \quad , \quad E_{2x} = E_2 \cos\theta_2 \quad , \quad E_{2y} = E_2 \sin\theta_2$$

- $E_{1x} = 3375\sqrt{2}$ N/C. It's positive because $\vec{\mathbf{E}}_1$ points to the right, not left.
- $E_{1y} = -3375\sqrt{2}$ N/C. It's negative because $\vec{\mathbf{E}}_1$ points downward, not upward.
- $E_{2x} = -3375\sqrt{2}$ N/C. It's negative because $\vec{\mathbf{E}}_2$ points to the left, not right.
- $E_{2y} = -3375\sqrt{2}$ N/C. It's negative because $\vec{\mathbf{E}}_2$ points downward, not upward.

(G) Add the respective components together.

$$E_x = E_{1x} + E_{2x} \quad , \quad E_y = E_{1y} + E_{2y}$$

- $E_x = 0$ and $E_y = -6750\sqrt{2}$ N/C.
- If you study the diagrams from part (A), you should be able to see why $E_x = 0$.

(H) Apply the Pythagorean theorem.

$$E = \sqrt{E_x^2 + E_y^2}$$

The magnitude of the net electric field is $E = 6750\sqrt{2}$ N/C. Magnitudes can't be negative.

(I) Take an inverse tangent.

$$\theta_E = \tan^{-1}\left(\frac{E_y}{E_x}\right)$$

Although the argument is undefined, $\theta_E = 270°$ since $E_x = 0$ and $E_y < 0$. Since $\vec{\mathbf{E}}$ lies on the negative y-axis, the angle is 270°.

10. Begin by sketching the forces that q_1 and q_2 exert on q_3.

(A) Recall that opposite charges attract, while like charges repel.

- Since q_1 and q_3 have opposite signs, q_1 pulls q_3 to the left. Thus, we draw $\vec{\mathbf{F}}_1$ towards q_1 (to the left).
- Since q_2 and q_3 are both positive, q_2 pushes q_3 down and to the right. Thus, we draw $\vec{\mathbf{F}}_2$ directly away from q_2 (down and to the right).

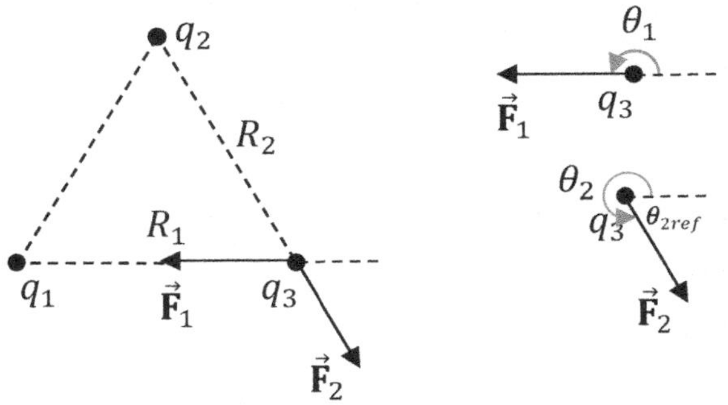

(B) Since each charge lies on the vertex of an equilateral triangle, each R equals the edge length: $R_1 = R_2 = 2.0$ m.

(C) Recall that the reference angle is the smallest angle between the vector and the horizontal axis. Since $\vec{\mathbf{F}}_1$ lies on the horizontal axis, its reference angle is $\theta_{1ref} = 0°$. The second vector lies $\theta_{2ref} = 60°$ from the horizontal. (**Equilateral** triangles have 60° angles.)

(D) Use the reference angles to determine the direction of each force counterclockwise from the $+x$-axis. Recall from trig that 0° points along $+x$, 90° points along $+y$, 180° points along $-x$, and 270° points along $-y$.

- $\vec{\mathbf{F}}_1$ lies on the negative x-axis:

$$\theta_1 = 180° - \theta_{1ref}$$

- $\vec{\mathbf{F}}_2$ lies in Quadrant IV:

$$\theta_2 = 360° - \theta_{2ref}$$

- You should get $\theta_1 = 180°$ and $\theta_2 = 300°$.

(E) First convert the charges from μC to C: $q_1 = -3.0 \times 10^{-6}$ C, $q_2 = 3.0 \times 10^{-6}$ C, and $q_3 = 4.0 \times 10^{-6}$ C.

- Use the following equations.

$$F_1 = \frac{k|q_1||q_3|}{R_1^2} \quad , \quad F_2 = \frac{k|q_2||q_3|}{R_2^2}$$

- Note that $10^9 10^{-6} 10^{-6} = 10^{-3}$ and that $27 \times 10^{-3} = 0.027$.
- The magnitudes of the forces are $F_1 = 0.027$ N $= 27$ mN and $F_2 = 0.027$ N $= 27$ mN.

(F) Use trig to determine the components of the forces.

$$F_{1x} = F_1 \cos\theta_1 \quad , \quad F_{1y} = F_1 \sin\theta_1 \quad , \quad F_{2x} = F_2 \cos\theta_2 \quad , \quad F_{2y} = F_2 \sin\theta_2$$

- $F_{1x} = -27$ mN. It's negative because $\vec{\mathbf{F}}_1$ points to the left, not right.
- $F_{1y} = 0$. It's zero because $\vec{\mathbf{F}}_1$ is horizontal.

- $F_{2x} = \frac{27}{2}$ mN. It's positive because $\vec{\mathbf{F}}_2$ points to the right, not left.
- $F_{2y} = -\frac{27}{2}\sqrt{3}$ mN. It's negative because $\vec{\mathbf{F}}_2$ points downward, not upward.

(G) Add the respective components together.

$$F_x = F_{1x} + F_{2x} \quad , \quad F_y = F_{1y} + F_{2y}$$

- $F_x = -\frac{27}{2}$ mN and $F_y = -\frac{27}{2}\sqrt{3}$ mN.
- If you study the diagrams from part (A), you should see why $F_x < 0$ and $F_y < 0$.

(H) Apply the Pythagorean theorem.

$$F = \sqrt{F_x^2 + F_y^2}$$

The magnitude of the net electric force is $F = 0.027$ N $= 27$ mN, where the prefix milli (m) stands for one-thousandth (10^{-3}).

(I) Take an inverse tangent.

$$\theta_F = \tan^{-1}\left(\frac{F_y}{F_x}\right)$$

The reference angle is 60° and the net force lies in Quadrant III because $F_x < 0$ and $F_y < 0$. The direction of the net electric force is $\theta_F = 240°$. In Quadrant III, $\theta_F = 180° + \theta_{ref}$.

---

11. Begin with a sketch of the electric fields in each region.
   - Imagine placing a positive "test" charge in each region. Since q_1 and q_2 are both positive, the positive "test" charge would be repelled by both q_1 and q_2. Draw the electric fields away from q_1 and q_2 in each region.
      I. Region I is left of q_1. $\vec{\mathbf{E}}_1$ and $\vec{\mathbf{E}}_2$ both point left. They won't cancel here.
      II. Region II is between q_1 and q_2. $\vec{\mathbf{E}}_1$ points right, while $\vec{\mathbf{E}}_2$ points left. They can cancel out in Region II.
      III. Region III is right of q_2. $\vec{\mathbf{E}}_1$ and $\vec{\mathbf{E}}_2$ both point right. They won't cancel here.

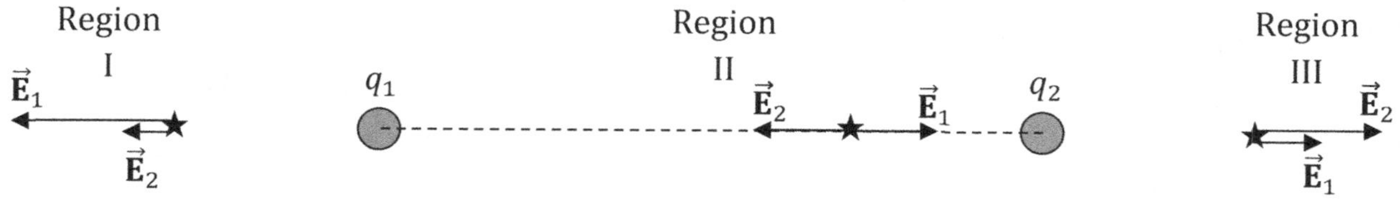

- The net electric field can only be zero in Region II.
- Set the magnitudes of the electric fields equal to one another.

$$E_1 = E_2$$

$$\frac{k|q_1|}{R_1^2} = \frac{k|q_2|}{R_2^2}$$

- Divide both sides by k (it will cancel out) and **cross-multiply**.

$$|q_1|R_2^2 = |q_2|R_1^2$$

- Plug in the values of the charges and simplify.

$$25R_2^2 = 16R_1^2$$

- Study the diagram to relate R_1 and R_2 to the distance between the charges, d.

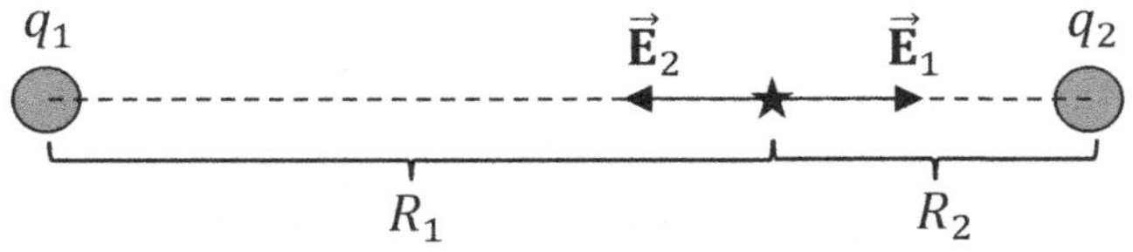

$$R_1 + R_2 = d$$

- Isolate R_2 in the previous equation: $R_2 = d - R_1$. Substitute this expression into the equation $25R_2^2 = 16R_1^2$, which we found previously.

$$25(d - R_1)^2 = 16R_1^2$$

- Squareroot both sides of the equation and simplify. Apply the rules $\sqrt{xy} = \sqrt{x}\sqrt{y}$ and $\sqrt{x^2} = x$ to write $\sqrt{25(d - R_1)^2} = \sqrt{25}\sqrt{(d - R_1)^2} = 5(d - R_1)$.

$$5(d - R_1) = 4R_1$$

- **Distribute** the 5 and **combine like terms**. Note that $5R_1 + 4R_1 = 9R_1$.

$$5d = 9R_1$$

- Divide both sides of the equation by 9.

$$R_1 = \frac{5d}{9}$$

- The net electric field is zero in Region II, a distance of $R_1 = 5.0$ m from the left charge (and therefore a distance of $R_2 = 4.0$ m from the right charge, since $R_1 + R_2 = d = 9.0$ m).

---

12. Begin with a sketch of the electric fields in each region.
    - Imagine placing a positive "test" charge in each region. Since q_1 is positive, $\vec{\mathbf{E}}_1$ will point away from q_1 in each region because a positive "test" charge would be repelled by q_1. Since q_2 is negative, $\vec{\mathbf{E}}_2$ will point towards q_2 in each region because a positive "test" charge would be attracted to q_2.

      IV. Region I is left of q_1. $\vec{\mathbf{E}}_1$ points left and $\vec{\mathbf{E}}_2$ points right. They can cancel here.

      V. Region II is between q_1 and q_2. $\vec{\mathbf{E}}_1$ and $\vec{\mathbf{E}}_2$ both point right. They won't cancel out in Region II.

      VI. Region III is right of q_2. $\vec{\mathbf{E}}_1$ points right and $\vec{\mathbf{E}}_2$ points left. Although $\vec{\mathbf{E}}_1$ and $\vec{\mathbf{E}}_2$ point in opposite directions in Region III, they won't cancel in this region because it is closer to the stronger charge, such that $|E_2| > |E_1|$.

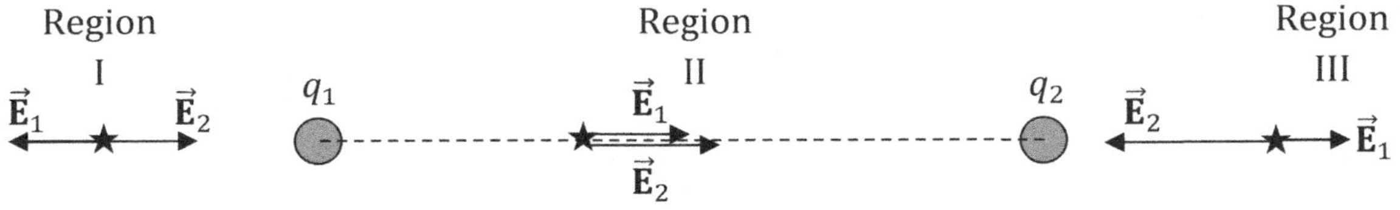

- The net electric field can only be zero in Region I.
- Set the magnitudes of the electric fields equal to one another.

$$E_1 = E_2$$
$$\frac{k|q_1|}{R_1^2} = \frac{k|q_2|}{R_2^2}$$

- Divide both sides by k (it will cancel out) and **cross-multiply**.

$$|q_1|R_2^2 = |q_2|R_1^2$$

- Plug in the values of the charges and simplify. Note the **absolute values**.

$$2R_2^2 = 8R_1^2$$

- Divide both sides of the equation by 2.

$$R_2^2 = 4R_1^2$$

- Study the diagram to relate R_1 and R_2 to the distance between the charges, d.

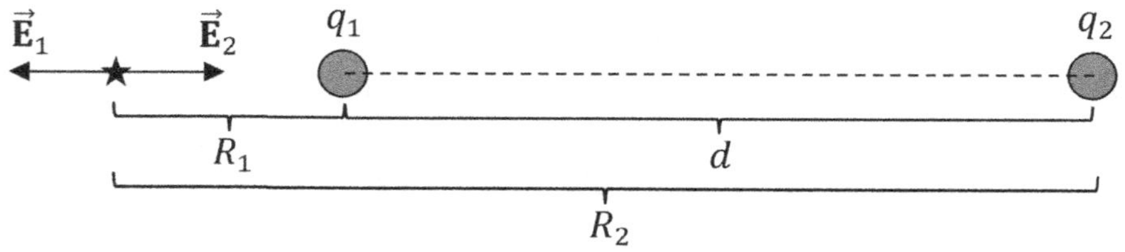

$$R_1 + d = R_2$$

- Substitute the above expression into the equation $R_2^2 = 4R_1^2$.

$$(d + R_1)^2 = 4R_1^2$$

- Squareroot both sides of the equation and simplify. Note that $\sqrt{x^2} = x$ such that $\sqrt{(d + R_1)^2} = d + R_1$. Also note that $\sqrt{4} = 2$.

$$d + R_1 = 2R_1$$

- **Combine like terms.** Note that $2R_1 - R_1 = R_1$.

$$d = R_1$$

- The net electric field is zero in Region I, a distance of $R_1 = 4.0$ m from the left charge (and therefore a distance of $R_2 = 8.0$ m from the right charge, since $R_1 + d = R_2$ and since $d = 4.0$ m).

Chapter 4: Electric Field Mapping

13. If a positive "test" charge were placed at points A, B, or C, it would be repelled by the positive sphere. Draw the electric field directly away from the positive sphere. Draw a shorter arrow at point B, since point B is further away from the positive sphere.

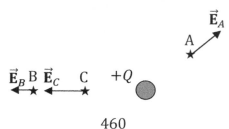

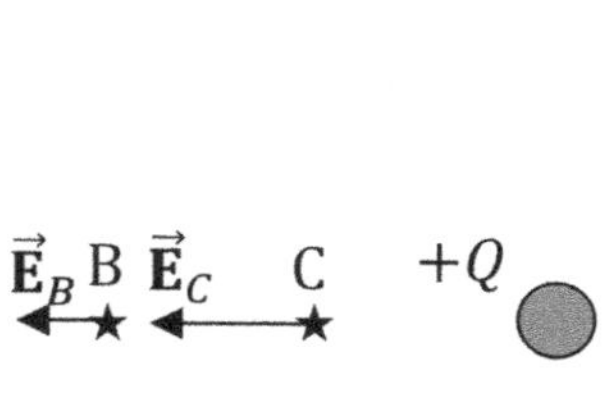

14. If a positive "test" charge were placed at points D, E, or F, it would be attracted to the negative sphere. Draw the electric field towards the negative sphere. Draw a longer arrow at point E, since point E is closer to the negative sphere.

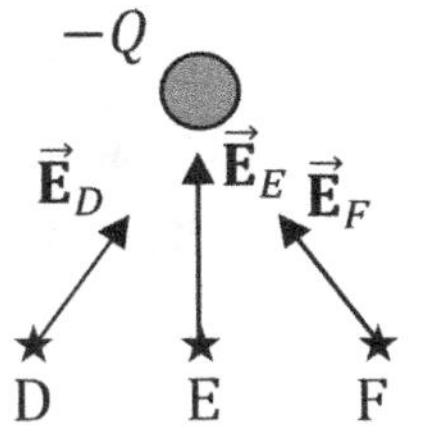

15. If a positive "test" charge were placed at any of these points, it would be repelled by the positive sphere and attracted to the negative sphere. First draw two separate electric fields (one for each sphere) and then draw the resultant of these two vectors for the net electric field. Move one of the arrows (we moved $\vec{\mathbf{E}}_R$) to join the two vectors ($\vec{\mathbf{E}}_L$ and $\vec{\mathbf{E}}_R$) tip-to-tail. The resultant vector, $\vec{\mathbf{E}}_{net}$, which is the net electric field at the specified point, begins at the tail of $\vec{\mathbf{E}}_L$ and ends at the tip of $\vec{\mathbf{E}}_R$. The length of $\vec{\mathbf{E}}_L$ or $\vec{\mathbf{E}}_R$ depends on how far the point is from the respective sphere.

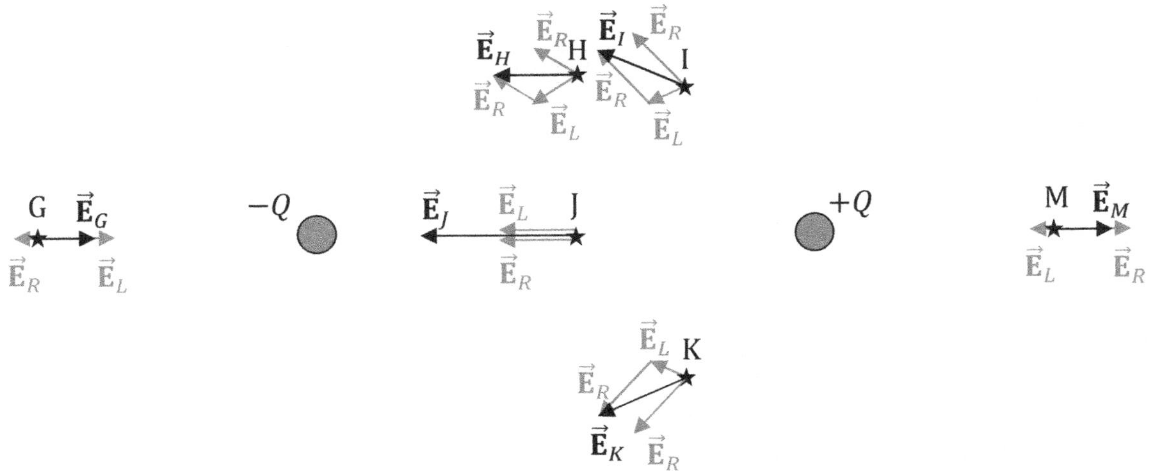

16. If a positive "test" charge were placed at any of these points, it would be attracted to each negative sphere. First draw two separate electric fields (one for each sphere) and then draw the resultant of these two vectors for the net electric field. Move one of the arrows (we moved $\vec{\mathbf{E}}_R$) to join the two vectors ($\vec{\mathbf{E}}_L$ and $\vec{\mathbf{E}}_R$) tip-to-tail. The resultant vector, $\vec{\mathbf{E}}_{net}$, which is the net electric field at the specified point, begins at the tail of $\vec{\mathbf{E}}_L$ and ends at the tip of $\vec{\mathbf{E}}_R$. The length of $\vec{\mathbf{E}}_L$ or $\vec{\mathbf{E}}_R$ depends on how far the point is from the respective sphere.

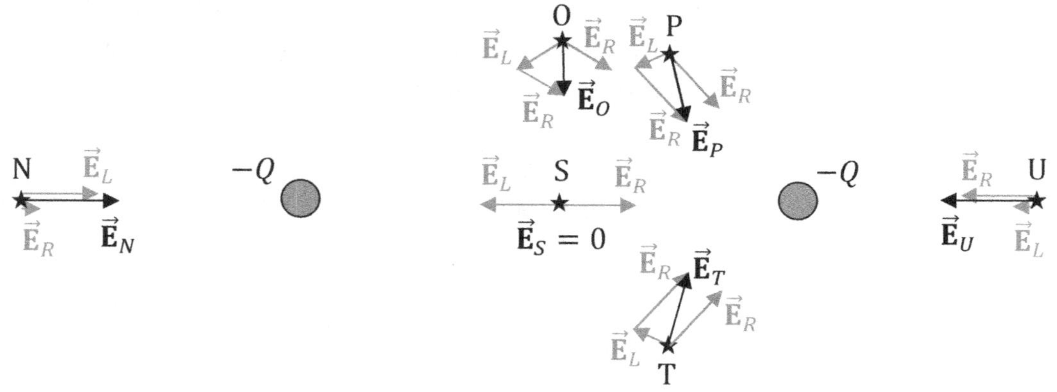

17. If a positive "test" charge were placed at point V, it would be repelled by the positive sphere and attracted to each negative sphere. First draw three separate electric fields (one for each sphere) and then draw the resultant of these three vectors for the net electric field. Move two of the arrows to join the three vectors tip-to-tail. The resultant vector, $\vec{\mathbf{E}}_{net}$, which is the net electric field at the specified point, begins at the tail of the first vector and ends at the tip of the last vector. Each of the three separate electric fields has the same length since point V is equidistant from the three charged spheres.

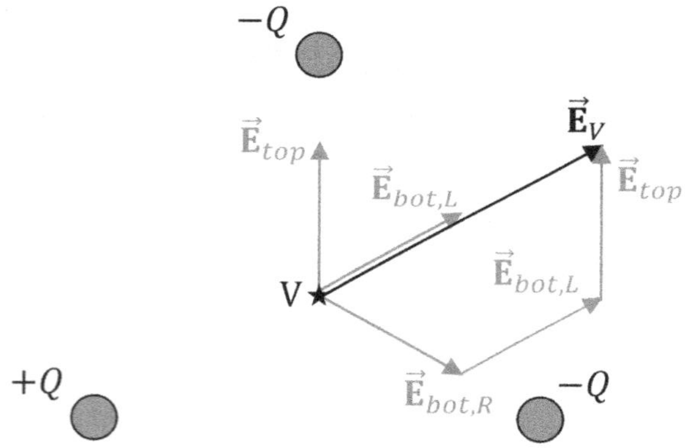

18. Make this map one step at a time:
- First, sample the net electric field in a variety of locations using the superposition strategy that we applied in the previous problems and examples.
  - The net electric field is <u>**zero**</u> in the <u>**center**</u> of the diagram (where a positive "test" charge would be repelled equally by both charged spheres). This point is called a **saddle point** in this diagram.
  - Along the horizontal line, the net electric field points to the left to the left of the left sphere and points to the right to the right of the right sphere.
  - The net electric field is somewhat **radial** (like the spokes of a bicycle wheel) **near either charge**, where the closer charge has the dominant effect.
- Beginning with the above features, draw smooth curves that leave each positive sphere. The lines aren't perfectly radial near either charge, but curve due to the influence of the other sphere.

- Check several points in different regions: The **tangent** line at any point on a line of force should match the direction of the net electric field from **superposition**.
- Draw smooth curves for the equipotential surfaces. Wherever an equipotential intersects a line of force, the two curves must be perpendicular to one another.

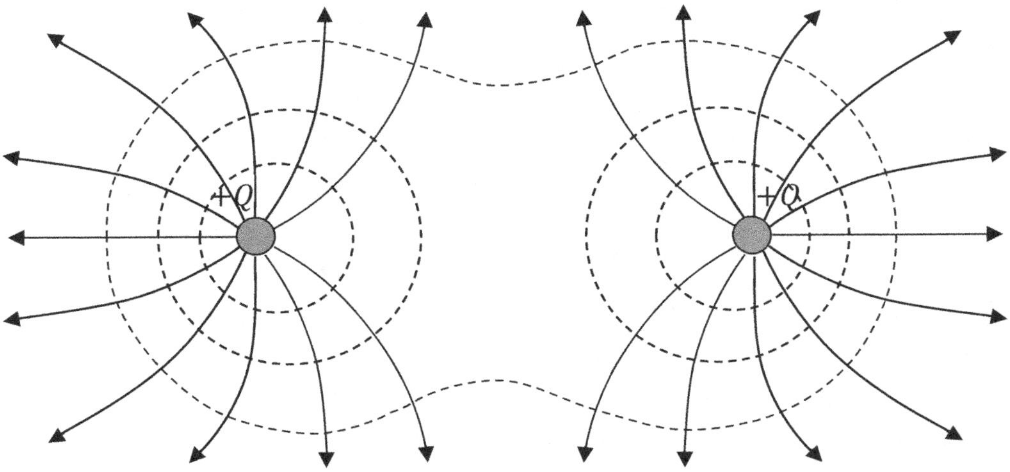

19. (A) Draw smooth curves that are **perpendicular** to the equipotentials wherever the lines of force intersect the equipotentials. Include arrows showing that the lines of force travel from higher electric potential (in Volts) to lower electric potential.

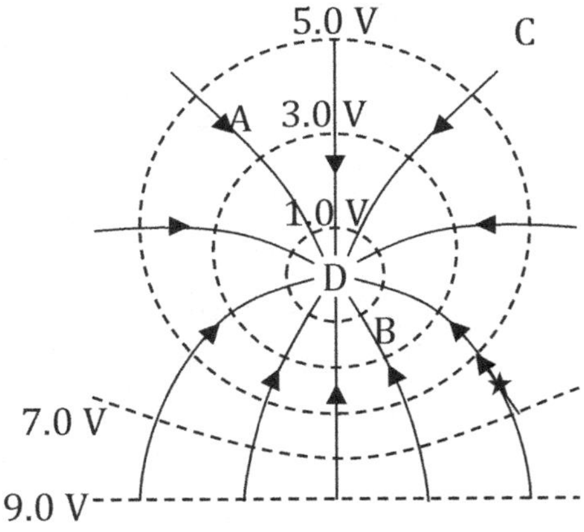

(B) There is a negative charge at point D where the lines of force **converge**.
(C) The lines of force are more dense at point B and less dense at point C: $E_B > E_A > E_C$.
(D) Draw an arrow through the star (★) that is on average roughly perpendicular to the neighboring equipotentials. The potential difference is $\Delta V = 7 - 5 = 2.0$ V. Measure the length of the line with a ruler: $\Delta R = 0.8$ cm $= 0.008$ m. Use the following formula.

$$E \approx \left|\frac{\Delta V}{\Delta R}\right| = \frac{2}{0.008} = 250\,\frac{\text{V}}{\text{m}} \text{ or } 250\,\frac{\text{N}}{\text{C}}$$

(E) It would be pushed **opposite** to the arrow drawn through the star above (because the arrow shows how a positive "test" charge would be pushed, and this question asks about a **negative** test charge). Note that this arrow isn't quite straight away from point D because there happens to be a positive horizontal rod at the bottom which has some influence.

Chapter 5: Electrostatic Equilibrium

20. Check your free-body diagrams (FBD). Note that the angle with the vertical is 30° (split the 60° angle at the top of the triangle in half to find this.)

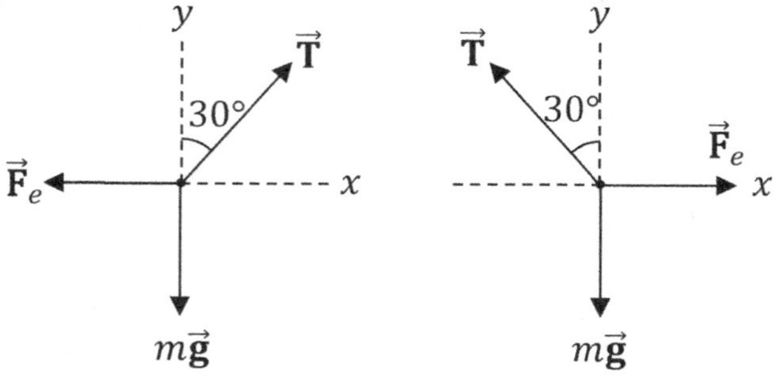

- Weight (mg) pulls straight down.
- Tension (T) pulls along the cord.
- The two bananas repel one another with an electric force (F_e) via Coulomb's law.
- Apply Newton's second law. In **electrostatic equilibrium**, $a_x = 0$ and $a_y = 0$.
- Since tension doesn't lie on an axis, T appears in both the x- and y-sums with trig. In the FBD, since y is adjacent to 30°, cosine appears in the y-sum.
- Since the electric force is horizontal, F_e appears only the x-sums with no trig.
- Since weight is vertical, mg appears only in the y-sums with no trig.

$$\sum F_{1x} = ma_x \quad , \quad \sum F_{1y} = ma_y \quad , \quad \sum F_{2x} = ma_x \quad , \quad \sum F_{2y} = ma_y$$

$$T\sin 30° - F_e = 0 \quad , \quad T\cos 30° - mg = 0 \quad , \quad F_e - T\sin 30° = 0 \quad , \quad T\cos 30° - mg = 0$$

(A) Solve for tension in the y-sum.

$$T = \frac{mg}{\cos 30°}$$

- The tension is $T \approx 180$ N. (If you don't round gravity, $T = 177$ N.)

(B) First solve for electric force in the x-sum.

$$F_e = T\sin 30°$$

- Plug in the tension from part (A). The electric force is $F_e \approx 90$ N.
- Apply Coulomb's law.

$$F_e = k\frac{|q_1||q_2|}{R^2}$$

- Set the charges equal to one another. Note that $|q||q| = q^2$.

$$F_e = k\frac{q^2}{R^2}$$

- Solve for q. Multiply both sides by R^2 and divide by k. Squareroot both sides.

$$q = \sqrt{\frac{F_e R^2}{k}} = R\sqrt{\frac{F_e}{k}}$$

- Note that $\sqrt{R^2} = R$. Coulomb's constant is $k = 9.0 \times 10^9 \frac{\text{N}\cdot\text{m}^2}{\text{C}^2}$.
- Plug in numbers. Note that $\frac{90}{9} = 10$ and that $\sqrt{\frac{10}{10^9}} = \sqrt{10^{-8}} = 10^{-4}$.
- The charge is $q = 2.0 \times 10^{-4}$ C, which can also be expressed as $q = 200$ μC using the metric prefix micro ($\mu = 10^{-6}$). Either both charges are positive or both charges are negative (there is no way to tell which without more information).

---

21. Check your free-body diagram (FBD). Verify that you labeled your angles correctly.

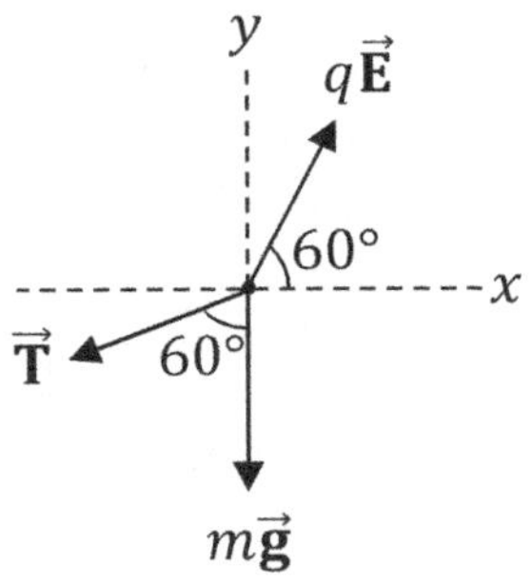

- Weight (mg) pulls straight down.
- Tension (T) pulls along the thread.
- The electric field ($\vec{\mathbf{E}}$) exerts an electric force ($\vec{\mathbf{F}}_e = q\vec{\mathbf{E}}$) on the charge, which is parallel to the electric field since the charge is positive.
- Apply Newton's second law. In **electrostatic equilibrium**, $a_x = 0$ and $a_y = 0$.
- Since tension doesn't lie on an axis, T appears in both the x- and y-sums with trig. In the FBD, since y is adjacent to 60° (for **tension**), cosine appears in the y-sum.
- Since the electric force doesn't lie on an axis, $|q|E$ appears in both the x- and y-sums with trig. Recall that $F_e = |q|E$ for a charge in an external electric field. In the FBD, since x is adjacent to 60° (for $q\vec{\mathbf{E}}$), cosine appears in the x-sum.
- Since weight is vertical, mg appears only in the y-sum with no trig.

$$\sum F_x = ma_x \quad , \quad \sum F_y = ma_y$$

$$|q|E\cos 60° - T\sin 60° = 0 \quad , \quad |q|E\sin 60° - T\cos 60° - mg = 0$$

- Simplify each equation. Note that $\cos 60° = \frac{1}{2}$ and $\sin 60° = \frac{\sqrt{3}}{2}$.

$$\frac{|q|E}{2} = \frac{T\sqrt{3}}{2} \quad , \quad \frac{|q|E\sqrt{3}}{2} - \frac{T}{2} = mg$$

(A) Each equation has two unknowns (E and T are both unknown). Make a substitution in order to solve for the unknowns. Isolate T in the first equation. Multiply both sides of the equation by 2 and divide both sides by $\sqrt{3}$.

$$T = \frac{|q|E}{\sqrt{3}}$$

- Substitute this expression for tension in the other equation.

$$\frac{|q|E\sqrt{3}}{2} - \frac{T}{2} = mg$$

$$\frac{|q|E\sqrt{3}}{2} - \frac{|q|E}{2\sqrt{3}} = mg$$

- The algebra is simpler if you multiply both sides of this equation by $\sqrt{3}$.

$$\frac{3|q|E}{2} - \frac{|q|E}{2} = mg\sqrt{3}$$

- Note that $\sqrt{3}\sqrt{3} = 3$ and $\frac{\sqrt{3}}{\sqrt{3}} = 1$. **Combine like terms**. The left-hand side becomes:

$$\frac{3|q|E}{2} - \frac{|q|E}{2} = \left(\frac{3}{2} - \frac{1}{2}\right)|q|E = |q|E$$

- Therefore, the previous equation simplifies to:

$$|q|E = mg\sqrt{3}$$

- Solve for the electric field. Divide both sides of the equation by the charge.

$$E = \frac{mg\sqrt{3}}{|q|}$$

- Convert the mass from grams (g) to kilograms (kg): $m = 60 \text{ g} = 0.060 \text{ kg} = \frac{3}{50} \text{ kg}$. Convert the charge from microCoulombs (μC) to Coulombs: $q = 300\ \mu\text{C} = 3.00 \times 10^{-4}$ C. Note that $\frac{\sqrt{3}}{5}10^4 = \frac{\sqrt{3}}{5}(10)(10^3)$ since $(10)(10^3) = 10^4$. Also note that $\frac{\sqrt{3}}{5}(10)(10^3) = 2\sqrt{3} \times 10^3 = 2000\sqrt{3}$.
- The magnitude of the electric field is $E = 2000\sqrt{3}\ \frac{\text{N}}{\text{C}}$.

(B) Plug $E = 2000\sqrt{3}\ \frac{\text{N}}{\text{C}}$ into $T = \frac{|q|E}{\sqrt{3}}$. The tension equals $T \approx \frac{3}{5}$ N $= 0.60$ N (if you round g).

Chapter 6: Integration Essentials

22. First find the anti-derivative using the formula $\int ax^b\, dx = \frac{ax^{b+1}}{b+1}$.
    - Compare $6x^2$ to the general form ax^b.
    - Note that $a = 6$ and $b = 2$.
    - The anti-derivative is $2x^3$. Evaluate $2x^3$ over the limits.
    - $[2x^3]_{x=2}^{3} = 2(3)^3 - 2(2)^3 = 54 - 16 = 38$.
    - The answer is 38.

---

23. Apply the rule $\frac{1}{x^m} = x^{-m}$ to write $\frac{8}{x^3}$ as $8x^{-3}$.
    - Find the anti-derivative using the formula $\int ax^b\, dx = \frac{ax^{b+1}}{b+1}$.
    - Compare $8x^{-3}$ to the general form ax^b.
    - Note that $a = 8$ and $b = -3$.

- Note that $b + 1 = -3 + 1 = -2$.
- The anti-derivative is $-4x^{-2}$. Evaluate $-4x^{-2}$ over the limits.
- $[-4x^{-2}]_{x=1}^{2} = -4(2)^{-2} - [-4(1)^{-2}]$.
- The two minus signs make a plus sign: $-4(2)^{-2} + 4(1)^{-2}$.
- Note that $2^{-2} = \frac{1}{2^2} = \frac{1}{4}$ and $1^{-2} = \frac{1}{1^2} = \frac{1}{1} = 1$. We get $-4\left(\frac{1}{4}\right) + 4(1) = -1 + 4 = 3$.
- The answer is 3.

---

24. Apply the rule $\sqrt{x} = x^{1/2}$ to write $3\sqrt{x}$ as $3x^{1/2}$.
    - Find the anti-derivative using the formula $\int ax^b\,dx = \frac{ax^{b+1}}{b+1}$.
    - Compare $3x^{1/2}$ to the general form ax^b.
    - Note that $a = 3$ and $b = \frac{1}{2}$.
    - Note that $b + 1 = \frac{1}{2} + 1 = \frac{1}{2} + \frac{2}{2} = \frac{1+2}{2} = \frac{3}{2}$.
    - Note that $\frac{a}{b+1} = \frac{3}{3/2} = 3\left(\frac{2}{3}\right) = 2$. (To divide by a fraction, multiply by its **reciprocal**.)
    - The anti-derivative is $2x^{3/2}$. Evaluate $2x^{3/2}$ over the limits.
    - $\left[2x^{3/2}\right]_{x=9}^{16} = 2(16)^{3/2} - 2(9)^{3/2}$.
    - Note that $(16)^{3/2} = \left(\sqrt{16}\right)^3 = (4)^3 = 64$ and $(9)^{3/2} = \left(\sqrt{9}\right)^3 = (3)^3 = 27$. (You can try entering 16^1.5 on your calculator, for example. Note that $\frac{3}{2} = 1.5$.)
    - You get $2(64) - 2(27) = 128 - 54$.
    - The answer is 74.

---

25. Apply the rule $\int (y_1 - y_2)\,dx = \int y_1\,dx - \int y_2\,dx$.
    - Find the anti-derivative of each term using the formula $\int ax^b\,dx = \frac{ax^{b+1}}{b+1}$.
    - The separate anti-derivatives are:

$$\int 8x^3\,dx = 2x^4 \quad , \quad \int 6x\,dx = 3x^2$$

    - Combine these anti-derivatives together to make $2x^4 - 3x^2$.
    - Evaluate $2x^4 - 3x^2$ over the limits.
    - $[2x^4 - 3x^2]_{x=1}^{3} = [2(3)^4 - 3(3)^2] - [2(1)^4 - 3(1)^2]$.
    - Distribute the minus sign. Two minus signs make a plus sign.
    - You should get $2(3)^4 - 3(3)^2 - 2(1)^4 + 3(1)^2$.
    - The final answer is 136.

---

26. First find the anti-derivative.
    - The anti-derivative is $\sin\theta$.
    - Evaluate the anti-derivative over the limits.

- $[\sin\theta]_{\theta=30°}^{90°} = \sin(90°) - \sin(30°)$.
- The answer is $\frac{1}{2}$. Note that $1 - \frac{1}{2} = \frac{1}{2}$.

---

27. First find the anti-derivative.
    - The anti-derivative is $-\cos\theta$.
    - Evaluate the anti-derivative over the limits.
    - $[-\cos\theta]_{\theta=45°}^{135°} = -\cos(135°) - [-\cos(45°)] = -\cos(135°) + \cos(45°)$.
    - The answer is $\sqrt{2}$. Note that $-\left(-\frac{\sqrt{2}}{2}\right) + \frac{\sqrt{2}}{2} = \frac{\sqrt{2}}{2} + \frac{\sqrt{2}}{2} = \frac{2\sqrt{2}}{2} = \sqrt{2}$.

---

28. Make the substitution $u = \frac{x}{4} + 1$.
    - Take an implicit derivative of both sides to get $du = \frac{dx}{4}$.
    - Solve for dx to get $dx = 4du$.
    - Plug the limits of x into the equation $u = \frac{x}{4} + 1$.
    - The new limits are from $u = 1$ to $u = 3$.
    - After making these substitutions, the integral becomes $\int_{u=1}^{3} 4u^3\,du$.
    - The anti-derivative is u^4.
    - Evaluate the anti-derivative over the limits.
    - $[u^4]_{u=1}^{3} = (3)^4 - (1)^4$.
    - The final answer is 80.

---

29. Make the substitution $u = 2x - 1$.
    - Take an implicit derivative of both sides to get $du = 2dx$.
    - Solve for dx to get $dx = \frac{du}{2}$.
    - Plug the limits of x into the equation $u = 2x - 1$.
    - The new limits are from $u = 1$ to $u = 9$.
    - After making these substitutions, the integral becomes $\int_{u=1}^{9} \frac{3}{2}u^{1/2}\,du$.
    - Note that $u^{1/2} = \sqrt{u}$ (see the hints to Problem 24). The anti-derivative is $u^{3/2}$.
    - Evaluate the anti-derivative over the limits.
    - $\left[u^{3/2}\right]_{u=1}^{9} = (9)^{3/2} - (1)^{3/2}$. Note that $(9)^{3/2} = \left(\sqrt{9}\right)^3 = (3)^3 = 27$.
    - The final answer is 26.

---

30. Make the substitution $x = 3\sin u$.
    - Take an implicit derivative of both sides to get $dx = 3\cos u\,du$.
    - Plug the limits of x into the equation $u = \sin^{-1}\left(\frac{x}{3}\right)$.
    - The new limits are from $u = 0°$ to $u = 90°$. This integral will only work in **radians**.

Convert to radians: $360° = 2\pi$ rad. The new limits are from $u = 0$ to $u = \frac{\pi}{2}$.

- After making these substitutions, you should get $\int_{u=0}^{\pi/2} 4\sqrt{9 - 9\sin^2 u}\ 3\cos u\, du$.
- Apply the trig identity $1 - \sin^2 u = \cos^2 u$. Simplify.
- You should get $36 \int_{u=0}^{\pi/2} \cos^2 u\, du$.
- Apply the trig identity $\cos^2 u = \frac{1+\cos 2u}{2}$.
- After this substitution, separate the integral into two integrals, one for each term.
- You should get $36 \int_{u=0}^{\pi/2} \frac{1}{2} du + 36 \int_{u=0}^{\pi/2} \frac{\cos 2u}{2} du = 18 \int_{u=0}^{\pi/2} du + 18 \int_{u=0}^{\pi/2} \cos 2u\, du$.
- The anti-derivatives are $18u$ and $9 \sin 2u$ (including the coefficients).
- Evaluate the anti-derivatives over the limits.
- The final answer is 9π. (Note that the second integral yields zero at both limits.)

---

31. Make the substitution $x = 4 \tan u$.
- Take an implicit derivative of both sides to get $dx = 4\sec^2 u\, du$.
- Plug the limits of x into the equation $u = \tan^{-1}\left(\frac{x}{4}\right)$.
- The new limits are from $u = 0°$ to $u = 45°$.
- After making these substitutions, the integral becomes $\int_{u=0°}^{45°} \frac{\sec^2 u}{\sqrt{1+\tan^2 u}} du$.
- Apply the trig identity $1 + \tan^2 u = \sec^2 u$. Simplify.
- You should get $\int_{u=0°}^{45°} \sec u\, du$.
- Find the anti-derivative of secant on page 70.
- The anti-derivative is $\ln|\sec u + \tan u|$.
- Evaluate the anti-derivative over the limits.
- $[\ln|\sec u + \tan u|]_{u=0°}^{45°} = \ln|\sec 45° + \tan 45°| - \ln|\sec 0° + \tan 0°|$.
- The last term vanishes because $\ln|1| = 0$.
- The final answer is $\ln|\sqrt{2} + 1|$.

---

32. First integrate over y (since the upper limit of y is x).
- Factor the $10x$ out of the y-integration: $\int_{x=0}^{3} 10x \left(\int_{y=0}^{x} y^2\, dy\right) dx$.
- The anti-derivative for the y-integration is $\frac{y^3}{3}$. Evaluate $\frac{y^3}{3}$ over the limits.
- The answer to the definite integral over y is $\frac{x^3}{3}$.
- Now you should have $\int_{x=0}^{3} \frac{10}{3} x^4\, dx$.
- The anti-derivative for the x-integration is $\frac{2x^5}{3}$. Evaluate $\frac{2x^5}{3}$ over the limits.
- The final answer to the double integral is 162.

33. First integrate over x (since the upper limit of x is y).

- Factor the y out of the x-integration: $\int_{y=0}^{3} 10y^2 \left(\int_{x=0}^{y} x\, dx\right) dy$.
- The anti-derivative for the x-integration is $\frac{x^2}{2}$. Evaluate $\frac{x^2}{2}$ over the limits.
- The answer to the definite integral over x is $\frac{y^2}{2}$.
- Now you should have $\int_{y=0}^{3} 5y^4\, dy$.
- The anti-derivative for the y-integration is y^5. Evaluate y^5 over the limits.
- The final answer to the double integral is 243.

---

34. You can do these integrals in any order. We will integrate over y first.

- Factor the x^3 out of the y-integration: $\int_{x=0}^{2} x^3 \left(\int_{y=0}^{3} y^2\, dy\right) dx$.
- The anti-derivative for the y-integration is $\frac{y^3}{3}$. Evaluate $\frac{y^3}{3}$ over the limits.
- The answer to the definite integral over y is 9.
- Now you should have $\int_{x=0}^{2} 9x^3\, dx$.
- The anti-derivative for the x-integration is $\frac{9x^4}{4}$. Evaluate it over the limits.
- The final answer to the double integral is 36.

---

35. First integrate over z: $\int_{x=0}^{4} \int_{y=0}^{x} 6\left(\int_{z=0}^{y} z\, dz\right) dy\, dx$.

- The anti-derivative for the z-integration is $\frac{z^2}{2}$. Evaluate $\frac{z^2}{2}$ over the limits.
- The answer to the definite integral over z is $\frac{y^2}{2}$.
- Now you should have $\int_{x=0}^{4} \left(\int_{y=0}^{x} 3y^2\, dy\right) dx$. Integrate over y.
- The anti-derivative for the y-integration is y^3. Evaluate y^3 over the limits.
- The answer to the definite integral over y is x^3.
- Now you should have $\int_{x=0}^{4} x^3\, dx$. Integrate over x.
- The anti-derivative for the x-integration is $\frac{x^4}{4}$. Evaluate $\frac{x^4}{4}$ over the limits.
- The final answer to the triple integral is 64.

---

36. Integrate over area: $A = \int dA$. Study the **example** on page 87.

- Express the differential area element using Cartesian coordinates: $dA = dxdy$.
- Find the equation of the line for the hypotenuse.
- The line for the hypotenuse has a slope of $-\frac{1}{2}$ and y-intercept of 4.
- The equation for the line is therefore $y = -\frac{x}{2} + 4$.
- Let x vary from 0 to 8. Then y will vary from 0 to $-\frac{x}{2} + 4$.

- The double integral is:

$$A = \int_{x=0}^{8} \int_{y=0}^{-\frac{x}{2}+4} dydx$$

- First integrate over y (since the upper limit of y is $-\frac{x}{2} + 4$).
- The anti-derivative for the y-integration is y. Evaluate y over the limits.
- The answer to the definite integral over y is $-\frac{x}{2} + 4$.
- Now you should have $\int_{x=0}^{8} \left(-\frac{x}{2} + 4\right) dx$. Find the anti-derivative of each term.
- The anti-derivative for the x-integration is $-\frac{x^2}{4} + 4x$. Evaluate this over the limits.
- The final answer to the double integral is $A = 16$.

---

37. Integrate over area: $A = \int dA$. Study the **example** on page 88.
- Express the differential area element using 2D polar coordinates: $dA = rdrd\theta$.
- The double integral is:

$$A = \int_{r=2}^{4} \int_{\theta=0}^{2\pi} rdrd\theta$$

- You can do these integrals in any order. We will integrate over θ first.
- Factor the r out of the θ-integation: $\int_{r=2}^{4} r \left(\int_{\theta=0}^{2\pi} d\theta\right) dr$.
- The anti-derivative for the θ-integration is θ. Evaluate θ over the limits.
- The answer to the definite integral over θ is 2π.
- Now you should have $\int_{r=2}^{4} 2\pi r\, dr$.
- The anti-derivative for the r-integration is πr^2. Evaluate πr^2 over the limits.
- The final answer to the double integral is $A = 12\pi$ m^2. Note that $4^2 - 2^2 = 12$.

Chapter 7: Electric Field Integrals

38. This problem is similar to the examples involving a charged rod. Study those examples and try to let them serve as a guide. If you need help, you can always return here.
- Begin with a labeled diagram. Draw a representative dq. Draw $\vec{\mathbf{R}}$ from the source, dq, to the field point $(0, -a)$. See the diagram on the next page.
- From the diagram on the following page, you should be able to see that $R = y + a$ (since dq is y units to the origin and another a units to the field point) and that $\hat{\mathbf{R}} = -\hat{\mathbf{y}}$ (since $\vec{\mathbf{R}}$ points downward).
- For a rod, write $dq = \lambda ds$ (Step 6 on page 94). For a rod on the y-axis, $ds = dy$.
- Since this rod has uniform charge density, you may pull λ out of the integral.

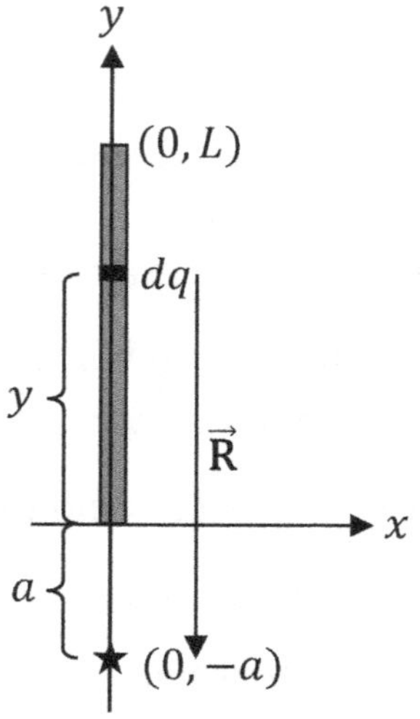

- Begin the math with the electric field integral.

$$\vec{\mathbf{E}} = k \int \frac{\hat{\mathbf{R}}}{R^2} dq$$

- Note that y varies from 0 to L (the endpoints of the rod).
- Substitute the expressions from the previous hints into the electric field integral. After simplifying, you should get:

$$\vec{\mathbf{E}} = -k\lambda\hat{\mathbf{y}} \int_{y=0}^{L} \frac{dy}{(y+a)^2}$$

- One way to perform the above integral is to make the following substitution.

$$u = y + a$$
$$du = dy$$

- Plug in the limits of y to find that $u(0) = a$ and $u(L) = L + a$. The integral becomes:

$$\vec{\mathbf{E}} = -k\lambda\hat{\mathbf{y}} \int_{u=a}^{L+a} \frac{du}{u^2}$$

- Apply the rule from algebra that $\frac{1}{u^2} = u^{-2}$.

$$\vec{\mathbf{E}} = -k\lambda\hat{\mathbf{y}} \int_{u=a}^{L+a} u^{-2}\, du$$

- The anti-derivative is $-u^{-1}$, which is the same as $-\frac{1}{u}$. Evaluate the anti-derivative over the limits. You should get $-\frac{1}{L+a} + \frac{1}{a}$. Make a **common denominator** of $a(L+a)$.
- Note that $-\frac{1}{L+a} + \frac{1}{a} = -\frac{a}{a(L+a)} + \frac{L+a}{a(L+a)} = \frac{-a+L+a}{a(L+a)} = \frac{L}{a(L+a)}$.
- Replace the previous definite integral with this expression.

$$\vec{\mathbf{E}} = -\frac{k\lambda L}{a(L+a)}\hat{\mathbf{y}}$$

- Integrate $Q = \int dq$ using the same substitutions as before.
- You should get $Q = \int \lambda\, ds = \lambda \int_{y=0}^{L} dy = \lambda L$. Solve for λ. You should get $\lambda = \frac{Q}{L}$.
- Plug λ into the previous equation for electric field. The final answer is:

$$\vec{\mathbf{E}} = -\frac{kQ}{a(L+a)}\hat{\mathbf{y}}$$

39. This problem is similar to the examples involving a charged rod. Study those examples and try to let them serve as a guide. If you need help, you can always return here.

- Begin with a labeled diagram. Draw a representative dq. Draw $\vec{\mathbf{R}}$ from the source, dq, to the field point $(0, p)$. See the diagram below.

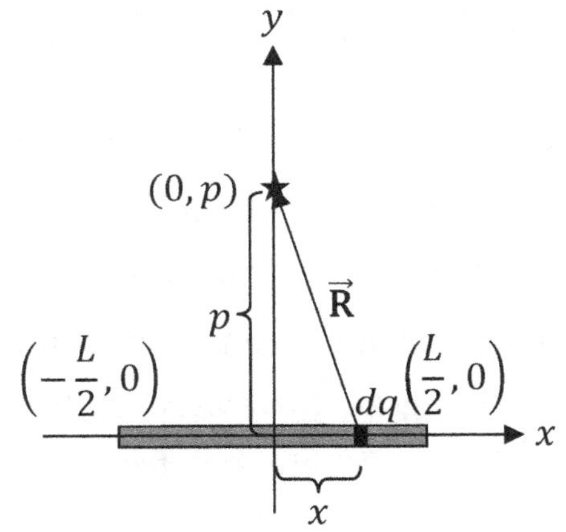

- From the diagram above, for the dq drawn, you should be able to see that $\vec{\mathbf{R}}$ extends x units to the left (along $-\hat{\mathbf{x}}$) and p units up (along $\hat{\mathbf{y}}$). Therefore, $\vec{\mathbf{R}} = -x\hat{\mathbf{x}} + p\hat{\mathbf{y}}$.
- Apply the Pythagorean theorem to find R. Divide $\vec{\mathbf{R}}$ by R to find $\hat{\mathbf{R}}$.
- You should get $R = \sqrt{x^2 + p^2}$ and $\hat{\mathbf{R}} = \frac{-x\hat{\mathbf{x}} + p\hat{\mathbf{y}}}{\sqrt{x^2+p^2}}$.
- For a rod, write $dq = \lambda ds$ (Step 6 on page 94). For a rod on the x-axis, $ds = dx$.
- Since the charge density is non-uniform, you may **not** pull λ out of the integral. Instead, plug in the expression $\lambda = \beta|x|$ that is given. You may pull β out of the integral.
- Begin the math with the electric field integral.

$$\vec{\mathbf{E}} = k \int \frac{\hat{\mathbf{R}}}{R^2} dq$$

- Note that x varies from $-\frac{L}{2}$ to $\frac{L}{2}$ (the endpoints of the rod).
- Substitute the expressions from the previous hints into the electric field integral. After simplifying, you should get:

$$\vec{\mathbf{E}} = k\beta \int_{x=-L/2}^{L/2} \frac{(-x\hat{\mathbf{x}} + p\hat{\mathbf{y}})|x|\,dx}{(x^2+p^2)^{3/2}}$$

- Note that $(x^2 + p^2)\sqrt{x^2 + p^2} = (x^2 + p^2)^1(x^2 + p^2)^{1/2} = (x^2 + p^2)^{3/2}$.
- Separate the integral into two terms. Note that $\frac{-x\hat{\mathbf{x}}+p\hat{\mathbf{y}}}{(x^2+p^2)^{3/2}} = \frac{-x\hat{\mathbf{x}}}{(x^2+p^2)^{3/2}} + \frac{p\hat{\mathbf{y}}}{(x^2+p^2)^{3/2}}$.

$$\vec{\mathbf{E}} = -k\beta \int_{x=-L/2}^{L/2} \frac{-x\hat{\mathbf{x}}|x|dx}{(x^2 + p^2)^{3/2}} + k\beta \int_{x=-L/2}^{L/2} \frac{p\hat{\mathbf{y}}|x|dx}{(x^2 + p^2)^{3/2}}$$

- The first integral is zero. One way to see this (without doing all the math!) is to look at the picture. By symmetry, the horizontal component of the electric field will vanish. Therefore, the $\hat{\mathbf{x}}$-integral is zero: $\vec{\mathbf{E}}$ will be along $\hat{\mathbf{y}}$ (straight up).
- You may pull $p\hat{\mathbf{y}}$ out of the second integral since they are constants.

$$\vec{\mathbf{E}} = k\beta p\hat{\mathbf{y}} \int_{x=-L/2}^{L/2} \frac{|x|dx}{(x^2 + p^2)^{3/2}}$$

- The integrand is an **even** function of x because $\frac{|x|}{(x^2+p^2)^{3/2}}$ is the same whether you plug in $+x$ or $-x$, since $|-x| = |x|$ and since $(-x)^2 = x^2$. Therefore, we would obtain the same result for the integral from $x = -\frac{L}{2}$ to 0 as we would for the integral from $x = 0$ to $\frac{L}{2}$. This means that we can change the limits to $x = 0$ to $\frac{L}{2}$ and add a factor of 2 in front of the integral. Now we may drop the absolute values from $|x|$, since x won't be negative when we integrate from $x = 0$ to $\frac{L}{2}$.

$$\vec{\mathbf{E}} = 2k\beta p\hat{\mathbf{y}} \int_{x=0}^{L/2} \frac{xdx}{(x^2 + p^2)^{3/2}}$$

- One way to perform the previous integral is to make the following substitution.

$$x = p \tan\theta$$
$$dx = p \sec^2\theta \, d\theta$$

- It's easier to ignore the new limits for now and deal with them later. After you make these substitutions and simplify, you should get:

$$\vec{\mathbf{E}} = 2k\beta\hat{\mathbf{y}} \int \sin\theta \, d\theta$$

- The anti-derivative is $-\cos\theta$. Draw a right triangle to express $\cos\theta$ in terms of x, and then we can simply evaluate the function over the old limits. Since $x = p\tan\theta$, which means that $\tan\theta = \frac{x}{p}$, we draw a right triangle with x opposite to θ and with p adjacent to θ.

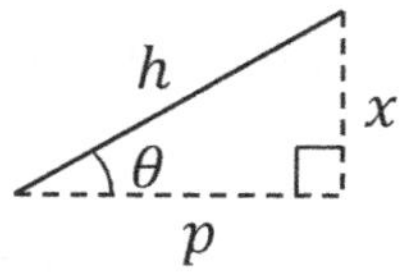

- Apply the Pythagorean theorem to find the hypotenuse: $h = \sqrt{x^2 + p^2}$.

- From the right triangle, you should see that $\cos\theta = \frac{p}{\sqrt{x^2+p^2}}$. The anti-derivative is thus $-\cos\theta = -\frac{p}{\sqrt{x^2+p^2}}$. Evaluate this over the limits from $x = 0$ to $\frac{L}{2}$.
- You should get $-\frac{p}{\sqrt{\frac{L^2}{4}+p^2}} + 1$. Replace the previous integral with this expression.

$$\vec{\mathbf{E}} = 2k\beta\hat{\mathbf{y}}\left(1 - \frac{p}{\sqrt{\frac{L^2}{4} + p^2}}\right)$$

- Note that $-\frac{p}{\sqrt{\frac{L^2}{4}+p^2}} + 1 = 1 - \frac{p}{\sqrt{\frac{L^2}{4}+p^2}}$.
- Integrate $Q = \int dq$ using the same substitutions as before.
- You should get $Q = \beta \int_{x=-L/2}^{L/2} |x|\, dx = 2\beta \int_{x=0}^{L/2} x\, dx = \frac{\beta L^2}{4}$.
- Solve for β. You should get $\beta = \frac{4Q}{L^2}$.
- Plug β into the previous equation for electric field. The final answer is:

$$\vec{\mathbf{E}} = \frac{8kQ}{L^2}\hat{\mathbf{y}}\left(1 - \frac{p}{\sqrt{\frac{L^2}{4} + p^2}}\right)$$

---

40. The "trick" to this problem is to apply the principle of superposition (see Chapter 3).
    - First, ignore the negative semicircle (as in the left diagram below) and find the electric field at the origin due to the positive semicircle. Call this $\vec{\mathbf{E}}_1$. This should be easy because this part of the solution is identical to one of the examples: If you need help with this, see the example on page 100 (but here the electric field has the opposite direction since this semicircle is **positive**). You should get $\vec{\mathbf{E}}_1 = -\frac{2kQ}{\pi a^2}\hat{\mathbf{y}}$.

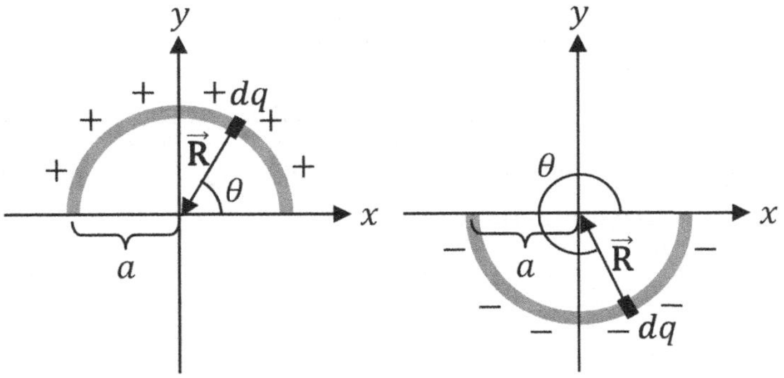

- Next, ignore the positive semicircle (as in the right diagram above) and find the electric field at the origin due to the negative semicircle. Call this $\vec{\mathbf{E}}_2$. This is very similar to the previous step. You should get $\vec{\mathbf{E}}_2 = -\frac{2kQ}{\pi a^2}\hat{\mathbf{y}}$.
- Note that $\vec{\mathbf{E}}_1$ and $\vec{\mathbf{E}}_2$ are both equal and both point downward: A positive "test"

charge at the origin would be repelled downward by the positive semicircle, and would also be attracted downward to the negative semicircle. That's why $\vec{\mathbf{E}}_1$ and $\vec{\mathbf{E}}_2$ both point downward.

- Add the vectors $\vec{\mathbf{E}}_1$ and $\vec{\mathbf{E}}_2$ to find the net electric field, $\vec{\mathbf{E}}_{net}$, at the origin. Since $\vec{\mathbf{E}}_1$ and $\vec{\mathbf{E}}_2$ are both equal (in magnitude and direction), the answer is simply two times $\vec{\mathbf{E}}_1$. The final answer is:

$$\vec{\mathbf{E}}_{net} = -\frac{4kQ}{\pi a^2}\hat{\mathbf{y}}$$

---

41. This problem is similar to the example involving a charged disc. Study that example and try to let it serve as a guide. If you need help, you can always return here.

- Begin with a labeled diagram. Draw a representative dq. Draw $\vec{\mathbf{R}}$ from the source, dq, to the field point $(0, 0, p)$. See the diagram below.

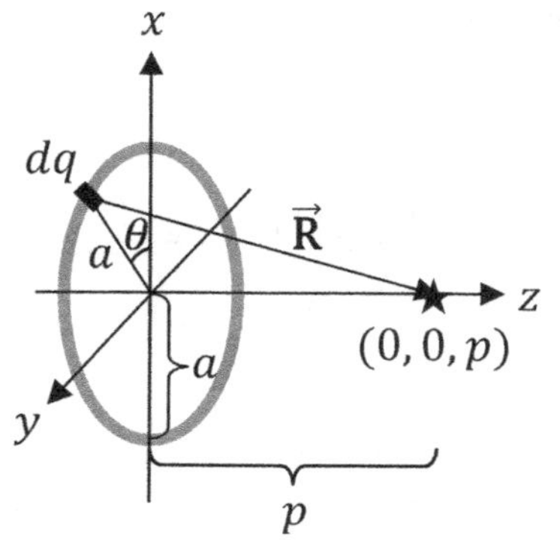

- From the diagram above, for the dq drawn, you should be able to see that $\vec{\mathbf{R}}$ extends a units inward, towards the z-axis (along $-\hat{\mathbf{r}}_c$), and p units along the z-axis (along $\hat{\mathbf{z}}$). Therefore, $\vec{\mathbf{R}} = -a\hat{\mathbf{r}}_c + p\hat{\mathbf{z}}$. Note that this is the $\hat{\mathbf{r}}_c$ of **cylindrical** coordinates.
- Apply the Pythagorean theorem to find R. Divide $\vec{\mathbf{R}}$ by R to find $\hat{\mathbf{R}}$.
- You should get $R = \sqrt{a^2 + p^2}$ and $\hat{\mathbf{R}} = \frac{-a\hat{\mathbf{r}}_c + p\hat{\mathbf{z}}}{\sqrt{a^2+p^2}}$.
- For a very thin ring, write $dq = \lambda ds$ (Step 6 on page 94). Since the ring is circular, $ds = a\, d\theta$. (This is $ds = r\, d\theta$, but with the radius of the ring equal to $r = a$.)
- Since this ring has uniform charge density, you may pull λ out of the integral.
- Begin the math with the electric field integral.

$$\vec{\mathbf{E}} = k \int \frac{\hat{\mathbf{R}}}{R^2} dq$$

- Note that θ varies from 0 to 2π (full circle).
- Substitute the expressions from the previous hints into the electric field integral. After simplifying, you should get:

$$\vec{\mathbf{E}} = k\lambda \int_{\theta=0}^{2\pi} \frac{(-a\hat{\mathbf{r}}_c + p\hat{\mathbf{z}})a}{(a^2 + p^2)^{3/2}} d\theta$$

- Note that $(a^2 + p^2)\sqrt{a^2 + p^2} = (a^2 + p^2)^1(a^2 + p^2)^{1/2} = (a^2 + p^2)^{3/2}$.
- Separate the integral into two terms. Note that $\frac{-a\hat{\mathbf{r}}_c + p\hat{\mathbf{z}}}{(a^2+p^2)^{3/2}} = \frac{-a\hat{\mathbf{r}}_c}{(a^2+p^2)^{3/2}} + \frac{p\hat{\mathbf{z}}}{(a^2+p^2)^{3/2}}$.

$$\vec{\mathbf{E}} = k\lambda \int_{\theta=0}^{2\pi} \frac{-a^2\hat{\mathbf{r}}_c}{(a^2 + p^2)^{3/2}} d\theta + k\lambda \int_{\theta=0}^{2\pi} \frac{pa\hat{\mathbf{z}}}{(a^2 + p^2)^{3/2}} d\theta$$

- The first integral is zero. One way to see this (without doing all the math!) is to look at the picture. By symmetry, the radial component (which extends outward away from the z-axsis) of the electric field will vanish. Therefore, the $\hat{\mathbf{r}}_c$-integral is zero: $\vec{\mathbf{E}}$ will be along $+\hat{\mathbf{z}}$ (horizontal, to the right, along the $+z$-axis).
- You may pull **<u>everything</u>** out of the second integral since they are all constants. None of those symbols depend on θ. (The only quantity that depended upon θ was $\hat{\mathbf{r}}_c$, but the first integral is zero and $\hat{\mathbf{r}}_c$ doesn't appear in the second integral.)

$$\vec{\mathbf{E}} = \frac{k\lambda pa}{(a^2 + p^2)^{3/2}} \hat{\mathbf{z}} \int_{\theta=0}^{2\pi} d\theta$$

- The remaining integral is trivial: The definite integral equals 2π.

$$\vec{\mathbf{E}} = \frac{2\pi k\lambda pa}{(a^2 + p^2)^{3/2}} \hat{\mathbf{z}}$$

- Integrate $Q = \int dq$ using the same substitutions as before.
- You should get $Q = \lambda \int_{\theta=0}^{2\pi} a\, d\theta = 2\pi\lambda a$.
- Solve for λ. You should get $\lambda = \frac{Q}{2\pi a}$.
- Plug λ into the previous equation for electric field. The final answer is:

$$\vec{\mathbf{E}} = \frac{kQp}{(a^2 + p^2)^{3/2}} \hat{\mathbf{z}}$$

---

42. This problem is similar to the example involving a charged disc. Study that example and try to let it serve as a guide. If you need help, you can always return here.

- Begin with a labeled diagram. Draw a representative dq. Draw $\vec{\mathbf{R}}$ from the source, dq, to the field point $(0, 0, p)$. See the diagram below.

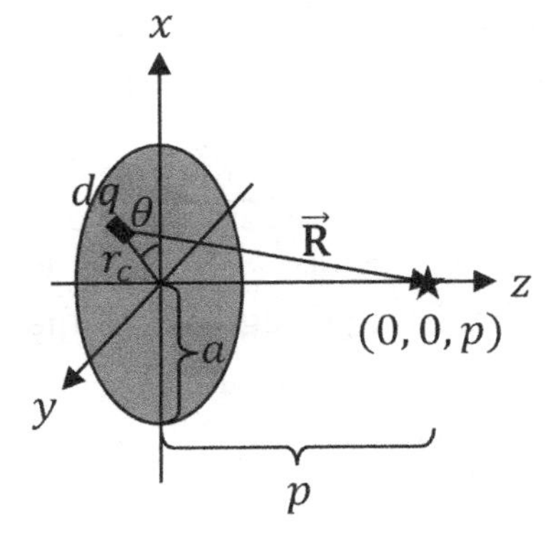

- From the previous diagram, for the dq drawn, you should be able to see that $\vec{\mathbf{R}}$ extends r_c units inward, towards the z-axis (along $-\hat{\mathbf{r}}_c$), and p units along the z-axis (along $\hat{\mathbf{z}}$). Therefore, $\vec{\mathbf{R}} = -r_c\hat{\mathbf{r}}_c + p\hat{\mathbf{z}}$. We are using **cylindrical** coordinates.
- Apply the Pythagorean theorem to find R. Divide $\vec{\mathbf{R}}$ by R to find $\hat{\mathbf{R}}$.
- You should get $R = \sqrt{r_c^2 + p^2}$ and $\hat{\mathbf{R}} = \frac{-r_c\hat{\mathbf{r}}_c + p\hat{\mathbf{z}}}{\sqrt{r_c^2+p^2}}$.
- For a solid disc, write $dq = \sigma dA$ (Step 6 on page 94). For a disc, $dA = r_c dr_c d\theta$.
- Since the charge density is non-uniform, you may **not** pull σ out of the integral. Instead, plug in the expression given in the problem: $\sigma = \beta r_c$. You may pull β out of the integral.
- Begin the math with the electric field integral.

$$\vec{\mathbf{E}} = k\int \frac{\hat{\mathbf{R}}}{R^2} dq$$

- Note that r_c varies from 0 to a and θ varies from 0 to 2π (full circle).
- Substitute the expressions from the previous hints into the electric field integral. After simplifying, you should get:

$$\vec{\mathbf{E}} = k\beta \int_{r_c=0}^{a}\int_{\theta=0}^{2\pi} \frac{(-r_c\hat{\mathbf{r}}_c + p\hat{\mathbf{z}})}{(r_c^2+p^2)^{3/2}} r_c^2 dr_c d\theta$$

- Note that one factor of r_c came from $dA = r_c dr_c d\theta$ and another factor came from $\sigma = \beta r_c$, and note that $(a^2+p^2)\sqrt{a^2+p^2} = (a^2+p^2)^1(a^2+p^2)^{1/2} = (a^2+p^2)^{3/2}$.
- Separate the integral into two terms. Note that $\frac{-r_c\hat{\mathbf{r}}_c+p\hat{\mathbf{z}}}{(r_c^2+p^2)^{3/2}} = \frac{-r_c\hat{\mathbf{r}}_c}{(r_c^2+p^2)^{3/2}} + \frac{p\hat{\mathbf{z}}}{(r_c^2+p^2)^{3/2}}$.

$$\vec{\mathbf{E}} = k\beta \int_{r_c=0}^{a}\int_{\theta=0}^{2\pi} \frac{-r_c\hat{\mathbf{r}}_c}{(r_c^2+p^2)^{3/2}} r_c^2 dr_c d\theta + k\beta \int_{r_c=0}^{a}\int_{\theta=0}^{2\pi} \frac{p\hat{\mathbf{z}}}{(r_c^2+p^2)^{3/2}} r_c^2 dr_c d\theta$$

- The first integral is zero. One way to see this (without doing all the math!) is to look at the picture. By symmetry, the radial component (which extends outward away from the z-axsis) of the electric field will vanish. Therefore, the $\hat{\mathbf{r}}_c$-integral is zero: $\vec{\mathbf{E}}$ will be along $+\hat{\mathbf{z}}$ (horizontal, to the right, along the $+z$-axis).
- You may pull $p\hat{\mathbf{z}}$ out of the second integral since they are constants.

$$\vec{\mathbf{E}} = k\beta p\hat{\mathbf{z}} \int_{r_c=0}^{a}\int_{\theta=0}^{2\pi} \frac{r_c^2 dr_c d\theta}{(r_c^2+p^2)^{3/2}} = k\beta p\hat{\mathbf{z}} \int_{r_c=0}^{a} \frac{r_c^2 dr_c}{(r_c^2+p^2)^{3/2}} \int_{\theta=0}^{2\pi} d\theta = 2\pi k\beta p\hat{\mathbf{z}} \int_{r_c=0}^{a} \frac{r_c^2 dr_c}{(r_c^2+p^2)^{3/2}}$$

- One way to perform the remaining integral is to make the following substitution.

$$r_c = p\tan\psi$$
$$dr_c = p\sec^2\psi\, d\psi$$

- It's easier to ignore the new limits for now and deal with them later. After you make these substitutions and simplify, you should get:

$$\vec{\mathbf{E}} = 2\pi k\beta p\hat{\mathbf{z}}\int \frac{\tan^2\psi \, d\psi}{\sec\psi}$$

- Note that a factor of $\sec^2\psi$ canceled between the numerator and denominator.
- Recall from trig that $\tan\psi = \frac{\sin\psi}{\cos\psi}$ and $\sec\psi = \frac{1}{\cos\psi}$. You should find that:

$$\frac{\tan^2\psi}{\sec\psi} = \frac{\sin^2\psi}{\cos\psi}$$

- Also recall from trig that $\sin^2\psi + \cos^2\psi = 1$, such that $\sin^2\psi = 1 - \cos^2\psi$. Thus:

$$\frac{\sin^2\psi}{\cos\psi} = \sec\psi - \cos\psi$$

$$\vec{\mathbf{E}} = 2\pi k\beta p\hat{\mathbf{z}}\int \sec\psi \, d\psi - 2\pi k\beta p\hat{\mathbf{z}}\int \cos\psi \, d\psi$$

- You can look up these anti-derivatives on page 70.

$$\vec{\mathbf{E}} = 2\pi k\beta p\hat{\mathbf{z}}(\ln|\sec\psi + \tan\psi| - \sin\psi)$$

- Draw a right triangle to express each trig function in terms of r_c, and then we can simply evaluate the function over the old limits. Since $r_c = p\tan\psi$, which means that $\tan\psi = \frac{r_c}{p}$, we draw a right triangle with r_c opposite and with p adjacent to ψ.

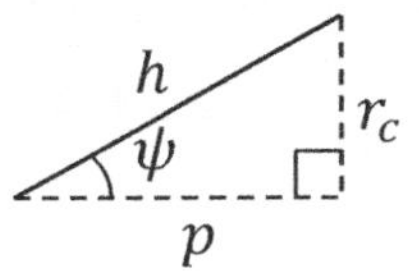

- Apply the Pythagorean theorem to find the hypotenuse: $h = \sqrt{r_c^2 + p^2}$.
- From the triangle, you should get $\sin\psi = \frac{r_c}{\sqrt{r_c^2+p^2}}$, $\sec\psi = \frac{\sqrt{r_c^2+p^2}}{p}$, and $\tan\psi = \frac{r_c}{p}$.
- Evaluate these functions over the limits from $r_c = 0$ to a.

$$\vec{\mathbf{E}} = 2\pi k\beta p\hat{\mathbf{z}}\left(\ln\left|\frac{\sqrt{a^2+p^2}}{p} + \frac{a}{p}\right| - \frac{a}{\sqrt{a^2+p^2}}\right)$$

- Integrate $Q = \int dq$ using the same substitutions as before.
- You should get $Q = \beta\int_{r_c=0}^{a}\int_{\theta=0}^{2\pi} r_c^2 \, dr_c d\theta = \frac{2\pi\beta a^3}{3}$.
- Solve for β. You should get $\beta = \frac{3Q}{2\pi a^3}$.
- Plug β into the previous equation for electric field. The final answer is:

$$\vec{\mathbf{E}} = \frac{3kQp}{a^3}\hat{\mathbf{z}}\left(\ln\left|\frac{\sqrt{a^2+p^2}}{p} + \frac{a}{p}\right| - \frac{a}{\sqrt{a^2+p^2}}\right)$$

GET A DIFFERENT ANSWER?

If you get a different answer and can't find your mistake even after consulting the hints and explanations, what should you do?

Please contact the author, Dr. McMullen.

How? Visit one of the author's blogs (see below). Either use the Contact Me option, or click on one of the author's articles and post a comment on the article.

www.monkeyphysicsblog.wordpress.com
www.improveyourmathfluency.com
www.chrismcmullen.wordpress.com

Why?

- If there happens to be a mistake (although much effort was put into perfecting the answer key), the correction will benefit other students like yourself in the future.
- If it turns out not to be a mistake, **you may learn something** from Dr. McMullen's reply to your message.

99.99% of students who walk into Dr. McMullen's office believing that they found a mistake with an answer discover one of two things:

- They made a mistake that they didn't realize they were making and learned from it.
- They discovered that their answer was actually the same. This is actually fairly common. For example, the answer key might say $t = \frac{\sqrt{3}}{3}$ s. A student solves the problem and gets $t = \frac{1}{\sqrt{3}}$ s. These are actually the same: Try it on your calculator and you will see that both equal about 0.57735. Here's why: $\frac{1}{\sqrt{3}} = \frac{1}{\sqrt{3}}\frac{\sqrt{3}}{\sqrt{3}} = \frac{\sqrt{3}}{3}$.

Two experienced physics teachers solved every problem in this book to check the answers, and dozens of students used this book and provided feedback before it was published. Every effort was made to ensure that the final answer given to every problem is correct.

But all humans, even those who are experts in their fields and who routinely aced exams back when they were students, make an occasional mistake. So if you believe you found a mistake, you should report it just in case. Dr. McMullen will appreciate your time.

43. First study the examples and the strategy to see if they can help you.

- Begin with a labeled diagram. Draw a representative dq. Draw $\vec{\mathbf{R}}$ from the source, dq, to the field point (at the origin). See the diagram below.

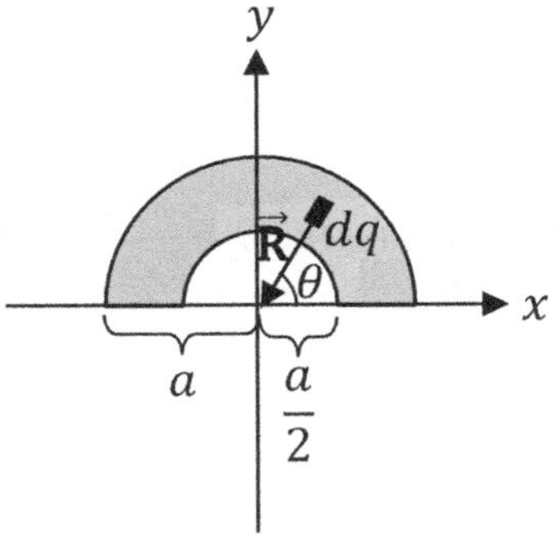

- For the dq drawn, you should be able to see that $\vec{\mathbf{R}}$ extends r units inward, towards the origin (along $-\hat{\mathbf{r}}$). Therefore, $R = r$ and $\hat{\mathbf{R}} = -\hat{\mathbf{r}}$ (using 2D polar coordinates).
- For a thick, solid ring, write $dq = \sigma dA$ (Step 6 on page 94) and $dA = r dr d\theta$.
- Since the charge density is non-uniform, you may **not** pull σ out of the integral. Instead, plug in the expression given in the problem: $\sigma = \beta r$. You may pull β out of the integral.
- Begin the math with the electric field integral.

$$\vec{\mathbf{E}} = k \int \frac{\hat{\mathbf{R}}}{R^2} dq$$

- Note that r varies from $\frac{a}{2}$ to a and θ varies from 0 to π (half circle).
- Substitute the expressions from the previous hints into the electric field integral. After simplifying, you should get:

$$\vec{\mathbf{E}} = -k\beta \int_{r=a/2}^{a} \int_{\theta=0}^{\pi} \hat{\mathbf{r}}\, dr d\theta$$

- Note that the r's from $dA = r dr d\theta$ and $\sigma = \beta r$ canceled the $\frac{1}{r^2}$ from $\frac{1}{R^2}$.
- Note that $\hat{\mathbf{r}}$ is a function of θ, since it points in a different direction for each dq. Thus, you may **not** pull $\hat{\mathbf{r}}$ out of the θ-integration. However, you may pull $\hat{\mathbf{r}}$ out of the r-integration, since $\hat{\mathbf{r}}$ is exactly one unit long (it's not a function of r).

$$\vec{\mathbf{E}} = -k\beta \int_{r=a/2}^{a} dr \int_{\theta=0}^{\pi} \hat{\mathbf{r}}\, d\theta = -\frac{k\beta a}{2} \int_{\theta=0}^{\pi} \hat{\mathbf{r}}\, d\theta$$

- Note that $\int_{r=a/2}^{a} dr = a - \frac{a}{2} = \frac{2a}{2} - \frac{a}{2} = \frac{2a-a}{2} = \frac{a}{2}$.
- Apply the equation $\hat{\mathbf{r}} = \hat{\mathbf{x}} \cos\theta + \hat{\mathbf{y}} \sin\theta$ from page 93.

$$\vec{\mathbf{E}} = -\frac{k\beta a}{2} \int_{\theta=0}^{\pi} (\hat{\mathbf{x}} \cos\theta + \hat{\mathbf{y}} \sin\theta)\, d\theta$$

- Separate the integral into two terms.

$$\vec{\mathbf{E}} = -\frac{k\beta a}{2}\int_{\theta=0}^{\pi} \hat{\mathbf{x}}\cos\theta\, d\theta - \frac{k\beta a}{2}\int_{\theta=0}^{\pi} \hat{\mathbf{y}}\sin\theta\, d\theta$$

- You may pull $\hat{\mathbf{x}}$ and $\hat{\mathbf{y}}$ out of the integrals since they are constants: They are exactly one unit long and each always points in the same direction, unlike $\hat{\mathbf{r}}$.

$$\vec{\mathbf{E}} = -\frac{k\beta a}{2}\hat{\mathbf{x}}\int_{\theta=0}^{\pi} \cos\theta\, d\theta - \frac{k\beta a}{2}\hat{\mathbf{y}}\int_{\theta=0}^{\pi} \sin\theta\, d\theta$$

- The anti-derivatives are $\sin\theta$ and $-\cos\theta$, respectively.
- Evaluate the anti-derivatives over the limits.

$$\vec{\mathbf{E}} = -k\beta a\hat{\mathbf{y}}$$

- Integrate $Q = \int dq$ using the same substitutions as before.
- You should get $Q = \beta\int_{r=a/2}^{a}\int_{\theta=0}^{\pi} r^2\, dr d\theta = 2\pi\beta\left(\frac{a^3}{3} - \frac{a^3}{24}\right) = \frac{7\pi\beta a^3}{24}$. Note that $\frac{1}{3}\left(\frac{a}{2}\right)^3 = \frac{a^3}{(3)(8)} = \frac{a^3}{24}$. Also note that $\frac{1}{3} - \frac{1}{24} = \frac{8}{24} - \frac{1}{24} = \frac{7}{24}$.
- Solve for β. You should get $\beta = \frac{24Q}{7\pi a^3}$.
- Plug β into the previous equation for electric field. The final answer is:

$$\vec{\mathbf{E}} = -\frac{24kQ}{7\pi a^2}\hat{\mathbf{y}}$$

Chapter 8: Gauss's Law

44. This problem is similar to the examples with a charged spherical insulator. Study those examples and try to let them serve as a guide. If you need help, you can always return here.

- Begin by sketching the electric field lines. See the diagram below.

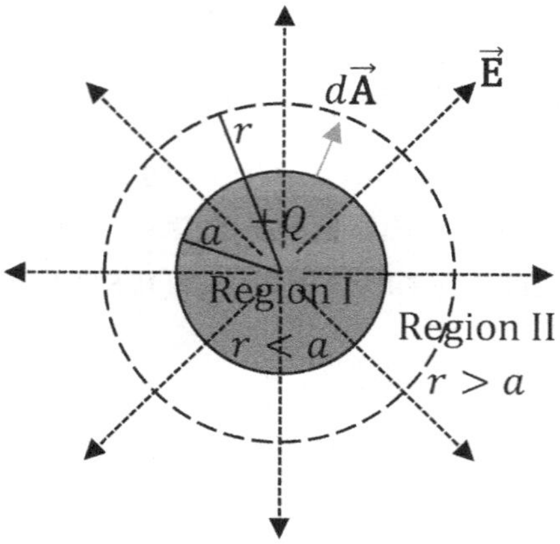

- Write down the equation for Gauss's law: $\oint_S \vec{\mathbf{E}}\cdot d\vec{\mathbf{A}} = \frac{q_{enc}}{\epsilon_0}$.
- Just as in the examples with spherical symmetry, Gauss's law reduces to $E4\pi r^2 = \frac{q_{enc}}{\epsilon_0}$. Isolate E and use $\epsilon_0 = \frac{1}{4\pi k}$ to get $E = \frac{kq_{enc}}{r^2}$.

- Integrate $q_{enc} = \int dq$ to find the charge enclosed in each region.
- For a solid sphere, $dq = \rho dV$ and $dV = r^2 \sin\theta \, dr d\theta d\varphi$ (see page 94).
- Since the charge density is non-uniform, you may **not** pull ρ out of the integral. Instead, plug in the equation given in the problem: $\rho = \beta r^2$. You may pull β out of the integral.
- After you make these substitutions and simplify, you should get:

$$q_{enc} = \beta \int_{r=0}^{r \text{ or } a} r^4 \, dr \int_{\theta=0}^{\pi} \sin\theta \, d\theta \int_{\varphi=0}^{2\pi} d\varphi$$

- The upper limit of the r-integration is r in region I ($r < a$) and a in region II ($r > a$).
- The anti-derivatives are $\frac{r^5}{5}$, $-\cos\theta$, and φ. Evaluate these over their limits.
- You should get $q_{enc}^I = \frac{4\pi\beta r^5}{5}$ in region I and $q_{enc}^{II} = Q = \frac{4\pi\beta a^5}{5}$ in region II.
- Divide these equations to see that $\frac{q_{enc}^I}{Q} = \frac{r^5}{a^5}$ (everything else cancels out).
- Multiply both sides by Q to get $q_{enc}^I = Q\frac{r^5}{a^5}$.
- Plug $q_{enc}^I = Q\frac{r^5}{a^5}$ and $q_{enc}^{II} = Q$ into the previous equation for electric field, $E = \frac{kq_{enc}}{r^2}$.
- Add $\hat{\mathbf{r}}$ for the direction. The answers are $\vec{\mathbf{E}}_I = \frac{kQr^3}{a^5}\hat{\mathbf{r}}$ and $\vec{\mathbf{E}}_{II} = \frac{kQ}{r^2}\hat{\mathbf{r}}$.

---

45. Each part of this problem was done in one of the examples.

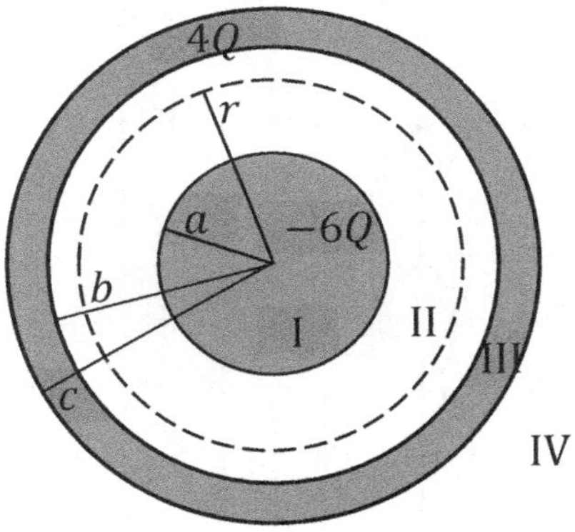

Region I: $r < a$.

- The conducting shell doesn't matter in region I. We solved the problem with a non-uniformly charged sphere with charge density $\rho = \beta r$ in one of the examples. Review that example, where the total charge of the sphere was Q and we found that the electric field was $\frac{kQr^2}{a^4}\hat{\mathbf{r}}$ **inside** of the sphere. Note that this is one of the alternate forms of the equation on page 128.
- The only differences with this problem are that the charge density is negative ($\rho = -\beta r$) and the total charge of the sphere is $-6Q$. The only difference this will make in the final answer is that Q will be replaced by $-6Q$. Thus, $\vec{\mathbf{E}}_I = -\frac{6kQr^2}{a^4}\hat{\mathbf{r}}$.

<u>**Region II**</u>: $a < r < b$.

- The conducting shell also doesn't matter in region II. Review the example that had a sphere with charge density $\rho = \beta r$, where the total charge of the sphere was Q and we found that the electric field was $\frac{kQ}{r^2}\hat{\mathbf{r}}$ **outside** of the sphere.
- The only differences with this problem are that the charge density is negative ($\rho = -\beta r$) and the total charge of the sphere is $-6Q$. The only difference this will make in the final answer is that Q will be replaced by $-6Q$. Thus, $\vec{\mathbf{E}}_{II} = -\frac{6kQ}{r^2}\hat{\mathbf{r}}$.

<u>**Region III**</u>: $b < r < c$.

- Inside the conducting shell, $\vec{\mathbf{E}}_{III} = 0$ for the same reasons discussed in the example that featured a conducting shell. See page 130.

<u>**Region IV**</u>: $r > c$.

- Outside of all of the spheres, the electric field is the same as that of a pointlike charge: It equals $\frac{kQ}{r^2}\hat{\mathbf{r}}$, except that in this problem the total charge isn't Q. The total charge enclosed by a Gaussian sphere in region IV is $-6Q + 4Q = -2Q$. Thus, we replace Q with $-2Q$ to obtain $\vec{\mathbf{E}}_{IV} = -\frac{2kQ}{r^2}\hat{\mathbf{r}}$.

---

46. This problem is similar to the examples with a line charge or cylinder. Study those examples and try to let them serve as a guide. If you need help, you can always return here.

- Begin by sketching the electric field lines. See the diagram below.

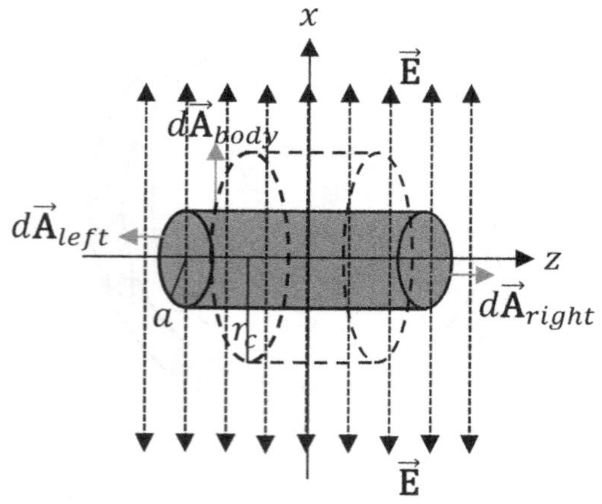

- Write down the equation for Gauss's law: $\oint_S \vec{\mathbf{E}} \cdot d\vec{\mathbf{A}} = \frac{q_{enc}}{\epsilon_0}$.
- Just as in the examples with cylindrical symmetry, Gauss's law reduces to $E 2\pi r_c L = \frac{q_{enc}}{\epsilon_0}$. Isolate E get $E = \frac{q_{enc}}{2\pi\epsilon_0 r_c L}$.
- Integrate $q_{enc} = \int dq$ to find the charge enclosed in each region.
- For a solid cylinder, $dq = \rho dV$ and $dV = r_c dr_c d\theta dz$ (see page 94). Note that we use ρdV for a solid cylindrical **insulator** (unlike the example with the **conductor**).
- Since the cylinder has uniform charge density, you may pull ρ out of the integral.

- After you make these substitutions and simplify, you should get:

$$q_{enc} = \rho \int_{r_c=0}^{r_c \text{ or } a} r_c \, dr_c \int_{\theta=0}^{2\pi} d\theta \int_{z=0}^{L} dz$$

- The upper limit of the r_c-integration is r_c in region I ($r_c < a$) and a in region II ($r_c > a$). The length of the Gaussian cylinder is L.
- The anti-derivatives are $\frac{r_c^2}{2}$, θ, and z. Evaluate these over their limits.
- You should get $q_{enc}^{I} = \rho\pi r_c^2 L$ in region I and $q_{enc}^{II} = \rho\pi a^2 L$ in region II.
- Plug $q_{enc}^{I} = \rho\pi r_c^2 L$ and $q_{enc}^{II} = \rho\pi a^2 L$ into the previous equation for electric field, $E = \frac{q_{enc}}{2\pi\epsilon_0 r_c L}$. Add $\hat{\mathbf{r}}_c$ for the direction. The answers are $\vec{\mathbf{E}}_I = \frac{\rho r_c}{2\epsilon_0}\hat{\mathbf{r}}_c$ and $\vec{\mathbf{E}}_{II} = \frac{\rho a^2}{2\epsilon_0 r_c}\hat{\mathbf{r}}_c$.

---

47. The "trick" to this problem is to apply the principle of superposition (see Chapter 3).
    - First, ignore the negative plane and find the electric field in each region due to the positive plane. Call this $\vec{\mathbf{E}}_1$. This should be easy because this part of the solution is identical to one of the examples: If you need help with this, see the example on page 119. You should get $E_1 = \frac{\sigma}{2\epsilon_0}$. To the left of the positive plane, $\vec{\mathbf{E}}_1$ points to the left, while to the right of the positive plane, $\vec{\mathbf{E}}_1$ points to the right (since a positive "test" charge would be **repelled** by the positive plane). The direction of $\vec{\mathbf{E}}_1$ in each region is illustrated below with <u>**solid**</u> arrows.

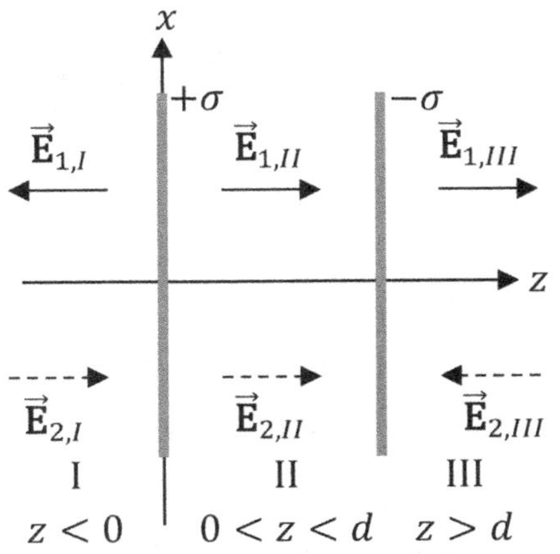

- Next, ignore the positive plane and find the electric field in each region due to the negative plane. Call this $\vec{\mathbf{E}}_2$. You should get $E_2 = \frac{\sigma}{2\epsilon_0}$. To the left of the negative plane, $\vec{\mathbf{E}}_2$ points to the right, while to the right of the negative plane, $\vec{\mathbf{E}}_2$ points to the left (since a positive "test" charge would be **attracted** to the negative plane). The direction of $\vec{\mathbf{E}}_2$ in each region is illustrated above with <u>**dashed**</u> arrows.
- Note that $\vec{\mathbf{E}}_1$ and $\vec{\mathbf{E}}_2$ are both equal in magnitude. They also have the same direction in region II, but opposite directions in regions I and III.

- Add the vectors $\vec{\mathbf{E}}_1$ and $\vec{\mathbf{E}}_2$ to find the net electric field, $\vec{\mathbf{E}}_{net}$, in each region. In regions I and III, the net electric field is zero because $\vec{\mathbf{E}}_1$ and $\vec{\mathbf{E}}_2$ are equal and opposite, whereas in region II the net electric field is twice $\vec{\mathbf{E}}_1$ because $\vec{\mathbf{E}}_1$ and $\vec{\mathbf{E}}_2$ are equal and point in the same direction (both point to the right in region II, as a positive "test" charge would be repelled to the right by the positive plane and also attracted to the right by the negative plane). Note that $\vec{\mathbf{E}}_{II} = 2\vec{\mathbf{E}}_{1,II} = 2\frac{\sigma}{2\epsilon_0}\hat{\mathbf{z}} = \frac{\sigma}{\epsilon_0}\hat{\mathbf{z}}$.

$$\vec{\mathbf{E}}_I = 0, \quad \vec{\mathbf{E}}_{II} = \frac{\sigma}{\epsilon_0}\hat{\mathbf{z}}, \quad \vec{\mathbf{E}}_{III} = 0$$

48. Apply the same strategy from the previous problem. Review the hints to Problem 47.
   - First, ignore the negative plane and find the electric field in the specified region due to the positive plane. Call this $\vec{\mathbf{E}}_1$. This should be easy because this part of the solution is identical to one of the examples: If you need help with this, see the example on page 119. You should get $\vec{\mathbf{E}}_1 = \frac{\sigma}{2\epsilon_0}\hat{\mathbf{z}}$. It points to the right since a positive "test" charge in the first octant would be **repelled** by the positive plane. The direction of $\vec{\mathbf{E}}_1$ in the specified region is illustrated below with a **solid** arrow.

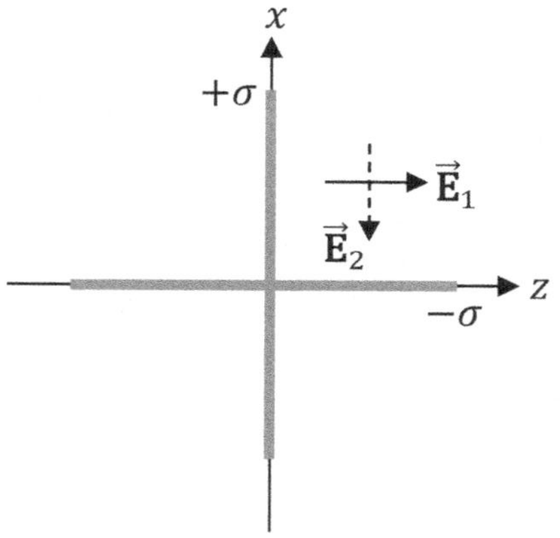

   - Next, ignore the positive plane and find the electric field in the first octant due to the negative plane. Call this $\vec{\mathbf{E}}_2$. You should get $\vec{\mathbf{E}}_2 = -\frac{\sigma}{2\epsilon_0}\hat{\mathbf{x}}$. It points to down since a positive "test" charge in the first octant would be **attracted** to the negative plane. The direction of $\vec{\mathbf{E}}_2$ in the specified region is illustrated above with a **dashed** arrow.
   - Note that the zx plane was drawn in the problem, with $+z$ pointing to the right and $+x$ pointing up. (The $+y$-axis comes out of the page, like most of the three-dimensional problems in this chapter.)
   - The vectors $\vec{\mathbf{E}}_1$ and $\vec{\mathbf{E}}_2$ have equal magnitude, but are perpendicular. When we add the vectors to get the net electric field, $\vec{\mathbf{E}}_{net}$, we get $\vec{\mathbf{E}}_{net} = \frac{\sigma}{2\epsilon_0}(\hat{\mathbf{z}} - \hat{\mathbf{x}})$, which has a magnitude of $E_{net} = \frac{\sigma}{2\epsilon_0}\sqrt{2}$ and points diagonally between $\vec{\mathbf{E}}_1$ and $\vec{\mathbf{E}}_2$.

49. This problem is similar to the example with an infinite slab. Study that example and try to let it serve as a guide. If you need help, you can always return here.

- Begin by sketching the electric field lines. See the diagram below.

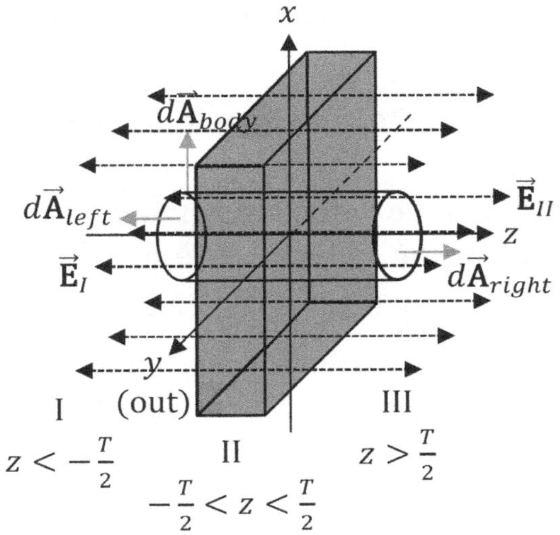

- Write down the equation for Gauss's law: $\oint_S \vec{\mathbf{E}} \cdot d\vec{\mathbf{A}} = \frac{q_{enc}}{\epsilon_0}$.
- Just as in the examples with an infinite plane or slab, Gauss's law reduces to $E2A_{end} = \frac{q_{enc}}{\epsilon_0}$, where A_{end} is the area of the end of the Gaussian cylinder. Isolate E get $E = \frac{q_{enc}}{2\epsilon_0 A_{end}}$.
- Integrate $q_{enc} = \int dq$ to find the charge enclosed in each region.
- For a solid slab, $dq = \rho dV$. For a Gaussian cylinder, $dV = r_c dr_c d\theta dz$ (see page 94).
- Since the charge density is non-uniform, you may **not** pull ρ out of the integral. Instead, plug in the expression given in the problem: $\rho = \beta|z|$. You may pull β out.
- To deal with the absolute values, multiply by 2 and begin the z integral from 0.

$$q_{enc} = 2\beta \int_{r_c=0}^{a} r_c \, dr_c \int_{\theta=0}^{2\pi} d\theta \int_{z=0}^{z \text{ or } T/2} z \, dz$$

- The limits of the z-integration are from 0 to z in region II ($-\frac{T}{2} < z < \frac{T}{2}$) and from 0 to $\frac{T}{2}$ for regions I and III ($|z| > \frac{T}{2}$). The radius of the Gaussian cylinder is a.
- Note that the double integral over r_c and θ equals the area of the end, which we have already called A_{end}. It will equal πa^2 (the area of a circle), but since it will cancel with A_{end} in a future step, there isn't any benefit to writing it as πa^2.
- The anti-derivative for the z integral is $\frac{z^2}{2}$. Evaluate this over the limits.
- Note that $\frac{1}{2}\left(\frac{T}{2}\right)^2 = \frac{1}{2}\left(\frac{T^2}{4}\right) = \frac{T^2}{8}$. (The overall factor of 2β will turn the $\frac{1}{8}$ into $\frac{1}{4}$.)
- You should get $q_{enc}^{I \text{ or } III} = \frac{\beta A_{end} T^2}{4}$ in region I or III and $q_{enc}^{II} = \beta A_{end} z^2$ in region II.

- Plug $q_{enc}^{I \text{ or } III} = \frac{\beta A_{end} T^2}{4}$ and $q_{enc}^{II} = \beta A_{end} z^2$ into the previous equation for electric field, $E = \frac{q_{enc}}{2\epsilon_0 A_{end}}$. Add $\pm\hat{\mathbf{z}}$ for the direction. The answers are $\vec{\mathbf{E}}_I = -\frac{\beta T^2}{8\epsilon_0}\hat{\mathbf{z}}$, $\vec{\mathbf{E}}_{II} = \pm\frac{\beta z^2}{2\epsilon_0}\hat{\mathbf{z}}$ (depending on whether or not $z > 0$), and $\vec{\mathbf{E}}_{III} = \frac{\beta T^2}{8\epsilon_0}\hat{\mathbf{z}}$.

---

50. This problem is very similar to the last example of this chapter, except that the shape is a cylinder instead of a sphere.

- First find the electric field due to the big cylinder (ignoring the cavity) at a distance r_c away from the axis of the cylinder. We did this in Problem 46. Review the solution to Problem 46, where $\vec{\mathbf{E}}_{big} = \frac{\rho r_c}{2\epsilon_0}\hat{\mathbf{r}}_c$ (for region I).
- Next find the electric field due to the small cylinder (the same size as the cavity) at a distance u_c away from the axis of this cylinder. We'll obtain the exact same result, except for the distance being smaller: $\vec{\mathbf{E}}_{small} = \frac{\rho u_c}{2\epsilon_0}\hat{\mathbf{u}}_c$.
- Just like we did in the last example of this chapter, find the electric field of the given shape (the cylinder with the cavity) through the principle of **superposition**. As shown in the last example of this chapter, we subtract: $\vec{\mathbf{E}}_{given} = \vec{\mathbf{E}}_{big} - \vec{\mathbf{E}}_{small}$.
- You should get $\vec{\mathbf{E}}_{given} = \frac{\rho}{2\epsilon_0}(r_c\hat{\mathbf{r}}_c - u_c\hat{\mathbf{u}}_c)$.
- Recall that any vector can be expressed as its magnitude times its direction. For example, $\vec{\mathbf{A}} = A\hat{\mathbf{A}}$. Therefore, $\vec{\mathbf{r}}_c = r_c\hat{\mathbf{r}}_c$ and $\vec{\mathbf{u}}_c = u_c\hat{\mathbf{u}}_c$. Thus, $\vec{\mathbf{E}}_{given} = \frac{\rho}{2\epsilon_0}(\vec{\mathbf{r}}_c - \vec{\mathbf{u}}_c)$.
- As explained in the last example of the chapter, $\vec{\mathbf{r}}_c - \vec{\mathbf{u}}_c = \vec{\mathbf{d}}$.
- The final answer is $\vec{\mathbf{E}}_{given} = \frac{\rho}{2\epsilon_0}\vec{\mathbf{d}}$.

Chapter 9: Electric Potential

51. We choose to call $q_1 = 7\sqrt{2}$ μC and $q_2 = -3\sqrt{2}$ μC.

(A) Use the distance formula.

$$R_1 = \sqrt{\Delta x_1^2 + \Delta y_1^2} \quad , \quad R_2 = \sqrt{\Delta x_2^2 + \Delta y_2^2}$$

- Check your distances: $|\Delta x_1| = |\Delta x_2| = 1.0$ m and $|\Delta y_1| = |\Delta y_2| = 1.0$ m.
- You should get $R_1 = R_2 = \sqrt{2}$ m.

(B) First convert the charges from μC to C: $q_1 = 7\sqrt{2} \times 10^{-6}$ C and $q_2 = -3\sqrt{2} \times 10^{-6}$ C.

- Use the equation for the electric potential of a system of two pointlike charges.

$$V_{net} = \frac{kq_1}{R_1} + \frac{kq_2}{R_2}$$

- Note that $\frac{\sqrt{2}}{\sqrt{2}} = 1$ and $10^9 10^{-6} = 10^3$. Note that the second term is **negative**.
- The net electric potential at (1.0 m, 0) is $V_{net} = 36$ kV, where the metric prefix kilo (k) is $10^3 = 1000$. It's the same as 36,000 V, 36×10^3 V, or 3.6×10^4 V.

---

52. This problem is similar to an example from Chapter 7, except that this problem is for electric potential instead of electric field. Study the example on page 100 and try to let it serve as a guide. If you need help, you can always return here.

- Begin with a labeled diagram. Draw a representative dq. Draw $\vec{\mathbf{R}}$ from the source, dq, to the field point at the origin. See the diagram below.

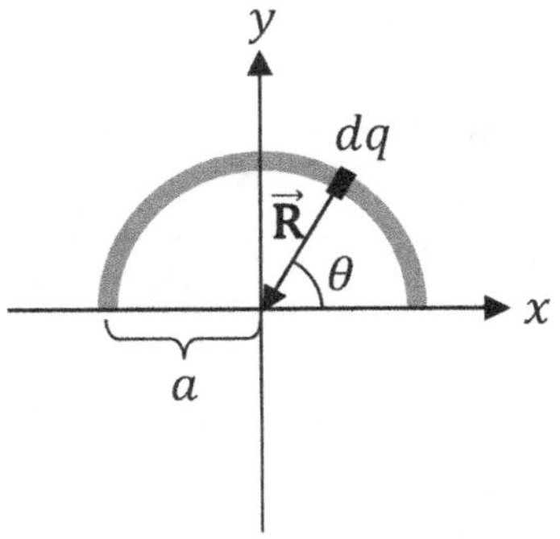

- The vector $\vec{\mathbf{R}}$ has the same length for each dq that makes up the semicircle. Its magnitude, R, equals the radius of the semicircle: $R = a$. (Note that this would **not** be the case for a thick ring or a solid semicircle.)
- For a very thin arc, write $dq = \lambda ds$ (Step 5 on page 155). Since the ring is circular, $ds = a\, d\theta$. (This is $ds = r\, d\theta$, but with the radius of the ring equal to $r = a$.)
- Since this semicircle has uniform charge density, you may pull λ out of the integral.
- Begin the math with the electric potential integral.

$$V = k \int \frac{dq}{R}$$

- Note that θ varies from 0 to π (half circle).
- Substitute the expressions from the previous hints into the electric potential integral. After simplifying, you should get:

$$V = k\lambda \int_{\theta=0}^{\pi} d\theta$$

- Note that **everything** is constant and comes out of the integral (none of the symbols depend upon θ). Also note that the $\frac{1}{a}$ from $\frac{1}{R} = \frac{1}{a}$ cancels the a from $ds = a\, d\theta$. The remaining integral equals π.

$$V = \pi k \lambda$$

- Integrate $Q = \int dq$ using the same substitutions as before.

- You should get $Q = \lambda \int_{\theta=0}^{\pi} a \, d\theta = \pi \lambda a$.
- Solve for λ. You should get $\lambda = \frac{Q}{\pi a}$.
- Plug λ into the previous equation for electric potential. The final answer is:

$$V = \frac{kQ}{a}$$

---

53. This problem is similar to the example involving a charged disc. Study that example and try to let it serve as a guide. If you need help, you can always return here.

- Begin with a labeled diagram. Draw a representative dq. Draw $\vec{\mathbf{R}}$ from the source, dq, to the field point $(0, 0, p)$. See the diagram below.

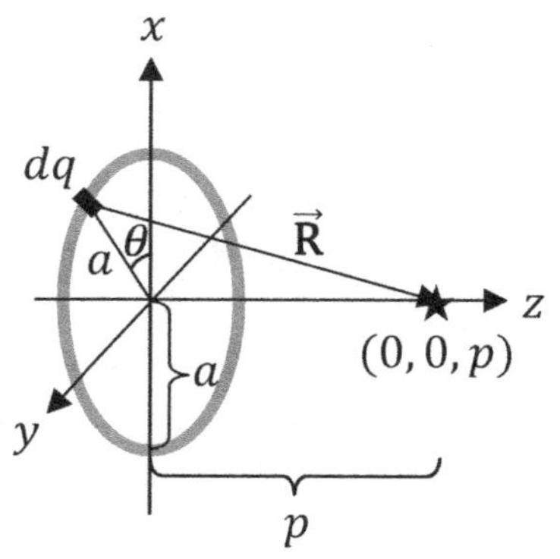

- From the diagram above, for the dq drawn, you should be able to see that $\vec{\mathbf{R}}$ extends a units inward, towards the z-axis (along $-\hat{\mathbf{r}}_c$), and p units along the z-axis (along $\hat{\mathbf{z}}$). Apply the Pythagorean theorem to find R. You should get $R = \sqrt{a^2 + p^2}$.
- For a very thin ring, write $dq = \lambda ds$ (Step 5 on page 155). Since the ring is circular, $ds = a \, d\theta$. (This is $ds = r \, d\theta$, but with the radius of the ring equal to $r = a$.)
- Since this ring has uniform charge density, you may pull λ out of the integral.
- Begin the math with the electric potential integral.

$$V = k \int \frac{dq}{R}$$

- Note that θ varies from 0 to 2π (full circle).
- Substitute the expressions from the previous hints into the electric potential integral. After simplifying, you should get:

$$V = \frac{k\lambda a}{\sqrt{a^2 + p^2}} \int_{\theta=0}^{2\pi} d\theta$$

- Note that **<u>everything</u>** is constant and comes out of the integral (none of the symbols depend upon θ). The remaining integral equals 2π.

$$V = \frac{2\pi k \lambda a}{\sqrt{a^2 + p^2}}$$

- Integrate $Q = \int dq$ using the same substitutions as before.

- You should get $Q = \lambda \int_{\theta=0}^{2\pi} a\, d\theta = 2\pi\lambda a$.
- Solve for λ. You should get $\lambda = \frac{Q}{2\pi a}$.
- Plug λ into the previous equation for electric potential. The final answer is:

$$V = \frac{kQ}{\sqrt{a^2 + p^2}}$$

Chapter 10: Motion of a Charged Particle in a Uniform Electric Field

54. This problem is similar to the first example of this chapter.

(A) First calculate the electric field between the plates with the equation $E = \frac{\Delta V}{d}$.

- Note that $d = 20$ cm $= 0.20$ m. You should get $E = 600$ N/C. (1 N/C = 1 V/m.)
- Draw a free-body diagram (FBD) for the tiny charged object.
  - The electric force ($q\vec{\mathbf{E}}$) pulls up. Since the charged object is positive, the electric force ($q\vec{\mathbf{E}}$) is parallel to the electric field ($\vec{\mathbf{E}}$), and since the electric field lines travel from the positive plate to the negative plate, $\vec{\mathbf{E}}$ points upward.
  - The weight ($m\vec{\mathbf{g}}$) of the object pulls straight down.

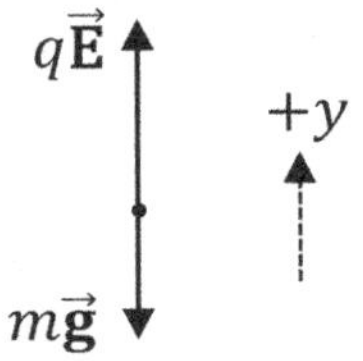

- Apply Newton's second law to the tiny charged object: $\sum F_y = ma_y$.
- You should get $|q|E - mg = ma_y$. Sign check: $q\vec{\mathbf{E}}$ is up (+) while $m\vec{\mathbf{g}}$ is down (−).
- Solve for the acceleration. You should get:

$$a_y = \frac{|q|E - mg}{m}$$

- Convert the charge from μ to C and the mass from g to kg.
- You should get $q = 1.5 \times 10^{-3}$ C and $m = 0.018$ kg.
- The acceleration is $a_y = 40$ m/s^2. The charge accelerates **upward**.

(B) Use an equation of one-dimensional uniform acceleration. First list the knowns.

- $a_y = 40$ m/s^2. We know this from part (A). It's **positive** because the object accelerates upward (and because we chose $+y$ to point upward).
- $\Delta y = 0.20$ m (the separation between the plates). It's **positive** because it finishes above where it started (and because we chose $+y$ to point upward).
- $v_{y0} = 0$. The initial velocity is zero because it starts from rest.
- We're solving for the final velocity (v_y).
- Choose the equation with these symbols: $v_y^2 = v_{y0}^2 + 2a_y\Delta y$.
- The final velocity is $v_y = 4.0$ m/s (heading upward).

Chapter 11: Equivalent Capacitance

55. (A) Redraw the circuit, simplifying it one step at a time by identifying series and parallel combinations. Try it yourself first, and then check your diagrams below.

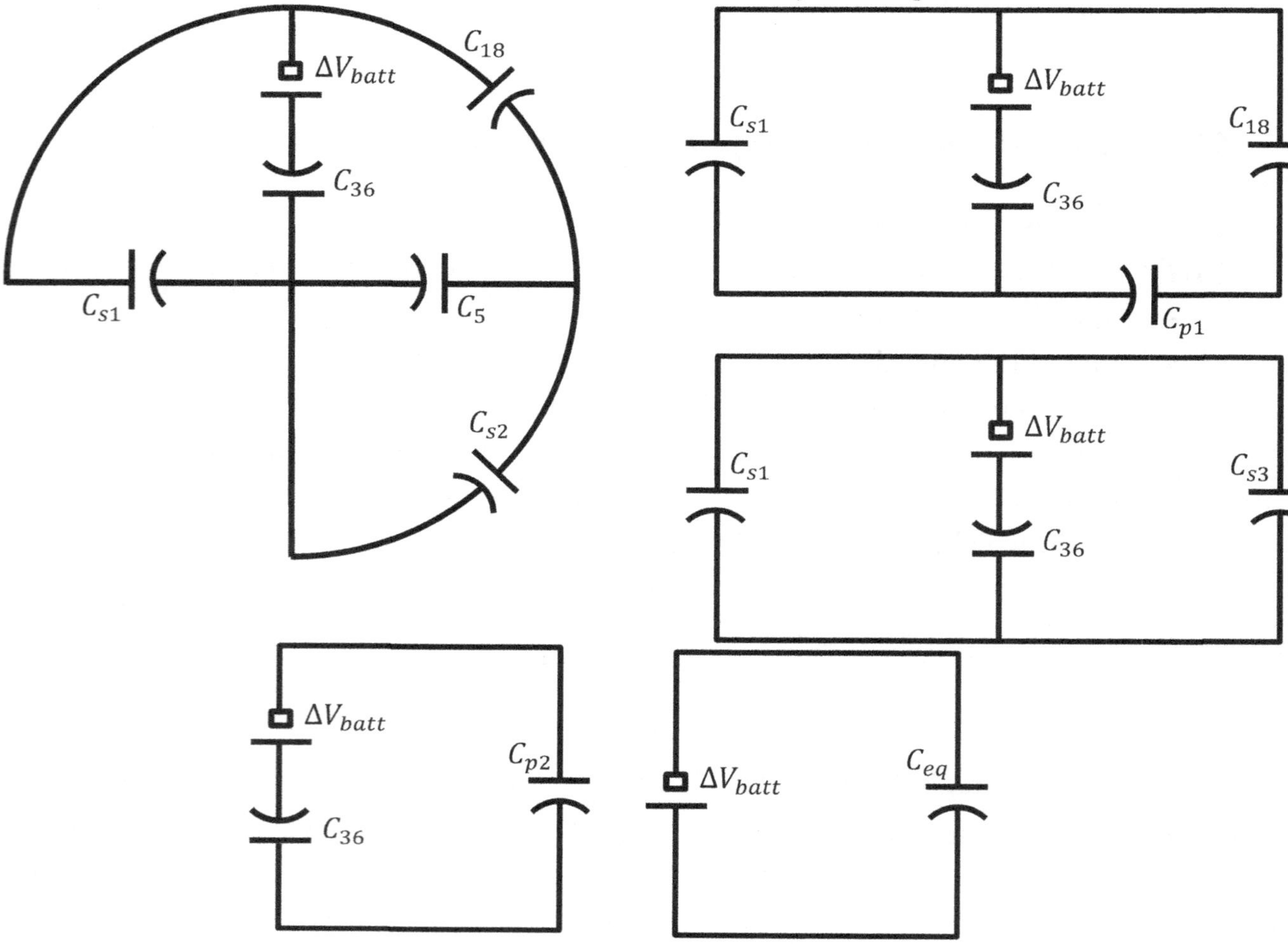

(B) Apply the formulas for series and parallel capacitors.

- The two 24-nF capacitors are in **series**. They became C_{s1}. The 6-nF capacitor is in **series** with the 12-nF capacitor. They became C_{s2}. In each case, an electron could travel from one capacitor to the other without crossing a junction. These reductions are shown in the left diagram above.
- Compute C_{s1} and C_{s2} using $\frac{1}{C_{s1}} = \frac{1}{C_{24}} + \frac{1}{C_{24}}$ and $\frac{1}{C_{s2}} = \frac{1}{C_6} + \frac{1}{C_{12}}$. Note that $\frac{1}{6} + \frac{1}{12} = \frac{2}{12} + \frac{1}{12} = \frac{3}{12} = \frac{1}{4}$. You should get $C_{s1} = 12.0$ nF and $C_{s2} = 4.0$ nF.
- In the top left diagram, C_5 and C_{s2} are in **parallel**. They became C_{p1}.
- Compute C_{p1} using $C_{p1} = C_5 + C_{s2}$. You should get $C_{p1} = 9.0$ nF.
- In the top right diagram, C_{p1} and C_{18} are in **series**. They became C_{s3}.
- Compute C_{s3} using $\frac{1}{C_{s3}} = \frac{1}{C_{p1}} + \frac{1}{C_{18}}$. You should get $C_{s3} = 6.0$ nF.

- Next, C_{s1} and C_{s3} are in **parallel**. They became C_{p2}.
- Compute C_{p2} using $C_{p2} = C_{s1} + C_{s3}$. You should get $C_{p2} = 18.0$ nF.
- Finally, C_{36} and C_{p2} are in **series**. They became C_{eq}.
- Compute C_{eq} using $\frac{1}{C_{eq}} = \frac{1}{C_{36}} + \frac{1}{C_{p2}}$. The answer is $C_{eq} = 12$ nF.

(C) Work your way backwards through the circuit one step at a time. Study the parts of the example that involved working backwards. Try to think your way through the strategy.

- Begin the math with the equation $Q_{eq} = C_{eq}\Delta V_{batt}$. You should get $Q_{eq} = 144$ nC.
- Going back one picture from the last, C_{36} and C_{p2} are in **series** (replaced by C_{eq}).
  - Charge is the same in series. Write $Q_{36} = Q_{p2} = Q_{eq} = 144$ nC.
  - Calculate potential difference: $\Delta V_{p2} = \frac{Q_{p2}}{C_{p2}}$. You should get $\Delta V_{p2} = 8.0$ V.
- Going back one more step, C_{s1} and C_{s3} are in **parallel** (replaced by C_{p2}).
  - ΔV's are the same in parallel. Write $\Delta V_{s1} = \Delta V_{s3} = \Delta V_{p2} = 8.0$ V.
  - Calculate charge: $Q_{s1} = C_{s1}\Delta V_{s1}$. You should get $Q_{s1} = 96$ nC.
- Going back one more step, C_{24} and C_{24} are in **series** (replaced by C_{s1}).
  - Charge is the same in series. Write $Q_{24} = Q_{24} = Q_{s1} = 96$ nC.
  - The answer is $Q_{24} = 96$ nC.

(D) We found $\Delta V_{s3} = 8.0$ V in part (C). Continue working backwards from here.

- Calculate the charge stored on C_{s3}: $Q_{s3} = C_{s3}\Delta V_{s3}$. You should get $Q_{s3} = 48$ nC.
- Going back one step from there, C_{18} and C_{p1} are in **series** (replaced by C_{s3}).
  - Charge is the same in series. Write $Q_{18} = Q_{p1} = Q_{s3} = 48$ nC.
  - Use the appropriate energy equation: $U_{18} = \frac{Q_{18}^2}{2C_{18}}$. Note that Q_{18} is squared.
  - The answer is $U_{18} = 64$ nJ.

---

56. (A) Redraw the circuit, simplifying it one step at a time by identifying series and parallel combinations. Try it yourself first, and then check your diagrams below.

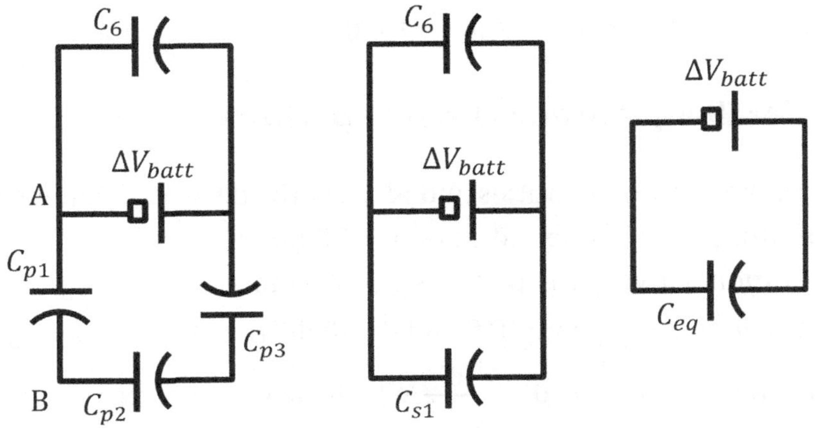

(B) Apply the formulas for series and parallel capacitors.

- Three pairs are in **parallel**: the 5 and 15, the 24 and 36, and the 12 and 18. These became C_{p1}, C_{p2}, and C_{p3}. These reductions are shown in the left diagram above.
- Compute C_{p1}, C_{p2}, and C_{p3}. For example, the formula for C_{p1} is $C_{p1} = C_5 + C_{15}$. You should get $C_{p1} = 20.0\ \mu\text{F}$, $C_{p2} = 60.0\ \mu\text{F}$, and $C_{p3} = 30.0\ \mu\text{F}$.
- In the left diagram above, C_{p1}, C_{p2}, and C_{p3} are in **series**. They became C_{s1}.
- Compute C_{s1} using $\frac{1}{C_{s1}} = \frac{1}{C_{p1}} + \frac{1}{C_{p2}} + \frac{1}{C_{p3}}$. You should get $C_{s1} = 10.0\ \mu\text{F}$. Note that $\frac{1}{20} + \frac{1}{60} + \frac{1}{30} = \frac{3}{60} + \frac{1}{60} + \frac{2}{60} = \frac{6}{60} = \frac{1}{10}$.
- Next, C_{s1} and C_6 are in **parallel**. They became C_{eq}.
- Compute C_{eq} using $C_{eq} = C_{s1} + C_6$. The answer is $C_{eq} = 16\ \mu\text{F}$.

(C) Work your way backwards through the circuit one step at a time. Study the parts of the example that involved working backwards. Try to think your way through the strategy.

- Begin the math with the equation $Q_{eq} = C_{eq}\Delta V_{batt}$. You should get $Q_{eq} = 64\ \mu\text{C}$.
- Going back one step, C_{s1} and C_6 are in **parallel** (replaced by C_{eq}).
  - ΔV's are the same in parallel. Write $\Delta V_{s1} = \Delta V_6 = \Delta V_{batt} = 4.0\text{ V}$.
  - Calculate charge: $Q_{s1} = C_{s1}\Delta V_{s1}$. You should get $Q_{s1} = 40\ \mu\text{C}$.
- Going back one more step, C_{p1}, C_{p2}, and C_{p3} are in **series** (replaced by C_{s1}).
  - Charge is the same in series. Write $Q_{p1} = Q_{p2} = Q_{p3} = Q_{s1} = 40\ \mu\text{C}$.
  - Calculate potential difference: $\Delta V_{p1} = \frac{Q_{p1}}{C_{p1}}$. You should get $\Delta V_{p1} = 2.0\text{ V}$.
- The answer is $\Delta V_{AB} = 2.0\text{ V}$ because $\Delta V_{AB} = \Delta V_{p1}$ (since C_{p1} is a single capacitor between points A and B).

(D) We found $\Delta V_{p1} = 2.0\text{ V}$ in part (C). Continue working backwards from here.

- Going back one step from there, C_5 and C_{15} are in **parallel** (replaced by C_{p1}).
  - ΔV's are the same in parallel. Write $\Delta V_5 = \Delta V_{15} = \Delta V_{p1} = 2.0\text{ V}$.
  - Use the appropriate energy equation: $U_5 = \frac{1}{2}C_5\Delta V_5^2$. Note that ΔV_5 is squared. The answer is $U_5 = 10\ \mu\text{J}$.

Chapter 12: Parallel-plate and Other Capacitors

57. Make a list of the known quantities and identify the desired unknown symbol:

- The capacitor plates have a radius of $a = 30$ mm.
- The separation between the plates is $d = 2.0$ mm.
- The dielectric constant is $\kappa = 8.0$ and the dielectric strength is $E_{max} = 6.0 \times 10^6\ \frac{\text{V}}{\text{m}}$.
- We also know $\epsilon_0 = 8.8 \times 10^{-12}\ \frac{\text{C}^2}{\text{N}\cdot\text{m}^2}$, which we will approximate as $\epsilon_0 \approx \frac{10^{-9}}{36\pi}\ \frac{\text{C}^2}{\text{N}\cdot\text{m}^2}$.
- The unknown we are looking for is capacitance (C).

- Convert the radius and separation to SI units.

$$a = 30 \text{ mm} = 0.030 \text{ m} = \frac{3}{100} \text{ m} \quad , \quad d = 2.0 \text{ mm} = 0.0020 \text{ m} = \frac{1}{500} \text{ m}$$

- Find the area of the circular plate: $A = \pi a^2$. Note that a is squared. Also note that $\left(\frac{1}{100}\right)^2 = \frac{1}{10{,}000}$. You should get $A = \frac{9\pi}{10{,}000}$ m^2. As a decimal, it's $A = 0.00283$ m^2.

(A) Use the equation $C = \frac{\kappa\epsilon_0 A}{d}$. Note that $\frac{1}{\left(\frac{1}{500}\right)} = 500$.

- The capacitance is $C = 0.10$ nF, which is the same as $C = 1.0 \times 10^{-10}$ F.

(B) First find the maximum potential difference (ΔV_{max}) across the plates.

- Use the equation $E_{max} = \frac{\Delta V_{max}}{d}$. Multiply both sides of the equation by d to solve for ΔV_{max}. You should get $\Delta V_{max} = 12{,}000$ V.
- Use the equation $Q_{max} = C\Delta V_{max}$. The maximum charge is: $Q = 1.2$ μC, which can also be expressed as $Q = 1.2 \times 10^{-6}$ C.

---

58. The first step is to apply Gauss's law to region II ($a < r < b$).

- We solved for the electric field in region II in a similar example in Chapter 8.

$$\vec{\mathbf{E}} = \frac{kQ}{r^2}\hat{\mathbf{r}}$$

- If you integrate outward, $\vec{\mathbf{E}} \cdot d\vec{\mathbf{s}} = E \cos 0° \, ds = Eds$.
- Integrate to find the potential difference between the conducting spheres.

$$\Delta V = V_f - V_i = -\int_i^f \vec{\mathbf{E}} \cdot d\vec{\mathbf{s}} = -\int_i^f E \, ds$$

- Note that $ds = dr$. After you plug in $E = \frac{kQ}{r^2}$ and simplify, you should get:

$$\Delta V = -kQ \int_{r_c=a}^{b} \frac{dr}{r^2}$$

- The anti-derivative is $-\frac{1}{r}$.
- Evaluate the anti-derivative over the limits. You should get:

$$\Delta V = -kQ\left(-\frac{1}{b} + \frac{1}{a}\right)$$

- Find a **common denominator**. You should get:

$$\Delta V = -kQ\left(\frac{-a+b}{ab}\right) = -kQ\left(\frac{b-a}{ab}\right)$$

- Note that $-a + b = b - a$.
- Plug the equation for ΔV into the capacitance formula: $C = \frac{Q}{|\Delta V|}$.
- The Q's will cancel. Note the absolute values. The answer is $C = \frac{ab}{k(b-a)}$.

Chapter 13: Equivalent Resistance

59. (A) Redraw the circuit, simplifying it one step at a time by identifying series and parallel combinations. Try it yourself first, and then check your diagrams below.

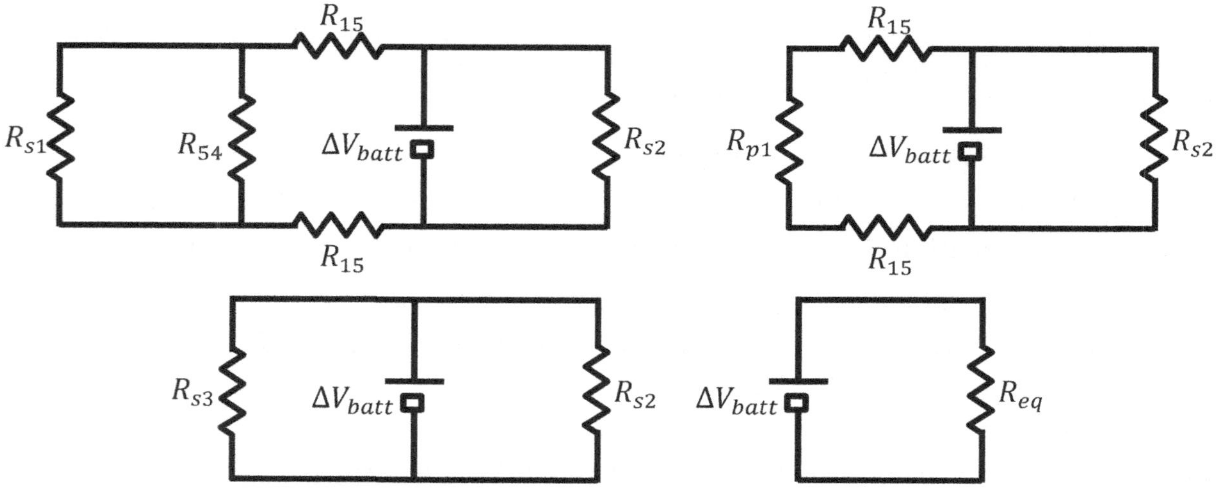

(B) Apply the formulas for series and parallel resistors.

- The 9.0-Ω and 18.0-Ω resistors are in **series**. They became R_{s1}. The 4.0-Ω and 12.0-Ω resistors are in **series**. They became R_{s2}. In each case, an electron could travel from one resistor to the other without crossing a junction. These reductions are shown in the left diagram above.
- Compute R_{s1} and R_{s2} using $R_{s1} = R_9 + R_{18}$ and $R_{s2} = R_4 + R_{12}$. You should get $R_{s1} = 27\ \Omega$ and $R_{s2} = 16\ \Omega$.
- In the top left diagram, R_{s1} and R_{54} are in **parallel**. They became R_{p1}.
- Compute R_{p1} using $\frac{1}{R_{p1}} = \frac{1}{R_{s1}} + \frac{1}{R_{54}}$. Note that $\frac{1}{27} + \frac{1}{54} = \frac{2}{54} + \frac{1}{54} = \frac{3}{54} = \frac{1}{18}$. You should get $R_{p1} = 18\ \Omega$.
- In the top right diagram, R_{p1} and the two R_{15}'s are in **series**. They became R_{s3}.
- Compute R_{s3} using $R_{s3} = R_{15} + R_{p1} + R_{15}$. You should get $R_{s3} = 48\ \Omega$.
- Finally, R_{s2} and R_{s3} are in **parallel**. They became R_{eq}.
- Compute R_{eq} using $\frac{1}{R_{eq}} = \frac{1}{R_{s2}} + \frac{1}{R_{s3}}$. The answer is $R_{eq} = 12\ \Omega$.

(C) Work your way backwards through the circuit one step at a time. Study the parts of the example that involved working backwards. Try to think your way through the strategy.

- Begin the math with the equation $I_{batt} = \frac{\Delta V_{batt}}{R_{eq}}$. You should get $I_{batt} = 20$ A.
- Going back one picture from the last, R_{s2} and R_{s3} are in **parallel** (replaced by R_{eq}).
  - ΔV's are the same in parallel. Write $\Delta V_{s2} = \Delta V_{s3} = \Delta V_{batt} = 240$ V. (Note that ΔV_{batt} is the potential difference across R_{eq}.)
  - Calculate current: $I_{s3} = \frac{\Delta V_{s3}}{R_{s3}}$. You should get $I_{s3} = 5.0$ A.

- Going back one more step, R_{p1} and the two R_{15}'s are in **series** (replaced by R_{s3}).
    - Current is the same in series. Write $I_{15} = I_{p1} = I_{15} = I_{s3} = 5.0$ A.
    - Use the appropriate power equation: $P_{15} = I_{15}^2 R_{15}$. Note that I_{15} is squared.
- The answer is $P_{15} = 375$ W.

---

60. (A) The first step is to redraw the circuit, treating the meters as follows:

- Remove the voltmeter and also remove its connecting wires.
- Remove the ammeter, patching it up with a line.

Redraw the circuit a few more times, simplifying it one step at a time by identifying series and parallel combinations. Try it yourself first, and then check your diagrams below.

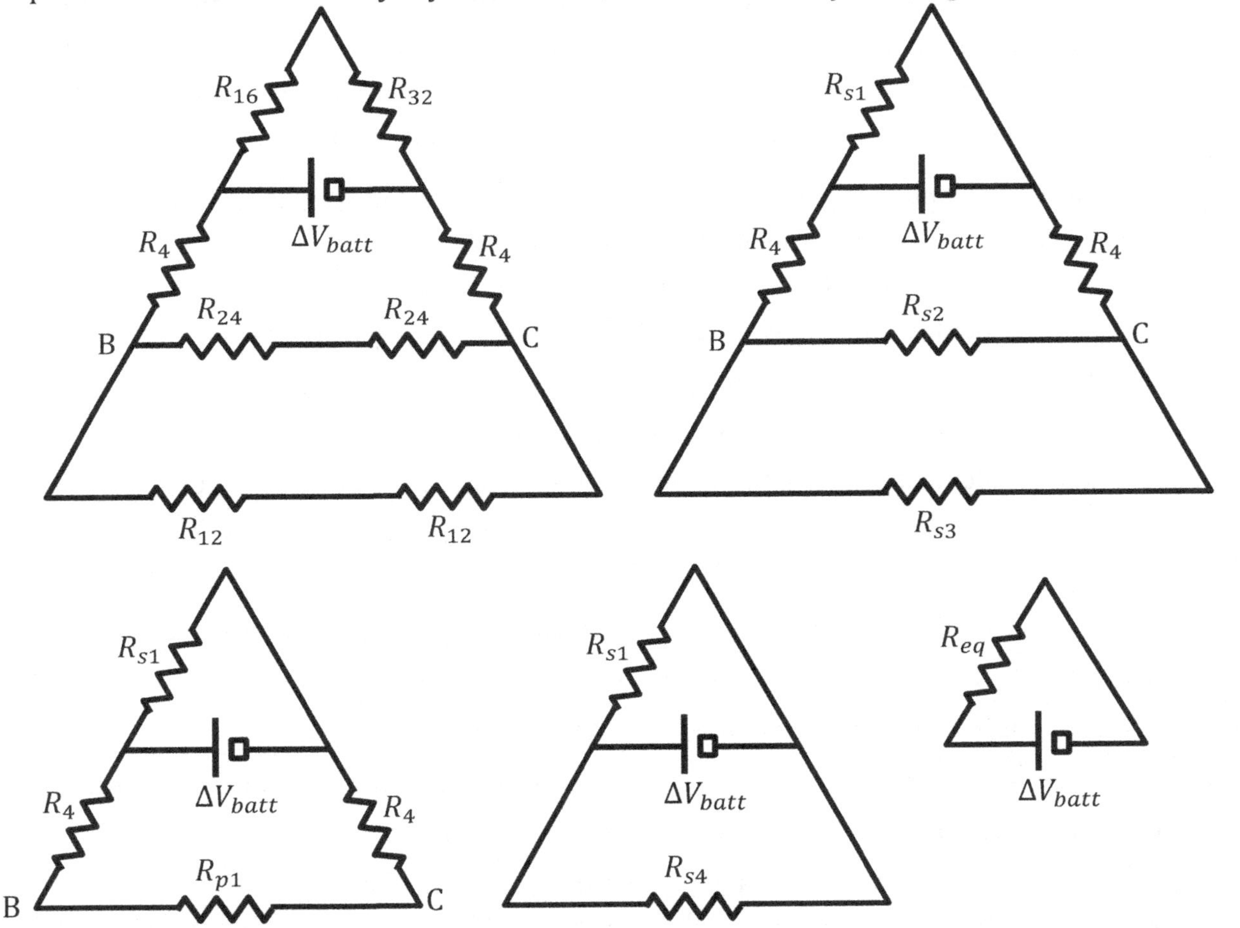

(B) Apply the formulas for series and parallel resistors.

- The 16.0-Ω and 32.0-Ω resistors are in **series**. They became R_{s1}. The two 24.0-Ω resistors are in **series**. They became R_{s2}. The two 12.0-Ω resistors are in **series**. They became R_{s3}. In each case, an electron could travel from one resistor to the other without crossing a junction. These reductions are shown above on the right.
- Compute R_{s1}, R_{s2}, and R_{s3} using $R_{s1} = R_{16} + R_{32}$, $R_{s2} = R_{24} + R_{24}$, and $R_{s3} = R_{12} + R_{12}$. You should get $R_{s1} = 48\ \Omega$, $R_{s2} = 48\ \Omega$, and $R_{s3} = 24\ \Omega$.

- In the top right diagram, R_{s2} and R_{s3} are in **parallel**. They became R_{p1}.
- Compute R_{p1} using $\frac{1}{R_{p1}} = \frac{1}{R_{s2}} + \frac{1}{R_{s3}}$. Note that $\frac{1}{48} + \frac{1}{24} = \frac{1}{48} + \frac{2}{48} = \frac{3}{48} = \frac{1}{16}$. You should get $R_{p1} = 16\ \Omega$.
- In the bottom left diagram, R_{p1} and the two R_4's are in **series**. They became R_{s4}.
- Compute R_{s4} using $R_{s4} = R_4 + R_{p1} + R_4$. You should get $R_{s4} = 24\ \Omega$.
- Finally, R_{s1} and R_{s4} are in **parallel**. They became R_{eq}.
- Compute R_{eq} using $\frac{1}{R_{eq}} = \frac{1}{R_{s1}} + \frac{1}{R_{s4}}$. The answer is $R_{eq} = 16\ \Omega$.

(C) Work your way backwards through the circuit one step at a time. Study the parts of the example that involved working backwards. Try to think your way through the strategy.

- An **ammeter** measures **current**. We need to find the current through the 12.0-Ω resistors because the ammeter is connected in series with the 12.0-Ω resistors.
- Begin the math with the equation $I_{batt} = \frac{\Delta V_{batt}}{R_{eq}}$. You should get $I_{batt} = 15$ A.
- Going back one picture from the last, R_{s1} and R_{s4} are in **parallel** (replaced by R_{eq}).
  - ΔV's are the same in parallel. Write $\Delta V_{s1} = \Delta V_{s4} = \Delta V_{batt} = 240$ V. (Note that ΔV_{batt} is the potential difference across R_{eq}.)
  - Calculate current: $I_{s4} = \frac{\Delta V_{s4}}{R_{s4}}$. You should get $I_{s4} = 10$ A.
- Going back one more step, R_{p1} and the two R_4's are in **series** (replaced by R_{s4}).
  - Current is the same in series. Write $I_4 = I_{p1} = I_4 = I_{s4} = 10$ A.
  - Calculate potential difference: $\Delta V_{p1} = I_{p1} R_{p1}$. You should get $\Delta V_{p1} = 160$ V.
- Going back one more step, R_{s2} and R_{s3} are in **parallel** (replaced by R_{p1}).
  - ΔV's are the same in parallel. Write $\Delta V_{s2} = \Delta V_{s3} = \Delta V_{p1} = 160$ V.
  - Calculate current: $I_{s3} = \frac{\Delta V_{s3}}{R_{s3}}$. You should get $I_{s3} = \frac{20}{3}$ A. Note that $\frac{160}{24} = \frac{20}{3}$.
- Going back one more step, the two R_{12}'s are in **series** (replaced by R_{s3}).
  - Current is the same in series. Write $I_{12} = I_{12} = I_{s3} = \frac{20}{3}$ A.
  - I_{12} is what the ammeter reads since the ammeter is in series with the R_{12}'s.
- The ammeter reads $I_{12} = \frac{20}{3}$ A. As a decimal, it's $I_{12} = 6.7$ A.

(D) Continue working your way backwards through the circuit.

- A **voltmeter** measures **potential difference**. We need to find the potential difference between points B and C. This means that we need to find ΔV_{p1}.
- This will be easy because we already found $\Delta V_{p1} = 160$ V in part (C). See the hints to part (C).
- The voltmeter reads $\Delta V_{p1} = 160$ V.

61. Redraw the circuit, simplifying it one step at a time by identifying series and parallel combinations. Try it yourself first, and then check your diagrams below.

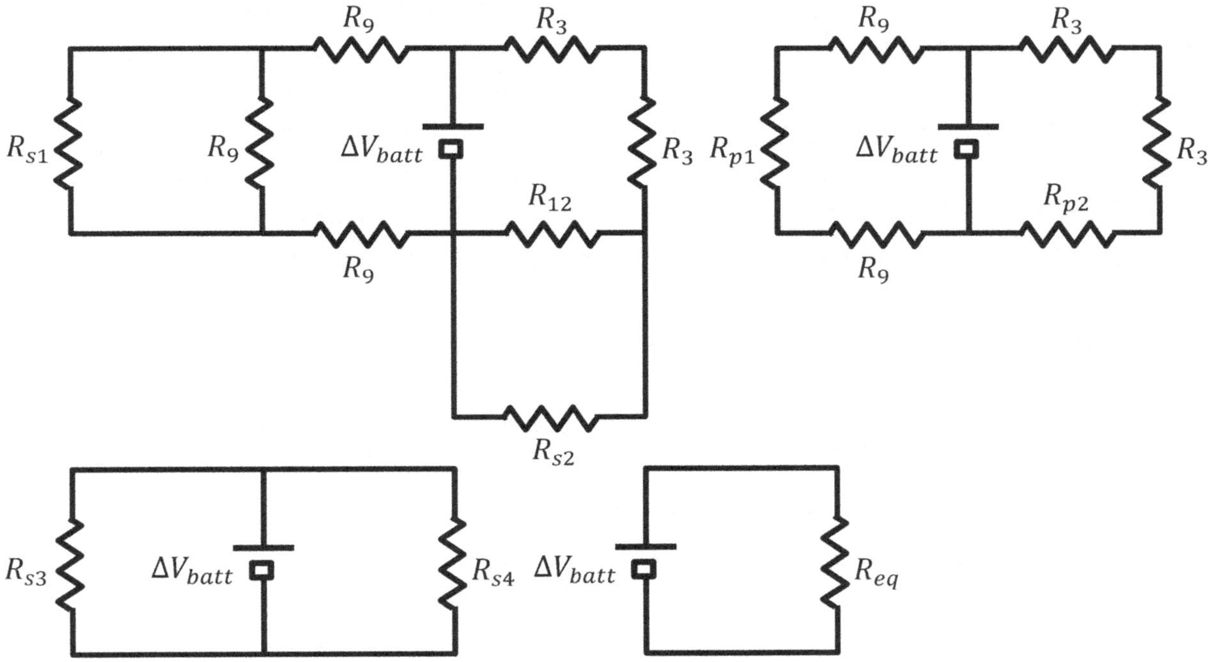

Apply the formulas for series and parallel resistors.

- The three 6.0-Ω resistors are in **series**. They became R_{s1}. The three 4.0-Ω resistors are in **series**. They became R_{s2}. (The two 3.0-Ω resistors are also in series, but it turns out to be convenient to save those for a later step. If you combine them now, it will be okay.)
- Compute R_{s1} and R_{s2} using $R_{s1} = R_6 + R_6 + R_6$ and $R_{s2} = R_4 + R_4 + R_4$. You should get $R_{s1} = 18\ \Omega$ and $R_{s2} = 12\ \Omega$.
- In the top left diagram, R_{s1} and R_9 are in **parallel**. They became R_{p1}. Also, R_{12} and R_{s2} are in **parallel**. They became R_{p2}.
- Compute R_{p1} and R_{p2} using $\frac{1}{R_{p1}} = \frac{1}{R_{s1}} + \frac{1}{R_9}$ and $\frac{1}{R_{p2}} = \frac{1}{R_{12}} + \frac{1}{R_{s2}}$. You should get $R_{p1} = 6.0\ \Omega$ and $R_{p2} = 6.0\ \Omega$.
- In the top right diagram, R_{p1} and the two R_9's are in **series**. They became R_{s3}. Also, R_{p2} and the two R_3's are in **series**. They became R_{s4}.
- Compute R_{s3} and R_{s4} using $R_{s3} = R_9 + R_{p1} + R_9$ and $R_{s4} = R_3 + R_{p2} + R_3$. You should get $R_{s3} = 24\ \Omega$ and $R_{s4} = 12\ \Omega$.
- Finally, R_{s3} and R_{s4} are in **parallel**. They became R_{eq}.
- Compute R_{eq} using $\frac{1}{R_{eq}} = \frac{1}{R_{s3}} + \frac{1}{R_{s4}}$. The answer is $R_{eq} = 8.0\ \Omega$.

Chapter 14: Circuits with Symmetry

62. Study the examples to try to understand how to identify points with the same electric potential and how to "unfold" the circuit.

- "Unfold" the circuit with point D at the bottom (call it the "ground") and point C at the top (call it the "roof"). This is how the battery is connected.
- Points B and H are each one step from point D (the "ground") and two steps from point C (the "roof"). Points B and H have the same electric potential.
- Points A and G are each two steps from point D (the "ground") and one step from point C (the "roof"). Points A and G have the same electric potential.
- Draw points B and H at the same height in the "unfolded" circuit.
- Draw points A and G at the same height in the "unfolded" circuit.
- Draw points B and H closer to point D (the "ground") and points A and G closer to point C (the "roof").
- Between those two pairs (B and H, and A and G), draw point F closer to point D (the "ground") and point E closer to point C (the "roof").
- Compare your attempt to "unfold" the circuit with the left diagram below.

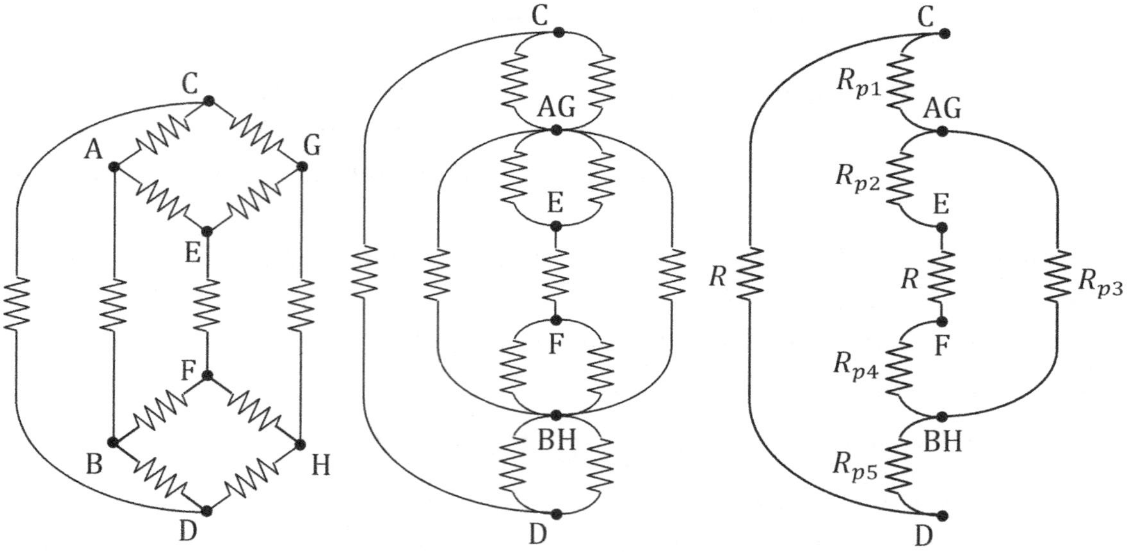

- Points B and H have the same electric potential. There are presently no wires connecting points B and H. Add wires to connect points B and H. Make these new wires so short that you have to move points B and H toward one another. Make it so extreme that points B and H merge into a single point, which we will call BH.
- Similarly, collapse points A and G into a single point called AG.
- With these changes, the diagram on the left above turns into the diagram in the middle above. In the middle diagram above, you should be able to find several pairs of resistors that are either in series or parallel.

- There are 5 pairs of parallel resistors in the middle diagram on the **previous** page:
  - The two resistors between C and AG form R_{p1}.
  - The two resistors between AG and E form R_{p2}.
  - The two resistors between AG and BH form R_{p3}.
  - The two resistors between F and BH form R_{p4}.
  - The two resistors between BH and D form R_{p5}.
- As usual when reducing a parallel combination, remove one of the paths and rename the remaining resistor. Compare the right two diagrams on the **previous** page.
- Calculate R_{p1} thru R_{p5} with formulas like $\frac{1}{R_{p1}} = \frac{1}{12} + \frac{1}{12}$.
- You should get $R_{p1} = R_{p2} = R_{p3} = R_{p4} = R_{p5} = 6.0\ \Omega$.
- R_{p2}, R, and R_{p4} are in series (forming R_{s1}) in the right diagram on the **previous** page.
- Calculate R_{s1} using $R_{s1} = R_{p2} + R + R_{p4}$. You should get $R_{s1} = 24.0\ \Omega$.

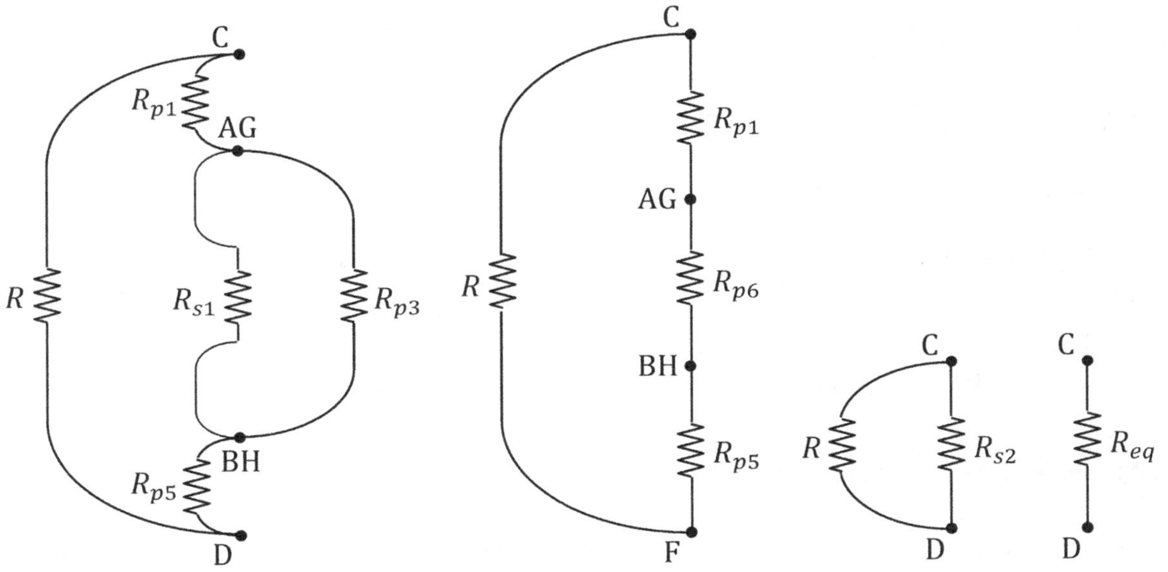

- R_{s1} and R_{p3} are in parallel in the left diagram above. They form R_{p6}.
- Calculate R_{p6} using $\frac{1}{R_{p6}} = \frac{1}{R_{s1}} + \frac{1}{R_{p3}}$. You should get $R_{p6} = \frac{24}{5}\ \Omega$ (or $4.8\ \Omega$). Note that $\frac{1}{24} + \frac{1}{6} = \frac{1}{24} + \frac{4}{24} = \frac{5}{24}$. (Then since $\frac{1}{R_{p6}} = \frac{5}{24}$, flip it to get $R_{p6} = \frac{24}{5}\ \Omega$.)
- R_{p1}, R_{p6}, and R_{p5} are in series (forming R_{s2}) in the second diagram above.
- Calculate R_{s2} using $R_{s2} = R_{p1} + R_{p6} + R_{p5}$. You should get $R_{s1} = \frac{84}{5}\ \Omega$ (or $16.8\ \Omega$). Note that $6 + \frac{24}{5} + 6 = \frac{30}{5} + \frac{24}{5} + \frac{30}{5} = \frac{84}{5}$. (This is series: **Don't** flip it.)
- R and R_{s2} are in parallel in the third diagram above. They form R_{eq}.
- Calculate R_{eq} using $\frac{1}{R_{eq}} = \frac{1}{R} + \frac{1}{R_{s2}}$. You should get $R_{eq} = 7.0\ \Omega$. Note that $\frac{1}{84/5} + \frac{1}{12} = \frac{5}{84} + \frac{1}{12} = \frac{5}{84} + \frac{7}{84} = \frac{12}{84} = \frac{1}{7}$. (This is parallel: Do flip it.)

63. Study the "Essential Concepts" section on page 210.

- Which points have the same electric potential?
- Point B is one-third the way from point A (the "ground," since it's at the negative terminal) to point D (the "roof," since it's at the positive terminal), since 10 is one third of 30 (the 30 comes from adding 10 to 20).
- Point C is also one-third the way from point A to point D, since 5 is one third of 15 (the 15 comes from adding 5 to 10).
- Therefore, points B and C have the same electric potential: $V_B = V_C$.
- The potential difference between points B and C is zero: $\Delta V_{BC} = V_C - V_B = 0$.
- From Ohm's law, $\Delta V_{BC} = I_{BC} R_{BC}$. Since $\Delta V_{BC} = 0$, the current from B to C must be zero: $I_{BC} = 0$.
- Since there is no current in the 8.0-Ω resistor, we may remove this wire without affecting the equivalent resistance of the circuit. See the diagram below.

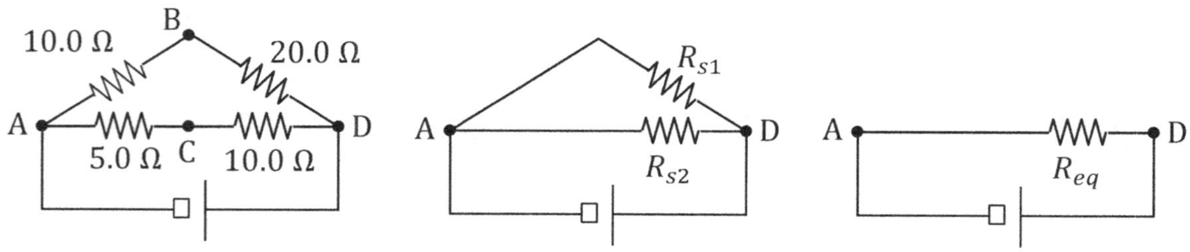

- Now it should be easy to solve for the equivalent resistance. Try it!
- The 10.0-Ω and 20.0-Ω resistors are in series. You should get $R_{s1} = 30.0\ \Omega$.
- The 5.0-Ω and 10.0-Ω resistors are in series. You should get $R_{s2} = 15.0\ \Omega$.
- R_{s1} and R_{s2} are in parallel. You should get $R_{eq} = 10.0\ \Omega$.

Chapter 15: Kirchhoff's Rules

64. Remove the voltmeter and its connecting wires. Draw the **currents**: See I_L, I_M, and I_R below – solid arrows (→). Draw the sense of **traversal** in each loop – dashed arrows (⇢).

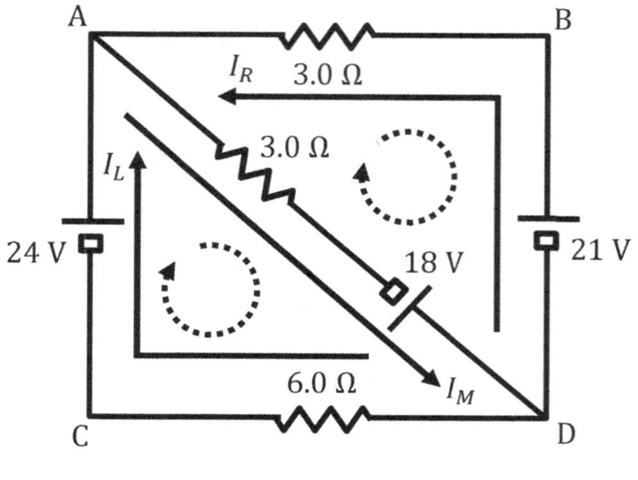

(A) Draw and label your currents the same way as shown on the previous page. That way, the hints will match your solution and prove to be more helpful. (If you've already solved the problem a different way, there isn't any reason that you can't get out a new sheet of paper and try it this way. If you're reading the hints, it appears that you would like some help, so why not try it?)

- Apply Kirchhoff's junction rule. We chose junction A. In our drawing, currents I_L and I_R enter A, whereas I_M exits A. Therefore, $I_L + I_R = I_M$. (If you draw your currents differently, the junction equation will be different for you.)
- Apply Kirchhoff's loop rule to the left loop. We chose to start at point A and traverse clockwise through the circuit. (If you draw your currents or sense of traversal differently, some signs will be different for you.)
$$-3\,I_M + 18 - 6\,I_L + 24 = 0$$
- Apply Kirchhoff's loop rule to the right loop. We chose to start at point A and traverse clockwise through the circuit. (If you draw your currents or sense of traversal differently, some signs will be different for you.)
$$+3\,I_R - 21 - 18 + 3\,I_M = 0$$
- **Tip**: If you apply the same method to perform the algebra as in the hints, your solution will match the hints and the hints will prove to be more helpful.
- Solve for I_R in the junction equation. (Why? Because you would have to plug in I_M twice, but will only have to plug in I_R once, so it's a little simpler.) You should get $I_R = I_M - I_L$.
- Replace I_R with $I_M - I_L$ the equation for the right loop. Combine **like terms**. Bring the constant term to the right. You should get $6\,I_M - 3\,I_L = 39$.
- Simplify the equation for the left loop. Combine **like terms**. Bring the constant term to the right. Multiply both sides of the equation by -1. (This step is convenient, as it will remove all of the minus signs.) You should get $3\,I_M + 6\,I_L = 42$.
- Multiply the equation $6\,I_M - 3\,I_L = 39$ by 2. This will give you $-6\,I_L$ in one equation and $+6\,I_L$ in the other equation. You should get $12\,I_M - 6\,I_L = 78$.
- Add the two equations ($3\,I_M + 6\,I_L = 42$ and $12\,I_M - 6\,I_L = 78$) together. You should get $15\,I_M = 120$.
- Divide both sides of the equation by 15. You should get $I_M = 8.0$ A.
- Plug this answer into the equation $3\,I_M + 6\,I_L = 42$. Solve for I_L.
- You should get $I_L = 3.0$ A. Plug I_L and I_M into the junction equation.
- You should get $I_R = 5.0$ A.

(B) Find the potential difference between the two points where the voltmeter had been connected. Study the example, which shows how to apply Kirchhoff's loop rule to find the potential difference between two points.

- Choose the route along the two batteries (the math is simpler).
- You should get $+18 + 21 = 39$ V (it's positive if you go **towards** point B).

(C) Study the last part of the example, which shows you how to **rank** electric potential.

- Set the electric potential at one point to zero. We chose $V_A = 0$.
- Apply the loop rule from A to B. Based on how we drew the currents, we get $V_B - V_A = +3\, I_R$. Plug in values for V_A and I_R. Solve for V_B. You should get $V_B = 15$ V.
- Apply the loop rule from B to D. You should get $V_D - V_B = -21$. Plug in the value for V_B. You should get $V_D = V_B - 21 = -6.0$ V.
- Apply the loop rule from D to C. You should get $V_C - V_D = -6\, I_L$. Plug in values for V_D and I_L. You should get $V_C = -24.0$ V. (Note that $V_C + 6 = -18$, since subtracting negative 6 equates to adding positive 6. Then subtracting 6 from both sides gives you the -24.)
- Check your answer by going from C to A. You should get $V_A - V_C = 24$. Plug in the value for V_C. You should get $V_A + 24 = 24$ (because subtracting negative 24 is the same as adding positive 24). Subtract 24 from both sides to get $V_A = 0$. Since we got the same value, $V_A = 0$, as we started out with, everything checks out.
- List the values of electric potential that we found: $V_A = 0$, $V_B = 15$ V, $V_D = -6.0$ V, and $V_C = -24.0$ V. Electric potential is highest at B, then A, then D, and lowest at C.

$$V_B > V_A > V_D > V_C$$

Note: Not every physics textbook solves Kirchhoff's rules problems the same way. If you're taking a physics course, it's possible that your instructor or textbook will apply a different (but equivalent) method.

Chapter 16: More Resistance Equations

65. Make a list of symbols. Choose the appropriate equation.
- The known symbols are $R = 5.0\ \Omega$, $L = \pi$ m, and $T = 0.80$ mm.
- Convert the thickness to meters: $T = 8.0 \times 10^{-4}$ m.
- The thicnkess equals the diameter of the wire. Solve for the radius of the wire.
- The radius of the wire is $a = \frac{D}{2} = \frac{T}{2}$. You should get $a = 4.0 \times 10^{-4}$ m.
- Find the cross-sectional area of the wire: $A = \pi a^2$. Note that $a^2 = (4 \times 10^{-4})^2 = (4)^2(10^{-4})^2 = 16 \times 10^{-8}\ \text{m}^2$. You should get $A = 16\pi \times 10^{-8}\ \text{m}^2$.
- Use the formula $R = \frac{\rho L}{A}$. Solve for ρ. You should get $\rho = \frac{RA}{L}$.
- The resistivity is $\rho = 8.0 \times 10^{-7}\ \Omega\cdot\text{m}$. It's the same as $\rho = 80 \times 10^{-8}\ \Omega\cdot\text{m}$.

66. Make a list of symbols. Choose the appropriate equation.
- The known symbols are $\alpha = 2.0 \times 10^{-3}$ /°C, $R_0 = 30\ \Omega$, $R = 33\ \Omega$, and $T_0 = 20$ °C.
- Use the formula $R \approx R_0(1 + \alpha \Delta T)$. Note that $\frac{33}{30} = 1.1$ and $1.1 - 1 = 0.1$.
- You should get $0.1 \approx 0.002(T - 20)$. Divide both sides by 0.002, then add 20.
- The answer is $T = 70$ °C.

67. Make a list of symbols. Choose the appropriate equation.
(A) The known symbols are $\varepsilon = 36$ V, $\Delta V = 32$ V, and $R = 8.0\ \Omega$.

- Apply Ohm's law: $\Delta V = IR$. Solve for the current: $I = \frac{\Delta V}{R}$.
- The current is $I = 4.0$ A.
- Use the formula $\varepsilon = I(R + r)$. Solve for the internal resistance (r).
- Note that $\frac{36}{4} = 9$ and $9 - 8 = 1$.
- The internal resistance is $r = 1.0\ \Omega$.

(B) Use the formula $P = I\Delta V$.

- The power dissipated in the resistor is $P = 128$ W.

---

68. Study the last example of this chapter and the strategy for deriving an equation for R.

- The first step is to apply Gauss's law to region II ($a < r < b$).
- We solved for the electric field in region II In a similar example in Chapter 8.

$$\vec{\mathbf{E}} = \frac{kQ}{r^2}\hat{\mathbf{r}}$$

- When you integrate outward to find potential difference, note that $\vec{\mathbf{E}} \cdot d\vec{\mathbf{s}} = E \cos 0°\, ds = Eds$ and $ds = dr$.

$$\Delta V = V_f - V_i = -\int_i^f \vec{\mathbf{E}} \cdot d\vec{\mathbf{s}} = -\int_i^f E\, ds$$

- After you plug in $E = \frac{kQ}{r^2}$ and simplify, you should get:

$$\Delta V = -kQ \int_{r_c=a}^{b} \frac{dr}{r^2}$$

- The anti-derivative is $-\frac{1}{r}$.
- Evaluate the anti-derivative over the limits. You should get:

$$\Delta V = -kQ\left(-\frac{1}{b} + \frac{1}{a}\right)$$

- Find a **common denominator**. You should get:

$$\Delta V = -kQ\left(\frac{-a+b}{ab}\right) = -kQ\left(\frac{b-a}{ab}\right)$$

- Note that $-a + b = b - a$.
- Since $\vec{\mathbf{J}} = \sigma\vec{\mathbf{E}}$, the integral for current is $I = \int \vec{\mathbf{J}} \cdot d\vec{\mathbf{A}} = \sigma \int \vec{\mathbf{E}} \cdot d\vec{\mathbf{A}}$.
- When you integrate over the surface area of a sphere, $\vec{\mathbf{E}} \cdot d\vec{\mathbf{A}} = E \cos 0°\, dA = EdA$ and $dA = r^2 \sin\theta\, d\theta d\varphi$. Recall the limits of θ and φ in spherical coordinates from Chapter 6. After you plug in $E = \frac{kQ}{r^2}$ and simplify, you should get:

$$I = \int \vec{\mathbf{J}} \cdot d\vec{\mathbf{A}} = \sigma \int \vec{\mathbf{E}} \cdot d\vec{\mathbf{A}} = kQ\sigma \int_{\theta=0}^{\pi} \int_{\varphi=0}^{2\pi} \sin\theta\, d\theta d\varphi$$

- Note that the r^2 from $dA = r^2 \sin\theta \, d\theta d\varphi$ cancels the $\frac{1}{r^2}$ in $E = \frac{kQ}{r^2}$.
- Note that the definite integral over the sine function is 2.
- You should get the following expression. Recall that $\sigma = \frac{1}{\rho}$.

$$I = 4\pi kQ\sigma = \frac{4\pi kQ}{\rho}$$

- Plug the equations for ΔV and I into the resistance formula: $R = \frac{|\Delta V|}{I}$.
- The k's and Q's will cancel. Note the absolute values. The answer is $R = \frac{\rho(b-a)}{4\pi ab}$.

Chapter 17: Logarithms and Exponentials

69. 5 raised to what power equals 625? The answer is 4 since $5^4 = 625$.

---

70. Divide both sides of the equation by 6. You should get $e^{-x/2} = \frac{1}{3}$.
    - Take the natural log of both sides of the equation. You should get $-\frac{x}{2} = \ln\left(\frac{1}{3}\right)$.
    - To see why, note that $e^{\ln(y)} = y$. Let $y = -\frac{x}{2}$.
    - Multiply both sides of the equation by -2. You should get $x = -2\ln\left(\frac{1}{3}\right)$.
    - Apply the rule $\ln\left(\frac{1}{y}\right) = -\ln(y)$. You should get $x = 2\ln(3)$.
    - The answer is $x = 2\ln(3)$, which is approximately $x = 2.197$.

---

71. Separate the integral into two terms: $\int_{x=0}^{2} dx - \int_{x=0}^{2} e^{-x/2}\, dx$.
    - The first definite integral equals 2.
    - Note that $\int e^{ax}\, dx = \frac{e^{ax}}{a}$. Compare $e^{-x/2}$ to e^{ax} to see that $a = -\frac{1}{2}$.
    - The anti-derivative for $\int_{x=0}^{2} e^{-x/2}\, dx$ is $-2e^{-x/2}$. Note that $\frac{1}{-1/2} = -2$.
    - Evaluate the anti-derivative over the limits. Note that $e^0 = 1$ and $e^{-2/2} = e^{-1} = \frac{1}{e}$.
    - The second definite integral works out to $\int_{x=0}^{2} e^{-x/2}\, dx = -\frac{2}{e} + 2$.
    - Figure in the overall minus sign: $-\int_{x=0}^{2} e^{-x/2}\, dx = -\left(-\frac{2}{e} + 2\right) = \frac{2}{e} - 2$.
    - The final answer is $\frac{2}{e}$. Note that when you add the two definite integrals together, the 2's cancel: $2 - \left(-\frac{2}{e} + 2\right) = 2 + \frac{2}{e} - 2 = \frac{2}{e}$.

---

72. The anti-derivative for $\int \frac{dx}{x}$ is $\ln(x)$.
    - Evaluate $\ln(x)$ over the limits.
    - You should get $\ln(2) - \ln(1)$. Note that $\ln(1) = 0$. The answer is $\ln(2)$.

Chapter 18: RC Circuits

73. Make a list of symbols. Choose the appropriate equation.
- The known symbols are $C = 5.0\ \mu\text{F}$, $Q_m = 60\ \mu\text{C}$, and $R = 20\ \text{k}\Omega$.
- Apply the metric prefixes: $\mu = 10^{-6}$ and $\text{k} = 10^3$.
- The known symbols are $C = 5.0 \times 10^{-6}$ F, $Q_m = 6.0 \times 10^{-5}$ C, and $R = 2.0 \times 10^4\ \Omega$.

(A) Use the equation for capacitance: $Q_m = C\Delta V_m$.
- Solve for ΔV_m. You should get $\Delta V_m = \frac{Q_m}{C}$.
- The initial potential difference across the capacitor is $\Delta V_m = 12.0$ V.

(B) Apply Ohm's law: $\Delta V_m = I_m R$. Solve for R. You should get $I_m = \frac{\Delta V_m}{R}$.
- The initial current is $I_m = 0.60$ mA, which is the same as 0.00060 A or 6.0×10^{-4} A.

(C) Use the equation $\tau = RC$.
- The time constant is $\tau = \frac{1}{10}$ s or $\tau = 0.10$ s.
- Use the equation $t_{1/2} = \tau \ln(2)$.
- The half-life is $t_{1/2} = \frac{\ln(2)}{10}$ s. If you use a calculator, $t_{1/2} = 0.069$ s.

(D) Use the equation $Q = Q_m e^{-t/\tau}$. Plug in $t = 0.20$ s.
- Note that $-\frac{0.20}{0.10} = -2$. Also note that $e^{-2} = \frac{1}{e^2}$.
- The charge stored on the capacitor is $Q = \frac{60}{e^2}\ \mu\text{C}$. If you use a calculator, $Q = 8.1\ \mu\text{C}$.

---

74. Make a list of symbols. Choose the appropriate equation.
- The known symbols are $C = 4.0\ \mu\text{F}$, $I_m = 6.0$ A, $I = 3.0$ A, and $t_{1/2} = 200$ ms.
- Apply the metric prefixes: $\mu = 10^{-6}$ and $\text{m} = 10^{-3}$.
- The capacitance is $C = 4.0 \times 10^{-6}$ F and the half-life is $t_{1/2} = 0.200\ \text{s} = \frac{1}{5}$ s.

(A) Use the equation $t_{1/2} = \tau \ln(2)$. Solve for τ. You should get $\tau = \frac{t_{1/2}}{\ln(2)}$.
- The time constant is $\tau = \frac{1}{5\ln(2)}$ s. If you use a calculator, $\tau = 0.29$ s.

(B) Use the equation $\tau = RC$. Solve for R. You should get $R = \frac{\tau}{C}$.
- The resistance is $R = \frac{50}{\ln(2)}\ \text{k}\Omega$. If you use a calculator, $R = 72\ \text{k}\Omega = 72{,}000\ \Omega$.

Chapter 19: Scalar and Vector Products

75. Identify the components of the given vectors.
- The components are: $A_x = 5$, $A_y = 2$, $A_z = 3$, $B_x = 4$, $B_y = -6$, and $B_z = -2$.
- Use the component form of the scalar product: $\vec{\mathbf{A}} \cdot \vec{\mathbf{B}} = A_x B_x + A_y B_y + A_z B_z$.
- You should get three terms: $20 - 12 - 6$.
- The scalar product is $\vec{\mathbf{A}} \cdot \vec{\mathbf{B}} = 2$.

- Use the determinant form of the vector product.

$$\vec{\mathbf{A}} \times \vec{\mathbf{B}} = \begin{vmatrix} \hat{\mathbf{x}} & \hat{\mathbf{y}} & \hat{\mathbf{z}} \\ A_x & A_y & A_z \\ B_x & B_y & B_z \end{vmatrix} = \hat{\mathbf{x}} \begin{vmatrix} A_y & A_z \\ B_y & B_z \end{vmatrix} - \hat{\mathbf{y}} \begin{vmatrix} A_x & A_z \\ B_x & B_z \end{vmatrix} + \hat{\mathbf{z}} \begin{vmatrix} A_x & A_y \\ B_x & B_y \end{vmatrix}$$

- The 2×2 determinants are $-4 + 18 = 14, -10 - 12 = -22$, and $-30 - 8 = -38$.
- Note the minus sign in the middle term. You will get $-(-22)\hat{\mathbf{y}} = +22\,\hat{\mathbf{y}}$.
- The vector product is $\vec{\mathbf{A}} \times \vec{\mathbf{B}} = 14\,\hat{\mathbf{x}} + 22\,\hat{\mathbf{y}} - 38\,\hat{\mathbf{z}}$.

---

76. Identify the components of the given vectors.
- The components are: $A_x = 3, A_y = -1, A_z = -4, B_x = 2, B_y = 0$, and $B_z = -1$.
- (Observe that $\vec{\mathbf{B}}$ doesn't have a $\hat{\mathbf{y}}$: That's why $B_y = 0$.)
- Use the component form of the scalar product: $\vec{\mathbf{A}} \cdot \vec{\mathbf{B}} = A_x B_x + A_y B_y + A_z B_z$.
- You should get two nonzero terms: $6 - 0 + 4$.
- The scalar product is $\vec{\mathbf{A}} \cdot \vec{\mathbf{B}} = 10$.
- Use the determinant form of the vector product.

$$\vec{\mathbf{A}} \times \vec{\mathbf{B}} = \begin{vmatrix} \hat{\mathbf{x}} & \hat{\mathbf{y}} & \hat{\mathbf{z}} \\ A_x & A_y & A_z \\ B_x & B_y & B_z \end{vmatrix} = \hat{\mathbf{x}} \begin{vmatrix} A_y & A_z \\ B_y & B_z \end{vmatrix} - \hat{\mathbf{y}} \begin{vmatrix} A_x & A_z \\ B_x & B_z \end{vmatrix} + \hat{\mathbf{z}} \begin{vmatrix} A_x & A_y \\ B_x & B_y \end{vmatrix}$$

- The 2×2 determinants are $1 - 0 = 1, -3 + 8 = 5$, and $0 + 2 = 2$.
- Note the minus sign in the middle term. You will get $-(5)\hat{\mathbf{y}} = -5\,\hat{\mathbf{y}}$.
- The vector product is $\vec{\mathbf{A}} \times \vec{\mathbf{B}} = \hat{\mathbf{x}} - 5\,\hat{\mathbf{y}} + 2\,\hat{\mathbf{z}}$.

---

77. Note that $A = 8, B = 5$, and $\theta = 60°$.

(A) Use the equation $\vec{\mathbf{A}} \cdot \vec{\mathbf{B}} = AB \cos\theta$.
- The scalar product is $\vec{\mathbf{A}} \cdot \vec{\mathbf{B}} = 20$.

(B) Use the equation $\|\vec{\mathbf{A}} \times \vec{\mathbf{B}}\| = AB \sin\theta$.
- The magnitude of the vector product is $\|\vec{\mathbf{A}} \times \vec{\mathbf{B}}\| = 20\sqrt{3}$.

Chapter 20: Bar Magnets

78. Sketch the magnetic field lines by rotating the diagram on page 265.

(A)

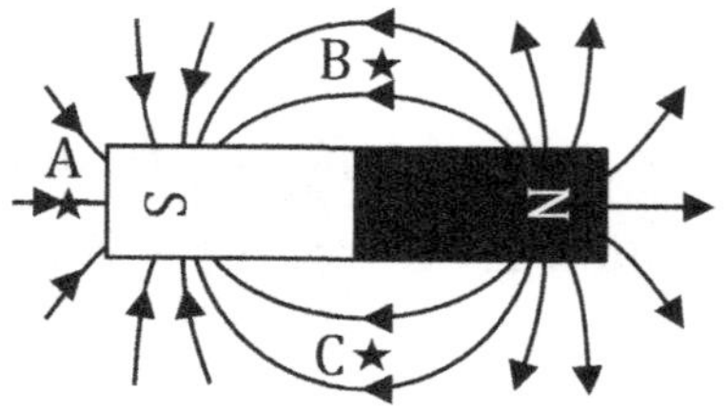

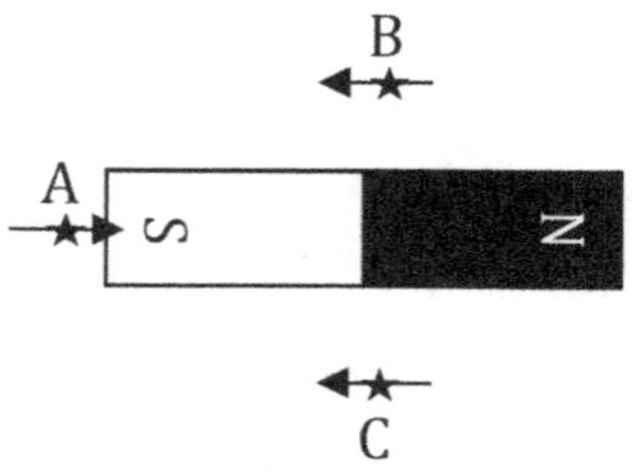

(B)

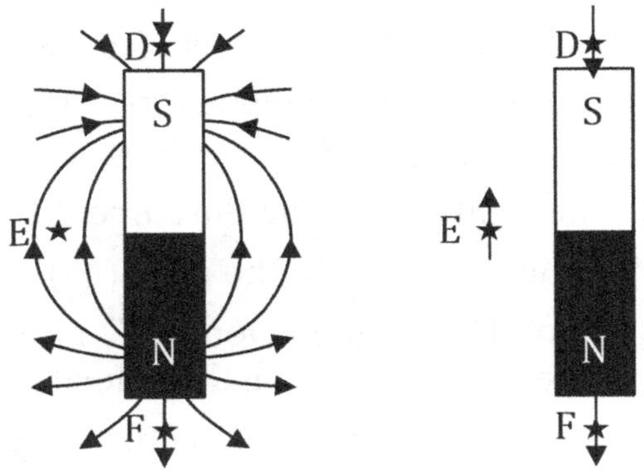

(C)

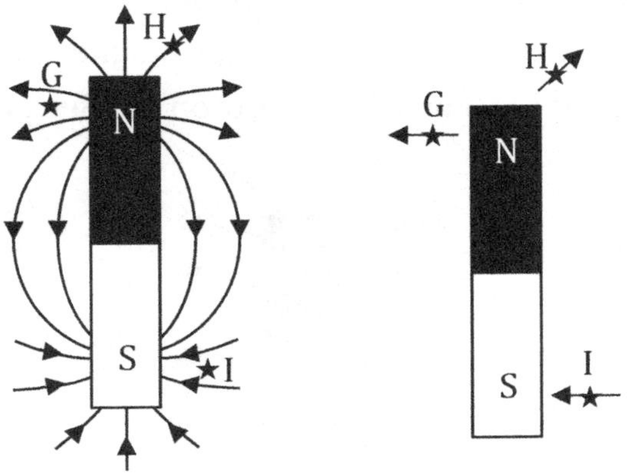

(D)

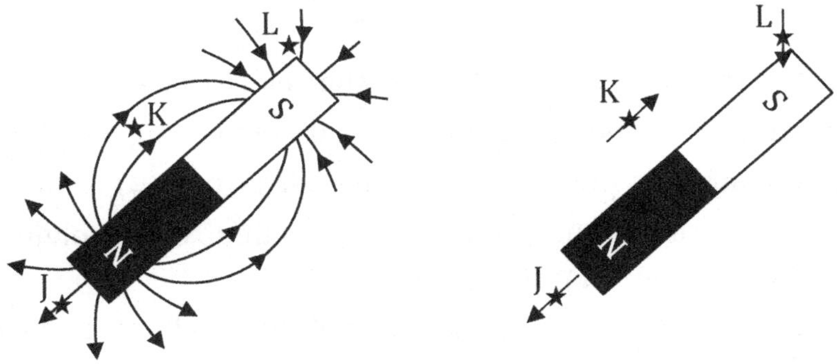

Chapter 21: Right-hand Rule for Magnetic Force

79. Study the right-hand rule on page 269 and think your way through the examples.

Tip: If you find it difficult to physically get your right hand into the correct position, try turning your book around until you find an angle that makes it more comfortable.

(A) Point your fingers down (↓), along the current (I).

- At the same time, face your palm to the right (→), along the magnetic field ($\vec{\mathbf{B}}$).
- Your thumb points out of the page: The magnetic force ($\vec{\mathbf{F}}_m$) is out of the page (⊙).

(B) Point your fingers up (↑), along the velocity ($\vec{v}$).

- At the same time, face your palm into the page (⊗), along the magnetic field ($\vec{\mathbf{B}}$).
- Your thumb points to the left: The magnetic force ($\vec{\mathbf{F}}_m$) is to the left (←).

(C) **Tip**: Turn the book to make it easier to get your right hand into the correct position.

- Point your fingers to the right (→), along the current (I).
- At the same time, face your palm out of the page (⊙), along the magnetic field ($\vec{\mathbf{B}}$).
- Your thumb points down: The magnetic force ($\vec{\mathbf{F}}_m$) is down (↓).

(D) Point your fingers out of the page (⊙), along the velocity ($\vec{v}$).

- At the same time, face your palm up (↑), along the magnetic field ($\vec{\mathbf{B}}$).
- Your thumb points to the left, but that's **not** the answer: The electron has **negative** charge, so the answer is **backwards**: The magnetic force ($\vec{\mathbf{F}}_m$) is to the right (→).

(E) First sketch the magnetic field lines for the bar magnet (see Chapter 20). What is the direction of the magnetic field lines where the proton is? See point A below: $\vec{\mathbf{B}}$ is down.

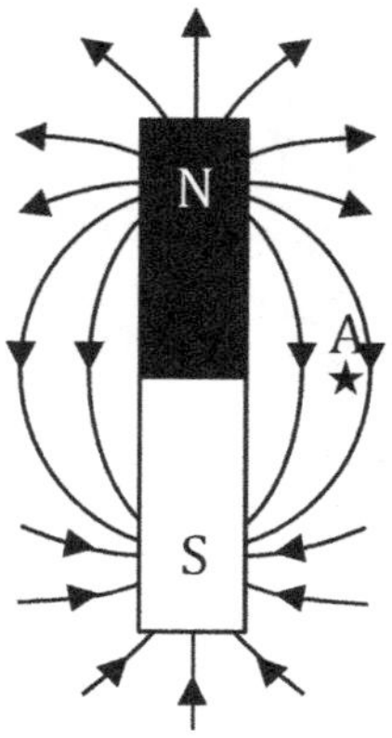

- Point your fingers to the right (→), along the velocity ($\vec{v}$).
- At the same time, face your palm down (↓), along the magnetic field ($\vec{\mathbf{B}}$).
- Your thumb points into the page: The magnetic force ($\vec{\mathbf{F}}_m$) is into the page (⊗).

(F) Invert the right-hand rule for magnetic force. This time, we're solving for $\vec{\mathbf{B}}$, **not** $\vec{\mathbf{F}}_m$.

- Point your fingers down (↓), along the current (I).
- At the same time, point your **thumb** (it's **not** your palm in this example) to the left (←), along the magnetic force ($\vec{\mathbf{F}}_m$).
- Your **palm** faces out of the page: The magnetic field ($\vec{\mathbf{B}}$) is out of the page (⊙).

---

80. Study the right-hand rule on page 269 and think your way through the examples.

(A) Point your fingers down (↓), along the velocity ($\vec{v}$).

- At the same time, face your palm to the left (←), along the magnetic field ($\vec{\mathbf{B}}$).
- Your thumb points into the page: The magnetic force ($\vec{\mathbf{F}}_m$) is into the page (⊗).

(B) Point your fingers into the page (⊗), along the current (I).

- At the same time, face your palm down (↓), along the magnetic field ($\vec{\mathbf{B}}$).
- Your thumb points to the left: The magnetic force ($\vec{\mathbf{F}}_m$) is to the left (←).

(C) This is a "trick" question. In this example, the current (I) is anti-parallel to the magnetic field ($\vec{\mathbf{B}}$). According to the strategy on page 271, the magnetic force ($\vec{\mathbf{F}}_m$) is therefore **zero** (and thus has no direction). In Chapter 24, we'll see that in this case, $\theta = 180°$ such that $F_m = ILB \sin 180° = 0$.

(D) First sketch the magnetic field lines for the bar magnet (see Chapter 20). What is the direction of the magnetic field lines where the proton is? See point A below: $\vec{\mathbf{B}}$ is down. (Note that the north pole is at the bottom of the figure. The magnet is upside down.)

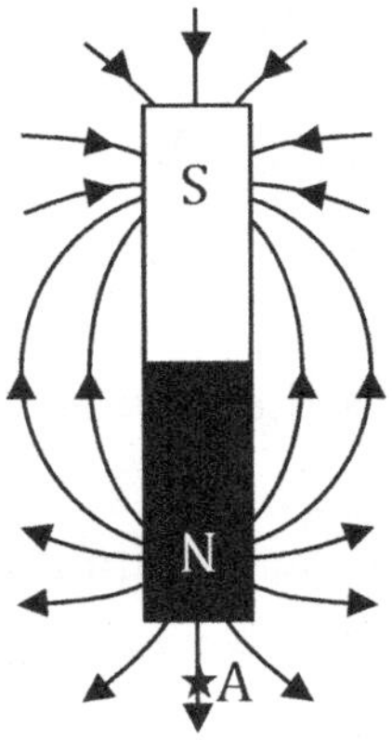

- Point your fingers to the left (←), along the velocity ($\vec{\boldsymbol{v}}$).
- At the same time, face your palm down (↓), along the magnetic field ($\vec{\mathbf{B}}$).
- Your thumb points out of the page: The magnetic force ($\vec{\mathbf{F}}_m$) is out of the page (⊙).

(E) Point your fingers diagonally up and to the left (↖), along the velocity ($\vec{\boldsymbol{v}}$).

- At the same time, face your palm out of the page (⊙), along the magnetic field ($\vec{\mathbf{B}}$).
- Your thumb points diagonally up and to the right (↗), but that's **not** the answer: The electron has **negative** charge, so the answer is **backwards**: The magnetic force ($\vec{\mathbf{F}}_m$) is diagonally down and to the left (↙).

(F) Invert the right-hand rule for magnetic force. This time, we're solving for $\vec{\mathbf{B}}$, **not** $\vec{\mathbf{F}}_m$.

- Point your fingers out of the page (⊙), along the current (I).
- At the same time, point your **thumb** (it's **not** your palm in this example) to the left (←), along the magnetic force ($\vec{\mathbf{F}}_m$).
- Your **palm** faces up: The magnetic field ($\vec{\mathbf{B}}$) is up (↑).

---

81. Study the right-hand rule on page 269 and think your way through the examples.

(A) Point your fingers diagonally down and to the right (↘), along the current (I).

- At the same time, face your palm diagonally up and to the right (↗), along the magnetic field ($\vec{\mathbf{B}}$).
- Your thumb points out of the page: The magnetic force ($\vec{\mathbf{F}}_m$) is out of the page (⊙).

(B) First sketch the magnetic field lines for the bar magnet (see Chapter 20). What is the direction of the magnetic field lines where the electron is? See point A below: $\vec{\mathbf{B}}$ is left.

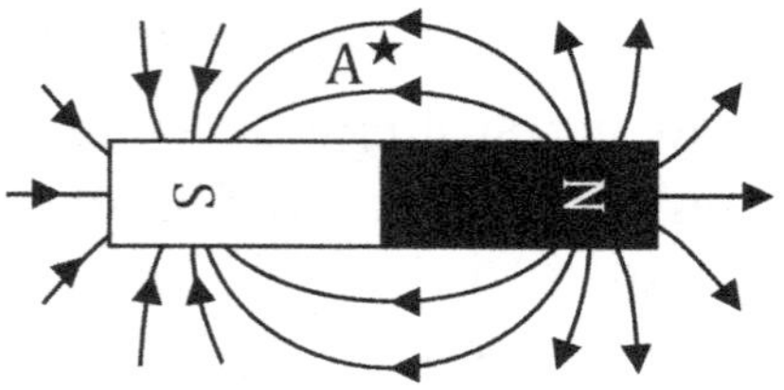

- Point your fingers into the page (⊗), along the velocity ($\vec{v}$).
- At the same time, face your palm to the left (←), along the magnetic field ($\vec{\mathbf{B}}$).
- Your thumb points up (↑), but that's **not** the answer: The electron has **negative** charge, so the answer is **backwards**: The magnetic force ($\vec{\mathbf{F}}_m$) is down (↓).

(C) The magnetic force ($\vec{\mathbf{F}}_m$) is a **centripetal** force: It pushes the proton towards the center of the circle. For the position indicated, the velocity ($\vec{v}$) is up (along a tangent) and the magnetic force ($\vec{\mathbf{F}}_m$) is to the left (toward the center). Invert the right-hand rule to find the magnetic field.

- Point your fingers up (↑), along the instantaneous velocity ($\vec{v}$).
- At the same time, point your **thumb** (it's **not** your palm in this example) to the left (←), along the magnetic force ($\vec{\mathbf{F}}_m$).
- Your **palm** faces into the page: The magnetic field ($\vec{\mathbf{B}}$) is into the page (⊗).

(D) It's the same as part (C), except that the electron has **negative** charge, so the answer is **backwards**: The magnetic field ($\vec{\mathbf{B}}$) is out of the page (⊙).

(E) First, apply the right-hand rule for magnetic force to each side of the rectangular loop:

- **Left side**: The current (I) points up (↑) and the magnetic field ($\vec{\mathbf{B}}$) also points up (↑). Since I and $\vec{\mathbf{B}}$ are parallel, the magnetic force ($\vec{\mathbf{F}}_{left}$) is **zero**.
- **Top side**: Point your fingers to the right (→) along the current (I) and your palm up (↑) along the magnetic field ($\vec{\mathbf{B}}$). The magnetic force ($\vec{\mathbf{F}}_{top}$) is out of the page (⊙).
- **Right side**: The current (I) points down (↓) and the magnetic field ($\vec{\mathbf{B}}$) points up (↑). Since I and $\vec{\mathbf{B}}$ are anti-parallel, the magnetic force ($\vec{\mathbf{F}}_{right}$) is **zero**.
- **Bottom side**: Point your fingers to the left (←) along the current (I) and your palm up (↑) along the magnetic field ($\vec{\mathbf{B}}$). The magnetic force ($\vec{\mathbf{F}}_{bot}$) is into the page (⊗).

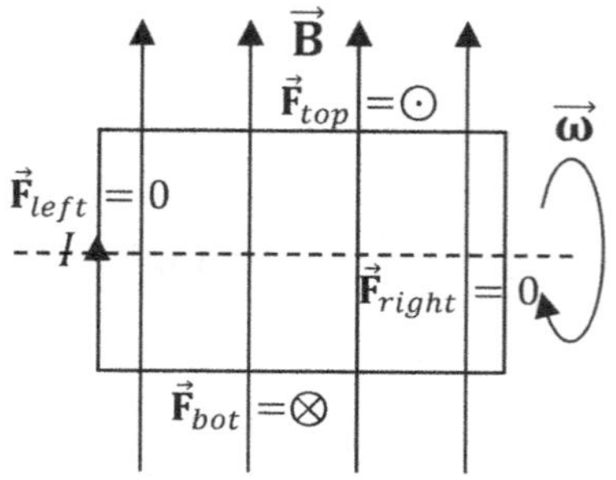

The top side of the loop is pulled out of the page, while the bottom side of the loop is pushed into the page. What will happen? The loop will rotate about the dashed axis.

(F) First, apply the right-hand rule for magnetic force to each side of the rectangular loop:

- **Bottom side**: Point your fingers to the right (→) along the current (I) and your palm out of the page (⊙) along the magnetic field ($\vec{\mathbf{B}}$). The magnetic force ($\vec{\mathbf{F}}_{bot}$) is down (↓).
- **Right side**: Point your fingers up (↑) along the current (I) and your palm out of the page (⊙) along the magnetic field ($\vec{\mathbf{B}}$). The magnetic force ($\vec{\mathbf{F}}_{right}$) is to the right (→).
- **Top side**: Point your fingers to the left (←) along the current (I) and your palm out of the page (⊙) along the magnetic field ($\vec{\mathbf{B}}$). The magnetic force ($\vec{\mathbf{F}}_{top}$) is up (↑).
- **Left side**: Point your fingers down (↓) along the current (I) and your palm out of the page (⊙) along the magnetic field ($\vec{\mathbf{B}}$). The magnetic force ($\vec{\mathbf{F}}_{left}$) is to the left (←).

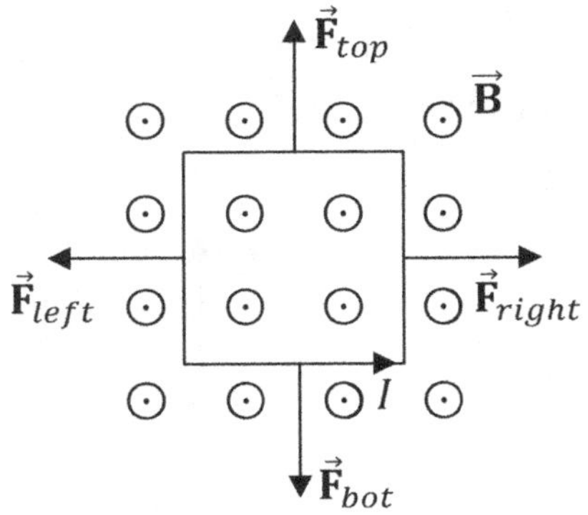

The magnetic force on each side of the loop is pulling outward, which would tend to make the loop try to expand. (That's the tendency: Whether or not it actually will expand depends on such factors as how strong the forces are and the rigidity of the materials.)

Chapter 22: Right-hand Rule for Magnetic Field

82. Study the right-hand rule on page 279 and think your way through the examples.

(A) Apply the right-hand rule for magnetic field:

- Grab the current with your thumb pointing to the down (↓), along the current (I).
- Your fingers make circles around the wire (toward your fingertips).
- The magnetic field ($\vec{\mathbf{B}}$) at a specified point is tangent to these circles, as shown in the diagram below on the right. At point A the magnetic field ($\vec{\mathbf{B}}_A$) points into the page (⊗), while at point C the magnetic field ($\vec{\mathbf{B}}_C$) points out of the page (⊙).

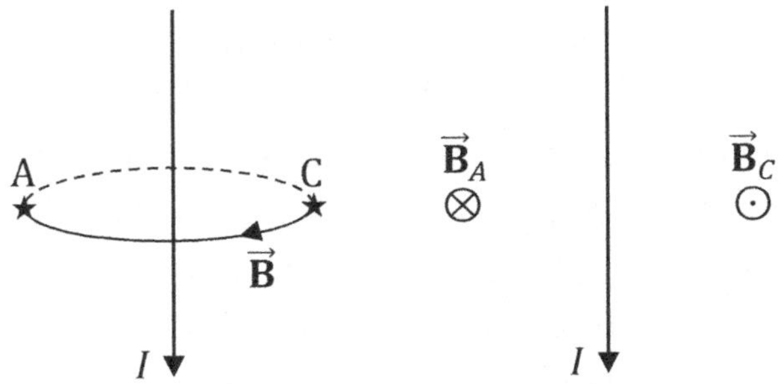

(B) Apply the right-hand rule for magnetic field:

- Grab the current with your thumb pointing into the page (⊗), along the current (I).
- Your fingers make **clockwise** (use the right-hand rule to see this) circles around the wire (toward your fingertips), as shown in the diagram below on the left.
- The magnetic field ($\vec{\mathbf{B}}$) at a specified point is **tangent** to these circles, as shown in the diagram below on the right. Draw tangent lines at points D, E, and F with the arrows headed **clockwise**. See the diagram below on the right. At point D the magnetic field ($\vec{\mathbf{B}}_D$) points up (↑), at point E the magnetic field ($\vec{\mathbf{B}}_E$) points left (←), and at point F the magnetic field ($\vec{\mathbf{B}}_F$) points diagonally down and to the right (↘).
- **Note**: The magnetic field is **clockwise** here, whereas it was counterclockwise in a similar example, because here the current is heading into the page, while in the example it was coming out of the page.

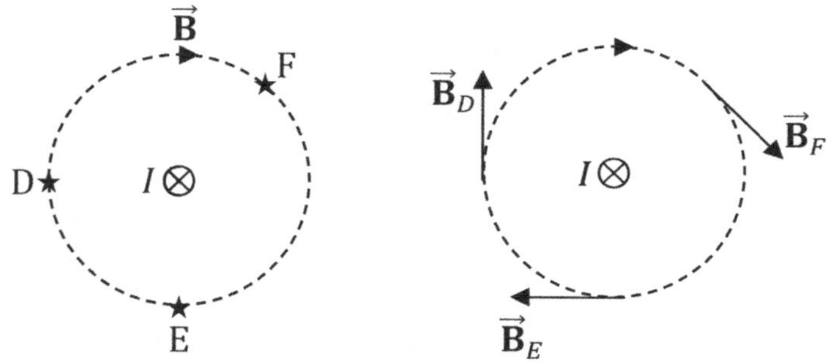

(C) Apply the right-hand rule for magnetic field:

- Grab the loop with your right hand, such that your thumb points **clockwise** (since that's how the current is drawn in the problem). No matter where you grab the loop, your fingers are going into the page (⊗) at point H. The magnetic field ($\vec{\mathbf{B}}_H$) points into the page (⊗) at point H.
- For point G, grab the loop at the leftmost point (that point is nearest to point G, so it will have the dominant effect). For point J, grab the loop at the rightmost point (the point nearest to point J). Your fingers are coming out of the page (⊙) at points G and J. The magnetic field ($\vec{\mathbf{B}}_G$ and $\vec{\mathbf{B}}_J$) points out of the page (⊙) at points G and J.
- **Note**: These answers are the opposite inside and outside of the loop compared to the similar example because the current is clockwise in this problem, whereas the current was counterclockwise in the similar example.

(D) This problem is essentially the same as part (C), except that the current is **counter-clockwise** instead of clockwise. The answers are simply the opposite of part (C)'s answers:

- The magnetic field ($\vec{\mathbf{B}}_L$) points out of the page (⊙) at point L.
- The magnetic field ($\vec{\mathbf{B}}_K$ and $\vec{\mathbf{B}}_M$) points into the page (⊗)at points K and M.

(E) First label the positive (+) and negative (−) terminals of the battery (the long line is the positive terminal) and draw the current from the positive terminal to the negative terminal. See the diagram on the following page on the left.

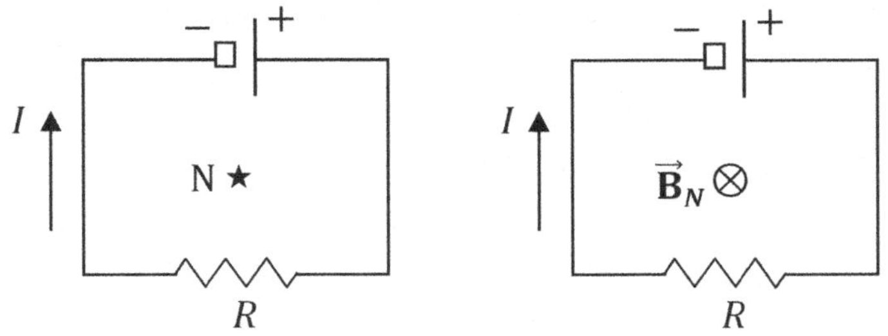

Now apply the right-hand rule for magnetic field:

- This problem is the same as point H from part (C), since the current is traveling through the loop in a **clockwise** path.
- The magnetic field ($\vec{\mathbf{B}}_N$) points into the page (⊗) at point N.

(F) Note that this loop (unlike the three previous problems) does **not** lie in the plane of the paper. This loop is a **horizontal** circle with the solid (—) semicircle in front of the paper and the dashed (---) semicircle behind the paper. It's like the rim of a basketball hoop. Apply the right-hand rule for magnetic field:

- Imagine grabbing the front of the rim of a basketball hoop with your right hand, such that your thumb points to your left (since in the diagram, the current is heading to the left in the front of the loop).
- Your fingers are going down (↓) at point O inside of the loop. The magnetic field ($\vec{\mathbf{B}}_O$) points down (↓) at point O.
- **Note:** This answer is the opposite of the similar example because the current is heading in the opposite direction in this problem compared to that example.

(G) This is similar to part (F), except that now the loop is vertical instead of horizontal. Apply the right-hand rule for magnetic field:

- Imagine grabbing the front of the loop with your right hand, such that your thumb points up (since in the diagram, the current is heading up in the front of the loop).
- Your fingers are going to the left (←) at point P inside of the loop. The magnetic field ($\vec{\mathbf{B}}_P$) points to the left (←) at point P.

(H) This is the same as part (G), except that the current is heading in the opposite direction. The answer is simply the opposite of part (G)'s answer. The magnetic field ($\vec{\mathbf{B}}_Q$) points to the right (→) at point Q.

(I) This solenoid essentially consists of several (approximately) horizontal loops. Each horizontal loop is just like part (F). Note that the current (I) is heading the same way (it is pointing to the left in the front of each loop) in parts (I) and (F). Therefore, just as in part (F), the magnetic field ($\vec{\mathbf{B}}_R$) points down (↓) at point R inside the solenoid.

Chapter 23: Combining the Two Right-hand Rules

83. First apply the right-hand rule from Chapter 22 to find the magnetic **field**. Next apply the right-hand rule from Chapter 21 to find the magnetic **force**.
(A) Draw the field point on I_2 since we want the force exerted on I_2.

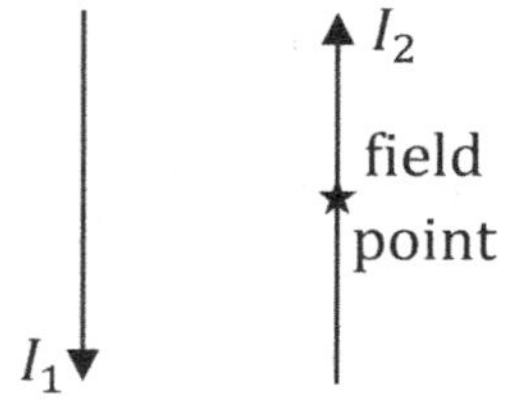

- Apply the right-hand rule for magnetic **field** (Chapter 22) to I_1. Grab I_1 with your thumb down (↓), along I_1, and your fingers wrapped around I_1. Your fingers are coming out of the page (⊙) at the field point (★). The magnetic field ($\vec{\mathbf{B}}_1$) that I_1 makes at the field point (★) is out of the page (⊙).
- Now apply the right-hand rule for magnetic force (Chapter 21) to I_2. Point your fingers up (↑), along I_2. At the same time, face your palm out of the page (⊙), along $\vec{\mathbf{B}}_1$. Your thumb points to the right (→), along the magnetic force ($\vec{\mathbf{F}}_1$).
- The left current (I_1) pushes the right current (I_2) to the **<u>right</u>** (→). (They repel.)

(B) Draw the field point on I_2 since we want the force exerted on I_2.

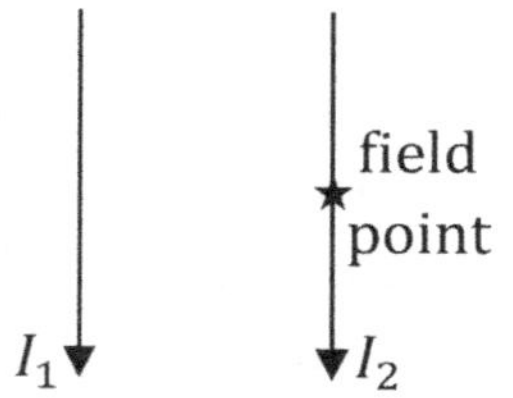

- Apply the right-hand rule for magnetic **field** (Chapter 22) to I_1. Grab I_1 with your thumb down (↓), along I_1, and your fingers wrapped around I_1. Your fingers are coming out of the page (⊙) at the field point (★). The magnetic field ($\vec{\mathbf{B}}_1$) that I_1 makes at the field point (★) is out of the page (⊙).
- Now apply the right-hand rule for magnetic force (Chapter 21) to I_2. Point your fingers down (↓), along I_2. At the same time, face your palm out of the page (⊙), along $\vec{\mathbf{B}}_1$. Your thumb points to the left (←), along the magnetic force ($\vec{\mathbf{F}}_1$).
- The left current (I_1) pulls the right current (I_2) to the **<u>left</u>** (←). (They attract.)

(C) Note that the **wording** of this problem is **different**: Note the difference in the **subscripts**.

- Draw the field point on I_1 (**<u>not</u>** on I_2) since we want the force exerted on I_1.
- Apply the right-hand rule for magnetic **field** (Chapter 22) to I_2 (**<u>not</u>** I_1). Grab I_2 with your thumb down (↓), along I_2, and your fingers wrapped around I_2. Your fingers are going into the page (⊗) at the field point (★). The magnetic field ($\vec{\mathbf{B}}_2$) that I_2 (it's **<u>not</u>** I_1 in this problem) makes at the field point (★) is into the page (⊗).

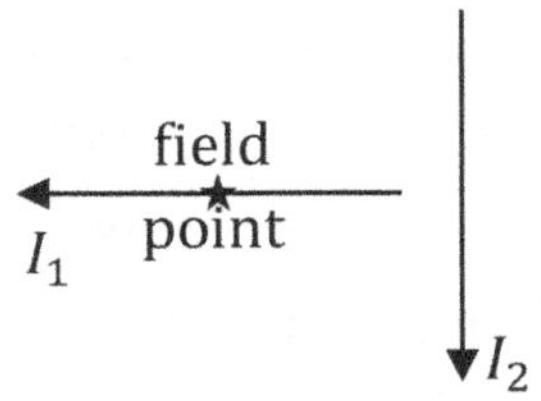

- Now apply the right-hand rule for magnetic force (Chapter 21) to I_1 (**not** I_2). Point your fingers to the left (←), along I_1 (it's **not** I_2 in this problem). At the same time, face your palm into the page (⊗), along $\vec{\mathbf{B}}_2$. Your thumb points down (↓), along the magnetic force ($\vec{\mathbf{F}}_2$).
- The right current (I_2) pushes the left current (I_1) **downward** (↓).
- Note that we found $\vec{\mathbf{B}}_2$ and $\vec{\mathbf{F}}_2$ in this problem (**not** $\vec{\mathbf{B}}_1$ and $\vec{\mathbf{F}}_1$) because this problem asked us to find the magnetic force exerted by I_2 on I_1 (**not** by I_1 on I_2).

(D) The field point is point A since we want the force exerted on I_2 at point A.

- Apply the right-hand rule for magnetic **field** (Chapter 22) to I_1. Grab I_1 with your thumb along I_1 and your fingers wrapped around I_1. Your fingers are going into the page (⊗) at the field point (point A). The magnetic field ($\vec{\mathbf{B}}_1$) that I_1 makes at the field point (point A) is into the page (⊗).
- Now apply the right-hand rule for magnetic force (Chapter 21) to I_2 at point A. Point your fingers up (↑), since I_2 runs upward at point A. At the same time, face your palm into the page (⊗), along $\vec{\mathbf{B}}_1$. Your thumb points to the left (←), along the magnetic force ($\vec{\mathbf{F}}_1$).
- The outer current (I_1) pushes the inner current (I_2) to the **left** (←) at point A. More generally, the outer current pushes the inner current inward. (They repel.)

(E) Draw the field point on I_2 since we want the force exerted on I_2.

- Label the positive (+) and negative (−) terminals of the battery: The long line is the positive (+) terminal. We draw the current from the positive terminal to the negative terminal: In this problem, the current (I_1) runs clockwise, as shown below.

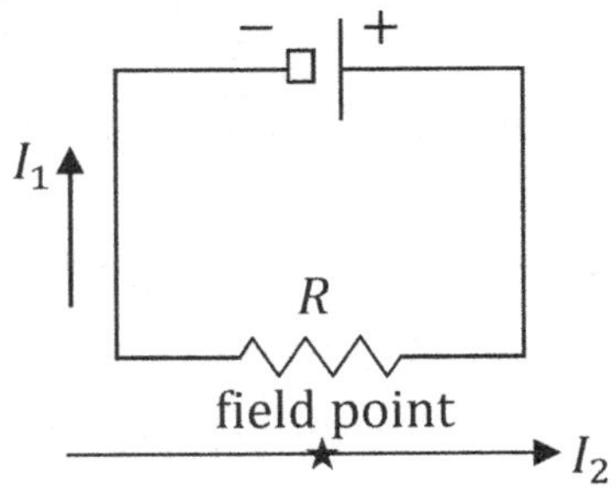

- Apply the right-hand rule for magnetic **field** (Chapter 22) to I_1. Grab I_1 at the **bottom** of the loop, with your thumb left (←), since I_1 runs to the left at the **bottom** of the loop. Your fingers are coming out of the page (⊙) at the field point (★). The magnetic field ($\vec{\mathbf{B}}_1$) that I_1 makes at the field point (★) is out of the page (⊙).

- Now apply the right-hand rule for magnetic force (Chapter 21) to I_2. Point your fingers to the right (→), along I_2. At the same time, face your palm out of the page (⊙), along $\vec{\mathbf{B}}_1$. **Tip**: Turn the book to make it more comfortable to position your hand as needed. Your thumb points down (↓), along the magnetic force ($\vec{\mathbf{F}}_1$).
- The top current (I_1) pushes the bottom current (I_2) **downward** (↓). (They repel.)

(F) Draw the field point in the center of the loop since we want the force exerted on I_2.

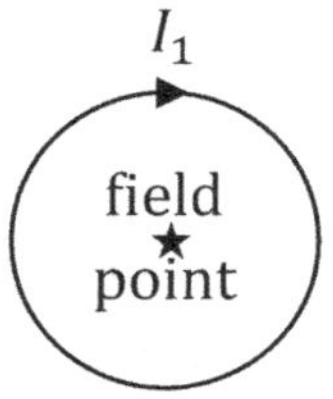

- Apply the right-hand rule for magnetic **field** (Chapter 22) to I_1. Grab I_1 with your thumb along I_1 and your fingers wrapped around I_1. Your fingers are going into the page (⊗) at the field point (★). The magnetic field ($\vec{\mathbf{B}}_1$) that I_1 makes at the field point (★) is into the page (⊗).
- Now apply the right-hand rule for magnetic force (Chapter 21) to I_2. The current (I_2) runs into the page (⊗), and the magnetic field ($\vec{\mathbf{B}}_1$) is also into the page (⊗). Recall from Chapter 21 that the magnetic force ($\vec{\mathbf{F}}_1$) is **zero** when the current (I_2) and magnetic field ($\vec{\mathbf{B}}_1$) are **parallel**.
- The outer current (I_1) exerts **no force** on the inner current (I_2): $\vec{\mathbf{F}}_1 = 0$.

(G) Draw the field point inside of the loop since we want the force exerted on the proton.

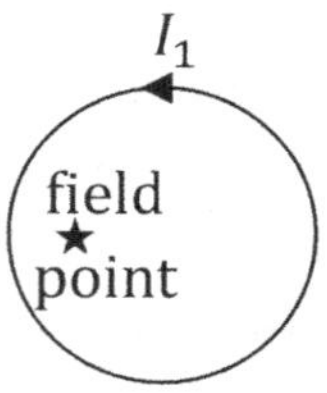

- Apply the right-hand rule for magnetic **field** (Chapter 22) to I_1. Grab I_1 with your thumb along I_1 and your fingers wrapped around I_1. Your fingers are coming out of the page (⊙) at the field point (★). The magnetic field ($\vec{\mathbf{B}}_1$) that I_1 makes at the field point (★) is out of the page (⊙).
- Note that the symbol p represents a proton, while the symbol $\vec{v}$ represents velocity.
- Now apply the right-hand rule for magnetic force (Chapter 21) to the proton (p). Point your fingers to the right (→), along $\vec{v}$. At the same time, face your palm out of the page (⊙), along $\vec{\mathbf{B}}_1$. **Tip**: Turn the book to make it more comfortable to position your hand as needed. Your thumb points down (↓), along the magnetic force ($\vec{\mathbf{F}}_1$).
- The current (I_1) pushes the proton **downward** (↓).

(H) Draw the field point on I_2 since we want the force exerted on I_2.

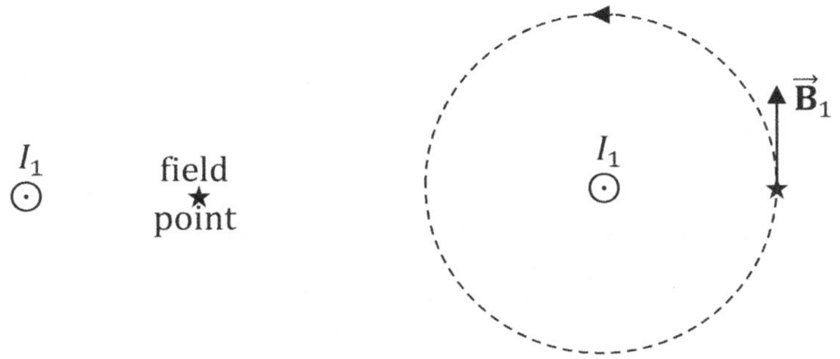

- Apply the right-hand rule for magnetic **field** (Chapter 22) to I_1. Grab I_1 with your thumb out of the page (⊙), along I_1, and your fingers wrapped around I_1. Your fingers are headed upward (↑) at the field point (★). The magnetic field ($\vec{\mathbf{B}}_1$) that I_1 makes at the field point (★) is upward (↑).
- **Note**: In the diagram above on the right, we drew a circular magnetic field line created by I_1. The magnetic field ($\vec{\mathbf{B}}_1$) that I_1 makes at the field point (★) is tangent to that circular magnetic field line, as we learned in Chapter 22.
- Now apply the right-hand rule for magnetic force (Chapter 21) to I_2. Point your fingers into the page (⊗), along I_2. At the same time, face your palm upward (↑), along $\vec{\mathbf{B}}_1$. Your thumb points to the right (→), along the magnetic force ($\vec{\mathbf{F}}_1$).
- The left current (I_1) pushes the right current (I_2) to the **right** (→). (They repel.)

(I) The field point is point C since we want the force exerted on I_2 at point C.

- Apply the right-hand rule for magnetic **field** (Chapter 22) to I_1. Grab I_1 at the **right** of the left loop, with your thumb upward (↑), since I_1 runs upward at the **right** of the left loop. (We're using the right side of the left loop since this point is nearest to the right loop, as it will have the dominant effect.) Your fingers are going into the page (⊗) at the field point (★). The magnetic field ($\vec{\mathbf{B}}_1$) that I_1 makes at the field point (★) is into the page (⊗).
- Now apply the right-hand rule for magnetic force (Chapter 21) to I_2 at point C. Point your fingers to the right (→), since I_2 runs to the right at point C. At the same time, face your palm into the page (⊗), along $\vec{\mathbf{B}}_1$. Your thumb points up (↑), along the magnetic force ($\vec{\mathbf{F}}_1$).
- The left current (I_1) pushes the right (I_2) **upward** (↑) at point C. More generally, the left loop pushes the right loop inward (meaning that at any point on the right loop, the magnetic force that I_1 exerts on I_2 is toward the center of the right loop).

Chapter 24: Magnetic Force

84. The magnetic field is $B = \frac{1}{2}$ T (or $\vec{\mathbf{B}} = \frac{1}{2}\,\hat{\mathbf{y}}$ in vector form) for all parts of this problem.

(A) Apply the equation $F_m = ILB \sin\theta$.

- $\theta = 90°$ because the current (I) is perpendicular to the magnetic field ($\vec{\mathbf{B}}$).
- The magnitude of the magnetic force is $F_m = 6.0$ N.
- One way to find the direction of the magnetic force ($\vec{\mathbf{F}}_m$) is to apply the right-hand rule for magnetic force (Chapter 21). Draw a right-handed, three-dimensional coordinate system with x, y, and z. For example, put $+x$ to the right, $+y$ up, and $+z$ out of the page, like the diagram on page 298. Then point your fingers into the page ($\otimes$), along the current (I). At the same time, face your palm upward ($\uparrow$), along the magnetic field ($\vec{\mathbf{B}}$). Your thumb points to the right ($\rightarrow$), along the magnetic force ($\vec{\mathbf{F}}_m$), which is along the $+x$-axis.
- Alternatively, write $\vec{\boldsymbol{L}} = -3\,\hat{\mathbf{z}}$ (since the length is 3.0 m and the current is heading along the $-z$-axis). Then use $\vec{\mathbf{F}}_m = I\vec{\mathbf{L}} \times \vec{\mathbf{B}}$ to get $\vec{\mathbf{F}}_m = 6\,\hat{\mathbf{x}}$.

(B) Apply the equation $F_m = |q|vB \sin\theta$.

- Convert the charge from μC to C using $\mu = 10^{-6}$. You should get $q = 2.00 \times 10^{-4}$ C.
- Convert the speed to SI units using k $= 1000$. You should get $v = 60{,}000$ m/s.
- $\theta = 120°$ because the velocity ($\vec{v}$) is 30° below the x-axis and the magnetic field ($\vec{\mathbf{B}}$) points along the $+y$-axis: $30° + 90° = 120°$. See the diagram below.

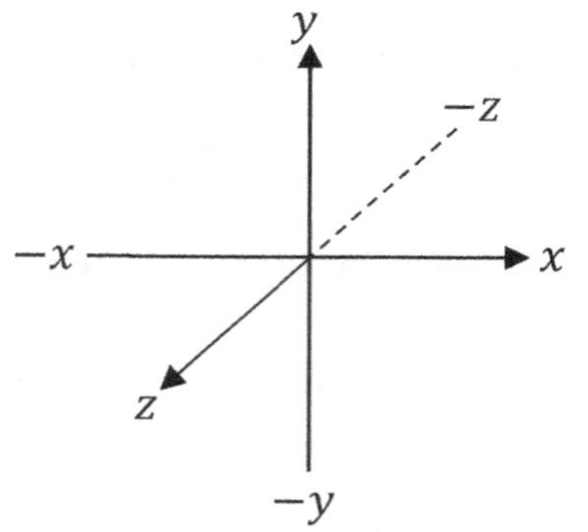

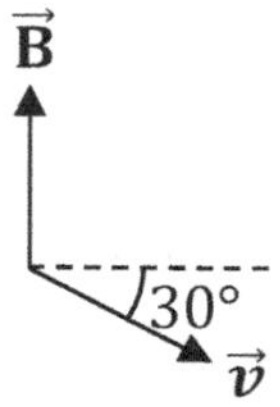

- Note that $10^{-4} \times 10^{3} = 10^{-1} = \frac{1}{10}$.
- The magnitude of the magnetic force is $F_m = 3\sqrt{3}$ N.
- One way to find the direction of the magnetic force ($\vec{\mathbf{F}}_m$) is to apply the right-hand rule for magnetic force (Chapter 21). Point your fingers right and downward ($\searrow$), along the velocity ($\vec{v}$). At the same time, face your palm upward ($\uparrow$), along the magnetic field ($\vec{\mathbf{B}}$). Your thumb points out of the page ($\odot$), along the magnetic force ($\vec{\mathbf{F}}_m$), which is along the $+z$-axis.

- Alternatively, write $\vec{v} = 30{,}000\sqrt{3}\,\hat{\mathbf{x}} - 30{,}000\,\hat{\mathbf{y}}$. This comes from $\vec{v} = v_x\,\hat{\mathbf{x}} - v_y\,\hat{\mathbf{y}}$, where $v_x = v\cos 330°$ and $v_y = v\sin 330°$. The angle 330° is equivalent to −30° (that is, 30° below the x-axis): Whether you go 330° counterclockwise from $+x$ or go 30° clockwise from $+x$, you arrive at the same place. Then use $\vec{\mathbf{F}}_m = q\vec{v} \times \vec{\mathbf{B}}$ to get $\vec{\mathbf{F}}_m = 3\sqrt{3}\,\hat{\mathbf{z}}$, applying the vector product according to Chapter 19.

(C) Apply the equation $F_m = |q|vB\sin\theta$.

- Which value(s) of θ make $\sin\theta = 0$?
- The answer is $\theta = 0°$ or $180°$, since $\sin 0° = 0$ and $\sin 180° = 0$.
- This means that the velocity ($\vec{v}$) must either be parallel or anti-parallel to the magnetic field ($\vec{\mathbf{B}}$). Since the magnetic field points along $+\hat{\mathbf{y}}$, the velocity must point along $+\hat{\mathbf{y}}$ or $-\hat{\mathbf{y}}$.

---

85. Apply the equation $\vec{\mathbf{F}}_m = q\vec{v} \times \vec{\mathbf{B}}$.

- Use the determinant form of the vector product.
- The components are: $v_x = 3$, $v_y = -2$, $v_z = -1$, $B_x = 5$, $B_y = 1$, and $B_z = -4$.

$$\vec{\mathbf{F}}_m = q\vec{v} \times \vec{\mathbf{B}} = q\begin{vmatrix} \hat{\mathbf{x}} & \hat{\mathbf{y}} & \hat{\mathbf{z}} \\ v_x & v_y & v_z \\ B_x & B_y & B_z \end{vmatrix} = q\left(\hat{\mathbf{x}}\begin{vmatrix} v_y & v_z \\ B_y & B_z \end{vmatrix} - \hat{\mathbf{y}}\begin{vmatrix} v_x & v_z \\ B_x & B_z \end{vmatrix} + \hat{\mathbf{z}}\begin{vmatrix} v_x & v_y \\ B_x & B_y \end{vmatrix}\right)$$

- The 2×2 determinants are $8 + 1 = 9$, $-12 + 5 = -7$, and $3 + 10 = 13$.
- Note the minus sign in the middle term. You will get $-(-7)\hat{\mathbf{y}} = +7\,\hat{\mathbf{y}}$.
- Remember to multiply each term by the charge: $q = 2.00 \times 10^{-4}$ C.
- The magnetic force is $\vec{\mathbf{F}}_m = 1.8 \times 10^{-3}\,\hat{\mathbf{x}} + 1.4 \times 10^{-3}\,\hat{\mathbf{y}} + 2.6 \times 10^{-3}\,\hat{\mathbf{z}}$.
- For example, $2 \times 10^{-4} \times 9\,\hat{\mathbf{x}} = 18 \times 10^{-4}\,\hat{\mathbf{x}} = 1.8 \times 10^{-3}\,\hat{\mathbf{x}}$.

---

86. (A) This is just like Problem 81, part (D) in Chapter 21 on page 278, which similarly has a negative charge. The only difference is that this problem has the charge moving clockwise instead of counterclockwise, so the answer is into the page (⊗).

(B) Apply Newton's second law to the particle. The acceleration is **centripetal**.

- Begin with $\sum F_{in} = ma_c$. Use the equations $F_m = |q|vB\sin\theta$ and $a_c = \frac{v^2}{R}$.
- You should get $|q|vB\sin\theta = \frac{mv^2}{R}$. Divide both sides of the equation by v.
- You should get $|q|B\sin\theta = \frac{mv}{R}$. Note that $\frac{v^2}{v} = v$. Also note that $\theta = 90°$.
- Multiply both sides of the equation by R and divide by $|q|B$.
- You should get $R = \frac{mv}{|q|B}$. Convert everything to SI units.
- You should get $|q| = 4.00 \times 10^{-4}$ C, $m = 0.00025$ kg $= \frac{1}{4000}$ kg, and $B = 20$ T.
- The radius is $R = 125$ m. (That may seem like a huge circle, but particles travel in much larger circles at some particle colliders, such as the Large Hadron Collider.)

87. First, apply the right-hand rule for magnetic force to each side of the rectangular loop:

- **Top side**: Point your fingers to the right (→) along the current (I) and your palm down (↓) along the magnetic field ($\vec{\mathbf{B}}$). The magnetic force ($\vec{\mathbf{F}}_{top}$) is into the page (⊗).
- **Right side**: The current (I) points down (↓) and the magnetic field ($\vec{\mathbf{B}}$) also points down (↓). Since I and $\vec{\mathbf{B}}$ are parallel, the magnetic force ($\vec{\mathbf{F}}_{right}$) is **zero**.
- **Bottom side**: Point your fingers to the left (←) along the current (I) and your palm down (↓) along the magnetic field ($\vec{\mathbf{B}}$). The magnetic force ($\vec{\mathbf{F}}_{bot}$) is out of the page (⊙).
- **Left side**: The current (I) points up (↑) and the magnetic field ($\vec{\mathbf{B}}$) points down (↓). Since I and $\vec{\mathbf{B}}$ are anti-parallel, the magnetic force ($\vec{\mathbf{F}}_{left}$) is **zero**.

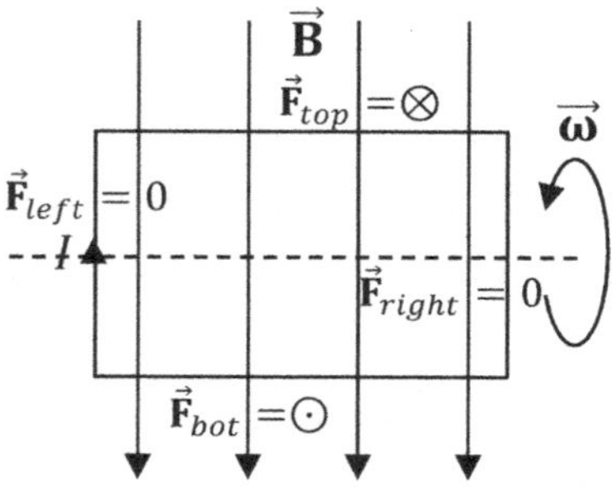

- The top side of the loop is pushed into the page, while the bottom side of the loop is pulled out of the page. The loop rotates about the dashed axis shown above.

(A) Find the force exerted on each side of the loop.

- Calculate the magnitudes of the two nonzero forces, F_{top} and F_{bot}, using the equation $F_m = ILB \sin\theta$. Note that $L = 50 \text{ cm} = \frac{1}{2}$ m (the correct length is the "width"), $B = 8000 \text{ G} = 0.80 \text{ T} = \frac{4}{5}$ T, and $\theta = 90°$. (Recall that $1 \text{ G} = 10^{-4}$ T.)
- You should get $F_{top} = F_{bot} = 12$ N.
- Since $\vec{\mathbf{F}}_{top}$ is into the page (⊗) and $\vec{\mathbf{F}}_{bot}$ is out of the page (⊙), and since the two forces have equal magnitudes, they **cancel** out: $F_{net} = 0$.

(B) **Don't** insert the net force, F_{net}, into the equation $\tau = rF \sin\theta$. Although the net force is zero, the net torque isn't zero. There are two ways to solve this problem.

- One way is to apply the formula $\tau_{net} = IAB \sin\theta$. Note that $\theta = 90°$.
- Note that $A = WH$ (width×height), where $W = 50 \text{ cm} = \frac{1}{2}$ m and $H = 25 \text{ cm} = \frac{1}{4}$ m.
- The net torque is $\tau_{net} = 3.0$ Nm.
- Another way to find the net torque is to apply the equation $\tau = rF \sin\theta$ to both F_{top} and F_{bot}, separately, using $r = \frac{H}{2}$ (see the dashed line in the figure above).
- Then add the two torques together: $\tau_{net} = \tau_1 + \tau_2$. Although the forces are opposite and cancel out, the torques are in the same direction because both torques cause the loop to rotate in the same direction. You should get $\tau_{net} = 3.0$ Nm.

Chapter 25: Magnetic Field

88. Draw the field point at the midpoint of the base, as shown below. Draw and label the relevant distances on the diagram, too.

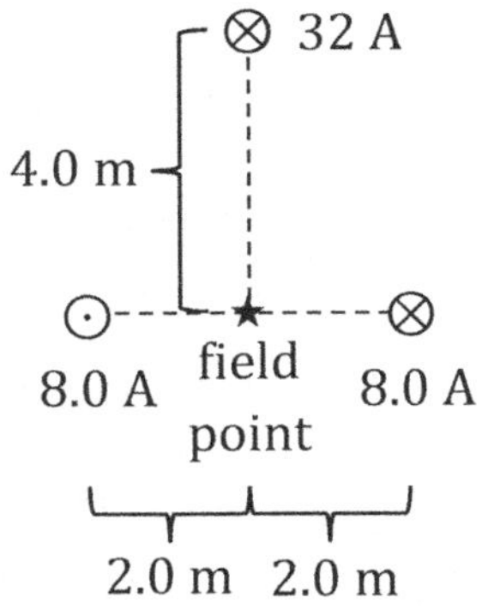

- There are three separate magnetic fields to find: One magnetic field is created by each current at the field point. Use the following equations.

$$B_{left} = \frac{\mu_0 I_{left}}{2\pi r_{left}} \quad , \quad B_{top} = \frac{\mu_0 I_{top}}{2\pi r_{top}} \quad , \quad B_{right} = \frac{\mu_0 I_{right}}{2\pi r_{right}}$$

- Note that $r_{left} = 2.0$ m, $r_{top} = 4.0$ m, and $r_{right} = 2.0$ m. (See the diagram above.)
- Recall that the permeability of free space is $\mu_0 = 4\pi \times 10^{-7} \frac{\text{T·m}}{\text{A}}$.
- Check that $B_{left} = 8.0 \times 10^{-7}$ T, $B_{top} = 16.0 \times 10^{-7}$ T, and $B_{right} = 8.0 \times 10^{-7}$ T.
- Apply the right-hand rule for magnetic **field** (Chapter 22) to find the direction of each magnetic field at the field point.
  - When you grab I_{left} with your thumb out of the page (⊙), along I_{left}, and your fingers wrapped around I_{left}, your fingers are going up (↑) at the field point (★). The magnetic field ($\vec{\mathbf{B}}_{left}$) that I_{left} makes at the field point (★) is straight up (↑).
  - When you grab I_{top} with your thumb into the page (⊗), along I_{top}, and your fingers wrapped around I_{top}, your fingers are going to the left (←) at the field point (★). The magnetic field ($\vec{\mathbf{B}}_{top}$) that I_{top} makes at the field point (★) is to the left (←).
  - When you grab I_{right} with your thumb into the page (⊗), along I_{right}, and your fingers wrapped around I_{right}, your fingers are going up (↑) at the field point (★). The magnetic field ($\vec{\mathbf{B}}_{right}$) that I_{right} makes at the field point (★) is straight up (↑).
- Since $\vec{\mathbf{B}}_{left}$ and $\vec{\mathbf{B}}_{right}$ both point straight up (↑), we add them:

$$B_y = B_{left} + B_{right}$$

- Check your intermediate answer: $B_y = 16.0 \times 10^{-7}$ T. You're not finished yet.

- Since $\vec{\mathbf{B}}_{top}$ is perpendicular to $\vec{\mathbf{B}}_{left}$ and $\vec{\mathbf{B}}_{right}$ (since $\vec{\mathbf{B}}_{top}$ points to the left, while $\vec{\mathbf{B}}_{left}$ and $\vec{\mathbf{B}}_{right}$ both point up), we use the Pythagorean theorem, with B_y representing the combination of $\vec{\mathbf{B}}_{left}$ and $\vec{\mathbf{B}}_{right}$.

$$B_{net} = \sqrt{B_{top}^2 + B_y^2}$$

- The magnitude of the net magnetic field at the field point is $B_{net} = 16\sqrt{2} \times 10^{-7}$ T. If you use a calculator, this comes out to $B_{net} = 23 \times 10^{-7}$ T $= 2.3 \times 10^{-6}$ T.
- You can find the direction of the net magnetic field with an inverse tangent.

$$\theta_B = \tan^{-1}\left(\frac{B_y}{B_x}\right)$$

- Note that $B_x = -B_{top}$ (since $\vec{\mathbf{B}}_{top}$ points to the left).
- We choose $+x$ to point right and $+y$ to point up, such that the answer lies in Quadrant II (since $\vec{\mathbf{B}}_{top}$ points left, while $\vec{\mathbf{B}}_{left}$ and $\vec{\mathbf{B}}_{right}$ point up). The reference angle is 45° and the Quadrant II angle is $\theta_B = 180° - \theta_{ref} = 135°$.
- The direction of the net magnetic field at the field point is $\theta_B = 135°$.

---

89. Draw a field point (★) at the location of I_2 (since the force specified in the problem is exerted on I_2), and find the magnetic field at the field point (★) created by I_1.

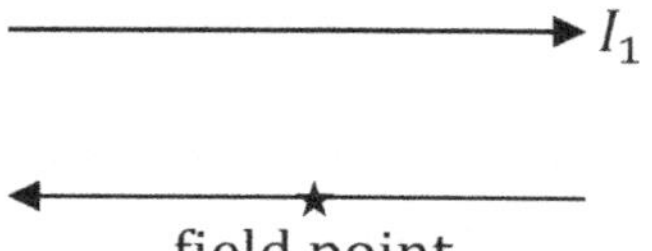

- First use the equation for the magnetic field created by a long straight wire, where $r_c = d = 0.050 \text{ m} = \frac{1}{20}$ m is the distance between the wires.

$$B_1 = \frac{\mu_0 I_1}{2\pi d}$$

- Check your intermediate answer: $B_1 = 320 \times 10^{-7}$ T $= 3.2 \times 10^{-5}$ T.
- Now use the equation for magnetic force: $F_1 = I_2 L_2 B_1 \sin\theta$.
- Note that $\theta = 90°$ because I_2 points left and B_1 is into the page (see below).
- The magnetic force that I_1 exerts on I_2 has a magnitude of $F_1 = 4.8 \times 10^{-4}$ N, which can also be expressed as $F_1 = 48 \times 10^{-5}$ N, $F_1 = 0.00048$ N, or $F_1 = 0.48$ mN.
- Apply the technique from Chapter 23 to find the direction of the magnetic force.
  - Apply the right-hand rule for magnetic **field** (Chapter 22) to I_1. Grab I_1 with your thumb to the right (→), along I_1, and your fingers wrapped around I_1. Your fingers are going into the page (⊗) at the field point (★). The magnetic field ($\vec{\mathbf{B}}_1$) that I_1 makes at the field point (★) is into the page (⊗).
  - Now apply the right-hand rule for magnetic **force** (Chapter 21) to I_2. Point

your fingers to the left (←), along I_2. At the same time, face your palm into the page (⊗), along $\vec{\mathbf{B}}_1$. Your thumb points down (↓), along the magnetic force ($\vec{\mathbf{F}}_1$).

- The top current (I_1) pushes the bottom current (I_2) straight **down** (↓). (They repel.)

90. Draw a field point (★) at the location of I_3 (since the force specified in the problem is exerted on I_3), and find the magnetic fields at the field point (★) created by I_1 and I_2.

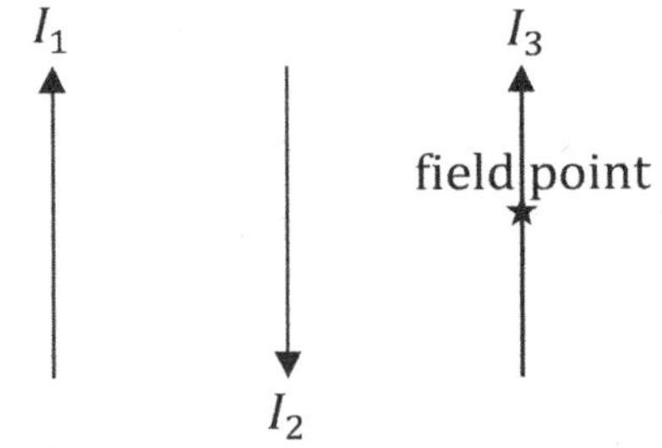

- First use the equation for the magnetic field created by a long straight wire.

$$B_1 = \frac{\mu_0 I_1}{2\pi d_1} \quad , \quad B_2 = \frac{\mu_0 I_2}{2\pi d_2}$$

- Note that $d_1 = 0.25 + 0.25 = 0.50$ m and $d_2 = 0.25$ m.
- Check your intermediate answers: $B_1 = 12 \times 10^{-7}$ T and $B_1 = 32 \times 10^{-7}$ T.
- Now use the equation for magnetic force: $F_1 = I_3 L_3 B_1 \sin\theta$ and $F_2 = I_3 L_3 B_2 \sin\theta$.
- Note that $\theta = 90°$ because I_3 is perpendicular to the magnetic fields.
- Check your intermediate answers: $F_1 = 360 \times 10^{-7}$ N and $F_2 = 960 \times 10^{-7}$ N.
- In order to determine how to combine the individual forces to find the net force, you must determine the direction of $\vec{\mathbf{F}}_1$ and $\vec{\mathbf{F}}_2$. Apply the strategy from Chapter 23, just as we did in the previous problem. Check your answers below:
  - $\vec{\mathbf{F}}_1$ is to the left (←), since I_1 is parallel to I_3 and parallel currents attract.
  - $\vec{\mathbf{F}}_2$ is to the right (→), since I_2 is anti-parallel to I_3 and anti-parallel currents repel.
- Since $\vec{\mathbf{F}}_1$ and $\vec{\mathbf{F}}_2$ have opposite directions, subtract them to find the net force.

$$F_{net} = |F_1 - F_2|$$

- The net magnetic force that I_1 and I_2 exert on I_3 has a magnitude of $F_{net} = 600 \times 10^{-7}$ N, which can also be expressed as $F_{net} = 6.0 \times 10^{-5}$ N.
- The direction of the net magnetic force that I_1 and I_2 exert on I_3 is to the right (→), since F_2 is greater than F_1 and since $\vec{\mathbf{F}}_2$ points to the right.

91. Draw a field point (★) at the location of the 2.0-A current (since the force specified in the problem is exerted on the 2.0-A current), and find the magnetic fields at the field point (★) created by the 4.0-A currents.

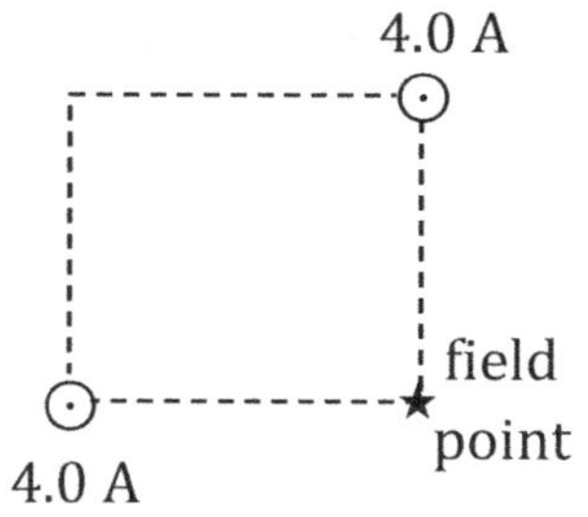

- First use the equation for the magnetic field created by a long straight wire.

$$B_{left} = \frac{\mu_0 I_{left}}{2\pi d_{left}} \quad , \quad B_{top} = \frac{\mu_0 I_{top}}{2\pi d_{top}}$$

- Note that $d_{left} = 0.25$ m and $d_{right} = 0.25$ m.
- Check your intermediate answers: $B_{left} = 32 \times 10^{-7}$ T and $B_{top} = 32 \times 10^{-7}$ T.
- Now use the equations $F_{left} = I_{br} L_{br} B_{left} \sin\theta$ and $F_{top} = I_{br} L_{br} B_{top} \sin\theta$.
- Note: The subscript *br* stands for "bottom right" (the 2.0-A current).
- Note that $\theta = 90°$ because I_{br} is perpendicular to the magnetic fields.
- Check your intermediate answers: $F_{left} = 192 \times 10^{-7}$ N and $F_{top} = 192 \times 10^{-7}$ N.
- In order to determine how to combine the individual forces to find the net force, you must determine the direction of $\vec{\mathbf{F}}_{left}$ and $\vec{\mathbf{F}}_{top}$. Apply the strategy from Chapter 23, just as we did in Problem 89. Check your answers below:
  - $\vec{\mathbf{F}}_{left}$ is to the right (→), since I_{left} is anti-parallel to I_{br} and anti-parallel currents repel.
  - $\vec{\mathbf{F}}_{top}$ is down (↓), since I_{top} is anti-parallel to I_{br} and anti-parallel currents repel.
- Since $\vec{\mathbf{F}}_{left}$ and $\vec{\mathbf{F}}_{top}$ are perpendicular, we use the Pythagorean theorem.

$$F_{net} = \sqrt{F_{left}^2 + F_{top}^2}$$

- The magnitude of the net magnetic force at the field point is $F_{net} = 192\sqrt{2} \times 10^{-7}$ N. If you use a calculator, this comes out to $F_{net} = 272 \times 10^{-7}$ N $= 2.7 \times 10^{-5}$ N.
- You can find the direction of the net magnetic force with an inverse tangent.

$$\theta_F = \tan^{-1}\left(\frac{F_y}{F_x}\right) = \tan^{-1}\left(\frac{-F_{top}}{F_{left}}\right)$$

- We choose $+x$ to point right and $+y$ to point up, such that the answer lies in Quadrant IV (since $\vec{\mathbf{F}}_{left}$ points right, while $\vec{\mathbf{F}}_{top}$ points down). The reference angle is 45° and the Quadrant IV angle is $\theta_F = 360° - \theta_{ref} = 315°$.

92. Study the last example from Chapter 25, since this problem is very similar to that example. The first step is to determine the direction of the magnetic force that I_1 exerts on each side of the rectangular loop, using the technique from Chapter 23.

- Apply the right-hand rule for magnetic **field** (Chapter 22) to I_1. Grab I_1 with your thumb along I_1 and your fingers wrapped around I_1. Your fingers are going out of the page ($\odot$) where the rectangular loop is. The magnetic field ($\vec{\mathbf{B}}_1$) that I_1 makes at the rectangular loop is out of the page ($\odot$).

Now apply the right-hand rule for magnetic force (Chapter 21) to each side of the loop.

- <u>Bottom side</u>: Point your fingers to the left ($\leftarrow$) along the current (I_2) and your palm out of the page ($\odot$) along the magnetic field ($\vec{\mathbf{B}}_1$). The magnetic force ($\vec{\mathbf{F}}_{bot}$) is up ($\uparrow$).
- <u>Left side</u>: Point your fingers up ($\uparrow$) along the current (I_2) and your palm out of the page ($\odot$) along the magnetic field ($\vec{\mathbf{B}}_1$). The magnetic force ($\vec{\mathbf{F}}_{left}$) is right ($\rightarrow$).
- <u>Top side</u>: Point your fingers to the right ($\rightarrow$) along the current (I_2) and your palm out of the page ($\odot$) along the magnetic field ($\vec{\mathbf{B}}_1$). The magnetic force ($\vec{\mathbf{F}}_{top}$) is down ($\downarrow$).
- <u>Right side</u>: Point your fingers down ($\downarrow$) along the current (I_2) and your palm out of the page ($\odot$) along the magnetic field ($\vec{\mathbf{B}}_1$). The magnetic force ($\vec{\mathbf{F}}_{right}$) is left ($\leftarrow$).

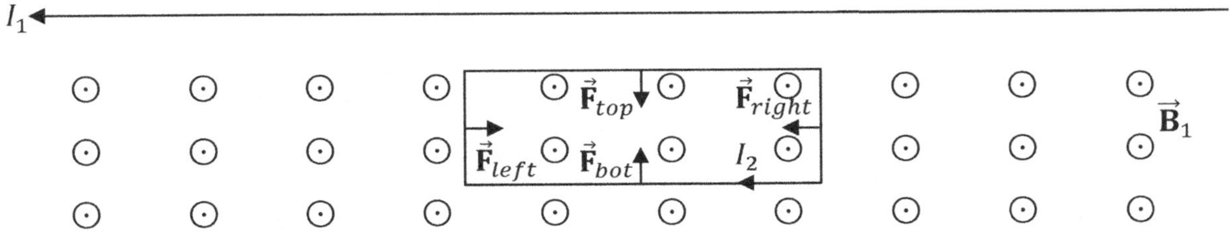

- Note that $\vec{\mathbf{F}}_{left}$ and $\vec{\mathbf{F}}_{right}$ cancel out in the calculation for the net magnetic force because they have equal magnitudes (since they are the same distance from I_1) and opposite direction. However, $\vec{\mathbf{F}}_{top}$ and $\vec{\mathbf{F}}_{bot}$ do <u>**not**</u> cancel.
- Begin the math with the following equations.

$$B_{top} = \frac{\mu_0 I_1}{2\pi d_{top}} \quad , \quad B_{bot} = \frac{\mu_0 I_1}{2\pi d_{bot}}$$

- Note that $d_{top} = 0.25$ m and $d_{bot} = 0.25 + 0.50 = 0.75$ m.
- Check your intermediate answers: $B_{top} = 48 \times 10^{-7}$ T and $B_{bot} = 16 \times 10^{-7}$ T.
- Now use the equations $F_{top} = I_2 L_2 B_{top} \sin\theta$ and $F_{bot} = I_2 L_2 B_{bot} \sin\theta$.
- Note that $\theta = 90°$ because the loop is perpendicular to the magnetic fields.
- Note that $L_2 = 1.5$ m (the width of the rectangular loop) since that is the distance that I_2 travels in the top and bottom sides of the rectangular loop.
- Check your intermediate answers: $F_{top} = 576 \times 10^{-7}$ N and $F_{bot} = 192 \times 10^{-7}$ N.
- Since $\vec{\mathbf{F}}_{top}$ and $\vec{\mathbf{F}}_{bot}$ have opposite directions, subtract them to find the net force.

$$F_{net} = |F_{top} - F_{bot}|$$

- The net magnetic force that I_1 exerts on I_2 has a magnitude of $F_{net} = 384 \times 10^{-7}$ N, which can also be expressed as $F_{net} = 3.84 \times 10^{-5}$ N.
- The direction of the net magnetic force that I_1 exerts on I_2 is down (↓), since F_{top} is greater than F_{bot} and since $\vec{\mathbf{F}}_{top}$ points down.

Chapter 26: The Law of Biot-Savart

93. This problem is similar to the first two examples from this chapter.
    - Divide the path up into 5 sections, as numbered in the diagram below.
    - The magnetic field at the origin is zero for the straight sections (1, 3, and 5) because $d\vec{\mathbf{s}}$ is either parallel or anti-parallel to $\hat{\mathbf{R}}$ for these sections, such that $d\vec{\mathbf{s}} \times \hat{\mathbf{R}} = 0$.
    - Just apply the law of Biot-Savart to the curved sections (2 and 4).
    - Draw $d\vec{\mathbf{s}}$ tangent to I and draw $\vec{\mathbf{R}}$ from the source, $d\vec{\mathbf{s}}$, to the field point (at the origin). See the diagram below on the right for section 2.

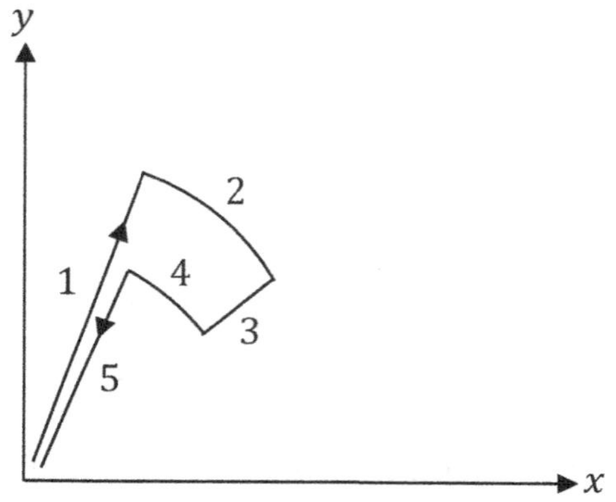

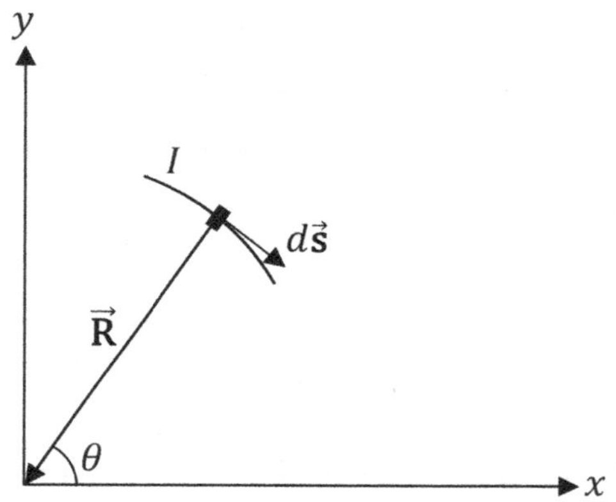

- From the diagram, you should see that $\hat{\mathbf{R}}$ points inward, along $-\hat{\mathbf{r}}$.
- Note that $R_2 = 3a$ for section 2 and $R_4 = 2a$ for section 4.
- Write $d\vec{\mathbf{s}}_2 = -3a\hat{\boldsymbol{\theta}}\, d\theta$ (clockwise) for section 2 and $d\vec{\mathbf{s}}_4 = 2a\hat{\boldsymbol{\theta}}\, d\theta$ (counter-clockwise) for section 4, since the radii of these arcs are $3a$ and $2a$, respectively.
- Apply the law of Biot-Savart to each arc: $\vec{\mathbf{B}} = \frac{\mu_0}{4\pi}\int \frac{I\, d\vec{\mathbf{s}} \times \hat{\mathbf{R}}}{R^2}$.
- Plug in the expressions for R, $\hat{\mathbf{R}}$, and $d\vec{\mathbf{s}}$. The limits are from $\theta = \frac{\pi}{4}$ to $\theta = \frac{\pi}{3}$.
- You should get $\vec{\mathbf{B}}_2 = \frac{\mu_0 I}{12\pi a}\int_{\theta=\pi/4}^{\pi/3} \hat{\boldsymbol{\theta}} \times \hat{\mathbf{r}}\, d\theta$ and $\vec{\mathbf{B}}_4 = -\frac{\mu_0 I}{8\pi a}\int_{\theta=\pi/4}^{\pi/3} \hat{\boldsymbol{\theta}} \times \hat{\mathbf{r}}\, d\theta$.
- Note, for example, that $R_2^2 = (3a)^2 = 9a^2$ and $R_4^2 = (2a)^2 = 4a^2$.
- Work out the vector product $\hat{\boldsymbol{\theta}} \times \hat{\mathbf{r}}$. You should get $\hat{\boldsymbol{\theta}} \times \hat{\mathbf{r}} = -\hat{\mathbf{z}}$.
- The integrals become $\vec{\mathbf{B}}_2 = -\frac{\mu_0 I}{12\pi a}\hat{\mathbf{z}}\int_{\theta=\pi/4}^{\pi/3} d\theta$ and $\vec{\mathbf{B}}_4 = \frac{\mu_0 I}{8\pi a}\hat{\mathbf{z}}\int_{\theta=\pi/4}^{\pi/3} d\theta$.
- Perform the integral $\int_{\theta=\pi/4}^{\pi/3} d\theta$. You should get $\frac{\pi}{3} - \frac{\pi}{4} = \frac{\pi}{12}$.
- The magnetic fields are $\vec{\mathbf{B}}_2 = -\frac{\mu_0 I}{144a}\hat{\mathbf{z}}$ and $\vec{\mathbf{B}}_4 = \frac{\mu_0 I}{96a}\hat{\mathbf{z}}$.

- Combine $\vec{\mathbf{B}}_2$ and $\vec{\mathbf{B}}_4$ together to form the net magnetic field, $\vec{\mathbf{B}}_{net}$.
- Subtract fractions with a **common denominator**: $-\frac{1}{144}+\frac{1}{96}=-\frac{2}{288}+\frac{3}{288}=\frac{1}{288}$.
- The net magnetic field at the origin is $\vec{\mathbf{B}}_{net}=\frac{\mu_0 I}{288a}\hat{\mathbf{z}}$.

---

94. This problem is very similar to the example involving an infinite straight wire. Study that example and try to let it serve as a guide. If you need help, you can always return here.

- Draw $d\vec{\mathbf{s}}$ along I and draw $\vec{\mathbf{R}}$ from the source, $d\vec{\mathbf{s}}$, to the field point $(a, 0, 0)$.

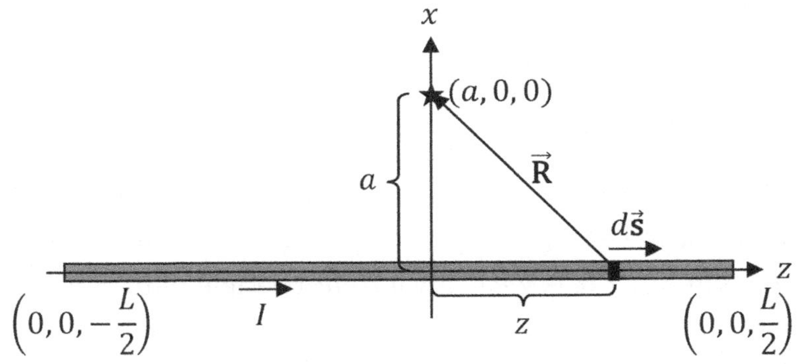

- From the diagram above, for the $d\vec{\mathbf{s}}$ drawn, you should be able to see that $\vec{\mathbf{R}}$ extends a units up (along $\hat{\mathbf{x}}$) and z units to the left (along $-\hat{\mathbf{z}}$). Therefore, $\vec{\mathbf{R}}=a\hat{\mathbf{x}}-z\hat{\mathbf{z}}$.
- Apply the Pythagorean theorem to find R. Divide $\vec{\mathbf{R}}$ by R to find $\hat{\mathbf{R}}$.
- You should get $R=\sqrt{a^2+z^2}$ and $\hat{\mathbf{R}}=\frac{a\hat{\mathbf{x}}-z\hat{\mathbf{z}}}{\sqrt{a^2+z^2}}$.
- For a straight filamentary current along the z-axis, write $d\vec{\mathbf{s}}=\hat{\mathbf{z}}\,dz$.
- Apply the law of Biot-Savart: $\vec{\mathbf{B}}=\frac{\mu_0}{4\pi}\int\frac{I\,d\vec{\mathbf{s}}\times\hat{\mathbf{R}}}{R^2}$.
- The limits of integration are from $z=-\frac{L}{2}$ to $z=\frac{L}{2}$ (the endpoints of the wire).
- Plug in the expressions for R, $\hat{\mathbf{R}}$, and $d\vec{\mathbf{s}}$.

$$\vec{\mathbf{B}}=\frac{\mu_0 I}{4\pi}\int_{z=-L/2}^{L/2}\frac{\hat{\mathbf{z}}\times(a\hat{\mathbf{x}}-z\hat{\mathbf{z}})}{(a^2+z^2)^{3/2}}dz$$

- Note that $(a^2+z^2)\sqrt{a^2+z^2}=(a^2+z^2)^1(a^2+z^2)^{1/2}=(a^2+z^2)^{3/2}$.
- Work out the vector product $\hat{\mathbf{z}}\times(a\hat{\mathbf{x}}-z\hat{\mathbf{z}})$ according to Chapter 19.

$$\hat{\mathbf{z}}\times(a\hat{\mathbf{x}}-z\hat{\mathbf{z}})=\begin{vmatrix}\hat{\mathbf{x}} & \hat{\mathbf{y}} & \hat{\mathbf{z}}\\ 0 & 0 & 1\\ a & 0 & z\end{vmatrix}=\hat{\mathbf{x}}\begin{vmatrix}0 & 1\\ 0 & z\end{vmatrix}-\hat{\mathbf{y}}\begin{vmatrix}0 & 1\\ a & z\end{vmatrix}+\hat{\mathbf{z}}\begin{vmatrix}0 & 0\\ a & 0\end{vmatrix}$$

- You should get $\hat{\mathbf{z}}\times(a\hat{\mathbf{x}}-z\hat{\mathbf{z}})=a\,\hat{\mathbf{y}}$. Plug this into the magnetic field integral.

$$\vec{\mathbf{B}}=\frac{\mu_0 I a}{4\pi}\int_{z=-L/2}^{L/2}\frac{\hat{\mathbf{y}}\,dz}{(a^2+z^2)^{3/2}}$$

- One way to perform the previous integral is to make the following substitution.

$$z = a \tan\theta$$
$$dz = a \sec^2\theta \, d\theta$$

- The denominator simplifies as follows, using the trig identity $1 + \tan^2\theta = \sec^2\theta$.

$$(a^2 + z^2)^{3/2} = [a^2 + (a\tan\theta)^2]^{3/2} = [a^2(1 + \tan^2\theta)]^{3/2} = (a^2\sec^2\theta)^{3/2} = a^3\sec^3\theta$$

- It's easier to ignore the new limits for now and deal with them later. After you make these substitutions and simplify, you should get:

$$\vec{\mathbf{B}} = \frac{\mu_0 I a}{4\pi}\hat{\mathbf{y}} \int \frac{d\theta}{a^2 \sec\theta}$$

- Recall that $\sec\theta = \frac{1}{\cos\theta}$ and note that $a\left(\frac{1}{a^2}\right) = \frac{1}{a}$. The integral becomes:

$$\vec{\mathbf{B}} = \frac{\mu_0 I}{4\pi a}\hat{\mathbf{y}} \int \cos\theta \, d\theta$$

- The anti-derivative is $\sin\theta$. Draw a right triangle to express $\sin\theta$ in terms of z, and then we can simply evaluate the function over the old limits. Since $z = a\tan\theta$, which means that $\tan\theta = \frac{z}{a}$, we draw a right triangle with z opposite to θ and with a adjacent to θ.

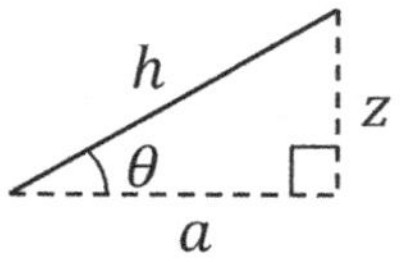

- Apply the Pythagorean theorem to find the hypotenuse: $h = \sqrt{z^2 + a^2}$.
- From the right triangle, you should see that $\sin\theta = \frac{z}{\sqrt{z^2+a^2}}$. Evaluate this over the limits from $z = -\frac{L}{2}$ to $z = \frac{L}{2}$.
- You should get $\frac{L/2}{\sqrt{\frac{L^2}{4}+a^2}} - \frac{-L/2}{\sqrt{\frac{L^2}{4}+a^2}} = \frac{L/2}{\sqrt{\frac{L^2}{4}+a^2}} + \frac{L/2}{\sqrt{\frac{L^2}{4}+a^2}} = \frac{L}{\sqrt{\frac{L^2}{4}+a^2}}$.
- Replace the previous integral with this expression.

$$\vec{\mathbf{B}} = \frac{\mu_0 I L}{4\pi a\sqrt{\frac{L^2}{4} + a^2}}\hat{\mathbf{y}}$$

---

95. You just need to apply the result from the third example, which involves a circular current loop and a field point at $(0, 0, p)$. You don't need to perform a new integral.

- Begin with the formula for the magnetic field a distance p along the axis of a circular loop of radius a, which we found in the third example (see page 332).

$$\vec{\mathbf{B}}_{one\ loop} = \frac{\mu_0 I a^2}{2(a^2 + p^2)^{3/2}}\hat{\mathbf{z}}$$

- The field point lies midway between the two loops, such that $p = \frac{a}{2}$ (since the two loops are a distance a apart in this example).

- Set $p = \frac{a}{2}$ in the previous expression for magnetic field. Note that $\left(\frac{a}{2}\right)^2 = \frac{a^2}{4}$.

$$\vec{\mathbf{B}}_{one\ loop} = \frac{\mu_0 I a^2}{2\left(a^2 + \frac{a^2}{4}\right)^{3/2}}\hat{\mathbf{z}}$$

- Add fractions with a common denominator: $a^2 + \frac{a^2}{4} = \frac{4a^2}{4} + \frac{a^2}{4} = \frac{5a^2}{4}$.

$$\vec{\mathbf{B}}_{one\ loop} = \frac{\mu_0 I a^2}{2\left(\frac{5a^2}{4}\right)^{3/2}}\hat{\mathbf{z}}$$

- Apply the rule from algebra that $\left(\frac{5a^2}{4}\right)^{3/2} = \frac{(5)^{3/2}(a^2)^{3/2}}{(4)^{3/2}}$.
- Note that $(a^2)^{3/2} = a^3$ according to the rule $(x^m)^n = x^{mn}$.
- Note that $(4)^{3/2} = \left(\sqrt{4}\right)^3 = 2^3 = 8$ (or enter 4^1.5 on your calculator).

$$\vec{\mathbf{B}}_{one\ loop} = \frac{\mu_0 I a^2}{2\frac{(5)^{3/2}a^3}{8}}\hat{\mathbf{z}}$$

- To divide by a fraction, multiply by its **reciprocal**. Note that $\frac{a^2}{a^3} = \frac{1}{a}$.

$$\vec{\mathbf{B}}_{one\ loop} = \frac{\mu_0 I a^2}{2}\hat{\mathbf{z}}\frac{8}{(5)^{3/2}a^3} = \frac{4\mu_0 I}{(5)^{3/2}a}\hat{\mathbf{z}}$$

- Both loops are equidistant from the field point, such that the magnitude of the magnetic field at the field point is the same for each loop.
- The direction of the magnetic field is also the same at the field point for both loops, which you can see by applying the right-hand rule for magnetic field (Chapter 22) to each loop.
- Since the magnitude and direction of both magnetic fields is the same at the field point, the net magnetic field is double the magnetic field created by one loop.
- The final answer is:

$$\vec{\mathbf{B}}_{net} = \frac{8\mu_0 I}{(5)^{3/2}a}\hat{\mathbf{z}}$$

---

96. This problem is very similar to the last example from this chapter. Study that example and try to let it serve as a guide. If you need help, you can always return here. Note that the mathematics of this problem is actually **<u>much simpler</u>** than the similar example, due to the location of the field point at the origin.

- Begin with a labeled diagram. Draw a representative dq. Draw $\vec{\mathbf{R}}$ from the source, dq, to the field point (at the origin). See the diagram that follows.

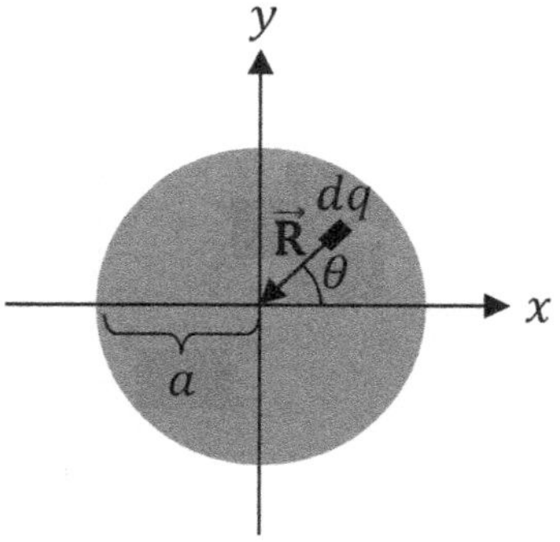

- For the dq drawn, you should be able to see that $\vec{\mathbf{R}}$ extends r units inward, towards the origin (along $-\hat{\mathbf{r}}$). Therefore, $R = r$ and $\hat{\mathbf{R}} = -\hat{\mathbf{r}}$ (using 2D polar coordinates).
- For a solid disc, write $dq = \sigma dA$ and $dA = r dr d\theta$ (Steps 6-7 on page 325).
- Apply the law of Biot-Savart for a rotating charged disc: $\vec{\mathbf{B}} = \frac{\mu_0}{4\pi}\int \frac{\vec{v}\, dq \times \hat{\mathbf{R}}}{R^2}$.
- Note that r varies from 0 to a and θ varies from 0 to 2π (full circle).
- Plug in the expressions for R, $\hat{\mathbf{R}}$, and dq. Note that $\frac{r}{r^2} = \frac{1}{r}$.

$$\vec{\mathbf{B}} = -\frac{\mu_0 \sigma}{4\pi} \int_{r=0}^{a} \int_{\theta=0}^{2\pi} \frac{\vec{v} \times \hat{\mathbf{r}}}{r^2} r dr d\theta$$

- Since the velocity ($\vec{v}$) of each dq is tangential to the circle of rotation, and since $\hat{\boldsymbol{\theta}}$ is tangential to a circle lying in the xy plane, we write $\vec{v} = v\,\hat{\boldsymbol{\theta}}$.
- For a rotating charged object, we write $v = r_{rot}\omega$, where r_{rot} is the radius of rotation of each dq. Based on how this solid disc is rotating, $r_{rot} = r$.
- Substitute $\vec{v} = v\,\hat{\boldsymbol{\theta}} = r\omega\hat{\boldsymbol{\theta}}$ into the magnetic field integral. Note that the r cancels.

$$\vec{\mathbf{B}} = -\frac{\mu_0 \sigma \omega}{4\pi} \int_{r=0}^{a} \int_{\theta=0}^{2\pi} \hat{\boldsymbol{\theta}} \times \hat{\mathbf{r}}\, dr d\theta$$

- Work out the vector product $\hat{\boldsymbol{\theta}} \times \hat{\mathbf{r}}$. You should get $\hat{\boldsymbol{\theta}} \times \hat{\mathbf{r}} = -\hat{\mathbf{z}}$.

$$\vec{\mathbf{B}} = \frac{\mu_0 \sigma \omega}{4\pi}\hat{\mathbf{z}} \int_{r=0}^{a} \int_{\theta=0}^{2\pi} dr\, d\theta$$

- This integral should be easy. You should get:

$$\vec{\mathbf{B}} = \frac{\mu_0 \sigma a \omega}{2}\hat{\mathbf{z}}$$

- Integrate $Q = \int dq$ using the same substitutions as before.
- You should get $Q = \sigma \int_{r=0}^{a} \int_{\theta=0}^{2\pi} r\, dr d\theta = \sigma \pi a^2$.
- Solve for σ. You should get $\sigma = \frac{Q}{\pi a^2}$.
- Plug σ into the previous equation for magnetic field. The final answer is:

$$\vec{\mathbf{B}} = \frac{\mu_0 Q \omega}{2\pi a}\hat{\mathbf{z}}$$

Chapter 27: Ampère's Law

97. This problem is similar to the example with the infinite solid cylindrical conductor. The only difference is that this current density is non-uniform.

- Begin by sketching the magnetic field lines. See the diagram below.

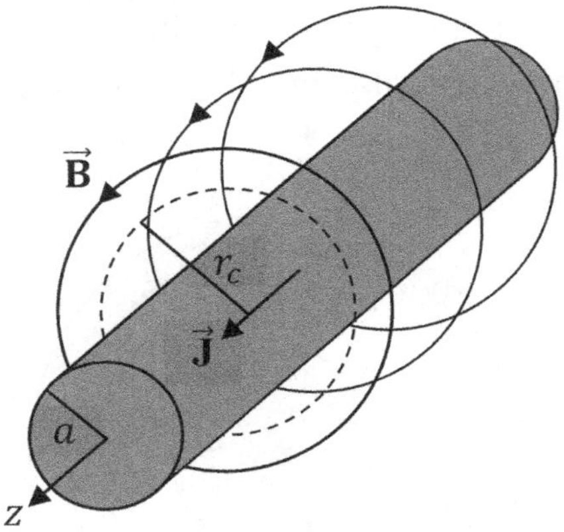

- Write down the equation for Ampère's law: $\oint_C \vec{\mathbf{B}} \cdot d\vec{\mathbf{s}} = \mu_0 I_{enc}$.
- Just as in the example with an infinite cylinder, Ampère's law reduces to $B2\pi r_c = \mu_0 I_{enc}$. Isolate B to get $B = \frac{\mu_0 I_{enc}}{2\pi r_c}$. See page 350.
- Integrate $I_{enc} = \int \vec{\mathbf{J}} \cdot d\vec{\mathbf{A}}$ to find the current enclosed in each region.
- Just as in the example with an infinite cylinder, this reduces to $I_{enc} = \int J\, dA$.
- Since the density is non-uniform, you may **not** pull J out of the integral. Instead, plug in the equation given in the problem: $J = \beta r_c$. You may pull β out of the integral.
- Work with cylindrical coordinates: $dA = r_c dr_c d\theta$ (see Chapter 6).
- After you make these substitutions and simplify, you should get:

$$I_{enc} = \beta \int_{r_c=0}^{r_c} \int_{\theta=0}^{2\pi} r_c^2\, dr_c\, d\theta$$

- The upper limit of the r_c-integration is r_c in region I ($r_c < a$) and is a in region II ($r_c > a$). The anti-derivatives are $\frac{r_c^3}{3}$ and θ. Evaluate these over their limits.
- You should get $I_{enc} = \frac{2\pi\beta r_c^3}{3}$ in region I and $I_{enc} = I = \frac{2\pi\beta a^3}{3}$ in region II.
- Divide these equations to see that $\frac{I_{enc}}{I} = \frac{r_c^3}{a^3}$ (everything else cancels out).
- Multiply both sides by I to get $I_{enc} = I\frac{r_c^3}{a^3}$.
- Substitute $I_{enc} = I\frac{r_c^3}{a^3}$ and $I_{enc} = I$ for the two regions into the previous equation for magnetic field, $B = \frac{\mu_0 I_{enc}}{2\pi r_c}$. Add $\hat{\boldsymbol{\theta}}$ for the direction.
- The answers are $\vec{\mathbf{B}}_I = \frac{\mu_0 I r_c^2}{2\pi a^3}\hat{\boldsymbol{\theta}}$ and $\vec{\mathbf{B}}_{II} = \frac{\mu_0 I}{2\pi r_c}\hat{\boldsymbol{\theta}}$.

98. This problem is similar to the example with the infinite solid cylindrical conductor.
Note: If you recall how we treated the conducting shell in the context of Gauss's law (Chapter 8), you'll want to note that the solution to a conducting shell problem in the context of Ampère's law is significantly different because current is **not** an electrostatic situation (since current involves a flow of charge). Most notably, the magnetic field is **not** zero in region III, like the electric field would be in an electrostatic Gauss's law problem. (However, as we will see, in this particular problem the magnetic field is zero in region IV.)

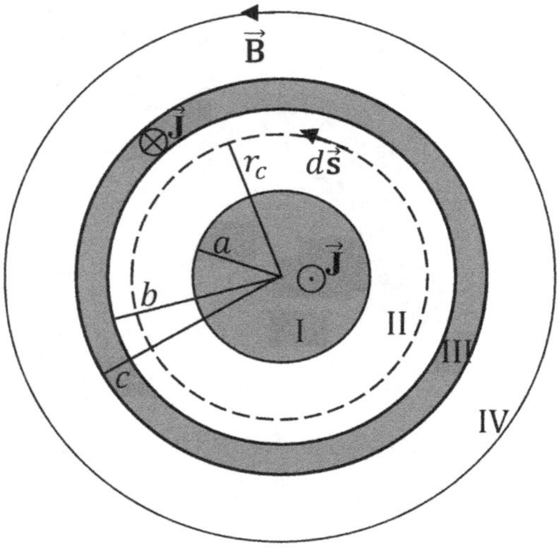

Region I: $r < a$.

- The conducting shell doesn't matter in region I. The answer is exactly the same as the example with the infinite solid cylinder: $\vec{\mathbf{B}}_I = \frac{\mu_0 I r_c}{2\pi a^2}\hat{\boldsymbol{\theta}}$. Note that this is one of the alternate forms of the equation on page 352.

Region II: $a < r < b$.

- The conducting shell also doesn't matter in region II. The answer is again exactly the same as the example with the infinite solid cylinder: $\vec{\mathbf{B}}_{II} = \frac{\mu_0 I}{2\pi r_c}\hat{\boldsymbol{\theta}}$.

Region III: $b < r < c$.

- There is a "trick" to finding the current enclosed in region III: You must consider both conductors for this region, plus the fact that the two currents run in opposite directions.
- The current enclosed in region III includes 100% of the solid cylinder's current, $+I$, which comes out of the page (⊙) **plus** a fraction of the cylindrical shell's current, $-I$, which goes into the page (⊗): $I_{enc} = I - I_{shell}$.
- Note that I_{shell} refers to the current enclosed within the shell in region III, which will be a fraction of the cylindrical shell's total current (I).
- The integral $I_{shell} = \int \vec{\mathbf{J}} \cdot d\vec{\mathbf{A}}$ gives the current enclosed within the shell in region III.
- Just as in the example with an infinite cylinder, this reduces to $I_{enc} = J \int dA$.
- Work with cylindrical coordinates: $dA = r_c dr_c d\theta$ (see Chapter 6).

- The limits of the r_c-integration are from $r_c = b$ to $r_c = r_c$ in region III ($b < r_c < c$).
- After you make these substitutions and simplify, you should get:

$$I_{shell} = J \int_{r_c=b}^{r_c} \int_{\theta=0}^{2\pi} r_c \, dr_c \, d\theta$$

- The anti-derivatives are $\frac{r_c^2}{2}$ and θ. Evaluate these over their limits.
- You should get $I_{shell} = \pi J [r_c^2]_{r_c=b}^{r_c} = \pi J (r_c^2 - b^2)$.
- Now we need an expression for the total current (I). If we do the same integral with the limits of the r_c-integration from $r_c = b$ to $r_c = c$, we will get the total current:

$$I = J \int_{r_c=b}^{c} \int_{\theta=0}^{2\pi} r_c \, dr_c \, d\theta$$

- You should get $I = \pi J [r_c^2]_{r_c=b}^{c} = \pi J (c^2 - b^2)$.
- Divide the two equations to get the following ratio. Note that the πJ cancels out.

$$\frac{I_{shell}}{I} = \frac{r_c^2 - b^2}{c^2 - b^2}$$

- Multiply both sides by I.

$$I_{shell} = I \left(\frac{r_c^2 - b^2}{c^2 - b^2} \right)$$

- Plug this expression for I_{shell} into the previous equation $I_{enc} = I - I_{shell}$. The minus **sign** represents that the shell's current runs **opposite** to the solid cylinder's current.

$$I_{enc} = I - I \left(\frac{r_c^2 - b^2}{c^2 - b^2} \right) = I \left(1 - \frac{r_c^2 - b^2}{c^2 - b^2} \right)$$

- We factored out the I. Subtract the fractions using a **common denominator**.

$$I_{enc} = I \left(\frac{c^2 - b^2}{c^2 - b^2} - \frac{r_c^2 - b^2}{c^2 - b^2} \right) = I \left(\frac{c^2 - b^2 - r_c^2 + b^2}{c^2 - b^2} \right) = I \left(\frac{c^2 - r_c^2}{c^2 - b^2} \right)$$

- Note that when you **distribute** the minus sign ($-$) to the second term, the two minus signs combine to make $+b^2$. That is, $-(r_c^2 - b^2) = -r_c^2 + b^2$. The b^2's cancel out.
- Plug this expression for I_{enc} into the equation for magnetic field from Ampère's law.

$$B_{III} = \frac{\mu_0 I_{enc}}{2\pi r_c} = \frac{\mu_0 I (c^2 - r_c^2)}{2\pi r_c (c^2 - b^2)}$$

- <u>**Check your answer**</u> for consistency: Verify that your solution matches your answer for regions II and IV at the boundaries. As r_c approaches b, the $c^2 - b^2$ will cancel out and the magnetic field approaches $\frac{\mu_0 I}{2\pi b}$, which matches region II at the boundary. As r_c approaches c, the magnetic field approaches 0, which matches region IV at the boundary. Our solution checks out.
- **Note**: It's possible to get a seemingly much different answer that turns out to be exactly the same (even if it doesn't seem like it's the same) if you approach the

enclosed current a different way. (Of course, if your answer is different, it could also just be **wrong**.) So if you obtained a different answer, see if it matches one of the alternate forms of the answer (you might not be wrong after all).

- A common way that students arrive at a different (yet equivalent) answer to this problem is to obtain the total current for the solid conductor instead of the conducting shell. We can find the total current for the solid cylinder by performing the following integral:

$$I = J \int_{r_c=0}^{a} \int_{\theta=0}^{2\pi} r_c \, dr_c \, d\theta = J\pi a^2$$

- In this case, find the shell current by dividing $I_{shell} = \pi J(r_c^2 - b^2)$ by $I = J\pi a^2$.

$$I_{shell} = I\left(\frac{r_c^2 - b^2}{a^2}\right)$$

- Recall that the current enclosed equals $I_{enc} = I - I_{shell}$.

$$I_{enc} = I - I\left(\frac{r_c^2 - b^2}{a^2}\right) = I\left(1 - \frac{r_c^2 - b^2}{a^2}\right) = I\left(\frac{a^2}{a^2} - \frac{r_c^2 - b^2}{a^2}\right) = I\left(\frac{a^2 - r_c^2 + b^2}{a^2}\right)$$

- In this case, the magnetic field in region III is:

$$B_{III} = \frac{\mu_0 I_{enc}}{2\pi r_c} = \frac{\mu_0 I(a^2 + b^2 - r_c^2)}{2\pi r_c a^2}$$

- You can verify that this expression matches region II as r_c approaches b, but it's much more difficult to see that it matches region IV as r_c approaches c. With this in mind, it's arguably better to go with our previous answer, which is the **same** as the answer shown above:

$$B_{III} = \frac{\mu_0 I(c^2 - r_c^2)}{2\pi r_c(c^2 - b^2)}$$

- As usual, we can turn this into a vector by adding the appropriate unit vector.

$$\vec{\mathbf{B}}_{III} = \frac{\mu_0 I(c^2 - r_c^2)}{2\pi r_c(c^2 - b^2)}\hat{\boldsymbol{\theta}}$$

- Another way to have a correct, but seemingly different, answer to this problem is to express your answer in terms of the current density (J) instead of the total current (I). For example, $B_{III} = \frac{\mu_0 J(c^2 - r_c^2)}{2r_c}$ or $B_{III} = \frac{\mu_0 J(a^2 - r_c^2 + b^2)}{2r_c}$. (There are yet other ways to get a correct, yet different, answer: You could mix and match our substitutions for I_{shell} and I, for example.)
- Two common ways for students to arrive at an **incorrect** answer to this problem are to integrate from $r_c = 0$ to $r_c = r_c$ for I_{shell} (when the lower limit should be $r_c = b$) or to forget to include the solid cylinder's current in $I_{enc} = I - I_{shell}$.

Region IV: $r > c$.

- Outside both of the conducting cylinders, the net current enclosed is zero: $I_{enc} = I - I = 0$. That's because one current equal to $+I$ comes out of page for the solid cylinder, while another current equal to $-I$ goes into page for the cylindrical shell. Therefore, the net magnetic field in region IV is zero: $\vec{\mathbf{B}}_{IV} = 0$.

---

99. This problem is similar to the example with an infinite current sheet. Study that example and try to let it serve as a guide. If you need help, you can always return here.

- Begin by sketching the magnetic field lines. See the diagram below.

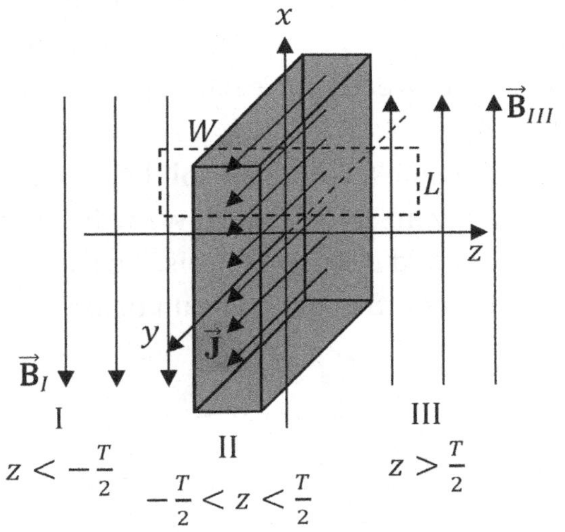

- Write down the equation for Ampère's law: $\oint_C \vec{\mathbf{B}} \cdot d\vec{\mathbf{s}} = \mu_0 I_{enc}$.
- Just as in the example with an infinite current sheet, Ampère's law reduces to $2BL = \mu_0 I_{enc}$. Isolate B to get $B = \frac{\mu_0 I_{enc}}{2L}$. See page 354.
- Integrate $I_{enc} = \int \vec{\mathbf{J}} \cdot d\vec{\mathbf{A}}$ to find the current enclosed in each region.
- This reduces to $I_{enc} = J \int dA$, where the integral is over the area of intersection between the Ampèrian rectangle (see above) and the infinite current sheet.
- For a rectangle lying in the zx plane, write $dA = dxdz$.
- After you make these substitutions and simplify, you should get:

$$I_{enc} = J \int_{x=-L/2}^{L/2} dx \int_{z=-z \text{ or } -T/2}^{z \text{ or } T/2} dz$$

- The limits of the z-integration are from $-z$ to z in region II ($-\frac{T}{2} < z < \frac{T}{2}$) and from $-\frac{T}{2}$ to $\frac{T}{2}$ for regions I and III ($|z| > \frac{T}{2}$).
- You should get $I_{enc} = JLT$ in regions I and III, and $I_{enc} = 2JLz$ in region II.
- Note that $\int_{x=-\frac{L}{2}}^{\frac{L}{2}} dx = [x]_{x=-\frac{L}{2}}^{\frac{L}{2}} = \frac{L}{2} - \left(-\frac{L}{2}\right) = \frac{L}{2} + \frac{L}{2} = L$, $\int_{z=-z}^{z} dx = [z]_{z=-z}^{z} = z -$

$(-z) = z + z = 2z$, and $\int_{z=-\frac{T}{2}}^{\frac{T}{2}} dz = [z]_{z=-\frac{T}{2}}^{\frac{T}{2}} = \frac{T}{2} - \left(-\frac{T}{2}\right) = \frac{T}{2} + \frac{T}{2} = T$.

- Substitute these expressions for the current enclosed in the three regions into the previous equation for magnetic field, $B = \frac{\mu_0 I_{enc}}{2L}$. Add $\pm\hat{\mathbf{x}}$ for the direction.
- The answers are $\vec{\mathbf{B}}_I = -\frac{\mu_0 JT}{2}\hat{\mathbf{x}}$, $\vec{\mathbf{B}}_{II} = \mu_0 Jz\,\hat{\mathbf{x}}$, and $\vec{\mathbf{B}}_{III} = \frac{\mu_0 JT}{2}\hat{\mathbf{x}}$.
- Note that in region II, when $z > 0$, $\vec{\mathbf{B}}_{II}$ points along $+\hat{\mathbf{x}}$, but when $z < 0$, $\vec{\mathbf{B}}_{II}$ points along $-\hat{\mathbf{x}}$.

100. There are a couple of "tricks" involved in this solution. The first "trick" is to apply the principle of superposition (vector addition), which we learned in Chapter 3. Another "trick" is to express the current correctly for each object (we'll clarify this "trick" when we reach that stage of the solution).

- Geometrically, we could make a complete solid cylinder (which we will call the "**big**" cylinder) by adding the given shape (which we will call the "**given**" shape, and which includes the cavity) to a solid cylinder that is the same size as the cavity (which we will call the "**small**" cylinder). See the diagram below.

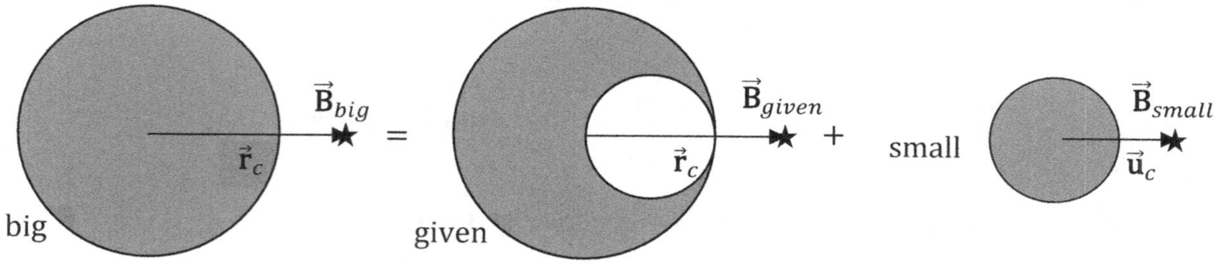

- First find the magnetic field due to the big cylinder (ignoring the cavity) at a distance $r_c = \frac{3a}{2}$ away from the axis of the cylinder. Since the field point $\left(\frac{3a}{2}, 0, 0\right)$ lies outside of the cylinder, use the equation for the magnetic field in region II from the example with the infinite cylinder. We will use one of the alternate forms from the bottom of page 352. $\vec{\mathbf{B}}_{big} = \frac{\mu_0 I_{big}}{2\pi r_c}\hat{\mathbf{y}} = \frac{\mu_0 I_{big}}{3\pi a}\hat{\mathbf{y}}$ (for region II with $r_c = \frac{3a}{2}$).
- Apply the right-hand rule for magnetic field (Chapter 22) to see that the magnetic field points up (↑) along $\hat{\mathbf{y}}$ at the field point $\left(\frac{3a}{2}, 0, 0\right)$.
- Next find the magnetic field due to the small cylinder (the same size as the cavity) at a distance $u_c = a$ away from the axis of this small cylinder. Note that the field point is closer to the axis of the small cylinder (which "fits" into the cavity) than it is to the axis of the big cylinder: That's why $u_c = a$ for the small cylinder, whereas $r_c = \frac{3a}{2}$ for the big cylinder. The equation is the same, except for the distance being smaller: $\vec{\mathbf{B}}_{small} = \frac{\mu_0 I_{small}}{2\pi u_c}\hat{\mathbf{y}} = \frac{\mu_0 I_{small}}{2\pi a}\hat{\mathbf{y}}$ (for region II with $u_c = a$).

- Note that the two currents are different. More current would pass through the big cylinder than would proportionately pass through the small cylinder: $I_{big} > I_{small}$.
- Find the magnetic field of the given shape (the cylinder with the cavity) through the principle of **superposition**. We subtract, as shown geometrically on the previous page.

$$\vec{\mathbf{B}}_{given} = \vec{\mathbf{B}}_{big} - \vec{\mathbf{B}}_{small}$$

- You should get $\vec{\mathbf{B}}_{given} = \frac{\mu_0}{\pi a}\hat{\mathbf{y}}\left(\frac{I_{big}}{3} - \frac{I_{small}}{2}\right)$. We factored out the $\frac{\mu_0}{\pi a}\hat{\mathbf{y}}$.
- Use the equation $I = \int \vec{\mathbf{J}} \cdot d\vec{\mathbf{A}}$ to find the current corresponding to each shape. Since the current density is uniform, this will simplify to $I = JA$, where A is the cross-sectional area (which has the shape of a circle). Check your intermediate answers:
  - $I_{big} = J\pi a^2$. (The big cylinder has a radius equal to a.)
  - $I_{small} = \frac{J\pi a^2}{4}$. (The small cylinder has a radius equal to $\frac{a}{2}$. When you square $\frac{a}{2}$ in the formula πr^2 for the area of a circle, you get $\frac{a^2}{4}$).
  - $I_{given} = \frac{3J\pi a^2}{4}$. (Get this by subtracting: $I_{big} - I_{small}$. Note that $1 - \frac{1}{4} = \frac{3}{4}$).
- Note the following. For example, you can divide equations to find that:

$$I_{big} = \frac{4}{3} I_{given}$$

$$I_{small} = \frac{1}{3} I_{given}$$

- The second "trick" to this solution is to realize that what the problem is calling the total current (I) refers to the current through the "**given**" shape (the cylinder with the cavity in it).
- Set $I_{given} = I$ in the above equations. You should get:

$$I_{big} = \frac{4}{3} I$$

$$I_{small} = \frac{1}{3} I$$

- Substitute these equations into the previous equation for $\vec{\mathbf{B}}_{given}$.
- Note that $\frac{1}{3}\frac{4}{3} - \frac{1}{2}\frac{1}{3} = \frac{4}{9} - \frac{1}{6} = \frac{8}{18} - \frac{3}{18} = \frac{5}{18}$.
- The final answer is $\vec{\mathbf{B}}_{given} = \frac{5\mu_0 I}{18\pi a}\hat{\mathbf{y}}$.
- **Note**: If you came up with $\frac{5}{24}$ instead of $\frac{5}{18}$, you probably made the "**mistake**" of setting I_{big} equal to the total current (I) instead of setting I_{given} equal to the total current (I).

Chapter 28: Lenz's Law

101. Apply the four steps of Lenz's law.

1. The **external** magnetic field ($\vec{\mathbf{B}}_{ext}$) is into the page (⊗). It happens to already be drawn in the problem.
2. The magnetic flux (Φ_m) is **increasing** because the problem states that the external magnetic field is increasing.
3. The **induced** magnetic field ($\vec{\mathbf{B}}_{ind}$) is out of the page (⊙). Since Φ_m is increasing, the direction of $\vec{\mathbf{B}}_{ind}$ is **opposite** to the direction of $\vec{\mathbf{B}}_{ext}$ from Step 1.
4. The **induced current** (I_{ind}) is counterclockwise, as drawn below. If you grab the loop with your thumb pointing counterclockwise and your fingers wrapped around the wire, inside the loop your fingers will come out of the page (⊙). Remember, you want your fingers to match Step 3 inside the loop: You <u>**don't**</u> want your fingers to match Step 1 or the magnetic field lines originally drawn in the problem.

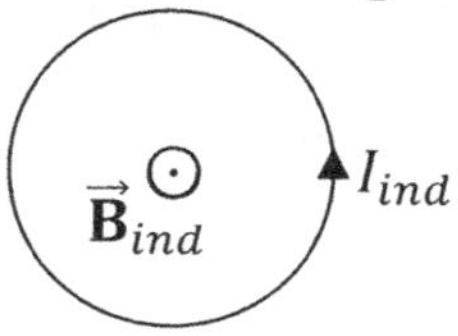

102. Apply the four steps of Lenz's law.

1. The **external** magnetic field ($\vec{\mathbf{B}}_{ext}$) is to the left (←). It happens to already be drawn in the problem.
2. The magnetic flux (Φ_m) is **decreasing** because the problem states that the external magnetic field is decreasing.
3. The **induced** magnetic field ($\vec{\mathbf{B}}_{ind}$) is to the left (←). Since Φ_m is decreasing, the direction of $\vec{\mathbf{B}}_{ind}$ is the **same** as the direction of $\vec{\mathbf{B}}_{ext}$ from Step 1.
4. The **induced current** (I_{ind}) runs up the front of the loop (and therefore runs down the back of the loop), as drawn below. Note that this loop is vertical. If you grab the front of the loop with your thumb pointing up and your fingers wrapped around the wire, inside the loop your fingers will go to the left (←).

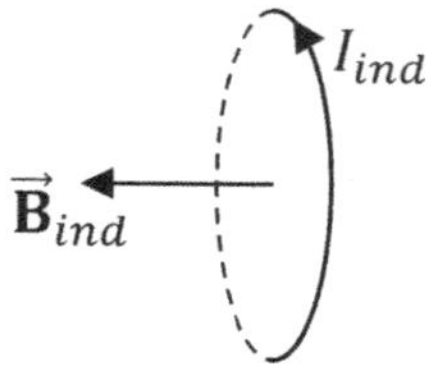

103. Apply the four steps of Lenz's law.

1. The **external** magnetic field ($\vec{\mathbf{B}}_{ext}$) is up (↑). It happens to already be drawn in the problem.

2. The magnetic flux (Φ_m) is **increasing** because the problem states that the external magnetic field is increasing.
3. The **induced** magnetic field ($\vec{\mathbf{B}}_{ind}$) is down (↓). Since Φ_m is increasing, the direction of $\vec{\mathbf{B}}_{ind}$ is **opposite** to the direction of $\vec{\mathbf{B}}_{ext}$ from Step 1.
4. The **induced current** (I_{ind}) runs to the left in the front of the loop (and therefore runs to the right in the back of the loop), as drawn below. Note that this loop is horizontal. If you grab the front of the loop with your thumb pointing to the left and your fingers wrapped around the wire, inside the loop your fingers will go down (↓). Remember, you want your fingers to match Step 3 inside the loop (**not** Step 1).

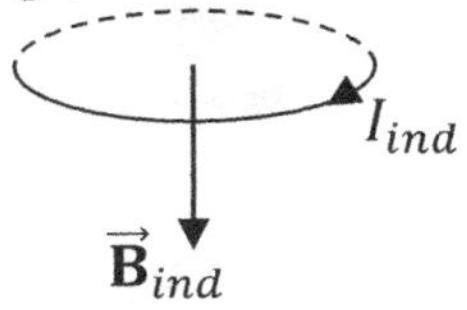

104. Apply the four steps of Lenz's law.

1. The **external** magnetic field ($\vec{\mathbf{B}}_{ext}$) is to the left (←). This is because the magnetic field lines of the magnet are going to the left, towards the south (S) pole, in the area of the loop as illustrated below. **Tip**: When you view the diagram below, ask yourself which way, on average, the magnetic field lines would be headed if you extend the diagram to where the loop is. (**Don't** use the velocity in Step 1.)

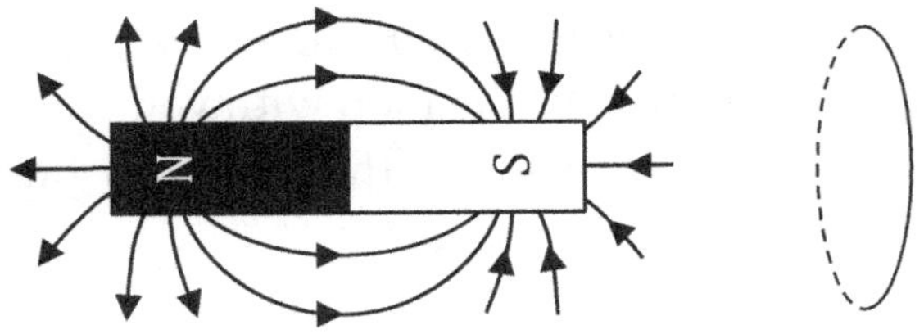

2. The magnetic flux (Φ_m) is **increasing** because the magnet is getting closer to the loop. This is the step where the direction of the velocity ($\vec{\boldsymbol{v}}$) matters.
3. The **induced** magnetic field ($\vec{\mathbf{B}}_{ind}$) is to the right (→). Since Φ_m is increasing, the direction of $\vec{\mathbf{B}}_{ind}$ is **opposite** to the direction of $\vec{\mathbf{B}}_{ext}$ from Step 1.
4. The **induced current** (I_{ind}) runs down the front of the loop (and therefore runs up in the back of the loop), as drawn below. Note that this loop is vertical. If you grab the front of the loop with your thumb pointing down and your fingers wrapped around the wire, inside the loop your fingers will go to the right (→). Remember, you want your fingers to match Step 3 inside the loop (**not** Step 1).

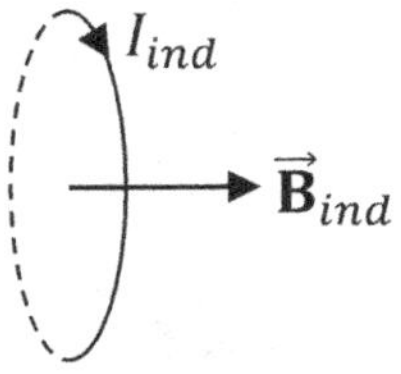

105. Apply the four steps of Lenz's law.

1. The **external** magnetic field ($\vec{\mathbf{B}}_{ext}$) is up (↑). This is because the magnetic field lines of the magnet are going up, away from the north (N) pole, in the area of the loop as illustrated below. Ask yourself: Which way, on average, would the magnetic field lines be headed if you extend the diagram to where the loop is?

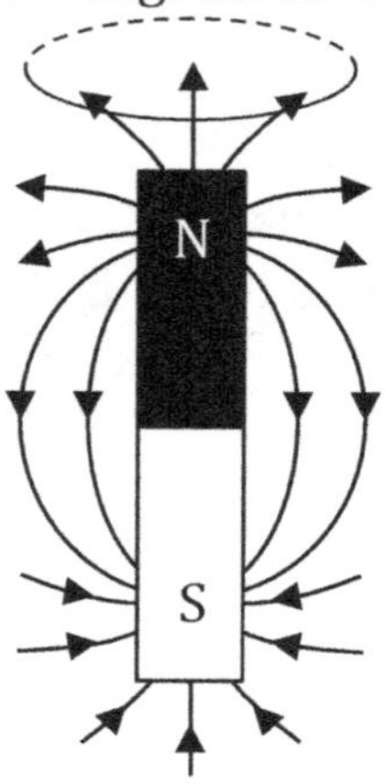

2. The magnetic flux (Φ_m) is **decreasing** because the magnet is getting further away from the loop. This is the step where the direction of the velocity ($\vec{v}$) matters.
3. The **induced** magnetic field ($\vec{\mathbf{B}}_{ind}$) is up (↑). Since Φ_m is decreasing, the direction of $\vec{\mathbf{B}}_{ind}$ is the **same** as the direction of $\vec{\mathbf{B}}_{ext}$ from Step 1.
4. The **induced current** (I_{ind}) runs to the right in the front of the loop (and therefore runs to the left in the back of the loop), as drawn below. Note that this loop is horizontal. If you grab the front of the loop with your thumb pointing right and your fingers wrapped around the wire, inside the loop your fingers will go up (↑).

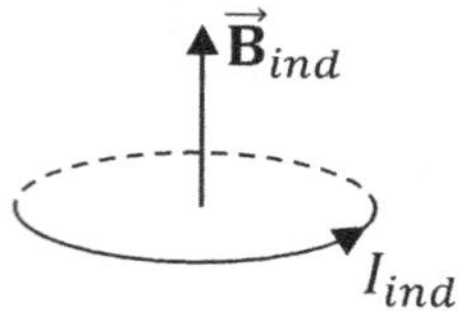

106. Apply the four steps of Lenz's law.

1. The **external** magnetic field ($\vec{\mathbf{B}}_{ext}$) is into the page (⊗). This is because the magnetic field lines of the magnet are going into the page (towards the loop), away from the north (N) pole, in the area of the loop as illustrated below. Ask yourself: Which way, on average, would the magnetic field lines be headed if you extend the diagram to where the loop is?

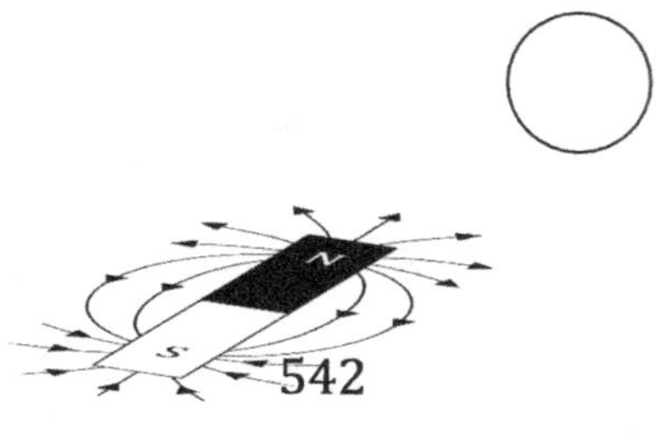

2. The magnetic flux (Φ_m) is **increasing** because the magnet is getting closer to the loop. This is the step where the direction of the velocity ($\vec{v}$) matters.
3. The **induced** magnetic field ($\vec{\mathbf{B}}_{ind}$) is out of the page ($\odot$). Since Φ_m is increasing, the direction of $\vec{\mathbf{B}}_{ind}$ is **opposite** to the direction of $\vec{\mathbf{B}}_{ext}$ from Step 1.
4. The **induced current** (I_{ind}) is counterclockwise, as drawn below. If you grab the loop with your thumb pointing counterclockwise and your fingers wrapped around the wire, inside the loop your fingers will come out of the page ($\odot$). Remember, you want your fingers to match Step 3 inside the loop (<u>not</u> Step 1).

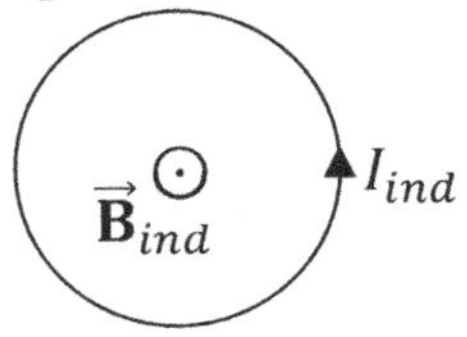

107. Apply the four steps of Lenz's law.

1. The **external** magnetic field ($\vec{\mathbf{B}}_{ext}$) is out of the page ($\odot$). This is because the outer loop creates a magnetic field that is out of the page in the region where the inner loop is. Get this from the right-hand rule for magnetic field (Chapter 22). When you grab the outer loop with your thumb pointed counterclockwise (along the given current) and your fingers wrapped around the wire, inside the outer loop (because the inner loop is inside of the outer loop) your fingers point out of the page.

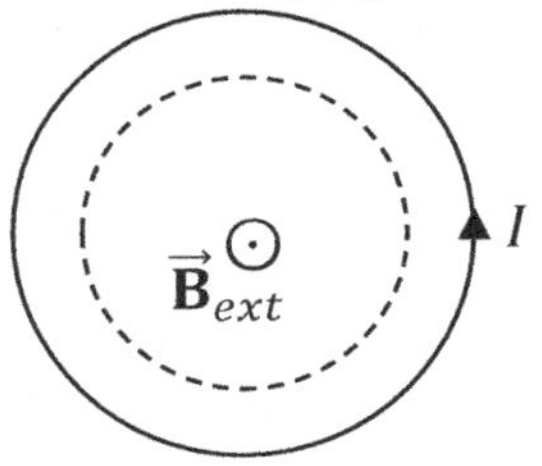

2. The magnetic flux (Φ_m) is **increasing** because the problem states that the given current is increasing.
3. The **induced** magnetic field ($\vec{\mathbf{B}}_{ind}$) is into the page ($\otimes$). Since Φ_m is increasing, the direction of $\vec{\mathbf{B}}_{ind}$ is **opposite** to the direction of $\vec{\mathbf{B}}_{ext}$ from Step 1.
4. The **induced current** (I_{ind}) is clockwise, as drawn below. If you grab the loop with your thumb pointing clockwise and your fingers wrapped around the wire, inside the loop your fingers will go into the page ($\otimes$). Remember, you want your fingers to match Step 3 inside the loop (<u>not</u> Step 1).

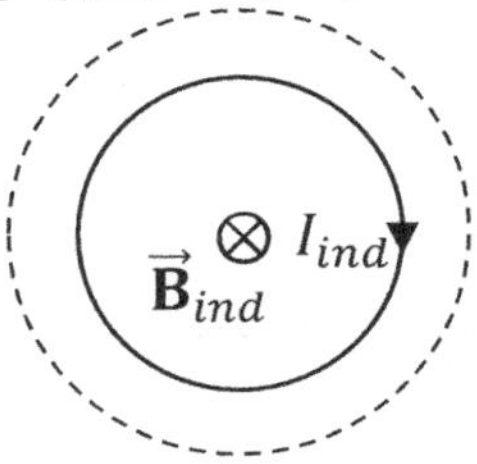

108. Apply the four steps of Lenz's law.

1. The **external** magnetic field ($\vec{\mathbf{B}}_{ext}$) is out of the page ($\odot$). This is because the given current creates a magnetic field that is out of the page in the region where the loop is. Get this from the right-hand rule for magnetic field (Chapter 22). When you grab the given current with your thumb pointed to the left and your fingers wrapped around the wire, below the given current (because the loop is below the straight wire) your fingers point out of the page.

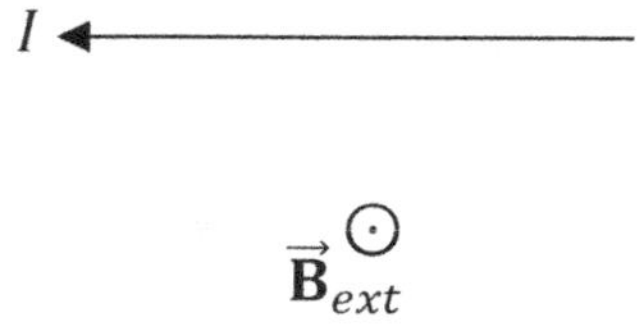

2. The magnetic flux (Φ_m) is **decreasing** because the problem states that the given current is decreasing.
3. The **induced** magnetic field ($\vec{\mathbf{B}}_{ind}$) is out of the page ($\odot$). Since Φ_m is decreasing, the direction of $\vec{\mathbf{B}}_{ind}$ is the **same** as the direction of $\vec{\mathbf{B}}_{ext}$ from Step 1.
4. The **induced current** (I_{ind}) is counterclockwise, as drawn below. If you grab the loop with your thumb pointing counterclockwise and your fingers wrapped around the wire, inside the loop your fingers will come out of the page ($\odot$).

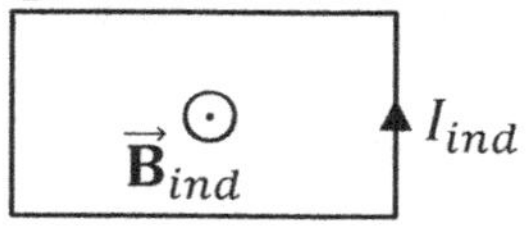

---

109. Apply the four steps of Lenz's law.

1. The **external** magnetic field ($\vec{\mathbf{B}}_{ext}$) is out of the page ($\odot$). First of all, an external current runs clockwise through the left loop, from the positive (+) terminal of the battery to the negative (−) terminal, as shown below. Secondly, the left loop creates a magnetic field that is out of the page in the region where the right loop is. Get this from the right-hand rule for magnetic field (Chapter 22). When you grab the **right side** of the **left loop** (because that side is nearest to the right loop) with your thumb pointed down (since the given current runs down at the right side of the left loop), outside of the left loop (because the right loop is outside of the left loop) your fingers point out of the page.

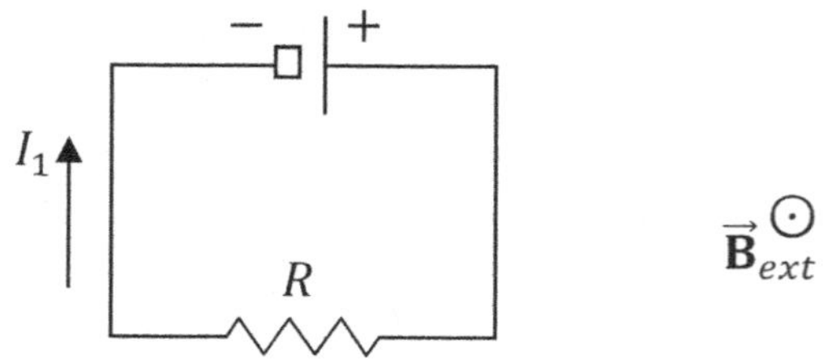

3. The magnetic flux (Φ_m) is **increasing** because the problem states that the potential difference in the battery is increasing, which increases the external current.
4. The **induced** magnetic field ($\vec{\mathbf{B}}_{ind}$) is into the page (⊗). Since Φ_m is increasing, the direction of $\vec{\mathbf{B}}_{ind}$ is **opposite** to the direction of $\vec{\mathbf{B}}_{ext}$ from Step 1.
5. The **induced current** (I_{ind}) is clockwise, as drawn below. If you grab the loop with your thumb pointing clockwise and your fingers wrapped around the wire, inside the loop your fingers will go into the page (⊗). Remember, you want your fingers to match Step 3 inside the loop (<u>not</u> Step 1).

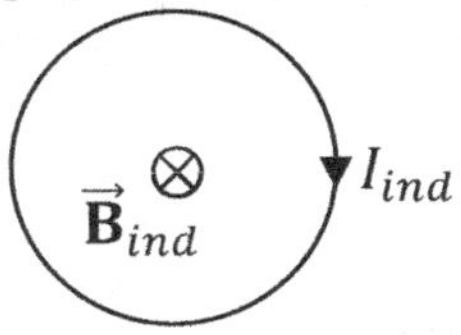

110. Apply the four steps of Lenz's law.

1. The **external** magnetic field ($\vec{\mathbf{B}}_{ext}$) is out of the page (⊙). It happens to already be drawn in the problem.
2. The magnetic flux (Φ_m) is **decreasing** because the area of the loop is getting smaller as the conducting bar travels to the right. Note that the conducting bar makes electrical contract where it touches the bare U-channel conductor. The dashed (---) line below illustrates how the area of the loop is getting smaller as the conducting bar travels to the right.

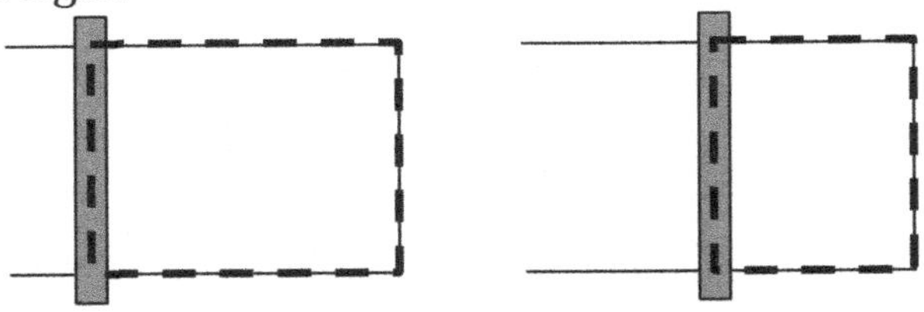

3. The **induced** magnetic field ($\vec{\mathbf{B}}_{ind}$) is out of the page (⊙). Since Φ_m is decreasing, the direction of $\vec{\mathbf{B}}_{ind}$ is the **same** as the direction of $\vec{\mathbf{B}}_{ext}$ from Step 1.
4. The **induced current** (I_{ind}) is counterclockwise, as drawn below. If you grab the loop with your thumb pointing counterclockwise and your fingers wrapped around the wire, inside the loop your fingers will come out of the page (⊙). Since the induced current is counterclockwise in the loop, the induced current runs down (↓) the conducting bar.

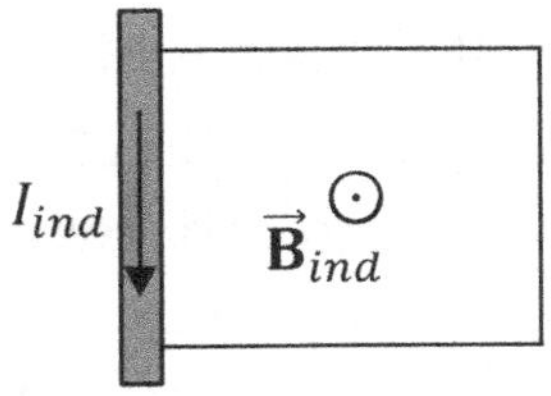

111. Apply the four steps of Lenz's law.

1. The **external** magnetic field ($\vec{\mathbf{B}}_{ext}$) is into the page (⊗). It happens to already be drawn in the problem.
2. The magnetic flux (Φ_m) is **increasing** because the area of the loop is getting larger as the conducting bar travels to the left. Note that the conducting bar makes electrical contract where it touches the bare U-channel conductor. The dashed (---) line below illustrates how the area of the loop is getting larger as the conducting bar travels to the left.

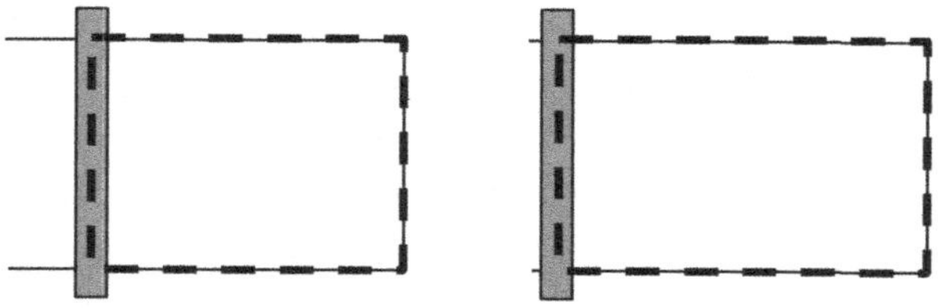

3. The **induced** magnetic field ($\vec{\mathbf{B}}_{ind}$) is out of the page (⊙). Since Φ_m is increasing, the direction of $\vec{\mathbf{B}}_{ind}$ is **opposite** to the direction of $\vec{\mathbf{B}}_{ext}$ from Step 1.
4. The **induced current** (I_{ind}) is counterclockwise, as drawn below. If you grab the loop with your thumb pointing counterclockwise and your fingers wrapped around the wire, inside the loop your fingers will come out of the page (⊙). Since the induced current is counterclockwise in the loop, the induced current runs down (↓) the conducting bar. Remember, you want your fingers to match Step 3 inside the loop (<u>**not**</u> Step 1).

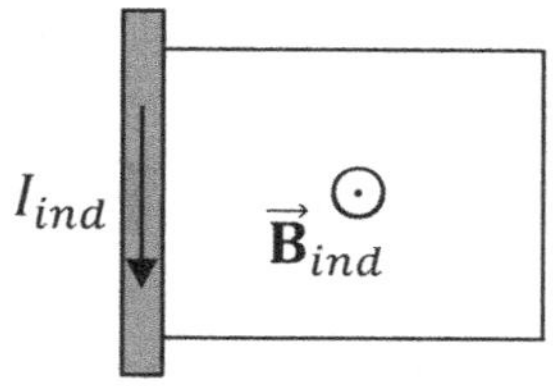

112. Apply the four steps of Lenz's law.

1. The **external** magnetic field ($\vec{\mathbf{B}}_{ext}$) is out of the page (⊙). It happens to already be drawn in the problem.
2. The magnetic flux (Φ_m) is **decreasing** because the area of the loop is getting smaller as the vertex of the triangle is pushed down from point A to point C. The formula for the area of a triangle is $A = \frac{1}{2}bh$, and the height (h) is getting shorter.
3. The **induced** magnetic field ($\vec{\mathbf{B}}_{ind}$) is out of the page (⊙). Since Φ_m is decreasing, the direction of $\vec{\mathbf{B}}_{ind}$ is the **same** as the direction of $\vec{\mathbf{B}}_{ext}$ from Step 1.
4. The **induced current** (I_{ind}) is counterclockwise, as drawn on the next page. If you grab the loop with your thumb pointing counterclockwise and your fingers wrapped around the wire, inside the loop your fingers will come out of the page (⊙).

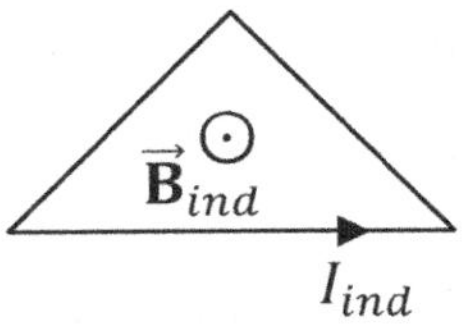

113. Apply the four steps of Lenz's law.

1. The **external** magnetic field ($\vec{\mathbf{B}}_{ext}$) is into the page ($\otimes$). It happens to already be drawn in the problem.
2. The magnetic flux (Φ_m) is **decreasing** because fewer magnetic field lines pass through the loop as it rotates. At the end of the described 90° rotation, the loop is vertical and the final magnetic flux is zero.
3. The **induced** magnetic field ($\vec{\mathbf{B}}_{ind}$) is into the page ($\otimes$). Since Φ_m is decreasing, the direction of $\vec{\mathbf{B}}_{ind}$ is the **same** as the direction of $\vec{\mathbf{B}}_{ext}$ from Step 1.
4. The **induced current** (I_{ind}) is clockwise, as drawn below. If you grab the loop with your thumb pointing clockwise and your fingers wrapped around the wire, inside the loop your fingers will go into the page ($\otimes$).

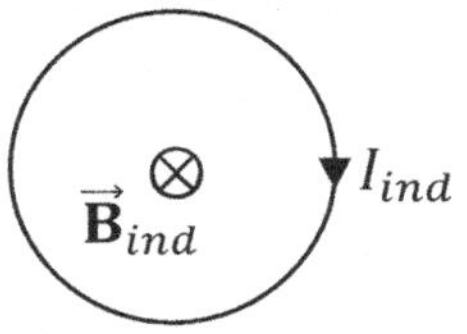

114. Apply the four steps of Lenz's law.

1. The **external** magnetic field ($\vec{\mathbf{B}}_{ext}$) is initially upward (↑). It happens to already be drawn in the problem.
2. The magnetic flux (Φ_m) is **constant** because the number of magnetic field lines passing through the loop doesn't change. The magnetic flux through the loop is zero at all times. This will be easier to see when you study Faraday's law in Chapter 29: The angle between the axis of the loop (which is perpendicular to the page) and the magnetic field (which remains in the plane of the page throughout the rotation) is 90° such that $\Phi_m = BA\cos\theta = BA\cos 90° = 0$ (since $\cos 90° = 0$).
3. The **induced** magnetic field ($\vec{\mathbf{B}}_{ind}$) is **<u>zero</u>** because the magnetic flux (Φ_m) through the loop is **constant**.
4. The **induced current** (I_{ind}) is also **<u>zero</u>** because the magnetic flux (Φ_m) through the loop is **constant**.

115. Apply the four steps of Lenz's law.

1. The **external** magnetic field ($\vec{\mathbf{B}}_{ext}$) is downward ($\downarrow$). It happens to already be drawn in the problem.
2. The magnetic flux (Φ_m) is **increasing** because more magnetic field lines pass through the loop as it rotates. At the beginning of the described 90° rotation, the initial magnetic flux is zero. This will be easier to see when you study Faraday's law in Chapter 29: The angle between the axis of the loop and the magnetic field is initially 90° such that $\Phi_m = BA\cos\theta = BA\cos 90° = 0$ (since $\cos 90° = 0$). As the loop rotates, θ decreases from 90° to 0° and $\cos\theta$ increases from 0 to 1. Therefore, the magnetic flux increases during the described 90° rotation.
3. The **induced** magnetic field ($\vec{\mathbf{B}}_{ind}$) is upward ($\uparrow$). Since Φ_m is increasing, the direction of $\vec{\mathbf{B}}_{ind}$ is **opposite** to the direction of $\vec{\mathbf{B}}_{ext}$ from Step 1.
4. The **induced current** (I_{ind}) runs to the right in the front of the loop (and therefore runs to the left in the back of the loop), as drawn below. Note that this loop is horizontal at the end of the described 90° rotation. If you grab the front of the loop with your thumb pointing right and your fingers wrapped around the wire, inside the loop your fingers will go up ($\uparrow$).

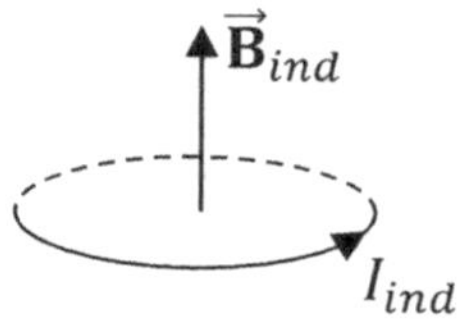

Chapter 29: Faraday's Law

116. First convert the diameter and magnetic field to SI units. Recall that $1\text{ G} = 10^{-4}$ T.

- You should get $D = 8.0$ cm $= 0.080$ m and $B = 2500$ G $= 0.25$ T $= \frac{1}{4}$ T.
- Find the **radius** from the diameter: $a = \frac{D}{2}$. You should get $a = 0.040$ m. (We're using a for radius to avoid possible confusion with resistance.)
- Find the area of the loop: $A = \pi a^2$. You should get $A = 0.0016\pi$ m^2, which works out to $A = 0.0050$ m^2 if you use a calculator.
- Use the equation $\Phi_m = BA\cos\theta$. Note that $\theta = 0°$ since the **axis** of the loop is the y-axis (because the axis of the loop is perpendicular to the loop).
- The magnetic flux through the area of the loop is $\Phi_m = 0.0004\pi$ T·m$^2 = 4\pi \times 10^{-4}$ T·m^2. If you use a calculator, it is $\Phi_m = 0.0013$ T·m$^2 = 1.3 \times 10^{-3}$ T·m^2.

---

117. Find the area of the loop: $A = L^2$. You should get $A = 4.0$ m^2.

- Use the equation $\Phi_m = BA\cos\theta$. Note that $\theta = 60°$ (**<u>not</u>** 30°) because the **axis** of the loop is perpendicular to the loop. See the diagram that follows.

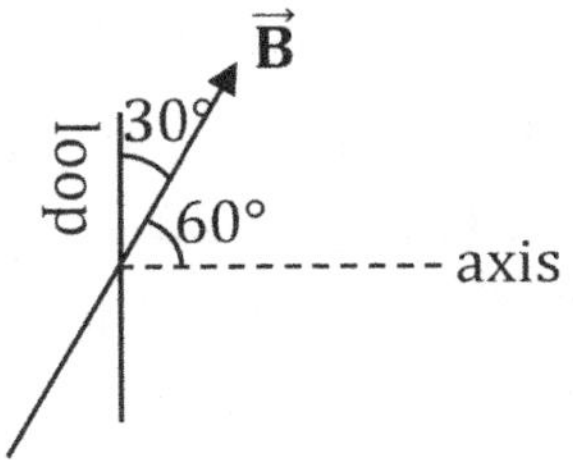

- Since the axis of the loop is perpendicular to the loop and since the loop is vertical, the **axis** of the loop is horizontal, as shown above. We need the angle that the magnetic field ($\vec{\mathbf{B}}$) makes with the axis of the loop, which is $\theta = 60°$.
- The magnetic flux through the area of the loop is $\Phi_m = 14$ T·m^2.

---

118. First convert the given quantities to SI units. Recall that 1 G $= 10^{-4}$ T.
- You should get $L = 0.50$ m, $W = 0.30$ m, $B_0 = 0.60$ T, $B = 0.80$ T, and $\Delta t = 0.500$ s.
- Find the area of the loop: $A = LW$. You should get $A = 0.15$ m^2.

(A) Use the equation $\Phi_{m0} = B_0 A \cos\theta$. Note that $\theta = 0°$.
- The initial magnetic flux is $\Phi_{m0} = 0.090$ T·m^2.

(B) Use the equation $\Phi_m = BA\cos\theta$.
- The final magnetic flux is $\Phi_m = 0.12$ T·m^2.

(C) Subtract your previous answers: $\Delta\Phi_m = \Phi_m - \Phi_{m0}$.
- Use the equation $\varepsilon_{ave} = -N\frac{\Delta\Phi_m}{\Delta t}$. Note that $N = 1$.
- The average emf induced in the loop is $\varepsilon_{ave} = -0.060$ V, which can alternatively be expressed as $\varepsilon_{ave} = -60$ mV, where the metric prefix milli (m) stands for 10^{-3}.

(D) Use the equation $I_{ave} = \frac{\varepsilon_{ave}}{R_{loop}}$.
- The average current induced in the loop is $I_{ave} = -0.015$ A, which can alternatively be expressed as $I_{ave} = -15$ mA, where the metric prefix milli (m) stands for 10^{-3}.

---

119. First convert the given quantities to SI units. Recall that 1 G $= 10^{-4}$ T.
- You should get $b = 0.50 \text{ m} = \frac{1}{2}$ m (where b is the **base** of the triangle), $h_0 = 0.25 \text{ m} = \frac{1}{4}$ m, $h = 0.50 \text{ m} = \frac{1}{2}$ m, $B = 0.40 \text{ T} = \frac{2}{5}$ T, and $\Delta t = 0.250 \text{ s} = \frac{1}{4}$ s.
- Find the initial area: $A_0 = \frac{1}{2}bh_0$. You should get $A_0 = 0.0625 \text{ m}^2 = \frac{1}{16}$ m^2.
- Find the final area: $A = \frac{1}{2}bh$. You should get $A = 0.125 \text{ m}^2 = \frac{1}{8}$ m^2.

(A) Use the equation $\Phi_{m0} = BA_0\cos\theta$. Note that $\theta = 0°$.
- The initial magnetic flux is $\Phi_{m0} = 0.025 \text{ T·m}^2 = \frac{1}{40}$ T·m^2.

(B) Use the equation $\Phi_m = BA\cos\theta$.
- The final magnetic flux is $\Phi_m = 0.050 \text{ T·m}^2 = \frac{1}{20}$ T·m^2.

(C) Subtract your previous answers: $\Delta\Phi_m = \Phi_m - \Phi_{m0}$.

- Use the equation $\varepsilon_{ave} = -N\frac{\Delta\Phi_m}{\Delta t}$. Note that $N = 200$.
- The average emf induced in the loop is $\varepsilon_{ave} = -20$ V.

(D) Apply the four steps of Lenz's law (Chapter 28).

1. The **external** magnetic field ($\vec{\mathbf{B}}_{ext}$) is into the page ($\otimes$). It happens to already be drawn in the problem.
2. The magnetic flux (Φ_m) is **increasing** because the area of the loop is getting larger as the vertex of the triangle is pushed up. The formula for the area of a triangle is $A = \frac{1}{2}bh$, and the height (h) is getting taller.
3. The **induced** magnetic field ($\vec{\mathbf{B}}_{ind}$) is out of the page ($\odot$). Since Φ_m is increasing, the direction of $\vec{\mathbf{B}}_{ind}$ is **opposite** to the direction of $\vec{\mathbf{B}}_{ext}$ from Step 1.
4. The **induced current** (I_{ind}) is counterclockwise, as drawn below. If you grab the loop with your thumb pointing counterclockwise and your fingers wrapped around the wire, inside the loop your fingers will come out of the page ($\odot$). Remember, you want your fingers to match Step 3 inside the loop (<u>**not**</u> Step 1).

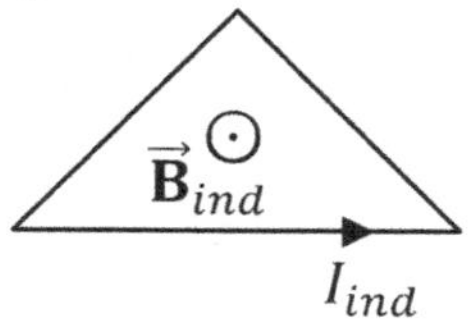

---

120. First convert the magnetic field to Tesla using 1 G = 10^{-4} T.

- You should get $B = \frac{\sqrt{3}}{2}$ T, which is $B = 0.87$ T if you use a calculator.
- Find the initial area (for the square): $A_0 = L^2$. You should get $A_0 = 4.0 \text{ m}^2$.
- Find the final area (for the rhombus). You don't need to look up the formula for the area of a rhombus. Divide the rhombus into four equal right triangles, as illustrated below. Note that the hypotenuse of each right triangle equals the length of each side of the square: $c = 2.0$ m. The height of each right triangle is one-half of the final distance between points L and N: $h = \frac{2}{2} = 1.0$ m.

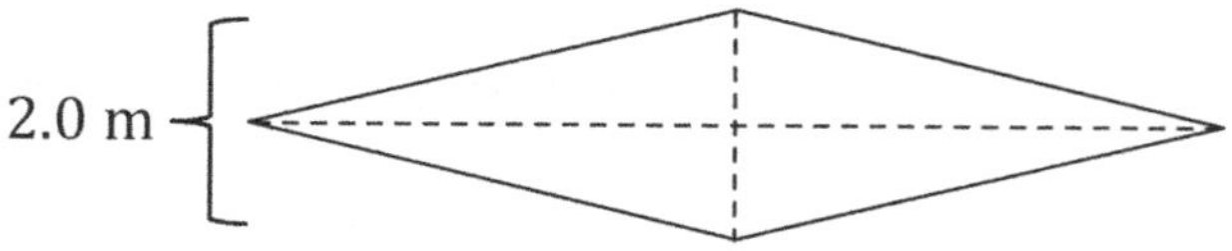

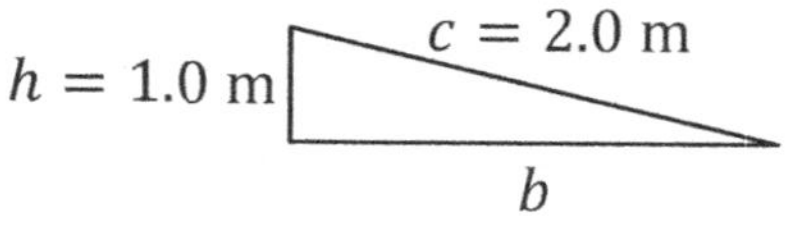

- Apply the Pythagorean theorem to solve for the base of each right triangle.
- Starting with $b^2 + h^2 = c^2$, solve for the base. You should get $b = \sqrt{c^2 - h^2}$.
- The base of each right triangle comes out to $b = \sqrt{3}$ m.
- Find the area of each right triangle: $A_{tri} = \frac{1}{2}bh$. You should get $A_{tri} = \frac{\sqrt{3}}{2} \text{ m}^2$.
- The area of the rhombus is 4 times the area of one of the right triangles: $A = 4A_{tri}$.
- The final area (for the rhombus) is $A = 2\sqrt{3} \text{ m}^2$. Using a calculator, $A = 3.5 \text{ m}^2$.

(A) Use the equation $\Phi_{m0} = BA_0 \cos\theta$. Note that $\theta = 0°$.

- The initial magnetic flux is $\Phi_{m0} = 2\sqrt{3}$ T·m^2. Using a calculator, $\Phi_{m0} = 3.5$ T·m^2.

(B) Use the equation $\Phi_m = BA\cos\theta$.

- The final magnetic flux is $\Phi_m = 3.0$ T·m^2.

(C) Subtract your previous answers: $\Delta\Phi_m = \Phi_m - \Phi_{m0}$.

- Use the equation $\varepsilon_{ave} = -N\frac{\Delta\Phi_m}{\Delta t}$. Note that $N = 1$.
- The average emf induced in the loop is $\varepsilon_{ave} = \sqrt{3}$ V. Using a calculator, $\varepsilon_{ave} = 1.7$ V.
- **Note**: If you're not using a calculator, there is a "trick" to the math. See below.

$$\varepsilon_{ave} = -N\frac{\Delta\Phi_m}{\Delta t} = -(1)\frac{3 - 2\sqrt{3}}{2 - \sqrt{3}} = \frac{2\sqrt{3} - 3}{2 - \sqrt{3}}\left(\frac{\sqrt{3}}{\sqrt{3}}\right) = \frac{(2\sqrt{3} - 3)(\sqrt{3})}{2\sqrt{3} - 3} = \sqrt{3}\text{ V}$$

- Note that we distributed the minus sign: $-(1)(3 - 2\sqrt{3}) = -3 + 2\sqrt{3} = 2\sqrt{3} - 3$. Then we multiplied by $\frac{\sqrt{3}}{\sqrt{3}}$. We then distributed the $\sqrt{3}$ in the denominator only: $(2 - \sqrt{3})\sqrt{3} = 2\sqrt{3} - 3$ (since $\sqrt{3}\sqrt{3} = 3$). Finally, $(2\sqrt{3} - 3)$ cancels out.
- Use the equation $I_{ave} = \frac{\varepsilon_{ave}}{R_{loop}}$. Note that $R_{loop} = 5 + 5 + 5 + 5 = 20\ \Omega$ (since the loop has 4 sides and each side has a resistance of 5.0 Ω).
- The average current induced in the loop is $I_{ave} = \frac{\sqrt{3}}{20}$ A. Using a calculator, $I_{ave} = 0.087$ A, which can also be expressed as $I_{ave} = 87$ mA since the metric prefix milli (m) stands for m $= 10^{-3}$.

(D) Apply the four steps of Lenz's law (Chapter 28).

1. The **external** magnetic field ($\vec{\mathbf{B}}_{ext}$) is into the page (⊗). It happens to already be drawn in the problem.
2. The magnetic flux (Φ_m) is **decreasing** because the area of the loop is getting smaller. One way to see this is to compare $A = 2\sqrt{3}$ m^2 = 3.5 m^2 to $A_0 = 4.0$ m^2. Another way is to imagine the extreme case where points L and N finally touch, for which the area will equal zero. Either way, the area of the loop is getting smaller.
3. The **induced** magnetic field ($\vec{\mathbf{B}}_{ind}$) is into the page (⊗). Since Φ_m is decreasing, the direction of $\vec{\mathbf{B}}_{ind}$ is the **same** as the direction of $\vec{\mathbf{B}}_{ext}$ from Step 1.
4. The **induced current** (I_{ind}) is clockwise, as drawn below. If you grab the loop with your thumb pointing clockwise and your fingers wrapped around the wire, inside the loop your fingers will go into the page (⊗).

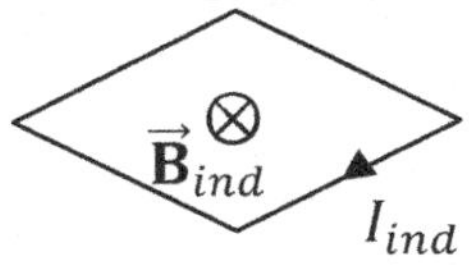

121. First convert the radius and magnetic field to SI units. Recall that 1 G $= 10^{-4}$ T.

- You should get $a = 0.040$ m and $B = 50$ T. (We're using a for radius to avoid possible confusion with resistance.)
- Note that the length of the solenoid (18 cm) is **irrelevant** to the solution. Just ignore it. (In the lab, you can measure almost anything you want, such as the temperature, so it's a valuable skill to be able to tell which quantities you do or don't need.)
- Find the area of each loop: $A = \pi a^2$. You should get $A = 0.0016\pi$ m^2, which works out to $A = 0.0050$ m^2 if you use a calculator.
- Convert $\Delta\theta$ from 30° to radians using $180° = \pi$ rad. You should get $\Delta\theta = \frac{\pi}{6}$ rad.

(A) Use the equation $\varepsilon_{ind} = -N\frac{d\Phi_m}{dt}$. For **instantaneous** emf, this involves a **derivative**.

- Plug in the expression $\Phi_m = BA\cos\theta$. Note that A and B are constants.
- You should get $\varepsilon_{ind} = -NBA\frac{d\cos\theta}{dt}$. Apply the **chain rule**, $\frac{df}{dx} = \frac{df}{du}\frac{du}{dx}$, with $f = \cos\theta$, $u = \theta$, and $x = t$. You should get $\varepsilon_{ind} = -NBA\frac{d\cos\theta}{d\theta}\frac{d\theta}{dt}$. Note that $\omega = \frac{d\theta}{dt}$. Plugging in $\omega = \frac{d\theta}{dt}$, you should get $\varepsilon_{ind} = -NBA\omega\frac{d\cos\theta}{d\theta}$. The derivative is $\frac{d\cos\theta}{d\theta} = -\sin\theta$. This should give you $\varepsilon_{ind} = NBA\omega\sin\theta$. The two minus signs make a plus sign.
- Plug numbers into $\varepsilon_{ind} = NBA\omega\sin\theta$. Note that $N = 300$, $B = 50$ T, $\omega = 20$ rad/s, and we previously found $A = 0.0016\pi$ m^2 $= 0.0050$ m^2.
- The **instantaneous** emf induced when θ reaches 30° is $\varepsilon_{ind} = 240\pi$ V. Using a calculator, this comes out to $\varepsilon_{ind} = 754$ V.

(B) Use the equation $\varepsilon_{ave} = -N\frac{\Delta\Phi_m}{\Delta t}$. For **average** emf, this involves a **ratio**.

- Use the equation $\Phi_{m0} = BA\cos\theta_0$. Note that $\theta_0 = 0°$.
- The initial magnetic flux is $\Phi_{m0} = 0.08\pi$ T·m^2. Using a calculator, $\Phi_{m0} = 0.251$ T·m^2.
- Use the equation $\Phi_m = BA\cos\theta$. Note that $\theta = 30°$.
- The final magnetic flux is $\Phi_m = 0.04\pi\sqrt{3}$ T·m^2. Using a calculator, $\Phi_m = 0.218$ T·m^2.
- Subtract: $\Delta\Phi_m = \Phi_m - \Phi_{m0}$. You should get $\Delta\Phi_m = 0.04\pi(\sqrt{3} - 2)$ T·m^2 $= -0.04\pi(2 - \sqrt{3})$ T·m^2. Using a calculator, $\Delta\Phi_m = -0.034$ T·m^2. (You could be off by a little round-off error.)
- The angular speed is $\omega = 20$ rad/s. Since the angular speed is constant, $\omega = \frac{\Delta\theta}{t}$. Multiply both sides by t to get $\omega t = \Delta\theta$. Divide both sides by ω to get $t = \frac{\Delta\theta}{\omega}$.
- Plug $\Delta\theta = \frac{\pi}{6}$ rad and $\omega = 20$ rad/s into $\Delta t = \frac{\Delta\theta}{\omega}$. You should get $\Delta t = \frac{\pi}{120}$ s. Using a calculator, this comes out to $\Delta t = 0.026$ s.
- Plug numbers into $\varepsilon_{ave} = -N\frac{\Delta\Phi_m}{\Delta t}$. Note that $N = 300$, $\Delta\Phi_m = -0.04\pi(2 - \sqrt{3})$ T·m^2 $= -0.034$ T·m^2, and $\Delta t = \frac{\pi}{120}$ s $= 0.026$ s.
- The **average** emf induced from 0° to 30° is $\varepsilon_{ave} = 1440(2 - \sqrt{3})$ V. Using a calculator, this comes out to $\varepsilon_{ave} = 386$ V. Note that $\frac{1}{\Delta t} = \frac{120}{\pi}$ in units of $\frac{1}{s}$.

(C) Consider the instantaneous emf again, which we found in part (A): $\varepsilon_{ind} = NBA\omega \sin\theta$.

- In part (A), we found that $\varepsilon_{ind} = 754$ V when θ reaches 30°.
- According to the equation $\varepsilon_{ind} = NBA\omega \sin\theta$, the instantaneous emf is initially **zero**: When θ is 0°, we get $\varepsilon_{ind} = 0$ (since $\sin 0° = 0$).
- Think about this: The **instantaneous** emf grows from 0 to 754 as θ varies from 0° to 30°. We should expect the **average** to be about $\frac{0+754}{2} = \frac{754}{2} = 357$ V.
- Indeed, the average emf turned out to be $\varepsilon_{ave} = 386$ V in part (B).
- (The reason it isn't "exactly" 357 V is because the emf is a **non-linear** function of θ.)

---

122. First convert the length to meters. You should get $\ell = 0.12$ m.

(A) The "easy" way to solve this problem is to use the **motional emf** equation: $\varepsilon_{ind} = -B\ell v$.

- If you prefer to do it the "long" way (or if that's the way you already did it), then you should follow the third example from this chapter closely, as that example solves a very similar problem the "long" way. See pages 384-385.
- Plug $B = 25$ T, $\ell = 0.12$ m, and $v = 3.0$ m/s into the equation $\varepsilon_{ind} = -B\ell v$.
- The emf induced across the ends of the conducting bar is $\varepsilon_{ind} = -9.0$ V.

(B) Use the equation $I_{ind} = \frac{\varepsilon_{ind}}{R_{loop}}$. Note that $R_{loop} = 3.0\ \Omega$.

- The current induced in the loop is $I_{ind} = -3.0$ A

(C) Apply the four steps of Lenz's law (Chapter 28).

1. The **external** magnetic field ($\vec{\mathbf{B}}_{ext}$) is into the page (⊗). It happens to already be drawn in the problem.
2. The magnetic flux (Φ_m) is **increasing** because the area of the loop is getting larger as the conducting bar travels to the right. Note that the conducting bar makes electrical contract where it touches the bare U-channel conductor. The dashed (---) line below illustrates how the area of the loop is getting larger as the conducting bar travels to the right.

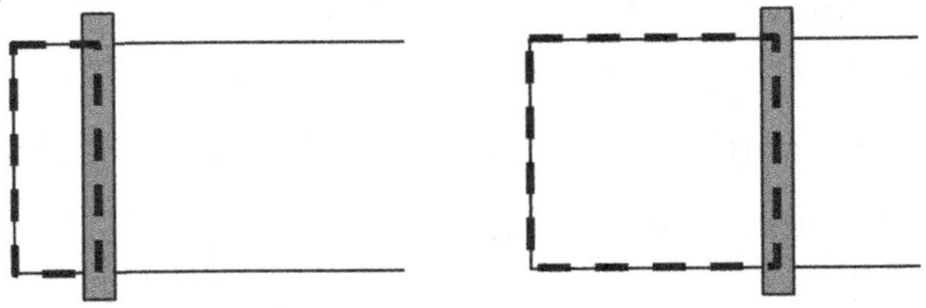

3. The **induced** magnetic field ($\vec{\mathbf{B}}_{ind}$) is out of the page (⊙). Since Φ_m is increasing, the direction of $\vec{\mathbf{B}}_{ind}$ is **opposite** to the direction of $\vec{\mathbf{B}}_{ext}$ from Step 1.
4. The **induced current** (I_{ind}) is counterclockwise, as drawn on the next page. If you grab the loop with your thumb pointing counterclockwise and your fingers wrapped around the wire, inside the loop your fingers will come out of the page (⊙). Since the induced current is counterclockwise in the loop, the induced current runs up (↑) the conducting bar. Remember, you want your fingers to match Step 3 inside the loop (**not** Step 1).

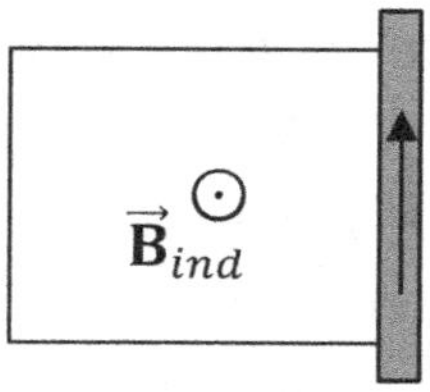

123. First convert the length and magnetic field to SI units. Recall that $1 \text{ G} = 10^{-4}$ T.

- You should get $\ell = 0.080$ m and $B = 4.0$ T.
- Use the **motional emf** equation: $\varepsilon_{ind} = -B\ell v$.
- The emf induced across the ends of the conducting bar is $\varepsilon_{ind} = -8.0$ V.

124. Use subscripts 1 and 2 to distinguish between the two different solenoids.

- $N_1 = 3000$, $L_1 = 0.50$ m, and $a_1 = 0.0050$ m for the first solenoid.
- $N_2 = 2000$, $L_2 = 0.20$ m, and $a_2 = 0.0050$ m for the second solenoid.
- $I_{10} = 3.0$ A and $I_1 = 7.0$ A for the initial and final current in the first solenoid.
- $R_2 = 5.0\ \Omega$ for the resistance of the second solenoid.
- $\Delta t = 0.016$ s since the metric prefix milli (m) stands for 10^{-3}.

(A) Note that the problem specifies magnetic **field** (B). We're **not** finding magnetic **flux** yet.

- Use the equation $B_{10} = \frac{\mu_0 N_1 I_{10}}{L_1}$ to find the initial magnetic field created by the first solenoid. (This equation was at the top of page 379, and is also in Chapter 25.)
- Recall that the permeability of free space is $\mu_0 = 4\pi \times 10^{-7}\ \frac{\text{T·m}}{\text{A}}$.
- The initial magnetic field created by the first solenoid is $B_{10} = 72\pi \times 10^{-4}$ T. Using a calculator, $B_{10} = 0.023$ T.

(B) Use the equation $\Phi_{2m0} = B_{10} A_2 \cos\theta$, where $\theta = 0°$ and A_2 is for the **second** solenoid.

- The initial magnetic flux is $\Phi_{2m0} = 18\pi^2 \times 10^{-8}$ T·m². Using a calculator, $\Phi_{2m0} = 1.8 \times 10^{-6}$ T·m². Check that your area is $A_2 = 25\pi \times 10^{-6}\ \text{m}^2 \approx 7.85 \times 10^{-5}\ \text{m}^2$. Note that $a_2 = 0.0050$ m is **squared** in $A_2 = \pi a_2^2$.

(C) Use the equation $\Phi_{2m} = B_1 A_2 \cos\theta$, where $\theta = 0°$ and A_2 is for the **second** solenoid.

- First find $B_1 = \frac{\mu_0 N_1 I_1}{L_1}$ using $I_1 = 7.0$ A. Check that $B_1 = 168\pi \times 10^{-4}\ \text{T} \approx 0.053$ T.
- The final magnetic flux is $\Phi_{2m} = 42\pi^2 \times 10^{-8}$ T·m². If you use a calculator, it is $\Phi_{2m} = 4.1 \times 10^{-6}$ T·m².

(D) Subtract your previous answers: $\Delta\Phi_{2m} = \Phi_{2m} - \Phi_{2m0}$.

- Use the equation $\varepsilon_{ave} = -N_2 \frac{\Delta\Phi_{2m}}{\Delta t}$. Be sure to use $N_2 = 2000$ (**not** $N_1 = 3000$). That's because we want the emf induced in the **second** solenoid.
- The average emf induced in the loop is $\varepsilon_{ave} = -0.03\pi^2$ V. Using a calculator, $\varepsilon_{ave} = -0.296$ V. (Since $\pi^2 \approx 10$, that's essentially how the decimal point moved.)
- Use the equation $I_{ave} = \frac{\varepsilon_{ave}}{R_{loop}}$. The average current induced in the loop is $I_{ave} = -6\pi^2 \times 10^{-3}\ \text{A} = -6\pi^2$ mA. Using a calculator, $I_{ave} = -0.059$ A.

125. First convert the distances to meters.

- You should get $L = 1.50$ m and $W = 0.80$ m.
- Find the area of the loop: $A = LW$. You should get $A = 1.2 \text{ m}^2$.

(A) Use the equation $\varepsilon_{ind} = -N\frac{d\Phi_m}{dt}$. For **instantaneous** emf, this involves a **derivative**.

- Plug in the expression $\Phi_m = BA\cos\theta$. Note that we can use the equation $\Phi_m = BA\cos\theta$ instead of $\Phi_m = \int_S \vec{\mathbf{B}} \cdot d\vec{\mathbf{A}}$ since B only depends on time (not x or y).
- You should get $\varepsilon_{ind} = -NA\cos\theta \frac{dB}{dt}$. Note that A and θ are constants.
- The derivative is $\frac{dB}{dt} = \frac{d}{dt}(15e^{-t/3}) = -5e^{-t/3}$ (see Chapter 17).
- Plug $\frac{dB}{dt} = -5e^{-t/3}$ into the equation from Faraday's law.
- You should get $\varepsilon_{ind} = 5NA\cos\theta\, e^{-t/3}$. The two minus signs make a plus sign.
- Plug numbers into this equation. Note that $N = 1$, $t = 3.0$ s, $\theta = 0°$, and we previously found $A = 1.2 \text{ m}^2$.
- The **instantaneous** emf induced at $t = 3.0$ s is $\varepsilon_{ind} = \frac{6}{e}$ V. Using a calculator, this comes out to $\varepsilon_{ind} = 2.2$ V. Note that $(5)(1.2) = 6$.

(B) Use the equation $\varepsilon_{ave} = -N\frac{\Delta\Phi_m}{\Delta t}$. For **average** emf, this involves a **ratio**.

- Use the equation $\Phi_{m0} = B_0 A\cos\theta$. Note that $\theta = 0°$.
- Plug $t = 0$ into $B = 15e^{-t/3}$ to find the initial magnetic field (B_0).
- You should get $B_0 = 15$ T. Note that $e^0 = 1$ (see Chapter 17).
- The initial magnetic flux is $\Phi_{m0} = 18 \text{ T·m}^2$. Note that $(15)(1.2) = 18$.
- Use the equation $\Phi_m = BA\cos\theta$. Note that $\theta = 0°$.
- Plug $t = 3.0$ s into $B = 15e^{-t/3}$ to find the final magnetic field (B).
- You should get $B = \frac{15}{e}$ T. Using a calculator with $e \approx 2.718$, you get $B = 5.52$ T.
- The final magnetic flux is $\Phi_m = \frac{18}{e} \text{ T·m}^2$. Using a calculator, $\Phi_m = 6.6 \text{ T·m}^2$.
- Subtract: $\Delta\Phi_m = \Phi_m - \Phi_{m0}$. You should get $\Delta\Phi_m = -\left(18 - \frac{18}{e}\right) \text{ T·m}^2$. Using a calculator, $\Delta\Phi_m = -11.4 \text{ T·m}^2$. (You could be off by a little round-off error.)
- Plug numbers into $\varepsilon_{ave} = -N\frac{\Delta\Phi_m}{\Delta t}$.
- The **average** emf induced from $t = 0$ to $t = 3.0$ s is $\varepsilon_{ave} = \left(6 - \frac{6}{e}\right)$ V. Using a calculator, this comes out to $\varepsilon_{ave} = 3.8$ V.

(C) Consider the instantaneous emf, which we found in part (A): $\varepsilon_{ind} = 5NA\cos\theta\, e^{-t/3}$.

- In part (A), we found that $\varepsilon_{ind} = 2.2$ V when $t = 3.0$ s.
- Plug $t = 0$ into $\varepsilon_{ind} = 5NA\cos\theta\, e^{-t/3}$ in order to find the initial emf: $\varepsilon_0 = 6.0$ V.
- Think about this: The **instantaneous** emf drops from 6.0 V to 2.2 V as t increases from 0 to 3.0 s. We should expect the **average** to be about $\frac{6+2.2}{2} = \frac{8.2}{2} = 4.1$ V.
- The average emf turned out to be $\varepsilon_{ave} = 3.8$ V in part (B), which is just 7% off.
- (The reason it isn't "exactly" 4.1 V because B is a **non-linear** function of t.)

Chapter 30: Inductance

126. First convert the inductance to Henry using $\text{m} = 10^{-3}$. You should get $L = 0.080$ H.

- Use the equation $\varepsilon_L = -L\frac{dI}{dt}$. Solve for $\frac{dI}{dt}$. (You don't need to apply calculus.)
- The rate at which the current increases is $\frac{dI}{dt} = 9.0$ A/s.

127. First convert the inductance to Henry using $\text{m} = 10^{-3}$. You should get $L = 0.020$ H.

- Use the equation $\tau = \frac{L}{R}$. The time constant is $\tau = 0.00040$ s $= 0.40$ ms.

128. First convert to SI units. You should get $L = 0.0160$ H and $C = 4.0 \times 10^{-5}$ F.

- Use the equation $\omega = \frac{1}{\sqrt{LC}}$. The angular speed is $\omega = 1250$ rad/s.
- It may help to write LC as $LC = 64 \times 10^{-8}$ such that $\sqrt{LC} = \sqrt{64 \times 10^{-8}}$. Use the rule from algebra that $\sqrt{ax} = \sqrt{a}\sqrt{x}$ to get $\sqrt{LC} = \sqrt{64} \times \sqrt{10^{-8}} = 8.0 \times 10^{-4}$.
- Note that $\frac{1}{8\times 10^{-4}} = \frac{10^4}{8} = \frac{10{,}000}{8} = 1250$.

129. Note that the square cross section (compared to a toroid with a circular cross section) actually makes the integral for magnetic flux simpler.

- First find the magnetic field inside of the toroidal coil. The answer for this step is the same as for a toroidal coil with circular cross section. We found this using Ampère's law in Chapter 27: $\vec{\mathbf{B}} = \frac{\mu_0 NI}{2\pi r_c}\hat{\boldsymbol{\theta}}$. See page 358.
- Substitute this into the integral for magnetic flux: $\Phi_m = \int_S \vec{\mathbf{B}} \cdot d\vec{\mathbf{A}}$.
- The differential area element is along $\hat{\boldsymbol{\theta}}$ (which is perpendicular to each square loop), such that $d\vec{\mathbf{A}} = \hat{\boldsymbol{\theta}}\, dA$.
- The integral is over the area of one of the square loops.
- For a square lying in the $r_c z$ plane, we write $dA = dr_c dz$ and $d\vec{\mathbf{A}} = \hat{\boldsymbol{\theta}}\, dr_c dz$. Note that r_c varies outward from $r_c = a$ to $r_c = a + p$, while z varies upward from $z = 0$ to $z = p$ (if you put the bottom of the toroidal coil in the xy plane).
- (Note that it's **incorrect** to write $r_c dr_c d\theta$ for this problem, even though we ordinarily write dA this way for a cylindrically shaped object. That's because we're integrating over a square loop that extends outward along r_c and upward along z, whereas when we write $r_c dr_c d\theta$, we would be integrating over the area of a circle.)
- After making these substitutions and simplifying, you should get:

$$\Phi_m = \frac{\mu_0 NI}{2\pi} \int_{r_c=a}^{a+p} \int_{z=0}^{p} \frac{\hat{\boldsymbol{\theta}}}{r_c} \cdot \hat{\boldsymbol{\theta}}\, dr_c\, dz$$

- Note that $\hat{\boldsymbol{\theta}} \cdot \hat{\boldsymbol{\theta}} = 1$. That's true for any unit vector dotted into itself.

- You should get:

$$\Phi_m = \frac{\mu_0 NI}{2\pi} \int_{r_c=a}^{a+p} \frac{dr_c}{r_c} \int_{z=0}^{p} dz$$

- Recall from Chapter 17 that $\int \frac{dx}{x} = \ln(x)$.
- When you evaluate $\ln(r_c)$ from $r_c = a$ to $r_c = a + p$, you get $\ln(a + p) - \ln(a)$.
- Use the logarithm identity $\ln(x) - \ln(y) = \ln\left(\frac{x}{y}\right)$ to get $\ln\left(\frac{a+p}{a}\right)$.
- The magnetic flux becomes:

$$\Phi_m = \frac{\mu_0 NIp}{2\pi} \ln\left(\frac{a+p}{a}\right)$$

- Substitute this into the equation for inductance: $L = \frac{N\Phi_m}{I}$.
- The final answer for the self-inductance is:

$$L = \frac{\mu_0 N^2 p}{2\pi} \ln\left(\frac{a+p}{a}\right)$$

Chapter 31: AC Circuits

130. Identify the given symbol. Which quantity represents the amplitude of the current?
- You are given $I_m = 2.0$ A. Use the equation $I_{rms} = \frac{I_m}{\sqrt{2}}$.
- The rms current is $I_{rms} = \sqrt{2}$ A, which is the same as $I_{rms} = \frac{2}{\sqrt{2}}$ A. Multiply the numerator and denominator by $\sqrt{2}$ in order to **rationalize** the denominator of the fraction: $\frac{2}{\sqrt{2}} = \frac{2}{\sqrt{2}}\frac{\sqrt{2}}{\sqrt{2}} = \frac{2\sqrt{2}}{2} = \sqrt{2}$. If you use a calculator, $I_{rms} = 1.4$ A.

---

131. You are given $f = 50$ Hz and $L = 30$ mH.
- Convert the inductance to Henry. You should get $L = 0.030$ H.
- Use the equation $\omega = 2\pi f$. You should get $\omega = 100\pi$ rad/s.
- Use the equation $X_L = \omega L$. The inductive reactance is $X_L = 3\pi\ \Omega$. If you use a calculator, $X_L = 9.4\ \Omega$.

---

132. You are given $f = 100$ Hz and $C = 2.0\ \mu$F.
- Convert the capacitance to Farads. You should get $C = 2.0 \times 10^{-6}$ F.
- Use the equation $\omega = 2\pi f$. You should get $\omega = 200\pi$ rad/s.
- Use the equation $X_C = \frac{1}{\omega C}$. The capacitive reactance is $X_C = \frac{2500}{\pi}\ \Omega$. If you use a calculator, $X_C = 796\ \Omega$, which is 800 Ω or 0.80 kΩ to two significant figures.

133. Convert the capacitance to Farads. You should get $C = 8.0 \times 10^{-4}$ F.

- First use the equations $X_L = \omega L$ and $X_C = \frac{1}{\omega C}$.
- You should get $X_L = 150\ \Omega$ and $X_C = 50\ \Omega$.

(A) Use the equation $Z = \sqrt{R^2 + (X_L - X_C)^2}$.

- The impedance is $Z = 200\ \Omega$. Note that $\left(100\sqrt{3}\right)^2 = 30{,}000$ and $\sqrt{40{,}000} = 200$.

(B) Use the equation $\varphi = \tan^{-1}\left(\frac{X_L - X_C}{R}\right)$.

- The phase angle is $\varphi = 30°$. Note that $\frac{100}{100\sqrt{3}} = \frac{1}{\sqrt{3}} = \frac{1}{\sqrt{3}}\frac{\sqrt{3}}{\sqrt{3}} = \frac{\sqrt{3}}{3}$ and $\tan^{-1}\left(\frac{\sqrt{3}}{3}\right) = 30°$.

---

134. Convert the capacitance to Farads. You should get $C = 1.0 \times 10^{-4}$ F.

- First use the equations $X_L = \omega L$ and $X_C = \frac{1}{\omega C}$.
- You should get $X_L = 150\ \Omega$ and $X_C = 200\ \Omega$.
- Use the equation $Z = \sqrt{R^2 + (X_L - X_C)^2}$.
- You should get $Z = 100\ \Omega$. Note that $\left(50\sqrt{3}\right)^2 = 7500$ and $\sqrt{10{,}000} = 100$.

(A) Use the equation $\Delta V_{rms} = I_{rms} Z$. Divide both sides by Z. You should get $I_{rms} = \frac{\Delta V_{rms}}{Z}$.

- (Alternatively, you could first use the equation $I_m = \frac{\Delta V_m}{Z}$ and then use the equations $I_{rms} = \frac{I_m}{\sqrt{2}}$ and $\Delta V_{rms} = \frac{\Delta V_m}{\sqrt{2}}$.)
- An AC ammeter would measure $I_{rms} = 2.0$ A.

(B) Use the equation $\Delta V_{Rrms} = I_{rms} R$.

- An AC voltmeter would measure $\Delta V_{Rrms} = 100\sqrt{3}$ V across the resistor. Using a calculator, $\Delta V_{Rrms} = 173$ V $= 0.17$ kV.

(C) Use the equation $\Delta V_{Lrms} = I_{rms} X_L$.

- An AC voltmeter would measure $\Delta V_{Lrms} = 300$ V across the inductor.

(D) Use the equation $\Delta V_{Crms} = I_{rms} X_C$.

- An AC voltmeter would measure $\Delta V_{Crms} = 400$ V across the capacitor.

(E) Subtract the previous two answers. $\Delta V_{LCrms} = |\Delta V_{Lrms} - \Delta V_{Crms}|$.

- The reason for this is that the phasors for the inductor and capacitor point in **opposite** directions: The phase angle for the inductor is 90°, while the phase angle for the capacitor is −90°.
- An AC voltmeter would measure $\Delta V_{LCrms} = 100$ V across the inductor-capacitor combination.

(F) Use the equation $\Delta V_{rms} = \sqrt{\Delta V_{Rrms}^2 + (\Delta V_{Lrms} - \Delta V_{Crms})^2}$.

- You should get $\Delta V_{rms} = 200$ V, which agrees with the value given in the problem.
- Note that $\left(100\sqrt{3}\right)^2 = 30{,}000$ and $\sqrt{40{,}000} = 200$.

135. Convert the inductance and capacitance to SI units.

- You should get $L = 0.080$ H and $C = 5.0 \times 10^{-5}$ F.

(A) Use the equation $\omega_0 = \frac{1}{\sqrt{LC}}$.

- The resonance angular frequency is $\omega_0 = 500$ rad/s.
- Note that $\sqrt{0.4 \times 10^{-5}} = \sqrt{4 \times 10^{-6}} = \sqrt{4}\sqrt{10^{-6}} = 2 \times 10^{-3}$.
- Also note that $\frac{1}{2\times10^{-3}} = \frac{10^3}{2} = \frac{1000}{2} = 500$.

(B) Use the equation $\omega_0 = 2\pi f_0$.

- Divide both sides of the equation by 2π. You should get $f_0 = \frac{\omega_0}{2\pi}$.
- The resonance frequency in Hertz is $f_0 = \frac{250}{\pi}$ Hz. Using a calculator, it is $f_0 = 80$ Hz.

---

136. As explained in part (I) of the first example in this chapter, the answer is **not** what we have called I_m. In our notation, the symbol I_m is the maximum value of the instantaneous current for a given frequency (the current oscillates between I_m and $-I_m$). See page 427.

(A) In our notation, I_{max} is the largest possible value of I_m that can be obtained by varying the frequency, and it occurs at the **resonance** frequency.

- However, note that I_{max} is the amplitude of the current when the frequency equals the resonance frequency. That's not quite what the question asked for.
- This problem asked for the maximum value of the **rms** current. So what we're really looking for is $I_{rms,max}$, where $I_{rms,max} = \frac{I_{max}}{\sqrt{2}}$.
- The equation $\Delta V_m = I_m Z$ reduces to $\Delta V_m = I_{max} R$ at the **resonance** frequency, since at resonance $X_L = X_C$ such that Z reduces to R.
- Divide both sides of the equation $\Delta V_m = I_{max} R$ by $\sqrt{2}$ in order to rewrite this equation in terms of rms values: $\Delta V_{rms} = I_{rms,max} R$. Divide both sides by R to get $I_{rms,max} = \frac{\Delta V_{rms}}{R}$. Plug numbers into this equation. Note that $\Delta V_{rms} = 120$ V.
- The maximum possible rms current that the AC power supply could provide to this RLC circuit is $I_{rms,max} = 4.0$ A.

(B) The maximum current and minimum impedance both occur at **resonance**.

- Less **impedance** results in more **current** (for a fixed rms power supply voltage).
- As we discussed in part (A), at resonance, the impedance equals the **resistance**.
- There is no math to do (but if this happens to be an assigned homework problem for you, then you should explain your answer).
- The minimum possible impedance that could be obtained by adjusting the frequency is $Z_{min} = R = 30\ \Omega$.

137. Convert the potential difference and capacitance to SI units.

- You should get $\Delta V_{rms} = 3000$ V and $C = 8.0 \times 10^{-5}$ F.
- First use the equations $X_L = \omega L$ and $X_C = \frac{1}{\omega C}$.
- You should get $X_L = 2000\ \Omega$ and $X_C = 500\ \Omega$.
- Use the equation $Z = \sqrt{R^2 + (X_L - X_C)^2}$.
- You should get $Z = 1000\sqrt{3}\ \Omega$.
- Note that $\left(500\sqrt{3}\right)^2 = 750{,}000$ and $\sqrt{3{,}000{,}000} = \sqrt{3}\sqrt{1{,}000{,}000} = 1000\sqrt{3}$.

(A) Use the equation $\Delta V_{rms} = I_{rms} Z$. Divide both sides by Z. You should get $I_{rms} = \frac{\Delta V_{rms}}{Z}$.

- (Alternatively, you could first use the equation $I_m = \frac{\Delta V_m}{Z}$ and then use the equations $I_{rms} = \frac{I_m}{\sqrt{2}}$ and $\Delta V_{rms} = \frac{\Delta V_m}{\sqrt{2}}$.)
- The rms current for this RLC circuit is $I_{rms} = \sqrt{3}$ A. Using a calculator, $I_{rms} = 1.7$ A.
- Note that $\frac{3000}{1000\sqrt{3}} = \frac{3}{\sqrt{3}} = \frac{3}{\sqrt{3}}\frac{\sqrt{3}}{\sqrt{3}} = \frac{3\sqrt{3}}{3} = \sqrt{3}$. Multiply by $\frac{\sqrt{3}}{\sqrt{3}}$ in order to **rationalize** the denominator. Note that $\sqrt{3}\sqrt{3} = 3$.

(B) Use the equation $P_{av} = I_{rms}^2 R$.

- The average power that the AC power supply delivers to the circuit is $P_{av} = 1500\sqrt{3}$ W. If you use a calculator, this comes out to $P_{av} = 2.6$ kW, where the metric prefix kilo (k) stands for k = 1000.
- Note that $\sqrt{3}\sqrt{3} = 3$ such that $\left(\sqrt{3}\sqrt{3}\right)\left(500\sqrt{3}\right) = (3)\left(500\sqrt{3}\right) = 1500\sqrt{3}$.
- Note that you would get the same answer using the equation $P_{av} = I_{rms}\Delta V_{rms}\cos\varphi$ (provided that you do the math correctly).

(C) Use the equation $\omega_0 = \frac{1}{\sqrt{LC}}$.

- The resonance angular frequency is $\omega_0 = \frac{25}{2}$ rad/s $= 12.5$ rad/s (or 13 rad/s if you round to two significant figures, as you should in this problem).
- Note that $\sqrt{640 \times 10^{-5}} = \sqrt{64 \times 10^{-4}} = \sqrt{64}\sqrt{10^{-4}} = 8.0 \times 10^{-2}$.
- Also note that $\frac{1}{8 \times 10^{-2}} = \frac{10^2}{8} = \frac{100}{8} = \frac{25}{2} = 12.5$.

(D) Use the same equation and reasoning that we applied in Problem 136, part (A).

- The equation you need is $I_{rms,max} = \frac{\Delta V_{rms}}{R}$.
- The maximum possible rms current that the AC power supply could provide to this RLC circuit is $I_{rms,max} = 2\sqrt{3}$ A. Using a calculator, $I_{rms,max} = 3.5$ A.
- Note that $\frac{3000}{500\sqrt{3}} = \frac{6}{\sqrt{3}} = \frac{6}{\sqrt{3}}\frac{\sqrt{3}}{\sqrt{3}} = \frac{6\sqrt{3}}{3} = 2\sqrt{3}$. Multiply by $\frac{\sqrt{3}}{\sqrt{3}}$ in order to **rationalize** the denominator. Note that $\sqrt{3}\sqrt{3} = 3$.

138. Convert the inductance and capacitance to SI units.

- You should get $L = 0.030$ H and $C = 1.2 \times 10^{-5}$ F.
- First use the equation $\omega_0 = \frac{1}{\sqrt{LC}}$.
- The resonance angular frequency is $\omega_0 = \frac{5000}{3}$ rad/s.
- Use the equation $Q_0 = \frac{\omega_0 L}{R}$.
- The quality factor for this RLC circuit is $Q_0 = 25$.

139. Read off the needed values from the graph:

- The resonance angular frequency is $\omega_0 = 550$ rad/s. This is the angular frequency for which the curve reaches its peak.
- The maximum average power is $P_{av,max} = 40$ W. This is the maximum vertical value of the curve.
- One-half of the maximum average power is $\frac{P_{av,max}}{2} = \frac{40}{2} = 20$ W.
- Study the pictures on pages 417 and 430.
- Draw a horizontal line on the graph where the vertical value of the curve is equal to 20 W, which corresponds to $\frac{P_{av,max}}{2}$.
- Draw two vertical lines on the graph where the curve intersects the horizontal line that you drew in the previous step. See the right figure on page 430. These two vertical lines correspond to ω_1 and ω_2.
- Read off ω_1 and ω_2 from the graph: $\omega_1 = 525$ rad/s and $\omega_2 = 575$ rad/s.
- Subtract these values: $\Delta\omega = \omega_2 - \omega_1 = 575 - 525 = 50$ rad/s. This is the **full-width at half max**. Recall that the resonance frequency is $\omega_0 = 550$ rad/s.
- Use the equation $Q_0 = \frac{\omega_0}{\Delta\omega}$. The quality factor for this graph is $Q_0 = 11$.

140. Identify the given quantities: $\Delta V_{in} = 240$ V, $N_p = 400$, and $N_s = 100$.

- Use the equation $\frac{\Delta V_{out}}{\Delta V_{in}} = \frac{N_s}{N_p}$. Cross multiply to get $\Delta V_{out} N_p = \Delta V_{in} N_s$.
- Divide both sides of the equation by N_p. You should get $\Delta V_{out} = \frac{\Delta V_{in} N_s}{N_p}$.
- The output voltage is $\Delta V_{out} = 60$ V.

141. Identify the given quantities: $\Delta V_{in} = 40$ V, $N_p = 200$, and $\Delta V_{out} = 120$ V.

- Use the equation $\frac{\Delta V_{out}}{\Delta V_{in}} = \frac{N_s}{N_p}$. Cross multiply to get $\Delta V_{out} N_p = \Delta V_{in} N_s$.
- Divide both sides of the equation by ΔV_{in}. You should get $N_s = \frac{\Delta V_{out} N_p}{\Delta V_{in}}$.
- The secondary has $N_s = 600$ turns (or loops).

WAS THIS BOOK HELPFUL?

A great deal of effort and thought was put into this book, such as:

- Breaking down the solutions to help make physics easier to understand.
- Careful selection of examples and problems for their instructional value.
- Multiple stages of proofreading, editing, and formatting.
- Two physics instructors worked out the solution to every problem to help check all of the final answers.
- Dozens of actual physics students provided valuable feedback.

If you appreciate the effort that went into making this book possible, there is a simple way that you could show it:

Please take a moment to post an honest review.

For example, you can review this book at Amazon.com or BN.com (for Barnes & Noble).

Even a short review can be helpful and will be much appreciated. If you're not sure what to write, following are a few ideas, though it's best to describe what's important to you.

- Were you able to understand the explanations?
- Did you appreciate the list of symbols and units?
- Was it easy to find the equations you needed?
- How much did you learn from reading through the examples?
- Did the hints and intermediate answers section help you solve the problems?
- Would you recommend this book to others? If so, why?

Are you an international student?

If so, please leave a review at Amazon.co.uk (United Kingdom), Amazon.ca (Canada), Amazon.in (India), Amazon.com.au (Australia), or the Amazon website for your country.

The physics curriculum in the United States is somewhat different from the physics curriculum in other countries. International students who are considering this book may like to know how well this book may fit their needs.

THE SOLUTIONS MANUAL

The solution to every problem in this workbook can be found in the following book:

100 Instructive Calculus-based Physics Examples

Fully Solved Problems with Explanations

Volume 2: Electricity and Magnetism

Chris McMullen, Ph.D.

ISBN: 978-1-941691-13-7

If you would prefer to see every problem worked out completely, along with explanations, you can find such solutions in the book shown below. (The workbook you are currently reading has hints, intermediate answers, and explanations. The book described above contains full step-by-step solutions.)

VOLUME 3

If you want to learn more physics, volume 3 covers additional topics.

Volume 3: Waves, Fluids, Sound, Heat, and Light

- Sine waves
- Simple harmonic motion
- Oscillating springs
- Simple and physical pendulums
- Characteristics of waves
- Sound waves
- The decibel system
- The Doppler effect
- Standing waves
- Density and pressure
- Archimedes' principle
- Bernoulli's principle
- Pascal's principle
- Heat and temperature
- Thermal expansion
- Ideal gases
- The laws of thermodynamics
- Light waves
- Reflection and refraction
- Snell's law
- Total internal reflection
- Dispersion
- Thin lenses
- Spherical mirrors
- Diffraction
- Interference
- Polarization
- and more

ABOUT THE AUTHOR

Chris McMullen is a physics instructor at Northwestern State University of Louisiana and also an author of academic books. Whether in the classroom or as a writer, Dr. McMullen loves sharing knowledge and the art of motivating and engaging students.

He earned his Ph.D. in phenomenological high-energy physics (particle physics) from Oklahoma State University in 2002. Originally from California, Dr. McMullen earned his Master's degree from California State University, Northridge, where his thesis was in the field of electron spin resonance.

As a physics teacher, Dr. McMullen observed that many students lack fluency in fundamental math skills. In an effort to help students of all ages and levels master basic math skills, he published a series of math workbooks on arithmetic, fractions, algebra, and trigonometry called the Improve Your Math Fluency Series. Dr. McMullen has also published a variety of science books, including introductions to basic astronomy and chemistry concepts in addition to physics textbooks.

Dr. McMullen is very passionate about teaching. Many students and observers have been impressed with the transformation that occurs when he walks into the classroom, and the interactive engaged discussions that he leads during class time. Dr. McMullen is well-known for drawing monkeys and using them in his physics examples and problems, applying his creativity to inspire students. A stressed-out student is likely to be told to throw some bananas at monkeys, smile, and think happy physics thoughts.

Author, Chris McMullen, Ph.D.

PHYSICS

The learning continues at Dr. McMullen's physics blog:

www.monkeyphysicsblog.wordpress.com

More physics books written by Chris McMullen, Ph.D.:

- An Introduction to Basic Astronomy Concepts (with Space Photos)
- The Observational Astronomy Skywatcher Notebook
- An Advanced Introduction to Calculus-based Physics
- Essential Calculus-based Physics Study Guide Workbook
- Essential Trig-based Physics Study Guide Workbook
- 100 Instructive Calculus-based Physics Examples
- 100 Instructive Trig-based Physics Examples
- Creative Physics Problems
- A Guide to Thermal Physics
- A Research Oriented Laboratory Manual for First-year Physics

SCIENCE

Dr. McMullen has published a variety of **science** books, including:

- Basic astronomy concepts
- Basic chemistry concepts
- Balancing chemical reactions
- Creative physics problems
- Calculus-based physics textbook
- Calculus-based physics workbooks
- Trig-based physics workbooks

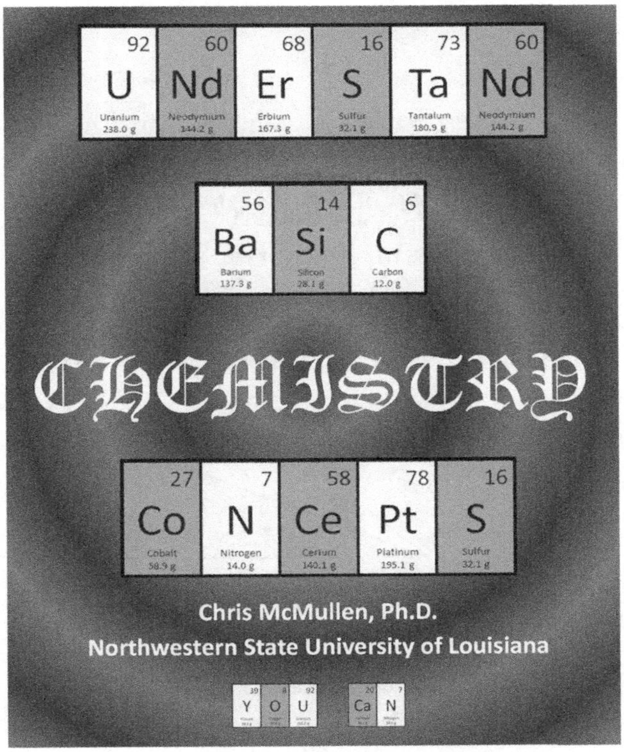

MATH

This series of math workbooks is geared toward practicing essential math skills:

- Algebra and trigonometry
- Fractions, decimals, and percents
- Long division
- Multiplication and division
- Addition and subtraction

www.improveyourmathfluency.com

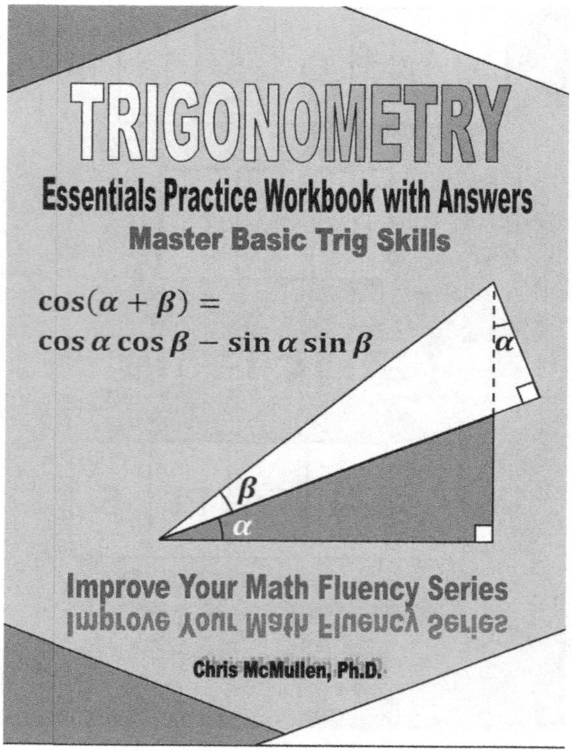

PUZZLES

The author of this book, Chris McMullen, enjoys solving puzzles. His favorite puzzle is Kakuro (kind of like a cross between crossword puzzles and Sudoku). He once taught a three-week summer course on puzzles. If you enjoy mathematical pattern puzzles, you might appreciate:

300+ Mathematical Pattern Puzzles

Number Pattern Recognition & Reasoning

- pattern recognition
- visual discrimination
- analytical skills
- logic and reasoning
- analogies
- mathematics

VErBAl ReAcTiONS

Chris McMullen has coauthored several word scramble books. This includes a cool idea called **VErBAl ReAcTiONS**. A VErBAl ReAcTiON expresses word scrambles so that they look like chemical reactions. Here is an example:

$$2\,C + U + 2\,S + Es \rightarrow S\,U\,C\,C\,Es\,S$$

The left side of the reaction indicates that the answer has 2 C's, 1 U, 2 S's, and 1 Es. Rearrange CCUSSEs to form SUCCEsS.

Each answer to a **VErBAl ReAcTiON** is not merely a word, it's a chemical word. A chemical word is made up not of letters, but of elements of the periodic table. In this case, SUCCEsS is made up of sulfur (S), uranium (U), carbon (C), and Einsteinium (Es).

Another example of a chemical word is GeNiUS. It's made up of germanium (Ge), nickel (Ni), uranium (U), and sulfur (S).

If you enjoy anagrams and like science or math, these puzzles are tailor-made for you.

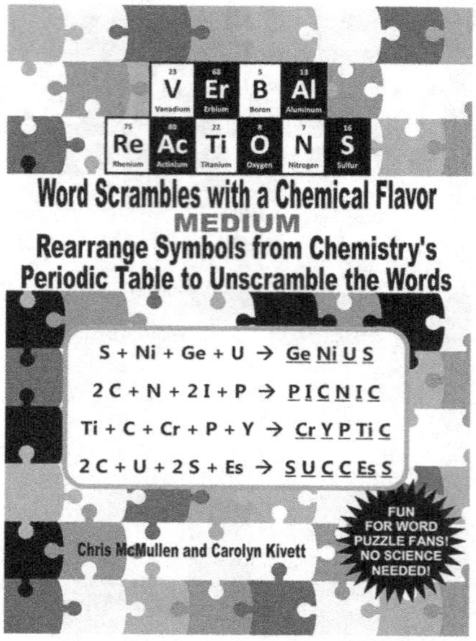

BALANCING CHEMICAL REACTIONS

$$2\ C_2H_6 + 7\ O_2 \rightarrow 4\ CO_2 + 6\ H_2O$$

Balancing chemical reactions isn't just chemistry practice.

These are also **fun puzzles** for math and science lovers.

Balancing Chemical Equations Worksheets

Over 200 Reactions to Balance

Chemistry Essentials Practice Workbook with Answers

Chris McMullen, Ph.D.

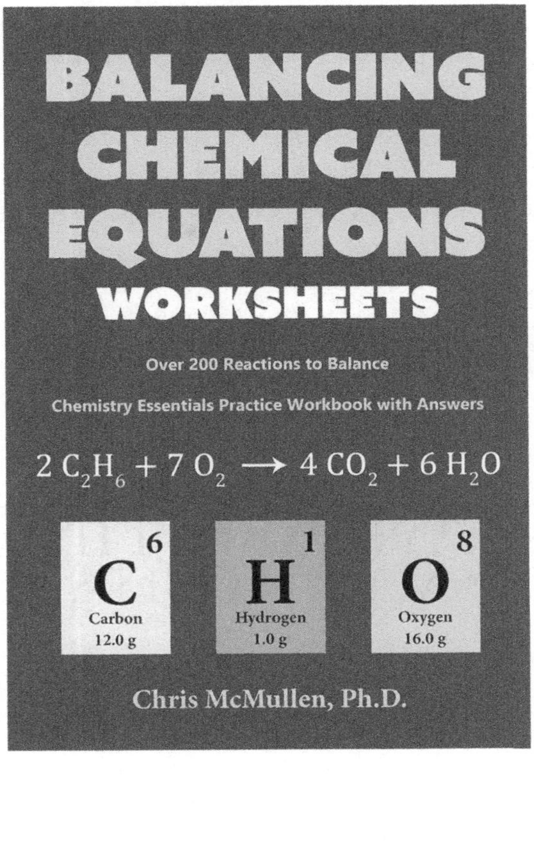

CURSIVE HANDWRITING

for... MATH LOVERS

Would you like to learn how to write in cursive?

Do you enjoy math?

This cool writing workbook lets you practice writing math terms with cursive handwriting. Unfortunately, you can't find many writing books oriented around math.

Cursive Handwriting for Math Lovers
by Julie Harper and Chris McMullen, Ph.D.

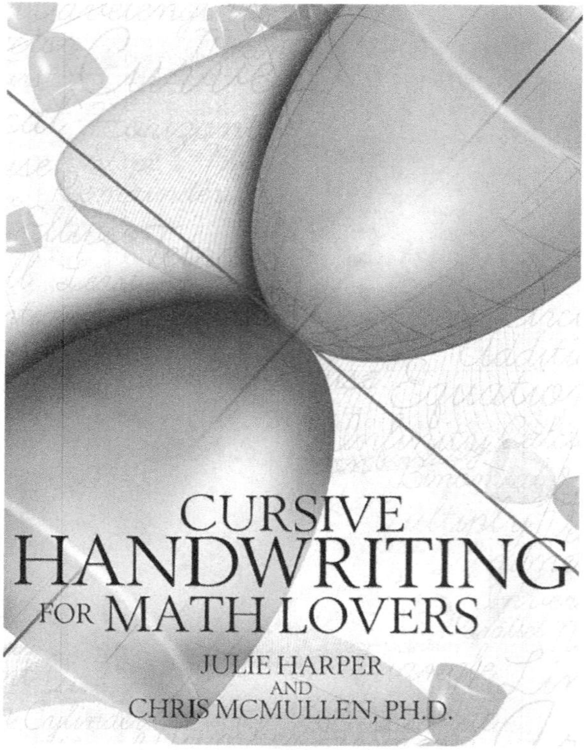

Made in the USA
Las Vegas, NV
11 March 2024

87045891R00315